ULTRAFAST LASERS

OPTICAL ENGINEERING

Founding Editor

Brian J. Thompson

University of Rochester
Rochester, New York

Editorial Board

ULTRAFAST LASERS

Technology and Applications

edited by

Martin E. Fermann
IMRA America, Inc.
Ann Arbor, Michigan, U.S.A.

Almantas Galvanauskas
University of Michigan
Ann Arbor, Michigan, U.S.A.

Gregg Sucha
IMRA America, Inc.
Ann Arbor, Michigan, U.S.A.

CRC Press
Taylor & Francis Group
Boca Raton London New York

CRC Press is an imprint of the
Taylor & Francis Group, an **informa** business

First published 2003 by Marcel Dekker, Inc.

Published 2019 by CRC Press
Taylor & Francis Group
6000 Broken Sound Parkway NW, Suite 300
Boca Raton, FL 33487-2742

© 2003 by Taylor & Francis Group, LLC
CRC Press is an imprint of Taylor & Francis Group, an Informa business

First issued in paperback 2019

No claim to original U.S. Government works

ISBN 13: 978-0-367-44696-3 (pbk)
ISBN 13: 978-0-8247-0841-2 (hbk)

Visit the Taylor & Francis Web site at
http://www.taylorandfrancis.com

and the CRC Press Web site at
http://www.crcpress.com

Foreword

For over a decade, the ultrafast (sub-100 fsec) dye laser served as the workhorse of research laboratories from its invention in 1981 until the advent of the Ar^+-pumped Ti:sapphire laser in 1992. As a graduate student working in the early 1990s in the Picosecond Ultrasonics Group under Humphrey Maris at Brown University, I learned first-hand the benefits that this solid-state laser brought to our pump-probe experiments. We used ultrashort light pulses to generate sub-THz acoustic waves and study their propagation in various materials. Relative to the dye lasers available at that time, the Ti:sapphire possessed lower noise, higher power, and greater environmental stability, and this opened up several new directions for ultrafast research in our group.

Second-generation Ti:sapphire technology further led to and enabled commercialization of the picosecond ultrasonics technique in 1998. Teams at Rudolph Technologies Inc. (RTEC) and Coherent, Inc., worked closely to develop the Vitesse™, a turnkey solid-state-pumped Ti:sapphire laser specifically designed for use in RTEC's MetaPULSE™ thin-film metrology system. The MetaPULSE is a highly automated system that has been widely adopted by manufacturers of microprocessors, DRAM, and magnetic storage for on-product thickness measurement of metal films ranging from several microns down to 20 angstroms in thickness. In 2001, RTEC announced the sale of its 100th MetaPULSE, representing a laser market of over $3 million per year since the product's introduction.

Three characteristics of the solid-state-pumped Ti:sapphire laser have proved instrumental to its industrial success: simple facilities requirements, compact size, and excellent stability. These improvements all result largely from replacing the Ar^+ pump with a Nd:vanadate laser. The all solid-state pump laser is highly efficient, permitting it to be plugged into a normal electrical outlet and resulting in a rate of heat generation lower than that of the average household lightbulb. Compact size is a key factor in product success in the semiconductor-manufacturing environment because of the high costs associated with building and maintaining a Class 1 cleanroom space. Laser stability and long lifetime are also essential in the semiconductor industry, in which equipment uptime in excess of 95% is standard in a 24 hours/7 days mode of operation. Coherent, Inc., designed the Vitesse with a highly stable, folded Ti:sapphire cavity, actively PZT-steered pump beam, and diode current servo loop for reliable hands-off laser operation over months or years. The latest product design refinements and increasing industry competition will have the result of extending the hands-off Ti:sapphire lifetime to greater than 10,000 hours by 2003, representing a stability improvement of nearly five orders of magnitude as compared with the ultrafast dye lasers of the early 1990s.

Thin-film semiconductor process metrology represents just one of many potential high-volume industrial applications for ultrafast lasers. The high market value of microprocessors and memory devices, at least in the near term, can support the typical $100,000 Ti:sapphire price tag over a limited number of process control tools (6–10 per manufacturing line). The same cannot be said for other potential markets: one timely example within the semiconductor industry is integrated metrology. In contrast to the stand-alone metrology model, in the integrated approach the metrology tool is "on board" with each process tool; in the case of an integrated Meta-PULSE, a metal film deposition or polishing system would be the most likely target for integration. While integrated metrology has several potential pitfalls and it is still too early to accurately predict the future market size, the approach may improve process tool uptime and the semiconductor product yield by shortening the loop, or adding feedback, between the process step and its control measurement. For a typical semiconductor process step, a standalone metrology tool may serve around five process tools; thus the market for integrated MetaPULSE units is potentially up to five times larger than the standalone market. Market analysis shows, however, that to be accepted and cost-effective, integrated metrology will likely require the equipment cost (including the laser) to decrease by at least the same factor (or down to $10,000– $20,000 or less). In addition, when metrology is placed inside a process tool, this requires a higher demand on laser compactness and reliability. Hands-off laser lifetimes greater than 10,000 hours will become an even greater necessity because failure of an integrated metrology tool leads to downtime of the process tool, thereby impeding the production line. These cost, compactness, and reliability considerations may lead to greater adoption of fiber-based ultrafast lasers in the near future as the market develops for integrated metrology.

Laser price performance points (e.g., pulse duration, power, wavelength, and noise) will, in general, be highly application-dependent; therefore, the industrial market for ultrafast lasers will most likely be split between different technologies and suppliers. For example, laser micromachining is already an established market with laser power, pulse duration, and wavelength requirements significantly different from those of nondestructive testing. At five years old the industrial ultrafast laser market is still in its infancy, and a rapidly growing market is plausible over the next decade as new industrial applications continue to emerge.

Christopher J. Morath
Rudolph Technologies, Inc.
Flanders, New Jersey

Preface

With the wide availability of commercial ultrafast laser systems, their applications to many areas of science and industry are blossoming. Indeed, ultrafast technology has diversified to such an extent that a comprehensive review of the whole field is nearly impossible. On the one hand, ultrafast lasers have entered the world of solid commerce; on the other, they continue to lead to breathtaking advances in areas of fundamental science.

The high degree of maturity that ultrafast optics has achieved in many areas inspired us to compile this book on selected topics in some of the most prominent areas. In selecting research topics, we concentrated on areas that have indeed already experienced a transformation to the commercial side of things, or where such a transformation is currently evolving or can at least be anticipated in the not too distant future, although, of course, we do not claim to predict where the field of ultrafast optics is heading. The only certainty is that ultrafast lasers will continue to play a rapidly growing role in technology with the so far elusive mass applications looming on the horizon.

Ultrafast Lasers is intended for researchers, engineers, and graduate students who are interested in a review of ultrafast optics technology. It consists of two parts, with the first describing some of the most widely used ultrafast laser systems and the second describing some of the best-known applications. On the laser systems side we cover ultrafast solid-state fiber, and diode lasers, and on the applications side we provide a general overview before presenting specific topics ranging from biology, electronics, optical communications, and mechanical engineering to integrated optics.

This book addresses the reader who is interested in a summary of the unique capabilities of ultrafast lasers. We hope that by providing this broad overview, we can contribute to the rapid advancement of a truly exciting technology.

Martin E. Fermann
Almantas Galvanauskas
Gregg Sucha

Contents

Contributors

A. Apolonski   Technische Universität Wien, Vienna, Austria
Mark Brezinski   Massachusetts Institute of Technology, Cambridge, Massachusetts, U.S.A.
Qin Chen   Rensselaer Polytechnic Institute, Troy, New York, U.S.A.
Peter J. Delfyett   University of Central Florida, Orlando, Florida, U.S.A.
Wolfgang Drexler   Massachusetts Institute of Technology, Cambridge, Massachusetts, U.S.A.
Martin E. Fermann   IMRA America, Inc., Ann Arbor, Michigan, U.S.A.
James G. Fujimoto   Massachusetts Institute of Technology, Cambridge, Massachusetts, U.S.A.
Almantas Galvanauskas   University of Michigan, Ann Arbor, Michigan, U.S.A.
José F. García   Harvard University, Cambridge, Massachusetts, U.S.A.
T. W. Hänsch   Max-Planck-Institut für Quantenoptik, Garching, Germany
Ingmar Hartl   Massachusetts Institute of Technology, Cambridge, Massachusetts, U.S.A.
R. Holzwarth   Max-Planck-Institut für Quantenoptik, Garching, Germany
Alan O. Jamison   Harvard University, Cambridge, Massachusetts, U.S.A.
Tibor Juhasz   University of Michigan, Ann Arbor, Michigan, U.S.A.
Franz Kärtner   Massachusetts Institute of Technology, Cambridge, Massachusetts, U.S.A.
Ursula Keller   Swiss Federal Institute of Technology, Zurich, Switzerland
F. Krausz   Technische Universität Wien, Vienna, Austria
Ronald M. Kurtz   University of California at Irvine, Irvine, California, U.S.A.
Xingde Li   Massachusetts Institute of Technology, Cambridge, Massachusetts, U.S.A.
Eric Mazur   Harvard University, Cambridge, Massachusetts, U.S.A.
Uwe Morgner   Massachusetts Institute of Technology, Cambridge, Massachusetts, U.S.A.
Michiel Müller   University of Amsterdam, Amsterdam, The Netherlands
Masataka Nakazawa   Tohoku University, Miyagi-ken, Japan

Stefan Nolte Friedrich Schiller University, Jena, Germany

Rüdiger Paschotta Swiss Federal Institute of Technology, Zurich, Switzerland

François Salin Université Bordeaux I, Talence, France

Melvin A. Sarayba University of California at Irvine, Irvine, California, U.S.A.

Chris B. Schaffer Harvard University, Cambridge, Massachusetts, U.S.A.

Jeff Squier Colorado School of Mines, Golden, Colorado, U.S.A.

Gregg Sucha IMRA America, Inc., Ann Arbor, Michigan, U.S.A.

G. Tempea Technische Universität Wien, Vienna, Austria

J. F. Whitaker University of Michigan, Ann Arbor, Michigan, U.S.A.

K. Yang University of Michigan, Ann Arbor, Michigan, U.S.A.

X.-C. Zhang Rensselaer Polytechnic Institute, Troy, New York, U.S.A.

ULTRAFAST LASERS

1

Ultrafast Solid-State Lasers

Rüdiger Paschotta and Ursula Keller
Swiss Federal Institute of Technology, Zurich, Switzerland

1.1 INTRODUCTION

Since 1990 we have observed tremendous progress in picosecond and femto-second pulse generation using solid-state lasers. Today ultrashort-pulse lasers are normally based on solid-state lasers, which provide sufficiently high average powers to efficiently generate many other frequencies with nonlinear frequency conversion schemes such as second harmonic to high harmonic generation, optical parametric oscillation, and amplification. The emphasis of this chapter is to give an updated review of the progress in pulsed solid-state lasers during the last ten years. Our goal is to give also to the non-expert an efficient starting point to enter into this field without providing all the detailed derivations. Relevant and useful references for further information are provided throughout the chapter.

For ultrashort-pulse generation we usually rely on continuous-wave (cw) mode locking, where the laser gain medium is pumped continuously and many axial laser modes are locked together in phase to form a short pulse. A homogeneously broadened laser would often lase in only one axial mode. In a modelocked laser, additional energy is transferred with the correct phase to adjacent modes with either active loss or a phase modulator (for active modelocking) or passive self-amplitude modulation (SAM) (for passive modelocking) inside the laser cavity. In the time domain, the many phase-locked axial modes result in a short pulse of a duration inversely proportional to the spectral width of all the phase-locked axial modes. Generally, passive modelocking generates shorter pulses and is much simpler than active modelocking. For reliable passive cw modelocking, the pulse generation starts from normal laser noise within less than approximately 1 ms. The

1

required SAM is typically obtained with a saturable absorber in the laser cavity. A saturable absorber is a device that has a lower loss for higher pulse intensities or energies. This can occur, for example, when the initial states of the absorbing transition are emptied or when the final states are occupied. The interaction of a short pulse with such a saturable absorber then produces a self-amplitude modulation due to the intensity-dependent absorption. For modelocking, the resulting pulses are typically much shorter than the cavity round-trip time, and the pulse repetition rate (typically between a few tens of megahertz and tens of gigahertz) is determined by the cavity round-trip time. Under certain conditions, the pulse repetition rate can be some integer multiple of the fundamental repetition rate; this is called harmonic modelocking (Becker et al., 1972).

Modelocking was first demonstrated in the mid-1960s using a HeNe laser (Hargrove et al., 1964), a ruby laser (Mocker and Collins, 1965), and an Nd:glass laser (De Maria et al., 1966). However, at the time passively modelocked solid-state lasers were also Q-switched. In this regime, called Q-switched modelocking (Fig. 1), the modelocked picosecond or femtosecond pulses are modulated with a much longer Q-switched pulse envelope (typically in the microsecond regime), which occurs at a much lower repetition rate (typically in the kilohertz regime). This continued to be a problem for most passively modelocked solid-state lasers until in 1992 the first intracavity saturable absorber was designed correctly to prevent self-Q-switching instabilities in solid-state lasers with microsecond or even millisecond upperstate lifetimes (Keller et al., 1992a).

For some time, the success of ultrafast dye lasers in the 1970s and 1980s diverted research interest away from solid-state lasers. Q-switching instabilities are not a problem for dye lasers. In 1974 the first subpicosecond passively modelocked dye lasers (Shank and Ippen, 1974; Ruddock and Bradley, 1976; Diels et al., 1978) and in 1981 the first sub-100 fs colliding pulse modelocked (CPM) dye lasers (Fork et al., 1981) were demonstrated. The CPM dye laser was the "workhorse" all through the 1980s for ultrafast laser spectroscopy in physics and chemistry. This CPM dye laser produced pulses as short as 27 fs with a typical average output power of about 20 mW (Valdmanis et al., 1985). The wavelength was centered around 630 nm and was not tunable. Shorter pulse durations, down to 6 fs, were achieved only through additional amplification and external pulse compression at much lower repetition rates (Fork et al., 1987).

The development of diode lasers with higher average powers in the 1980s again stimulated strong interest in solid-state lasers. Diode laser pumping provides dramatic improvements in efficiency, lifetime, size, and other important laser characteristics. For example, actively modelocked diode-pumped Nd:YAG (Maker and Ferguson, 1989a) and Nd:YLF

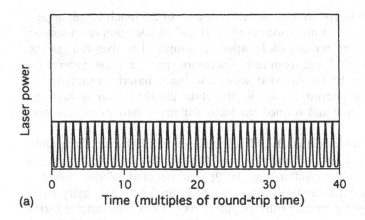

(a)

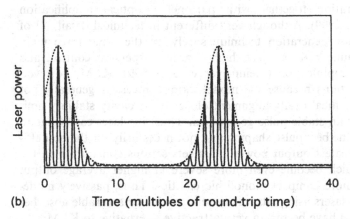

(b)

Figure 1 Schematic illustration of (a) continuous-wave (cw) modelocking and (b) Q-switched modelocking.

(Maker and Ferguson, 1989b; Keller et al., 1990a; Weingarten et al., 1990; Juhasz et al., 1990) lasers generated 7–12 ps pulse durations for the first time. In comparison, flashlamp-pumped Nd:YAG and Nd:YLF lasers typically produced pulse durations of $\sim$100 ps and $\sim$30 ps, respectively. Before 1992, however, all attempts to passively modelock diode-pumped solid-state lasers resulted in Q-switching instabilities that at best produced stable modelocked pulses within longer Q-switched macropulses, as mentioned above.

The breakthrough of ultrafast solid-state lasers happened with the discovery of the Ti:sapphire lase medium (Moulton, 1986), which was the first solid-state laser medium that was able to support ultrashort pulses without cryogenic cooling. The existing passive modelocking techniques, primarily developed for dye lasers, were inadequate because of the much longer upper-state lifetime (i.e., in the microsecond regime) and smaller gain cross section (i.e., in the $10^{-19}\,cm^2$ regime) of Ti:sapphire compared to dyes (i.e., in the nanosecond and $10^{-16}\,cm^2$ regimes). Therefore, passive pulse generation techniques had to be re-evaluated with new laser material properties in mind. The strong interest in an all-solid-state ultrafast laser technology was the driving force and formed the basis for many new inventions and discoveries.

Kerr lens modelocking (KLM) (Spence et al., 1991) of Ti:sapphire lasers was discovered in 1991 and produced the shortest laser pulses—pulses of less than 6 fs duration (Sutter et al., 1999; Morgner et al., 1999a, 1999b)—directly from the laser cavity without any additional external cavity pulse compression. Slightly shorter sub-5 fs pulses have been demonstrated with external pulse compression (Baltuska et al., 1997; Nisoli et al., 1997) and continuum generation together with parametric optical amplification (Shirakawa et al., 1999). Although very different in technical detail, all of these sub-6 fs pulse generation techniques rely on the same three main components: nonlinear Kerr effect, higher order dispersion control, and ultrabroadband amplification (Steinmeyer et al., 1999). KLM, however, has serious limitations because the modelocking process is generally not self-starting and critical cavity alignment close to the cavity stability limit is required to obtain stable pulse generation. Thus, the laser cavity has to be optimized for the best pulse shaping and not necessarily for the best efficiency and the greatest output power. This sets serious constraints on the cavity design, which become even more severe at higher average output powers and for more compact monolithic cavities. Thus, passively mode-locked solid-state lasers using intracavity semiconductor saturable absorber mirrors (SESAMs) have become a very attractive alternative to KLM.

In 1992, semiconductor saturable absorber mirrors (SESAMs) (Keller et al., 1992a; 1996) allowed for the first time self-starting and stable passive modelocking of diode-pumped solid-state lasers with an intracavity satur-able absorber. The design freedom of SESAMs allowed for a systematic investigation of the stability regimes of passive Q-switching and passive cw modelocking with an improved understanding and modeling of Q-switching (Hönninger et al., 1999a) and multiple pulsing instabilities (Aus der Au et al., 1997; Kärtner et al., 1998). Simple laser and SESAM design guidelines allowed us to continue to push the frontiers of ultrafast solid-state lasers during the last ten years. Presently, the frontiers in average output power of diode-pumped solid-state lasers are set in the picosecond regime with Nd:YAG lasers (27 W average output power and 19 ps pulse duration) (Spühler et al., 2000) and in the femtosecond regime with Yb:YAG lasers (16 W, 730 fs) (Aus der Au et al., 2000). This basically means that microjoule-level pulse energies in both the picosecond and femtosecond regimes are available directly from compact solid-state lasers without cavity dumping or further pulse amplification. In addition, the frontier in pulse repetition rate has been recently pushed to nearly 80 GHz by the use of quasi-monolithic miniature Nd:YVO$_4$ laser cavities (Krainer et al., 2000). Today SESAMs are well established for ultrafast all-solid-state lasers. The main reason for this device's utility is that both the linear and nonlinear optical properties can be engineered over a wide range, allowing for more freedom in the specific laser cavity design. The main absorber parameters such as operation wavelength, modulation depth, saturation fluence, and absorber lifetime can be custom designed for stable cw modelocking or Q-switching. Initially, semiconductor saturable absorber mirrors were used in coupled cavities (Keller et al., 1990b), because they introduced too much loss inside solid-state lasers with small gain cross sections (i.e., 10^{-19} cm^2 and smaller). Two years later, in 1992, this work resulted in a new type of intracavity saturable absorber mirror, the antiresonant Fabry-Perot saturable absorber (A-FPSA) (Keller et al., 1992a), where the absorber was integrated inside a Fabry-Perot structure of which the bottom reflector was a high reflector (with nearly 100% reflectivity). The Fabry-Perot was operated at antiresonance to obtain broad bandwidth and low loss. The A-FPSA mirror was based on a semiconductor Bragg mirror, an absorber layer, and a dielectric top reflector and therefore allowed for a large variation of the absorber parameters. The result was a much better understanding of the absorber and laser design necessary to obtain stable passive mode locking or Q-switching of many different solid-state lasers. Scaling of the A-FPSA design resulted in a semiconductor Bragg mirror with only one single-quantum-well saturable absorber layer (Brovelli et al., 1995; Jung et al., 1995a). This design was later also referred to as a saturable Bragg reflector (SBR) (Tsuda et al., 1995). In 1995, it was further realized that the intracavity saturable absorber can be integrated into a more general mirror structure that allows for both saturable absorption and negative dispersion control, which is now generally referred to as a SESAM. In a general sense, then, we can reduce the design problem of a SESAM to the analysis of multilayer interference filters for a given desired nonlinear reflectivity response for both the amplitude and phase. Thus, if required, the SESAM provides not only saturable absorption but also negative dispersion compensation. The A-FPSA, and SBR, and the dispersive saturable absorber mirror (D-SAM) (Kopf et al., 1996a) are therefore special examples of SESAM designs. A more detailed summary of different SESAM designs is given in a recent book chapter (keller, 1999).

A new modelocking technique, referred to as soliton modelocking (Kärtner and Keller, 1995; Jung et al., 1995b; Kärtner et al., 1996), totally removes the cavity design constraints of KLM lasers and substantially relaxes the requirements of the SESAM performance in the femtosecond regime. The pulse formation is basically done by the soliton effects, and the saturable absorber is required only to start and stabilize the modelocking. Thus, even a SESAM with a relatively long recovery time can support very short pulses. For example, a SESAM with a recovery time of 10 ps was sufficient to stabilize soliton pulses of only 300 fs duration in a Ti:sapphire laser (Jung et al., 1995b). In contrast to KLM, no critical cavity alignment is necessary and modelocking is self-starting. Pulses as short as 13 fs have been produced with soliton modelocked Ti:sapphire lasers (Kärtner et al., 1996).

Today, a large variety of reliable and compact all-solid-state ultrafast lasers are available with pulse durations ranging from picoseconds to less than 100 fs. Table 1 summarizes the results achieved for various solid-state lasers and various modelocking (ML) techniques. For each technique we provide the first demonstration and the best results. A more detailed table with all results is provided elsewhere (Keller, 2001). For each laser material we emphasize the best results in terms of pulse duration, average outpur power, and pulse repetition rates with bold letters. For completeness we also include the coupled-cavity modelocking (CCM) techniques even though they have become less relevant today because SESAM-based ultrafast lasers are much simpler and more compact. We distinguish between two different CCM techniques: Additive pulse modelocking (APM) is a coupled-cavity modelocking technique for which the coupled cavity generates a nonlinear phase shift that adds constructively at the peak of the pulse inside the main cavity and destructively in the wings, thus shortening the pulse duration inside the main cavity. This was first discovered by Kean et al. (1989) and later explained in this simple model by Ippen et al. (1989). However, APM requires interferometric cavity length stabilization of the coupled cavity. Resonant passive modelocking (RPM) (Keller et al., 1990b; Haus et al., 1991) removed this problem by using an amplitude nonlinearity inside the coupled cavity.

This chapter is organized as follows. In Section 1.2 we discuss the demands on gain media for ultrashort-pulse generation and give an overview on available media as well as the corresponding achievements. Sections 1.3 and 1.4 are devoted to technical issues of particular importance, namely the effects of dispersion and nonlinearities in laser cavities and different modelocking techniques. Several examples for ultrafast lasers are then discussed in Section 1.5. Although ultrafast pulses are usually generated with modelocked lasers, in Section 1.6 we also discuss a kind of Q-switched laser that can generate picosecond pulses. Finally, we give a short summary and outlook in Section 1.7.

Table 1. Continuous-Wave Modelocked Solid-State Lasers Using Different Modelocking Techniques[a]

Laser material	ML technique	λ_0	τ_p	$P_{av,out}$	f_{rep}	Remarks	Ref.
Ti:sapphire							
	Active AOM	814 nm	150 fs	600 mW	80.5 MHz		Curley and Ferguson, 1991
	CCM-APM		1.4 ps	300 mW		Highly chirped 1.4 ps output pulses, externally compressed down to 200 fs	Goodberlet et al., 1989
	CCM-RPM Dye sat. absorber	860 nm 750 nm	2 ps 140 fs	90 mW	250 MHz	KLM started with dye-saturable absorber (not understood—assumed to have a CPM Ti:sapphire laser)	Keller et al, 1990b Sarukura et al., 1991
	KLM	880 nm	60 fs	300 mW		First demonstration of KLM (but KLM not understood)	Spence et al., 1991
		~800 nm	<6 fs	300 mW	85 MHz	Fused silica prisms and double-chirped mirrors, KLM is self-starting with SESAM	Sutter et al., 1999

(*Continued*)

Laser material	ML technique	λ_0	τ_p	$P_{av,out}$	f_{rep}	Remarks	Ref.
		~800 nm	<6 fs	200 mW	90 MHz	CaF$_2$ prisms, double-chirped mirrors	Morgner et al., 1999b
		780 nm	8.5 fs	1 W	75 MHz	1.5 MW peak, focused intensity 5×10^{13} W/cm^2	Xu et al., 1998
		850 nm	13 fs	1.5 W	110 MHz	1 MW peak, 13 nJ out	Beddard et al., 1999
		800 nm	16.5 fs	170 mW	15 MHz	0.7 MW peak	Cho et al., 1999
		782 nm	23 fs	300 mW	2 GHz	Ring laser, self-starting modelocking due to feedback from external mirror	Bartels et al., 1999
	Soliton-SESAM	840 nm	34 fs	140 mW	98.9 MHz	First SESAM design with a single quantum well absorber in a Bragg reflector	Brovelli et al., 1995
		810 nm	13 fs	80 mW	85 MHz	Shortest pulse with soliton modelocking and no KLM	Kärtner et al., 1996

Cr:LiSAF	KLM	800–880 nm	150 fs	50 mW	82 MHz	First modelocked femtosecond Cr:LiSAF laser, Kr-pumped	Miller et al., 1992
			300 ps	~1 mW		**First** diode-pumped modelocked Cr:LiSAF laser, AOM or RPM for starting KLM	French et al., 1993
			220 fs	~10 mW		KLM started by SESAM	Mellish et al., 1994
	Soliton-SESAM	850 nm	**12 fs**	23 mW	200 MHz	ML not self-starting	Uemura et al., 1999
		840 nm	98 fs	50 mW	120 MHz	**First** soliton mode-locking, no KLM required	Kopf et al., 1994a
		850 nm	45 fs	105 mW	176 MHz		Kopf et al., 1995a
		875 nm	110 fs	**500 mW**	150 MHz	Low-brightness 0.9 cm wide, 15 W diode laser array	Kopf et al., 1997
Cr:LiCAF	KLM	800 nm	170 fs	100 mW	90 MHz	Kr-pumped	LiKamWa et al.,1992
		820 nm	**20 fs**	13 mW	95 MHz		Gabel et al., 1998
Cr:LiSGaF	KLM	830 nm	100 fs (50 fs)	35 mW	71 MHz		Yanovsky et al., 1995
		895 nm	**14 fs**	100 mW	70 MHz	Kr ion laser pumped, chirped mirror	Sorokina et al., 1997
	Soliton-SESAM	839 nm	61 fs	78 mW	119 MHz		Loesel et al., 1997

(*Continued*)

Laser material	ML technique	λ_0	τ_p	$P_{av,out}$	f_{rep}	Remarks	Ref.
Cr:LiSCAF							
	KLM	860 nm	90 fs	100 mW	140 MHz	Kr-ion-laser pumped	Wang et al., 1994
Cr:Forsterite							
	KLM	1.23 μm (1.21–1.27 μm)	48 fs	380 mW	81 MHz	Nd:YAG laser pumped, KLM self-starting with AOM	Sennaroglu et al., 1993
		1.3 μm	**14 fs**	80 mW	100 MHz	Double chirped mirrors	Chudoba et al., 2000
	Soliton-SESAM	1.29 μm	40 fs	60 mW		Nd:YVO$_4$ laser pumped	Zhang et al., 1997a
		1.29 μm	36 fs	60 mW		Nd:YVO$_4$ laser pumped	Zhang et al., 1997b
		1.26 mW	78 fs	**800 mW**	83 MHz	Nd:YAG laser pumped	Petrov et al., 1998
Cr:YAG							
	KLM	1.52 μm	120 fs	360 mW	81 MHz	Nd:YAG laser pumped, regen. AOM for starting KLM	Sennaroglu et al., 1994
		1.54 μm	**43 fs**	200 mW	70 MHz	Nd:YVO$_4$ laser pumped	Tong et al., 1997
		1.52 μm	55 fs		**1.2 GHz**	Nd:YVO$_4$ laser pumped	Tomaru et al., 2000
	Soliton-SESAM	1.541 μm	110 fs	70 mW		Nd:YVO$_4$ laser pumped	Collings et al., 1996
		1.52 μm	**44 fs**	65 mW		Nd:YVO$_4$ laser pumped	Zhang et al., 1999

Nd:YAG

Method	Wavelength	Pulse duration	Power	Rep. rate	Comments	Reference
Active AOM	1.064 μm	25 ps			Lamp-pumped: pulse shortening due to intracavity etalon	Roskos et al., 1986
Active FM	1.064 μm	12 ps	65 mW	350 MHz		Maker et al., 1989a
Active EOM	1.32 μm	8 ps	240 mW	1 GHz		Zhou et al., 1991
Active AOM	1.32 μm	53 ps	1.5 W	200 MHz	Lamp-pumped and harmonically modelocked	Keller et al., 1988
CCM-APM	1.064 μm	1.7 ps	25 mW	136 MHz		Goodberlet et al., 1990
KLM	1.32 μm	10 ps	700 mW	100 MHz	Lamp-pumped	Liu et al., 1990a
KLM	1.064 μm	8.5 ps	1 W	100 MHz		Liu et al., 1992
KLM	1.064 μm	6.7 ps	675 mW	106 MHz	Ti:sapphire laser-pumped	Henrich and Beigang, 1997
SESAM	1.064 μm	8.7 ps	100 mW	100 MHz		Weingarten et al., 1993
SESAM	1.064 μm	6.8 ps	400 mW	217 MHz		Keller, 1994
SESAM	1.064 μm	19 ps	27 W	55 MHz	Three side-pumped laser heads	Spühler et al., 2000
Polarization switching in nonlinear crystal	1.064 μm	23 ps	4 W	150 MHz	Lamp-pumped, KTP crystal	Kubecek et al., 1999

(Continued)

Laser material / ML technique	λ_0	τ_p	$P_{av,out}$	f_{rep}	Remarks	Ref.
Nd:YLF						
Active AOM	1.053 μm	37 ps	6.5 W	100 MHz	Laser-pumped	Bado et al., 1987
	1.053 μm	18 ps	12 mW	230 MHz		Maker and Ferguson, 1989c
	1.047 μm	7 ps	135 mW	2 GHz		Weingarten et al., 1990
	1.047 μm	6.2 ps	20 mW	1 GHz	Ti:sapphire laser-pumped	Walker et al., 1990
Active FM	1.053 μm	4.5 ps	400 mW	2.85 GHz		Weingarten et al., 1992
KLM	1.3 μm	8 ps	240 mW	1 GHz		Zhou et al., 1991
	1.047 μm	3 ps	250 mW			Lincoln and Ferguson, 1994
	1.047 μm	2.3 ps	800 mW	82 MHz	Lamp-pumped, microdot mirror	Ramaswamy et al., 1993
SESAM	1.047 μm	3.3 ps	700 mW	220 MHz	Ti:sapphire laser-pumped	Keller et al., 1992a
CCM-APM	1.047 μm	2.8 ps	460 mW	220 MHz		Keller, 1994
	1.3 μm	5.7 ps	130 mW	98 MHz		Fluck et al., 1996
	1.053 μm	3.7 ps	7 W	76 MHz	Lamp-pumped	Liu and Chee, 1990b
	1.053 μm	1.7 ps		103 MHz		Jeong et al., 1999
	1.047 μm	**1.5 ps**	20 mW	123 MHz		Malcolm et al., 1990
CCM-RPM	1.047 μm	3.7 ps	550 mW	250 MHz	Ti:sapphire laser-pumped	Keller et al., 1992b

Material		Wavelength	Pulse	Power		Notes	Reference
Nd:YVO₄	SESAM	1.064 μm	7 ps	4.5 W	84 MHz		Ruffing et al., 1997
		1.064 μm	21 ps	20 W	90 MHz		Burns et al., 2000
		1.064 μm	**2.7 ps**	65 mW	**77 GHz**	Ti:sapphire laser pumped	Krainer et al., 2000
	Nonlinear mirror ML	1.3 μm	4.6 ps	50 mW	93 MHz		Fluck et al., 1996
		1.064 μm	7.9 ps	1.35 W	150 MHz		Agnesi et al., 1997
	Intensity-dependent polarization rotation	1.064 μm	**2.8 ps**	670 mW	130 MHz		Couderc et al., 1999
Nd:LSB	SESAM	1.062 μm	**1.6 ps**	210 mW	240 MHz	Ti:sapphire laser-pumped	Braun et al., 1996
		1.062 μm	2.8 ps	400 mW	177 MHz		Braun et al., 1996
Nd:BEL	Active-AOM	1.070 μm	7.5 ps	230 mW	250 MHz		Li et al., 1991
	Active-Fm	1.070 μm	**2.9 ps**	30 mW	238 MHz	Harmonic mode locking	Godil et al., 1991
		1.070 μm	3.9 ps	30 mW	**20 GHz³**	Harmonic mode locking	Godil et al., 1991
Nd:glass **Nd:phosphate**	Active-AOM	1.054 μm	7 ps	20 mW		Ar ion laser-pumped	Yan et al., 1986
		1.054 μm	~10 ps	30 mW			Basu and Byer, 1988
		1.054 μm	9 ps	30 mW	240 MHz		Hughes et al., 1992

(Continued)

Laser material	ML technique	λ_0	τ_p	$P_{av,out}$	f_{rep}	Remarks	Ref.
		1.063 µm	310 fs	70 mW	240 MHz	Ti:sapphire laser-pumped, regeneratively actively modelocked	Kopf et al., 1994b
Nd:phosphate glass	Active-FM	1.054 µm	9 ps	14 mW	235 MHz		Hughes et al., 1991
Nd:phosphate glass	CCM-APM	1.054 µm	122 fs	200 mW		Kr ion laser-pumped	Spielmann et al., 1991
Nd:phosphate glass	Soliton-SESAM	1.054 µm	150 fs	110 mW	180 MHz		Kopf et al., 1995b
Nd:phosphate glass		1.054 µm	120 fs	30 mW	150 MHz	Single prism for dispersion compensation	Kopf et al., 1996b
		1.054 µm	275 fs	1.4 W	74 MHz		Paschotta et al., 2000a
Nd:fluoro-phosphate glass		1.065 µm	60 fs	84 mW	114 MHz		Aus der Au et al., 1997
Nd:silicate glass		1.064 µm	130 fs	80 mW	180 MHz		Kopf et al., 1995b
Yb:YAG	Soliton-SESAM	1.03 µm	540 fs	100 mW	81 MHz	First passively modelocked Yb:YAG laser	Hönninger et al., 1995

Material	Modelocking	λ	Pulse duration	Power	Rep. rate	Reference
Yb:KGW		1.03 μm	**340 fs**	170 mW		Hönninger et al., 1999b
		1.03 μm	730 fs	**16.2 W**	35 MHz	Aus der Au et al., 2000
	Soliton-SESAM	1.037 μm	176 fs (112 fs)	1.1 W (0.2 W)	86 MHz	Brunner et al., 2000
Yb:GdCOB	Soliton-SESAM	1.045 μm	90 fs	40 mW	100 MHz	Druon et al., 2000
Yb:glass Yb:phosphate glass	Soliton-SESAM	1.025–1.065 μm	58 fs	65 mW	112 MHz	Hönninger et al., 1998
Yb:silicate glass	Soliton-SESAM	1.03–1.082 μm	61 fs	53 mW	112 MHz	Hönninger et al., 1998

[a]"Best" means in terms of pulse duration, highest average output power, highest pulse repetition rate etc. The result for which "best" applies is in bold letters. The lasers are assumed to be diode-pumped, if not stated otherwise (except Ti:sapphire laser). ML = modelocking; CCM = coupled-cavity modelocking; APM = additive pulse modelocking; RPM = resonant passive modelocking; KLM = Kerr lens modelocking; SESAM = semiconductor saturable absorber mirror; soliton-SESAM = soliton modelocking with a SESAM; SESAM = saturable absorber modelocking using SESAMs; AOM = acousto-optic modulator; EOM = electro-optic phase modulator; λ_0-center lasing wavelength; τ_p-measured pulse duration; $P_{av,out}$-average output power; f_{rep}-pulse repetition rate.

1.2 GAIN MEDIA FOR ULTRASHORT-PULSE GENERATION

Gain media for ultrafast lasers have to meet a number of conditions. We first list those criteria that apply to continuous-wave (cw) lasers as well. Obviously the gain medium should have a laser transition in the desired wavelength range and a pump transition at a wavelength where a suitable pump source is available. Several factors are important to achieve good power efficiency: a small quantum defect, the absence of parasitic losses, and a high gain ($\sigma\tau$ product) are desirable. The latter allows for the use of an output coupler with relatively high transmission, which makes the laser less sensitive to intracavity losses. For high-power operation, we prefer media with good thermal conductivity, a week (or even negative) temperature dependence of the refractive index (to reduce thermal lensing), and a week tendency to undergo thermally induced stress fracture.

For ultrafast lasers, in addition we require a broad emission bandwidth, because ultrashort pulses intrinsically have a large bandwidth. More precisely, we need a large range of wavelengths in which a smooth gain spectrum is obtained for a fixed inversion level. The latter restrictions explain why the achievable modelocked bandwidth is in some cases [e.g., some Yb^{3+}-doped media (Hönninger et al., 1999b)] considerably smaller than the tuning range achieved with tunable cw lasers, particularly for quasi-three-level gain media. A less obvious requirement is that the laser cross sections should be high enough. Although the requirement of a reasonably small pump threshold can be satisfied even with low laser cross sections if the fluorescence lifetime is large enough, it can be very difficult to overcome Q-switching instabilities (see Sec. 1.4.3) in a passively modelocked laser based on a gain material with small laser cross sections. Unfortunately, many broadband gain media tend to have small laser cross sections, which can significantly limit their usefulness for passive modelocking, particularly at high pulse repetition rates and in cases where the pump beam quality is poor, necessitating a large mode area in the gain medium. Finally, a short pump absorption length is desirable because it permits the use of a short path length in the medium, which allows for operation with a small mode area in the gain medium and also limits the effects of dispersion and Kerr nonlinearity.

Most ultrafast lasers belong to one of two groups. The first group is based on gain media that have quite favorable properties for diode-pumped high-power cw operation but cannot be used for femtosecond pulse generation because of their relatively small amplification bandwidth. Typical examples are Nd^{3+}:YAG and Nd^{3+}:YVO$_4$. With high-power laser diodes, one or several conventional end-pumped or side-pumped laser rods and a SESAM (Sec. 1.4.3) for modelocking, up to 27 W or average power in 20 ps pulses has been achieved with Nd^{3+}:YAG (Spühler et al., 2000) and

20 W in 20 ps pulses with Nd^{3+}:YVO_4 (Burns et al., 2000). Significantly shorter pulse durations have been achieved at lower output powers, down to 1.5 ps with 20 mW (Malcolm et al., 1990), using the technique of additive pulse modelocking (APM, Sec. 1.4.3). For all these Nd^{3+}-doped media the relatively large laser cross sections usually make it relatively easy to achieve stable modelocked operation without Q-switching instabilities. See Section 1.5.1 for typical cavity setups.

The second group of gain media are characterized by a much broader amplification bandwidth, typically allowing for pulse durations well below 0.5 ps but also usually by significantly poorer thermal properties and smaller laser cross sections. Ti^{3+}:sapphire (Moulton, 1986) is a notable exception, combining nearly all desired properties for powerful ultrafast lasers, except that the short pump wavelength excludes the use of high-power diode pump lasers and the quantum defect is large. Using an argon ion laser or a frequency-doubled solid-state laser as a pump source, Ti^{3+}:sapphire lasers have been demonstrated to generate pulses with durations below 6 fs and a few hundred milliwatts of average power (Sutter et al., 1999; Morgner et al., 1999b). For these pulse durations, KLM (Sec. 1.4.3) is required, and self-starting may be achieved with a SESAM in addition (Sutter et al., 1999). With a SESAM alone, 13 fs pulses with 80 mW have been demonstrated. If significantly longer pulse durations are acceptable, several watts of average power can be generated with a commercially available Ti^{3+}:sapphire laser, usually pumped with an argon ion laser. Recently, Cr^{2+}:ZnSe (Page et al., 1997) was identified as another very interesting gain material that is in various ways similar to Ti^{3+}:sapphire but emits at mid-infrared wavelengths, around 2.2–2.8 μm. This very broad bandwidth should allow for pulse durations below 20 fs, although to date the shortest achieved pulse duration is much longer, ≈ 4 ps (Carrig et al., 2000).

Diode-pumped femtosecond lasers can be built with crystals such as Cr^{3+}:LiSAF, Cr^{3+}:LiSGaF, or Cr^{3+}:LiSCAF, which can be pumped at longer wavelengths. However, these media have much poorer thermal properties and thus cannot compete with Ti^{3+}:sapphire in terms of output power. Cr^{3+}:LiSAF lasers have generated pulses as short as 12 fs (Uemura and Torizuka, 1999), but only with 23 mW of output power, using KLM without self-starting ability. The highest achieved modelocked power was 0.5 W in 110 fs pules (Kopf et al., 1997).

Cr^{4+}:forsterite emits around 1.3 μm and is suitable for pulse durations down to 14 fs with 80 mW (Chudoba et al., 2000), or 78 fs pulses with 800 mW (Petrov et al., 1998). Normally, an Nd-doped laser (which may be diode-pumped) is used for pumping of Cr^{4+}:forsterite. The same holds for Cr^{4+}:YAG, which emits around 1.4–1.5 μm and has allowed the generation of pulses of 43 fs duration at 200 mW (Tong et al., 1997).

Other broadband gain materials are phosphate or silicate glasses doped with rare earth ions such as Nd^{3+} or Yb^{3+}, for pulse durations down to $\approx 60\,fs$ (Aus der Au et al., 1997; Hönninger et al., 1998) and output powers of a few hundred milliwatts. The relatively poor thermal properties make high-power operation challenging. Up to 1.4 W of average power in 275 fs pulses (Paschotta et al., 2000a) or 1 W in 175 fs pulses (Aus der Au et al., 1998) have been obtained from Nd^{3+}:glass by using a specially adapted elliptical mode pumping geometry (Paschotta et al., 2000b). Here, a strongly elliptical pump beam and laser mode allow the use of a fairly thin gain medium that can be efficiently cooled from both flat sides. The resulting nearly one-dimensional heat flow reduces the thermal lensing compared to cylindrical rod geometries if the aspect ration is large enough.

Yb^{3+}:YAG has thermal properties similar to those of Nd^{3+}:YAG and at the same time a much larger amplification bandwidth. Another favorable property is the small quantum defect. However, challenges arise from the quasi-three-level nature of this medium and from the small laser cross sections, which favor Q-switching instabilities. High pump intensities help in both respects. An end-pumped laser based on a Yb^{3+}:YAG rod has generated 340 fs pulses with 170 mW (Hönninger et al., 1995). As much as 8.1 W in 2.2 ps pulses was obtained from an elliptical mode Yb^{3+}:YAG laser (Aus der Au et al., 1999). Recently, the first Yb^{3+}:YAG thin disk laser (Giesen et al., 1994) was passively modelocked, generating 700 fs pulses with 16.2 W average power (Aus der Au et al., 2000). The concept of the passively modelocked thin disk laser appears to be power-scalable, so that even much higher powers should become possible in the near future.

A few Yb^{3+}-doped crystalline gain materials have been developed that combine a relatively broad amplification bandwidth (sufficient for pulse durations of a few hundred femtoseconds) with thermal properties that are better than those of other broadband materials, although not as good as, for example, those of YAG or sapphire. Examples are Yb^{3+}:YCOB (Valentine et al., 2000), Yb^{3+}:YGdCOB (Druon et al., 2000), Yb^{3+}:SFAP (Gloster et al., 1997), and Yb^{3+}:KGW (Brunner et al., 2000). With an end-pumped Yb^{3+}:KGW rod, 1.1 W of average power has recently been achieved in 176 fs pulses (Brunner et al., 2000). Yb^{3+}:KGW and Yb^{3+}:KYW may be applicable in a thin disk laser, possibly generating tens of watts in pulses with $<200\,fs$ duration, but this remains to be demonstrated. Another new class of materials of particular importance are the Yb^{3+}-doped sesquioxides (Larionov et al., 2001) such as Y_2O_3, Sc_2O_3, and Lu_2O_3, which appear to be very suitable for high-power operation.

Color center crystals can also be used for femtosecond pulse generation (Blow and Nelson, 1988; Yakymyshyn et al., 1989; Islam et al.,

1989), but we do not discuss them here because they need cryogenic conditions.

1.3 DISPERSION AND NONLINEARITIES

1.3.1 Dispersion

When a pulse travels through a medium, it acquires a frequency-dependent phase shift. A phase shift that varies linearly with the frequency corresponds to a time delay, without any change of the temporal shape of the pulse. Higher order phase shifts, however, tends to modify the pulse shape and are thus of relevance for the formation of short pulses. The phase shift can be expanded in a Taylor series around the center angular frequency ω_0 of the pulse:

$$\varphi(\omega) = \varphi_0 + \frac{\partial \varphi}{\partial \varphi}(\omega - \omega_0) + \frac{1}{2}\frac{\partial^2 \varphi}{\partial \omega^2}(\omega - \omega_0)^2 + \frac{1}{6}\frac{\partial^3 \varphi}{\partial \omega^3}(\omega - \omega_0)^3 + \cdots \quad (1)$$

Here the derivatives are evaluated at $\omega_0 \cdot \partial \varphi / \partial \omega$ is the group delay T_g, $\partial^2 \varphi / \partial \omega^2$ the group delay dispersion (GDD), and, $\partial^3 \varphi / \partial \omega^3$ the third-order dispersion (TOD). The GDD describes a linear frequency dependence of the group delay and thus tends to separate the frequency components of a pulse: For positive GDD, e.g., the components with higher frequencies are delayed with respect to those with lower frequencies, which results in a positive "chirp" ("up-chirp") of the pulse. Higher orders of dispersion generate more complicated distortions.

The broader the bandwidth of the pulse (i.e., the shorter the pulse duration), the more terms of this expansion are significant. GDD, which acts on an initially unchirped Gaussian pulse with full width at half maximum (FWHM) pulse duration τ_0, increases the pulse duration according to

$$\tau = \tau_0 \left[1 + \left(4 \ln 2 \frac{\text{GDD}}{\tau_0^2} \right)^2 \right]^{1/2} \quad (2)$$

It is apparent that the effect of GDD becomes strong if $\text{GDD} > \tau_0^2$. Similarly, TOD becomes important if $\text{TOD} > \tau_0^3$.

1.3.2 Dispersion Compensation

If no dispersion compensation is used, the net GDD for one cavity round-trip is usually positive, mainly because of the dispersion in the gain medium.

Other components such as mirrors may also contribute to this. However, in lasers with > 10 ps pulse duration the dispersion effects can often be ignored, because the total GDD in the laser cavity is typically at most a few thousand square femtoseconds, much less than the pulse duration squared. For shorter pulse durations, the GDD has to be considered, and pulse durations below about 30 fs usually necessitate the compensation of TOD or even higher orders of dispersion. In most cases, the desired total GDD is not zero but negative, so that soliton formation (see Sec. 1.3.4) can be exploited. Usually, one requires sources of negative GDD and in addition appropriate higher order dispersion for shorter pulses. The most important techniques for dispersion compensation are discussed in the following subsections.

Dispersion from Wavelength-Dependent Refraction

If the intracavity laser beam hits a surface of a transparent medium with non-normal incidence, the wavelength dependence of the refractive index can cause wavelength-dependent refraction angles. In effect, different wavelength components will travel on slightly different paths, and this in general introduces an additional wavelength dependence to the round-trip phase and thus a contribution to the overall dispersion. The most frequently used application of this effect is to insert a prism pair in the cavity (Fork et al., 1984), where the different wavelength components travel in different directions after the first prism and along parallel but separated paths after the second prism. The wavelength components can be recombined simply on the way back after reflection at a plane end mirror (of a standing wave cavity) or by a second prism pair (in a ring cavity). Spatial separation of different wavelengths occurs only in part of the cavity. The negative GDD obtained from the geometric effect is proportional to the prism separation, and an additional (usually positive) GDD contribution results from the propagation in the prism material. The latter contribution can be easily adjusted via the prism insertion, so that the total GDD can be varied over an appreciable range. Some higher order dispersion is also generated, and the ratio of TOD and GDD can for a given prism material be varied only in a limited range by using different combinations of prism separation and insertion. Some prism materials with lower dispersion (e.g., fused quartz instead of SF10 glass) can help to reduce the amount of TOD generated together with a given value of GDD but also require greater prism separation. The prism angles are usually not used as optimization parameters but rather are set to be near Brewster's angle in order to minimize reflection losses. The small losses and the versatility of the prism pair technique are the reasons prism pairs are very widely used in ultrafast lasers.

In Ti^{3+}:sapphire lasers, pulse durations around 10 fs can be reached with negative dispersion only from a prism pair.

More compact geometries for dispersion compensation make use of a single prism only (Ramaswamy-Paye and Fujimoto, 1994; Kopf et al., 1996b). In this case, the wavelength components are spatially separated in the whole resonator, not just part of it. Even without any additional prisms, refraction at a Brewster interface of the gain medium can generate negative dispersion. In certain configurations, where the cavity is operated near a stability limit, the refraction effect can be strongly increased (Paschotta et al., 1999a), so significant negative GDD can be generated in a compact cavity. The amount of GDD may then also strongly depend on the thermal lens in the gain medium and on certain cavity dimensions.

Grating Pairs

Compared to prism pairs, pairs of diffraction gratings can generate higher dispersion in a compact setup. However, because of the limited diffraction efficiency of gratings, the losses of a grating pair are typically higher than acceptable for use in a laser cavity, except in cases with a high gain (e.g., in fiber lasers). For this reason, grating pairs are normally used only for external pulse compression.

Gires-Tournois Interferometers (GTIs)

A compact device to generate negative GDD (even in large amounts) is the Gires-Tournois interferometer (GTI) (Gires and Tournois, 1964), which is a Fabry-Perot interferometer operated in reflection. Because the rear mirror is highly reflective, the GTI as a whole is highly reflective over the whole wavelength range, whereas the phase shift varies nonlinearly by 2π for each free spectral range, calculated as $\Delta v = c/2nd$, where n and d are refractive index and the thickness of the spacer material, respectively. Within each free spectral range, the GDD oscillates between two extremes the magnitude of which is proportional to d^2 and also depends on the front mirror reflectivity. Ideally, the GTI is operated near a minimum of the GDD, and the usable bandwidth is some fraction (e.g., one-tenth) of the free spectral range, which is proportional to d^{-1}. Tunable GDD can be achieved if the spacer material is a variable air gap, which, however, must be carefully stabilized to avoid unwanted drifts. More stable but not tunable GDD can be generated with monolithic designs, based, e.g., on thin films of dielectric media such as TiO_2 and SiO_2, particularly for use in femtosecond lasers. The main drawbacks of GTI are the fundamentally limited bandwidth (for a given amount of GDD) and the limited amount of control of higher order dispersion.

Dispersive Mirrors

Dielectric Bragg mirrors with regular $\lambda/4$ stacks have a fairly small dispersion when operated well within their reflection bandwidth, but increasing dispersion at the edges of this range. Modified designs can be used to obtain well-controlled dispersion over a large wavelength range. One possibility already discussed is the use of a GTI structure (see preceding subsection). Another broad range of designs are based on the concept of the chirped mirror: If the Bragg wavelength is appropriately varied within a Bragg mirror design, longer wavelengths can be reflected deeper in the structure, thus acquiring a larger phase change, which leads to negative dispersion. However, the straightforward implementation of this idea leads to strong oscillations of the GDD (as a function of frequency), which render such designs useless. These oscillations can be greatly reduced by numerical optimizations that introduce complicated (and not analytically explainable) deviations from the simple chirp law. A great difficulty is that the figure of merit to optimize is a complicated function of many layer thickness variables; it typically has a large number of local extremes and thus is quite difficult to optimize. Nevertheless, refined computing algorithms led to designs with respectable performance that were realized with precision growth of dielectric mirrors. Such mirrors can compensate for the dispersion in Ti^{3+}:sapphire lasers for operation with pulse durations well below 10 fs (Jung et al., 1997). Further analytical studies led to the design of double-chirped mirrors, where both the Bragg wavelength and the coupling strength (determined by the ratio of thickness for low- and high-index layers) are varied, and a broadband high quality antireflection coating on the front surface is used in addition (Kärtner et al., 1997). Because such analytically found designs are already close to optimum, a slight refinement by local numerical optimization is sufficient to arrive at very broadband designs. Mirrors based on such designs have been used in Ti^{3+}:sapphire lasers to generate pulses with durations of <6 fs (Morgner et al., 1999b; Sutter et al., 2000), which are the shortest pulses ever generated directly from a laser.

Dispersive SESAMs

Negative dispersion can also be obtained from semiconductor saturable absorber mirrors (SESAMs) (see Sec. 1.4.3) with specially modified designs. The simplest option is to use a GTI-like structure (see earlier in this section) (Kopf et al., 1996a). A double-chirped dispersive semiconductor mirror has also been demonstrated (Paschotta et al., 1999b), and a saturable absorber could be integrated into such a device.

1.3.3 Kerr Nonlinearity

Because of the high intracavity intensities, the Kerr effect is relevant in most ultrafast lasers. The refractive index. of, e.g., the gain medium is modified according to

$$n(I) = n_0 + n_2 I \tag{3}$$

where I is the laser intensity and n_2 a material-dependent coefficient that also weakly depends on the wavelengths. (The pump intensity is normally ignored because it is far smaller than the peak laser intensity.) These nonlinear refractive index changes have basically two consequences. The first is a transverse index gradient resulting from the higher intensities on the beam axis compared to the intensities in the wings of the transverse beam profile. This leads to a so-called Kerr lens with an intensity-dependent focusing effect (for positive n_2) that can be exploited for a passive modelocking mechanism as discussed in Section 1.4.3.

The second consequence of the Kerr effect is that the pulses experience self-phase modulation (SPM): The pulse center is delayed more (for positive n_2) than the temporal wings. For a freely propagating Gaussian transverse beam profile of radius w (defined so that at this radius we have $1/e^2$ times the peak intensity), the nonlinear coefficient γ_{SPM}, related the on-axis phase change φ to the pulse power P according to $\varphi = \gamma_{SPM} P$, is given by

$$\gamma_{SPM} = \frac{2\pi}{\lambda} n_2 \left(\frac{\pi w^2}{2} \right)^{-1} L = \frac{4 n_2 L}{\lambda w^2} \tag{4}$$

where L is the propagation length in the medium. Note that the peak intensity of a Gaussian beam is $I = P/(\pi w^2/2)$, and the on-axis phase change (and not an averaged phase change) is relevant for freely propagating beams. For guided beams, an averaged phase change has to be used, which is two times smaller.

The most important consequences of SPM in the context of ultrafast lasers is the possibility of soliton formation (Sec. 1.3.4). Another important aspect is that SPM can increase or decrease the bandwidth of a pulse, depending on the original phase profile of the pulse. In general, positive SPM tends to increase the bandwidth of positively chirped (i.e., up-chirped) pulses, whereas negatively chirped pulses can be spectrally compressed. An originally unchirped pulse traveling through a nonlinear a nondispersive medium will experience an increase of bandwidth only to second order of the propagation distance, whereas a chirp grows in first order.

1.3.4 Soliton Formation

If a pulse propagates through a medium with both second-order dispersion (GDD) and a Kerr nonlinearity, the two effects can interact in complicated ways. A special case is that the intensity has a sech^2 temporal profile

$$P(t) = P_p \, \text{sech}^2\left(\frac{t}{\tau_S}\right) = \frac{P_p}{\cosh^2(t/\tau_S)} \tag{5}$$

with the peak power P_p and the FWHM pulse duration $\tau_{\text{FWHM}} \approx 1.76\tau_S$. If such a pulse is unchirped and fulfills the condition

$$\tau_S = \frac{2|\text{GDD}|}{|\gamma_{\text{SPM}}|E_p} \tag{6}$$

where GDD and γ_{SPM} have opposite signs and are calculated for the same propagation distance, and

$$E_p \approx 1.13 P_p \tau_{\text{FWHM}} \tag{7}$$

is the pulse energy, then we have a so-called fundamental soliton. Such a pulse propagates in the medium with constant temporal and spectral shape and acquires only an overall nonlinear phase shift. Higher order solitons, where the peak power is higher by a factor that is the square of an integral number, do not preserve their temporal and spectral shape but evolve in such a way that the original shape is restored after a certain propagation distance, the so-called soliton period in the case of a second-order soliton.

Solitons are remarkably stable against various kinds of distortions. In particular, stable soliton-like pulses can be formed in a laser cavity even though dispersion and Kerr nonlinearity occur in discrete amounts and the pulse energy varies due to amplification in the gain medium and loss in other elements. As long as the soliton period amounts to many (at least about five to ten) cavity round-trips, the solition simply "sees" the average GDD and Kerr nonlinearity, and this "average solition" behaves in the same way as in a homogeneous medium. The soliton period in terms of the number of cavity round-trips is

$$N_S = \frac{\pi \tau_S^2}{2|\text{GDD}|} \approx \frac{\tau_{\text{FWHM}}^2}{2|\text{GDD}|} \tag{8}$$

where GDD is calculated for one cavity round-trip. N_S is typically quite large in lasers with pulse durations of $> 100\,\text{fs}$, so the average soliton is a good approximation. Once N_S becomes less than about 10, the soliton is significantly disturbed by the changes of dispersion and nonlinearity during a round-trip, and this may lead to pulse break-up. In Ti^{3+}:sapphire lasers for pulse durations below 10 fs, the regime of small N_S is unavoidable and can be stabilized only by using a fairly strong saturable absorber. On the other hand, in cases with very large values of N_S it can be beneficial to decrease N_S by increasing both |GDD| and γ_{SPM}, because stronger soliton shaping can stabilize the pulse shape and spectrum and make the pulse less dependent on other influences.

Note that soliton effects can fix the pulse duration at a certain value even if other cavity elements (most frequently the laser gain with its limited bandwidth) tend to reduce the pulse bandwidth. The pulse will then acquire a positive chirp (assuming $n_2 > 0$ for the Kerr medium), and under these conditions SPM can generate the required extra bandwidth.

1.4 MODELOCKING

1.4.1 General Remarks

Ultrashort light pulses are in most cases generated by modelocked lasers. In this regime of operation, usually a single short pulse propagates in the laser cavity and generates an output pulse each time it hits the output coupler mirror. The generated pulses are usually quite short compared to the round-trip time.

In the frequency domain, modelocking means operation of the laser on a number of axial cavity modes, whereby all these modes oscillate in phase (or at least with nearly equal phases). In this case, the mode amplitudes interfere constructively only at certain times, which occur with the period of the round-trip time of the cavity. At other times, the output power is negligibly small. The term "modelocking" resulted from the observation that a fixed phase relationship between the modes has to be maintained in some way to produce short pulses. The achievable pulse duration is then inversely proportional to the locked bandwidth, i.e., to the number of locked modes time their frequency spacing.

It is obvious in the domain description that modelocking cannot be achieved if a significant amount of the laser power is contained in higher order transverse modes of the cavity, because these usually have different resonance frequencies, so that the periodic recurrence of constructive addition of all mode amplitudes is not possible. Therefore, laser operation on the fundamental transverse cavity mode (TEM_{00}) is usually a prerequisite to stable modelocking.

In some cases, a modelocked laser is operated with several pulses circulating in the cavity. This mode of operation, called "harmonic modelocking," can be attractive for the generation of pulse trains with higher repetition rates. The main difficulty is that the timing between the pulses has to be maintained in some way, either by some kind of interaction between the pulses or with the aid of externally applied timing information. A number of solutions have been found, but in this chapter we concentrate on fundamental modelocking, with a single pulse in the laser cavity, as it occurs in most ultrafast lasers.

So far we have assumed that the modelocked laser operates in a steady state, where the pulse energy and duration may change during a round-trip (as an effect of gain and loss, dispersion, nonlinearity, etc.) but always return to the same values at a certain position in the cavity. This regime is called cw (continuous-wave) modelocking and is indeed very similar to ordinary cw (not modelocked) operation, because the output power in each axial cavity mode is constant over time. Another important regime is Q-switched modelocking (Sec. 1.4.3), where the modelocked pulses are contained in periodically recurring bunches that have the envelope of a Q-switched pulse. It is in some cases a challenge to suppress an unwanted tendency for Q-switched modelocking.

In addition to Q-switching tendencies, a number of other mechanisms can destabilize the modelocking behavior of a laser. In particular, short pulses have a broader bandwidth than, e.g., a competing cw signal and thus tend to experience less gain than the latter. For the pulses to be stable, some mechanism is required that increases the cavity loss for cw signals more than for pulses. Also, a modelocking mechanism should ideally give a clear loss advantage for shorter pulses compared to any other mode of operation. We discuss various stability criteria in the following sections.

Mechanisms for modelocking are grouped into active and passive schemes and hybrid schemes that utilize a combination of the two. Active modelocking (Sec. 1.4.2) is achieved with an active element (usually an acousto-optic modulator) generating a loss modulation that is precisely synchronized with the cavity round-trips. Passive schemes (Sec. 1.4.3) rely on a passive loss modulation in some type of saturable absorber. This passive loss modulation can occur on a much faster time scale, so that passive modelocking typically allows for the generation of much shorter pulses. Passive schemes are also usually simpler, because they do not rely on driving circuits and synchronization electronics. The pulse timing is then usually not externally controlled. Synchronization of several lasers is more easily achieved with active modelocking schemes.

Modelocked lasers are in most cases optically pumped with a cw source, e.g., one or more laser diodes. This requires a gain medium that can store the excitation energy over a time of more than one cavity round-trip (which typically takes a few nanoseconds). Typical solid-state gain materials fulfill this condition very well, even for rather low repetition rates, because the lifetime of the upper laser level is usually at least a few microseconds, in some cases even more than a millisecond. Synchronous pumping (with a modelocked source), as is often used for dye lasers, is therefore rarely applied to solid-state lasers and is not discussed in this chapter.

In the following sections we discuss some details of active and passive modelocking. The main emphasis is on passive modelocking, because such techniques clearly dominate in ultrafast optics.

1.4.2 Active Modelocking

An actively modelocked laser contains some kind of electrically controlled modulator, in most cases of acousto-optic and sometimes of electro-optic type. The former uses a standing acoustic wave with a frequency of tens or hundreds of megahertz, generated with a piezoelectric transducer in an acousto-optic medium. In this way one obtains a periodically modulated refractive index pattern at which the laser light can be refracted. The refracted beam is normally eliminated from the cavity, and the refraction loss oscillates with twice the frequency of the acoustic wave because the refractive index change depends only on the modulus of the pressure deviation.

A pulse circulating in the laser cavity is hardly affected by the modulator provided that it always arrives at those times when the modulator loss is at its minimum. Even then, the temporal wings of the pulse experience a somewhat higher loss than the pulse center. This mechanism tends to shorten the pulses. On the other hand, the limited gain bandwidth always tends to reduce the pulse bandwidth and thus to increase the pulse duration. Other effects such as dispersion or self-phase modulation (Sec. 1.3) can have additional influences on the pulse. Usually, a steady state is reached where the mentioned competing effects exactly cancel each other for every complete cavity round-trip.

The theory of Kuizenga and Siegman (Kuizenga and Siegman, 1970) describes the simplest situation with a modulator (in exact synchronism with the cavity) and gain filtering but no dispersion or self-phase modulation. The gain is assumed to be homogeneously broadened and of Lorentzian shape. The simple result is the steady-state FWHM pulse duration

$$\tau \approx 0.45 \left(\frac{g}{M}\right)^{1/4} (f_m \, \Delta f_g)^{-1/2} \tag{9}$$

where g is the power gain at the center frequency, M is the modulation strength ($2M$ = peak-to-peak variation of power transmission), f_m is the modulation frequency, and Δf_g is the FWHM gain bandwidth. The peak gain g is slightly higher than the cavity loss l (for minimum modulator loss) according to

$$g = l + \frac{M}{4} (2\pi f_m \tau)^2 \tag{10}$$

so that the total cavity losses (including the modulator losses) are exactly balanced for the pulse train. Note that the losses would be higher than the gain for either a longer pulse (with higher modulator loss) or a shorter pulse (with stronger effect of gain filtering).

It is important to note that for non-Lorentzian gain spectra the quantity Δf_g should be defined such that the corresponding Lorentzian reasonably fits the gain spectrum within the pulse bandwidth, which may be only a small fraction of the gain bandwidth; only the "curvature" of the gain spectrum within the pulse bandwidth is important. This means that even weak additional filter effects, caused, e.g., by intracavity reflections (etalon effects), can significantly reduce the effective bandwidth and thus lead to longer pulses. Such effects are usually eliminated by avoiding any optical interfaces that are perpendicular to the incident beam. Also note that the saturation of inhomogeneously broadened gain can distort the gain spectrum and thus affect the pulse duration.

With SPM in addition (but no dispersion), somewhat shorter pulses can be generated, because SPM can generate additional bandwidth. However, with too much SPM the pulses become unstable. It has been shown with numerical simulations (Haus et al., 1991) that a reduction of the pulse duration by a factor on the order of 2 is possible. The resulting pulses are then chirped, so that some further compression with extracavity negative dispersion is possible.

The situation is quite different if positive SPM occurs together with negative GDD, because in this case soliton-like pulses (Sec. 1.3.4) can be formed. If the soliton length corresponds to at least about 10 but less than 100 cavity round-trips, soliton-shaping effects can be much stronger than the pulse-shaping effect of the modulator. The pulse duration is then determined only by the soliton equation, Eq. (6). The solition experiences only a very small loss at the modulator. However, owing to its bandwidth it also experiences less gain than, e.g., a long background pulse, the so-called continuum, which can have a small bandwidth. The shorter the soliton pulse duration, the smaller the soliton gain compared to the continuum gain and the higher the modulator strength M must be to keep the soliton stable against growth of the continuum. Using results from Kärtner et al. (1995a), it can be shown that the minimum pulse duration is proportional to

$$\frac{g^{1/3}}{\Delta f_g^{2/3}} \left(\frac{N_S}{M} \right)^{1/6} \tag{11}$$

where N_S is the soliton period in terms of cavity round-trips. This indeed shows that a smaller modulator strength is sufficient if N_S is kept small, but also that the cavity loss should be kept small (to keep g small) and the gain bandwidth is the most important factor.

With gain media like Nd^{3+}:YAG or Nd^{3+}:YLF, typical pulse durations of actively modelocked lasers are a few tens of picoseconds. The minimum appears to be 6.2 ps (Walker et al., 1990). Using Nd^{3+}:glass and the regime of soliton formation, pulse durations of around 7 ps have been achieved (Yan et al., 1986). For significantly shorter pulses, active modelocking is not effective because the time window of low modulator loss would be much longer than the pulse duration.

1.4.3 Passive Modelocking

The Starting Mechanism

Passive modelocking relies on the use of some type of saturable absorber, which favors the generation of a train of short pulses against other modes of operation such as cw emission. Starting from a cw regime, the saturable absorber will favor any small noise spikes, so those can grow faster than the cw background. Once these noise spikes contain a significant part of the circulating energy, they saturate the gain so that the cw background starts to decay. Later on, the most energetic noise spike, which experiences the least amount of saturable absorption, will eliminate all the others by saturating the gain to a level where these experience net loss in each round-trip. In effect we obtain a single circulating pulse. Owing to the action of the saturable absorber, which favors the peak over the wings of the pulse, the duration of the pulse is then reduced further in each cavity round-trip, until broadening effects (e.g., dispersion) become strong enough to prohibit further pulse shortening. Note that other shortening effects, e.g., soliton-shaping effects, can also become effective.

The described start-up can be prevented if strong pulse-broadening effects are present in an early phase. In particular, the presence of spurious intracavity reflections can be significant, because those tend to broaden (or split up) pulses even before they have acquired a significant bandwidth. A significantly stronger saturable absorber may then be required to get self-starting modelocking. These effects are usually difficult to quantify but can be suppressed in most bulk solid-state lasers by using suitable designs. For example, even antireflection-coated surfaces in the cavity should be slightly tilted against the laser beam so that beams resulting from residual reflections are eliminated from the cavity.

Note that a saturable absorber with long recovery time (low saturation intensity) is most effective for fast self-starting modelocking, although a short recovery time may allow the generation of shorter pulses. Other techniques to facilitate self-starting modelocking include the use of optical feedback from a moving mirror (Smith, 1967), which tends to increase the intracavity fluctuations in cw operation.

Parameters of Fast and Slow Saturable Absorbers

An important parameter of a saturable absorber is its recovery time. In the simplest case, we have a so-called fast saturable absorber, which can recover on a faster time scale than the pulse duration. In this case the state of the absorber is largely determined by the instantaneous pulse intensity, and strong shaping of the leading edge as well as the trailing edge of the pulse takes place. However, for short enough pulse durations we have the opposite situation of a slow absorber, with the absorber recovery occurring on a long time scale compared to the pulse duration. This regime is frequently used in ultrafast lasers, because the choice of fast saturable absorbers is very limited for pulse durations below 100 fs. We discuss in the following subsection why it is possible that pulses with durations well below 100 fs can be generated even with absorbers that are much slower.

Concrete types of saturable absorbers are compared in the following subsection. Here we discuss some parameters that can be used to quantitatively characterize the action of saturable absorbers.

The intensity loss* q generated by a fast saturable absorber depends only on the instantaneous intensity I, the incident power divided by the mode area. Simple absorber models lead to a function

$$q(I) = \frac{q_0}{1 + I/I_{\text{sat}}} \qquad (12)$$

which is a reasonable approximation in many cases. The response is then characterized by the parameter I_{sat}, called the saturation intensity, and the unsaturated loss q_0. For pulses with a given peak intensity I_p, an average value of q can be calculated that represents the effective loss for the pulse. Figure 2 shows this quantity as a function of the normalized peak intensity for Gaussian and solition (sech^2) pulses, together with $q(I_p)$. We see that the pulse form has little influence on the average loss.

The behavior of a slow saturable absorber is described by the differential equation

---

*Note that some authors define q as the amplitude (instead of intensity) loss coefficient, which is one-half the value used here.

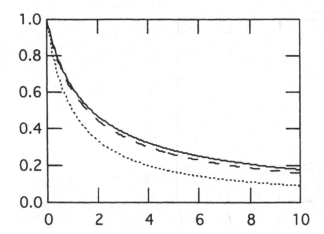

Figure 2 Solid curve: Effective loss for a soliton pulse on a fast saturable absorber, as a function of the ratio of peak intensity to the saturation intensity. Dashed curve: Same as for a Gaussian pulse. Dotted curve: Loss for the peak intensity.

$$\frac{dq}{dt} = -\frac{q - q_0}{\tau} - \frac{I}{F_{sat}} q \tag{13}$$

with the recovery time τ, the unsaturated loss q_0, and the saturation fluence F_{sat}. If the recovery is so slow that we can ignore the first term, then the value of q after a pulse with fluence F_{sat} is $q_0 \exp(-F_p/F_{sat})$ if the pulse hits an initially unsaturated absorber. The effective loss for the pulse is (independent of the pulse form)

$$q_p(F_p) = q_0 \left[1 - \exp\left(\frac{-F_p}{F_{sat}}\right) \right] \left(\frac{F_{sat}}{F_p}\right) = q_0 \frac{1 - \exp(-S)}{S} \tag{14}$$

with the saturation parameter $S := F_p/F_{sat}$. For strong saturation (as usual in modelocked lasers) ($S > 3$), the absorbed pulse fluence is $\approx F_{sat} \Delta R$, and we have

$$q_p(F_p) \approx q_0/S \tag{15}$$

Figure 3 shows a plot of this function, compared to the loss after the pulse. It is important to observe that the loss after the pulse gets very small for $S > 3$, whereas the average loss q_p for the pulse is still significant then.

Another important parameter for any saturable absorber is the damage threshold in terms of the applicable pulse fluence or intensity. Note that the absolute value of the damage threshold is actually less relevant than the ratio of the damage fluence and the saturation fluence, because the latter determines the typical operating parameters.

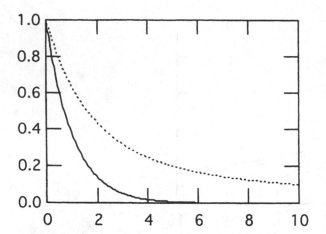

Figure 3 Dotted curve: Loss for a pulse on a slow saturable absorber, as a function of the saturation parameter (pulse energy divided by saturation energy). Solid curve: Loss after the pulse.

Passive Modelocking with Fast and Slow Saturable Absorbers

Here we give some guidelines as to what pulse durations can be expected from passively modelocked lasers. First we consider cases with a fast saturable absorber and without significant influence of dispersion and self-phase modulation. We also assume that significant gain saturation does not occur during a pulse. For this situation, which can occur particularly in picosecond solid-state lasers modelocked with SESAMs (see later), analytical results have been obtained (Haus et al., 1991). These calculations are strictly valid only for weak absorber saturation, which is not the desired case, but numerical simulations show that for a fully saturated absorber the obtained pulse duration can be estimated from

$$\tau \approx \frac{0.9}{\Delta f_g} \sqrt{\frac{g}{\Delta R}} \tag{16}$$

where Δf_g is th FWHM gain bandwidth (assuming that a Lorentzian shape fits the gain spectrum well within the range of the pulse spectrum; see Sec. 1.4.2), g is the power gain coefficient (which equals the overall cavity losses), and ΔR is the modulation depth (maximum reflectivity change) of the absorber. Compared to the equation from the analytical results, we have increased the constant factor by about 10% because the analytical calculations are not accurate for a fully saturated gain and numerical simulations (which do not need to approximate the gain saturation with a linear function) result in typically $\approx 10\%$ longer pulse durations. Note that significantly weaker or stronger absorber saturation results in longer pulses. If we introduce some self-phase modulation, the pulse duration can be somewhat reduced because this effect helps to increase the pulse bandwidth. However, the dynamics become unstable if too much self-phase modulation occurs. The same holds for phase changes that can arise in an absorber when it is saturated.

Note that an increase in ΔR also increases the required gain g, so that a value of ΔR greater than the linear cavity losses does not significantly reduce the pulse duration. The limit for the pulse duration is on the order of $1/\Delta f_g$.

With a slow saturable absorber, somewhat longer pulse durations are obtained. Without significant influence of dispersion and self-phase modulation, we can estimate the pulse duration (Paschotta and Keller, 2001) with

$$\tau \approx \frac{1.07}{\Delta f_g} \sqrt{\frac{g}{\Delta R}} \tag{17}$$

This equation, which is very similar to Eq. (16), is an empirical fit to results from numerical simulations. It holds if the absorber is operated at roughly 3–5 times the saturation fluence. For significantly weaker or stronger absorber saturation, the pulse duration gets somewhat longer. In contrast to the situation for fast absorbers, the influence of self-phase modulation always makes the pulses longer, apart from the instability occurring when the effect is too strong. Particularly in femtosecond lasers, it can be difficult to make the nonlinearity weak enough. In this case, soliton mode locking (see below) is a good solution. Here, a significant nonlinear phase shift is even desirable for good stability.

The pulse duration obtained with a slow saturable absorber can be significantly shorter than the absorber recovery time; but in such cases the stability and pulse quality can be somewhat reduced because a weak pedestal can grow behind the pulse. This result may seem quite surprising because a slow absorber can clean up only the leading part of the pulse, not the trailing part. However, only a little energy can be fed into the trailing part as long as dispersion effects are weak. Also, the pulse is constantly delayed because the absorber attenuates the leading part; in effect, the pulse tends to "eat up" the trailing part, which is not delayed (Paschotta and Keller, 2001).

In the femtosecond regime, many absorbers (including SESAMs) become slow compared to the pulse duration. The question arises as to how stable modelocked operation can be achieved with a pulse duration much shorter than the absorber recovery time. In dye lasers, it was found that gain saturation during each pulse can very much shorten the time window with effective gain: After the passage of the pulse, the absorber stays saturated for a while, but the gain is also exhausted and may need some longer time to recover. For this reason, dye lasers can be used to generate pulses with durations down to $\approx 27\,\text{fs}$ (Valdmanis et al., 1985) [with external compression, even 6 fs (Fork et al., 1987)], which is much less than the absorber recovery time (typically on the order of 0.1 ns to a few nanoseconds for dye absorbers). In ion-doped solid-state gain media, this principle cannot be used because such gain media have relatively small laser cross sections, and the gain saturation fluence is much larger than the achievable pulse fluence. Therefore, gain saturation occurs only on a time scale on the order of the fluorescence decay time and is caused by the integral effect of many pulses, whereas gain saturation by a single pulse is very weak. Nevertheless it has been shown that even from solid-state media one can obtain pulses that are much shorter than the absorber recovery time. This is particularly the case if the mechanism of soliton modelocking (Kärtner and Keller, 1995) is employed. In the regime of negative intracavity dispersion, soliton effects (see Sec. 1.3.4) can fix the pulse duration at a certain value. The saturable absorber is then required only to start the modelocking process and to stabilize the solitons. The latter means that two types of instability must be suppressed. First, a competing cw background (or long pulse pedestal) can have less bandwidth and thus experience more gain than the soliton, but this can be suppressed by the saturable absorber because this introduces a higher loss for the background. Second, pulse breakup could occur because a breakup into two solitons with, e.g., half the energy would increase the gain (because of the reduced pulse bandwidth); the absorber must prevent this by having a higher loss for two pulses compared to the single pulse. It turns out that the limit for the pulse duration depends only weakly on the modulation depth and recovery time of the absorber.

Note that apart from the shortest achievable pulse duration, selfstarting modelocking is also a desirable goal. In this respect, a slow absorber is usually superior, because it has a lower saturation intensity, which facilitates the modelocking process in an early phase, as discussed earlier in this section.

Saturable Absorbers for Passive Modelocking

In this section we discuss a number of different saturable absorbers (or artificial saturable absorbers) for passive modelocking, concentrating on the most important techniques for ultrashort-pulse generation.

Semiconductor Absorbers. A semiconductor can absorb light if the photon energy is sufficient to excite carriers from the valence band to the conduction band. Under conditions of strong excitation, the absorption is saturated because possible initial states of the pump transition are depleted while the final states are partially occupied. Within typically 60–300 fs after the excitation, the carriers in each band thermalize, and this leads to a partial recovery of the absorption. On a longer time scale—typically between a few picoseconds and a few nanoseconds—they will be removed by recombination and trapping. Both processes can be used for modelocking of lasers.

A few percent of the incident radiation can be absorbed in a semiconductor layer of only a few nanometers thickness. Such a structure is called a quantum well because for structures with such small dimensions the density of states is modified by quantum effects. Quantum well absorbers or somewhat thicker absorbers (called bulk absorbers) are usually integrated into a microstructure with different surrounding semiconductor materials that are not absorbing due to a larger band gap. Typically, such structures are grown on a semiconductor wafer ≈ 0.5 mm thick (of GaAs, e.g.) and contain a Bragg mirror structure so they can be operated in reflection. In some cases, the Bragg mirror is replaced by a metallic mirror with broader bandwidth (Jung et al., 1997). In any case, such a structure has been called a semiconductor saturable absorber mirror (SESAM) since its first use for modelocking in 1992 (Keller et al., 1992). This acronym includes all types of saturable semiconductor absorbers that are operated in reflection. Most SESAMs contain a Bragg mirror, and if the absorber is incorporated into the top part of this Bragg mirror, the term saturable Bragg reflector (SBR) introduced in 1995 (Tsuda et al., 1995), is also sometimes used.

Most SESAMs that are used nowadays consist of a semiconductor Bragg mirror and a quantum well or bulk absorber, usually embedded between two nonabsorbing semiconductor layers grown on top of this mirror. The semiconductor air interface has a reflectivity of $\approx 30\%$, which together with the Bragg mirror forms a Fabry-Perot structure. In most cases, the SESAM is designed so that this Fabry-Perot is antiresonant at the operation wavelengths. In this regime, the largest possible bandwidth is achieved, and the saturation fluence is on the order of $100 \, \mu J/cm^2$. (The absorber layer itself would have a saturation fluence on the order of $30 \, \mu J/cm^2$.) We call such a device a low-finesse SESAM. A higher finesse of the Fabry-Perot can be obtained with a dielectric top mirror. Such devices were initially used for modelocking of lasers (Keller et al., 1992) and were termed antiresonant Fabry-Perot saturable absorbers (A-FPSAs). Compared to the same SESAM without a top reflector, the high-finesse SESAM has a greater saturation fluence ($> 100 \, \mu J/cm^2$) and an accordingly smaller modulation depth (which can often be compensated for with a thicker absorber layer, if required). The increased damage fluence is no advantage because such samples have to be operated at accordingly higher fluence values. For use in high-power lasers, low-finesse SESAMs operated with an accordingly larger laser spot size and lower pulse fluence are preferable (Paschotta et al., 2000a) because the generated heat is distributed over a larger area.

Most modelocked lasers are operated in wavelength regimes of around 0.71–0.9 μm (e.g., Ti^{3+}:sapphire) and 1.03–1.07 μm (e.g., Nd^{3+}:YAG, Nd^{3+}:glass, Yb^{3+}:YAG). For 0.8 μm, GaAs absorbers can be used together with Bragg mirrors made of AlAs and $Ga_{x}Al_{1-x}As$ (where the gadolinium content x is kept low enough to avoid absorption). For wavelengths around 1.03–1.064 μm, the Bragg mirrors can be made of GaAs and AlAs, and the absorber is made of $In_{x}Ga_{1-x}As$, where the indium content x is adjusted to obtain a small enough band gap. Because $In_{x}Ga_{1-x}As$ cannot be grown lattice-matched to GaAs, increasing the indium content leads to increasing strain, which can lead to surface degradation by cracking and thus limits the absorber thickness. Nevertheless, such samples have been applied for mode-locking of lasers operating around 1.3 μm (Fluck et al., 1996) and even 1.5 μm (Spühler et al., 1999a). SESAMs for 1.3 and 1.5 μm have also been developed on the basis of InGaAsP grown on InP substrates.

SEASAMs can be produced either with molecular beam epitaxy (MBE) or with metal-organic chemical vapor deposition (MOCVD). The MOCVD process is faster and thus appears to be most suitable for mass production. It also leads to relatively small nonsaturable losses, so that such SESAMs are well suited for high-power operation. Similarly low losses can be achieved with MBE. However, MBE gives us the additional flexibility to grow semiconductors at lower temperatures, down to ≈200°C, whereas MOCVD is usually done at ≈600°C. Lower growth temperatures lead to microscopic structural defects that act as traps for excited carriers and thus reduce the recovery time, which can be beneficial for use in ultrafast lasers. However, the nonsaturable losses of such samples also increase with decreasing growth temperature. This compromise between speed of recovery and quality of the surfaces can be improved by optimized annealing procedures or by beryllium doping (Haiml et al., 1999a, 1999b).

The presence of two different time scales in the SESAM recovery, resulting from intraband thermalization and carrier trapping and/or recombination, can be rather useful for modelocking. The longer time constant results in a reduced saturation intensity for part of the absorption, which facilitates self-starting modelocking, whereas the faster time constant is more effective in shaping subpicosecond pulses. Therefore, SESAMs make it possible to easily obtain self-starting modelocking in most cases.

SESAMs are quite robust devices. The ratio of the damage fluence to the saturation fluence is typically on the order of 100, so that a long device lifetime can be achieved with the usual parameters of operation (pulse fluence ≈ 3–10 times the saturation fluence). For high-power operation, the thermal load can be substantial, but the relatively good heat conductivity, e.g., of GaAs and the geometry of a thin disk allow for efficient cooling through the back side, which is usually soldered to a copper heat sink. The latter may be water-cooled in multiwatt lasers. Once the mode diameter is larger than the substrate thickness (typically 0.5 mm), further power scaling by increasing the mode area and keeping the pulse fluence constant will not substantially increase the temperature rise (Paschotta et al., 2000a). The temperature rise can be limited to below 100 K by the use of SESAM designs with low saturation fluence (100 μJ/cm^2 or less), because these allow operation with a larger mode area. SESAM damage can thus be avoided by using suitable designs and operation parameters, and further power scaling is not prohibited by such issues.

Recently, a new type of semiconductor-based saturable absorber, based on InAs semiconductor nanoparticles in silica, was demonstrated (Bilinsky et al., 1999a, 1999b). The fabrication technique may be well suitable for cheap mass production. Using such an absorber, modelocking of a Ti^{3+}:sapphire laser with 25 fs pulse duration was obtained. A disadvantage of these absorber devices is their high saturation fluence (25 mJ/cm^2), at least about two orders of magnitude higher than for typical SESAMs. For this reason, strong focusing on the absorber is necessary, and high powers can probably not be handled.

Kerr Lens Modelocking. Very fast effective saturable absorbers, suitable for the generation of pulses with durations below 10 fs, can be implemented using the Kerr effect. In typical gain media, the Kerr effect (i.e., the dependence of the refractive index on the light intensity) has a time constant on the order of at most a few femtoseconds. The intensity gradients across the transverse mode profile, e.g., in the gain medium lead to a Kerr lens with intensity-dependent focusing power. This can be combined with an aperture to obtain an effective saturable absorber, which can be used for Kerr lens modelocking (KLM) (Keller et al., 1991; Negus et al., 1991; Salin et al., 1991). For example, a pinhole at a suitable location in the laser cavity leads to significant losses for cw operation but reduced losses for short pulses for which the beam radius at the location of the pinhole is reduced by the Kerr lens. This is called hard aperture KLM. Another possibility [soft aperture KLM (Piché et al., 1993)] is to use a pump beam that has a significantly smaller beam radius than the laser beam in the gain medium. In this situation, the effective gain is higher for short pulses if the Kerr lens reduces their beam radius in the gain medium, because then the pulses have a better spatial overlap with the pumped region.

In any case, KLM requires the use of a laser cavity that reacts relatively strongly to changes in the focusing power of the Kerr lens. This is usually achieved only by operating the laser cavity near one of the stability limits, possibly with detrimental effects on the long-term stability.

Detailed modeling of the KLM action requires sophisticated three-dimensional simulation codes, including both the spatial dimensions and the time variable (Cerullo et al., 1996; Christov et al., 1995; Rothenberg, 1992). The basic reason for this is that the strength of the Kerr lens varies within the pulse duration, which leads to complicated spatiotemporal behavior. However, a reasonable understanding can be obtained simply by calculating the transverse cavity modes with a constant Kerr lens, using some averaged intensity value.

Self-starting modelocking with KLM can be achieved with specially optimized laser cavities (Chen et al., 1991; Krausz et al., 1991; Cerullo et al., 1994a, 1994b; Lai, 1994). Unfortunately, this optimization is in conflict with the goal of generating the shortest possible pulses. This is because the laser intensity changes by several orders of magnitude during the transition from cw to mode-locked operation, and the Kerr effect for very short pulses is too strong when the cavity is optimized to react sensitively to long pulses during the start-up phase. Basically one would face a similar problem with any fast saturable absorber, whereas a slow saturable absorber is less critical in this respect because its action depends on pulse energy and not intensity.

Generation of very short pulses—below 6 fs duration—with a self-starting Ti^{3+}:sapphire laser has become possible by combining a SESAM with a cavity set up for KLM (Jung et al., 1997a). The SESAM is then responsible for fast self-starting and the pulse shaping in the femtosecond domain is mainly done by the Kerr effect.

Additive Pulse Modelocking. Before the invention of Kerr lens mode-locking, the Kerr effect was used in a different technique called "additive pulse modelocking" (APM) (Kean et al., 1989; Ippen et al., 1989). A cavity containing a single-mode glass fiber is coupled to the main laser cavity. In this cavity, the Kerr effect introduces a larger nonlinear phase shift for the temporal pulse center compared to the wings. When the pulses from both cavities meet again at the coupling mirror, they interfere in such a way that the pulse center is constructively enhanced in the main cavity while its wings are reduced by destructive interference. A prerequisite for this to happen is, of course, that the cavity lengths be equal and stabilized inter-ferometrically. Because other methods (e.g., KLM or the use of saturable absorbers) do not need such stabilization, they are nowadays usually preferred to APM, although APM has been proven to be very effective, particularly in picosecond lasers. For example, the shortest pulses (1.7 ps) from a Nd^{3+}:YAG laser have been achieved with APM (Goodberlet et al., 1990).

A variant of APM that is no longer used is resonant passive modelocking (RPM) (Keller et al., 1990b; 1992b). As with APM, a coupled cavity is used, but this contains a saturable absorber (i.e., an intensity nonlinearity) instead of a Kerr nonlinearity. With this methods, 2 ps pulses were generated with a Ti:sapphire laser (Keller et al., 1990b) and 3.7 ps pulses in a Nd:YLF laser (Keller et al., 1992b).

Nonlinear Mirror Modelocking. Effective saturable absorbers can also be constructed by using $\chi^{(2)}$ nonlinearities (Stankov, 1988). A nonlinear mirror based on this principle consists of a frequency-doubling crystal and a dichroic mirror. For short pulses, part of the incident laser light is converted to the second harmonic, for which the mirror is highly reflective, and converted back to the fundamental wave if an appropriate relative phase shift for fundamental and second harmonic light is applied. On the other hand, unconverted fundamental light experiences a significant loss at the mirror. Thus the device has a higher reflectivity at higher intensities. This has been used for modelocking with, e.g., up to 1.35 W of average output power in 7.9 ps pulses from a Nd^{3+}:YVO_4 laser (Agnesi et al., 1997). The achievable pulse duration is often limited by group velocity mismatch between fundamental and second harmonic light.

Q-Switching Instabilities

The usually desired mode of operation is cw modelocking, where a train of pulses with constant parameters (energy, duration, shape) is generated. However, the use of saturable absorbers may also lead to Q-switched modelocking (QML), where the pulse energy oscillates between extreme values (see Fig. 1). The laser then emits bunches of modelocked pulses, which may or may not have a stable Q-switching envelope.

The reason for the QML tendency is the following. Starting from the steady state of cw modelocking, any small increase in pulse energy will lead to stronger saturation of the absorber and thus to a positive net gain. This will lead to exponential growth of the pulse energy until this growth is stopped by gain saturation. In a solid-state laser, which usually exhibits a large gain saturation fluence, this may take many cavity round-tips. The pulse energy will then drop even below the steady-state value. A damped oscillation around the steady state (and finally stable cw modelocking) is obtained only if gain saturation sets in fast enough.

The transition between the regimes of cw modelocking and QML has been investigated in detail (Hönninger et al., 1999; Kärtner et al., 1995b) for slow saturable absorbers. If the absorber is fully saturated, if it always fully recovers between two cavity round-trips, and if soliton shaping effects do not occur, a simple condition for stable cw modelocking can be found (Hönninger et al., 1999a):

$$E_p^2 > E_{L,\text{sat}} E_{A,\text{sat}} \Delta R \tag{18}$$

where E_p is the intracavity (not output) pulse energy and $E_{L,\text{sat}}$, $E_{A,\text{sat}}$ are the gain and absorber saturation energies, respectively. Using the saturation parameter $S := E_p / E_{A,\text{sat}}$, we can rewrite this to obtain

$$E_p > E_{L,\text{sat}} \frac{\Delta R}{S} \tag{19}$$

This explains why passively modelocked lasers often exhibit QML when weakly pumped and stable cw modelocking for higher pump powers. Normally, modelocking is fairly stable even for operation only slightly above the QML threshold, so there is no need for operation far above this threshold. The QML threshold is high when $E_{L,\text{sat}}$ is large (laser medium with small laser cross sections and/or large mode area in the gain medium, enforced, e.g., by poor pump beam quality or by crystal fracture), when E_p cannot be made large (limited power of pump source, high repetition rate, or large intracavity losses), or when a high value of ΔR is needed for some reason. We thus find the following prescriptions to avoid QML:

> Use a gain medium with small saturation fluence and optimize the pumping arrangement for a small mode area in the gain medium.
> Minimize the cavity losses so that a high intracavity pulse energy can be achieved.
> Operate the saturable absorber in the regime of strong saturation, although this is limited by the tendency for pulse breakup or by absorber damage.
> Do not use a larger modulation depth ΔR than necessary.
> Use a cavity with low repetition rate.

Another interesting observation (Hönninger et al., 1999) is that in soliton modelocked lasers the minimum intracavity pulse energy for stable cw modelocking is lower by typically a factor on the order of 4. The reason for this is that a soliton acquires additional bandwidth if its energy increases for some reason. This reduces the effective gain, so that we have a negative feedback mechanism, which tends to stabilize the pulse energy. Thus the use of soliton formation in a laser can help not only to generate shorter pulses but also to avoid QML.

Finally we note that two-photon absorption (which may occur in a SESAM) can modify the saturation behavior in such a way that the QML tendency is reduced (Schibli et al., 2000). In this case, the two-photon absorption acts as an all-optical power limiter (Walker et al., 1986).

Passive Modelocking at High Repetition Rates

The repetition rates of typical modelocked solid-state lasers are on the order of 30–300 MHz. Pulse trains with much higher repetition rates— > 1 GHz or even > 10 GHz—are required for some applications, e.g., in data transmission or for optical clocking of electronic circuits.

It is clear that there are two basic approaches to realizing such high repetition rates: Either one must use a very compact laser cavity, which has a short round-trip time for the circulating pulse, or one must arrange for several pulses to circulate in the cavity with equal temporal spacing. The latter technique, called harmonic modelocking, requires some mechanism to stabilize the pulse spacing. It is widely used in fiber lasers but rarely in solid-state bulk lasers as discussed in this chapter. A Cr^{4+}:YAG laser has been demonstrated with three pulses in the cavity (i.e., harmonic modelocking), resulting in a repetition rate of 2.7 GHz (Collings et al., 1997), which is to our knowledge the highest value achieved with this technique applied to diode-pumped solid-state lasers. A fundamentally modelocked laser with 1.2 GHz repetition rate was also demonstrated (Tomaru et al., 2000).

In the following we concentrate on fundamentally (i.e., not harmonically) modelocked bulk lasers, which have generated repetition rates of up to 77 GHz (Krainer et al., 2000a). It is not difficult to construct a sufficiently compact laser cavity—a cavity length of a few millimeters (or slightly more than 1 mm for 77 GHz) is easily obtained when a compact modelocker such as a SESAM (see earlier) is used. The main challenge is to overcome the tendency for Q-switched modelocking (see previous subsection), which becomes very strong at high repetition rates. Because this issue formerly constituted a challenge even at repetition rates on the order of 1 GHz, specially optimized laser cavities had to be developed for repetition rates > 10 GHz. First it was realized that Nd^{3+}:YVO_4 is a laser medium with a particularly high gain cross section of 114×10^{-20} cm^2 (Peterson et al., 2002). This leads to a small gain saturation energy, which can be further minimized by using a laser cavity with a rather small mode radius of 30 μm or less. Another important factor is to minimize the intracavity losses, so that operation with a small-output coupler transmission and accordingly high intracavity pulse energy is achieved, and to use a SESAM with optimized parameters.

Since the first milestone at 13 GHz (Krainer et al., 1999a), based on a cavity with a Brewster-cut Nd^{3+}:YVO_4 crystal, a SESAM, and an air

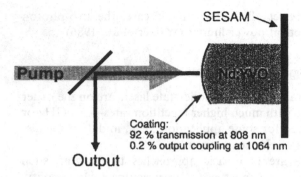

Figure 4 Quasi-monolithic setup as used for Nd^{3+}:YVO_4 lasers with repetition rate above 20 GHz. SESAM: semiconductor saturable absorber mirror.

spacing, quasi-monolithic Nd^{3+}:YVO_4 lasers have been used for higher repetition rates of 29 GHz (Krainer et al., 1999b) and later 59 GHz (Krainer et al., 2000b). Here the laser crystal has a reflective coating on a curved side, and the SESAM is attached to the other end (Fig. 4). In this regime, the repetition rate is limited not only by the Q-switching tendency but also because of the finite pulse duration (typically 5–6 ps), which leads to an increasing overlap of consecutive pulses. However, both limits have been pushed further by employing the effect of soliton modelocking which reduced the pulse duration to 2.7 ps and also has a stabilizing effect, counteracting Q-switching instabilities. The negative dispersion required for soliton modelocking was obtained from a GTI effect (Sec. 1.3.2) generated at a small air gap between the laser crystal and the SESAM. In this way, a repetition rate of 77 GHz was achieved (Krainer et al., 2000a), and further increases even beyond 100 GHz may be possible with this scheme. Note that the average output powers are at least a few tens of milliwatts in all cases, sometimes a few hundred milliwatts. Powers above 1 W at repetition rates of tens of gigahertz appear to be possible with further optimization and are well beyond the foreseeable potential of other approaches, except with external amplification.

Unfortunately, the choice of laser media for high repetition rate at other wavelengths, e.g., 1.5 µm as required for telecom applications, is very limited. Erbium-doped gain media have rather low laser cross sections, which greatly limits their use for passive modelocking at high repetition rates. Cr^{4+}:YAG is more favorable in this respect but suffers from higher crystal losses and a quite variable crystal quality. Repetition rates greater than 10 GHz may still be possible but remain to be demonstrated.

Although we concentrate on lasers based on ion-doped crystals in this chapter, we note that semiconductor gain media are also very interesting for obtaining high pulse repetition rates. The low gain saturation fluence of such a medium basically eliminates the problem of Q-switching instabilities in passively modelocked lasers. The large amplification bandwidth allows for subpicosecond pulse duration, and such devices can be developed for different laser wavelengths. With modelocked edge-emitting semiconductor lasers, repetition rates of up to 1.5 THz (Arahira et al., 1996) have been obtained, but typically at very low power levels and often not with transform-limited pulses. Surface-emitting semiconductor lasers allow for larger mode areas and thus for higher powers, and a good spatial mode quality can be achieved by using an external cavity. Recently, it was recognized that optically pumped surface-emitting semiconductor lasers have a great potential for passive modelocking with multiwatt output power, multi-gigahertz repetition rates, and possibly subpicosecond pulse durations at the same time. The first passively modelocked lasers of this kind were recently demonstrated (Hoogland et al., 2000), using SESAMs as modelockers. The maximum average output power obtained so far is ≈ 0.2 W, and the pulse durations are typically a few picoseconds, but greatly enhanced performance in terms of output power and pulse duration appears to be feasible in the near future.

Summary: Requirements for Stable Passive Modelocking

Here we briefly summarize a number of conditions that must be met to obtain stable passive modelocking:

> The laser must have a good spatial beam quality, i.e., it must operate in the fundamental transverse mode (see Sec. 1.4.1).
>
> The saturable absorber must be strong and fast enough that the loss difference between the pulses and a cw background is sufficient to compensate for the smaller effective gain of the short pulses.
>
> The saturable absorber must be strong enough to guarantee stable self-starting. (A longer absorber recovery times helps in this respect, see earlier in this section.)
>
> The absorber should not be too strongly saturated, because this would introduce a tendency for double pulses, particularly in the soliton regime. The effective gain is larger for double pulses (with decreased energy and bandwidth), and a strongly saturated absorber cannot sufficiently favor the more energetic single pulses.
>
> Nonlinear phase shifts from self-phase modulation must not be too strong, particularly in cases with zero or positive dispersion. In the soliton mode-locked regime, significantly stronger self-phase modulation can be tolerated and is even desirable.

To avoid Q-switching instabilities, the condition discussed earlier in this section must be fulfilled.

In some cases where inhomogeneous gain saturation occurs [e.g., induced by spatial hole burning (Braun et al., 1995, Kärtner et al., 1995c, Paschotta et al., 2001)], spectral stability may not be achievable. A spectral hole in the gain, generated by the current lasing spectrum, can lead to a situation where a shift of the gain spectrum to longer or shorter wavelengths increases the gain. Such a situation is unstable and must be avoided, possibly by the use of a suitable intracavity filter.

1.5 DESIGNS OF MODELOCKED LASERS

In the preceding sections we discussed the most important physical effects that are relevant for the operation of modelocked lasers. Here we give an overview on laser designs, with emphasis on those designs that are currently used and/or promise to find widespread application in the foreseeable future. We discuss a number of practical aspects for the design and operation of such lasers.

1.5.1 Picosecond Lasers

The setups of picosecond lasers typically do not differ very much from those of lasers for continuous-wave operation. Some modelocker is installed, which might be either an acousto-optic modulator (AOM) for active modelocking (Sec. 1.4.2) or, e.g., a SESAM (Sec. 1.4.3) for passive modelocking. Also, the cavity design needs to fulfill a few additional demands. As an example, we refer to Figure 5, which shows the setup of a high-power Nd^{3+}:YAG laser (Spühler et al., 1999b), passively modelocked with a SESAM. The cavity design must provide appropriate beam radii both in the laser head (where the fundamental Gaussian mode should just fill the

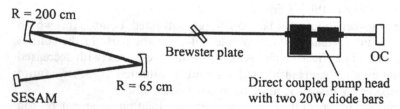

Figure 5 Setup of passively modelocked high-power Nd^{3+}: YAG laser, containing a DCP laser head, two curved mirrors, a SESAM, and an output coupler mirror (OC) with 7% transmission.

usable area) and on the SESAM, so as to use an appropriate degree of saturation. The latter depends on a number of factors: the output power, the output coupler transmission, the cavity length, and the saturation fluence of the SESAM. Obviously the cavity length must be chosen to obtain the desired repetition rate. The equations given in Section 1.4.3 can be used to ensure that the chosen design will not suffer from Q-switching instabilities. The laser head is side-pumped in the mentioned example, but end-pumped laser heads can also be used, where the pump light is typically injected through a folding mirror that is highly reflective for the laser light.

The SESAM should typically be used as an end mirror, not as a folding mirror. Otherwise a tendency for the generation of multiple pulses, which would meet on the SESAM, might be induced.

Similar setups can be used for actively modelocked lasers, where the SESAM is replaced by an AOM. The AOM should be placed close to an end mirror, for reasons similar to those discussed above for the SESAM.

1.5.2 High-Power Thin-Disk Laser

By far the highest average powers in the subpicosecond domain can be obtained from thin-disk Yb^{3+}:YAG lasers, passively modelocked with a SESAM: 16.2 W in 700 fs pulses has been demonstrated (Aus der Au et al., 2000). The thin-disk laser head (Giesen et al., 1994) is power-scalable because of the nearly one-dimensional heat flow in the beam direction: Thermal effects (such as thermal lensing) do not become more severe if the mode area is scaled up in proportion to the power level. The same is true for the SESAM (Sec. 1.4.3), which also has the geometry of a thin disk. Thus the whole concept of the passively modelocked thin-disk laser is power-scalable, and average power levels above 50 W are expected for the near future.

Because the thin disk must be cooled from one side, a reflecting coating on this side is used, and the laser beam is always reflected at the disk. For modelocked operation, the disk is preferably used as a folding mirror rather than as an end mirror, because the two ends of the standing-wave cavity are then available for the output coupler mirror and the SESAM. An appropriate amount of negative dispersion for soliton modelocking (Sec. 1.4.3) is generated with a GTI (Sec. 1.3.2) or with a dispersive mirror. Prism pairs (Sec. 1.3.2) appear to be unsuitable because either they absorb too highly (and thus exhibit thermal lensing), as, e.g., for SF10 glass, or they do not provide sufficient dispersion (as for silica). Soliton modelocked operation is essential because otherwise the pulse duration would be longer and the modelocking operation would be destabilized by the action by spatial hole burning (Paschotta et al., 2001). Also, it helps to suppress Q-switching instabilities (Sec. 1.4.3) (see also Hönninger et al., 1999).

1.5.3 Typical Femtosecond Lasers

Most femtosecond lasers are based on an end-pumped laser setup, with a broadband laser medium such as Ti^{3+}:sapphire, Cr^{3+}:LiSAF, or Yb^{3+}:glass (see Sec. 1.2 for an overview). In the case of Ti^{3+}:sapphire, the pump source can be either an argon ion laser or a frequency-doubled solid-state laser. In any case, one typically uses a few watts of pump power in a beam with good transverse beam quality, because the mode radius in the Ti^{3+}:sapphire rod is usually rather small. Other gain media such as Cr^{3+}:LiSAF or Yb^{3+}:glass are typically pumped with high-brightness diode lasers, delivering a few watts with beam quality M^2 factor on the order of 10 in one direction and <5 in the other direction.

Typically laser cavities (see Fig. 6 as an example) contain two curved mirrors within a distance of a few centimeters on opposite sides of the gain medium. The pump power is usually injected through one or both of these mirrors, which also focus the intracavity laser beam to an appropriate beam waist. One of the two "arms" of the cavity ends with the output coupler mirror, and the other one may be used for a SESAM as a passive mode-locker. One arm typically contains a prism pair (Sec. 1.3.2) for dispersion compensation, which is necessary for femtosecond pulse generation. In practically all cases, femtosecond lasers operate in the regime of negative overall intracavity dispersion so that soliton-like pulses are formed.

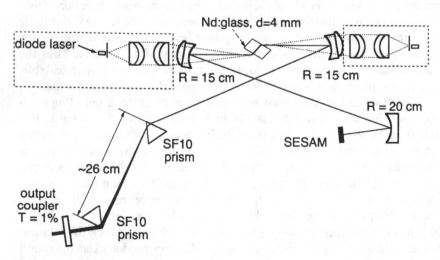

Figure 6 Setup of a femtosecond Nd:glass laser (Aus der Au et al., 1997). The gain medium is pumped with two laser diodes. A prism pair is used for dispersion compensation, and a SESAM as modelocker. This laser generated 60 fs pulses with 84 mW average power.

Instead of a SESAM, or in addition to it, the Kerr lens in the gain medium can be used for modelocking (Sec. 1.4.3). In most cases, soft-aperture KLM is used. Here, the cavity design is made such that the Kerr lens reduces the mode area for high intensities and thus improves the overlap with the (strongly focused) pump beam. This is achieved by operating the laser cavity near one of the stability limits of the cavity, which are found by varying the distance between the above-mentioned curved folding mirrors or some other distance in the cavity.

For the shortest pulse durations around 5–6 fs (Morgner et al., 1999b; Sutter et al., 2000), the strong action of KLM as an effective fast saturable absorber is definitely required. Also double-chirped mirrors (Sec. 1.3.2) are required for precise dispersion compensation over a very broad bandwidth. Typically, several dispersive mirrors are used in the laser cavity, and additional mirrors are used for further external compression. A SESAM allows for self-starting modelocking (Jung et al., 1997a).

Higher pulse energies and peak powers have been generated by using laser setups with reduced repetition rates of, e.g., 15 MHz (Cho et al., 1999). The long cavity length required for such repetition rates is achieved by inserting a multipass cell. However, the limiting factor to the pulse energy is ultimately not the practically achievable cavity length but rather the non-linearity of the gain crystal—at least in the sub-30 fs domain: If self-phase modulation becomes too strong, this destabilizes the modelocking process.

1.5.4 Lasers with High Repetition Rates

For ultrashort pulses with multigigahertz repetition rate, laser designs are required that are very different from those described in the previous sections. We discussed such designs with up to 77 GHz repetition rate in Section 1.4.3. Another type of laser in this domain is described in the next section.

1.5.5 Passively Modelocked Optically Pumped Semiconductor Lasers

Here we discuss a novel type of modelocked laser that is based not on an ion-doped crystal or glass as gain medium, but on a semiconductor medium. This special case of a modelocked semiconductor laser is described here because it allows for much higher output powers than other modelocked semiconductor lasers and can thus be compared directly with ion-doped lasers. The gain medium is a surface-emitting semiconductor chip, with a Bragg mirror and one or several amplifying quantum wells grown on a wafer (made, e.g., of GaAs). The diffraction-limited laser mode is formed in an external cavity that also contains a SESAM (Sec. 1.4.3) for passive modelocking. Optical pumping of the gain medium allows the use of a rather large mode with correspondingly high output power. If the cooling (e.g., through the GaAs substrate, which may be thinned down) is sufficient, the output power can be scaled by increasing the mode area so that the temperature is not increased. Thus we have a power-scalable concept that should allow for multiwatt average output powers. Q-switching instabilities (Sec. 1.4.3) do not occur because of the small saturation fluence of the semiconductor gain medium, so very high repetition rates can be achieved. The gain bandwidth is sufficient for the generation of subpicosecond pulses.

Since the first demonstration of this concept (Hoogland et al., 2000), the average output power has been increased to $> 200\,\text{mW}$ (Häring et al., 2001), and further increases should be possible soon. The pulse durations have so far been at least a few picoseconds, but shorter pulses should also become possible. We envisage that this concept will lead to lasers with multi-gigahertz repetition rates, multiwatt output powers, and subpicosecond pulse durations. This combination of properties would be very difficult to achieve with ion-doped gain media.

1.6 PASSIVELY Q-SWITCHED MICROCHIP LASERS

Ultrashort pulses are in most cases generated with modelocked lasers of some kind. Q-switching is a method that is typically used to generate much longer pulses ($>1\,\text{ns}$ pulse duration). This is because the pulse duration achievable with Q-switching is longer than the cavity round-trip time, which is typically a few nanoseconds. The achievable pulse energy is significantly higher than with modelocking, because the repetition rate is relatively low. Another feature is that the output linewidth can be rather small; in some cases the emission is limited to a single cavity mode.

A very compact laser setup can be obtained using the principle of a microchip laser (Zayhowski et al., 1989). In such a setup (Fig. 7), a thin disk of a gain medium (and optionally a modulator) is sandwiched between two mirrors. Because the round-trip time can be far lower than 1 ns, Q-switched pulses from such a laser can be much shorter than pulses from conventional Q-switched lasers.

The first Q-switching result for a microchip laser was obtained in 1992 with an electro-optic modulator in the cavity. In this way, 115 ps pulses were generated (Zayhowski et al., 1992). With a Cr^{4+}:YAG absorber for passive Q-switching, the shortest pulse duration was 218 ps (Zayhowski, 1996). Further shortening of the cavity, and consequently a pulse duration to 56 ps, were achieved in 1997 by passive Q-switching with a SESAM (Sec. 1.4.3) (Braun et al., 1997). The cavity length is then basically determined by the thickness of the laser crystal, because the penetration depth in the

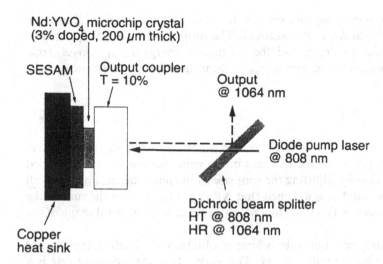

Figure 7 Setup of a Nd : YVO₄ microchip laser that is passively Q-switched with a SESAM. The Nd:YVO₄ crystal is sandwiched between the SESAM and the output coupler mirror. A dichroic mirror is used to separate the output beam from the pump beam.

SESAM is only a few micrometers. Optimization of various parameters of a Nd^{3+}:YVO₄ laser led to the shortest pulses of 37 ps duration (Spühler et al., 1999c). Although such pulse durations are comparable to those of many modelocked lasers, other parameters are quite different. In particular, the repetition rate of Q-switched microchip lasers is much lower; it can be varied in a wide range from a few hundred hertz to a few megahertz and depends on the pumping conditions (pump power and spot size). The pulse energy can well exceed a microjoule in some cases. Such pulses are very suitable for range finding, for example, where subnanosecond pulse durations at a low repetition rate are desired and some fluctuations of the repetition rate can be tolerated. If operation on a single transverse cavity mode can be achieved (as is the case for a thin enough gain medium), these fluctuations can be fairly small (a few percent or less).

For calculation of the main operation parameters, a simple set of equation has been derived (Spühler et al., 1999c). These equations are based on a number of approximations, which are typically well fulfilled in real lasers and have been tested experimentally. Here we briefly summarize the results. According to

$$\tau_p \approx 3.5 \frac{T_r}{\Delta R} \tag{20}$$

the pulse duration τ_p depends only on the cavity round-trip time T_r and the modulation depth ΔR of the SESAM. The output pulse energy E_p depends on the modulation depth and the saturation energy $E_{L,\text{sat}} = h\nu_L A_L / \sigma_{\text{em}}$ (with the emission cross section σ_{em}, assuming a four-level system) of the laser medium:

$$E_p \approx E_{L,\text{sat}} \Delta R \tag{21}$$

The laser mode area A_L is defined by the pumping conditions, which determine the strength of the thermal lens in the gain medium. A_L can be varied in a certain range by adjusting the spot size of the pump beam. For Eqs. (20) and (21) to be valid, it is assumed that ΔR is not larger than the sum of the liner cavity losses, which leads to a nearly symmetrical temporal shape of the pulses.

The pulse repetition rate is simply obtained by dividing the average output power by the pulse energy. This shows that the repetition rate is a linear function of the pump power, provided that the laser mode area stays approximately constant.

To generate rather short pulses, a thin gain crystal has to be combined with a SESAM with high modulation depth. It is thus important to choose a highly doped crystal with a high gain per unit length and short pump absorption length. Nd^{3+}:YVO_4 has proven to be very suitable for this purpose.

For high pulse energies, gain media with large saturation fluence are ideal. An Yb^{3+}:YAG microchip laser generated $1.1\,\mu J$ pulses of $0.53\,ns$ duration at $1030\,nm$ (Spühler et al., 2001), and up to $4\,\mu J$ in subnanosecond pulses in the eye-safe wavelength regime around $\approx 1.5\,\mu m$ have been obtained with Er^{3+}:Yb^{3+}:glass (Fluck et al., 1998). In this case, the limits are set by SESAM damage, because the obtained intracavity pulse fluence is much higher than the SESAM saturation fluence, even with optimized high-finesse SESAM designs. On the other hand, the laser mode area cannot be easily scaled up because of the requirement that the thermal lens form a stable cavity mode.

An interesting feature of the passively Q-switched microchip laser is that the pulse parameters (energy, duration, shape) are very stable, because they are determined only by fixed properties of the cavity, the gain medium, and the saturable absorber. The pump power basically influences only the pulse repetition rate, because the formation of a pulse is always triggered when a certain fixed level of stored energy (and thus gain) is reached. Apart from this, only weak fluctuations of the pulse energy may be expected if fluctuations of the pump power translate into slight changes of the laser mode area. The situation is different for actively Q-switched microchip lasers; here the repetition rate is electronically determined, whereas the pulse parameters (energy, duration, shape) depend on the energy stored in the gain medium at the moment the pulse is triggered. Therefore, the pulse parameters are affected by low-frequency pump power fluctuations.

1.7 SUMMARY AND OUTLOOK

We have shown that the technology of ultrafast lasers has become very refined and is now suitable for application in many areas. Points of particular importance in this respect are:

The transition from dye lasers to solid-state lasers, which can be compact, powerful, efficient, and long-lived. It has been shown that solid-state lasers can generate pulses that are even shorter than those generated in dye lasers.

The development of diode lasers for direct pumping of solid-state lasers. This has led not only to very efficient and compact lasers, but also to modelocked lasers with tens of watts of output power.

The development of semiconductor saturable absorbers that can be optimized for operation in very different parameter regimes concerning laser wavelength, pulse duration, and power levels.

Within the next few years we expect the following developments in the filed:

New solid-state gain media will be used. In particular, Cr^{2+}:ZnSe (see Sec. 1.2) appears to be suitable for the generation of pulses of 20 fs duration or less in a new spectral region around 2.7 μm.

Very high power levels (tens of watts or even $> 100\,W$ of average power) should become possible with passively modelocked thin-disk lasers. Pulse durations just below 1 ps are already feasible in this power regime, but with new materials the regime of 200 fs or even below should become accessible with similarly high powers. Amplifier devices for lower repetition rates will become important for material processing.

Nonlinear frequency conversion stages (based on second harmonic generation, sum frequency mixing, or parametric oscillation) will be pumped with high-power modelocked lasers to generate short, powerful pulses at other wavelengths. This will be of interest, e.g., for application in large-screen RGB display systems.

As an alternative to ion-doped gain media, optically pumped semiconductor lasers with an external cavity and a SESAM for passive modelocking (Sec. 1.5.5) will generate higher output powers and shorter pulses in the regime of multigigahertz pulse repetition rates.

We thus believe that the development of ultrafast laser sources has not come to its end but will continue to deliver new devices with superior properties for many applications.

ACKNOWLEDGMENTS

We thank T. Südmeyer for help with the preparation of the manuscript.

REFERENCES

Agnesi A, Pennacchio C, Reali GC, Kubecek V. (1997). High-power diode-pumped picosecond Nd:YVO$_4$ laser. Opt Lett 22:1645–1647.

Aus der Au J, Kopf D, Morier-Genoud F, Moser M, Keller U. (1997). 60-fs pulses from a diode-pumped Nd:glass laser. Opt Lett 22:307–309.

Aus der Au J, Loesel FH, Morier-Genoud F, Moser M, Keller U. (1998). Femtosecond diode-pumped Nd:glass laser with more than 1-W average output power. Opt Lett 23:271–273.

Aus der Au J, Schaer SF, Paschotta R, Hönninger C, Keller U, Moser M. (1999). High-power diode-pumped passively modelocked Yb:YAG lasers. Opt Lett 24:1281–1283.

Aus der Au J, Spühler GJ, Südmeyer T, Paschotta R, Hövel R, Moser M, Erhard S, Karszewski M, Giesen A, Keller U. (2000). 16.2 W average power from a diode-pumped femtosecond Yb:YAG thin disk laser. Opt Lett 25:859.

Bado P, Bouvier M, Coe JS. (1987). Nd:YLF mode-locked oscillator and regenerative amplifier. Opt Lett 12:319–321.

Baltuska A, Wei Z, Pshenichnikov MS, Wiersma DA, Szipöcs R. (1997). All-solid-state cavity dumped sub-5-fs laser. Appl Phys B 65:175–188.

Bartels A, Dekorsky T, Kurz H. (1999). Femtosecond Ti:sapphire ring laser with 2-GHz repetition rate and its application in time-resolved spectroscopy. Opt Lett 24:996–998.

Basu S, Byer RL. (1988). Continuous-wave mode-locked Nd:glass laser pumped by a laser diode. Opt Lett 13:458–460.

Becker MF, Kuizenga KJ, Siegman AE. (1972). Harmonic mode locking of the Nd:YAG laser. IEEE J Quantum Electron QE-8:687–693.

Beddard T, Sibbett W, Reid DT, Garduno-Mejia J, Jamasbi N, Mohebi M. (1999). High-average-power, 1-MW peak-power self-mode-locked Ti:sapphire oscillator. Opt Lett 24:163–165.

Blow KJ, Nelson BP. (1988). Improved modelocking of an F-center laser with a nonlinear nonsoliton external cavity. Opt Lett 13:1026–1028.

Braun B, Hönninger C, Zhang G, Keller U, Heine F, Kellner T, Huber G. (1996). Efficient intracavity frequency doubling of a passively modelocked diode-pumped Nd:LSB laser. Opt Lett 21:1567–1569.

Brovelli LR, Jung ID, Kopf D, Kamp M, Moser M, Kärtner FX, Keller U. (1995). Self-starting soliton modelocked Ti:sapphire laser using a thin semiconductor saturable absorber. Electron Lett 31:287–289.

Brunner F, Spühler GJ, Aus der Au J, Krainer L, Morier-Genoud F, Paschotta R, Lichtenstein N, Weiss S, Harder C, Lagatsky AA, Abdolvand A, Kuleshov NV, Keller U. (2000). Diode-pumped femtosecond Yb:KGd(WO$_4$)$_2$ laser with 1.1-W average power. Opt Lett 25:1119–1121.

Burns D, Hetterich M, Ferguson AI, Bente E, Dawson MD, Davies JI, Bland SW. (2000). High-average-power (> 20 W) Nd:YVO$_4$ lasers mode locked by strain-compensated saturable Bragg reflectors. J Opt Soc Am B 17:919–926.

Carrig TJ, Wagner GJ, Sennaroglu A, Jeong JY, Pollock CR. (2000). Mode-locked Cr^{2+}:ZnSe laser. Opt Lett 25:168–170.

Cerullo G, Dienes A, Magni V. (1996). Space-time coupling and collapse threshold for femtosecond pulses in dispersive nonlinear media. Opt Lett 21:65–67.

Cho SH, Bouma BE, Ippen EP, Fujimoto JG. (1999). Low-repetition-rate high-peak-power Kerr-lens mode-locked Ti:Al$_2$O$_3$ laser with a multiple-pass cell. Opt Lett 24:417–419.

Christov IP, Stoev VD, Murnane MM, Kapteyn HC. (1995). Mode locking with a compensated space-time astigmatism. Opt Lett 20:2111–2113.

Chudoba C, Fujimoto JG, Ippen EP, Haus HA, Morgner U, Kärtner FX, Scheuer V, Angelow G, Tschudi T. (2000). All-solid-state Cr:forsterite laser generating 14-fs pulses at 1.3 μm. Conf. on Lasers and Electro-Optics, 2000, Postdeadline paper CPD4.

Collings BC, Stark JB, Tsuda S, Knox WH, Cunningham JE, Jan WY, Pathak R, Bergman K. (1996). Saturable Bragg reflector self-starting passive modelocking of a Cr^{4+}:YAG laser pumped with a diode-pumped Nd:YVO$_4$ laser. Opt Lett 21:1171–1173.

Couderc V, Louradour F, Barthelemy A. (1999). 2.8 ps pulses from a modelocked diode pumped Nd:YVO$_4$ laser using quadratic polarization switching. Opt Commun 166:103–111.

Curley PF, Ferguson AI. (1991). Actively modelocked Ti:sapphire laser producing transform-limited pulses of 150 fs duration. Opt Lett 16:1016–1018.

De Maria AJ, Stetser DA, Heynau H. (1966). Self mode-locking of lasers with saturable absorbers. Appl phys Lett 8:174–176.

Diels JC, Stryland EWV, Benedict G. (1978). Generation and measurement of pulses of 0.2 ps duration. Opt Commun 25:93.

Druon F, Balembois F, Georges P, Brun A, Courjaud A, Hönninger C, Salin F, Aron A, Mougel F, Aka G, Vivien D. (2000). Generation of 90-fs pulses from a modelocked diode-pumped Yb:Ca$_4$GdO(Bo$_3$)$_3$ laser. Opt Lett 25:423–425.

Fluck R, Zhang G, Keller U, Weingarten KJ, Moser M. (1996). Diode-pumped passively mode-locked 1.3 μm Nd:YVO$_4$ and Nd:YLF lasers by use of semiconductor saturable absorbers. Opt Lett 21:1378–1380.

Fork RL, Cruz CHB, Becker PC, Shank CV. (1987). Compression of optical pulses to six femtoseconds by using cubic phase compensation. Opt Lett 12:483–485.

Fork RL, Greene BI, Shank CV. (1981). Generation of optical pulses shorter than 0.1 ps by colliding pulse modelocking. Appl Phys Lett 38:617–619.

Fork RL, Martinez OE, Gordon JP. (1984). Negative dispersion using pairs of prisms. Opt Lett 9:150–152.

French PMW, Mellish R, Taylor JR, Delfyett PJ, Florez LT. (1993). Modelocked all-solid-state diode-pumped Cr:LiSAF laser. Opt Lett 18:1934–1936.

Gabel KM, Russbuldt P, Lebert R, Valster A. (1998). Diode-pumped Cr^{3+}:LiCAF fs-laser. Opt Commun 157:327–334.

Giesen A, Hügel H, Voss A, Wittig K, Brauch U, Opower H. (1994). Scalable concept for diode-pumped high-power solid-state lasers. Appl Phys B 58:363–372.

Gires F, Tournois P. (1964). Interferometre utilisable pour la compression d'impulsions lumineuses modulees en frequence. CR Acad Sci Paris 258:6112–6115.

Gloster LAW, Cormont P, Cox AM, King TA, Chain BHT. (1997). Diode-pumped Q-switched Yb:S-FAP laser. Opt Commun. 146:177–180.

Godil AA, Hou A, Auld BA, Bloom DM. (1991). Harmonic mode locking of a Nd: BEL laser using a 20-GHZ dielectric resonator/optical modulator. Opt Lett 16:1765–1767.

Goodberlet J, Jacobson J, Fujimoto JG, Schulz PA, Fan TY. (1990). Self-starting additive pulse modelocked diode-pumped Nd:YAG laser. Opt Lett 15:504–506.

Goodberlet J, Wang J, Fujimoto JG, Schulz PA. (1989). Femtosecond passively mode-locked Ti:sapphire laser with a nonlinear external cavity. Opt Lett 14:1125–1127.

Haiml M, Siegner U, Morier-Genoud F, Keller U, Luysberg M, Lutz RC, Specht P, Weber ER. (1999b). Optical nonlinearity in low-temperature grown GaAs: microscopic limitations and optimization strategies. Appl Phys Lett 74:3134–3136.

Haiml M, Siegner U, Morier-Genoud F, Keller U, Luysberg M, Specht P, Weber ER. (1999a). Femtosecond response times and high optical nonlinearity in beryllium doped low-temperature grown GaAs. Appl Phys Lett 74:1269–1271.

Hargrove LE, Fork RL, Pollack MA. (1964). Locking of HeNe laser modes induced by synchronous intracavity modulation. Appl Phys Lett 5:4.

Haus HA, Keller U, Knox WH. (1991). A theory of coupled cavity modelocking with resonant nonlinearity. J Opt Soc Am B 8:1252–1258.

Henrich B, Beigang R. (1997). Self-starting Kerr-lens mode locking of a Nd:YAG-laser. Opt Commun 135:300–304.

Hönninger C, Morier-Genoud F, Moser M, Keller U, Brovelli LR, Harder C. (1998). Efficient and tunable diode-pumped femtosecond Yb:glass lasers. Opt Lett 23:126–128.

Hönninger C, Paschotta R, Graf M, Morier-Genoud F, Zhang G, Moser M, Biswal S, Nees J, Braun A, Mourou GA, Johannsen I, Giesen A, Seeber W, Keller U. (1999b). Ultrafast ytterbium-doped bulk lasers and laser amplifiers. Appl Phys B 69:3–17.

Hönninger C, Zhang G, Keller U, Giesen A. (1995). Femtosecond Yb:YAG laser using semiconductor saturable absorbers. Opt Lett 20:2402–2404.

Hönninger C, Paschotta R, Morier-Genoud F, Moser M, Keller U. (1999a). Q-switching stability limits of continuous-wave passive mode locking. J Opt Soc Am B 16:46–56.

Hughes DW, Barr JRM, Hanna DC. (1991). Mode locking of a diode-laser-pumped Nd:glass laser by frequency modulation. Opt Lett 16:147–149.

Hughes DW, Phillips MW, Barr JRM, Hanna DC. (1992). A laser-diode-pumped Nd:glass laser: mode-locked, high-power, and single frequency performance. IEEE J Quantum Electron 28:1010–1017.

Ippen EP, Haus HA, Liu LY. (1989). Additive pulse modelocking. J Opt Soc Am B 6:1736–1745.

Islam MN, Sunderman ER, Soccolich CE, Bar-Joseph I, Sauer N, Chang TY, Miller BL. (1989). Color center lasers passively mode locked by quantum wells. IEEE J Quantum Electron 25:2454–2463.

Jeong TM, Kang EC, Nam CH. (1999). Temporal and spectral characteristics of an additive-pulse mode-locked Nd:YLF laser with Michelson-type configuration. Opt Commun 166:95–102.

Juhasz T, Lai ST, Pessot MA. (1990). Efficient short-pulse generation from a diode-pumped Nd: YLF laser with a piezoelectrically induced diffraction modulator. Opt Lett 15:1458–1460.

Jung ID, Brovelli LR, Kamp M, Keller U, Moser M. (1995a). Scaling of the anti-resonant Fabry-Perot saturable absorber design toward a thin saturable absorber. Opt Lett 20:1559–1561.

Jung ID, Kärtner FX, Brovelli LR, Kamp M, Keller U. (1995b). Experimental verification of soliton modelocking using only a slow saturable absorber. Opt Lett 20:1892–1894.

Jung ID, Kärtner FX, Matuschek N, Sutter DH, Morier-Genoud F, Zhang G, Keller U, Scheuer V, Tilsch M, Tschudi T. (1997). Self-starting 6.5 fs pulses from a Ti:sapphire laser. Opt Lett 22:1009–1011.

Kärtner FX, Aus der Au J, Keller U. (1998). Modelocking with slow and fast saturable absorbers: what's the difference? IEEE J Selected Topics Quantum Electron 4:159–168.

Kärtner FX, Jung ID, Keller U. (1996). Soliton modelocking with saturable absorbers. Special Issue on Ultrafast Electronics, Photonics and Optoelectronics, IEEE J Selected Topics Quantum Electron (JSTQE) 2:540–556.

Kärtner FX, Keller U. (1995). Stabilization of soliton-like pulses with a slow saturable absorber. Opt Lett 20:16–18.

Kärtner FX, Matuschek N, Schibli T, Keller U, Haus HA, Heine C, Morf R, Scheuer V, Tilsch M, Tschudi T. (1997). Design and fabrication of double-chirped mirrors. Opt Lett 22:831–833.

Kean PN, Zhu X, Crust DW, Grant RS, Landford N, Sibbett W. (1989). Enhanced modelocking of color center lasers. Opt Lett 14:39–41.

Keller U. (1994). Ultrafast all-solid-state laser technology. Appl Phys B 58:347–363.

Keller U. (1999). Semiconductor nonlinearities for solid-state laser modelocking and Q-switching, in Nonlinear Optics in Semiconductors, A. Kost and E. Garmire (eds.), Boston: Academic Press, pp. 211–286.

Keller U. (2001). Short and Ultra short Pulse Generation, in Laser Physics and Application, G. Herzinfer, R. Poprawe, H. Weber (eds.), Springer, 2002.

Keller U, Chiu TH. (1992b). Resonant passive modelocked Nd: YLF laser. IEEE J Quantum Electron 28:1710–1721.

Keller U, Knox WH, Roskos H. (1990b). Coupled-cavity resonant passive modelocked (RPM) Ti:sapphire laser. Opt Lett 15:1377–1379.

Keller U, Li KD, Khuri-Yakub BT, Bloom DM, Weingarten KJ, Gerstenberger DC. (1990a). High-frequency acousto-optic modelocker for picosecond pulse generation. Opt Lett 15:45–47.

Keller U, Miller DAB, Boyd GD, Chiu TH, Ferguson JF, Asom MT. (1992a). Solid-state low-loss intracavity saturable absorber for Nd:YLF lasers: an antiresonant semiconductor Fabry-Perot saturable absorber. Opt Lett 17:505–507.

Keller U, Valdmanis JA, Nuss MC, Johnson AM. (1988). 53 ps pulses at 1.32 µm from a harmonic mode-locked Nd:YAG laser. IEEE J Quantum Electron 24:427–430.

Keller U, Weingarten KJ, Kärtner FX, Kopf D, Braun B, Jung ID, Fluck R, Hönninger C, Matuschek N, Aus der Au J. (1996). Semiconductor saturable absorber mirrors (SESAMs) for femtosecond to nanosecond pulse generation in solid-state lasers. IEEE J Selected Topics Quantum Electron 2:435–453.

Kopf D, Kärtner FX, Weingarten KJ, Keller U. (1995b). Diode-pumped mode-locked Nd:glass lasers using an A-FPSA. Opt Lett 20:1169–1171.

Kopf D, Kärtner F, Weingarten KJ, Keller U. (1994b). Pulse shortening in a Nd:glass laser by gain reshaping and soliton formation. Opt Lett 19:2146–2148.

Kopf D, Spühler GJ, Weingarten KJ, Keller U. (1996b). Mode-locked laser cavities with a single prism for dispersion compensation. Appl Opt 35:912–915.

Kopf D, Weingarten KJ, Brovelli LR, Kamp M, Keller U. (1994a). Diode-pumped 100-fs passively mode-locked Cr:LiSAF laser with an antiresonant Fabry-Perot saturable absorber. Opt Lett 19:2143–2145.

Kopf D, Weingarten KJ, Brovelli LR, Kamp M, Keller U. (1995a). Sub-50-fs diode-pumped mode-locked Cr:LiSAF with an A-FPSA. Conference on Lasers and Electro-Optics.

Kopf D, Weingarten KJ, Zhang G, Moser M, Emanuel MA, Beach RJ, Skidmore JA, Keller U. (1997). High-average-power diode-pumped femtosecond Cr:Li-SAF lasers. Appl Phys B 65:235–243.

Kopf D, Zhang G, Fluck R, Moser M, Keller U. (1996a). All-in-one dispersion-compensating saturable absorber mirror for compact femtosecond laser sources. Opt Lett 21:486–488.

Krainer L, Paschotta R, Moser M, Keller U. (2000). 77 GHz soliton modelocked Nd:YVO$_4$ laser. Electron Lett 36:1846–1848.

Kubecek V, Couderc V, Bourliaguet B, Louradour F, Barthelemy A. (1999). 4-W and 23-ps pulses from a lamp-pumped Nd:YAG laser passively mode-locked by polarization switching in a KTP crystal. Appl Phys B 69:99–102.

Larionov M, Gao J, Erhard S, Giesen A, Contag K, Peters V, Mix E, Fornasiero L, Petermann K, Huber G, Aus der Au J, Spühler GJ, Paschotta R, Keller U,

Lagatsky AA, Abdolvand A, Kuleshov NV. (2001). Thin disk laser operation and spectroscopic characterization of Yb-doped sesquioxides and potassium tungstates. Advanced Solid State Lasers. Vol. 50, P. 625.

Li KD, Sheridan JA, Bloom DM. (1991). Picosecond pulse generation in Nd:BEL with a high-frequency acousto-optic mode locker. Opt Lett 16:1505–1507.

LiKamWa P, Chai BHT, Miller A. (1992). Self-mode-locked Cr^{3+}:LiCaAlF$_6$ laser. Opt Lett 17:1438–1440.

Lincoln JR, Ferguson AI. (1994). All-solid-state self-mode locking of a Nd:YLF laser. Opt Lett 19:2119–2121.

Liu JM, Chee JK. (1990b). Passive modelocking of a cw Nd:YLF laser with a non-linear external coupled cavity. Opt Lett 15:685–689.

Liu KX, Flood CJ, Walker DR, Driel HMv. (1992). Kerr lens mode locking of a diode-pumped Nd:YAG laser. Opt Lett 17:1361–1363.

Liu LY, Huxley JM, Ippen EP, Haus HA. (1990a). Self-starting additive pulse mode-locking of a Nd:YAG laser. Opt Lett 15:553–555.

Loesel FH, Horvath C, Grasbon F, Jost M, Niemz MH. (1997). Selfstarting femto-second operation and transient dynamics of a diode-pumped Cr:LiSGaF laser with a semiconductor saturable absorber mirror. Appl Phys B 65:783–787.

Maker GT, Ferguson AI. (1989a). Frequency-modulation mode locking of diode-pumped Nd:YAG laser. Opt Lett 4:788–790.

Maker GT, Ferguson AI. (1989b). Electron Lett 25:1025.

Maker GT, Ferguson AI. (1989c). Modelocking and Q-switching of a diode laser pumped neodymium-doped yttrium lithium fluoride laser. Appl Phys Lett 54:403–405.

Malcolm GPA, Curley PF, Ferguson AI. (1990). Additive pulse modelocking of a diode pumped Nd:YLF laser. Opt Lett 15:1303–1305.

Mellish PM, French PMW, Taylor JR, Delfyett PJ, Florez LT. (1994). All-solid-state femtosecond diode-pumped Cr:LiSAF laser. Electron Lett 30:223–224.

Miller A, LiKamWa P, Chai BHT, Stryland EWV. (1992). Generation of 150-fs tunable pulses in Cr:LiSAF. Opt Lett 17:195–197.

Mocker HW, Collins RJ. (1965). Mode competition and self-locking effects in a Q-switched ruby laser. Appl Phys Lett 7:270–273.

Morgner U, Kärtner FX, Cho SH, Chen Y, Haus HA, Fujimoto JG, Ippen EP, Scheuer V, Angelow G, Tschudi T. (1999a). Sub-two-cycle pulses from a Kerr-lens mode-locked Ti:sapphire laser: addenda. Opt Lett 24:920.

Morgner U, Kärtner FX, Cho SH, Chen Y, Haus HA, Fujimoto JG, Ippen EP, Scheuer V, Angelow G, Tschudi T. (1999b). Sub-two-cycle pulses from a Kerr-lens mode-locked Ti:sapphire laser. Opt Lett 24:411–413.

Moulton PF. (1986). Spectroscopic and laser characteristics of Ti:Al$_2$O$_3$. J Opt Soc Am B 3:125–132.

Nisoli M, Stagira S, Silvestri SD, Svelto O, Sartania S, Cheng Z, Lenzner M, Spielmann C, Krausz F. (1997). A novel high-energy pulse compression system: generation of multigigawatt sub-5-fs pulses. Appl Phys B 65:189–196.

Page RH, Schaffers KI, DeLoach LD, Wilke GD, Patel FD, Tassano JB, Payne SA, Krupke WF, Chen K-T, Burger A. (1997). Cr^{2+}-doped zinc chalcogenides as efficient, widely tunable mid-infrared lasers. IEEE J Quantum Electron 33:609–619.

Paschotta R, Aus der Au J, Keller U. (1999a). Strongly enhanced negative dispersion from thermal lensing or other focussing elements in femtosecond laser cavities. J Opt Soc Am B 17:646–651.

Paschotta R, Aus der Au J, Keller U. (2000b). Thermal effects in high power end-pumped lasers with elliptical mode geometry. J Selected Topics Quantum Electron 6:636–642.

Paschotta R, Aus der Au J, Spühler GJ, Morier-Genoud F, Hövel R, Moser M, Erhard S, Karszewski M, Giesen A, Keller U. (2000a). Diode-pumped passively mode-locked lasers with high average power. Appl Phys B 70:S25–S31.

Paschotta R, Spühler GJ, Sutter DH, Matuschek N, Keller U, Moser M, Hövel R, Scheuer V, Angelow G, Tschudi T. (1999b). Double-chirped semiconductor mirror for dispersion compensation in femtosecond lasers. Appl Phys Lett 75:2166–2168.

Petrov V, Shcheslavskiy V, Mirtchev T, Noack F, Itatani T, Sugaya T, Nakagawa T. (1998). High-power self-starting femtosecond Cr:forsterite laser. Electron Lett 34:559–561.

Ramaswamy M, Gouveia-Neto AS, Negus DK, Izatt JA, Fujimoto JG. (1993). 2.3-ps pulses from a Kerr-lens mode-locked lamp-pumped Nd: YLF laser with a microdot mirror. Opt Lett 18:1825–1827.

Ramaswamy-Paye M, Fujimoto JG. (1994). Compact dispersion-compensating geometry for Kerr-lens mode-locked femtosecond lasers. Opt Lett 19:1756–1758.

Roskos H, Robl T, Seilmeier A. (1986). Pulse shortening to 25 ps in a cw mode-locked Nd:YAG laser by introducing an intracavity etalon. Appl Phys B 40:59–65.

Rothenberg JE. (1992). Space-time focusing: breakdown of slowly varying envelope approximation in the self-focusing of femtosecond pulses. Opt Lett 17:1340–1342.

Ruddock IS, Bradley DJ. (1976). Bandwidth-limited subpicosecond pulse generation in modelocked cw dye lasers. Appl Phys Lett 29:296–297.

Ruffing B, Nebel A, Wallenstein R. (1997). A 20-W KTA-OPO synchronously pumped by a cw modelocked Nd:YVO$_4$ oscillator-amplifier system, Conference on Lasers and Electro-Optics. OSA Tech Dig Ser Vol II, CWB2. New York: Opt Soc Am: p 199.

Sarukura N, Ishida Y, Nakano H. (1991). Generation of 50 fs pulses from a pulse compressed, cw, passively modelocked Ti:sapphire laser. Opt Lett 16:153–155.

Sennaroglu A, Pollock CR, Nathel H. (1993). Generation of 48-fs pulses and measurement of crystal dispersion by using a regeneratively initiated self-mode-locked chromium-doped forsterite laser. Opt Lett 18:826–828.

Sennaroglu A, Pollock CR, Nathel H. (1994). Continuous-wave self-modelocked operation of a femtosecond Cr^{4+}:YAG laser. Opt Lett 19:390–392.

Shank CV, Ippen EP. (1974). Subpicosecond kilowatt pulses from a modelocked cw dye laser. Appl Phys Lett 24:373–375.

Shirakawa A, Sakane I, Takasaka M, Kobayashi T. (1999). Sub-5-fs visible pulse generation by pulse-front-matched noncollinear optical parametric amplification. Appl Phys Lett 74:2268–2270.

Sorokina IT, Sorokin E, Wintner E, Cassanho A, Jenssen HP, Szipöcs R. (1997). 14-fs pulse generation in Kerr-lens modelocked prismless Cr:LiSGaF and Cr:LiSAF lasers: observation of pulse self-frequency shift. Opt Lett 22:1716–1718.

Spence DE, Kean PN, Sibbett W. (1991). 60-fsec pulse generation from a self-mode-locked Ti:sapphire laser. Opt Lett 16:42–44.

Spielmann C, Krausz F, Brabec T, Wintner E, Schmidt AJ. (1991). Femtosecond passive modelocking of a solid-state laser by dispersively balanced nonlinear interferometer. Appl Phys Lett 58:2470–2472.

Spühler GJ, Paschotta R, Kullberg MP, Graf M, Moser M, Mix E, Huber G, Harder C, Keller U. (2001). A passively Q-switched Yb:YAG microchip laser. Appl Phys B 72:285–287.

Spühler GJ, Südmeyer T, Paschotta R, Moser M, Weingarten KJ, Keller U. (2000). Passively mode-locked high-power Nd:YAG lasers with multiple laser heads. Appl Phys B 71:19–25.

Steinmeyer G, Sutter DH, Gallmann L, Matuschek N, Keller U. (1999). Frontiers in ultrasoft pulse generation: pushing the limits in linear and nonlinear optics. Science 286:1507–1512.

Sutter DH, Gallmann L, Matuschek N, Morier-Genoud F, Scheuer V, Angelow G, Tschudi T, Steinmeyer G, Keller U. (2000). Sub-6-fs pulses from a SESAM-assisted Kerr-lens modelocked Ti:sapphire laser: at the frontiers of ultrashort pulse generation. Appl Phys B 70:S5–S12.

Sutter DH, Steinmeyer G, Gallmann L, Matuschek N, Morier-Genoud F, Keller U, Scheuer V, Angelow G, Tschudi T. (1999). Semiconductor saturable-absorber mirror-assisted Kerr-lens mode-locked Ti:sapphire laser producing pulses in the two-cycle regime. Opt Lett 24:631–633.

Tomaru T, Petek H. (2000). Femtosecond Cr^{4+}:YAG laser with an L-fold cavity operating at a 1.2 GHz repetition rate. Opt Lett 25:584–586.

Tong YP, French PMW, Taylor JR, Fujimoto JO. (1997). All-solid-state femtosecond sources in the near infrared. Opt Commun 136:235–238.

Tsuda S, Knox WH, De Souza EA, Jan WY, Cunningham JE. (1995). Low-loss intracavity AlAs/AlGaAs saturable Bragg reflector for femtosecond mode locking in solid-state lasers. Opt Lett 20:1406–1408.

Uemura S, Torizuka K. (1999). Generation of 12-fs pulses from a diode-pumped Kerr-lens mode-locked Cr:LiSAF laser. Opt Lett 24:780–782.

Valdmanis JA, Fork RL, Gordon JP. (1985). Generation of optical pulses as short as 27 fs directly from a laser balancing self-phase moduation, group-velocity dispersion, saturable absorption, and saturable gain. Opt Lett 10:131–133.

Valentine GJ, Kemp AJ, Birkin DJL, Burns D, Balembois F, Georges P, Bernas H, Aron A, Aka G, Sibbett W, Brun A, Dawson MD, Bente E. (2000). Femtosecond Yb:YCOB laser pumped by narrow-stripe laser diode and passively modelocked using ion implanted saturable-absorber mirror. Electron Lett 36:1621–1623.

Walker SJ, Avramopoulos H, Sizer T. (1990). Compact mode-locked solid-state lasers at 0.5- and 1-GHz repetition rates. Opt Lett 15:1070–1072.

Walker AC, Kar AK, Ji W, Keller U, Smith SD. (1986). All-optical power limiting of CO_2 laser pulses using cascaded optical bistable elements. Appl Phys Lett 48:683–685.

Wang HS, LikamWa P, Lefaucheur JL, Chai BHT, Miller A. (1994). Cw and self-mode-locking performance of a red pumped Cr:LiSCAF laser. Opt Commun 110:679–688.

Weingarten KJ, Godil AA, Gifford M. (1992). FM modelocking at 2.85 GHz using a microwave resonant optical modulator. IEEE Photon Technol Lett 4:1106–1109.

Weingarten KJ, Keller U, Chiu TH, Ferguson JF. (1993). Passively mode-locked diode-pumped solid-state lasers that use an antiresonant Fabry-Perot saturable absorber. Opt Lett 18:640–642.

Weingarten KJ, Shannon DC, Wallace RW, Keller U. (1990). Two gigahertz repetition rate, diode-pumped, mode-locked, Nd:yttrium lithium fluoride (YLF) laser. Opt Lett 15:962–964.

Xu L, Tempea G, Spielmann C, Krausz F. (1998). Continuous-wave mode-locked Ti:sapphire laser focusable to $5 \times 10^{13}\,W/cm^2$. Opt Lett 23:789–791.

Yakymyshyn P, Pinto JF, Pollock CR. (1989). Additive pulse modelocked NaCl:OH-laser. Opt Lett 14:621–623.

Yan L, Ling JD, Ho P-T, Lee CH. (1986). Picosecond-pulse generation from a continuous-wave neodymium: phosphate glass laser. Opt Lett 11:502–503.

Yanovsky VP, Wise FW, Cassanho A, Jenssen HP. (1995). Kerr-lens mode-locked diode-pumped Cr:LiSGAF laser. Opt Lett 20:1304–1306.

Zhang Z, Nakagawa T, Torizuka K, Sugaya T, Kobayashi K. (1999). Self-starting modelocked Cr^{4+}:YAG laser with a low-loss broadband semiconductor saturable-absorber mirror. Opt Lett 24:1768–1770.

Zhang Z, Torizuka K, Itatani T, Kobayashi K, Sugaya T, Nakagawa T. (1997a). Self-starting mode-locked femtosecond forsterite laser with a semiconductor saturable-absorber mirror. Opt Lett 22:1006–1008.

Zhang Z, Torizuka K, Itatani T, Kobayashi K, Sugaya T, Nakagawa T. (1997b). Femtosecond Cr:forsterite laser with modelocking initiated by a quantum well saturable absorber. IEEE J Quantum Electron 33:1975–1981.

Zhou F, Malcolm GPA, Ferguson AI. (1991). 1-GHz repetition-rate frequency-modulation mode-locked neodymium lasers at 1.3 µm. Opt Lett 16:1101–1103.

2
Ultrafast Solid-State Amplifiers

François Salin
Université Bordeaux I, Talence, France

2.1 INTRODUCTION

Soon after the first demonstration of ultrashort pulses, it appeared that although the pulse duration was very short, its peak power was still in the kilowatt range, too small to be used in most nonlinear optics experiments. Amplifiers were designed to increase the energy of the pulse, and multigigawatt pulses were obtained in 1982 (Fork et al., 1982; Migus et al., 1982). At that time, femtosecond lasers were based on dyes, and the pulses were amplified by simply sending them into a series of dye cells pumped by a nanosecond green laser. The efficiency was in general below 1%, and the pulse was hidden in a large amplified spontaneous emission (ASE) pedestal. In the meantime people were trying to decrease the duration of the pulse produced by a large-scale Nd:glass system. The main difficulties were linked to damage in the amplifier chain due to the very high peak power of the amplified short pulses.

A new era started when Mourou and coworkers introduced the concept of chirped pulse amplification in optics (Strickland and Mourou, 1985; Maine et al., 1988). This idea had been developed during World War II for radar but had never been applied to lasers. The peak power of short pulses quickly increased to the terawatt level. The final stage of this evolution was reached when femtosecond pulses were obtained in Ti:sapphire (Spence et al., 1991), opening the way for ultrashort and ultra-intense pulses. The first amplification of 100 fs pulses in a solid-state amplifier (Squier et al., 1991) and the first demonstration of a kilohertz repetition rate amplifier (Salin et al., 1991) followed this discovery by only a few months. Although great progress has been made in the performance of these systems (see Fig. 1), nowadays high-intensity laser chains are still based on concepts developed

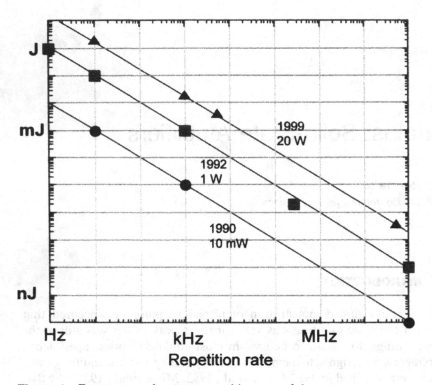

Figure 1 Energy per pulse versus repetition rate of the system.

in the early 1990s. Surprisingly, the best performance of various systems at a given time correspond to the same average power. Over the years, we came from 10 mW average power with dye systems to over 20 W produced by oscillators at a 100 MHz repetition rate as well as by 10 Hz amplifiers.

This chapter describes the basic phenomena involved in the amplification of ultrashort pulses, with some emphasis on the specific difficulties linked to the very short duration and very high peak power achieved. We first give a simplified theoretical background of pulse amplification and then introduce the concept of pulse stretching and compression. Finally the chapter concludes with some technological hints for the design of complete solid-state ultrafast amplifers.

2.2 AMPLIFICATION

Amplification of short pulses is basically similar to that of any optical radiation. A laser medium is pumped by an external source of radiation, which in most cases is a laser. If the excited state lifetime of the amplifier medium is assumed to be longer than any other time involved in the process, one can easily calculate the small signal gain of the amplifier (Siegman, 1986):

$$g_0 = \frac{J_{sto}}{J_{sat}} \tag{1}$$

where J_{sto} is the pump fluence (= energy density in joules per square centimeter) stored in the medium and J_{sat} is an intrinsic parameter called the saturation fluence (J/cm^2).

J_{sat} is given by the inverse of the gain cross section σ_e:

$$J_{sat} = \frac{h\nu}{\sigma_e}$$

where h is the Planck constant and ν the frequency of the transition.

J_{sto} can be deduced from experimental values:

$$J_{sto} = \frac{E_p \alpha}{S} \left(\frac{\lambda_p}{\lambda_L} \right) \tag{2}$$

where E_p is the pump energy incident on the amplifier, α is the total absorption coefficient of the pump radiation, and S is the pump beam cross section. λ_p and λ_L are respectively the pump and laser wavelengths. A pulse with a small energy E_{in} will be amplified by passing through the amplifier, and its output energy will be

$$E_{out} = E_{in} \exp(g_0) \tag{3}$$

This very simple formula is valid for only small input fluences. As soon as the input fluence reaches a level comparable to that of the saturation fluence J_{sat} of the amplifier, one must use a more complicated expression derived by Frantz and Nodvick (1963) that is valid only for a four-level system:

$$J_{out} = E_{out}/S = J_{sat} \log[g_0[\exp(J_{in}/J_{sat}) - 1] + 1] \tag{4}$$

where J_{out} is the output fluence of the amplified pulse and the other parameters are as defined above. It is highly recommended to spend some time on this formula, which is the basis of pulse amplification. In particular, one can see that the amplifier gain J_{out}/J_{in} decreases while the extraction efficiency $(J_{out} - J_{in})/J_p$ increases as J_{in} increases. Amplifying light pulses is thus always a trade-off between gain and efficiency. Figure 2 shows

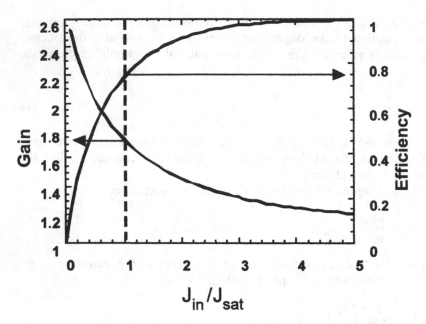

Figure 2 Gain and extraction efficiency versus input fluence normalized to the saturation fluence of the gain medium. The stored fluence is equal to the saturation fluence.

the gain and the extraction efficiency as a function of the input fluence normalized to the saturation fluence. The stored fluence has been arbitrarily set to the saturation fluence. The small signal gain per pass in the amplifier is 2.8 but quickly decreases toward 1. Conversely, the extraction efficiency slowly increases toward its maximum value. It is important to note that even for very large input fluences the output energy can reach only a fraction of the pump energy. This is due to the so-called quantum defect, which is simply the difference in the energy of the pump and signal photons. Because one pump photon can give rise to only one excited atom, which can in turn give only one signal photon, the maximum efficiency (extracted energy divided by pump energy) is given by the ratio of the pump to signal wavelengths: $\eta_{max} = \lambda_p/\lambda_s$.

Figure 2 can be split arbitrarily into two regions. For low input fluence the gain is high and can be used for the first stages of amplification to quickly increase the energy of pulses produced by oscillators. The price to be paid is low efficiency. On the other hand, for high enough input fluence, large extraction efficiency can be obtained at the expense of single-pass gain. A typical amplifier chain will include several stages working in different states. The first stages work in the small-signal regime and are used to bring the pulse energy to a value high enough to enter the extraction regime for the final stages. As a general rule, the goal is to reach an input fluence close to the saturation fluence, which gives an interesting trade-off between gain and extraction.

The role of the parameter J_{sat} in the design of an amplifier is obvious and will dictate the precise technology that one can use with a particular laser medium. A few examples of typical saturation fluences are given in Table 1. One can see in this table that two groups can be found: on the one hand, liquids and gases with low J_{sat}, and on the other hand, solids with large J_{sat}. This distinction leads naturally to two different types of amplifiers.

Dye amplifiers were the first to be used for femtosecond pulses (Fork et al., 1982; Migus et al., 1982; Knox et al., 1984). Since then, solid-state materials have surpassed them, and we will not describe their technology in detail. The typical structure of a dye amplifier is a series of dye cells, in general four or five, that are pumped transversely by a nanosecond blue or green laser (excimer or Nd:YAG/2ω). The gain per stage can be higher than 1000, with gain values after four stages reaching 10^6. The major drawback of these amplifiers is the high gain. It is very difficult to avoid having photons from the fluorescence of the first stage become amplified in the chain. When nanosecond pulses are used to pump the amplifier, the output of the amplifier consists of a short amplified pulse on the top of a very broad pedestal. This nanosecond pedestal is called amplified spontaneous emission (ASE) and corresponds to noise photons following the same path as the femtosecond pulse and hence seeing the same gain. Several techniques are used to reduce the amount of ASE—for example, saturable absorbers or spatial filtering—but the ASE accounts in general for 5–10% of the total

Table 1 Saturation Fluence of Typical Amplifier Media Used in Short-Pulse Amplification

Amplifier medium	J_{sat}
Dyes	$\sim 1\,\mathrm{mJ/cm^2}$
Excimers	$\sim 1\,\mathrm{mJ/cm^2}$
Nd:YAG	$0.5\,\mathrm{J/cm^2}$
Ti:Al$_2$O$_3$	$1\,\mathrm{J/cm^2}$ at 800 nm
Cr:LiSAF	$5\,\mathrm{J/cm^2}$ at 830 nm
Nd:glass	$5\,\mathrm{J/cm^2}$
Yb:glass	$100\,\mathrm{J/cm^2}$
Alexandrite	$22\,\mathrm{J/cm^2}$

output energy. The efficiency of a dye amplifier pumped by the second harmonic of a Nd:YAG laser is low, typically 0.1–0.3%. This kind of amplifier has been adapted to several kinds of pump lasers with various characteristics. High repetition rate, low energy pulses have been obtained using copper vapor lasers (Know et al., 1984). Direct amplification in an excimer amplifier has been the subject of several studies (Szatmari et al., 1987). Excimers, owing to their low saturation fluence, suffer the same advantages (high gain) and drawback (ASE) as dyes with the complication of working in the UV. They are still used in some high-energy systems because they can be relatively easily scaled up to large amplifier diameters and hence large energies. Except for very particular applications, dye amplifiers have now been abandoned and are replaced by their solid-state counterpart.

As can be seen in Table 1, solid-state materials in general have high saturation fluence leading to much smaller gain per pass in the amplifier than in dyes or excimers. In order to extract the energy stored in the medium, one must use multiple passes in the amplifier. Two techniques have been developed: regenerative (Murray et al., 1980) and multipass amplification (Georges et al., 1991). Multipass amplification is based on bow-tie types of amplifiers in which the different passes in the amplifier are separated geometrically (see Fig. 3). The number of passes is limited by the geometrical complexity of the design and by the increasing difficulty of focusing all the passes in a single spot on the crystal. The typical number of passes is four to eight. For higher gains, several multipass amplifiers can be cascaded (Chambaret et al., 1996). This technique is simple and cheap but requires long and tedious adjustments, and the crystal is used close to the damage threshold in order to keep the gain high enough to be compatible with the small number of passes. It is nevertheless the only technique that can be used to produce energies above 50 mJ.

In regenerative amplification (see Fig. 4) a pulse is trapped in a laser resonator. Regardless of the gain per pass, the pulse is kept in the resonator

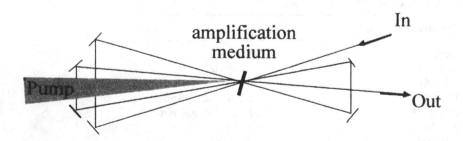

Figure 3   Generic multipass amplifier setup.

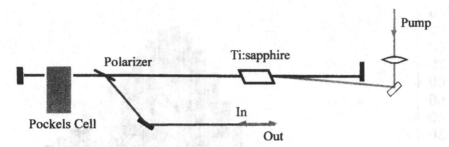

Figure 4 Generic regenerative amplifier setup.

until it has extracted all the energy stored in the amplification medium. Trapping and dumping the pulse is done using a Pockels cell and a broad-band polarizer. The Pockels cell is initially set to be equivalent to a quarter-wave plate either by slightly rotating it or by applying a $V_{\lambda/4}$ bias voltage. A vertically polarized input pulse reflected off the polarizer becomes horizontally polarized after a double pass through the Pockels cell and is transmitted by the polarizer. After passing twice through the gain medium, it is transmitted again by the polarizer. If no voltage (besides the bias voltage) is applied to the Pockels cell, a double pass through the cell again rotates the polarization and the pulse is reflected off the polarizer and leaves the cavity. If, however, a $\lambda/4$ voltage step is applied to the Pockels cell when the pulse is in the right part of the cavity, the Pockels cell becomes equivalent to a half-wave plate and a double pass through it does not change the polarization. In that case the horizontally polarized pulse present in the cavity at that moment will not be affected by the Pockels cell and is trapped in the reso-nator. The pulse can stay in the cavity until it reaches saturation, and a second $\lambda/4$ voltage step is applied to the Pockels cell to extract the pulse. From a physical point of view the only difference between multipass and regenerative amplification is the way input and output beams are distin-guished: In a multipass amplification this is done geometrically, and in a regenerative amplifier polarization is used. In both cases the repetition rate of the amplifier is given by that of the pump laser.

The propagation in a regenerative or multipass amplifier can be easily simulated by using a Frantz–Nodvick model (Frantz and Nodvick, 1963). Equation (4) is applied at each pass, taking into account the depletion of the excited state population due the gain process. One typically obtains the picture given in Figure 5. From this figure it is obvious that the pulse should be extracted from the amplifier after 20 passes in this particular example. This figure also shows that the gain per pass tends to decrease when the energy reaches its maximum. This is due not to the fact that

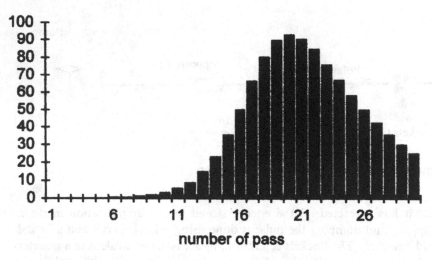

Figure 5 Theoretical evolution of pulse energy versus the number of passes in a regenerative amplifier.

the pulse fluence gets close to the saturation fluence, but to the depletion of the excited state population, in other words, to the extraction of the energy. In contrast to single-pass amplifiers described above, a multipass amplifier or a regenerative amplifier can extract most of the energy stored in the gain medium without reaching the saturation fluence.

Multipass amplifiers are used in two different applications. The first one is for amplifying weak (typically nanojoule) pulses to the millijoule level. In that case a gain of 10^6–10^8 is required. Since, practically speaking, the maximum number of passes is six to eight, the gain per pass has to be around 10. For solid-state gain media with relatively high saturation fluence, this translates into a small pump spot on the gain medium. In order to match the signal beam size on the gain medium to that of the pump at each pass, the multipass amplifier must include focusing optics. A typical design is given in Figure 6.

Two spherical mirrors with an aperture in the middle to let the pump beam reach the gain medium are used in a confocal geometry (Georges et al., 1991). The radii of curvature of the two mirrors are slightly different, and the beam walks slowly from the center to the side of the system. A major drawback of this design is the fact that it is equivalent to a confocal resonator with a very large Fresnel number. The diffraction losses are very small, and spontaneous emission can be easily amplified. One solution to reduce the ASE energy is to limit the effective diameter of the optics with

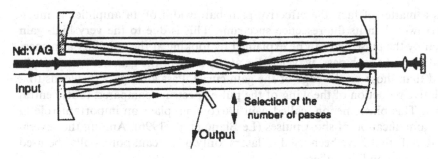

Figure 6 Multipass amplifier with a refocusing geometry.

an aperture and include a Pockels cell, set in general after the pulse has traveled half the total number of passes.

A second typical use of a multipass amplifier is to extract a large amount of energy. Because a regenerative amplifier is based on a laser cavity, the beam size is limited to that of the TEM_{00} mode supported by the cavity. Resonators with TEM_{00} modes larger than 1 mm are either complicated or cumbersome, and the energy is typically limited to 10 mJ by the damage fluence of the optical element. Although some new schemes have been demonstrated, based on gain guiding or phase plates (Salin et al., 1992), typical regenerative amplifiers are limited to energies in the range of 10 mJ. To produce greater energies one must use a larger beam size, and multipass amplifiers in which the beam diameter can be made arbitrarily large are used. For beam diameters larger than 1 mm, divergence is no longer a problem, and a very simple scheme based on flat mirrors can be used (see Fig. 7). In order to limit the pump fluence, the gain medium is generally pumped on both sides.

The amplifiers presented here can be used with any kind of pulses and any kind of laser media and are not specific to ultrashort pulses. When femtosecond pulses are to be amplified, new problems arise. First of all the amplifier bandwidth must be broad enough to support the pulse spectrum.

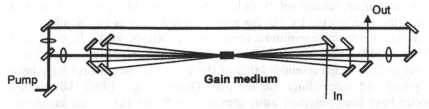

Figure 7 Typical high-energy multipass amplifier. Depending upon the pump energy, this design can produce pulses with energy anywhere from 1 mJ to 10 J.

As a matter of fact, the effective gain bandwidth of an amplifier is much narrower than its fluorescence spectrum. This is due to the very high gain seen by the pulse and can be explained by looking at the amplifier as a spectral filter. At each pass, wavelengths at the peak of the gain are more amplified than those on the edge of the gain curve. Pass after pass this leads to a relative reduction of the wing of the pulse spectrum compared to its central part. This phenomenon, called gain narrowing, plays an important role in the amplification of short pulses (Le Blanc et al., 1996). Among the several materials that have been used in lasers, only a few can, potentially, be used in femtosecond amplifiers:

Ti:sapphire (Ti:Al$_2$O$_3$)	650–1100 nm
Alexandrite (Cr:Be$_2$O$_3$)	700–820 nm
Colquirites (Cr:LISAF, Cr:LICAF, ...)	800–1000 nm
Fosterite	1250–1300 nm
Yb-doped materials	1030–1080 nm
Nd:glasses	1040–1070 nm

Figure 8 shows an example of the gain curve of Ti:sapphire as well as the spectrum of a 10 fs pulse. The 65 nm spectrum of this very short pulse seems to fit easily in the gain curve, but a simple calculation shows that this spectrum is reduced to a mere 32 nm after amplification by a factor of 10^6. Indeed, the actual gain of the amplifier (ratio between the output and the input) is not the parameter of importance. Gain narrowing is related to the total gain delivered by the amplifier even if part of this gain has been used to cancel the optical loss of the amplifier.

The total gain of a regenerative amplifier with a 10% loss per pass and a small signal gain of 2 is typically 10^7 for an amplification factor of 10^6. In that sense, multipass amplifiers have higher gain per pass and lower loss and do exhibit lower gain narrowing than regenerative amplifiers for the same amplification factor.

Because gain narrowing was a severe limitation to the amplification of ultrashort pulses, several solutions have been proposed. The basic idea is to flatten the gain curve. By far the most effective way is to include in the amplifier a filter whose minimum transmission corresponds to the maximum of the amplifier gain curve (including the mirror's reflectivity curve) and whose spectral shape resembles that of the gain. Different setups have been proposed and successfully implemented (Barty et al., 1996). Ultrabroad spectra have been obtained using thin air gap Fabry-Perot interferometers (see Fig. 9). The main difficulty is to stabilize these etalons against thermal drift. A more robust solution consists of a thin birefringent filter (Bagnoud

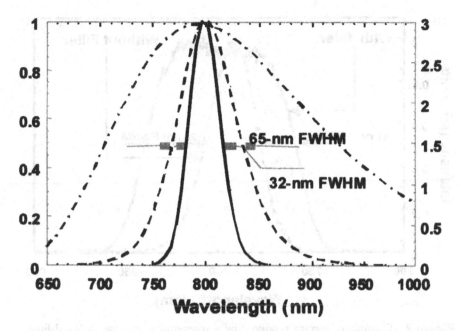

Figure 8 Gain narrowing in a Ti:sapphire regenerative amplifier. The dashed line gives the gain cross section of Ti:sapphire (right-hand scale). The dot-dashed line gives the spectrum of a 10 fs input pulse, and the solid line is the spectrum of the sample pulse after amplification from 1 nJ to 1 mJ.

and Salin, 2000). These filters are easily implemented in regenerative amplifiers, and their use has been called regenerative gain filtering. This technique has made it possible to produce spectra as broad as 120 nm, and further improvement is now limited by the bandwidth of mirrors and polarizers.

2.3 PULSE COMPRESSION

2.3.1 Dispersion Control

As shown in the introduction, chirped pulse amplification (CPA) has become the common technique for avoiding nonlinear effects during amplification of short pulses in solid-state material (Strickland and Mourou, 1985). Furthermore, efficient amplification with minimal material path lengths requires that the amplifier be operated at or above the saturation fluence (1 J/cm^2 at 800 nm for Ti:sapphire) of the gain medium. Most coatings and materials cannot be operated much above 5 GW/cm^2, which is generally the rough rule of thumb used in designing the upper tolerable intensity

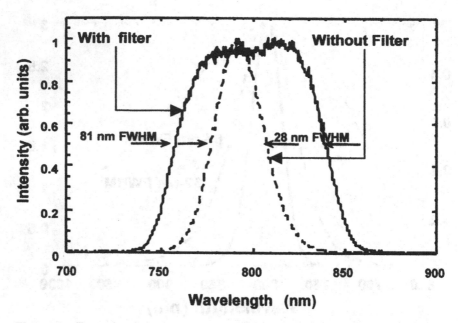

Figure 9 Example of spectra produced by a regenerative amplifier (dashed line), and the same amplifier including a filter to flatten the gain profile (solid line).

limit within a laser (Barty et al., 1995). This number, in conjunction with the desire to run at or above the gain saturation of the amplifier, gives the minimum pulse duration that one would like to use in short-pulse amplifiers. An initial 100 fs pulse cannot be amplified directly without risking severe damage in the amplifier. The first solution was to increase the amplifier diameter, but the limit was rapidly reached and the maximum peak power produced by short-pulse lasers saturated until Mourou and coworkers realized that direct amplification could not be the way to go. Their idea is sketched schematically in Figure 10. The upper part of the figure (above the damage limit line) is a forbidden area in which the pulse must travel only in a low density medium (vacuum or gas). In order to reach the goal depicted by the upper stars, one must first increase the pulse duration. Then the energy can be safely increased until the saturation fluence for efficient energy extraction is reached. Finally, the pulse duration is brought back to its initial value but must avoid passing through any dense material.

This need to avoid nonlinear effects and damage while reaching the saturation fluence of the gain medium places the stretched pulse duration at or above 200 ps for materials such as Ti:sapphire. This means that a 20 fs pulse must be stretched by a factor of 10,000. Systems that use

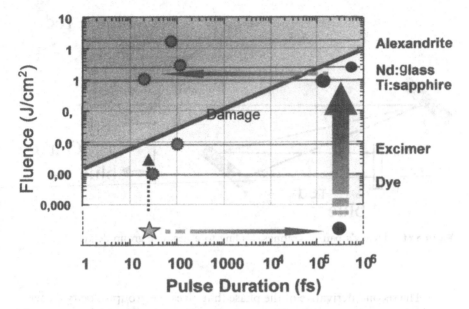

Figure 10 Illustration of the chirped pulse amplification strategy.

stretching factors significantly below this damage/saturation limit will produce less high-order distortion and thus more easily achieve accurate recompression (Backus et al., 1997; Lenzner et al., 1995). However, this type of system will ultimately be limited in its ability to efficiently extract all stored energy by the onset of optical damage to the amplifier and its optical components.

The need to stretch and then recompress the pulse was solved by using dispersive delay lines. In 1969 Treacy (Treacy, 1969) showed that a pair of gratings introduces a delay that depends upon the wavelength. In other words, the blue path is shorter than the red one, and an initial "white pulse" will exit a pair of gratings with a chirp and a much longer duration (see Fig. 11). A single pass through the pair of gratings produces a spatially chirped beam, and a mirror is generally used to double pass the grating pair and in doing so double the dispersion introduced by the system. The spectral phase introduced by Treacy's pair of gratings is given by the equation

$$\Phi(\omega) = 2\frac{\omega}{c}\left(\frac{G}{\cos\theta_i}\right)\cos(\theta_i - \theta_r(\omega))$$

where G is the slant distance the gratings, and θ_i and θ_r are the incident and diffracted angles, respectively.

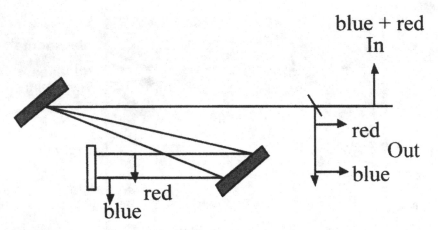

Figure 11   Treacy's pair of gratings, leading to a negative group delay.

The second derivative of the phase that gives the group velocity dispersion is then

$$\frac{d^2\Phi}{d\omega^2} = \frac{-4\pi^2 Gc}{\omega^3 d^2 \cos^3(\theta_r(\omega))} \tag{5}$$

where d is the gratings' groove density. This equation shows that the group delay is not perfectly linear with frequency. It depends upon the angle of incidence on the grating and the groove density and linearly increases with the distance between gratings.

This grating pair makes it possible to chirp a short pulse or to compress a positively chirped pulse, but, as Eq. (5) shows, the sign of the dispersion is negative regardless of the parameters. In 1987 Oscar Martinez introduced a system (Martinez, 1987) based on a pair of gratings including a -1 magnification telescope that provides a positive dispersion (Fig. 12). Coupling this stretcher to a Treacy's grating pair made it possible to stretch a short pulse and then recompress it at the end of the amplifier. All the elements to build a CPA system were ready, and the design of these systems has not changed much since that time.

It is important to note, however, that in chirped pulse amplification systems the higher the stretching factor, the harder it is to achieve perfect recompression. In addition, for pulses below 100 fs, the dispersion introduced by the amplification material and optics also has to be taken into consideration. Furthermore, compressing a 20 fs pulse to its transform-limited

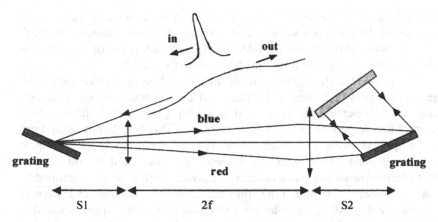

Figure 12 Generic stretcher as first described by Martinez (1987).

duration means that the final residual group delay of the pulse must be almost constant over a 100 nm bandwidth.

Nowadays oscillators can produce ultrashort pulses, so an optimized chirped pulse amplifier should be capable of amplifying the pulses without increasing their duration. This translates into the fact that the total dephasing along the amplifier chain must be identically null over the whole spectral range:

$$\Phi_s(\omega) + \Phi_a(\omega) + \Phi_c(\omega) = 0 \qquad \text{for all } \omega$$

where Φ_s, Φ_a, and Φ_c are, respectively, the dephasing introduced by the stretcher, the amplifier material, and the compressor.

The spectral phase of any component $\Phi(\omega)$ can be expanded in a Taylor series as

$$\phi(\omega) = \phi(\omega_0) + \phi_1(\omega - \omega_0) + \phi_2(\omega - \omega_0)^2 + \phi_3(\omega - \omega_0)^3$$
$$+ \phi_4(\omega - \omega_0)^4 + \cdots \qquad (6)$$

where $\phi_1(\omega)$ is the group delay of the pulse, and $\phi_2(\omega)$, $\phi_3(\omega)$, and $\phi_4(\omega)$ are, respectively, the second-, third-, and fourth-order dispersion terms.

For a system compensated through fourth order, the condition to obtain near-transform-limited pulses is to eliminate one by one (or globally) the terms introduced by the stretcher, the amplifiers, and the compressor:

$$\phi_i^{\text{stretcher}} + \phi_i^{\text{amplifiers}} + \phi_i^{\text{compressor}} = 0 \qquad \text{with } i = 2, 3, 4 \qquad (7)$$

To compensate for dispersion terms we need at least as many variables as the order of the phase we want to counterbalance. Some common "knobs" used to achieve this balance include grating separation (second-order control) and grating angle of incidence (third-order control).

Two separate strategies for designing a CPA system have been used. The first consists in finding a stretcher and a compressor design with perfectly equal dispersion terms. In that case, Eq. (7) can be fulfilled only if the amplifier dispersion is as small as possible. This first family of systems uses amplifiers with minimum dispersive materials such as multipass amplifiers associated with specially designed stretchers that exhibit a group delay (derivative of the phase) exactly opposed to that of the compressor. It can be shown (Martinez, 1987) that the stretcher presented in Figure 12 introduces a dephasing given by the same function as a compressor with a distance between the compressor gratings equal to

$$G_{eq} = 2f - S_1 - S_2 \tag{8}$$

This relation, unfortunately, is true only if the incidence angles are the same in the stretcher and compressor and if the optical system in the stretcher is perfect. The first problem comes from chromatic aberration introduced by the lenses. This aberration makes it impossible to use lens-based stretchers with pulses shorter than 150 fs, and most modern stretchers include mirrors. Even with mirrors, however, special designs must be used to cancel all the geometrical aberrations.

The aberration-free stretcher developed by Chériaux et al. (1996) and simultaneously by Du et al. (1995) is one of the new generations of stretcher designs (see Fig. 13). Using spherical mirrors in an Offner triplet design, this system operates essentially aberration-free, which means that it is well matched to a Treacy compressor. The main advantage of this configuration is that not only is it symmetrical and exhibits no chromatic aberration, but also the precise relationship between the mirror ROCs allows the spherical aberration and astigmatism of one mirror to be compensated for by the other, thereby offering the elimination of on-axis coma. However, when the amplifier material is added to the system, the compressor grating separation must be increased. This results in a net mismatch of the overall dispersion of the system. Adjusting the angle and the separation in the compressor can compensate for second- and third-order dispersion. On the other hand, this operation leaves a residual fourth-order and higher dispersion that cannot be compensated for. The shortest pulse is then obtained by balancing second-order against fourth-order, leading to a global minimization of the residual phase distortion (Bagnoud and Salin, 1998). For amplifiers with just a few passes and a minimal path length, the residual phase distortion is very small, and pulses as short as 25 fs have been produced.

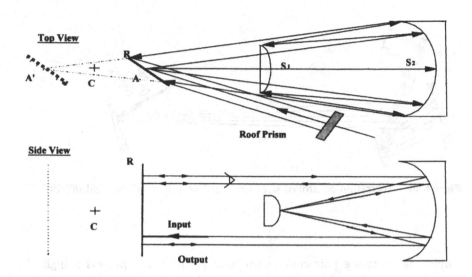

Figure 13 Aberration-free stretcher based on an Offner triplet. C is the common center of curvature, A is the position of the grating, and A' is its image through the Offner triplet.

For large material path lengths, as in the case of regenerative amplifiers, or for large stretched pulse duration, other dispersion compensation systems have to be considered. One of these, which has made possible the compression of pulses as short as 18 fs, is a dispersion-balancing expander (Lemoff and Barty, 1993) (see Fig. 14). This system is able to control the phase delay over a bandwidth of more than 100 nm, leaving only a residual quintic phase. Because it uses an off-axis cylindrical telescope, the stretcher introduces spherical aberration onto the diffracted beam, which in turn results in a modification of the spectral dispersion of the expander. This spectral dispersion change dramatically alters the fourth-order dispersion term. The right amount of aberration can be introduced by the stretcher, such that the combination of stretcher, material path length, and compressor results in only a few femtoseconds of group delay error over a bandwidth of 100 nm for pulses stretched to 1 ns (a stretching/compression ratio >100,000).

In addition to the cylindrical mirror expander system, Kane and Squier (1997b) theoretically demonstrated that it is possible to use an aberration-free stretcher and to compensate for large material path lengths in a compressor whose grating groove density differs from those found in the stretcher. Their calculations indicate that an aberration-free CPA system can also provide quintic-phase-limited operation simply by

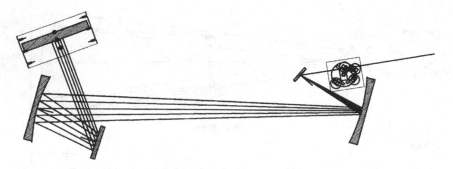

Figure 14 Aberration-controlled stretcher based on off-axis cylindrical mirrors.

employing a grating pair compressor with gratings that possess a higher groove density than those in the stretcher. Experiments by Squier et al. (1998) demonstrated that even with a very high groove density grating $(2000 \, \text{mm}^{-1})$ compressor, 50 fs pulse duration can be obtained. Using a $1200 \, \text{mm}^{-1}$ stretcher and $1400 \, \text{mm}^{-1}$ compressor grating combination, Bagnoud and Salin (2000) showed that it is possible to obtain near-transform-limited pulses as short as 18 fs. Note additionally that this system has the advantage of ease of computation. No ray tracing code is necessary to determine phase compensation due to the aberration-free nature of the stretcher. The system is also straightforward to align. With this mixed grating system, near-transform-limited pulses can be obtained even with large amounts of material as in regenerative amplifiers. It should also be noted that in general the diffraction efficiency of compressor gratings has been reported to be higher for 1400 or $1500 \, \text{mm}^{-1}$ gratings than for $1200 \, \text{mm}^{-1}$ gratings; thus mixed grating systems may also be more efficient than those using lower groove density gratings (Britten et al., 1996).

Even though several designs have been able to produce short high-power pulses, the quality of the amplified pulses is still an issue. The limitation in the contrast ratio of short pulses stretched by the above-mentioned systems and recompressed by a standard compressor have been studied using a three-dimensional ray tracing program. It was found that the most important limitation on the contrast comes from beam clipping on the edges of the mirrors in the stretcher. This leads to a rule of thumb that says that, for a sech^2 pulse, in order to keep the contrast ratio higher than 10^6 at delays of $\pm 5 \times$ pulse duration, the stretcher bandwidth must be larger than 4 times the pulse bandwidth. Both experiments and calculations show that the uncompensated dispersion left when both the angle of incidence and the length of the compressor have been optimized does not affect the pulse contrast ratio for pulses as short as 30 fs. Conversely, a dramatic dependence of the pulse contrast ratio on the mirror quality was found experimentally.

Figure 15 shows a set of autocorrelation traces obtained using the same stretcher–compressor setup with stretcher mirrors of increasing quality. This figure clearly indicates the fundamental importance of mirror quality to the pulse quality. This can be easily understood because spatial deformations in or near the Fourier plane directly translate into spectral dephasing and hence into pulse wings or pre-pulses as shown in Figure 15. In order to match the limitation set by the spectral clipping or uncompensated dispersion in the amplifier, the stretcher mirrors must be specified at $\lambda/20$ or better. This specification depends on the number of reflections on the mirror but not on the design of the stretcher (spherical or cylindrical mirrors, Offner or classical configuration, etc.). For ultrashort pulses and optimized stretcher–compressor configuration the mirror quality is the limitation of the system. To increase the contrast, several groups are

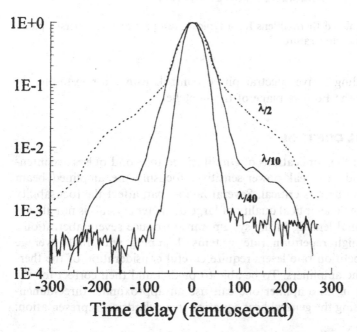

Figure 15 Autocorrelation trace of an initial 20 fs pulse after passing through a stretcher set at zero dispersion for increasing mirror quality.

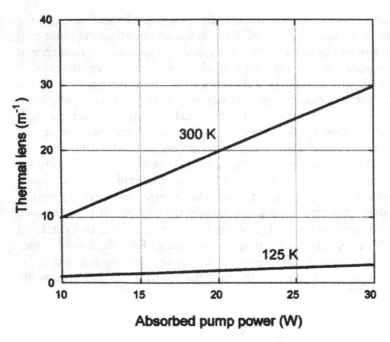

Figure 16 Calculated thermal lens for a typical multipass amplifier versus pump power and cooling temperature.

considering adding active spectral phase control, using, for example, an optical valve in the Fourier plane of the stretcher.

2.4 THERMAL CONTROL

Because most of the applications of amplified femtosecond pulses are intensity-sensitive and not peak-power-sensitive, focusing the amplified beam over a very small area is critical. Several factors can affect the focusability of the final beam. The optical quality of large diameter crystals is not always good, and thermal lensing in the system can introduce severe aberrations, especially for high repetition rate systems. Pumping with high average power, high repetition rate lasers require careful consideration of the thermal effects in the amplifiers. To be able to predict and then correct for the thermal lensing in an amplifier one can use an approximate three-dimensional model, using the general heat equation in a cylindrical representation:

$$\frac{1}{r}\frac{\partial}{\partial r}\left(r\frac{\partial T}{\partial r}\right) + \frac{q}{k} = 0 \tag{9}$$

where T is temperature, q the heat source in W/cm^3, and k the thermal conductivity in $W/(m \cdot K)$. We consider, for instance, a Ti:sapphire rod as a cylinder in contact with a copper block cooled with water. The Ti:sapphire crystal is pumped in the center, with a pump laser providing the heat. With this model, we find a quadratic dependence on the temperature in the pumped region:

$$T_1(r) = -\frac{q}{4k}r^2 + C_1 \tag{10}$$

and a logarithmic behavior out of the pumped region in the Ti:sapphire crystal:

$$T_2(r) = C_2 \ln r + C_3 \tag{11}$$

with C_1, C_2, C_3 constants that depend on the pump power, thermal conductivity, and radius of the system.

The index of refraction follows the temperature profile as

$$n(r) = n_0 + \Delta T \frac{dn}{dT} \tag{12}$$

which implies that it is parabolic in the pumped region and logarithmic outside the pumped region. A classical spherical lens with the appropriate focal length can compensate for the parabolic part of the index. The logarithmic part of the index profile induces aberrations that cannot be compensated for by classical spherical optics and are detrimental to the beam quality.

This model can give an estimate of the output beam profile depending upon the ratio between the pump beam diameter and the amplified beam diameter, for instance, and allows to estimate the influence of thermal aberrations on the focusability of the beam. The aberrations induced by the non quadratic portion of the thermal lens increase as the ratio of the amplified beam radius to the pump beam radius. As an example, with a 20 W average pump power and a 1 mm pump beam diameter, the aberrations are so strong for a 2 mm diameter input beam that the focused far-field beam profile is increased by a factor of 3, resulting in a factor of 10 reduction in peak intensity. For any input beam diameter that is smaller than the pump laser, we calculate that the aberrations are, in fact, negligible. Consequently, the thermal aberrations are quite strong outside the pumped area, and the best solution is therefore to work with pump and laser beams of the same size, to simultaneously maximize extraction efficiency and beam quality.

The thermal lens can be partially compensated by difference methods. In multipass amplifiers, one can use corrective optics such as diverging lenses. However, working with strong thermal lenses requires the addition of corrective optics for each pass in the amplifier. Liquid nitrogen cooling of the Ti:sapphire crystal in order to increase the thermal conductivity has also been used (Backus et al., 1997). This reduced the thermal lens by a factor of 12 but suffers from the fact that it involves a liquid nitrogen–cooled cell for crystal cooling. Alternatively, this severe thermal lens can be used to one's advantage. Using the calculated thermal lens based on the spherical portion of the thermal index gradient, a multipass amplifier can be constructed such that the thermal lens forms a stable lens guide, or eigenmode. For a particular combination of amplifier length and beam diameter, the crystal re-images the beam on itself as in an optical resonator (Salin et al., 1997).

The ultimate solution consists in introducing a deformable mirror into the amplifier chain that is fed back by a beam spatial phase measurement to compensate for any phase front deformation. Although still complicated, these deformable mirrors have shown excellent performance and are the best and most efficient way to obtain a diffraction-limited beam on target (Albert et al., 2000).

2.5 TYPICAL FEMTOSECOND SOLID-STATE AMPLIFIERS

As we saw in the previous section, a typical high-power femtosecond chain includes four fundamental elements: an oscillator, a stretcher, an amplifier, and a compressor. An example of such a system based here on Ti:sapphire is depicted in Figure 17. An oscillator produces 100 fs pulses at a repetition rate of typically 100 MHz. After passing through the stretcher the pulse duration increases up to 400 ps and can be amplified. Depending on the pump laser the Ti:sapphire amplifier can work at repetition rates from 10 HZ (Squier et al., 1991) to 1 kHz (Salin et al., 1991). Similar techniques can be used with other wavelengths and laser media. The gain medium is a Ti:sapphire crystal pumped by the second harmonic of a Nd:YAG laser running at 10 Hz. The very same design is used at 1000 Hz by replacing the Nd:YAG laser by a Q-switch Nd:YLF laser. Once trapped in the cavity the pulse is amplified at each pass in the crystal with a typical gain per pass of 2. After 15–20 round-trips the pulse saturates the gain and reaches its maximum energy. It is extracted from the amplifier and sent into a multipass power amplifier and finally into the compressor. There it recovers its initial duration, and pulses as short as 30 fs with energies in the joule range can be obtained (Chambaret et al., 1996). The efficiency of typical amplifiers can be in excess of 10%.

Recently, pulses as powerful as 100 TW have been obtained with this kind of system. Figure 18 shows an example of a very high energy

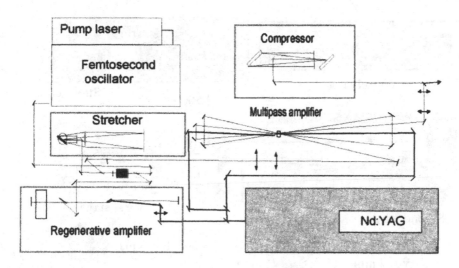

Figure 17 Typical CPA amplifier setup including a stretcher, a regenerative amplifier, a multipass amplifier, and a compressor.

femtosecond amplifier built at the Laboratoire d'Optique Appliquée (Palaiseau, France). This system uses only multipass amplifiers in order to increase the contrast ratio (no pre-pulse) and to limit the effect of gain narrowing. Without any spectral compensation this system can produce 30 fs pulses with energies in excess of 2.4 J. For very high energy extraction and good beam quality, a cooling system has been installed in the last amplifier (not shown in the figure), and extra pump energy is used at this stage. An alternative system has been built at JEARY (Japan), using a regenerative amplifier as the first stage. Including a Fabry-Perot etalon in the resonator made it possible to produce very large spectra, and 20 fs, 2 J (100 TW) pulses have been produced at 10 Hz (Yamakawa et al., 1998).

Although less powerful in terms of peak power, high repetition rate systems have reached the terawatt level (Bagnoud and Salin, 2000). The laser system built at CELIA (Université Bordeaux I, France) and producing 18.5 fs, 20 mJ (1 TW) pulses is shown in Figure 19. This system also uses regenerative pulse shaping as well as different gratings [the so-called mixed grating scheme (Kane and Squier, 1997b)] in the compressor and stretcher. Excellent beam quality is ensured despite strong thermal lensing by using thermal eigenmode propagation through the amplifiers (Salin et al., 1997).

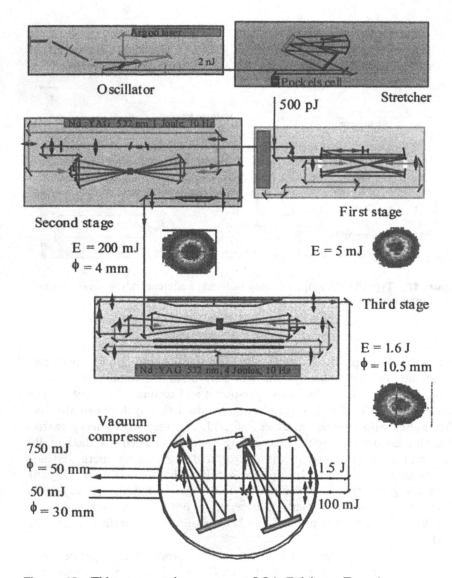

Figure 18 Thirty terawatt laser system at LOA (Palaiseau, France).

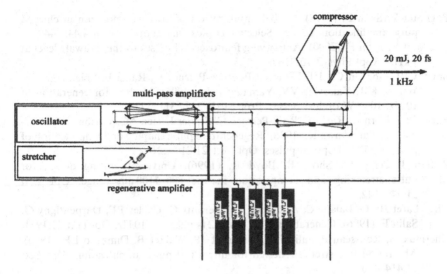

Figure 19 One terawatt laser system running at 1 kHz. (After Bagnoud and Salin, 2000.)

2.6 CONCLUSION

The past 20 years have brought femtosecond pulses from the nanojoule to the multijoule level. The technology of chirped pulse amplifiers has been pushed to the limit, and little is left for the next generation. The duration of amplified pulses has reached 15 fs, and a mere factor of 2 or 3 better can be expected. Energies over 1 J have been obtained, and much higher energies at reasonable repetition rates would require very large budgets. Most of the progress in short-pulse amplifiers can be expected in the change of technology through diode pumping. This shift will require finding new laser materials and even abandoning classical CPA for fiber amplifiers or other new schemes. These changes are necessary to bring high-energy pulses to real-world applications, and this is definitively the next challenge.

REFERENCES

Albert O, Wang H, Liu D, Chang Z, Mourou G. (2000). Generation of relativistic intensity pulses at a kilohertz repetition rate. Opt Lett 25:1125–1127.

Backus S, Durfee CG III, Mourou G, Kapteyn HC, Murnane MM. (1997). 0.2 TW laser system at 1 kHz. Opt Lett 22:1256–1258.

Bagnoud V, Salin F. (1998). Global optimization of pulse compression in chirped pulse amplification. IEEE J Selected Topics Quantum Electron 4:445–448.

Bagnoud V, Salin F. (2000). Amplifying femtosecond pulses to the terawatt level at 1 kHz. Appl Phys B 70:S165.

Barty CPJ, Gordon CL III, Lemoff BE, Rose-Petruck C, Raski F, Spielmann Ch, Wilson KR, Yakovlev VV, Yamakawa K. (1995). Method for generation of 10 Hz, 100 TW optical pulses. Proc SPIE 2377:311–322.

Barty CPJ, Korn G, Raski F, Rose-Petruck C, Squier J, Tien AC, Wilson K, Yakovlev VV, Yamakawa K. (1996). Regenerative pulse shaping and amplification of ultrabroadband optical pulses. Opt Lett 21:219–221.

Britten J, Perry M, Shore B, Boyd R. (1996). Universal grating design for pulse stretching and compression in the 800–1100 nm range. Opt Lett 21:540–542.

Chambaret JP, Le Balanc C, Antonetti A, Cheriaux G, Curley PF, Darpentigny G, Salin F. (1996). Generation of 25 TW, 32 fs pulses at 10 Hz. Opt Lett 21:1921.

Cheriaux G, Rousseau P, Salin F, Chambaret JP, Walker B, Dimauro LF. (1996). Aberration free stretcher design for ultrashort-pulse amplification. Opt Lett 21:414–416.

Du D, Squier J, Kane S, Korn G, Mourou G, Pang Y, Cotton C. (1995). Terawatt Ti:sapphire laser with spherical reflective optic pulse-expander. Opt Lett 20:2114–2116.

Fork RL, Shank CV, Yen RT. (1982). Amplification of 70 fs optical pulses to gigawatt powers. Appl Phys Lett 41:223.

Frantz LM, Nodvick JS. (1963). Theory of pulse propagation in a laser amplifier. J Appl Phys 34:2346–2349.

Georges P, Estable F, Salin F, Poizat JP, Grangier P, Brun A. (1991). High efficiency multipass Ti:sapphire amplifiers for continuous wave single mode laser. Opt Lett 16:144–146.

Kane S, Squier J. (1997a). Grism pair stretcher compressor system for simultaneous second and third order dispersion compensation in chirped pulse amplification. J Opt Soc Am B 14:661.

Kane S, Squier J. (1997b). Fourth-order dispersion limitations of aberration free chirped pulse amplification systems. J Opt Soc Am B 14:1237–1244.

Knox WH, Downer MC, Fork RL, Shank CV. (1984). Amplified fs optical pulses and continuum generation at 5 kHz repetition rate. Opt Lett 9:552.

Le Blanc C, Curley P, Salin F. (1996). Gain narrowing and gain shifting of ultrashort pulses in Ti:sapphire amplifiers. Opt Commun 131:391–398.

Lemoff BE, Barty CPJ. (1993). Quintic-phase-limited, spatially uniform expansion and recompression of ultrashort optical pulses. Opt Lett 18:1651–1653.

Lenzner M, Spielmann Ch, Wintner E, Krausz F, Schmidt AJ. (1995). Sub-20-fs, kilohertz-repetition rate Ti:sapphire amplifier. Opt Lett 20:1397–1399.

Maine P, Strickland D, Bado P, Pessot M, Mourou G. (1988). Generation of ultrahigh peak power pulse by chirped pulse amplification. IEEE J Quantum Electron 24:398–403.

Martinez OE. (1987). 3000 times grating compressor with positive group velocity dispersion: application to fiber compensation in the 1.3–1.6 μm region. IEEE J Quantum Electron QE-23:59.

Migus A, Shank CV, Ippen EP, Fork RL. (1982). Amplification of subpicosecond optical pulses: theory and experiment. IEEE J Quantum Electron QE-18:101.

Murray JE, Lowdermilk WH. (1980). Nd:YAG regenerative amplifier. J Appl Phys 51:3548.

Salin F, Squier J. (1992). Gain guiding in solid-state laser media. Opt Lett 17:1352–1354.

Salin F, Squier J, Mourou G, Vaillancourt G. (1991). Multikilohertz Ti:Al$_2$O$_3$ amplifier for high power femtosecond pulses. Opt Lett 16:1964–1966.

Salin F, Le Blanc C, Squier J, Barty CPJ. (1997). Thermal eigenmode amplifiers for diffraction limited amplification of ultrashort pulses. Opt Lett 22:718–720.

Siegman AE. (1986). Lasers, Mill Valley, Univ Sci Books.

Spence DE, Kean PN, Sibbett W. (1991). 60-fs pulse generation from a self-mode-locked Ti-sapphire laser. Opt Lett 16:42–44.

Squier J, Salin F, Mourou G, Harter D. (1991). 100 femtosecond pulse generation and amplification in Ti-sapphire. Opt Lett 16:324–326.

Squier J, Le Blanc C, Salin F, Barty CPJ, Kane S. (1998). Using mismatched grating pairs in chirped pulse amplification systems. Appl Opt

Strickland D, Mourou G. (1985). Compression of amplified chirped optical pulses. Opt Commun 56:219.

Szatmari S, Schafer FP, Muller-Horsche E, Muckenheim W. (1987). Opt Commun 63:305.

Treacy EB. (1969). Optical pulse compression with diffraction gratings. IEEE J Quantum Electron 5:454.

Yamakawa K, Aoyoma M, Matsuoka S, Kase T, Akahane Y, Takuma H. (1998). 100 TW sub-20 fs Ti-sapphire laser system operating at a 10 Hz repetition rate. Opt Lett 23:1468–1470.

3
Ultrafast Fiber Oscillators

Martin E. Fermann
IMRA America, Inc., Ann Arbor, Michigan, U.S.A.

3.1 INTRODUCTION

As ultrafast laser technology has begun to be deployed in industrial applications, the availability of commercial ultrafast lasers has greatly proliferated. The most widely used ultrafast laser systems in the market are based mainly on modelocked bulk solid-state and fiber lasers, with a bias toward bulk Ti:sapphire and erbium (Er) fiber oscillators, respectively. Indeed, several companies offer commercial hands-off oscillators based on these systems that promise a level of unprecedented reliability for any potential customer. The preference for the above two lasers arises from their favorable material properties, such as wide bandwidth and superior thermal properties for the case of Ti:sapphire. Erbium fiber lasers, on the other hand, can be pumped by telecom-compatible pump diodes, do not require any water cooling, and allow for straightforward excitation of soliton pulses using standard optical fibers. Moreover, when operating at Er fiber–compatible wavelengths, nonlinear pulse compression techniques in standard fibers can be employed to produce ultrashort femtosecond pulses without resorting to any actual ultrafast oscillators, i.e., by appropriately shaping an initial small perturbation from a continuous-wave (cw) signal.

In this chapter the bias is toward the unique capabilities of modelocked ultrafast Er fiber oscillators with their potential for miniaturization and integration, though we also review some of the work on other rare earth fiber oscillators and on nonlinear pulse compression. Indeed, fiber oscillators based on alternative rare earth ions such as thulium (Tm), neodymium, (Nd), or ytterbium (Yb) are now also available and have operation characteristics very similar to those of Er fiber lasers. The large family of rare earth ions allows ultrashort pulse generation at many different wavelengths with very high optical efficiencies. Moreover, modelocked fiber lasers have been constructed that produce pulses with widths from around 30 fs to 1 ns at repetition rates ranging from less than 1 MHz to 200 GHz. This versatility is unique in laser technology and opens up applications of fiber lasers ranging from telecommunications to optical machining. For example, ultrafast fiber lasers have been used in all-optical scanning [1], nonlinear frequency conversion [2], injection-seeding [3], multiphoton microscopy [4], terahertz generation [5], and optical telecommunications [6]. A much larger number of applications are pursued behind closed doors, ensuring continued growth for this technology for years to come.

3.2 OVERVIEW OF ULTRAFAST FIBER PULSE SOURCES

Modelocked fiber lasers are the most popular fiber-based ultrafast pulse sources and are the only fiber-based ultrafast pulse sources that are commercially available to date. Traditionally, modelocked fiber lasers are classified as being either actively or passively modelocked. Actively modelocked oscillators incorporate intracavity optical modulators operating at a given modulation frequency. Passively modelocked oscillators operate without modulators but incorporate components that act on the amplitude of the oscillating pulses. Present-day actively modelocked fiber lasers operate mainly in the soliton regime at gigahertz repetition rates, producing picojoule pulses with a width of a few picoseconds without the need for dispersion compensation. In contrast, passively modelocked fiber lasers typically operate at repetition rates of around 100 MHz and are often designed with internal dispersion compensation, allowing for the generation of sub-100 fs pulses with nanojoule energies. Passively modelocked fiber lasers mostly resort to semiconductor saturable absorbers for passive amplitude modulation and to enable reliable operation. Passively modelocked fiber lasers can be constructed with both single-mode and multimode fibers providing pulses with energies up to 5–10 nJ. The operation of passively modelocked fiber lasers in the similariton regime (Sec. 3.7.6) may increase the available pulse energies even further.

Modelocked fiber lasers represent only a small fraction of available fiber-based ultrafast pulse sources. As an alternative to modelocking, short pulses have been generated in optical fibers by direct modulation of a cw signal. In this, a variety of amplification and pulse compression stages allow the generation of high quality femtosecond pulses without the use of an optical cavity [7–10]. Ultrashort optical pulses have also been generated by using gain-switched diode lasers with subsequent pulse compression and pulse cleanup in additional fiber amplifier stages [11,12]. Sub-100 fs pulses with energies in the picojoule regime have been so produced. Direct modulation of cw signals is used mainly for the generation of pulses with gigahertz repetition rates, though in principle directly addressable pulses can also be generated. The triggering of short optical pulses with respect to an external signal source has been an important motivation for the development of pulse-compressed gain-switched laser diodes.

To increase the pulse energies from fiber systems to the 10 nJ to 1 μJ range, a variety of techniques have been implemented. Direct amplification of Raman solitons has produced pulse energies up to 10 nJ [13]; in-line chirped pulse amplification systems have generated pulse energies up to 100 nJ [14]. Nonlinear amplification of parabolic pulses has generated pulse energies up to 250 nJ [15], and chirped pulse amplification of chirped pulses generated with directly modulated diode lasers has produced pulses with energies up to a few microjoules [16].

Pulse energies in the 100 μJ to 1 mJ regime have been generated using chirped pulse amplification in Yb amplifiers [17] as well as parametric chirped pulse amplification in periodically poled $LiNbO_3$ [18]. Parametric chirped pulse amplification can potentially allow even further increases in available pulse energies from fiber sources, because it can overcome the restrictions in modal size and energy storage of optical fibers with diffraction-limited outputs. Ultrahigh-power fiber amplifiers producing pulse energies in the microjoule regime are discussed in Chapter 4 and are only mentioned here.

The performance achieved with several exemplary fiber-based pulse sources is shown in Table 1. System 1 represents the performance of a

Table 1 Typical Performance Data for a Variety of Devices Based on Modelocked Fiber Lasers

Pulse source	Repetition rate (GHz)	Average power (mW)	Pulse energy (nJ)	Pulse width (ps)	Wavelength	Ref.
1 Er actively modelocked	10	34	3.4×10^{-3}	1.4	1.55	19
2 Diode laser seeded erbium amplifier	2	1200	0.6	0.02–0.08	1.6	12
3 Modulated cw laser	10	650	0.065	1.0	1.55	8
4 Stretched pulse fiber laser	0.032	85	2.7	0.10	1.56	24
5 Multimode soliton fiber laser	0.067	300	4.5	0.36	1.535	25
6 Yb amplifier	0.050	6500	115	0.10	1.05	26

typical actively modelocked Er fiber laser operating at a repetition rate of 10 GHz [19]. Such systems are also readily commercially available. In some variations, repetition rates up to 20 GHz [20] and, using rational modelocking, repetition rates as high as 200 GHz [21] have been generated. Actively modelocked fiber lasers are particularly attractive due to the low pulse jitter that is achievable, which can be as low as 10 fs in a time period of 10 ms [22].

System 2 is based on a gain-switched diode laser that is pulse-compressed in a variety of nonlinear pulse-shaping fibers [12]. Though the present system operates at repetition rates of 2 GHz, the repetition rates are freely electronically selectable, and the pulses can even be freely triggered. For the generation of stable spectra, seeding of the nonlinear fiber amplifiers and pulse compression fibers with shorter pulses such as are obtainable from modelocked diode lasers can also be readily implemented [6]. This can produce wide-bandwidth higher quality frequency combs that are useful for dense wavelength-division multiplexing.

System 3 is based on a cw laser modulated by an electroabsorption modulator with subsequent fiber pulse compression stages [8]. In early versions of such pulse sources, typically the beat signal from two single-frequency lasers was used to produce an initial modulated signal, from which a high quality, high repetition rate pulse train was generated by the implementation of a variety of nonlinear pulse compression stages. Repetition rates as high as 70 GHz were indeed generated [23].

Though systems 2 and 3 and their derivatives can produce pulses at very high, and moreover adjustable, repetition rates, they are rather complex, relying on a range of special fibers for pulse compression.

The component count of ultrashort fiber pulse sources can be minimized by implementation of passively modelocked fiber lasers such as systems 4 and 5. In system 4, high-power operation is achieved by implementing a stretched pulse laser design [24], whereas in system 5 a multimode fiber laser allows for the generation of high-energy pulses [25,26].

System 6 is representative of on an oscillator–amplifier system comprising a passively modelocked fiber laser and a fiber amplifier. In system 6, similariton formation readily produces high-energy sub-100 fs pulses with pulse energies up to 250 nJ in conjunction with simple dispersive delay lines at the output of the amplifier [15].

3.3 LINEAR PULSE PROPAGATION IN OPTICAL FIBERS

The linear propagation characteristics of optical fibers are of prime importance in the design of ultrafast fiber oscillators, and some of the most important parameters are briefly reviewed here. We discuss an appropriate handling of fiber gain, fiber polarization, mode coupling, fiber dispersion, and group velocity. In the construction of actual fiber lasers, splice losses as well as diffraction losses due to non-fiber cavity elements also need to be considered.

3.3.1 Fiber Gain

Erbium fiber lasers operate on the three-level transition $^4I_{13/2} \rightarrow {}^4I_{15/2}$ with a bandwidth that is dependent on the average inversion inside the amplifier [27]. For most considerations, an approximate bandwidth of $\Delta\lambda = 45\,\mathrm{nm}$ can be assumed, where the lineshape function can be approximated with a parabolic shape. For high-gain Er fiber lasers, the distribution of the pump light and the signal inside the fiber laser as a function of fiber length have to be taken into account; these can be obtained from numerical solutions of the rate equations [28] while neglecting the contribution of amplified stimulated emission on the saturation characteristics of the amplifier. For the case of the technologically important cladding-pumped Er/Yb fiber lasers [25,26,29–31], energy transfer rates between Yb and Er also have to be considered, further complicating the numerical solution of the exact signal distribution in such systems. As an approximation, the fiber can be divided into a number of sections with the same small signal gain g_0 and varying saturated gain g_i.

$$g_i = \frac{g_0}{1 + \bar{P}/P_{\mathrm{sat}}} \tag{1}$$

where $\bar{P}$ is averaged signal power and P_{sat} is the saturation power. The small signal gain and the saturation power can be assumed to be constant if only a small fraction of the pump light is absorbed inside the fiber. For greater pump light absorption the variation in small signal gain and saturation power must also be incorporated.

3.3.2 Fiber Polarization

Linear as well as nonlinear polarization evolution in optical fibers as relevant to short-pulse fiber lasers is reviewed in Ref. 32. Most present-day ultrafast fiber oscillators operate on only one fiber polarization axis or incorporate means for the compensation of polarization drifts inside the cavity. To enable operation on one polarization axis, polarization-maintaining fibers must be used. The polarization-holding ability of optical fibers is generally characterized by the polarization beat length of the fiber at the operation wavelength. The beat length is given by

$$\Lambda_b = \frac{2\pi}{\delta\beta} \tag{2}$$

where $\delta\beta$ is the difference in propagation constants of the fundamental mode along the two fiber axes. Because most fiber lasers have relatively short lengths, a high degree of polarization holding can be achieved
with moderate beat lengths of the order of 10 cm at a wavelength of
1.5 μm. For a given beat length, the ability of a fiber to hold a single
polarization is affected by polarization mode coupling and depends on
external fiber perturbations. The influence of external fiber perturbations
is reduced by the implementation of large outside fiber diameters as discussed in Section 3.3.3.

A large amount of birefringence and a high degree of polarization
holding can also be induced by tension coiling of near-isotropic fibers
[33,34]. In this case the induced beat length for a fiber with a cylindrical
shape is given by [33]

$$\Lambda_{tc} \approx 2\frac{\lambda R}{\varepsilon r} \tag{3}$$

where R is the radius of curvature of the fiber coil, r is the outer radius of the
fiber, and ε is the mean axial strain. A constant strain can be easily applied
to a fiber by attaching a fixed weight to one end of the fiber and securing the
other end. For a fiber with an outside diameter of 125 μm, an applied weight
of 1 kg corresponds to an axial strain of $\varepsilon = 0.0124$ and 80 kg/mm^2 of tension. To avoid fiber breakage over long periods of time, ε should be < 0.01.
For a fiber with a coil radius $R = 5$ mm, this corresponds to a maximum
weight of 800 g and a fiber beat length of 12 cm at a wavelength of
1.55 μm. Note, however, that some special fibers may tolerate only much
smaller values of axial strain without breaking. Thus one has to be careful
when using this technique.

Rather than using polarization-maintaining fibers, near isotropic
fibers can also be implemented in fiber cavities. Any polarization drifts
can then be compensated for by incorporating a Faraday rotator mirror
inside a cavity as shown in Figure 1 [35]. A Faraday rotator mirror ensures
that in reflection any polarization changes are exactly compensated for
except for a rotation of the incident polarization state by 90°. Thus a single-polarization, polarization-insensitive fiber cavity can be constructed by
the incorporation of two Faraday rotators and a polarizer. The cavity is
based on an Er-doped fiber amplifier (EDFA) as the gain medium. It also
includes a half-wave plate and a quarter-wave plate for the optimization
of nonlinear polarization evolution (Sec. 3.4.4) to obtain the shortest
possible pulses. The pump light is injected through the wavelength-division
multiplexing (WDM) coupler.

In the presence of polarization compensation or polarization-maintaining fibers, it is generally sufficient to ignore the vector nature (Sec.

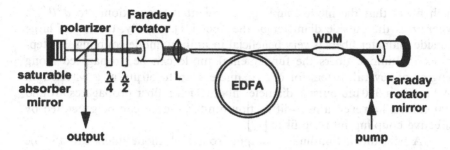

Figure 1 Experimental setup for an environmentally stable erbium-doped fiber laser. The waveplates are used to control both the linear and nonlinear polarization evolution in the fiber cavity. (From Ref. 35.)

3.4.4) of the pulses generated inside the fiber. Otherwise one has to resort to a rather involved set of Jones matrices and coupled nonlinear equations to describe linear and nonlinear polarization evolution, respectively [32].

3.3.3 Modes in Optical Fibers and Mode Coupling

The consideration of mode coupling is important to establish the limits of polarization holding for single-mode fibers and the allowable core size for multimode fibers. Most ultrafast optical applications require diffraction-limited laser beams. Therefore the modal quality of the output generated by an optical fiber needs to be accurately controlled, particularly for multimode fibers. However, diffraction-limited output beams can be readily obtained from multimode fibers by appropriate coupling and filtering of the fundamental mode inside the multimode fiber provided mode coupling is sufficiently small. Because the fundamental mode size in optical fibers is proportional to the core size, the use of large core diameters is very attractive, particularly for high-power applications. Various fiber designs have been suggested that allow an increase in fundamental mode size beyond the limits of conventional single-mode telecommunication fibers. Generally, it can be assumed that the susceptibility of a given fiber to mode coupling between two modes is governed by the difference in their propagation constants $\delta\beta = \beta_1 - \beta_2$.

The simplest large-mode fiber design is a standard step-index multimode fiber. For fibers that support a large number of modes, $\delta\beta$ is proportional to λ/d^2, where λ is the operation wavelength and d is the core diameter. Thus the susceptibility to mode coupling is independent of the numerical aperture of the fiber. Moreover, it has been shown for such fibers that the mode-coupling coefficient is proportional to $d^8/b^6\lambda^4$, where b is the outside diameter of the fiber [36]. Clearly, the use of large outside diameter fibers is very beneficial in limiting mode coupling. In step-index multimode fibers the fundamental mode can be propagated along lengths of several meters for core diameters up to 50 μm at a wavelength of 1 μm in a 300 μm outside diameter fiber. Such a fiber propagates around 100 modes; however, a near-diffraction-limited output can be generated by effective coupling into the fiber [17].

A near-diffraction-limited output from multimode fibers can also be obtained by tightly bending the fiber. In this case, the increased bending loss of higher order modes is exploited to preferentially transmit the fundamental mode in the fiber [37]. However, owing to the reduced lifetime of tightly coiled fibers this technique is preferable for fibers with diameters smaller than 200 μm.

Alternative methods for increasing the fundamental mode in optical fibers comprise the use of very low NA fibers, which increases the size of the fundamental mode without the complication of guiding additional higher order modes. The great susceptibility of such fibers to bending losses can be reduced by incorporating additional raised index rings outside the core [38]. Fibers with core diameters of up to 50 μm can also be used in this case.

Photonic crystal fibers have also been shown to allow the use of large fundamental mode sizes [39]. The limits to the size of the fundamental mode set by bending losses are similar to those of conventional step-index fibers. However, the influence of mode coupling has not yet been fully analyzed.

Another possibility for an increase in mode size is the use of multicore fibers [40]. If the cores are sufficiently close to each other, coherent coupling between the various fiber cores can occur, leading to the establishment of a single near-diffraction-limited supermode inside the fiber. However, the stability of a supermode is also affected by mode coupling, and it is uncertain whether any increase in fundamental mode size compared to the other techniques can be obtained.

3.3.4 Fiber Dispersion

Current optical fibers offer many options for accurate dispersion control. Moreover, fiber gratings allow dispersion manipulation in the whole transparency regime of optical fibers [41,42]. Even without fiber gratings, positive and negative dispersion fibers are available at wavelengths anywhere from around 600 nm to the 2 μm regime. In the normal dispersion regime of silica glass for wavelengths less than 1.3 μm, negative dispersion fibers can be constructed by the implementation of photonic crystal fibers designed with anomalous waveguide dispersion [43]. In addition, not only do optical fibers allow the control of second-order dispersion with corresponding quadratic spectral phase distortions, but also fibers with well-defined third-order and fourth-order dispersion terms can be manufactured [44]. In ultrafast fiber oscillators, it is mainly second-order dispersion that determines the operation regime of the laser. In the presence of negative dispersion, soliton generation can occur [45]. Alternating sections of positive and negative dispersion fibers give rise to the oscillation of Gaussian pulses [24,46]. The presence of positive dispersion fiber and gain in the same fiber can give rise to similariton generation and the formation of pulses with reduced pulse wings [47].

Dispersion arises from the frequency dependence of the refractive index. The dispersion parameter β_2 at a frequency ω_0 is obtained by expanding the mode-propagation constant β in a Taylor series about the center frequency:

$$\beta(\omega) = \beta_0 + \beta_1(\omega - \omega_0) + \frac{1}{2}\beta_2(\omega - \omega_0)^2 + \frac{1}{6}\beta_3(\omega - \omega_0)^3$$

$$+ \frac{1}{24}\beta_4(\omega - \omega_0)^4 + \cdots \tag{4}$$

Here $1/\beta_1 = v_g$ is the group velocity at frequency ω_0; β_2, β_3, β_4 characterize second-, third-, fourth-order dispersion.

Dispersion produces a phase modulation of the pulses and leads to dispersive pulse broadening. The dispersively broadened pulses acquire a frequency chirp, which can be linear or nonlinear depending on the dominance of the dispersion terms inside the laser. For an initially unchirped sech^2-shaped pulse with power $P(t) = P_0 \, \mathrm{sech}^2(t/\tau)$ and a small propagation length, the phase delay induced by second-order dispersion can be written as $\phi_d(t) = (-\beta_2/\tau^2)z \, \mathrm{sech}^2(t/\tau)$. We can thus define a dispersive length Z_d as the length over which the accumulated dispersive (second-order) peak phase delay for a sech^2-shaped pulse is unity, i.e., $Z_d = \tau^2/\beta_2$. A more common interpretation [48] of the dispersive length is the length it takes for the FWHM pulse width of an initially unchirped Gaussian pulse to broaden by a factor of $\sqrt{2}$.

The most dramatic dispersion manipulation can be obtained with photonic crystal fibers [43,49], where an arrangement of air holes surrounding a central section of the fiber produces the fiber core as sketched in Figure 2. The dispersion of the photonic crystal fiber can be approximated with a step-index fiber with an equivalent glass–air index step. The dispersion calculated for a 3 μm diameter silica rod in air is shown in Figure 3 and is compared to the dispersion of a standard telecom fiber.

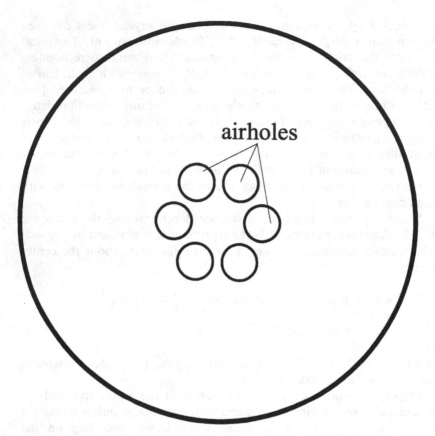

Figure 2 Cross section of photonic crystal fiber.

Early work on ultrafast fiber lasers used positive-dispersion Nd-doped fiber operating at 1.06 μm, and the therefore bulk diffraction gratings [48,50] or prism pairs [51,52] had to be used inside the laser to minimize the amount of dispersion inside the cavity to obtain the shortest possible pulses. As a recent development, fibers with negative dispersion [53] or chirped fiber gratings [54] can be used for dispersion control even in the 1 μm wavelength region [48], though the use of intracavity bulk diffraction gratings is still popular [55]. The dispersion of a chirped fiber grating can be written as [41]

$$\beta_2 = \frac{n}{\pi}\left(\frac{\lambda_0^2}{\delta\lambda}\right)\left(\frac{L_g}{c^2}\right) \tag{5}$$

where n is the core refractive index, λ_0 is the center wavelength, $\delta\lambda$ is the grating bandwidth, c is the velocity of light, and L_g is the grating length.

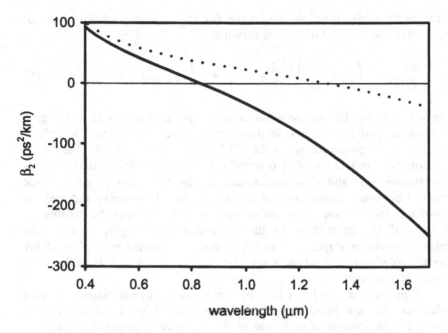

Figure 3 Dispersion of 3 μm diameter glass rod in air (solid line) compared to dispersion of a standard telecom fiber (dotted line).

The dispersion of a 2 mm length fiber grating with a bandwidth of 16 nm operating at the 1.05 μm wavelength region is thus $\beta_2 = 0.7$ ps^2. Such a grating could, for example, be written with a phase mask with a chirp rate of 80 nm/cm. It can further be shown that such a grating would have a reflectivity of around 10%. A dispersion of around 1 ps^2 represents roughly the minimal fiber grating dispersion that can be obtained using phase mask writing techniques due to the limited possible refractive index modulation of silica fibers.

Hence when used in conjunction with fiber oscillators, fiber gratings generally significantly overcompensate for the fiber dispersion, producing cavity designs with high values of intra-cavity dispersion. However, such cavities can be used for efficient high power picosecond soliton generation in the whole transparency range of silica glass [54,56].

3.4 NONLINEAR PULSE PROPAGATION IN OPTICAL FIBERS

Nonlinear pulse propagation in optical fibers and fiber amplifiers is effectively described with a modified nonlinear Schrödinger equation. For a parabolic gain profile, we may write for the evolution of the temporal pulse envelope A as a function of distance

$$\frac{\partial A}{\partial z} = -\frac{i}{2}\left(\beta_2 + i\frac{g}{\Omega_g^2}\right)\frac{\partial^2 A}{\partial t^2} + i\gamma|A|^2A + \frac{1}{2}(g-l)A \tag{6}$$

where g and l are the intensity gain and loss per unit length, Ω_g is the gain bandwidth, and β_2 is the second-order dispersion coefficient. Self-phase modulation is governed by $\gamma = 2\pi n_2/\lambda F$, where $n_2 = 3.2 \times 10^{-20}\,\text{m}^2/\text{W}$ is the nonlinear refractive index in standard silica fiber [48], λ is the operation wavelength, and F is approximately the core area in single-mode fibers. For small propagation distances z, the phase delay induced by self-phase modulation on a pulse with power $P(t)$ can be written as $\phi_{nl}(t) = \gamma P(t)z$. In analogy to the dispersive fiber length, we can thus define a nonlinear length z_{nl} over which the accumulated peak phase delay due to self-phase modulation is one, i.e., $z_{nl} = 1/\gamma P_0$, where P_0 is the peak power of the pulse.

Equations of the form Eq. (6) can produce several classes of pulse solutions that are important in the operation of ultrashort-pulse fiber lasers. In the absence of gain and in the presence of negative dispersion, Eq. (6) reduces to the standard nonlinear Schrödinger equation with stationary sech^2-shaped pulse solutions referred to as solitons [57,58]. Stationary chirped sech^2 pulse solutions can further be found in the presence of bandwidth-limited gain and are referred to as gain-guided solitary pulses [59]. In the presence of alternating sections of positive and negative dispersion fiber, quasi-stationary Gaussian-shaped pulse solutions exist [60]. Finally, in the presence of gain and no bandwidth limitation, nonstationary self-similar parabolic pulse solutions can be found [61]. All of the above-mentioned pulse shapes have been observed experimentally.

3.4.1 Solitons and Gain-Guided Solitary Pulses

In the absence of gain and loss, stationary solitons form when the dispersive phase delay $\phi_d(t)$ of a sech^2 pulse is opposite to the nonlinear phase delay induced by self-phase modulation $\phi_{nl}(t)$, i.e., $\phi_d(t) = \phi_{nl}(t)$. Hence the dispersive length has to be equal to the nonlinear length. The fundamental soliton power inside the optical fiber is thus obtained as

$$P_s = \frac{|\beta_2|}{\gamma\tau^2} = \frac{3.11|\beta_2|}{\gamma\,\Delta\tau^2} \tag{7}$$

where $\Delta\tau$ is the full width at half maximum (FWHM) pulse width; for a sech^2 pulse $\Delta\tau = 1.763\tau$. In turn, the energy of a soliton can be calculated as

$$E_s = \frac{3.53|\beta_2|}{\gamma\,\Delta\tau} \tag{8}$$

It is also instructive to introduce the soliton period $z_s = (\pi/2)z_d$, which is the oscillation period of a higher order soliton in the fiber. The wavevector of the soliton is given by $k_s = 2\pi/8z_s$, which is independent of frequency. Thus in one soliton period a soliton accumulates a nonlinear phase delay of $\pi/4$.

In the presence of bandwidth-limited gain with $(g - l) > 0$ in Eq. (6), chirped solitary pulses can be found as stationary solutions. The chirp is linear near the center of the pulse and reaches a limit for large time. In the limit of large negative dispersion and small gain dispersion, weakly solitary sech^2-shaped pulses with a pulse width

$$\tau = \frac{1}{\Omega_g}\left(\frac{g}{3(g-l)}\right)^{1/2} \tag{9}$$

are obtained. In this regime the pulse width decreases with an increase in excess gain. In the presence of positive fiber dispersion and a limited gain bandwidth, stationary soliton-like pulse propagation is also possible. Here stable pulse propagation is obtained by a balance of nonlinear pulse broadening and amplitude gain on the one side and limited gain bandwidth on the other side. For positive values of fiber dispersion, the pulse chirp greatly increases, producing pulse solutions with greatly increased energies compared to basic soliton pulses.

The increase in chirp and pulse energies for modelocked fiber oscillators operating in the positive dispersion regime is readily observable [62,63]. Experimentally, the reproducibility of the chirp as well as pulse stability can be problematic when operating short-fiber lasers with large values of positive dispersion.

3.4.2 Dispersion Management

Solitons and gain-guided solitary pulses are invariant with fiber length. In real optical fiber systems truly invariant pulse solutions play only a very limited role, because fiber losses or periodic fiber cavity losses nearly always need to be considered. For example, in optical communication systems, fiber losses become appreciable only after a few kilometers propagation. The combination of loss and optical amplification produces pulses with

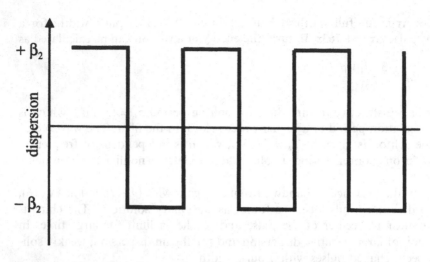

Figure 4 Fiber dispersion as a function of length for a typical dispersion-managed fiber transmission line.

periodically varying amplitudes. For transmission systems based on negative dispersion fibers, average soliton pulses can then still be defined [64], where the soliton pulse width from Eq. (6) is obtained approximately from the path-averaged pulse power.

Recent communication systems go even further and incorporate dispersion management, i.e., an alteration of positive and negative dispersion fiber along the transmission line [65]. An example of a distribution of fiber dispersion for a transmission line with balanced amounts of positive and negative dispersion fiber is shown in Figure 4. The pulse width in these transmission lines is not constant and varies along each fiber segment. For such dispersion-managed transmission systems, stable average pulse solutions can still be found [65,66]. For the case of Figure 4, the stable pulse solutions comprise Gaussian-shaped pulses.

The existence of Gaussian pulse solutions can be understood from a balance of self-phase modulation and dispersion in such systems just as for soliton pulses. For Gaussian pulses the average quadratic phase perturbations due to dispersion (if the amounts of negative and positive dispersion fiber are slightly different) are balanced by the average quadratic phase perturbations due to self-phase modulation. It can indeed be shown that in dispersion-managed transmission lines, self-phase modulation along the fiber segments can be lumped together as a single parabolic phase modulation at the point of shortest pulse width [60]. Just as in active modelocking, parabolic phase modulation calls for

Gaussian-shaped pulse solutions. This simple picture leaves out the influence of gain dispersion, which can, however, also be readily accounted for. The most important feature of pulse propagation in dispersion-managed fiber systems is that higher energy pulses can be transmitted than in soliton transmission systems. In dispersion management the pulses are compressed to a minimal pulse width for only a small fraction of the total propagation length, which greatly reduces the absolute amount of self-phase modulation for a given pulse energy. Similarly, when implemented in fiber laser designs, dispersion management allows a substantial increase in oscillating pulse energy compared to soliton laser designs. Relatively high energy pulses can also be transmitted (or generated) for systems with overall zero dispersion, where the energy of a soliton goes to zero [Eq. (8)]. Whereas the implementation of dispersion management in fiber lasers is relatively straightforward (Secs. 3.6 and 3.7.5), dispersion management allows for large variations in the design of optical transmission systems, which cannot be completely covered here. For a review of the pulse propagation phenomena in communication systems, the reader is referred to Ref. 67.

3.4.3 Similaritons

Whereas solitons and Gaussians can provide stationary solutions for pulse propagation processes, more generally nonstationary pulse propagation in optical fiber amplifiers needs to be considered. Of particular interest is self-similar pulse propagation in high-gain fiber amplifiers in the presence of positive dispersion [68]. In such systems seed pulses with a bandwidth much smaller than the gain bandwidth of the amplifier can evolve asymptotically into linearly chirped parabolic pulses independent of the exact shape of the seed pulses [61]. These parabolic pulses propagate self-similarly inside the amplifier; i.e., their parabolic pulse form is preserved while their amplitude and width grow exponentially. Because of the unique stability of these pulse solutions, they are sometimes referred to as similaritons. Similariton formation in highly nonlinear fiber amplifiers can be adequately described by Eq. (6) under the assumption of an infinite gain bandwidth and negligible gain saturation. This approximation is sufficiently accurate as long as the generated pulse bandwidth is significantly smaller than the gain bandwidth. The asymptotic self-similar parabolic pulse solutions can be described as

$$A(z, t) = A_0(z)\{1 - [t/\tau_0(z)]^2\}^{1/2} \exp[i(\varphi(z, t)] \qquad (10)$$

where $A_0(z)$ is the pulse amplitude and $\tau_0(z)$ is the width parameter as a function of z. τ_0 is related to the FWHM pulse width as $\Delta\tau = \sqrt{2}\tau_0$. The quadratic phase is given by

$$\varphi(z,t) = \varphi_0 + 3\gamma(2g)^{-1}A_0^2(z) - g(6\beta_2)^{-1}t^2 \tag{11}$$

where φ_0 is an arbitrary constant. The corresponding linear chirp is given by $\delta\omega(t) = \partial\varphi/\partial t = gt/3\beta_2$. The amplitude and pulse width scale exponentially with length as

$$A_0(z) = 0.5(gE_{in})^{1/3}\frac{\gamma\beta_2}{2}\exp\left(\frac{gz}{3}\right) \tag{12a}$$

and

$$\tau_0(z) = 3g^{2/3}(\gamma\beta_2)^{1/3}\exp\left(\frac{gz}{3}\right) \tag{12b}$$

where E_{in} is the seed pulse energy to the amplifier. The asymptotic pulse solutions depend only on the seed pulse energy and not on the initial pulse width or phase. The product of pulse width and power scales exponentially with g, which demonstrates the absence of gain saturation. Gain saturation, a length-dependent gain, and a limited gain bandwidth can readily be incorporated into Eq. (6), and pulse propagation can be calculated numerically to evaluate the limits of similariton formation in fiber amplifiers [69] and to obtain a more exact estimate of the pulse quality. Generally, the pulse quality of a similariton is best evaluated after pulse compression. Numerically, a linearly chirped dispersive delay line [48] can be assumed for pulse compression; in a real system, higher order dispersion in the dispersive delay line also needs to be accounted for [44]. One can then show that high quality compressed pulses can be obtained for integrated values of self-phase modulation inside a similariton amplifier in excess of 10π [44,70].

3.4.4 Nonlinear Polarization Evolution

Nonlinear polarization evolution in optical fibers is reviewed in Ref. 32. When using non-polarization-maintaining fibers, generally the full nonlinear coupled wave equations have to be solved to evaluate nonlinear pulse propagation in optical fibers. Indeed, the early work on modelocked fiber lasers heavily relied on the exploitation of nonlinear polarization evolution for the construction of stable modelocked fiber lasers [51,52]. Even now, nonlinear polarization evolution is exploited in stretched pulse fiber lasers to generate short pulses with pulse widths comparable to the bandwidth of the gain medium [24,46].

Nonlinear coupling of the two polarization eigenmodes in an optical fiber can also be used to allow stable oscillation of vector solitons inside a fiber cavity. Vector solitons become relevant when intracavity polarizers are omitted from the cavity designs, allowing for the simultaneous oscillation of both polarization eigenmodes in the fiber. Vector solitons as relevant to the construction of modelocked fiber lasers were recently classified into two groups [71,72]. A first group of vector solitons comprises group-velocity-locked vector solitons that propagate with a common group velocity for the two polarization eigenmodes, though their phase velocity is generally different. A second group comprises polarization-locked vector solitons, where the polarization eigenmodes are locked in both group and phase velocity. Phase locking of the two polarization eigenmodes results in a stationary polarization profile during pulse propagation. Polarization-locked vector solitons generally have an elliptical polarization state, with the polarization component along the fast axis possessing the greater intensity.

The use of polarization-locked vector solitons in actual fiber laser designs is not straightforward, because the oscillation of stable vector solitons requires a sensitive balance of birefringence and nonlinearity in the fiber. In general, small linear polarization perturbations can break up this balance and lead to the oscillation of unstable group-velocity-locked vector solitons [73] or even Q-switching instabilities. When the group-velocity walkoff between the two polarization eigenmodes in a fiber is large compared to the pulse width, the stable oscillation of two soliton pulses along the two polarization axes with two different repetition rates can also be observed.

To overcome all these instabilities and to enable actual predictable manufacturing processes in the construction of modelocked fiber lasers, polarization-maintaining fibers can be implemented into fiber laser cavity designs. Moreover, polarization-maintaining fibers allow single polarization operation without the implementation of intracavity polarizers by simple bending of the fiber [74]. Due to the nondegeneracy of the bend loss in polarization-maintaining fiber, bending leads to the preferential attenuation of one polarization eigenmode over the other.

Even without the availability of polarization-maintaining fiber, polarization compensated cavity designs can be implemented [35] that do not necessarily rely on nonlinear polarization evolution for their stable operation. The disadvantage of such cavity designs is the increased component count.

3.5 MODELING OF MODELOCKED FIBER LASERS

Stable operation regimes of modelocked fiber lasers are defined by the stability of the pulse energy against a growth of relaxation oscillations and the stability of the generated ultrafast pulse form. In principle, a complete set of equations can be written that accounts for both processes by accounting for the gain dynamics of the laser on a time scale of the relaxation oscillations of the system as well as for pulse propagation inside the cavity on a time scale of the cavity round-trip time. Due to the large difference between these time scales for typical rare-earth-doped fiber lasers, it is preferable to investigate the two processes separately. The gain dynamics of the laser can be obtained from a system of rate equations [75,76] with the pulse energy as an input. For lasers operating at the fundamental cavity frequency, the pulse form inside the laser can thus be evaluated with the assumption of time-invariant gain; moreover, because of the long carrier lifetime of rare-earth-doped fibers, gain saturation by individual pulses can be neglected. To establish the stability limits for harmonically modelocked lasers, more complicated models are generally required, where the oscillation of more than one pulse inside the cavity needs to be assumed (see Sec. 3.6).

    The time-averaged gain inside the laser can then be described by Eq. (1), whereas the steady-state pulse form can be obtained by solving Eq. (6) with added terms corresponding to any additional non-fiber cavity components. Due to the recursive nature of pulse propagation inside a cavity, recursive relations [77] are conveniently applied to find solutions to equations of the form of Eq. (6). In such a model, the envelope function of the pulse amplitude $A_n(t)$ is evaluated at successive round trips $n, n + 1, \ldots$ through the cavity, accounting for nonlinear propagation inside the fiber and pulse shaping by the intracavity elements. For a steady-state pulse solution $A_n(t)$ it holds that

$$A_{n+1}(t) = e^{i\varphi} A_n(t) \tag{13}$$

where φ is a constant phase shift between round-trips. For a highly nonlinear fiber laser it is also convenient to divide the fiber into k sections of length l_j and evaluate nonlinear pulse propagation inside each section separately. We can then write down the following operator equation governing pulse propagation inside a typical fiber laser cavity:

$$A^{n+1}(t) = e^{i\varphi} \hat{S} \hat{F} \left[\prod_{j=1}^{k} \hat{N} \hat{D} \hat{G}_j \right] A^n(t) \tag{14}$$

In Eq. (14), $R^{1/2}A^n(t)$ is the pulse amplitude reflected back into the cavity at the output coupler; R is the reflectivity of the output coupler; $\hat{N}, \hat{D}$, and $\hat{G}_j$ are differential operators, where $\hat{N} = i\gamma l_j|A|^2$ represents self-phase modulation inside the fiber,

$$\hat{D} = -\frac{i}{2}\left(\beta_2 l_j + i\frac{g_j}{\Omega_g^2}\right)\frac{\partial^2}{\partial t^2}$$

accounts for fiber dispersion and a finite bandwidth of the gain material, and $\hat{G}_j$ is the saturated amplitude gain in a fiber section j given by $\hat{G}_j = \exp(g_j/2)$, where g_j is given by Eq. (1). $\hat{F}$ accounts for a filter with optical bandwidth B, and $\hat{S}$ represents a modulation mechanism, which can comprise an optical modulator, a Kerr-type all-optical switching mechanism, a saturable absorber, or more complicated optical modulating elements.

For the case of an active amplitude modulator such as an acousto- or electro-optic modulator, $\hat{S}$ is given by $\hat{S} = \exp(-\delta_a\omega_m^2 t^2/2)$, where $\omega_m = 2\pi f_m$, f_m is the modulation frequency, and δ_a is the amplitude modulation coefficient.

For the case of a passive amplitude modulator such as a Kerr-type all-optical switch or a semiconductor saturable absorber, we can write $\hat{S} = \exp\{-\delta_{ns} + \delta_s(t)\}$, where δ_{ns} is the nonsaturable and δ_s the saturable loss of the passive modulator. In the presence of a Kerr-type all-optical switching mechanism we can assume an instantaneous response of the passive amplitude modulator to the pulse intensity. Assuming that the modulation mechanism saturates with a saturation power P_A, we can write

$$\delta_s(t) = \frac{-\delta_{s0}}{1 + |A(t)|^2/P_A}$$

where $A(t)$ is the pulse amplitude and δ_{s0} is the saturable loss induced by the amplitude modulator.

For the case of a semiconductor saturable absorber, the time dependence of $\delta_s(t)$ can be obtained from [78]

$$\frac{\partial\delta}{\partial t} = -\frac{\delta_s - \delta_{s0}}{t_s} - \frac{|A(t)|^2}{E_A}\delta_s \tag{15}$$

where δ_{s0} is the saturable loss of the unexcited absorber, t_s is the carrier lifetime, and E_A is the absorber saturation energy. Analytical solutions of Eq. (15) are described in Ref. 79.

Equations (13)–(15) provide for a full set of equations for finding stable pulse solutions for a given modelocked laser system, assuming the laser is stable aganist the onset of relaxation oscillations. Conveniently, the pulse solutions are found numerically by using a standard fast Fourier transform algorithm [48].

Note, however, that other nonlinear effects sometimes need to be considered to fully model a laser. Such higher order nonlinear effects comprise two-photon absorption, self-focusing, and thermal processes inside the absorber, just to mention a few phenomena. In the presence of pulse-shaping elements that only weakly perturb the intracavity pulse per round-trip, an averaged form of Eq. (14) referred to as the "master equation" with corresponding analytical solutions can be readily obtained [60,66,80–82].

3.6 ACTIVELY MODELOCKED FIBER LASERS

Active modelocking of fiber lasers is currently the method of choice for generating ultralow-jitter pulses at high repetition rates. Actively modelocked fiber lasers can be operated in the presence of relatively high amounts of self-phase modulation, allowing for the formation of pulses with subpicosecond pulse widths. Moreover, further amplification and pulse compression in optical fibers can produce high quality sub-100 fs pulses [83] as well as high quality frequency combs [6].

Actively modelocked fiber lasers were first demonstrated in the mid-1980s by the use of bulk acousto-optic modulators [84,85]. In Nd-doped fibers, bandwidth-limited pulses with widths ranging from 8 to 20 ps have so been generated [86–88], where cavity modulation at the fundamental cavity frequency was typically employed. Current communications-oriented applications rely on integrated $LiNbO_3$ amplitude or phase modulators to produce modelocked pulses at repetition rates of tens of gigahertz with highly integrated cavity designs [19–22]. Because of the long cavity lengths of such lasers (several hundred meters), these systems operate at very high harmonics of the fundamental cavity frequency. Recently, commercial systems also became available.

Most commonly, actively modelocked fiber lasers operate in a linear regime with negligible self-phase modulation [89], as nonlinear soliton lasers [45,90,91], as nonlinear solitary [92] lasers, and with dispersion management [19].

In the linear regime, actively modelocked fiber lasers can be described by the modelocking theory for homogeneously broadened gain media [93], where we can distinguish amplitude modulation (AM) and frequency modulation (FM) modelocking. Amplitude modulation modelocking uses an intracavity amplitude modulator, and FM modelocking uses a phase

modulator. Typically, the time-varying modulation exerted by the intracavity modulator may be cast in the form

$$S(t) = \exp[-(\delta_a - i\delta_p)\omega_m^2 t^2/2] \tag{16}$$

where $\omega_m = 2\pi f_m$ and f_m is the optical modulation frequency and δ_a and δ_p
are the amplitude and phase modulation indices, respectively. δ_p is simply
the exerted peak phase retardation in the cavity. For pure FM modelocking,
the predicted output is a train of chirped Gaussian-shaped pulses with a
FWHM pulse width of

$$\Delta\tau = 0.45 \left(\frac{2g}{\delta_p}\right)^{1/4} \left(\frac{1}{f_m \Delta f_a}\right)^{1/2} \tag{17}$$

where Δf_a is the 3 dB bandwidth of the gain medium and g is the saturated
intensity gain in the cavity. The time–bandwidth product in this case is
$\Delta\tau \Delta f = 0.63$, where $\Delta\tau$ is the FWHM pulse width and Δf is the FWHM
spectral bandwidth. The time–bandwidth product for bandwidth-limited
Gaussian pulses is $\Delta\tau \Delta f = 0.44$, and therefore the chirp produced by FM
modelocking is relatively small. To describe the case of pure AM modelocking, δ_p has to be replaced by δ_a in Eq. (17). The resulting pulses also have a
Gaussian shape and are bandwidth-limited. Equation (17) is valid for both
fundamental and higher harmonic modelocking. Typical modulation indices
for standard electro-optic and acousto-optic modulators are of the order of 1.

In a weakly nonlinear actively modelocked fiber laser cavity with overall positive dispersion, self-phase modulation can lead to pulse-shortening
by up to a factor of 2 compared to Eq. (17) before instabilities set in [94],
as demonstrated in early work in fiber lasers [86,87]. As a unique feature
of nonlinear actively modelocked fiber lasers operating in the positive dispersion regime, passive cavity length stabilization can be achieved [92]. In
this, any cavity length fluctuations are automatically compensated for by
corresponding wavelength fluctuations of the laser. The pulses in such lasers
are strongly chirped, and pulse stability is obtained by an interplay of self-
phase modulation, dispersion, the limited bandwidth of the gain medium,
and amplitude modulation. Dispersive broadening of the pulses is limited
by the amplitude modulator and the finite bandwidth of the gain medium,
just as in solitary pulse propagation, suggesting the name solitary laser for
such systems.

In a negative dispersion cavity with approximately uniform dispersion
maps, i.e., a cavity incorporating constant negative dispersion fiber, nonlinear actively modelocked fiber lasers can produce high quality soliton pulses
[45,90,91]. An excessive amount of non linearity in the cavity leads to pulse

breakup and unstable pulse formation, however. In a soliton laser, the corresponding pulses have a sech^2 shape with an FWHM pulse width given by Eq. (6). In general the soliton pulses are accompanied by a weak continuum that arises from the perturbations of the solitons in the cavity. For the pulses to be stable the loss of the continuum should be higher than the loss of the soliton. From this condition it can be shown [95] that soliton pulses with a width decreased by a factor R below the value given by standard linear active modelocking theory can be sustained in the cavity, where R is given by

$$R \leq 1.37 \left(\frac{\beta_2 L}{g/(2\pi \, \Delta f_a)^2} \right)^{1/4} \tag{18}$$

For the pulses to remain stable, the amount of self-phase modulation in the cavity should be limited to a fraction of π to reduce the amount of continuum generation. Typically, pulse widths of the order of 1 ps are obtainable at repetition rates of up to 20 GHz, where the pulse energies are of the order of 1 pJ [90,91].

The development of actively modelocked fiber lasers has culminated in the demonstration of nonlinear dispersion-managed lasers that produce significantly higher pulse energies than soliton lasers and also lower pulse jitter. A recent example of such a dispersion-managed actively modelocked fiber laser is shown in Figure 5 [19]. The round-trip fiber length in this particular

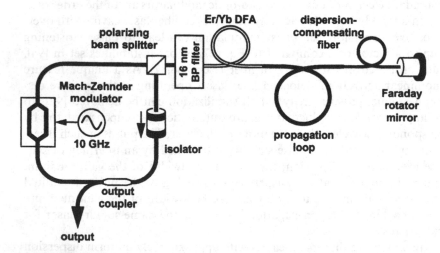

Figure 5 Sketch of an actively modelocked dispersion-managed Er fiber laser. The laser produces pulses with a repetition rate of 10 GHz and an average power of 34 mW. The cavity length is ≈ 100 m. (From Ref. 19.)

device was nearly 200 m, and the laser was modelocked at approximately the 10,000th harmonic of the fundamental cavity frequency. The laser has a sigma cavity to compensate for any polarization drifts in the non-polarization-maintaining fiber sections, allowing for environmentally stable operation. Dispersion management (Sec. 3.4.3) is used with the cavity design; the total dispersion inside the cavity is slightly negative and is adjusted by inserting appropriate lengths of standard and dispersion-compensating fiber. Mode competition between counterpropagating signals inside the loop section is avoided by the insertion of an isolator. A phase-locked loop is used to lock the repetition rate of the fiber laser to an external clock, where the cavity length is continuously adjusted with a fiber stretcher (see also Sec. 3.8). The laser uses an Er/Yb-doped fiber amplifier (Er/Yb DFA) as the gain medium and generates stable pulses with an FWHM width of 1.4 ps and an average power of 34 mW, corresponding to a pulse energy of 3.4 pJ. Dispersion management increases the energy content of the pulses compared to a simple soliton laser, leading also to a smaller amount of amplified spontaneous emission in the cavity. With low phase noise master clocks, these cavity designs have generated a pulse jitter as low as 10 fs in a 10 ms time window [22]. The dispersion-managed laser described here indeed exhibits three distinguishable operation modes, related to linear, soliton, and dispersion-managed operation. In this particular example, only high-power operation in the dispersion-managed regime produces stable pulses.

However, simply modulating a fiber laser at a harmonic of the fundamental cavity frequency or increasing the pulse energy does not generally ensure pulse stability. Due to the long relaxation times of typical trivalent rare earths, individual pulses of the harmonically modelocked pulse train cannot saturate the gain of the laser and the fiber laser saturates only with respect to the average laser power. Thus there is no mechanism to ensure that the pulses inside the pulse train have equal energy. Hence harmonically modelocked pulses can suffer from considerable pulse-to-pulse instabilities, which show up in the RF spectrum of the laser as side-bands at multiples of the fundamental cavity frequency [19,90].

In the experimental setup shown in Figure 5, a large degree of sideband suppression was achieved by implementing nonlinear polarization evolution [96] as well as an optical bandpass filter [97,98] as an optical limiting mechanism. When operated as an all-optical limiter, nonlinear polarization evolution in the cavity is set to produce a minimal cavity loss at a certain pulse power. Hence an increase or a decrease in pulse power increases the cavity loss. The peak power of the pulses in the cavity is thus fixed, and the pulse energies are equalized accordingly because the pulse widths are relatively insensitive to power fluctuations. This phenomenon produces better amplitude stabilization in soliton lasers than in dispersion-managed lasers [19]. The reason is that the pulse peak power in a dispersion-managed laser is only weakly dependent on pulse energy. Similarly, the narrow-band-pass intracavity filter (BP filter) in Figure 5 limits the spectral width of pulses and thus the optical power. A filter is also slightly more effective as an optical limiter in a soliton laser than in a dispersion-managed laser [99]. Compared to nonlinear polarization evolution, a filter has the advantage that it is compatible with an all-polarization-maintaining cavity. Equally compatible with a polarization-maintaining fiber cavity is the introduction of a semiconductor two-photon absorber [100], though such an element adds to component count.

Pulse evolution in actively modelocked soliton and dispersion-managed lasers can in principle be modeled using a formalism as described in Eq. (14) with the operator $\hat{S}$ accounting for the optical modulator. However, such a model can predict only the expected pulse quality and energy. Because no pulse interactions apart from an optical continuum are included when solving Eq. (14), pulse-to-pulse instabilities such as pulse drop-outs, pulse-to-pulse amplitude variations, and noise cannot be described. Recently, a more detailed model using multipulse propagation [99] in a harmonically modelocked laser was developed that can successfully describe pulse-to-pulse instabilities. The model comprises a small number of individual pulses as well as a fictitious superpulse that contains the remaining intracavity pulse energy. With the help of the superpulse the influence of amplified spontaneous emission and gain saturation can be more accurately accounted for in the numerical simulations.

3.7 PASSIVE MODELOCKING

For applications in instrumentation and as seed sources for high-power amplifiers, passively modelocked fiber lasers are the system of choice. Currently three passively modelocked fiber lasers are commercially available: a wavelength-tunable Er fiber laser from Calmar [101], a high-power soliton fiber laser from IMRA America Inc. [102], and a system from Clark MXR [103] based on a dispersion-managed fiber laser design. All these commercially available lasers are diode pumped and produce sub-picosecond pulses with picojoule to nanojoule pulse energies. Options for pulse compression down to around 100 fs also exist. The first two systems comprise some form of saturable absorber as the modelocking element, whereas the Clark laser uses an all-optical switching mechanism based on nonlinear polarization switching in a ring cavity to induce modelocking [24,46]. As an example, the package of an IMRA femtosecond fiber laser is shown in Figure 6.

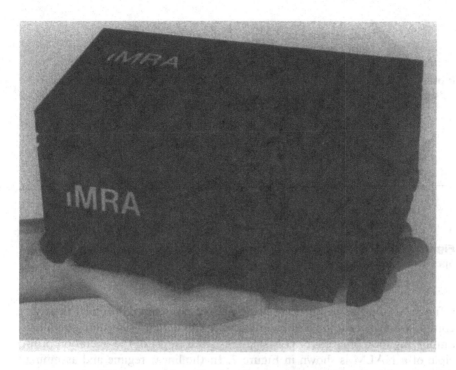

Figure 6 Photograph of a commercially available frequency-doubled femtosecond Er-doped fiber oscillator–amplifier system. (Courtesy of IMRA America, Inc.)

Generally, the classification of passively modelocked fiber lasers can follow the classification of actively modelocked fiber lasers, and we can distinguish between soliton lasers, solitary lasers, and dispersion-managed lasers. Solitary lasers have not received a lot of attention to date and are not separately discussed here. Because of the strong pulse shaping that is possible in passively modelocked lasers, it was recently suggested that similariton lasers can also be constructed [47]. The strong pulse shaping in passively modelocked lasers as well as the generated short pulse widths also allow for the construction of stable passively modelocked multimode fiber lasers [25,56]. Short-pulse generation in passively modelocked fiber lasers is induced with passive amplitude modulators, such as all-optical intracavity fiber switches or saturable absorbers.

3.7.1 Fiber-Based All-Optical Switches

All-optical switches come in a variety of forms; their common feature is a nonlinear interferometer that provides an intracavity loss that decreases as

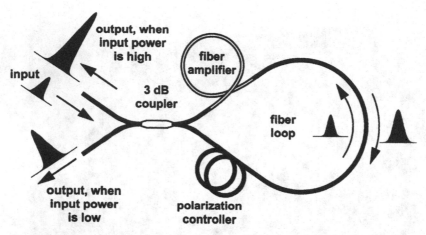

Figure 7 Principle of operation of the nonlinear amplifying loop mirror (NALM) operated in reflection.

a function of pulse power. Early examples of nonlinear all-optical switches comprised nonlinear optical loop mirrors (NOLMs) [104] and nonlinear amplifying loop mirrors (NALMs) [105]. An example of the operation principle of a NALM is shown in Figure 7. In the linear regime and assuming the polarization state is preserved inside the fiber loop, the NALM is equivalent to a Sagnac interferometer with an output amplified by the intraloop amplifier section. Pulses are injected into the NALM in port 1 and split equally by the fiber coupler, producing clockwise and counterclockwise propagating pulses in the Sagnac ring. In the linear regime the clockwise and counterclockwise propagating pulses accumulate the same linear phase delay across the loop, are recombined at the coupler, and are reflected back from port 1. Hence here the NALM is completely reciprocal and acts as an amplifying mirror.

When the pulse energy is increased, self-phase modulation comes into play and causes an imbalance in the optical path lengths for the two counterdirectonal pulses inside the loop, because the clockwise direction gets amplified first, whereas the counterclockwise direction gets amplified last. The reflectivity of the NALM thus becomes power-dependent. By also manipulating the linear polarization state inside the NALM, a linear phase delay Φ_0 can be induced between the two counterpropagating signals in the loop. The reflectivity of the NALM can thus be freely adjusted to either decrease or increase as a function of pulse power. It can be shown that the reflectivity R of the NALM varies sinusoidally as a function of pulse power P injected into port 1:

$$R = \cos^2\left\{\frac{1}{2}\left[\Phi_0 + \gamma(g-1)L\frac{P}{2}\right]\right\} \qquad (19)$$

where g is the gain of the amplifier, γ is the nonlinearity parameter of the fiber in the loop, L is the fiber loop length, and we neglected the length of the amplifier. Early passively modelocked fiber lasers indeed implemented the NALM as an all-optical switch to produce femtosecond pulses [50,106].

The required polarization control inside the NALM complicates its operation characteristics beyond the simple explanation given above [32]. Thus for current passive modelocking applications, nonlinear interferometers based on the nonlinear interference of the two polarization eigenmodes inside a fiber are most popular, because they require the least number of optical components. The principle of nonlinear polarization evolution [51] in a slightly birefringent isotropic fiber such as that used in a Fabry-Perot cavity is explained in Figure 8. For simplicity only one high reflector is shown. The output coupling mirror is located just in front of the polarizer and is omitted. Here the nonlinear interferometer consists simply of the two linear polarization eigenmodes of the fiber, which are excited

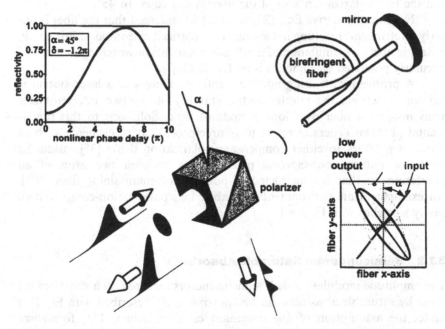

Figure 8 Nonlinear polarization evolution in a weakly birefringent fiber. At high power the polarization state at the polarizer rotates into the axis of the polarizer. (From Ref. 52.)

by the linearly polarized light coming from the polarizer. The angle of the polarizer with respect to the major polarization axis of the fiber is α. After one round-trip through the fiber, the two eigenmodes accumulate a differential linear phase delay $2\Phi_0$ and finally interfere again at the intracavity polarizer.

The nonlinear reflectivity of the interferometer, i.e., the backreflection through the polarizer, may then be written to first order as [52]

$$R = R_0 + \kappa P \tag{20}$$

where $R_0 = \cos^4 \alpha + \sin^4 \alpha$ is the linear reflectivity, P is the pulse power injected from the polarizer into the birefringent fiber, and κ is given by

$$\kappa = \frac{-\gamma L}{6} \sin 4\alpha \, \sin 2\alpha \, \sin 2\Phi_0 \tag{21}$$

κ is maximized when $\delta = -\pi/4 (\mathrm{mod}\, n2\pi)$. A nonlinear switching action is obtained by exciting the two polarization eigenmodes with different intensities, i.e., by rotating the polarizer such that the angle between the polarizer and the two polarization axes of the fiber is not equal to 45°.

Note that to derive Eqs. (20) and (21) we assumed that the fiber is linearly birefringent, resulting in two linearly polarized eigenmodes. For nearly isotropic fibers a formulation of nonlinear polarization switching in terms of circularly polarized eigenmodes is preferred [32].

A problem with using nonlinear interferometers in a laser system is that of preserving the interferometric stability of the two interferometer arms inside the fiber over long periods of time. Solutions to this fundamental problem generally resort to nonreciprocal cavity elements such as Faraday rotators to either compensate polarization drifts [35] inside the cavity or induce nonreciprocal phase delays between two arms of an interferometer that is constructed of polarization-maintaining fiber [107]. An example of an environmentally stable, i.e., polarization-compensated, cavity was shown in Figure 1.

3.7.2 Semiconductor Saturable Absorbers

The amplitude modulation with semiconductors incurred on a short optical pulse by saturable absorbers can be approximately described with Eq. (15) under the assumption of one dominant carrier lifetime. This formalism can also be readily extended when more than one lifetime in the absorber needs to be considered [81,82]. However, equations of the form of Eqs. (14) and (15) generally predict only the steady-state pulse form. To evaluate the stability of the laser against Q-switching, a rate equation analysis of the gain dynamics inside the cavity is necessary [75,76].

Despite the relatively long relaxation times of semiconductor saturable absorbers (1 ps) compared to the response time of nonlinear fiber Kerr switches (3 fs), very short femtosecond soliton pulses can still be generated [108]. For soliton pulses dispersion and self-phase modulation are balanced, preventing pulse broadening in a cavity. In contrast, the continuum accompanying the soliton pulses as well as any noise that exists propagate linearly through the cavity and spread dispersively between round-trips. Thus the noise components inside the cavity can be effectively suppressed by a transmission window induced by the saturable absorber with a carrier lifetime significantly longer than the pulse width. Indeed, soliton pulses with a width 10–20 times shorter than the carrier lifetime of the saturable absorber can be readily produced with well-designed soliton lasers [108].

Femtosecond pulses can also be generated by using saturable absorbers in conjunction with dispersion management [55,109]. Generally, the stability of such systems is enhanced by providing for chirp linearization by minimizing self-phase modulation of the negative dispersion cavity elements and maximizing self-phase modulation in the positive dispersion cavity elements. This can be accomplished, for example, by using positive dispersion Nd or Yb fiber lasers in conjunction with bulk dispersive delay lines [50–52,55,109]. Pulse formation in the presence of a saturable absorber and negative or positive dispersion fiber was recently analyzed [81,82], though no detailed theory of the operation of saturable absorbers in dispersion-managed fiber systems or the similariton regime is not yet available.

A design of a saturable absorber mirror used for modelocking of Nd [109] and Yb fiber lasers is shown in Figure 9. It is a multiple quantum well design with 50 periods of 60 Å thick GaAs barriers and 62 Å InGaAs barriers grown on a GaAs substrate with a GaAs/AlAs dielectric mirror. The band gap of the absorber was selected to be around 1.06 μm, close to the operation wavelength of the laser. The saturation characteristics of the saturable absorber are determined by the carrier lifetime (1 ps to 30 ns) and a time constant of about 300 fs arising from exciton screening [110] and carrier thermalization [111]. The carrier lifetime can be effectively controlled by the growth temperature of the saturable absorber. Alternatively, ion bombardment can also be implemented to minimize the carrier lifetime [112]. To improve the long-term stability of saturable absorbers, thermal annealing can also be readily implemented [112].

For passive modelocking of Er-doped fiber lasers, bulk InGaAsP saturable absorbers [113] as well as saturable Bragg reflectors (SBRs) based on InGaAs/InP multiple quantum wells [81,82] have been success-

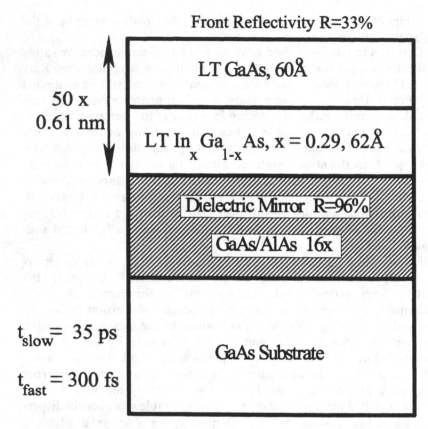

Figure 9 Design of a GaAs saturable absorber mirror for passive modelocking of Nd- and Yb-doped fiber lasers. (From Ref. 109.)

fully used. An example of an SBR used for passive modelocking of an Er fiber laser [81] is shown in Figure 10. Due to the broad band edge of bulk saturable absorbers compared to the narrow exciton absorption resonances of multiple quantum wells, bulk absorbers are the easiest to implement. In passively modelocked fiber lasers, bulk absorbers are operated either at the band edge or in the tail of the absorption band. Typical values for the saturation energy density of bulk InGaAsP at the band edge due to long-lived carriers are of the order of $100 \, \mu J/cm^2$. For the two fast mechanisms, saturation energy densities smaller by a factor of 3 are typically assumed. These values are inversely proportional to the linear absorption and increase linearly with the ratio of the absorption at the band edge to the absorption at the band tail.

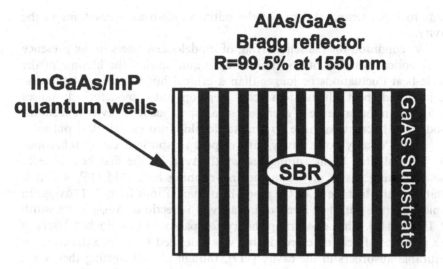

Figure 10 Design of a saturable Bragg reflector (SBR) for passive modelocking of Er- or Er/Yb-doped fiber. (From Ref. 82.)

3.7.3 Start-Up of Passively Modelocked Fiber Lasers

Fiber lasers modelocked by semiconductor saturable absorbers generally self-start from a Q-switching instability. Only when the laser power exceeds a certain threshold power is stable cw modelocking obtained. The stability ranges obtainable with saturable absorber modelocked lasers are well understood [75,76] and have recently been shown to apply also to fiber soliton lasers [114]. To minimize the possibility of optical damage of saturable absorbers during the Q-switching period, optical limiting via two-photon absorption in a semiconductor operated above half band gap can be implemented [114].

In contrast, Fabry-Perot fiber lasers modelocked by intracavity all-optical switches can start from random noise fluctuations. The random noise fluctuations present in lasers prior to modelocking arise from the presence of several axial modes in the cavity that have randomly fluctuating amplitudes and phases and therefore produce long seed pulses with a certain lifetime. The lifetime of the initial seed pulses is limited by intracavity reflections, which produce an uneven axial mode spacing. A physical explanation for this phenomenon was given by Brabec et al. [115]. The mode spacing is given by the ratio of the phase velocity of light to the round-trip cavity length, so an uneven mode spacing can be described as an uneven phase velocity for the axial cavity modes. An uneven phase velocity thus leads to rapid temporal decay of the initial mode-beat fluctuations in the cavity.

A condition for the self-starting of modelocked lasers in the presence of all-optical switches can be derived by requiring that the lifetime of the mode-beat fluctuations be longer than a critical buildup time, so that the initial mode-beat fluctuations can develop into modelocked pulses [115,116]. In the presence of even very small spurious intracavity reflections, mode-beat fluctuations have no time to develop into modelocked pulses.

The "quality" of a cavity with respect to spurious cavity reflections can be evaluated by measuring the width Δv_{3dB} of the first beat note of the oscillating cavity modes in the free-running laser [115,116], which is related to the lifetime τ_c of the mode-beat fluctuations by $\tau_c = 1/\Delta v_{3dB}$. In typical cavities with low levels of intracavity reflections, Δv_{3dB} has a width of 1 kHz [116]. The self-starting theory for passively modelocked lasers in the presence of beat-note broadening was extended to include the effect of saturable absorbers in the cavity [117], though no self-starting theory yet exists that accounts for the start-up of modelocking from Q-switching instabilities. Due to the long lifetime of rare earth fiber gain media, Q-switching instabilities can indeed be expected to play a role in the start-up dynamics of any fiber laser design beyond the simple model described in Refs. 115 and 116.

Generally, unidirectional ring cavities are less sensitive than a Fabry-Perot cavity to spurious reflections, because two reflectors inside a ring are required to produce coherent coupling of the reflected light to the circulating intracavity light [118]. Because a Fabry-Perot cavity is bidirectional, any one reflection can couple coherently to the intracavity light and lead to a frequency shift of the axial intracavity modes via mode pulling. Thus generally Fabry-Perot cavities with intracavity all-optical switches do not self-start, and additional "start" modulators [50] or cavity perturbations [119] have to be implemented to induce the modelocking process.

3.7.4 Soliton Fiber Lasers

The first passively modelocked all-fiber soliton laser was demonstrated by Duling [106]. It comprises a uniform dispersion map, an Er-doped gain fiber, and a NALM [105] as an all-optical switching element, and because of its layout it is referred to as a figure-of-eight laser (F8L). An example of an experimental setup for an F8L is shown in Figure 11. Here the NALM is operated in transmission. To ensure unidirectional propagation in the loop outside the NALM an isolator is also required. In addition to Er-based F8Ls, Pr-based F8Ls were also demonstrated [120] delivering subpicosecond near-soliton pulses in the 1.3 μm region.

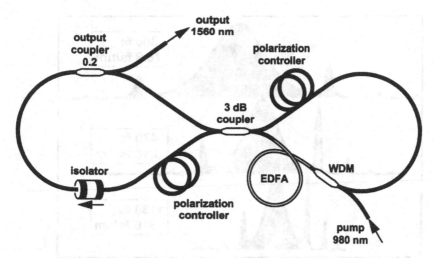

Figure 11 Typical setup for an all-fiber, self-starting figure eight laser (F8L) made with Er-doped fiber. (From Ref. 106.)

The nonlinear intracavity phase delay in the F8L is of the order of π [106], which in the NALM also corresponds to the switching power. From the NALM switching power and the expression for the soliton pulse width, the FWHM width $\Delta\tau$ of the soliton pulses generated in the F8L is calculated as [106]

$$\Delta\tau^2 = 0.49|\beta_2|L(g-1) \qquad (22)$$

where L is the length of the fiber in the NALM loop and g is the gain of the NALM. In Figure 12, we show a spectrum of a typical F8L [121]. Characteristic of the F8L as well as all (passively modelocked) soliton lasers in general are the spectral resonances (so-called Kelly side bands [122]), which occur because the soliton and the weak continuum arising from soliton perturbations couple coherently inside the cavity. For energy to couple between the soliton and the continuum wave, their phase delay per round-trip needs to be a multiple of 2π, from which the separation $\Delta\Omega_n$ of the side bands from the central frequency can be calculated as [122]

$$\Delta\Omega_n = \pm\frac{1}{\tau}\left(-1 + \frac{8nz_s}{L}\right)^{1/2} \qquad (23)$$

where τ and z_s are the soliton pulse width and period, respectively. From a measurement of the pulse width and the frequency spacing of the side bands, the dispersion of the cavity can be calculated.

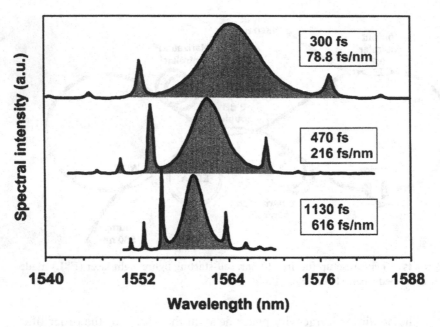

Figure 12 Spectra of pulses obtained with different cavities of figure eight lasers. The data listed next to each curve are the pulse durations and the total cavity dispersion. (From Ref. 121.)

The minimal FWHM pulse width $\Delta\tau_{min}$ obtainable from the F8L (and other soliton lasers) is determined by the amount of intracavity dispersion. It was shown [121,123] that $\Delta\tau_{min}$ is given by

$$\Delta\tau_{min} \geq 0.75\sqrt{\overline{\beta_2}L} \qquad (24)$$

where $\overline{\beta_2}L$ is the total dispersion in the cavity. Good agreement of Eq. (24) with the measured pulse width was observed for pulse widths down to 100 fs. In Er-doped fiber lasers, pulses shorter than 100 fs are in general affected by the limited bandwidth of Er as well as higher order dispersion.

To minimize the component count compared to F8Ls, and to ensure more reliable self-starting operation, soliton fiber ring lasers have been constructed [118]. In these structures, nonlinear polarization evolution is used as an all-optical switching mechanism. Both F8L and soliton fiber ring lasers require control of the linear polarization state inside the cavity for optimum operation. Because generally the polarization state in non-polarization-maintaining fibers is environmentally unstable, it has been suggested that nonreciprocal cavity elements such as Faraday rotators be implemented to minimize any long-term polarization drifts in the cavity. Nonreciprocal cavity elements can in principle also be introduced into F8Ls, but their introduction into polarization-switched Fabry-Perot cavities is the most convenient [35].

A limitation of such cavities is that modelocking is not self-starting and therefore semiconductor saturable absorbers are generally also included as a starting mechanism [32]. Indeed such soliton lasers have been commercially available since 1998 [102] and are finding widespread applications as absolutely turnkey ultrafast laser sources in industrial environments.

Despite the excellent industrially proven reliability of environmentally stable cavities, alternative soliton laser cavity designs have been investigated to reduce the component count. Owing to the stability of soliton pulse formation in the presence of saturable absorbers with a carrier lifetime significantly longer than the generated pulse width, femtosecond soliton fiber lasers can also be constructed without resorting to nonlinear polarization evolution. Indeed, single-polarization fiber laser cavities can be constructed with polarization-maintaining fiber and an intracavity polarizer aligned with one of the fiber polarization axes [113]. The intracavity polarizer can be eliminated by the use of tightly coiled highly birefringent Er-doped fiber, which produces a differential loss for the two polarization axes in the fiber. As a result a completely integrated single-polarization fiber soliton laser can be constructed as shown in Figure 13 [74].

Another example of completely integrated fiber soliton lasers are vector soliton lasers, which operate with near-isotropic E-doped fiber [71,72]. A limitation of such systems is that long-term environmental stability cannot be easily ensured owing to drifts in the polarization state in the near-isotropic fiber. Moreover, the output polarization is equally sensitive to any polarization drifts inside the fiber [73].

It should also be noted that saturable absorbers were introduced into soliton fiber ring lasers early on [124]. However, the lack of deterministic polarization control in the early fiber laser designs was an issue. Subsequently, the incorporation of polarization-maintaining fiber into the cavity solved this problem [125], though polarization-maintaining fiber ring lasers remain difficult to construct.

One disadvantage of integrated fiber laser designs discussed above is the limited adjustability of the cavity, and as a result the obtainable pulse energies are in the range of only a few hundred picojoules. Though the generated soliton pulses can be amplified in subsequent fiber amplifiers, this may be undesirable because of the increased component count. To increase the output power of soliton fiber lasers to the nanojoule regime, large-core fibers can be implemented [126] in conjunction with maximization of the

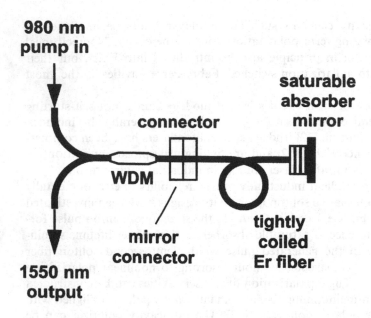

Figure 13 Experimental setup for a single-polarization integrated fiber soliton laser. (From Ref. 74.)

output coupling fraction to 99% and higher [25,26]. Both these improvements minimize the nonlinearity of the oscillator fiber. Because the average power of such high-power soliton lasers can exceed 100 mW [26], the implementation of high-power cladding pumping of double-clad Er/Yb-doped fibers provides further improvement in laser design [25,26,127,128]. Pulses with an average power of 300 mW at a repetition rate of 67 MHz and a pulse width as short as 300 fs were so obtained in large-core multimode Er/Yb-doped cladding pumped fibers [26].

The maximum sustainable pulse energy in a passively modelocked soliton single-mode (or multimode) fiber soliton laser can also be raised by greatly increasing the negative cavity dispersion by the incorporation of a chirped fiber grating into the cavity, as shown in Figure 14 [56]. In the presence of the grating the dispersion of a cavity of length L incorporating just a few meters of fiber is completely governed by the dispersion D_2 of the grating. The soliton pulse energy then scales with $\sqrt{|D_2|}/L$, whereas the pulse width scales with $\sqrt{|D_2|}$. Using this technique, pulses with widths of 4 ps and energies of up to 10 nJ have been obtained directly from single-mode fiber oscillators [129]. With chirped fiber Bragg gratings, high-power picosecond soliton fiber lasers can be built in the whole transparency range of silica glass [54,56].

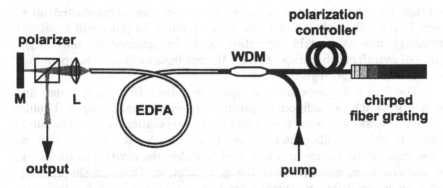

Figure 14 Experimental setup for an erbium-doped fiber laser containing a chirped fiber grating to produce a high-power picosecond pulse source. (From Ref. 56.)

3.7.5 Dispersion-Managed Fiber Lasers

The concept of dispersion management for the control of ultrafast pulse propagation goes back to the design of early femtosecond dye lasers that readily incorporated dispersion-compensating elements into the cavity [130]. The amount of dispersion compensation required in such systems was relatively small due to the short length of the gain medium. Passively modelocked solid-state lasers incorporated comparatively longer gain media and also relied on dispersion compensation for the generation of the shortest possible pulses [131].

The first dispersion-compensated ultrafast fiber laser was demonstrated by Kafka et al. [132]. The system was based on a synchronously pumped fiber Raman laser with a length of 100 m operating at a wavelength of 1100 nm. The system produced pulses with a minimum FWHM width of 800 fs by compensating for the fiber dispersion with a bulk negative dispersion grating pair. The magnitude of intracavity pulse width variations in the system by Kafka was not published.

The cavity design pioneered by Kafka was later extended to the construction of the first dispersion-compensated rare earth fiber lasers [86]. In synchronously pumped Raman fiber lasers as well as in rare earth fiber lasers, the amount of dispersion compensation was orders of magnitude greater than in dye and bulk lasers. Therefore the term "pulse compression modelocking" (PCM) was introduced to describe the pulse formation processes in such cavities [133]. Pulse compression modelocking is based on the idea that the interplay of positive dispersion fiber and self-phase modulation produces linearly chirped broad-bandwidth pulses inside the cavity that are intracavity pulse-compressed by a negative bulk dispersion grating or prism pair. However, the early work in dispersion-compensated fiber lasers [86,133] showed only a small 30% variation in pulse width within the cavity; moreover, stable operation could be obtained for only large values of overall negative dispersion [133]; thus these cavities were operating mainly in the soliton regime.

The first truly dispersion-managed fiber laser [50] incorporated an all-optical switch in addition to positive dispersion gain fiber and bulk negative dispersion elements, resulting in the generation of a 100 fs pulse from a modelocked fiber laser for the first time. A substantial variation in the pulse width by up to a factor of 3 within the cavity was observed for the first time, and moreover the generation of Gaussian-shaped [50] pulses in a passively modelocked laser was also observed for the first time. In this early work, the term "additive pulse compression modelocking" (APCM) was used to describe the pulse formation processes in the cavity.

The cavity design from Ref. 50 is shown in Figure 15. The system used an Nd-doped fiber amplifier (Nd DFA) as the gain element, and a NALM [105] operated in reflection as the all-optical switch. Here a Faraday rotator was employed to compensate for the 90° polarization rotation inside the NALM, which was implemented to achieve a fully adjustable linear phase delay between the clockwise and counterclockwise propagating pulses. The system also incorporated an acousto-optic modulator to initiate passive modelocking. The intracavity optical fiber provided positive dispersion at the operation wavelength of 1.06 μm. The dispersion of the fiber was compensated for with the bulk intracavity grating pair [48].

To simplify the cavity setup the NALM in Ref. 50 was later replaced with a nonlinear polarization switch [51], which increased the variation in the intracavity pulse width by up to a factor of 8. The obtained pulse width

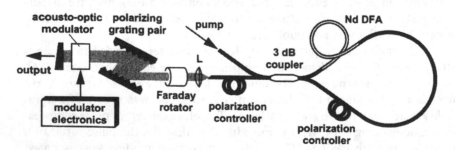

Figure 15 Dispersion-managed passively modelocked neodymium-doped fiber laser using a nonlinear amplifying loop mirror in reflection. The laser delivered pulses with a width of 100 fs. (From Ref. 50.)

was 70 fs, and further optimization allowed the generation of 37 fs pulses [52], the shortest pulses produced with passively modelocked fiber lasers to date. An autocorrelation of a 37 fs pulse from Ref. 52 is shown in Figure 16. These early dispersion-managed laser systems were generally not self-starting and relied on an intracavity start modulator or some other cavity perturbation [119] to initiate modelocking. Detailed measurements of the operation characteristics of these early dispersion-managed fiber lasers can also be found in Ref. 52.

The concept of intracavity pulse compression with bulk optic dispersion-compensating elements was later also used in conjunction with Er-doped fibers [134]. A great simplification in the design of dispersion-managed fiber lasers was achieved by the implementation of an all-fiber cavity consisting of highly nonlinear positive dispersion Er-doped fiber and weakly nonlinear negative dispersion undoped fiber [46]. Due to the large amount of

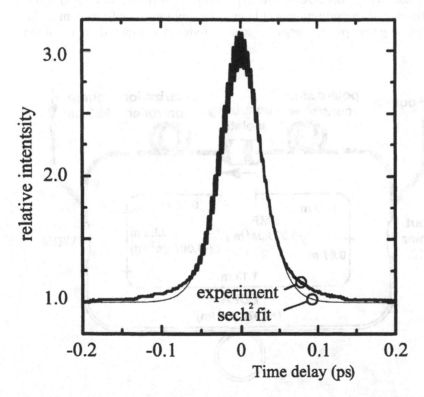

Figure 16 Autocorrelation of the shortest pulses generated from fiber laser oscillators to date. The pulse width is 37 fs, assuming a sech^2 shape. (From Ref. 52.)

intracavity pulse width variations in this cavity, which can exceed a factor of 10, the term "stretched pulse laser" was introduced for this cavity design. Stretched pulse lasers rely on nonlinear polarization evolution as an intracavity all-optical switch. A primary benefit of the stretched pulse laser is that a ring design is easily implemented and the laser can therefore start from noise [46]. Stretched pulse lasers reliably produce sub-100 fs pulses with nanojoule pulse energies [24] in the important Er wavelength region at the fundamental cavity frequency. The pulse formation processes in a stretched pulse laser are reviewed in Ref. 135. Stretched pulse cavity designs were also extended to enable modelocking of Tm- [136] and Yb-doped [137] fibers.

An example of an experimental setup of a stretched pulse laser is shown in Figure 17 [46]. The cavity is dispersion-balanced with a slightly positive overall cavity dispersion, which produces an increased output pulse energy [63]. The polarizing isolator ensures unidirectional operation, and the polarization controllers optimize nonlinear polarization evolution within the cavity. Because of the large amount of intracavity pulse width variations in dispersion-managed lasers, the amount of self-phase modulation for a given pulse energy is greatly reduced compared to a soliton

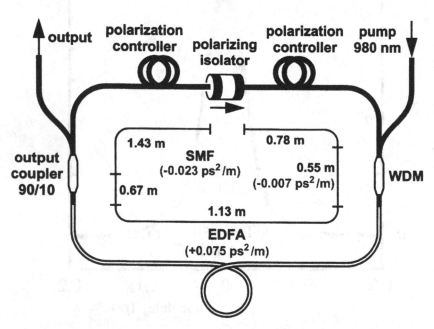

Figure 17 Experimental setup for a dispersion-managed, stretched pulse erbium-doped fiber ring laser. (From Ref. 46.)

laser, and as a result significantly higher pulse energies can be generated. Indeed, the maximum pulse energy from a single-mode stretched pulse laser [24] was around 3 nJ for a 100 fs pulse, which compares to a maximum pulse energy of 0.6 nJ for a 200 fs pulse as generated with a single-mode soliton laser [2].

The use of nonlinear polarization evolution in the early dispersion-managed fiber lasers complicates the construction of the cavity, because generally the optimum setting of the polarization controllers cannot be determined a priori. In typical fibers the fiber birefringence is temperature- and pressure-dependent, which leads to a time-varying linear and nonlinear phase delay between the eigenmodes of the fiber and a limited temperature range over which stable modelocked operation can be achieved. Moreover, long-term drifts in the polarization state can also not be excluded. Therefore, environmentally stable [35] stretched pulse lasers were also constructed, though, owing to the resulting complications in the cavity design, environmentally stable stretched pulse lasers relied on an intracavity optical modulator for pulse start-up [138].

Most of the demonstrated passively modelocked dispersion-managed fiber lasers producing pulses with a bandwidth comparable to the bandwidth of the gain medium rely on the interplay of gain, positive dispersion, and self-phase modulation in a single element, just as in similariton formation [61]. Dispersion-managed Er fiber lasers incorporating negative dispersion Er-doped fiber can also be constructed, but the pulses generated are generally significantly longer, and only 200 fs was obtained [126]. Moreover, the pulse energy was reduced to 100 pJ, whereas conventional stretched pulse lasers [24] readily produce nanojoule pulse energies. Clearly, with gain, positive dispersion, and self-phase modulation combined in one cavity element, the formation of an optimal pulse-forming process is ensured. Indeed, even in the presence fo relatively large amounts of self-phase modulation, the chirp in the positive dispersion gain fiber can be linearized, producing stable pulses at the bandwidth limit of the gain medium in conjunction with intra-cavity dispersion compensation. Moreover, the combination of gain, positive dispersion, and self-phase modulation can in principle also allow the generation of similariton pulses.

The basic operational characteristics of dispersion-managed lasers can be analyzed by an analytic model based on the master equation [60]. The key to the analytical model is the assumption of an effective dispersion corresponding to the dispersion imbalance in the cavity and the assumption of an effective parabolic phase and amplitude modulation due to self-phase modulation and an all-optical switching mechanism, respectively. With a chirped Gaussian pulse Ansatz, two readily solvable complex equations for the pulse width and chirp are produced. The analytical model is insensitive to the location of the cavity elements and therefore does not produce the evolution of the pulse parameters inside the cavity. These higher order effects need to be calculated using numerical solutions to Eqs. (13)–(15).

3.7.6 Similariton Lasers

In contrast to all other laser systems discussed so far similariton lasers rely on both strong linear and nonlinear intracavity pulse shaping. That is, not only the temporal width but also the spectral width of such lasers is expected to vary significantly during one round-trip through the cavity. Similariton lasers therefore rely on the presence of large amounts of intracavity self-phase modulation (up to 10π) and have so far been predicted only theoretically [47]. However, the expected great increase in obtainable pulse energies from similariton lasers compared to standard fiber lasers justifies a brief discussion at this point.

An experimental setup for a previously proposed similariton laser is shown in Figure 18. The laser comprises a positive dispersion gain fiber and some optical bulk dispersion-compensating (DC) element similar to an APCM laser setup [50]. Optimally the dispersion-compensating elements compensate for about 50% of the positive dispersion of the gain fiber. The laser comprises also a narrow-bandwidth filter inserted after the output coupler. Generally, a similariton laser requires strong intracavity amplitude modulation to ensure pulse stability. Therefore an intracavity saturable absorber with a lifetime shorter than the maximum pulse width inside the cavity is required, where the linear saturable loss should be of the order of 50–80%. At this time such large saturable losses are not obtainable with

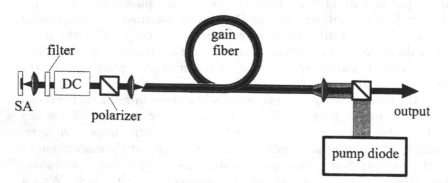

Figure 18 Theoretical concept for the design of a similariton laser. DC = optional dispersion compensation element; SA = saturable absorber.

conventional semiconductor saturable absorbers; therefore, additional all-optical switching elements would be needed for similariton generation.

Figure 19 shows the calculated output pulse intensity as well as the pulse spectrum from a similariton laser incorporating a 100 m length of EDFA and a 3 nm bandwidth intracavity filter. For simplicity a polarization-maintaining fiber was assumed, and the pulse solutions were obtained from a numerical solution of Eqs. (13)–(15). Parabolic pulses are readily generated, where the intracavity pulse energy can be of the order of 1 nJ. The parabolic pulse shape enables self-similar propagation inside the cavity, which results in a large linear chirp across the pulse, but no pulse breakup. Owing to the strong intracavity pulse shaping by self-phase modulation, the bandwidth of the output spectrum is around seven times broader than that of the intracavity filter. The pulse spectrum generated in the similariton laser is very similar to pulse spectra generated in single-pass similariton amplifiers [61,70], which demonstrates that pulse shaping is indeed dominated by self-similar nonlinear propagation inside the amplifier. The output pulses are also linearly chirped and are compressible to an FWHM pulse width of 500 fs.

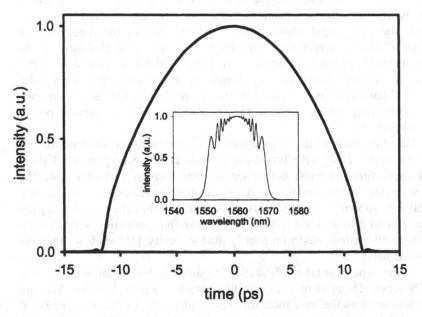

Figure 19 Calculated parabolic pulse and corresponding pulse spectrum generated with a similariton laser. (From Ref. 47.)

The energy of a similariton, proportional to $|A_0(z)|^2 \tau_0(z)$ in Eqs. (12) and (13), increases in proportion to $g^{4/3}$ with a reduction in amplifier length because the gain per unit length g has to increase in proportion to the amplifier length reduction to keep the overall gain $\exp(gz)$ the same. Hence a reduction in amplifier length by a factor of 10 increases the similariton energy by a factor of $10^{4/3} \approx 20$. Therefore, similariton fiber lasers generating pulses with an energy in the 10 nJ regime are feasible for 10 m intracavity fiber lengths. The generation of sub-100 fs pulses can also be expected with oscillator fiber lengths of 10 m or less. Particularly attractive is the implementation of a similariton laser with a double-clad Yb fiber amplifier, where it can be shown that similariton pulses with energies similar to the pulse energies obtained with single-pass similariton fiber amplifiers [61,70] can be generated.

3.7.7 Passively Modelocked Multimode Fiber Lasers

Of all modelocked fiber laser designs, multimode fiber lasers have generated the highest average powers to date. The reason is the greater core diameter achievable with multimode fibers, which minimizes the nonlinearity of the fiber. The increase in fiber core size possible in multimode fibers is particularly beneficial in cladding-pumped oscillator designs, because the pump light absorption is proportional to the core/cladding area ratio. Thus a larger core with respect to the same cladding significantly increases pump light absorption. Hence relatively short fiber oscillators can be used, which in turn allows for the generation of high pulse energies. Moreover, large pump powers are readily available from broad-area diode pump lasers, enabling the construction of high average power fiber oscillators.

Cladding-pumped multimode oscillators can be either end-pumped [25] or side-pumped [128] with broad-area diode lasers. Side-pumped Fabry-Perot oscillators are most conveniently implemented and allow for the highest degree of miniaturization. An example of a side-pumped multimode oscillator is shown in Figure 20. It is remarkable that the complete optics package fitted into a $14 \times 9 \times 2.5$ cm box, even in conjunction with a polarization-compensated, environmentally stable cavity [35] such as the one employed here.

In the experimental configuration depicted in Figure 20, a multimode Er/Yb-doped 16 μm core diameter fiber with a length between 1.4 and 2.4 m was used as the gain medium. The multimode Er/Yb fiber had a V value of 6.5 at the operational wavelength of 1.56 μm. The nonlinearity parameter of the Er/Yb fiber was calculated as $\gamma = 0.001$. Self-starting

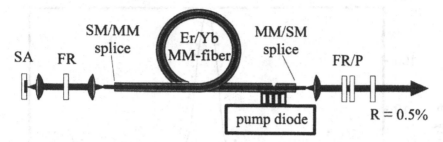

Figure 20 Experimental setup for a modelocked Er/Yb multimode fiber laser. The fiber has a double cladding and is pumped through the side. [MM = multimode; SM = single-mode.]

stable passive modelocking was ensured by employing a near-resonant [114] InGaAsP Fabry-Perot saturable absorber with a carrier lifetime of 1 ps. The output coupling fraction was typically chosen to be 99.5% by using an AR-coated wedge for output coupling. Only 0.5% of the output light was coupled back into the cavity, which further minimizes the nonlinearity of the cavity and increases the possible laser output power. The shown intra-cavity filter was used for gain shaping, enabling lasing at either 1.545 or 1.56 µm.

The fundamental mode in the multimode fiber was selected by two short single-mode fiber pigtails spliced onto both ends of the multimode fiber, though fiber tapers could also be used for fundamental mode selection [139]. A maximum cw modelocked power of 185 mW and a pulse energy of 2.9 nJ were obtained at 1.545 µm at a repetition rate of 64 MHz. The generated FWHM pulse width was of the order of 300 fs assuming a sech^2 pulse shape. The resulting pulse peak power was thus 10 kW, about 4 times higher than what is achievable in typical passively modelocked single-mode Er fiber lasers [2]. Though the pulses were not quite bandwidth-limited, they closely resembled solitons, as shown in Figure 21. The small satellite pulses observable in the autocorrelation trace in Figure 21 result from the small excitation of higher order modes in the fiber.

The stable operation of a passively modelocked multimode fiber oscillator is possible only under certain conditions, which eventually limits the maximum achievable output power. First of all, mode coupling should be minimized and the fundamental mode in the multimode fiber should be excited with high efficiency. Further, to avoid modal interference at the fiber splices and to obtain bend-insensitive operation of a multimode fiber oscillator, the generated pulse width should be significantly shorter than the single-pass group-velocity walkoff time between the fundamental and

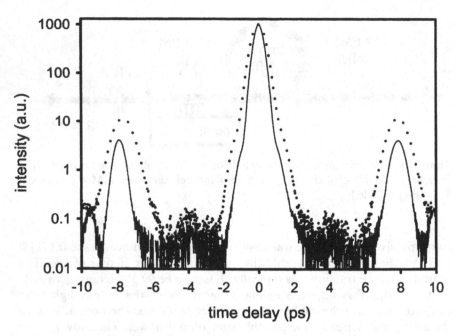

Figure 21 Autocorrelation of pulses generated in a passively modelocked multi-mode Er/Yb fiber laser. Solid line: full laser power; dotted line: 76% of full laser power. (From Ref. 26.)

the next higher order mode. Therefore, significant bandwidth restrictions of modelocked multimode fibers lead to pulse instabilities. Moreover, active modelocking of multimode fiber lasers tends to be equally unstable because of the longer intracavity pulses involved.

Stable operation regimes of passively modelocked multimode fiber lasers can be conveniently modeled using Eqs. (13)–(15) with an additional scattering operator $\hat{O}_s(A(t)) = A(t) + \eta A(t - \delta t)$ in Eq. (14) arising from the limited excitation $(1 - \eta)$ of the fundamental LP_{01} mode at the ends of the multimode fiber, where $\eta \ll 1$. It is sufficient to account only for scattering into the next higher order mode [36], which results in an excitation η of the LP_{11} mode. At the fiber ends the LP_{01} and LP_{11} modes are thus excited with amplitudes $(1 - \eta)^{1/2}A(t)$ and $\eta^{1/2}A(t)$, respectively. At the opposite fiber end, the LP_{11} mode is scattered back into the fundamental mode, resulting to first order in a scattering operator $\hat{O}_s$ as shown above.

For values of η of the order of a few percent, it can then be shown that modelocking stability is ensured when the single-pass group delay between the LP_{01} and the LP_{11} modes is about twice as large as the saturable absorber lifetime, in reasonable agreement with experimental observation. This may be understood from the requirement that the injection signal [140] due to intramode scattering must be smaller than the injection signal produced by the saturable absorber to allow for stable modelocking. Moreover, the model can also account for the large intracavity gain, predicting that pulse stability can be obtained for an nonlinear intracavity phase delay of up to 1.5π. As a result, at maximum power the laser does not produce true solitons, but rather slightly distorted soliton pulses.

Finally, the model can also predict that owing to the small temporal separation between the main pulse and the satellite pulses, the main pulse produces an injection signal for the growth of the satellite pulses; the injection signal arises from the wings of the main pulse and thus increases with an increase in pulse width. Therefore, the relative magnitude of the satellite pulses increases with a reduction in intracavity pulse energy [from Eq. (8) the pulse width is inversely proportional to the pulse energy for soliton pulses]. This is further demonstrated in Figure 21, which shows the measured autocorrelation trace of the multimode laser at full output power and at 73% of the full power. The relative increase in the magnitude of the satellite pulse is clearly seen.

The requirement that the saturable absorber lifetime be significantly shorter than the single-pass group delay between the fundamental and the next higher order mode and the requirement that any satellite pulses be well separated from the main pulse clearly set a limit to the minimum fiber length and the maximum core diameter that can be implemented in such a cavity design. However, with saturable absorbers with lifetimes of the order of 1 ps, passive modelocking of multimode fibers with core diameters of up to 25 μm should be possible, though the fiber length would have to be appropriately adjusted to ensure good temporal separation between the main and satellite pulses for a given pulse width.

Multimode fiber oscillators operating at 1.56 μm are attractive because they can be frequency-doubled to produce an output at around 800 nm, which is a preferred emission wavelength for many currently considered applications of ultrafast optics. Using a 1.2 mm length of antireflection-coated periodically poled $LiNbO_3$ (PPLN) with a poling period of 17.8 μm for frequency doubling [2], a doubling efficiency greater than 60% was obtainable from a multimode Er/Yb fiber oscillator, compared to a doubling efficiency of only about 25% obtainable with single-mode Er fiber oscillators [2]. A measurement of the frequency-doubled pulse power versus input power is shown in Figure 22. With an oscillator power of 165 mW (corresponding to a pulse energy of 4 nJ for this case), a frequency-doubled power of 100 mW was obtained.

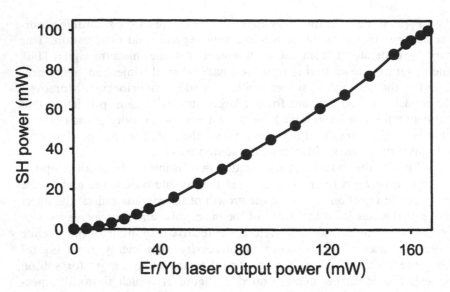

Figure 22   Second harmonic (SH) power at 773 nm versus fundamental power at 1546 nm obtained with a passively modelocked multimode Er/Yb soliton fiber laser operating at 40 MHz.

3.7.8 Passive Harmonic Modelocking

Whereas the output of passively modelocked fiber lasers is a train of phase-locked pulses at the fundamental cavity round-trip time, passively harmonically modelocked fiber lasers produce additional pulses located in between the train of phase-locked pulses. An example of the pulse train generated with passive modelocking and passive harmonic modelocking is shown in Figure 23. Passively harmonically modelocked pulses are typically not phase-locked and jitter around their average positions. At repetition rates of 2.5 GHz, a pulse jitter as small ±15 ps has been measured in an integrated cavity [141].

Passively harmonically modelocked lasers are very useful in scaling up the repetition rates of femtosecond fiber lasers while still preserving a very simple cavity setup. Though the timing jitter of passively harmonically mode locked fiber lasers is relatively high, making their use in nonlinear optical devices problematic, some applications may be able to tolerate this amount of timing jitter.

Moreover, passively harmonically modelocked lasers sometimes allow the repetition rate to be adjusted by simply changing the pump power level to the fiber cavity [141–146]. An example of a measurement of the number of

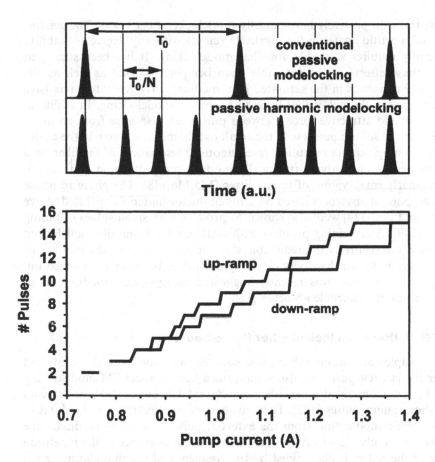

Figure 23 Top: Conceptual drawing of pulse locations in a passively modelocked laser operating at the fundamental cavity frequency and a passively harmonically modelocked fiber laser operating at four times the fundamental cavity frequency. Bottom: Number of pulses inside the cavity of a passively harmonically modelocked fiber laser as a function of current applied to the pump laser. (From Ref. 142.)

pulses N as a function of pump current in a cladding-pumped fiber laser cavity [142] is also shown in Figure 23. In passive harmonic modelocking, interactions between the intracavity pulses distribute all pulses approximately evenly across the cavity, producing an adjustable repetition rate Nf_0 in steps of multiples N of the fundamental cavity frequency f_0.

Passive harmonic modelocking is generally observed in soliton fiber lasers with overall negative group velocity dispersion [143]. A number of different processes have been suggested that lead to the self-stabilization of the pulse trains in passively harmonically modelocked fiber lasers. The generation of repetition rates in the gigahertz regime with any degree of stability generally requires saturable absorber modelocking. It has been suggested that phase effects in the saturable absorber [142,144,145] as well as the recovery dynamics in the saturated gain medium [147] generate a repulsive force between the pulses that leads to harmonic modelocking. In addition, repulsive and attractive forces between pulses can also arise from an interaction of the soliton pulses with the small oscillating continuum in the cavity [145]. A subgigahertz repetition rates, acoustic resonances of the fiber tend to further stabilitze the harmonically modelocked pulse train and lead to particularly small values of timing jitter [145,146,148]. The phase response of two-photon absorbers based on semiconductor materials operated above half band gap [114] were also shown to produce very small values of timing jitter [149]. A remaining problem with passively harmonically modelocked lasers is the relatively unpredictable start-up behavior; i.e., the number of pulses generated in the cavity can vary each time the laser is switched on. A possible solution was recently suggested using optical modulation of the intracavity saturable absorber [150].

3.7.9 Other Modelocked Fiber Pulse Sources

For completeness a few other pulse sources that have been demonstrated over the last few years are also summarized here. Optical FM modelocking has been demonstrated using either an external data stream [151] or a stream of short pump pulses [152]. Here instead of an electro-optic modulator, cross-phase modulation from the external pulses is used to modulate the phase of the fiber laser cavity. Just as in active modelocking, the repetition rate of the pulses is determined by the frequency of the modulator, or the frequency of the external data stream in this case. It is interesting to note that grey solitons, i.e., repetitive pulses with an intensity minimum in an optical background wave with a constant amplitude, have also been generated by optically induced modelocking [153].

A phase-insensitive passively additive pulse modelocked (APM) laser using a coupled fiber laser cavity with three fiber gratings was also constructed [154]. Whereas NALMs [105] constitute a nonlinear interferometeric optical switch, where the length of the interferometer arms is automatically balanced due to the Sagnac configuration, in principle other types of interferometers (such as Mach-Zehnder or Michelson configurations) can also be used to construct a nonlinear switch. In such interferometers the exact balancing of the length of the interferometer arms is typically a problem, but fiber gratings allow the construction of nonlinear interferometers, where the interferometer phase for optimum modelocking is automatically set by way of passive selection of the oscillating wavelength [154].

Another method for the generation of short optical pulses is the use of a sliding frequency soliton fiber laser. In sliding frequency lasers a continuous frequency shifter and a wavelength filter are incorporated into the cavity [155,156]. By continuously frequency-shifting the oscillating wave inside the cavity, the formation of axial modes is suppressed, and in conjunction with the wavelength filter a loss mechanism for continuous oscillation is introduced. However, owing to the ability of soliton pulses to follow small frequency perturbations, the lower loss of soliton pulses can lead to the formation of a soliton pulse train. Soliton pulses at repetition rates of several gigahertz have been generated in this manner. All-fiber versions have also been constructed by the incorporation of all-fiber frequency modulators [157]. Sliding frequency fiber lasers have also been operated in the positive dispersion regime in cladding-pumped Yb fiber lasers [158]. High power 4–5 ps widely wavelength-tunable pulses could be so generated. Sliding frequency fiber lasers are relatively insensitive to variations in polarization state and have also allowed modelocking of a photonic crystal Yb fiber, where 15 ps pulses were generated [53].

3.8 TIMING STABILIZATION

Most applications of modelocked lasers do not require only the presence of pulses of a certain short width; very often, the timing of the pulses is just as critical. In an ideal case the position of the modelocked pulses in the time domain would always be known to within a fraction of an optical half-cycle, which corresponds to a time of 2.5 fs at 1.55 μm. In fact, because the pulse train of lasers modelocked at the fundamental cavity round-trip time consists of a number of phase-locked axial modes, such timing accuracy exists between successive pulses in the pulse train. However, owing to a variety of physical effects, the pulses are subject to timing jitter, and the pulse arrival times after many round-trips inside the cavity will be subject to statistical deviations. However, one of the advantages of fiber lasers over bulk and semiconductor lasers is that they provide pulse sources with the lowest timing jitter presently available.

3.8.1 Physical Origins of Timing Jitter

The timing of the pulses in modelocked fiber lasers is affected by cavity length fluctuations, ASE, pump laser fluctuations, and acoustic interactions in the fiber. With present diode laser technology the effect of pump power fluctuations can generally be neglected. Equally, acousto-optic interactions are very weak and play a role only in passively harmonically modelocked fiber lasers (see Sec. 3.7.8).

Cavity length fluctuations arise mainly from temperature-induced refractive index fluctuations inside the optical fiber. With the temperature dependence of the refractive index of silica $\partial n/\partial T = 1.1 \times 10^{-5}$, a temperature change of just 1°C produces an optical path length change of 11 μm in 1 m of fiber.

Theories analyzing the jitter of modelocked fiber lasers due to ASE noise have been developed for both soliton [159] and stretched pulse lasers [160]. For a passively modelocked soliton laser the expected timing jitter $\langle \Delta t \rangle$ in a time window T due to ASE was evaluated as [159]

$$\langle \Delta t \rangle^2 = 3D^2 \frac{\theta}{g} \Omega_g^4 \tau^2 \frac{1}{N_{ph}} \left(\frac{T}{T_R} \right) \tag{25}$$

where D is the cavity dispersion, θ is the noise enhancement factor due to imperfect inversion in a three-level system, g is the intracavity gain, Ω_g is the gain bandwidth in circular frequency, N_{ph} is the number of photons per pulse, and T_R is the cavity round-trip time. N_{ph} is given by $N_{ph} = E/h\nu$, where E is the pulse energy. The rms timing jitter can be obtained by taking the square root of Eq. (25). A related expression can also be obtained for stretched pulse lasers [160]. As expected for a random walk process, Eq. (25) predicts an rms timing jitter proportional to $T^{1/2}$. Note, however, that direct measurements of the evolution of timing jitter have shown an increase in timing jitter proportional to T [161], which indicates that other noise sources in addition to ASE determined the jitter of the modelocked fiber laser in Ref. 161.

The timing jitter due to ASE in erbium-doped fiber lasers was evaluated as a few hundred femtoseconds and less than 100 fs in a time interval of 100 ms for soliton [159] and stretched pulse lasers [160], respectively.

3.8.2 Methods for Timing Stabilization of Modelocked Lasers

To ensure the stability of actively modelocked pulses and to lock the repetition rate of passively modelocked lasers to an external clock, phase-locked loops (PLLs) are generally used. PLLs can be implemented in various fashions. For example, the cavity length can be controlled by mounting a cavity mirror onto a moving piezoelectric transducer. Feedback bandwidths up to about 1 kHz, limited by the response time of the piezoelectric transducer, can thus be achieved [162]. Integrated cavity length adjusters can also be constructed by coiling the intracavity fiber onto a piezoelectric drum with an electronically controlled diameter [19,22]. A third technique does not require any piezoelectric transducers and relies on acousto-optically deflecting a beam onto different parts of a diffraction grating. In this the deflection angle from the acousto-optic modulator is controlled by varying the modulation frequency. Though the cavity length adjustment goes along with a wavelength change, much higher feedback bandwidths compared to piezoelectric transducers can be obtained. Feedback bandwidths up to 10 kHz have indeed been implemented, limited by the response time of the drive electronics for the acousto-optic modulator [163].

The above three techniques can be used for both actively and passively modelocked fiber lasers. The repetition rate of actively modelocked lasers can also be stabilized by a feedback loop that minimizes relaxation oscillations inside the cavity [164]; the magnitude of relaxation oscillations is then measured in the frequency domain and minimized by matching the cavity length to the cavity modulation frequency.

For actively modelocked lasers, passive cavity length adjustment techniques can also be employed. For example, in an AM-modelocked fiber laser cavity with a relatively large amount of positive group velocity dispersion, small cavity length drifts can be self-adjusting. In this the large amount of cavity dispersion is sufficient to compensate for cavity length drifts by a change in wavelength [92].

To eliminate the use of cavity length controlling elements, the regenerative modelocking technique can be used in active modelocking [165,166]. In regenerative modelocking the RF beat-note (see Sec. 3.7.3) of the initially free-running laser is detected, amplified, and fed back to an intracavity modulator. Provided the phase setting of the feed back loop is correct, the modelocking process starts from noise, and the correct modelocking frequency is continuously regeneratively generated from the laser signal itself, eliminating the need for cavity length control. However, the repetition rate of the laser is now drifting in unison with the cavity length, which is a problem when the laser must operate at a fixed clock frequency. Regenerative modelocking can also be used to start the modelocking process in passively modelocked lasers.

To minimize cavity length drifts between two fiber lasers, it is also useful to coil both fiber lasers onto a single drum and ensure that they are subjected to the same environmental conditions, i.e., to environmentally couple the two lasers. A system of two environmentally coupled fiber lasers operating at 5 MHz has been electronically stabilized with relative timing jitters as low as 5 ps over a time period of 10 min [167], corresponding to a phase error of only 30 arcsec. These phase errors are in general limited by long-term drifts in the PLL circuits, which are temperature-dependent and have a limited phase accuracy.

Note that here we refer to the pulse jitter as phase noise with respect to the fundamental *RF frequency* corresponding to the cavity round-trip time. Recently, the absolute locking of the optical phase between two lasers was also achieved by minimizing the difference between the two *optical frequency* combs generated by the lasers [168]. In applications of fiber laser pairs to all-optical scanning [1], the improvement in pulse jitter by optical phase-locking techniques cannot be used, because a physical separation of the pulses is required. Optical phase-locking techniques, on the other hand, rely on producing a temporal overlap between the two pulses generated by the lasers. However, in the generation of stable frequency combs for wavelength-division multiplexing, absolute optical phase locking could be invaluable, as discussed in the next section.

3.9 RECENT DEVELOPMENTS

Though ultrafast fiber lasers are readily commercially available, further research on improved fiber pulse sources is ongoing. One direction of research deals with the generation of very high energy pulses from fiber amplifiers, which is discussed in Chapter 4. Though ultrafast fiber amplifiers can readily compete with bulk solid-state lasers in many areas, one limitation of current ultrafast fiber lasers is their relatively narrow bandwidth, which presents many challenges for the generation of sub-100 fs pulses.

However, the application of nonlinear pulse compression techniques using fibers with controllable higher order dispersion terms has allowed the generation of optical pulses with widths as short as 20 fs, albeit with pulses with energies in the nanojoule range [12]. Fifty femtosecond pulses with energies in the 10 nJ range have been generated with Yb fiber similariton amplifiers [44]; in conjunction with higher power fiber amplifiers, the generation of sub-10 fs pulses should also be possible.

Another area of activity comprises the incorporation of photonic crystal fibers into actual fiber laser designs, with the promise of demonstrating ultracompact soliton lasers operating in the normal dispersion regime of silica fibers at wavelengths of less than 1.3 μm [53].

The tuning range of ultrafast fiber lasers has been greatly improved by the application of all-fiber wavelength conversion techniques [70,169–171]. These sources employ the soliton self-frequency shift [172,173] to enable Raman shifting of the wavelength of ≈100 fs soliton pulses in a spectral range spanning up to 500 nm. Wavelength tuning is achieved by injecting a variable fraction of the ultrafast pulses from a fixed wavelength fiber oscillator into a separate Raman-shifting fiber.

In conjunction with standard step-index fibers, femtosecond pulses from an Er fiber laser could be continuously Raman shifted in a range of 1.55–2.00 µm [169]. Further extensions of the tuning range from 1.3 to 1.5 µm [174] and from 2.00 to 2.13 µm [70] have also been possible. By frequency-doubling in periodically poled LiNbO$_3$, femtosecond pulses in the important 1.064 µm wavelength range could be generated using a Raman-shifted Er soliton laser. Pulses tunable in the 1.05–1.5 µm wavelength range [171] have been generated using an ytterbium fiber femtosecond laser in conjunction with Raman shifting in a photonic crystal fiber.

A novel area of activity comprises the generation of supercontinua [49,175,176] and coherent frequency combs [6,177,178]. Because of the small core sizes of photonic crystal fibers, supercontinuum generation in optical fibers is possible, with nanojoule pulses allowing for the construction of compact all-fiber supercontinuum sources [179]. Coherent frequency combs with exact frequency spacings of 25 GHz have been generated with actively modelocked semiconductor lasers [6] in conjunction with a variety of pulse compression fibers. Such signal sources are of great interest for dense wavelength division multiplexing systems in optical communications. The research into all these areas is just beginning, offering great rewards for years to come.

3.10 CONCLUSIONS

Over the last decade, the use of ultrafast fiber laser systems has greatly proliferated. Whenever ultrafast pulses are required in real industrial environments, the use of fiber lasers is one of the top choices. Industrial laser systems in particular benefit from the reduced complexity of ultrafast fiber laser systems, their superior durability, and their uniquely small size. Ultrafast fiber laser systems are readily implemented in advanced instrumentation applications that are not easily compatible with conventional ultrafast solid-state laser technology. It even seems feasible that ultrafast fiber lasers can potentially reach the consumer market. Indeed, some of the ultrafast fiber lasers described in this chapter readily lend themselves to mass production, and it can be considered only a matter of time until automated assembly techniques will be fully installed in conjunction with the manufacture of ultrafast fiber lasers.

The recent advances in ultrafast laser technology, such as precision frequency metrology [180], absolute optical phase control [181], absolute optical phase locking [168] and coherent frequency comb generation

[6,178], are fully compatible with ultrafast fiber lasers and offer novel avenues for research and development. The use of ultrafast fiber lasers, not only in instrumentation devices but also in actual Telcordia-certified optical communication systems, could be a result of these endeavors.

REFERENCES

1. Sucha G, Fermann ME, Harter DJ, Hofer M. A new method for rapid temporal scanning of ultrafast lasers. IEEE J Selected Topics Quantum Electron 2:605–621 (1996).
2. Arbore MA, Fejer MM, Fermann ME, Hariharan A, Galvanauskas A, Harter D. Frequency-doubling of femtosecond erbium-fiber soliton lasers in periodically poled lithium niobate. Opt Lett 22:13–15 (1997).
3. Hariharan A, Fermann ME, Stock ML, Harter DJ, Squier J. Alexandrite-pumped alexandrite regenerative amplifier for femtosecond pulse amplification. Opt Lett 21:128–130 (1996).
4. Millard AC, Wiseman PW, Fittinghoff DN, Wilson KR, Squier J, Müler M. Third-harmonic generation microscopy by use of a compact, femtosecond fiber laser source. Appl Opt 38:7393–7397 (1999).
5. Li M, Zhang X-C, Sucha G, Harter D. Portable THz systems and their applications. Proc SPIE 3616:126–135 (1999).
6. Yamada E, Takara H, Ohara T, Sato K, Morioka T, Jinguji K, Itoh M, Ishii M. A high SNR, 150 ch supercontinuum cw optical source with precise 25 GHz spacing for 10 Gbit/s DWDM systems. Optical Fiber Communication Conf., 2001, OFC paper ME2.
7. Mamyshev PV, Chernikov SV, Dianov EM. Generation of fundamental soliton trains for high-bit-rate optical fiber communication lines. IEEE J Quantum Electron 27:2347–2355 (1991).
8. Reeves-Hall PC, Lewis SAE, Chernikov SV, Taylor JR. Picosecond soliton pulse-duration-selectable source based on adiabatic compression in Raman amplifier. Electron Lett 36:622–623 (2000).
9. Reeves-Hall PC, Taylor JR. Wavelength and duration tunable sub-picosecond source using adiabatic Raman compression. Electron Lett 37:417–418 (2001).
10. Hansryd J, Andrekson PA. Wavelength tunable 40 GHz pulse source based on a fiber optical parametric amplifier. Electron Lett 37:584–585 (2001).
11. Miyamoto M, Tsuchiya M, Liu HF, Kamiya T. Generation of ultrashort (~65 fs) pulses from 1.55 µm gain-switched distributed feedback (DFB) laser with soliton compression by dispersion arrangement. Jpn J Appl Phys 35:L1330–L1332 (1996).
12. Matsui Y, Pelusi MD, Suzuki A. Generation of 20-fs optical pulses from a gain-switched laser diode by a four-stage soliton compression technique. IEEE Photon Tech Lett 11:1217–1219 (1999).

13. Hofer M, Fermann ME, Galvanauskas A, Harter D, Windeler RS. High-power 100 fs pulse generation by frequency doubling of an erbium-ytterbium fiber master oscillator power amplifier. Opt Lett 23:1840–1842 (1998).

14. Galvanauskas A, Harter D, Arbore MA, Chou MH, Fejer MM. Chirped-pulse-amplification circuits for fiber amplifiers, based on chirped-period quasi-phase-matching gratings. Opt Lett 23:1695–1697 (1998).

15. Galvanauskas A, Fermann ME. 13 W average power ultrafast fiber laser. Conf. on Lasers and Electro-Optics, CLEO, 2000, Paper CPD-3.

16. Galvanauskas A, Fermann ME, Blixt P, Tellefson JA. Jr, Harter D. Hybrid diode-laser fiber-amplifier source of high-energy ultrashort pulses. Opt Lett 19:1043–1045 (1994).

17. Galvanauskas A, Sartania Z, Bischoff M. Millijoule femtosecond all-fiber system. Conf. on Lasers and Electro-Optics, CLEO, 2001, Paper CMA-1.

18. Galvanauskas A, Hariharan A, Raksi F, Wong KK, Harter D, Imeshev G, Fejer MM. Generation of diffraction limited femtosecond beams using spatially multimode nanosecond pump sources in parametric chirped pulse amplification systems. Conf. on Lasers and Electro-Optics, CLEO, 2001, Paper CTHB4.

19. Carruthers TF, Duling IN. III, Horowitz M, Menyuk CR. Dispersion management in a harmonically modelocked fiber soliton laser. Opt Lett 25: 153–155 (2000).

20. Takara H, Kawanishi S, Saruwatari M, Noguchi K. Generation of highly stable 20 GHz transform-limited optical pulses from actively modelocked Er^{3+}-doped fiber lasers with an all-polarization maintaining ring cavity. Electron Lett 28:2095–2096 (1992).

21. Yoshida E, Nakazawa M. 80–200 GHz erbium doped fibre laser using a rational harmonic modelocking technique. Electron Lett 32:1371–1372 (1996).

22. Clark TR, Carruthers TF, Duling IN. III, Matthews PJ. Phase noise measurements of ultrastable 10 GHz harmonically modelocked fibre laser. Electron Lett 35:720–721 (1999).

23. Chernikov SV, Richardson DJ, Laming RI, Dianov EM, Payne DN. 70 Gbit/s fibre based source of fundamental solitons at 1550 nm. Electron Lett 28:1210–1212 (1992).

24. Nelson LE, Fleischer SB, Lens G, Ippen EP. Efficient frequency-doubling of a femtosecond fiber laser. Opt Lett 21:1759–1761 (1996).

25. Fermann ME, Hofer M, Windeler RS. Multimode fiber soliton laser. Trends Opt Photon 1:389–393 (1999).

26. Fermann ME, Galvanauskas A, Hofer M. Ultrafast pulse sources based on multimode optical fibers. Appl Phys B B70:S13–S23 (2000).

27. Wysocki PF, Simpson JR, Lee D. Prediction of gain peak wavelength for Er-doped fiber amplifiers and amplifier chains. IEEE Photonics Tech Lett 6:1098–1100 (1994).

28. Desurvire E. Erbium Doped Fiber Amplifiers: Principles and Applications. New York: Wiley, 1994.

29. Grubb SG, Humer WF, Cannon RS, Windhorn TH, Vendetta SW, Sweeney KL, Leilabady PA, Barnes WL, Jedrzejewski KP, Townsend JE. +21dBm erbium power amplifier pumped by a diode-pumped Nd:YAG laser. IEEE Photonics Tech Lett 4:553–555 (1992).

30. Po H, Snitzer E, Tumminelli R, Zenteno L, Hakimi F, Cho NM, Haw T. Doubly-clad high brightness Nd fiber laser pumped by GaAlAs phased array. Proc Optical Fiber Communication Conf OFC Houston, TX, 1989, Paper PD7.

31. Minelly JD, Barnes WL, Laming RI, Morkel P, Townsend JE, Grubb S, Payne DN. Diode-array pumping of Er^{3+}/Yb^{3+} co-doped fiber lasers and amplifiers. IEEE Photonics Tech Lett 5:301–303 (1993).

32. Fermann ME. Nonlinear polarization evolution in passively modelocked fiber laser. In: IN Duling, ed. Compact Ultrafast Pulse Sources. New York: Cambridge Univ. Press, 1995.

33. Rashleigh SC, Ulrich R. High-birefringence in tension-coiled single-mode fibers. Opt Lett 5:354–356 (1980).

34. Koplow JP, Golberg L, Moeller RP, Kliner DAV. Polarization-maintaining double-clad fiber amplifier employing externally applied stress-induced birefringence. Opt Lett 25:387–389 (2000).

35. Fermann ME, Yang LM, Stock ML, Andrejco MJ. Environmentally stable Kerr-type modelocked erbium fiber laser producing 360 fsec pulses. Opt Lett 19:43–45 (1994).

36. Fermann ME. Single-mode excitation of multimode fibers with ultrashort pulses. Opt Lett 23:52–54 (1998).

37. Koplov JP, Kliner DAV, Goldberg L. Single-mode operation of a coiled multimode fiber amplifier. Opt Lett 25:442–444 (2000).

38. Offerhaus HL, Broderick NG, Richardson DJ, Sammut R, Caplen J, Dong L. High-energy single-transverse-mode Q-switched fiber laser based on a multimode large-mode-area erbium doped fiber. Opt Lett 23:1683–1685 (1998).

39. Knight JC, Birks TA, Cregan RF, Russell PStJ, Sandro de J-P. Large mode area photonic crystal fiber. Electron Lett 43:1347–1348 (1998).

40. Cheo P. Clad-pumped eye-safe and multi-core phase-locked fiber lasers. US Patent 6,031,850.

41. Oulette F. Dispersion cancelation using linearly chirped Bragg grating filters in optical waveguides. Opt Lett 12:847–849 (1987).

42. Kashap R. Fiber Bragg Gratings. New York: Academic Press, 1999.

43. Knight JC, Arriaga J, Birks T, Ortigosa-Blanc A, Wadsworth WJ, Russell PStJ. Anomalous dispersion in photonic crystal fiber. IEEE Photon Tech Lett 12:807–809 (2000).

44. Fermann ME, Stock ML, Galvanauskas A, Cho GC, Thomsen BC. Third-order dispersion control in ultrafast Yb fiber amplifiers. Opt Soc Am Topical Meeting on Advanced Solid State Lasers. ASSL, 2001, Paper TuA3.

45. Kafka JD, Bear T. Modelocked erbium-doped fiber laser with soliton pulse shaping. Opt Lett 14:1269–1271 (1989).

46. Tamura K, Ippen EP, Haus HA, Nelson LE, 77 fs pulse generation from a stretched-pulse modelocked all-fiber ring laser. Opt Lett 18:1080–1082 (1993).

47. Peacock AC, Kruglov VI, Thomsen BC, Harvey J, Fermann ME, Sucha G, Harter D, Dudley JM. Generation and interaction of parabolic pulses in high gain fiber amplifiers and oscillators. Conf on Optical Fiber Communication, OFC, 2001, Paper WP4.

48. Agrawal GP. Nonlinear Fiber Optics. San Diego: Academic Press, 1989.

49. Ranka JK, Windeler RS, Stentz AJ. Optical properties of high-delta air-silica microstructure optical fibers. Opt Lett 25:796–798 (2000).

50. Fermann ME, Hofer M, Haberl F, Ober MH, Schmidt AJ. Additive-pulse-compression mode locking of a neodymium fiber laser. Opt Lett 16:244–246 (1991).

51. Hofer M, Fermann ME, Haberl F, Ober MH, Schmidt AJ. Modelocking with cross-phase and self-phase modulation. Opt Lett 16:502–504 (1991).

52. Hofer M, Ober MH, Haberl F, Fermann ME. Characterization of ultra-short pulse formation in passively modelocked fiber lasers. IEEE J Quantum Electron 28:720–728 (1992).

53. Furusawa K, Monro TM, Petropoulos PPW, Richardson DJ. Modelocked laser based on ytterbium doped holey fiber laser. Electron Lett 37: 560–561 (2001).

54. Hofer M, Ober MH, Hofer R, Fermann ME, Sucha G, Harter D, Sugden K, Bennion I, Mendonca CAC, Chiu T. High-power neodymium soliton fiber laser that uses a chirped fiber grating. Opt Lett 20:1701–1703 (1995).

55. Price JHV, Lefort L, Richardson DJ, Spuhler GJ, Paschotta R, Keller U, Barty C, Fry A, Weston J. A practical, low-noise, stretched pulse Yb^{3+} doped fiber laser. Conf on Lasers and Electro-Optics, CLEO, 2001, Paper CtuQ6.

56. Fermann ME, Sugden K, Bennion I. High-power soliton fiber laser based on pulse width control with chirped fiber Bragg gratings. Opt Lett 20:172–174 (1995).

57. Hasegawa A, Tappert F. Transmission of stationary nonlinear optical pulses in dispersive dielectric fibers. Appl Phys Lett 23:142–144 (1973).

58. Mollenauer LF, Stolen RH, Gordon JP. Experimental observation of pico-second pulse narrowing and solitons in optical fibers. Phys Rev Lett 45:1095–1098 (1980).

59. Bélanger PA, Gagnon L, Paré C. Solitary pulses in an amplified nonlinear dispersive medium, Opt Lett 14:943–945 (1989).

60. Haus HA, Tamura K, Nelson LE, Ippen EP. Stretched pulse additive pulse modelocking in fiber ring lasers:theory and experiment. IEEE J Quantum Electron 31:591–598 (1995).

61. Fermann ME, Kruglov VI, Thomsen BC, Dudley JM, Harvey JD. Self-similar propagation and amplification of parabolic pulses in optical fibers. Phys Rev Lett 84:6010–6013 (2000).

62. Hofer M, Ober MH, Haberl F, Fermann ME, Taylor ER, Jedrzejewski KP. Regenerative Nd glass amplifier seeded with a Nd fiber laser. Opt Lett 17:807–809 (1992).

63. Tamura K, Nelson LE, Haus HA, Ippen EP. Soliton versus nonsoliton operation of fiber ring lasers. Appl phys Lett 64:149–151 (1994).

64. Hasegawa A, Kodoma Y. Guiding-center soliton. Phys Rev Lett 66:161–164 (1991).
65. Smith NJ, Doran NJ, Forysiak W, Knox FM. Soliton transmission using periodic dispersion compensation. J Lightw Technol. 15:1808–1822 (1997).
66. Jones DJ, Haus HA, Nelson LE, Ippen EP. Stretched pulse generation and propagation. IEICE Trans Electron E81-C:180–187 (1997).
67. Turitsyn SK, Shapiro EG, Mezentsev. Dispersion-managed solitons and optimization of the dispersion management. Opt Fiber Technol 4:384–412 (1998).
68. Tamura K, Nakazawa M. Pulse compression by nonlinear pulse evolution with reduced optical wave breaking in erbium-doped fiber amplifiers. Opt Lett 21:68–70 (1996).
69. Kruglov VI, Peacock AC, Dudley JM, Harvey JD. Self-similar propagation in high-power parabolic pulses in optical fiber amplifiers. Opt Lett 25:1753–1755 (2000).
70. Fermann ME, Galvanauskas A, Stock ML, Wong KK, Harter D, Goldberg L. Ultrawide tunable Er Soliton fiber laser amplified in Yb-doped fiber. Opt Lett 24:1428–1430 (1999).
71. Soto-Crespo JM, Akhmediev NN, Collings BC, Cundiff ST, Bergman K, Knox WH. Polarization-locked temporal vector solitons in a fiber laser: theory. J Opt Soc Am B 17:366–372 (2000).
72. Collings BC, Cundiff ST, Akhmediev NN, Soto-Crespo JM, Bergman K, Knox WH. Polarization-locked temporal vector solitons in a fiber laser: experiment. J Opt Soc Am B 17:354–365 (2000).
73. Hofer M, Ober MH, Hofer R, Reider GA, Sugden K, Bennion I, Fermann ME, Sucha G, Harter D, Mendonca CAC, Chiu TH. Monolithic polarization-insensitive passively modelocked fiber laser. Conf on Optical Fiber Communication, OFC, 1996.
74. Fermann ME, Harter DJ. Integrated passively modelocked fiber laser and method for constructing the same. US Patent 6,072,811, 2000.
75. Kärtner FX, Brovelli LR, Kopf D, Kamp M, Calasso I, Keller U. Control of solid-state laser dynamics by semiconductor devices. Opt Eng 34:2024–2036 (1995).
76. Hönninger C, Paschotta R, Morier-Genoud F, Moser M, Keller U. Q-switching stability limits of continuous-wave passive mode locking. J Opt Soc Am B 16:46–56 (1999).
77. Sucha G, Chemla DS, Bolton SR. Effects of cavity topology on the nonlinear dynamics of additive-pulse modelocked lasers. J Opt Soc Am B 15:2847–2853 (1998).
78. Agrawal GA, Olson NA. Amplification and compression of weak picosecond optical pulses by using semiconductor-laser amplifiers. Opt Lett 14:500–502 (1989).
79. Akhmediev NN, Ankiewicz A, Lederer MJ, Luther-Davis B. Ultrashort pulses generated by modelocked lasers with either a slow or a fast saturable-absorber response. Opt Lett 23:280–282 (1998).

80. Haus HA, Ippen EP, Tamura K. Additive pulse modelocking in fibre laser. IEEE J Quantum Electron 30:200–208 (1994).

81. Collings BC, Bergman K, Cundiff ST, Tsuda S, Kutz JN, Cunnigham JE, Jan WY, Koch M, Knox WH. Short cavity erbium/ytterbium fiber lasers modelocked with saturable Bragg reflectors. IEEE J Selected Topics Quantum Electron 3:1–11 (1997).

82. Kutz JN, Collings B, Bergman K, Tsuda S, Cundiff ST, Knox WH, Holmes P, Weinstein M. Modelocking pulse dynamics in a fiber laser with a saturable Bragg reflector. J Opt Soc Am B14:2681–2690 (1997).

83. Tamura KR, Nakazawa M. 54-fs, 10 GHz soliton generation from a polarization-maintaining dispersion-flattened dispersion-decreasing fiber pulse compressor. Opt Lett 26:762–764 (2001).

84. Alcock IP, Ferguson AI, Hanna DC, Tropper AC. Modelocking of a neodymium-doped monomode fibre laser. Electron Lett 22:268–269 (1986).

85. Duling IN, Goldberg L, Weller JF. High-power, modelocked Nd:fibre laser pumped by an injection-locked diode array. Electron Lett 24:1333–1334 (1988).

86. Hofer M, Fermann ME, Haberl F. Active modelocking of a neodymium-doped fiber laser using intra-cavity pulse compression. Opt Lett 15:1467–1469 (1990).

87. Phillips MW, Ferguson AI, Hanna DC. Frequency modulation mode locking of a Nd^{3+}-doped fiber laser. Opt Lett 14:219–221 (1989).

88. Phillips MW, Ferguson AI, Kino GS, Paterson DB. Modelocked fiber laser with a fiber phase modulator. Opt Lett 14:680–682 (1989).

89. Harvey GT, Mollenauer LF. Harmonically modelocked fiber ring laser with an internal Fabry-Perot stabilizer for soliton transmission. Opt Lett 18:107–109 (1993).

90. Carruthers T, Duling IN III. 10 GHz, 1.3 ps erbium fiber laser employing soliton pulse shortening. Opt Lett 21:1927–1929 (1996).

91. Jones DJ, Haus HA, Ippen EP. Subpicosecond solitons in an actively modelocked fiber laser. Opt Lett 21:1818–1820 (1996).

92. Tamura K, Nakazawa M. Dispersion-tuned harmonically modelocked fiber ring laser for self-synchronization to an external clock. Opt Lett 21:1984–1986 (1996).

93. Kuizenga DJ, Siegmann AE. FM and AM modelocking of the homogeneous laser-part I: theory. IEEE J Quantum Electron QE-6:694–715 (1970).

94. Haus HA, Silberberg Y. Laser mode locking with addition of nonlinear index. IEEE J Quantum Electron QE-22:325–331 (1986).

95. Kärtner FX, Kopf D, Keller U. Solitary-pulse stabilization and shortening in actively modelocked lasers. J Opt Soc Am B 12(3):486–497 (1995).

96. Doerr CR, Haus HA, Ippen EP, Shirasaki M, Tamura K. Additive pulse limiting. Opt Lett 19:1747–1749 (1994).

97. Tamura K, Nakazawa M. Pulse energy equalization in harmonically FM modelocked lasers with slow gain. Opt Lett 21:1930–1932 (1996).

98. Nakazawa M, Tamura K, Yoshida E. Supermode noise suppression in a harmonically modelocked fiber laser by self-phase modulation and spectral limiting. Electron Lett 32:461–462 (1996).

99. Horowitz M, Menyuk CR, Carruthers TF, Duling IN III. Theoretical and experimental study of harmonically modelocked fiber lasers for optical communication systems. J Lightw Technol 18:1565–1574 (2000).

100. Thoen ER, Grein ME, Koontz EM, Ippen EP, Haus HA, Kolodziejski LA. Stabilization of an active harmonically modelocked fiber laser using two-photon absorption. Opt Lett 25:948–950 (2000).

101. www.calmaropt.com

102. www.imra.com

103. www.cmxr.com

104. Doran N, Wood D. Nonlinear optical loop mirror. Opt Lett 13:56–58 (1988).

105. Fermann ME, Haberl F, Hofer M, Hochreiter H. Nonlinear amplifying loop mirror. Opt Lett 15:752–754 (1990).

106. Duling IN III. All-fiber ring soliton laser mode locked with a nonlinear mirror. Opt Lett 16:539–541 (1991).

107. Lin H, Donald DK, Sorin WV. Optimizing polarization states in a figure-8 laser using nonreciprocal phase shifter. J Lightw Technol 12:1121–1128 (1994).

108. Kärtner FX, Keller U. Stabilization of solitonlike pulses with a slow saturable absorber. Opt Lett 20:16–18 (1995).

109. Ober MH, Hofer M, Chiu TH, Keller U. Self-starting diode-pumped femtosecond Nd fiber laser. Opt Lett 18:1532–1534 (1993).

110. Chemla DS, Miller DAB. Room-temperature excitonic nonlinear-optical effects in semiconductor quantum-well structures. J Opt Am B 1155–1173 (1985).

111. Islam MN, Sunderman ER, Soccolich CE, Bar-Joseph I, Sauer N, Chang TY, Miller BI, Color Center Lasers passively mode locked by quantum wells. IEEE J Quantum Electron QE-25:2454–2463 (1989).

112. Lederer MJ, Kolev V, Luther-Davis B, Tan HH, Jagadish C. Ion-implanted InGaAs single quantum well semiconductor saturable absorber mirrors for passive modelocking, To be published.

113. DeSouza EA, Islam MN, Soccolich CE, Pleibel W, Stolen RH, Simpson JR, DiGiovanni DJ. Saturable absorber modelocked polarization maintaining erbium-doped fibre laser. Electron Lett 29:447–449 (1993).

114. Jiang M, Sucha G, Fermann ME, Harter D, Dagenais M, Fox S, Hu Y. Nonlinearly limited saturable absorber modelocking of an erbium fiber laser. Opt Lett 24:1074–1076 (1999).

115. Brabec T, Kelly SMJ, Krausz F. Passive modelocking in solid state lasers. In: IN Duling III, ed., Compact Sources of Ultrashort Pulses, Cambridge Studies Mod Opt London, UK:Cambridge Univ Press, 1995.

116. Krausz F, Brabec T, Spielmann C. Self-starting passive modelocking. Opt Lett 16:235–237 (1991).

117. Chou YF, Wang J, Liu HH, Kuo NP. Measurements of the self-starting threshold of kerr-lens modelocking lasers. Opt Lett 19:566–568 (1994).
118. Tamura K, Jacobson J, Ippen EP, Haus HA, Fujimoto JG. Unidirectional ring resonator for self-starting passively mode locked lasers. Opt Lett 18:220–222 (1993).
119. Ober MH, Hofer M, Fermann ME. 42 fsec pulse generation from a fiber laser started with a moving mirror. Opt Lett 18:367–369 (1993).
120. Guy MJ, Noske DU, Boskovic A, Taylor JR. Femtosecond soliton generation in a praseodymium fluoride fiber laser. Opt Lett 19:828–830 (1994).
121. Dennis M, Duling IN III. Role of dispersion in limiting pulse widths in fiber lasers. Appl Phys Lett 62:2911–2913 (1993).
122. Kelly SMJ. Characteristic sideband instability of periodically amplified average soliton. Electron Lett 28:806–807 (1992).
123. Matsas VJ, Loh WH, Richardson DJ. Self-starting, passively modelocked Fabry-Perot fiber soliton laser using nonlinear polarization evolution. IEEE Photonics Tech Lett 5:492–494 (1993).
124. Zirngibl M, Stulz LW, Stone J, Hugi J, DiGiovanni D, Hansen PB. 1.2 ps pulses from passively modelocked laser diode pumped Er-doped fibre ring laser. Electron Lett 27:1734–1735 (1991).
125. Hong L, Donald DK, Chang KW, Newton SA. Colliding pulse modelocked lasers using Er-doped fiber and a semiconductor saturable absorber. Conf. on Lasers and Electro-Optics, CLEO, 1995, Paper JtuE1.
126. Broderick NGR, Offerhaus HL, Richardson DJ, Sammut RA. Power scaling in passively modelocked laser area fiber lasers. IEEE Photon Technol Lett 10:1718–1720 (1998).
127. Fermann ME, Harter DJ, Minelly JD, Vienne GG. Cladding-pumped passively modelocked fiber laser generating femtosecond and picosecond pulses. Opt Lett 21:967–969 (1996).
128. Hofer M, Fermann ME, Goldberg L. High power side-pumped passively modelocked Er/Yb fiber laser. IEEE Photon Technol Lett 10:1247–1249 (1998).
129. Fermann ME, Sugden K, Bennion I. Generation of 10 nJ picosecond pulses from a modelocked fiber laser. Electron Lett 31:194–195 (1995).
130. Valdmanis JA, Fork RL, Gordon JP. Generation of optical pulses as short as 27 femtoseconds directly from a laser balancing self-phase modulation, group-velocity dispersion, saturable absorption, and saturable gain. Opt Lett 10:131–133 (1985).
131. Lamb K, Spence DE, Hong J, Yelland C, Sibbett W. All-solid-state self-modelocked Ti:sapphire laser. Opt Lett 19:1864–1866 (1994).
132. Kafka JD, Head DF, Baer T. Dispersion Compensated Fiber Raman Oscillator. In: Ultrafast Phenomena V. Springer Ser Chem Phys 46. New York: Springer-Verlag, 1986, pp 51–53.
133. Fermann ME, Hofer M, Haberl F, Craig-Ryan SP. Femtosecond fiber laser. Electron Lett 26:1737–1738 (1990).

134. Fermann ME, Andrejco MJ, Silberberg Y, Weiner AM. Generation of pulses shorter than 200 fsec from a passively modelocked Er fiber laser. Opt Lett 18:48–50 (1993).

135. Nelson LE, Jones DJ, Tamura K, Haus HA, Ippen EP. Ultrashort-pulse fiber ring lasers. Appl Phys B, Lasers Opt 65:277–294 (1997).

136. Nelson LE, Tamura K, Ippen EP, Haus HA. Broadly tunable sub-500 fs pulses from an additive-pulse modelocked thulium-doped fiber ring laser. Appl Phys Lett 67:19–21 (1995).

137. Cautaeris V, Richardson DJ, Paschotta R, Hanna DC. Stretched pulse Yb^{3+}:silica fiber laser. Opt Lett 22:316–318 (1997).

138. Jones DJ, Nelson LE, Haus HA, Ippen EP. Environmentally stable stretched pulse fiber laser generating 120 fs pulses. Conf on Lasers and Electro-Optics, CLEO, 1997, Paper CTUY3.

139. Alvarez-Chavez JA, Grudinin AB, Nilsson J, Turner PW, Clarkson WA. Mode selection in high power cladding pumped fibre lasers with tapered sections. Conf on Lasers and Electro-Optics, CLEO, 1999, Paper CWE 7.

140. Haus HA, Ippen EP. Self-starting of passively modelocked lasers. Opt Lett 16:1331 (1991).

141. Collings BC, Bergman K, Knox WH. Stable multigigahertz pulse-train formation in a short-cavity passively harmonic modelocked erbium/ytterbium fiber laser. Opt Lett 23:123–125 (1998).

142. Fermann ME, Minelly JD. Cladding-pumped passive harmonically modelocked fiber laser. Opt Lett 21:970–972 (1996).

143. Grudinin AB, Richardson DJ, Payne DN. Passive harmonic modelocking of a fiber soliton ring laser. Electron lett 29:1860–1861 (1993).

144. Gray S, Grudinin AB. Soliton fiber laser with a hybrid saturable absorber. Opt Lett 21:207–209 (1996).

145. Grudinin AB, Gray S. Passive harmonic mode locking in soliton fiber lasers. J Opt Soc Am B 14:144–154 (1997).

146. Gray S, Grudinin AB, Loh WH, Payne DN. Femtosecond harmonically modelocked fiber laser with timing jitter below 1 ps. Opt Lett 20:189–191 (1995).

147. Kutz JN, Collings BC, Bergman K, Knox WH. Stabilized pulse spacing in soliton lasers due to gain depletion and recovery. IEEE J Quantum Electron 43:1749–1757 (1998).

148. Pilipetskii AN, Golovchenko EA, Menyuk CR. Acoustic effect in passively modelocked fiber ring lasers. Opt Lett 20:907–909 (1995).

149. Hofer M, unpublished data.

150. Bonadeo NH, Knox WH, Roth JM, Bergman K. Passive harmonic modelocked soliton fiber laser stabilized using an optically pumped saturable Bragg reflector. Conf on Lasers and Electro-Optics CLEO, 2000, Paper CMP8.

151. Greer EJ, Smith K. All-optical FM modelocking of fiber laser. Electron Lett 28:1741–1743 (1992).

152. Stock ML, Yang LM, Andrejco MJ, Fermann ME. Synchronous mode locking using pump-induced phase modulation. Opt Lett 18:1529–1531 (1993).

153. Pataca DM, Rocha ML, Kashap R, Smith K. Bright and dark pulse generation in an optically modelocked fibre laser at 1.3 μm. Electron Lett 31:35–36 (1995).
154. Cheo PK, Wang L, Ding M. Low-threshold, self-tuned and passively modelocked coupled-cavity all-fiber lasers. IEEE Photonics Tech Lett 8:66–68 (1996).
155. Fontana F, Bossalini L, Franco P, Midrio M, Romagnoli M, Wabnitz S. Self-starting sliding frequency fibre soliton laser. Electron Lett 30:321–322 (1994).
156. Romagnoli M, Wabnitz S, Franco P, Midrio M, Bossalini L, Fontana F. Role of dispersion in pulse emission from a sliding-frequency fiber laser. J Opt Soc Am B12:938–944 (1995).
157. Culverhouse DO, Richardson DJ, Birks TA, Russell PStJ. All-fiber sliding-frequency Er^{3+}/Yb^{3+} soliton fiber laser. Opt Lett 20:2381–2383 (1995).
158. Porta J, Grudinin AB, Chen ZJ, Minelly JD, Traynor NJ. Environmentally stable picosecond ytterbium fiber laser with a broad tuning range. Opt Lett 23:615–617 (1998).
159. Namiki S, Yu CX, Haus HA. Observation of nearly quantum-limited timing jitter in an all-fiber ring laser. J Opt Soc Am B13:2817–2823 (1996).
160. Yu CX, Namiki S, Haus HA. Noise of the stretched pulse fiber laser: Part II. Experiments. IEEE J Quantum Electron 33:649–659 (1997).
161. Sucha G, Fermann ME, Harter DJ. Time-domain jitter measurements of ultrafast lasers. Conf on Lasers and Electro-Optics, San Francisco, 1998, Paper CTu01.
162. Haberl F, Ober MH, Hofer M, Fermann ME, Wintner E, Schmidt AJ. Low-noise operation modes of a passively modelocked fiber laser. IEEE Photon Tech Lett 3:1071–1073 (1991).
163. Jiang M, Sha W, Rahman L, Barnett BC, Anderson JK, Islam MN, Reddy KV. Synchronization of two passively modelocked erbium-doped fiber lasers by an acousto-optic modulator and grating scheme. Opt Lett 21:809–811 (1996).
164. Takara H, Kawanishi S, Saruwatari M. Stabilisation of a modelocked Er-doped fibre laser by suppressing the relaxation oscillation frequency component. Electron Lett 31:292–293 (1995).
165. Huggett GR. Modelocking of CW lasers by regenerative RF feedback. Appl Phys Lett 13:186–187 (1968).
166. Nakazawa M, Yoshida E, Kimura Y. Ultrastable harmonically and regeneratively modelocked polarization-maintaining erbium fiber ring laser. Electron Lett 30:1603–1605 (1994).
167. Sucha G, Hofer M, Fermann ME, Harter D. Synchronization of environmentally coupled, passively modelocked fiber lasers. Opt Lett 21:1570–1572 (1996).
168. Shelton RK, Ma LS, Hall JL, Kapteyn HC, Murnane MM. Coherent pulse synthesis from two (formerly) independent passively modelocked Ti:sapphire oscillators. Conf on Lasers and Electro-Optics, CLEO, 2001, Paper CDP10.

169. Nishizawa N, Goto T. Compact system of wavelength-tunable femtosecond pulse generation using optical fibers. IEEE Photon Tech Lett 11:325–327 (1999).

170. Chandalia JK, Eggleton BJ, Windeler RS, Kosinski SG, Liu X, Xu C. Adiabatic coupling in tapered air-silica microstructured optical fiber. IEEE Photon Tech Lett 13:52–54 (2001).

171. Price JHV, Furusawa K, Monro TM, Lefort L, Richardson DJ. A tuneable, femtosecond pulses source operating in the range 1.06–1.33 microns based on a Yb doped holey fiber amplifier. Conf on Lasers and Electro-Optics, CLEO, 2001, Paper CPD1.

172. Mitschke FM, Mollenauer LF. Discovery of the soliton self-frequency shift. Opt Lett 11:659–661 (1986).

173. Gordon JP. The theory of the soliton self-frequency shift. Opt Lett 11:662–664 (1986).

174. Nishizawa, Okamura R, Goto M. Widely wavelength tunable ultrashort soliton pulses and anti-Stokes pulse generation for wavelengths of 1.32–1.75 μm. Jpn J Appl Phys 39:409–411 (2000).

175. Morioka T, Mori K, Saruwatari M. More than 100-wavelength-channel picosecond optical pulse generation from single laser source using supercontinuum in optical fibers. Electron Lett 29:862–864 (1993).

176. Ranka J, Windeler RS, Stentz AJ. Visible continuum generation in air silica microstructure optical fibers with anomalous dispersion at 800 nm. Opt Lett 25:25–27 (2000).

177. Kubota H, Tamura K, Nakazawa M. Analyses of coherence-maintained ultrashort optical pulse trains and supercontinuum generation in the presence of soliton-amplified spontaneous-emission interaction. J Opt Soc B 16:2223–2232 (1999).

178. Takara H, Yamada E, Ohara T, Sato K, Jinguji K, Inoue Y, Shibata T, Morioka T. 106 × 10 Gbit/s, 25 GHz-spaced, 640 km DWDM transmission employing a single supercontinuum multi-carrier source, Conf on Lasers and Electro-Optics, CLEO, 2001, Paper CPD11.

179. Fermann ME, Stock ML, Harter D, Birks T, Wadsworth WJ, Russell PStJ, Fujimoto J. Wavelength-tunable soliton generation in the 1400–1600 nm region using an Yb fiber laser. Conf on Opt Fiber Commun, OFC, 2001, Paper TuI2.

180. Udem T, Reichert J, Holzwarth R, Hänsch TW. Absolute optical frequency measurement of the cesium D_1 line with a mode-locked laser. Phys Rev Lett 82:3568–3571 (1999).

181. Apolinski A, Poppe A, Tempea G, Spielmann C, Udem T, Holzwarth R, Hänsch TW, Krausz F. Controlling the phase evolution of few-cycle light pulses. Phys Rev Lett 85:740–743 (2000).

4

Ultrashort-Pulse Fiber Amplifiers

Almantas Galvanauskas
University of Michigan, Ann Arbor, Michigan, U.S.A.

4.1 INTRODUCTION

Rapid development of ultrafast technology within the last several years has established a number of important practical applications and, consequently, produced a strong demand for practical turnkey femtosecond laser systems. This demand has been the dominant force driving the recent development of ultrashort-pulse lasers. An important part of this development has been an emergence of a new generation of ultrashort pulse lasers, based on the use of optical fibers as the gain medium. Technological advantages of fibers are apparent in their intrinsic robustness and compactness, high diode-pumping efficiencies, and resilience to thermal effects, thus making fiber-based systems prime candidates from developing highly manufacturable and reliable systems.

For many emerging applications, the availability of high average power and high pulse energies are of critical importance. This poses a significant challenge for ultrashort-pulse fiber technology, because the very attributes that make fibers attractive, such as a confined mode and long propagation length, constitute a principal limitation for reaching the high peak powers associated with high energy ultrashort pulses. Nevertheless, over the last few years a number of technological developments proved that this difficulty can be overcome and high pulse energies as well as powers can be reached using fiber amplifiers. This development has led to fiber amplifier systems that produce femtosecond pulses in the millijoule energy range and average powers of more than 10 W, essentially at the same level as from conventional solid-state lasers. The basic solution has been to scale down the peak intensities inside the fiber core by a variety of techniques so that nonlinear effects can be kept under control even at high peak powers.

The objective of this chapter is to review the technological development that has produced these achievements, and is continuing to lead to further advances. The chapter consists of two principal parts. Section 4.2 describes fundamental aspects of amplifying ultrashort pulses in fiber amplifiers, and include a description of a variety of pulse amplification techniques using optical fibers as well as discussion of the basic limitations caused by nonlinear effects. Section 4.3 presents an overview of the variety of ultrashort-pulse fiber amplifiers, grouped into high average power and high pulse energy systems.

4.2 FUNDAMENTALS OF AMPLIFYING ULTRASHORT PULSES IN RARE-EARTH FIBERS

4.2.1 Techniques for Ultrashort-Pulse Amplification in Fibers

The essential part of amplifying ultrashort pulses in a fiber amplifier is the control of nonlinear effects in the fiber core, such that ultrashort pulses after amplification would still maintain good quality and short duration. This chapter reviews four basic approaches of such control. Peak intensity in the amplified pulse can be reduced though temporal scaling using chirped pulse amplification technique and through spatial scaling by increasing fiber core size even beyond the single-mode limit. The latter can be either accomplished while still maintaining diffraction-limited output from a multimode core or by employing nonlinear beam-cleaning techniques. Alternatively, nonlinear effects can be exploited rather than avoided during pulse amplification in a fiber amplifier, with the addition advantage of further compressing the duration of such nonlinearly amplified optical pulses.

Temporal Scaling: Chirped Pulse Amplification in Fibers

Broad spectral bandwidth, characteristic of ultrashort optical pulses, makes it relatively easy to stretch and compress such pulses by large factors of up to 10^3–10^4 by using spectrally dependent spatial effects, one example of which is angular dispersion [1] in diffraction-grating based pulse stretching [2] and compression [3] arrangements. This gives a very efficient method for reducing peak power of ultrashort pulses in an optical amplifier by stretching seed pulses prior to amplification and recompressing then afterward—so-called chirped pulse amplification (CPA) [4]. Such "temporal scaling" can increase the maximum achievable energies from an optical amplifier by factors comparable to the stretching ratio. However, because ultrashort-pulse stretching by these large factors fundamentally relies on spatial effects, the longest stretched pulse durations are limited by the largest practically acceptable size of a stretcher/compressor arrangement to about 1 ns, thus setting a limit for temporal scaling of amplified pulse energies.

Currently, CPA is a well-established method for high-energy amplification in solid-state laser systems, relying on diffraction gratings for pulse stretching and recompression. Such a technique is not quite compatible with the targeted compactness, robustness, and manufacturability of fiber-based systems, because diffraction grating arrangements (stretchers in particular) are typically large and require rather complicated alignment. This brings us to the need to develop new pulse stretching/compression technologies that would provide practical alternatives to the conventional technique. Recently, a number of new devices have emerged for implementing compact femtosecond-pulse stretchers and compressors, such as the chirped fiber Bragg gratings [5], chirped QPM gratings [6] and even an initial demonstration of volume Bragg gratings [7], which will be described in more detail in subsequent sections of this chapter. The development of such compact stretching/compression techniques is an essential part of the general development of high-power and high-energy fiber-based ultrashort pulse amplification systems.

It is necessary to note, however, that with respect to extracting the highest saturation-fluence-limited energies from a fiber-based system, none of the compact devices demonstrated to date can fully replace conventional diffraction grating arrangements. The reasons for that become clear from an inspection of Table 1, where a comparative summary of the main characteristics of different pulse-compressing devices are presented.

Table 1 Characteristics of Pulse-Compressing Devices

	Diffraction grating	Fiber/ diffraction grating	Fiber Bragg grating	Quasi-phase- matched	Volume Bragg grating
Maximum $\Delta T_{\text{stretched}}$	$\sim 1\,\text{ns}$	$\sim 1\,\text{ns}$	$\sim 1\,\text{ns}$	$< 50\,\text{ps}$	$< 50\,\text{ps}$ $(\sim 1\,\text{ns})^a$
Maximum E_{recomp}	Scalable	Scalable	$\ll 1\,\mu\text{J}$	Scalable	Scalable
Monolithic	No	No	Yes	Yes	Yes
Recompressible pulse duration[b]	BWL	Non-BWL	BWL	BWL	BWL
Engineerability of device dispersion	No	No	Yes	Yes	Yes

[a]Possible, but has not been demonstrated yet.
[b]BWL = bandwidth-limited.

The maximum stretched pulse duration $\Delta T_{stretched}$ determines the maximum extent of the temporal scaling and consequently the maximum energies extractable from a fiber amplifier. The maximum recompressible energy E_{recomp} characterizes the energy scalability of the compressor itself, which essentially is dependent on whether the beam size in a particular device is transversely scalable. Not all stretching/compression arrangements permit reaching bandwidth-limited (BWL) durations for the recompressed pulses. Additionally, a combination of a few different devices can be used for achieving a specific technological goal, which would require matching the dispersion between different devices up to the higher order terms. Therefore, the ability to engineer dispersion properties is very important.

Conventional diffraction-grating arrangements for both stretching and compressing optical pulses in CPA are represented in Table 1. They provide long stretched pulse durations and are energy-scalable but are not monolithic devices. One step towards more compact system is to use a combination of a fiber- or a linearly chirped fiber grating stretcher and a conventional diffraction grating compressor. The main difficulty encountered with this combined system is that currently it is not possible to match all dispersion orders between a fiber or a fiber grating stretcher and a diffraction grating compressor, thus hindering achieving bandwidth-limited femtosecond pulses after recompression (see Ref. 8 and also Sec. 4.3.2). The remaining columns represent fully monolithic designs based on a chirped fiber Bragg grating (FBG), a chirped quasi-phase-matched second harmonic compressor (QPM), or a volume chirped period Bragg grating (volume Bragg), respectively. Currently the drawback of the monolithic devices is either in the limited stretched-pulse duration or in the limited recompressible energy. The most promising monolithic device in this respect appears to be a volume Bragg grating, which potentially can provide ~ 1 ns stretched pulses and also be scalable in the transverse dimension for handling high recompressed pulse energies. So far, however, only stretched pulses of < 50 ps have been achieved [7], and further significant technological development is necessary in this direction.

Spatial Scaling: Fiber Core-Size Scaling

Fiber core size scaling beyond the single-mode limit while still maintaining diffraction-limited output rely on the fact that in a highly doped optical fiber amplifier pulse propagation length can be made significantly shorter than the characteristic distance for power coupling from fundamental to higher-order modes. Such spatial scaling can increase effective mode area by ten to hundred times, correspondingly complementing scalability of chirped pulse amplification or nonlinear amplification approaches.

In principle, a single-mode fiber core can be made arbitrarily large by reducing the refractive index difference between the core and the cladding. The practical limit, however, for such straightforward mode-size scaling is set by the bending loss increase accompanying mode-diameter increases [9]. An alternative path is to increase fiber core size without reducing the refractive index difference between the core and the cladding. This, however, leads to fiber core dimensions supporting more than one transverse mode. Traditionally it was believed that such mode-size scaling was unacceptable, because the maximum usable fiber core size is limited by the requirement of allowing only the fundamental mode to exist.

Conventional wisdom maintained that multimode core fibers were not practical for providing diffraction-limited output beams for two reasons. First, it is difficult to launch only the fundamental mode into a highly multi-mode fiber. Second, if only the fundamental mode were initially launched into such a fiber, it would be difficult to prevent strong higher-mode excitation as the beam travels along. However, in reality it is the quantitative measure of these factors that is important in determining whether multimode core fibers can be used in practice.

Coupling the fundamental mode into a multimode core fiber does not constitute a principal problem. For example, by fabricating an adiabatic taper at the injection end, fundamental mode excitation can be robustly achieved. More generally, experimental practice has proven that the single fundamental mode can be excited in a multimode core by accurate mode matching between the fiber mode and the focused aberration-free input beam. It is the mode scattering during propagation in the fiber that poses a principal problem.

In an ideal, perfectly flawless multimode core fiber the fundamental mode could propagate without being scattered into higher order modes. In a real fiber, imperfections in the core provide the mechanism of coupling between the modes [10] and consequently can cause unacceptable degradation of beam quality and ultrashort pulse shape. The ultrashort pulse shape degrades when propagating in a multimode beam, because every mode travels at a different group velocity, producing multiple pulses at the exit of a fiber.

Fiber core imperfections can be caused by a number of factors such as glass density fluctuations, microscopic random bends caused by stress, diameter variations, and defects such as cracks, microvoids, or dust particles trapped during the manufacturing process. The effect of the these imperfections can be described as a random perturbation δn in the refractive index profile of the fiber core $\delta n^2(r, \theta, z) = n^2(r, \theta, z) - n^2(r)$, where $n(r)$ is the ideal index profile and r, θ, and z are cylindrical coordinates representing the radius, azimuthal angle, and distance along fiber axis. The random perturbation is characterized by a power spectrum in the spatial frequency domain $2\pi/\Lambda_i$, where Λ_i corresponds to a particular spatial period component of this perturbation. Coupling between any two modes, for example between the fundamental LP01 and the next higher order mode LP11, can occur only if this random perturbation spectrum contains a spatial period component Λ_i such that the phase-matching condition for the propagation constants of these modes is met: $2\pi/\Lambda_i = |\beta_{LP01} - \beta_{LP11}|$.

This coupling between two modes i and j can be described by a coupling coefficient d_{ij}, which by definition represents a fraction of the ith-mode power scattered into the jth-mode per unit length (or, equivalently, scattering probability) and can be measured in units of inverse meters (m^{-1}) [10]. In the linear case $d_{ij} = d_{ji}$. It has been found that this coupling occurs predominantly between adjacent modes because, for majority of random perturbations, the power spectrum is a strongly decreasing function of the spatial frequency [11,12]. Therefore, the coupling coefficient d_{ij} is typically a strongly decreasing function of the difference between mode propagation constants $\Delta\beta$: $d(\Delta\beta)$.

Consequently, the value of $d(\Delta\beta)$ is determined by the separation between the propagation constants of the modes and by the flawlessness of the fiber core. Generally, $\Delta\beta$ decreases with increasing fiber core diameter. Flawlessness of the fiber should be strongly dependent on the fabrication conditions, geometry, and material of the fiber. Indeed, it was shown as early as 1975 by Gambling et al. [13] that, for example, by resorting to liquid-core fibers this scattering coefficient could be reduced by at least two orders of magnitude compared to that of all-glass multimode core fibers. It is straightforward to estimate the magnitude of $d(\Delta\beta)$ required to achieve diffraction-limited operation of a fiber amplifier. Because the typical length of a highly doped ultrashort-pulse fiber amplifier is approximately 1–2 m, obtaining less than $\sim 1\%$ of total power in the higher order modes at the amplifier output requires $d(\Delta\beta) < 0.01\,m^{-1}$.

For the analysis of beam quality evolution along the multimode core fiber amplifier it is sufficient to consider interaction between only two modes: the fundamental LP_{01} and the closest higher order mode LP_{11}. Indeed, we are interested only in determining the conditions at which unacceptable beam degradation occurs. Beam quality could be considered to be acceptable as long as the power in the higher order mode LP_{11} is not more than approximately 1% of the total beam power. The power flow between the two modes when traveling along the fiber can be described with this set of equations [11]:

$$\frac{dP_{LP01}}{dz} = d(\Delta\beta)(P_{LP11} - P_{LP01}) - \alpha_{LP01}P_{LP01} \tag{1}$$

$$\frac{dP_{LP11}}{dz} = d(\Delta\beta)(P_{LP01} - P_{LP11}) - \alpha_{LP11}P_{LP11} \tag{2}$$

Here the terms $d(\Delta\beta)P_{LP01}$ and $d(\Delta\beta)P_{LP11}$ give the rate of power flow from the corresponding mode to the neighboring mode, and $\alpha_{LP01}P_{LP01}$ and $\alpha_{LP11}P_{LP11}$ describe the power loss ($\alpha > 0$) or gain ($\alpha < 0$) for each mode due to absorption or gain in the fiber. Note that in general modal loss/gain coefficients α_{LPij} are different for each transverse mode. Because we are interested in the case when power in the LP_{11} mode is negligible compared to the power in the LP_{01} mode, power flow from the LP_{11} mode due to scattering off the core imperfections, $d(\Delta\beta)P_{LP11}$, is also negligible and can be omitted from the right-hand side of each equation. The resulting system of two differential equations can be solved analytically, and the power evolution in the LP_{11} mode along the fiber z can be expressed as

$$P_{LP11}(z) = P_{LP11}(0)\,\exp(-\alpha_{LP11}z) + d\,P_{LP01}(0)\,\exp(-\alpha_{LP11}z)$$

$$\times \int_0^z \exp[(\alpha_{LP11} - \alpha_{LP01} - d)x]\,dx \tag{3}$$

This solution can be significantly simplified for the case of $P_{LP11}(0) = 0$, which corresponds to the excitation of only the fundamental LP_{01} mode at the fiber input:

$$P_{LP11}(z) = dP_{LP01}(0)\left(\frac{1}{\alpha_{LP11} - \alpha_{LP01} - d}\right)$$

$$\times \{\exp[(\alpha_{LP11} - \alpha_{LP01} - d)z] - 1\}\,\exp(-\alpha_{LP11}z) \tag{4}$$

After some simple algebraic transformations, evolution of the power ratio between the LP_{11} and LP_{01} modes can be expressed as

$$\frac{P_{LP11}(z)}{P_{LP01}(Z)} = \frac{d(\Delta\beta)}{\Delta\alpha_{differential} - d(\Delta\beta)}\{1 - \exp[-(\Delta\alpha_{differential} - d(\Delta\beta)]z)\} \tag{5}$$

where $\Delta\alpha_{differential} - \alpha_{LP11} - \alpha_{LP01}$ is the differential modal gain/loss coefficient. Plots of this solution when $d(\Delta\beta) \ll 1$ and for different values of differential modal gain/loss coefficient are shown in Figure 1.

As we can see from this plot, when no differential gain or loss is present, power is transferred from LP_{01} to LP_{11} linearly with the propagation distance at these low scattering values and at the low level of power in LP_{11} mode. It is interesting to note that when differential gain or loss between the two modes is present, the situation is qualitatively different.

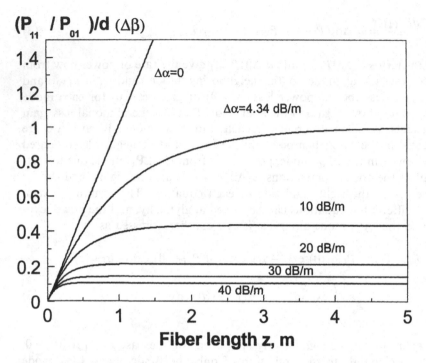

Figure 1 Evolution of power transfer between the modes LP_{01} and LP_{11} for different values of the differential gain/loss coefficient $\Delta\alpha$. Ratio between the mode powers in normalized to the value of the modal-coupling coefficient $d(\Delta\beta)$.

From the solution of Eq. (5) it is clear that after a sufficiently long propagation distance (limit $z \to \infty$) the power ratio between the LP_{11} and LP_{01} modes stops increasing approaches some steady-state value:

$$\frac{P_{LP11}}{P_{LP01}} \Longrightarrow \frac{d(\Delta\beta)}{\Delta\alpha_{\text{differential}} - d(\Delta\beta)} \tag{6}$$

This performance can be understood by considering that when the fundamental LP_{01} mode experiences less loss (or equivalently, more gain) than the LP_{11} mode ($\Delta\alpha_{\text{differential}} > 0$) the situation is equivalent to introducing relative power flow from LP_{11} to LP_{01}. Under the conditions assumed in this analysis this effective power flow acts in the direction opposite to the actual power flow caused by core imperfections, and its rate is increasing with increasing power in the LP_{11} mode as the beam travels along the fiber. After sufficiently long propagation it balances out the power flow from LP_{01} to LP_{11}, because the rate of the latter flow is approximately constant along the fiber.

Two main conclusions follow from this analysis. First, the presence of differential gain/loss stabilizes the ratio between the higher order and fundamental modes, allowing longer propagation lengths in a multimode core fiber. Second, the particular ratio that can be achieved is essentially determined by the mode-scattering coefficient d. Indeed, inspection of Figure 1 reveals that a large increase in $\Delta\alpha_{\text{differential}}$ causes only a relatively small reduction of the ratio between LP_{11} to LP_{01}. For example, increasing the differential gain/loss value by approximately four orders of magnitude (to $\sim 40\,dB/m$) brings only a factor of ~ 10 reduction in the mode ratio.

It is therefore necessary to estimate the actual values of the scattering coefficient d that can be achieved in a multimode core fiber. Apart from the earlier work on liquid core fibers [13], such analysis has been reported recently [14] for solid-core silica-glass fibers. The findings are summarized in Table 2

It has been found [14] that depending on the fiber fabrication procedure (rod-in-tube vs MCVD) and fiber size, the scattering coefficient d can vary over orders of magnitude. The manufacturing procedure affects defect concentration and residual stress in the core interface. Large-diameter fibers are more rigid and are less susceptible to micro-bending deformations and consequently have significantly lower mode-scattering coefficients. It is essential that scattering coefficients below $\sim 0.01\,m^{-1}$ be achievable, thus providing nearly diffraction-limited operation of multimode core fiber amplifiers over lengths of several meters.

A number of techniques for providing differential gain between LP_{01} and higher order modes have been investigated. Some techniques rely on confinement of the active ion doping profile in the fiber core for achieving the highest gain for the fundamental mode [15]. Other approaches exploit higher loss for the higher order mode induced by a tight bending radius of the multimode core fiber [16]. However, as it follows from the above analysis, such techniques alone could not compensate for the prohibitively large levels of mode scattering and are therefore useful only as additional means to optimize multimode core fiber performance. The proper choice of optical

Table 2 Estimated Scattering Coefficient d Achievable in Multimode Core Fiber

Fabrication method	Rod-in-tube	MCVD	MCVD
Core diameter, μm	50	50	50
Fiber diameter, μm	125	125	250
Scattering coefficient d, m^{-1}	~ 10	~ 1	~ 0.01

fiber geometry and fabrication conditions is critical for sustaining diffraction-limited operation in multimode core fiber amplifiers.

Nonlinear Pulse Amplification

Effects of nonlinear interactions in a fiber are not necessarily detrimental under all conditions for ultrashort-pulse amplification in rare-earth doped fiber amplifiers. In fact, when the pulse energy is relatively low, for certain conditions nonlinear pulse amplification can lead to relatively simple and compact fiber systems with little or no pulse stretching and requiring no compression or only a small amount of pulse compression after amplification. Such nonlinear amplification is particularly useful when the goal is to achieve high average powers at high pulse repetition rates rather than the highest pulse energies at the saturation fluence level. Additionally, such an approach can provide much shorter pulses than are supported by the amplification band of a fiber amplifier and even allows wavelength-tunable output to be obtained beyond this band.

Nonlinear pulse amplification is achieved by injecting a bandwidth-limited or nearly bandwidth-limited pulse into a fiber amplifier, where the interplay between the gain, self-phase modulation (SPM), intrapulse Raman scattering and group-velocity dispersion (GVD) spectrally and temporally reshapes the seed pulse, producing either a compressed or a chirped pulse at the output. Essentially, nonlinear pulse amplification is always associated with nonlinear pulse compression. Similarly to nonlinear pulse compression in non-amplifying fibers, two general types of such compression could be distinguished with respect to the sign of the fiber dispersion: soliton effect compression in negative-GVD fibers [17] and SPM-induced spectral broadening in positive-GVD fibers [18]. Negative-GVD nonlinear amplification generally produces compressed optical pulses, whereas positive-GVD nonlinear amplification results in chirped pulses, which have to be compressed in additional compressor after the amplifier stage.

For characterizing ultrashort-pulse nonlinear compression in an optical fiber it is useful to use the parameter N^2 given by

$$N^2 = \frac{\gamma P_0 T_0^2}{|\beta_2|} \tag{7}$$

Here P_0 is the pulse peak power, T_0 is its width, $\beta_2 = d^2\beta/d\omega^2$ is the fiber group-velocity dispersion coefficient, $\gamma = n_2\omega_0/cA_{\mathrm{eff}}$ is the fiber nonlinearity, n_2 is the nonlinear index coefficient, and $A_{\mathrm{eff}} = \pi r^2$ is the effective area of the mode with radius r [19]. In a negative dispersion fiber N represents a soliton order.

Theoretical analysis of soliton effect compression in fiber amplifiers is presented in Ref. 20. When the peak power of an amplified pulse becomes large enough, a fundamental soliton is formed with $N = 1$. Further increase in power leads to pulse duration decrease. If N exceeds 1 this compression is associated with higher order soliton formation. However, the soliton order can be maintained with the increase in peak power if the corresponding pulse-duration decrease occurs according to Eq. (7), which constitutes the condition for adiabatic amplification. The compression process eventually should stop owing to the finite gain bandwidth of a fiber amplifier [20]. Experimentally such compression was first demonstrated in Er-doped fibers [21]. The practical limitation of this method, however, is that because of the formation of higher order solitons, it typically produces short pulses riding on a broad pedestal. For properly chosen conditions, Raman soliton formation can be achieved, which can stabilize and "clean" the generated compressed pulses [22]. Due to the associated Raman soliton self-frequency shift [23], the compressed pulse and its pedestal become spectrally separated and the pedestal can be further removed through spectral filtering.

Positive-GVD nonlinear amplification is particularly important for high-power Yb-doped fiber amplifiers, which operate below the zero dispersion wavelength of $\sim 1.3\,\mu m$ for fused silica glass fibers. The principal distinction between amplifying and nonamplifying positive-GVD compression schemes is that the latter typically (for strong GVD) produces square pulses, whereas the former produces parabolic pulses with significantly better chirp linearity and consequently better compressed pulse quality [24]. Furthermore, positive-GVD nonlinear amplification should generally lead to significantly higher pulse energies compared to negative-GVD amplifiers. Positive dispersion tends to stretch the pulses, which are spectrally broadened through the action of SPM, thus reducing the peak power and consequently increasing the maximum achievable amplified-pulse energies. Also, the achieved quality of compressed pulses is, in the majority of cases, better for positive-GVD amplifiers than for negative-GVD amplifiers.

Fiber Mode Cleaning Using Parametric Chirped Pulse Amplification

Traditionally, nonlinear interactions have been exploited exclusively for wavelength conversion rather than for optical power and energy amplification. Currently, however, a new type of femtosecond energy amplifier is emerging that relies on parametric amplification of chirped (stretched) pulses in a nanosecond-pulse pumped nonlinear medium. The concept of parametric chirped pulse amplification (PCPA) is based on stretching femtosecond pulses to match the duration of the nanosecond pump pulse in an optical parametric amplifier (OPA) crystal in order to achieve efficient energy extraction from the pump [25,26]. In this, PCPA is distinguished from conventional chirped pulse amplification (CPA) [4], where pulse stretching is required solely for eliminating pulse distortions at high peak intensities. The principal advantage of the PCPA approach is the fact that parametric interaction can provide tens of decibels of single-pass gain, compared to < 10 dB in bulk solid-state media, thus offering a simple alternative to conventional regenerative and multiple-pass amplifiers.

The practical importance of using PCPA in conjunction with fiber technology is that the formation of spatial solitary waves during optical parametric amplification can efficiently convert a highly multimodal pump beam into a diffraction-limited amplified signal beam [27]. This enables to overcome core-size limitations in fiber-based systems and opens a new avenue of using multimode core fiber amplifiers for high-energy femtosecond pulse generation.

It is well established that stable two-transverse-dimensional spatial solitons can exist in frequency-doubling crystals. Because such solitons emerge due to $\chi^{(2)}$ nonlinearity, they are known as quadratic solitons and are stable cylindrically symmetric solutions of the coupled-mode equations describing the SHG process at some phase-mismatch condition [28]. Furthermore, at phase-matching conditions, stable spatial solitary waves can exist due to $\chi^{(2)}$ nonlinearity [29]. Such solitary waves also develop in the phase-matched three-wave optical parametric interaction, which can be described by a standard set of three coupled-wave equations:

$$\frac{\partial E_1}{\partial z} + \frac{i}{2k_1} \nabla_t^2 E_1 = i\kappa_1 E_3 E_2^* \exp(i\Delta kz)$$

$$\frac{\partial E_2}{\partial z} + \frac{i}{2k_2} \nabla_t^2 E_2 = i\kappa_2 E_3 E_1^* \exp(i\Delta kz) \qquad (8)$$

$$\frac{\partial E_3}{\partial z} + \frac{i}{2k_3} \nabla_t^2 E_3 = i\kappa_3 E_3 E_2 \exp(-i\Delta kz)$$

Here $i = 1, 2, 3$ refer to the signal, idler and pump beams, respectively, and Δk is the wave-vector mismatch. These equations fully describe spatial mode evolution in a parametric amplifier by including the effects of pump depletion and beam diffraction in the paraxial approximation.

In general, analysis of the solutions obtained shows that after initial beam reshaping, the three optical waves at pump, signal, and idler wavelengths can propagate locked together in space without any significant spatial diffraction at intensities for which the characteristic parametric gain length is shorter than the diffraction length. Furthermore, the obtained solutions are robust even for a wide variety of non-diffraction-limited pump beam inputs, resulting in stable Gaussian-shaped and diffraction-limited output signal beams.

A specific example of implementing such a fiber-based system can be found in Section 4.3.2. It is anticipated that ultimately such an approach can provide the highest energies achievable with a fiber-based system.

4.2.2 Limitations Caused by Nonlinear Effects in Pulsed Fiber Amplifiers

In a solid-state amplifier CPA is typically employed for avoiding optical damage and for eliminating the occurence of nonlinear effects. The optical path in a material of a typical solid-state amplifier is relatively short and the beam size is relatively large. Consequently, stretching pulses to $\sim$1 ns duration in majority of cases is sufficient in order to extract maximum energies that are limited not by damage or by nonlinear pulse distortions but by the saturation fluence value of the amplifier material. The situation is quite different for fiber amplifiers,where the fiber length is at least 100 times longer and the mode area is at least 100–1000 times smaller than for a solid-state amplifier. As a result, even for 1 ns pulses it is nonlinear optical effects that typically limit the maximum achievable energy from a fiber amplifier.

The two nonlinear effects that are the most important in this respect are self-phase modulation and Raman gain. Self-phase modulation can be detrimental in two ways. First, at high peak pulse intensities SPM affects the phase of the amplified stretched pulses, which after recompression manifest as pulse shape distortions affecting both pulse duration and quality [30,31,32]. Second, for multimode fibers high peak intensities can cause self-focusing of propagating beams [33,34] leading to mode-quality degradation and, ultimately, to optical damage in an optical fiber. At high peak intensities Raman gain leads to the "shedding" of optical power from the amplified pulse to long-wavelength Raman spectral components, thus manifesting itself as a significant power loss at the fiber-amplified pulse wavelength.

At very high pulse energies and high average powers it is also necessary to consider optical damage and thermal effects in an optical fiber.

Note that the strength of self-phase modulation, Raman gain and interactions leading to optical damage are inversely proportional to the optical wavelength [19]. This indicates that, generally, it is easier to achieve higher energies for longer wavelength pulses.

Self-phase Modulation in Chirped Pulse Fiber Amplifiers

The effect of self-phase modulation on the phase of the amplified chirped pulse is essentially determined by its spectral shape. The basic equation that describes a propagating optical pulse in a fundamental mode of a fiber amplifier core is given by [19]

$$i\frac{\partial A(z,T)}{\partial z} = \frac{i}{2}g(z)A(z,T) + \frac{1}{2}\beta_2\frac{\partial^2 A(z,T)}{\partial T^2} - \gamma |A(z,T)|^2 A(z,T) \qquad (9)$$

where $A(z,T)$ is the slowly varying amplitude of the pulse envelope and T is measured in a frame of reference moving with the pulse at the group velocity $v_g(T = t - z/v_g)$ in the fiber core. The first term on the right-hand side accounts for optical gain in a fiber, $G = \exp[g(z)z]$. Longitudinal variation of the gain coefficient $g(z)$ takes into account the fact that in a fiber amplifier inversion is not homogeneous along the fiber and is determined by the longitudinal distribution of pump absorption and saturation of the amplifier due to amplified signal and ASE. The second term on the right-hand side accounts for the linear group-velocity dispersion in a fiber. The last term accounts for the action of self-phase modulation. Fiber nonlinearity is characterized by $\gamma = n_2\omega_0/cA_{\text{eff}}$. Here n_2 is the nonlinear refractive index and $A_{\text{eff}} = \pi r^2$ is the fundamental mode area determined by the mode radius r.

Because in the case of chirped pulse amplification typically $\Delta T_{\text{stretched}} \gg \Delta t_{\text{bandwidth-limited}}$, the dispersion term can be ignored in the majority of case in analyzing the propagation of a stretched pulse in a chirped pulse fiber amplifier. This significantly simplifies the analysis allowing an analytical description of the SPM effects on the recompressed pulses.

Without the GVD term, Eq. (9) can be easily solved [19], and the SPM effect on the phase ϕ_{SPM} of a pulse $P(t)$ is described by

$$\phi_{\text{SPM}}(z,t) = P(t)\gamma z_{\text{eff}} = P_0\gamma z_{\text{eff}}|U(t)|^2 \qquad (10)$$

Here $\phi_{\text{SPM}}(z,t)$ is an instantaneous phase shift due to SPM after the pulse has propagated a distance z in a fiber, and γ is the fiber non linearity coefficient. $U(t)$ is the normalized pulse amplitude such that $U(0) = 1$ and P_0 is pulse peak power at $t = 0$. The effective propagation distance z_{eff} accounts for the gain g in a fiber amplifier:

$$z_{\text{eff}} = \frac{1}{g}[1 - \exp(-gz)] \qquad (11)$$

For simplicity, gain is here assumed to be constant along the fiber.

Further simplification can be obtained by noting that since in the CPA system typically $\Delta T_{\text{stretched}} \gg \Delta t_{\text{bandwidth-limited}}$, from the properties of Fourier transforms it follows that the temporal shape $U(t)$ of such a highly chirped pulse and its spectral shape [the Fourier transform $\hat{U}(\omega)$ of $U(t)$]

are similar to each other:

$$U(t) = \tilde{U}_0 \tilde{U}(\omega) = \tilde{U}_0 \tilde{U}(t/\beta_2^{\text{stretcher}}) \tag{12}$$

The main consequence of this condition is that it can be shown (as presented in the Appendix) that

$$\beta_n^{\text{SPM}} = P_0 \gamma z_{\text{eff}} \left. \frac{d^n |\tilde{U}_0 \tilde{U}(\omega)|^2}{d\omega^n} \right|_{\omega=0} \tag{13}$$

This shows that SPM induces a phase shift, which is equivalent to the action of a dispersive device with corresponding dispersion orders β_n^{SPM}. The magnitude of each dispersion order is completely determined by the maximum SPM phase shift at the peak of the pulse $\phi_{\text{peak}} = P_0 \gamma z_{\text{eff}}$ and nth order derivative of the pulse spectral shape.

The conclusion here is that the action of SPM is equivalent to the action of a dispersive device and its detrimental effect on the recompressed pulse can be described in the same way as in the case of dispersion mismatch between a stretcher and a compressor [35]. For example, the effect of SPM could be compensated by choosing $\beta_n^{\text{SPM}} = -\beta_n^{\text{Compressor}}$. Alternatively, SPM effects can be minimized or eliminated through the appropriate choice (or control) of the amplified pulse spectral shape.

In practice, however, it is very important to maintain smooth temporal and spectral profile of the amplified stretched pulses, since, according to the analysis above, these shape irregularities would transform into irregular phase modulation for the amplified pulse. After pulse recompression such phase modulation transforms into low intensity pedestal and satellite pulses, thus degrading quality of the recompressed amplified pulses significantly.

Raman Gain in Pulsed Fiber Amplifiers

In practice, Raman gain constitutes the main limitation for the achievable pulse peak power and energy in a fiber CPA system. Once the peak power in the amplified pulse reaches a certain critical level corresponding to the Raman threshold, the generated Stokes wave starts growing rapidly inside the fiber such that most of the pulse energy transfers into the Stokes component after a short propagation distance. Thus the amplified power in the pulse is lost in that it cannot be recompressed to the initial femtosecond duration. The threshold for the onset of such energy "shedding," to a good approximation, is given in a fiber amplifier by [36].

$$P_{\text{critical}} \approx \frac{16 A_{\text{mode}}}{g_{\text{Raman}} z_{\text{eff}}} \tag{14}$$

Here g_{Raman} is the Raman gain coefficient for the fiber glass and z_{eff} is the effective fiber length given by Eq. (11).

Clearly, increasing core sizes and using fiber amplifiers with high gain coefficients and relatively short lengths should maximize the Raman-limited pulse energies.

Self-Focusing and Optical Damage in Multimode Core Fibers

The ultimate limit for the peak power and hence for the maximum nanosecond stretched pulse energy achievable in a multimode core fiber is set by the threshold for self-focusing, the process occurring due to intensity dependence of the refractive index. Self-focusing is characterized by a critical power for the propagating beam, upon reaching which the beam can collapse, creating very high peak intensities and causing damage to the fiber. It is interesting to note that, as has been shown recently [34], this critical power for self-focusing in a multimode fiber core is exactly the same as in a bulk medium. The expression for the critical power P_{cr} has been shown to be of the form [37]

$$P_{cr} = \alpha \frac{\lambda^2}{4\pi n n_2} \qquad (15)$$

where α is a constant and $\alpha \approx 1.83$ according to Ref. [37]. For fused silica, $n_2 = 3.2 \times 10^{-20}$ m/W, and self-focusing is expected to occur in a multimode core optical fiber at the critical peak power $P_{cr} \sim 4$ MW corresponding to a nanosecond stretched pulse energy of ~ 4 mJ.

More detailed theoretical analyses of the self-focusing effects in multimode core fibers [33,34] indicate that even before a catastrophic collapse would occur, significant beam distortions could be induced at peak powers approaching the critical power level. However, no reliable experimental studies of these self-focusing phenomena in multimode fibers have been reported so far.

From practical perspective, it is also necessary to consider issues associated with the surface damage at the fiber output end of a high energy fiber amplifier. For nanosecond stretched pulses surface damage in fused silica occurs at fluencies of 30–40 J/cm^2 [38]. For a 15–30 μm diameter fundamental mode in a large-core fiber this would result in surface damage occurring at the unprotected fiber output end for stretched-pulse energies in the range of ~ 50–300 μJ. This is more than an order of magnitude lower than the bulk-damage threshold due to self-focusing. Fortunately, surface damage can be avoided by the proper treatment of the fiber ends. One solution is to expand the exciting beam before it reaches the output surface of a fiber, thus effectively reducing the fluence level at the surface.

In practice this technique has allowed to completely eliminate the surface damage problem.

Thermal Effects in High-Power Multimode Core Fibers

One of the principal technological advantages of fiber lasers and amplifiers is their relative immunity from detrimental thermal effects compared to other solid-state lasers. The main reason for this is their relatively large surface area corresponding to a pumped fiber-laser volume. If we approximate the pumped volume of any laser as a cylinder of radius R and length L, then the volume-to-surface ratio is $V/S = \pi R^2 L / 2\pi R L = R/2$. Typically, a double-clad fiber radius is at least an order of magnitude smaller than the pumped-volume radius of a solid-state laser, thus indicating that the volume-to-surface ratio and consequently the immunity to thermal effects in a fiber should be improved by at least a comparable factor. Nevertheless, thermal effects are not completely absent from fibers, and it is necessary to consider their effect for high-power multimode core fiber lasers and amplifiers.

Thermal lensing is the most common detrimental thermal effect in high-power bulk solid-state lasers and amplifiers. This effect occurs when a radial dependence on the index of refraction of the medium under consideration is induced by the radial temperature distribution in the pumped volume. This dependence can be expressed as [39]

$$\Delta n_T(r) = -\frac{P}{4\pi\kappa} n'(T) \left(\frac{r}{R_{\text{core}}}\right)^2 \qquad (16)$$

Here r is the radial coordinate, P is the dissipated pump power per unit length, R_{core} is the radius of the rare-earth-doped volume in the core, $n'(T)$ is the temperature coefficient of the index of refraction, and κ is the thermal conductivity of the fiberglass. Thermal self-focusing can occur in a multimode core fiber when the thermal lens numerical aperture $\sqrt{2n\,\Delta n_T}$ becomes equal to the diffraction angle $\theta_{\text{diff.}} = \lambda/\pi\omega_0$ for the fundamental mode beam of radius ω_0 [40].

A numerical example can be given using typical parameters for high-power large-core double-clad Yb-doped fibers used in most of the work described in the following sections of this chapter. For fused silica $n(T) = 2.5 \times 10^{-6}\,\text{K}^{-1}$ and $\kappa = 1.38\,\text{V m}^{-1}\text{K}^{-1}$ [39]. Typically $R_{\text{fiber}}/R_{\text{core}} \sim 10$ in multimode core double-clad Yb-doped fibers used in high energy and high-power fiber CPA systems, which usually have an $\sim 30\,\mu\text{m}$ diameter core and $\sim 300\,\mu\text{m}$ diameter pump cladding. Such double-clad fibers can be highly doped to $\sim 2.5\,\text{wt}\%$ of Yb and, therefore, can absorb estimated up to $10\,\text{W/m}$ of pump at 976 nm. By assuming that less than

10% of absorbed 976 nm pump power is dissipated in the form of heat, one can estimate that $P < 1\text{W/m}$. Equation (16) then yields $\Delta n_T < 1.44 \times 10^{-5}$, and the corresponding thermal lens numerical aperture $\sqrt{2n\,\Delta n_T} < 6.5 \times 10^{-3}$. The largest mode radius so far achieved in robust single-mode operation of multimode core fiber amplifier is approximately 15 μm corresponding to the numerical aperture of 22×10^{-3}.

This specific example indicates that although for 10–20 W output from a large-core fiber the dissipated pump power is still below the threshold for thermal self-focusing, further increases in fiber size, doping level, and pumping power can lead to significant thermal beam-distortions in high-power multimode core fiber lasers and amplifiers. Conversely, fiber immunity to thermal effects increases with the decrease in fiber structure transverse dimensions. More detailed analyses of thermal effects in high-power pumped fibers can be found in Refs. [39,41,42].

4.3 FIBER SYSTEMS FOR AMPLIFYING ULTRASHORT PULSES

4.3.1 High Average Power Systems

The distinction between *high average power* and *high pulse energy* fiber systems is much more significant than is merely implied by their respective names. Indeed, average power from a high-energy system can actually be higher than that from a high-power system. It is the compactness and robustness of the design achieved by the use of novel pulse-stretching and compressing devices in high-power systems that essentially distinguishes them from high-energy systems. The use of these novel stretching/compressing devices, however, currently comes at the cost of limiting the achievable pulse energy to significantly below the saturation fluence level of an optical amplifier (hence, high average power is the most appropriate attribute). In contrast, high-energy fiber systems rely on conventional diffraction-grating based pulse compressors, which allow achieving the maximum pulse energies at the saturation-fluence and stored-energy limits of a particular fiber amplifier, but at the cost of the large size and limited robustness of the overall system. Such pulse compressors (and stretchers) are essentially bulk devices and have a number of other attributes that place them technologically apart from the compact stretchers and compressors. This justifies the importance of the distinction between these two classes of ultrashort-pulse fiber amplifiers.

It is probable, however, that the current situation is only temporary. Further development might lead to new types of compact stretchers and compressors capable of supporting the maximum pulse energies in and from a fiber amplifier, thus diminishing the technological difference between the two types of systems in the future.

Compact Pulse Stretchers and Compressors

To fully exploit the technological advantages offered by optical fibers for ultrashort-pulse amplification, stretching and recompression of pulses in a fiber CPA should be accomplished by using devices whose compactness and robustness match that of the fiber-optic components of the rest of the system. In contrast, the conventional diffraction-grating-based stretchers and compressors used with solid-state amplifiers since the inception of the CPA concept [4] rely on angular dispersion [1] and are therefore typically associated with large separation between components, large gratings (e.g., ~ 1 ns stretching requires ~ 10 cm width), and critical spatial alignment. The search for and development of new types of stretchers and compressors has been one of the main areas in fiber CPA development.

Over the last several years the quest for compact pulse stretching/compressing devices for fiber CPA has led to the development of a number of novel technologies. It is interesting to note, however, that although various such devices involve completely different physical processes, they can all be attributed to the same broad class of longitudinally chirped grating stretchers and compressors.

The operation principle of such gratings is illustrated in Figure 2. Two different types can be distinguished: reflection and transmission gratings. For sufficiently large longitudinal chirp, coupling between the incident and output (either reflected or transmitted) beams for each particular wavelength is strongly localized in a relatively short fraction of the total grating length. Consequently, components of different wavelengths λ_1 and λ_2 of an optical wave incident into such a chirped grating will couple into an output wave at different longitudinal positions z_1 and z_2 respectively and, therefore, will experience different time delays. For a reflection grating this time-delay difference $\Delta\tau$ is the sum of the time delays for the incident and reflected waves propagating between z_1 and z_2:

$$\Delta\tau_{\text{reflection}} = \frac{\Delta L}{v_{g_{\text{incident}}}} + \frac{\Delta L}{v_{g_{\text{reflected}}}} \tag{17}$$

and for a transmission grating it is the difference between the delays for the incident and transmitted waves propagating between z_1 and z_2:

$$\Delta\tau_{\text{transmission}} = \frac{\Delta L}{v_{g_{\text{incident}}}} - \frac{\Delta L}{v_{g_{\text{transmitted}}}} \tag{18}$$

a)

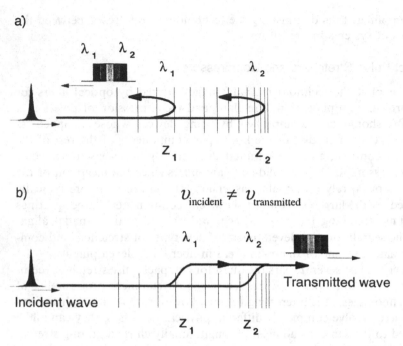

b)

Figure 2 Operation principle of (a) chirped-period reflection and (b) transmission grating based pulse stretchers and compressors. In transmission gratings group velocities of incident and transmitted waves are different. Incident and transmitted waves, for example, can be core and cladding modes in a fiber or different-wavelength modes in a QPM-grating SHG compressor.

Here v_g is the group velocity of light in the structure for a particular propagating wave and $\Delta L = z_2 - z_1$. In general, these group velocities are not necessarily equal, because the corresponding waves might be, for example, different waveguide modes or different polarization modes in a long-period fiber grating, or fundamental and second harmonic waves in a QPM SHG compressor.

An important example of a reflection grating for a compact CPA is a chirped fiber Bragg grating [5]. Such a grating can be written into a fiber core, exploiting the photosensitivity of certain optical glasses to UV light illumination [43]. Coupling for a specific wavelength λ between the incident and reflected waves occurs at the longitudinal position z where the local period $\Lambda(z)$ if this index-variation fulfils the Bragg condition $\lambda_{\text{Bragg}} = 2n_{\text{eff}}\Lambda(z)$. Here $\Delta\tau = 2\,\Delta L/v_g$ because both waves propagate in the fundamental mode of a fiber. The exact phase response of a chirped fiber grating can be calculated by using a variety of numerical simulation techniques [43].

An example of a transmission grating used for a fiber CPA is a chirped-period quasi-phase-matched second harmonic generator [44,45]. Such a grating is typically a periodic variation of the sign of the second-order nonlinear susceptibility χ_2 of a material, implemented through the periodic reversal of ferroelectric domains in a crystal [46]. This can be accomplished, for example, through electric field poling of various ferro-electric materials, such as $LiNbO_3$, $LiTaO_3$, KTP, RTA, and CTA. (See, for example, Ref. [46–49].) Coupling for a specific wavelength λ of the incident fundamental wave into the corresponding second-harmonic wavelength $\lambda_{SH} = \lambda/2$ of the transmitted wave occurs at the longitudinal position z where the local period Λ (z) of the poled domains fulfils the phase-matching condition $\Lambda(z) = \{n(\lambda_{SH})/\lambda_{SH} - 2n(\lambda)/\lambda\}^{-1}$ for second harmonic generation (SHG). Due to the different group velocities for the fundamental and second harmonic waves in a crystal $\Delta\tau = \Delta L(1/v_{g\,\text{fundamental}} - 1/v_{g\,SH})$. Methods of calculating the exact phase response of a chirped-period QPM compressor are described in Refs. [50,51]. It is interesting to note that although such QPM compressors are based on nonlinear optical interactions, the resulting action at low SHG conversion efficiencies is equivalent to the linear effect of group-velocity dispersion (GVD). At high conversion efficiency, however, it is necessary to also consider the effects produced by the cascaded nonlinear interactions [52], and the linear GVD-equivalent description becomes inaccurate.

The essential difference between reflection and transmission gratings is obvious from a comparison of Eqs. (17) and (18): Total delay for reflection is proportional to the group-velocity delay, whereas for transmission it is proportional to the difference between group-velocity delays. Consequently, reflection gratings typically provide orders-of-magnitude longer stretched pulse durations than transmission gratings. For example, a chirped 10 cm long fiber grating produces up to ~ 1 ns maximum delay, which is comparable to that of a typical diffraction grating arrangement. In comparison, a 6 cm long QPM grating in an electric field poled lithium niobate crystal (at approximately the technological maximum), for 1550 nm SHG produces only an 18 ps maximum delay. Also, a chirped long-period fiber grating for coupling between two orthogonally polarized modes in a single-mode bi-refringent fiber could produce maximum delays in the order of 1 to 2 ps, due to the negligibly small difference between propagation constants of these modes [53]. Consequently, such long-period fiber gratings are practically unsuitable for use in fiber CPA.

There is a significant technological difference between the fabrication of reflection and transmission gratings. Typically, phase matching in a reflection grating requires periods comparable to the optical wavelengths of the interacting waves (e.g., for a fiber grating Λ_{Bragg} ~ 250–400 nm), whereas in the transmission gratings these periods are significantly longer (e.g., for SHG in PP-LiNbO$_3$ at $\lambda = 1.55\,\mu\mathrm{m}\,\Lambda_{QPM} \sim 20\,\mu\mathrm{m}$. Obviously, such long periods are much easier to fabricate than subwavelength periods. For example, QPM for counterpropagating SHG [54] is essentially a reflection grating and therefore when chirped could provide delays comparable to those of a Bragg grating. But such first-order gratings have not been fabricated yet due to the current technological difficulties of achieving subwavelength period electric field poled domains.

An important practical consideration for using a compact pulse stretcher and compressor in a CPA systems is the transversal size of the acceptable beam: the larger the beam the higher the recompressed pulse energies the compressor can handle. In this respect, fiber gratings are at a clear disadvantage compared to QPM gratings, since they are limited to single-mode fibers. However, as described in Section 4.3.2, initial demonstrations of using a bulk Bragg grating compressor in fiber CPA [7] indicates the technological path for implementing high-energy compact pulse compressors.

Finally, properties of longitudinally chirped gratings can be engineered, thus providing significant technological advantage compared to conventional diffraction grating arrangements. The bandwidth $\Delta\lambda$ of a chirped grating is determined by the grating-period span $\Delta\Lambda$ and is therefore a fully engineerable parameter. Additionally, by tailoring the local period $\Lambda(z)$ one can also engineer the higher order dispersion terms of such a structure [55] (see also Sec. 4.3.2).

Chirped Fiber Grating CPA Circuits

Use of chirped fiber Bragg grating (FBG) pulse stretchers and compressors enable the construction of very compact, robust all-fiber CPA systems, which can essentially be considered as fiber-optic circuits for amplifying femtosecond pulses in a fiber [5]. FBG pulse stretchers/compressors are technologically advantageous owing to their compatibility with other fiber-optic components and their suitability for large-scale manufacturing. Chirped (as well as un-chirped) FBG's can be currently mass produced using various implementations of either interferometric [56,57] or phase-mask techniques [58,59]. Initially, chirped FBGs have been suggested [60] and then demonstrated [61,62,63] as components for dispersion compensation in optical fiber communication systems.

A fiber Bragg grating is inherently a single-mode device, the use of which does not permit complete elimination of detrimental peak-power related effects in all-fiber CPA circuits. Nevertheless, the use of chirped FBGs significantly increases the maximum pulse energies achievable from an ultrashort-pulse fiber amplifier because a typical chirped FBG compressor length of 1–10 cm is approximately 10^2–10^3 times shorter compared to a typical 1–10 m length of an ultrashort-pulse fiber amplifier. This leads to a comparable nonlinear interaction length reduction and consequently to the corresponding increase in obtainable recompressed pulse energy. Additionally, such long chirped FBGs provide long stretched pulses from tens of picoseconds to nanoseconds in length, sufficient to reduce phase distortions in a CPA fiber amplifier down to a negligible level.

Clearly, there is a certain design trade-off for constructing an all-fiber CPA circuit: reduction of the FBG length reduces the detrimental effects in the compressor but increases phase-distortions in a fiber amplifier. Optimization of every particular CPA circuit design requires careful consideration of the effects of nonlinear interactions in both the CPA fiber amplifier (described earlier in Sec. 4.2.2) and in a chirped FBG compressor.

Experimental and theoretical study of the nonlinear effects during ultrashort-pulse compression in chirped fiber Bragg gratings [64] has revealed that similar to the conventional case of nonlinear pulse propagation in a single-mode fiber, self-phase modulation plays the major role in pulse reshaping in a chirped FBG compressor. However, the effect of SPM is significantly modified by the presence of the chirped grating itself. The SPM for a pulse propagating in a fiber produces low frequencies at the leading edge and high frequencies at the trailing edge (positive chirping). In a chirped FBG, the compressed pulse is not just propagating but is being formed by continuously adding a consecutive reflected part of the pulse spectrum. The result is that, depending on the orientation of the grating (positive or negative FBG dispersion), one edge of the pulse becomes steeper while the other one becomes longer. Furthermore, some of the SPM-broadened part of the spectrum gets reflected from the grating, producing further steepening of the corresponding pulse edge and a low-intensity pedestal with some satellites on the lengthened edge of the pulse. Finally, the high intensities of the recompressed pulse instantaneously change the photonic band structure at the pulse location in the fiber grating through the nonlinear component of the refractive index, thus affecting pulse synthesis from the reflected pulse spectrum as well as the reflection of the SPM broadened spectral part. Such reshaping effects are illustrated in Figure 3, where both experimentally measured and numerically calculated autocorrelation traces are shown for various recompressed pulse energies.

The most important practical limitation resulting from nonlinear interactions is the energy-dependent loss occurring at high energies in a chirped FBG compressor. The experimentally measured (for conditions identical to those for Figure 3) dependence of the recompressed pulse energy reflected from an approximately 10 cm long linearly chirped and 18 nm bandwidth grating is shown in Figure 4. The bandwidth-limited duration

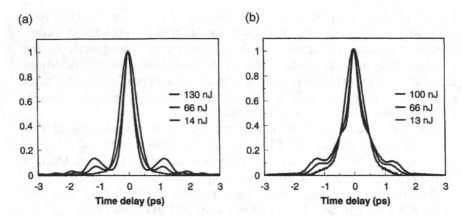

(a)   (b)

Figure 3 Calculated (a) and measured (b) autocorrelation traces of pulses recompressed with a 10 cm long chirped fiber Bragg grating for different recompressed pulse energies (from Ref. 64.). Fiber has 8 μm diameter core. Input stretched pulses correspond to 300 fs bandwidth limited duration at 1550 nm. Pulse reshaping due to nonlinear interactions in a grating compressor becomes evident as the energy of recompressed pulses increases.

of the incident stretched pulse corresponds to 300 fs. Significant pulse energy losses of 20%–30% appear in the energy range of ~ 50 nJ–100 nJ. Together with observed pulse distortions this has been determined to be the practical limit for the maximum recompressed pulse energy for the given parameters of a fiber grating, a fiber core, and a pulse.

From the perspective of constructing an all-fiber CPA circuit it is essential to understand how the recompressible pulse energies scale with the length L_{FBG} of an FBG compressor and with the bandwidth-limited duration ΔT of an incident stretched pulse. Nonlinear interactions in a chirped FBG compressor can be described with a set of two nonlinear Schrödinger equations, each for an optical wave propagating in one of the two directions in the fiber and mutually coupled through the periodic variation of the refractive index profile in the core [19]. Due to the fact that typical fiber grating compressors are relatively short (1–10 cm) and the spectral widths of the pulses in the majority of practical cases are relatively small, the effect of group-velocity dispersion in the fiber itself can be neglected. Consequently, similar to the case of dispersion-free pulse propagation in a fiber which is modeled with a single nonlinear Schrodinger equation [19], the solution of this set of two equations depends on a single dimensionless parameter Γ:

$$\Gamma = \frac{4\pi n_2 L_{FBG}}{\lambda A_{eff}} P_0 \tag{19}$$

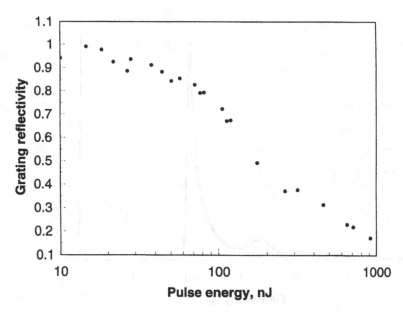

Figure 4 Measured reflectivity of a 10 cm long chirped-FBG pulse compressor as a function of recompressed pulse energy. Conditions of the experiment are identical to those corresponding to Figure 3. Reflectivity of the FBG compressor starts decreasing at pulse energies ∼ 50 nJ due to the change in the fiber core refractive index induced by high peak intensities of recompressed pulses in the grating.

Here A_{eff} is the effective mode cross section in the grating fiber, n_2 is the nonlinear refractive index, and P_0 is the effective peak power obtained by dividing the output pulse energy E_{compr} by the bandwidth-limited duration ΔT of the stretched pulse incident into the grating: $P_0 = E_{compr}/\Delta T$. Γ effectively describes the strength of SPM-induced effects in a fiber grating compressor.

In Figure 5 the calculated intensity profiles ([64]) of the compressed pulses from 1, 2, 5, and 10 cm long chirped fiber gratings are shown. All four traces were calculated for the same value of $\Gamma = 110$, at which there is strong nonlinear pulse compression in fiber gratings. Fixed Γ means that for every grating length L_{FBG} the corresponding effective peak power P_0 is proportionally adjusted according to the Eq. (19). All other parameters, such as the incident pulse shape spectrum and its detuning from the central wavelength of the grating spectrum, grating reflection spectrum, reflectivity and fiber parameters are kept unchanged. Indeed, the recompressed traces for all four grating lengths are identical, thus confirming that the solution of the set of nonlinear Schrodinger equations scales with the value of Γ.

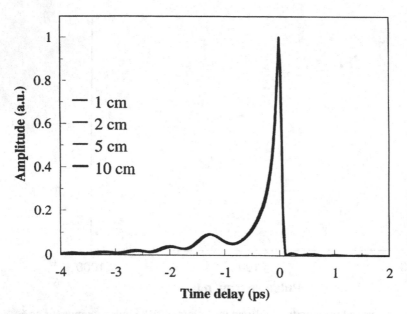

Figure 5 Calculated intensity profiles of the recompressed pulses from 1, 2, 5 and 10 cm long chirped fiber gratings. All four traces were calculated keeping $\Gamma = 110$ by adjusting pulse peak power [according to Eq. (19)] in order to compensate for the change in the grating length in each of the cases. All other parameters, such as incident-pulse shape and spectrum, were kept unchanged. All four calculated pulse profiles are identical, indicating the scalability of the coupled set of nonlinear Schrodinger equations with the parameter Γ.

This result establishes that the maximum obtainable pulse energy from an FBG compressor scales inversely with the FBG compressor length. Also, because $P_0 \sim 1/\Delta T$ one can assume that the maximum obtainable pulse energy should approximately scale proportionately to the bandwidth-limited duration of the stretched incident pulse ΔT.

An important distinction of a chirped FBG from a conventional bulk diffraction grating stretcher/compressor is the reversibility of its dispersion sign, simply achievable by using opposite directions of propagation in a grating. Indeed, as it is discussed in Section 4.3.2, reflection of different wavelength components of an optical wave incident on a chirped Bragg grating are localized at different longitudinal positions, experiencing the time delay described by Eq. (17). If the same optical wave would enter the grating from the opposite end this delay difference would have the same magnitude but reverse sign. These intuitive considerations can be confirmed with a rigorous analysis showing that when the bandwidth of a chirped grating is much larger than the bandwidth of an un-chirped grating of the same length and optical strength, the phase-responses for each propagation direction of such a chirped Bragg grating are exactly reciprocal to each other irrespective of the chirp profile shape. This enables to first stretch and subsequently to accurately recompress a femtosecond pulse using the same or identical grating without any complicated pulse compression alignment procedure. Of course, in order to achieve complete recompression of a femtosecond pulse on has to ensure that the net dispersion of the fibers and all other optical components located in the beam path between the input and output ports of the system is also compensated. This can be achieved by simply inserting an appropriate length of a dispersion-compensating fiber [5].

Two typical examples of implementing fiber grating CPA circuits for high-power fiber amplification are shown in Figure 6a and 6b. In Figure 6a, a completely all-fiber CPA circuits is represented, all the details of which can be found in Ref. 5. The key feature here is that pulses never leave the fiber medium until it reaches the output port, and all the manipulation of the pulse duration and its amplification is accomplished exclusively with standard single mode fiber optic components. The initial 300 fs pulses from a modelocked fiber oscillator were stretched to 30 ps in the first FBG and after amplification recompressed down to 408 fs. It has been verified experimentally that the recompressed pulses were bandwidth-limited and the longer duration of the recompressed pulses was solely due to the some spectral reshaping in the fiber amplifier. Both gratings were identical having 5 mm length and 15 nm reflection bandwidth, centered at 1561 nm to match the central wavelength of the fiber oscillator. The incident and reflected beams for each grating were separated using standard 50:50 fiber couplers. While such an arrangement permits only a limited energy throughput of a maximum of 25%, the use of other types of components such as fiber-pigtailed optical circulators should allow an increase in the transmitted energy to close to 100%, the only limit being the insertion losses of the components and the reflectivity of the grating itself. Note the presence of the dispersion-compensating fiber at the input of the system, which was used to ensure that the net GVD of the system was completely compensated. In this CPA system, the maximum energy of 6 nJ after amplification was limited by the nonlinear distortion threshold for 30 ps stretched pulses in the single-mode fiber amplifier rather than by nonlinear limitations in a fiber grating compressor.

For achieving higher energies larger stretching durations for the amplified pulses are required. In Figure 6b, a fiber CPA is shown producing 55 nJ pulses and 1 W of average power, that uses a single 10 cm long and 18 nm bandwidth FBG stretcher/compressor with up to 1 ns optical stretching to eliminate nonlinear distortions in the amplifier. One important practical

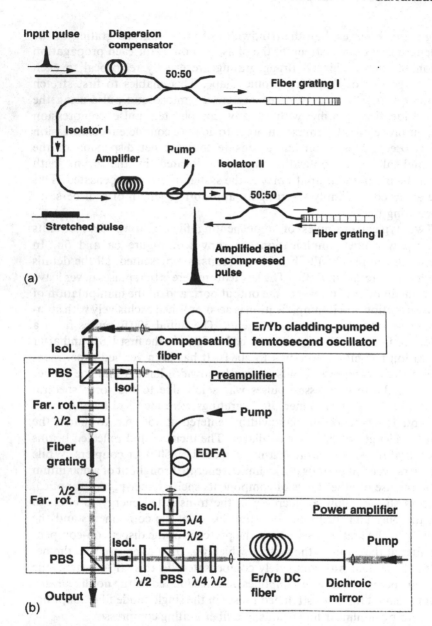

Figure 6 Examples of implementing fiber-grating CPA circuits in high-power fiber amplifiers: (a) an all-fiber CPA circuit using two separate gratings for pulse stretching and recompressing (from Ref. 5); (b) a fiber CPA system using the same grating for both pulse stretching and recompressing (from Ref. 65). Average power of 1 W has been obtained with the system (b) by using a double-clad Er/Yb-doped fiber in the power amplifier stage.

consideration associated with such long FBG stretchers and compressors is that it becomes difficult to ensure phase reciprocity of two separate and long fiber gratings. The reason for that is the accumulation of small errors in the refractive index profile during fabrication of each grating even under identical conditions. Such errors become increasingly more important with the length increase of the grating and the bandwidth increase of the pulses that are being stretched and recompressed. Although this problem can probably be addressed by refining the fabrication process, the configuration using a single grating for both stretching and compression allows to eliminate the problem altogether. A detailed description of the system can be found in Ref. 65. An important result is shown in Figure 7, where the autocorrelation traces of the measured 310 fs recompressed spectrum are compared. Nearly perfect overlap of these traces indicates the remarkable reciprocity of this stretching/compression scheme.

It is interesting to note that the system in Figure 6b represents another limiting case of a fiber CPA circuit design compared to the system shown in Figure 6a in that the recompressed pulse energy was close to the onset of nonlinear distortions for compressed pulses in the fiber grating rather than

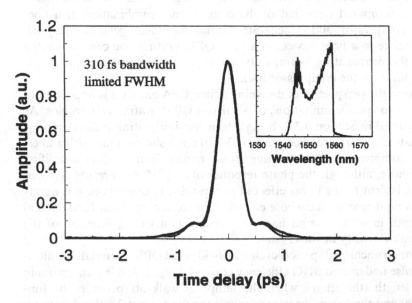

Figure 7 Measured autocorrelation trace (solid dark line) of 310 fs pulse generated by the system shown in Figure 6b. Light grey line is the calculated autocorrelation trace for the bandwidth-limited pulse corresponding to the spectrum at the output of the amplifier (shown in the insert at the top right of the chart).

for stretched pulses in the fiber amplifier. In fact, for ∼ 1 ns stretched pulses approximately ten times higher energies could have been obtained from this particular amplifier fiber. Clearly, by reducing the FBG compressor length, one could in this case increase the overall energy or, equivalently, average power output from such a fiber CPA system.

Chirped Quasi-Phase-Matched Grating CPA Circuits

Another compact CPA circuit for high-power femtosecond pulse amplification in an optical fiber is based on a chirped-period quasi-matching grating compressor [6]. The uniqueness of such a compressor is that it provides with second harmonic generation and frequency-chirp compensation simultaneously is one crystal. The use of a QPM compressor offers several important technological advantages. First, because it is a transmission device the CPA circuits based on it are much simpler with a significantly lower number of components that fiber grating based CPA circuits. Second, such a compressor directly provides second harmonic output, which is particularly useful in conjunction with Er-doped fiber systems for producing femtosecond pulses in the technologically important 780 nm wavelength region. The overall system efficiency for this wavelength region can be substantially increased compared with that of the conventional combination of a fiber grating compressor and a separate second harmonic generator crystal. Third, unlike in a fiber device, in a bulk QPM grating one can choose the size of the propagating beam, thus making it possible to generate high-energy pulses in the compressor itself.

From the perspective of designing a fiber CPA circuit it is important to consider two specific limitations of a chirped QPM grating compressor. As was discussed in Section 4.3.2, being a transmission grating it can compress only relatively short stretched pulses (10–100 times shorter than with a fiber-grating compressor), thus limiting pulse energy from a fiber amplifier. Furthermore, although the phase response of a QPM compressor is essentially equivalent to the linear effect of group-velocity dispersion, it is nevertheless a nonlinear device whose efficiency of conversion from fundamental wavelength input to second harmonic output is a strong function of the pulse peak intensity in the crystal.

For an unchirped (periodic) electric field poled QPM material, the ultrashort-pulse undepleted SHG efficiency is $\eta = \xi E_{fund}$, assuming an optimum crystal length (the length when the temporal walk-off between the fundamental and the second harmonic equals the fundamental pulse duration), confocal focusing, and no pump depletion [66]. The coefficient ξ is determined by the properties of a particular poled material and the pulse shape. Its values for several commonly used electric field poled ferroelectric materials are given in [50]. The SHG conversion efficiency in a linearly chirped QPM grating is reduced by the ratio between the durations τ_0 of the bandwidth-limited pulse and τ the stretched pulse: $\eta = \xi E_{\text{fund}}/(\tau/\tau_0)$ [50]. Note that this expression is equally valid for generating either stretched or compressed bandwidth-limited pulses.

In a QPM-based fiber CPA circuit, the peak power of the stretched and amplified pulses should be low enough to avoid phase distortions in a fiber amplifier due to SPM but should be high enough to provide efficient second harmonic conversion in the QPM compressor [6]. It has been shown in [6] that this requirement for the peak power P_{peak} can be expressed as $\eta/\tau_0\xi < P_{\text{peak}} < \pi/\gamma z_{\text{eff}}$. Here $\gamma = 2\pi n_2/\lambda A_{\text{eff}}$ is the (defined before) fiber nonlinearity coefficient, A_{eff} is the effective core area, n_2 is the fiberglass nonlinear index coefficient, and z_{eff} is the effective propagation distance in the amplifier. For the poled ferroelectric materials with the highest nonlinearity, such as electric field poled lithium niobate, this requirement can be achieved even using a single-mode core fiber amplifier [6].

Typical implementation of a QPM fiber CPA circuit is shown in Figure 8 [67]. The simplicity of such a system is apparent. It consists of a modelocked Er-doped fiber oscillator, Er-fiber preamplifier, high-power cladding-pumped Er/Yb co-doped fiber amplifier and a chirped period poled lithium niobate (CPPLN) pulse compressor.

A mode-locked fiber oscillator generated 180 fs, 100 pJ seed pulsed at 1560 nm center wavelength and 5 MHz repetition rate. These pulses

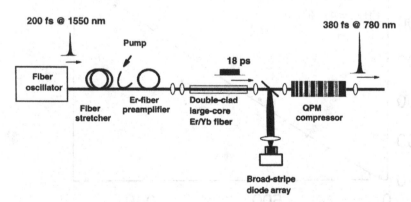

Figure 8 An example of implementing a fiber CPA circuit using a quasi-phase-matching SHG compressor. Femtosecond pulses from a mode-locked fiber oscillator are stretched to 18 ps in a fiber stretcher, amplified in an Er-doped fiber and a double-clad Er/Yb fiber amplifiers, and recompressed through second-harmonic generation in a 6 cm long chirped-period poled LiNbO$_3$ compressor to 380 fs at 780 nm. (Generated power and conversion efficiency are shown in Figure 9.)

were stretched to about 18 ps in 12 m of a positive-dispersion fiber ($\beta_2 = +0.108 \, ps^2/m$) and after amplification were launched into 6 cm long CPPLN crystal, producing 380 fs long compressed second harmonic pulses at 780 nm. The recompressed pulses were bandwidth-limited and the pulse duration increase occurred due to spectral gain narrowing in the Er/Yb co-doped fiber amplifier.

This particular system used a 24 μm diameter multi-mode core Er/Yb double-clad fiber for the power amplification stage. Up to 10 W at 976 nm from a broad-stripe diode array has been coupled into a 200-μm diameter and NA = 0.35 inner cladding of the fiber, producing 1.2 W maximum average power output at 1560 nm for 30 mW of injected preamplified signal. The CPPLN compressor power-conversion characteristic is shown in Figure 9, indicating that up to 0.5 W of 780 nm signal was generated with a corresponding SHG conversion efficiency of 42%. Such large core size, when operated in a single transverse mode, allowed extracting relatively high energies despite the limited duration of the stretched pulses. Indeed, at full 5 MHz repetition rate pulse output energies reached 100 nJ. The maximum energy extraction from such a system has been tested in a separate

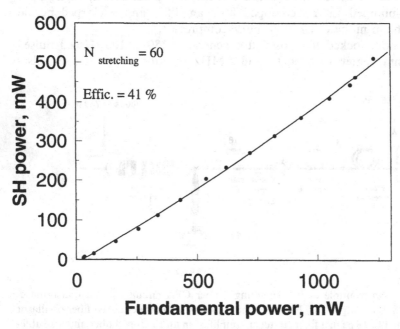

Figure 9 Chirped-period-poled lithium niobate pulse compressor power conversion characteristics. (Experimental results have been obtained with the fiber CPA system shown in Figure 8.)

measurement at a down-counted repetition rate yielding distortion-free pulses for up to 0.4 µJ at 1560 nm at the fiber amplifier output [67].

The other important result demonstrated with this system was that 100 nJ recompressed pulses obtained at 780 nm were sufficient to efficiently pump a PPLN optical parametric generator crystal. This allowed achieving up to 100 mW of wavelength-tunable output in the 1–3 µm range (including both signal and idler branches) from such a compact 5 MHz pulse repetition rate source.

Multimode Core Mode Scalable Nonlinear Amplifiers

Mode-scalable multimode core fibers are particularly useful for developing simple and therefore compact high average power systems based on non-linear pulse amplification at high pulse repetition rates. A larger mode size permits higher pulse energies, and a larger core size facilities coupling of higher pump powers. Indeed, the effective pump absorption coefficient α_{fiber} of a double-clad fiber is determined by the absorption in the fiber core α_{core} and the core-to-cladding area ratio: $\alpha_{fiber} = \alpha_{core} A_{core} / A_{cladding}$. The maximum pump power that can be coupled into the pump cladding is proportional to the cladding area: $P_{pump} \propto A_{cladding}$. Therefore, any increase in the core size allows a proportional increase in the pump-cladding size without sacrificing pump absorption in the fiber, and consequently allows a proportional increase in the maximum coupled pump power.

In a negative dispersion range at 1.55 µm (see Sec. 5.2.3) a multi-mode core Er/Yb fiber has been used for Raman-soliton nonlinear amplification [68], where use of a large-core allowed to increase power significantly compared to a single-mode fiber [22] producing 1.2 W of average power. However, in a positive-dispersion wavelength range at ~1 µm, the use of a mode-scalable large core Yb-doped fiber recently has led to the demonstration of the first femtosecond fiber system with an average power of ~10 W [69]. The importance of this result is that such a power level significantly exceeds average powers achievable with conventional mode-locked Ti:Sapphire lasers. The system is shown in Figure 10 and consists of an all-fiber ultrashort-pulse seed laser, a short length of single-mode fiber stretcher, a 4.3 m length of 25 µm diameter core and 300 µm diameter cladding Yb-doped fiber power-amplifier, and a pulse compressor. The core-to-cladding area ratio of 7×10^{-3} is sufficiently large to ensure efficient pump absorption in the short power-amplifier fiber length. The amplifier is pumped from both ends, with two fiber-bundle coupled laser diode sources operating at 976 nm. An estimated 14 W of pump power is coupled into each of the amplifier-fiber ends, of which 97% is absorbed. A maximum amplified output power of 13 W has been obtained (Fig. 11). At a pulse repetition rate of

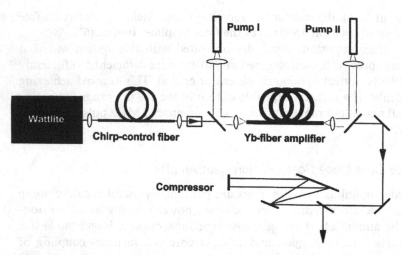

Figure 10 An example of a multimode core mode scalable nonlinear fiber amplifier. This particular system uses a 25 µm core doubled-clad Yb-doped fiber amplifier
to generate 13 W amplified pulses at 50 MHz repetition rate. (From Ref. 69).

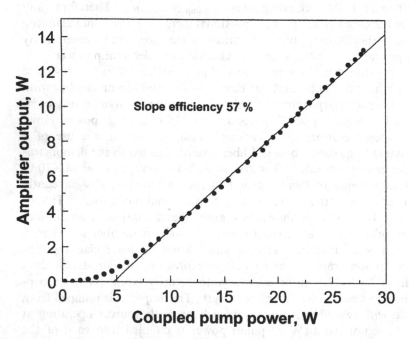

Figure 11 Amplified average power from a 25 µm core Yb-doped fiber for the system shown in Figure 10.

50 MHz, the resulting pulse energy is 0.26 µJ. The slope efficiency of this large-core fiber amplifier with respect to absorbed pump power is ~57%.

It is essential that the system shown in Figure 10 is similar to a CPA approach in that the width of the amplified pulses is increased while propagating along the amplifier due to the interplay between SPM and positive dispersion in the amplifier fiber. The combined action of large gain, self-phase modulation and positive dispersion ensures a linear pulse chirp at the output even in the presence of large spectral and temporal broadening, thus overcoming the nonlinear limitations of the fiber [70,71]. This is in contrary to the negative-dispersion case, where dispersion tends to compress spectrally broadened pulses and, thus, leads to soliton formation or modulation instability. The maximum achievable pulse energy in the latter case is related to soliton energy and is significantly smaller compared to the energy achievable with a positive-dispersion amplifier.

In the system shown in Figure 10 the pulse width and the peak power at the power amplifier output are 5 ps and ~52 kW, respectively. These pulses are recompressed down to 100 fs after passing through a small diffraction grating compressor consisting of a pair of 600 mm^{-1} gratings separated by only 10 cm. The pulse spectrum is centered at 1055 nm. The grating throughput efficiency is 40% giving 5 W of average power. The use of properly optimized gratings should provide significantly higher throughput (typically 60–70%).

This simple approach to generating high average power femtosecond pulses is suitable for further scaling to significantly higher power levels. Recently, a 50 W average power 10 ps system was demonstrated using similar fiber geometry [72].

4.3.2 High Pulse Energy Systems

The main technical target for a high pulse energy fiber system is to reach maximum achievable energies at the saturation fluence and stored energy limits of a fiber. This requires stretching amplified pulses to the longest recompressable durations, in the nanosecond range. At present such long and high-energy pulses can be only compressed using bulk diffraction-grating compressors. Therefore, all high-energy systems demonstrated so far have relied on bulk diffraction-grating based stretchers and compressors, as described in the following paragraphs. As noted in previous sections, from the perspective of compactness and ease of manufacture, such bulk-grating-based devices do not provide an optimum technical match to the rest of a fiber-based system. The primary goal of these demonstrations has been to show that high energies can indeed be reached using a fiber-based system.

The anticipation, however, is that current technical difficulties of implementing compact devices that can compress high-energy and nanosecond-long optical pulses will eventually be overcome. Efforts to develop such technology are ongoing, and some of the approaches will be briefly discussed in the following sections, including higher order dispersion compensation using nonlinearly chirped quasi-phase-matched structures and the first demonstration of the use of volume Bragg gratings for implementing compact high-energy CPA systems.

High Energy Er-Doped Fiber CPA Systems

Initial fiber CPA work started with Er-doped fibers because they were the most technologically developed for use in optical communication systems. At the very outset of this work pulse energies of greater than 1 μJ were reached [73], which constituted the approximate limit achievable with standard single-mode fibers. The introduction of cladding-pumping technology made it possible to significantly increase the average powers of fiber CPA systems, leading to the demonstration of the first femtosecond fiber systems with ~1 W output [65]. To further increase the achievable pulse energies, large-core single-mode fibers were developed with an approximately 16 μm diameter mode approaching the theoretical bending loss limit [74]. This led to the extraction of >10 μJ from a single-mode fiber CPA system [75], constituting an approximate technological limit for single-mode fiber CPA.

The next technological step came in 1997 with the first demonstration of generating ~100 μJ picosecond pulses using CPA based on multi-mode core cladding-pumped Er/Yb fiber [76]. This was the first experimental evidence that multi-mode core fiber amplifiers can be successfully used for nearly diffraction-limited operation. The experimental arrangement of this CPA system is shown in Figure 12, and its basic layout exemplifies a typical high-energy fiber CPA. The system comprises a femtosecond fiber oscillator, a pulse stretcher and a compressor based on standard arrangements of diffraction-gratings, and a few stages of amplification. The particular arrangement in Figure 12 consists of three main stages, although there is one additional preamplifier stage, not shown in the figure, which is the part of the oscillator arrangement. This multi-stage arrangement is necessary to achieve the required ~50-60 dB total gain for boosting pulse energies from an initial ~0.1-1 nJ to ~100 μJ. Acousto-optic modulators (AOMs) between the amplification stages are used as optical gates, allowing to control pulse repetition rate from an initial tens of megahertz from the mode-locked oscillator down to the kilohertz range required for achieving maximum pulse energies. Also, AOM gates prevent ASE buildup due to the feedback between different amplification stages. The mode size of the fibers comprising each amplifier is progressively increased to enable higher energy extraction from each stage. This increase in mode size helps to avoid pulse-phase distortions

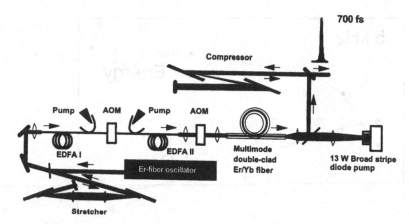

Figure 12   Experimental set-up of Er/Yb-doped 30 μm core fiber high-energy CPA system.

by nonlinear effects. In this particular system the first stage uses a standard in-core pumped (~200 mW at 980 nm) single-mode Er-doped fiber (~3 m long). The second is based on ~2 m long in-core pumped large-core single mode fiber with ~15 μm mode-field diameter (MFD) and is pumped with approximately 500 mW at 980 nm. The final power amplifier comprises a 1.5 m length of double-clad Er/Yb fiber with a 200 μm cladding and a 30 μm core. This fiber is heavily doped to achieve maximum gain over short lengths of fiber. The dopant levels in this particular case were approximately 15,000 ppm Yb^{3+} and 1200 ppm Er^{3+}.

Figure 13 shows the gain and amplified energy characteristics of the final stage measured versus the incident input energy at 5 kHz repetition rate. At low injection levels, the measured small-signal gain in approximately 19 dB. Also, the amplified energy reaches saturation at about 3 μJ of incident pulse energy. The maximum pulse energy of 250 μJ was reached at repetition rates below ~1 kHz. The stretched-pulse duration from a 1200 mm^{-1} diffraction grating stretcher in the Martinez configuration in this case was ~350 ps. The average output power at high repetition rates was up to 1.2 W. The launched pump power was 4 W at 976 nm and approximately 800 mW of this pump passed though the amplifier. Allowing for a measured bleaching power of 500 mW (absorbed), the average power extraction slope-efficiency is 38% with respect to the absorbed pump power.

After recompression, pulse energies of up to 100 μJ were achieved. The autocorrelation measurement, shown in Figure 14, corresponds to a pulse duration (after deconvolution) of ~0.7 ps. Some satellites in the wings of the trace indicate the presence of scattering into higher order modes

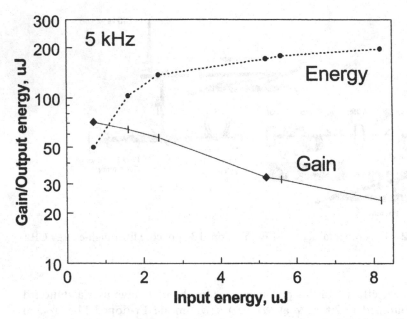

Figure 13 Gain and amplified energy from the 30 μm core Er/Yb-doped fiber measured as a function of the incident input energy at 5 kHz repetition rate.

in this first experimental demonstration of multi-mode core fiber CPA. Nevertheless, the measured $M^2 \sim 1.5$ of the output beam confirms a good quality of the recompressed output already with this first demonstration.

High Energy Yb-Doped Fiber CPA Systems

Multimode core fibers constitute a significant departure from standard communication grade optical fibers. This eliminates the technological advantages of working at $\sim 1.55\,\mu m$ as compared to other wavelength regions. In fact, the recent emergence of Yb-doped fibers provides a much more suitable medium for generating high power ultrashort pulses. The two main advantages of Yb-doped fibers compared to Er-doped ones are the significantly broader amplification band (50–100 nm compared to 10–30 nm, respectively) and significantly higher optical pumping efficiencies 60–80% compared to 30–40%, respectively). In fact, such high optical efficiency sets Yb-fibers favorably apart from any other rate-earth-doped fiber media. An additional important advantage of Yb doping is that the maximum achievable active-ion concentrations in the fiber core can be significantly higher than for Er doping, thus enabling a very high optical gain to be reached in a relatively short length of fiber.

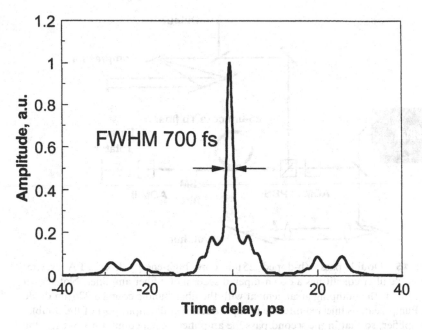

Figure 14   Measured autocorrelation trace of recompressed 100 μJ pulses.

Due to these advantages further advances in high-energy fiber CPA have been associated with Yb fibers. Recent work [77] demonstrated the generation of ~100 μJ femtosecond pulses at 1050 nm from an all-fiber based Yb-doped CPA. The results of this work immediately revealed the main improvements achievable with Yb-fibers. Higher pumping efficiencies led to significantly higher output average powers: up to 5.5 W at 1 MHz, before recompression. Broad amplification bandwidth provided with pulses as short as ~220 fs, compared to ~500–700 fs typical from high-energy Er-fiber CPA. The higher optical gain of Yb-fibers allowed the construction of CPA system similar to the one in Figure 12, but with only two amplification stages needed to obtain >100 μJ from ~0.5 nJ seed [77]. These improvements were achieved with an Yb-fiber, whose geometry and core size were similar to the corresponding Er/Yb doped fiber parameters and the stretched-pulse duratio was approximately the same as that used in Ref. 76.

Further exploitation of high Yb fiber gain allowed significant simplification of the basic fiber CPA design (shown in Fig. 12) by implementing a double-pass amplifier [78], which eliminates the need for multiple amplification stages and provides polarization-stable operation of the system using only simple round-core (nonpolarization preserving) fibers. The system

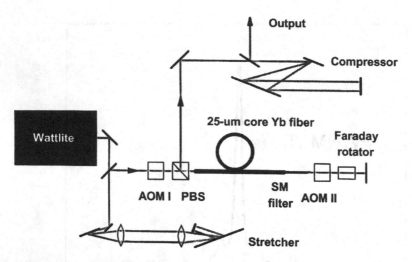

Figure 15 Double-pass Yb-doped 25 μm core high-energy fiber CPA system. Single-mode filter constitutes a down-tapered section of a fiber amplifier. Not shown in the figure is the pumping arrangement with the fiber-bundle coupled 976 nm diode array. Pump beam is injected into the pump cladding at the input port of the double-pass amplifier, so that in the second pass the amplified signal counter propagates the pump beam.

layout is shown in Figure 15. It principally consists of a pulse source, a pulse stretcher, a single amplifier stage and a pulse compressor. The pulse source in this system produces 0.5 nJ and 2 ps duration pulses, which are linearly chirped over 20 nm bandwidth and are compressible to ~150 fs duration. These seed pulses are stretched in a Martinez-type 1200 mm^{-1} diffraction-grating stretcher, which for 110 cm equivalent grating separation and 20 nm bandwidth provides 420 ps delay between the "red" and "blue" wings of the spectrum. The double-pass amplifier is constructed using ~2 m of 25 μm core Yb-doped fiber, which is pumped into 300 μm and NA = 0.35 inner cladding with a 15 W fiber-bundle coupled 976 nm laser diode array. The core has $V = 10.33$ and supports approximately 50 modes. Nevertheless, only fundamental mode excitation is achieved in the core by appropriate focusing conditions at the fiber input, as confirmed by the measured $M^2 = 1.1$ for the output beam.

Two important design features of this double-pass arrangement should be noted. First, polarization stability is achieved by exploiting the non-reciprocity of the Faraday rotator, which in the second pass of the beam through the amplifier fiber compensates a phase difference between the two orthogonal polarizations accumulated in the first pass. Second, robust fundamental-mode excitation in this double-pass configuration has been ensured by using a single-mode filter at the fiber end, implemented by tapering the 25 μm core fiber down to a size corresponding to a 6 μm core in which only a fundamental LP_{01} mode can propagate. The taper length has been kept sufficiently long in order to have an adiabatic transition in the taper [79] and thus to prevent scattering into higher order modes in the up-taper direction. The estimated fundamental mode power transmission through the taper is >75%.

Double-pass gain and amplified pulse energies as a function of pulse repetition rate are shown in Figure 16. The maximum net small-signal gain achieved at low injection power is 53 dB. Note, however, that the actual gain after two passes in the fiber is 60 dB, which is offset by a 7 dB loss in the back-reflecting arrangement. The highest output pulse energies of 100 μJ were achieved at repetition rates below 10 kHz.

The autocorrelation trace and spectrum of 220 fs recompressed pulses are shown in Figure 17. Transmission of ~50% through the compressor grating pair provided recompressed pulse energies of ~70 μJ and ~50 μJ for the single- and double-pass configurations respectively. The maximum obtained pulse energies were limited by the onset of stimulated Raman scattering.

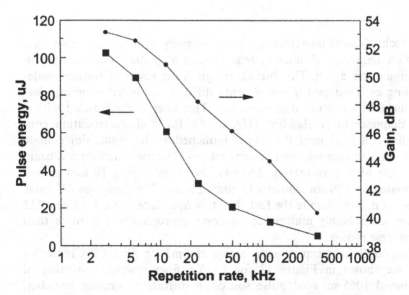

Figure 16 Measured double-pass gain and output energy as a function of pulse repetition rate for a double pass Yb-doped 25 μm core fiber amplifier.

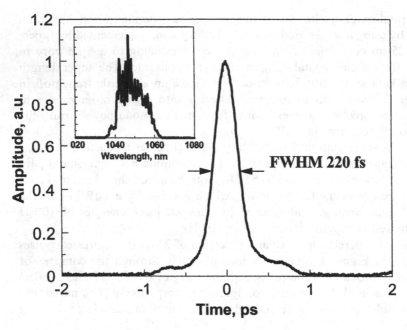

Figure 17 Autocorrelation trace and spectrum of amplified and recompressed pulses from a double-pass Yb-doped 25 µm core fiber CPA.

A technological breakthrough came recently with the demonstration of the first femtosecond fiber system generating pulses with energies in the millijoule range [80]. This breakthrough is the result of further mode-size scaling of Yb-doped fibers: Robust diffraction-limited operation has been achieved in a 50 µm diameter core. This fiber is double-clad with a 350 µm diameter inner cladding (NA = 0.4). Fiber characterization confirmed that if a fundamental mode is launched at the input, single-mode propagation is maintained over several meters. The measured output beam mode profile after propagating 2.6 m is shown in Figure 18 and is well approximated by a 30 µm diameter Gaussian beam. The measured M^2 value of the beam is 1.16, despite the fact that this fiber core has a V value of 15 and therefore is highly multimode supporting propagation of more than 100 transverse modes.

Structurally, the experimental setup of 1 mJ Yb fiber CPA is similar to the one shown in Figure 12 for Er/Yb fiber system, consisting of a fiber-based 1055 nm seed pulse source, a diffraction grating stretcher, a three-stage Yb fiber amplifier chain with two optical gates between the stages and a diffraction-grating compressor. The seed pulse source produces

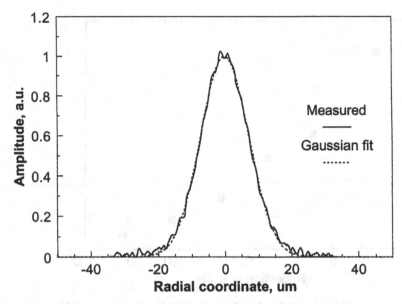

Figure 18 Measured fundamental mode profile of 50 μm core Yb-doped fiber. Measured mode-field diameter at $1/e^2$ intensity level is 30 μm. Measured M^2 of the beam is 1.16 indicating that the output is essentially diffraction-limited, despite the fact that this particular fiber can support propagation of up to ~100 transversal modes.

330 mW of 20 nm bandwidth and 2 ps linearly chirped pulses at 50 MHz repetition rate [71]. These pulses are further stretched to up to 800 ps. The first preamplifier uses a standard cladding-pumped single-mode core fiber and is operated at the full 50 MHz repetition rate. The second stage consists of a cladding-pumped 25 μm core fiber and is operated at 58 kHz repetition rate. The second amplifier produces ~3 μJ fixed-energy pulses for injection into the last stage. It is essential that at this repetition rate the second amplification stage is fully depleted, thus producing amplified pulses with no ASE background. The pulse repetition rate is varied only with the second acousto-optic modulator, which is placed after the second stage and can be opened at any sub-harmonic frequency of 58 kHz. This ensures identical seed pulses produced at any used repetition rate. The last amplification stage uses 2.6 m of the above-described 50 μm core double-clad fiber pumped with a 20 W laser diode array. It was important to counter-propagate pump and signal beams in a fiber amplifier, since this significantly reduces the effective propagation length for the amplified pulses to a small fraction of the actual amplifier length.

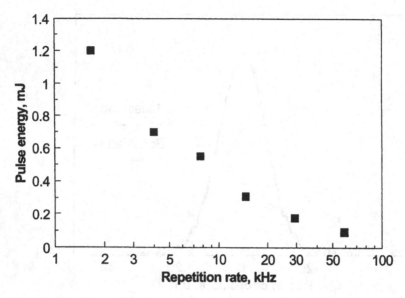

Figure 19 Measured pulse energy as a function of pulse repetition rate from Yb-doped 50 μm core fiber CPA.

The measured pulse energy as a function of the seed repetition rate is shown in Figure 19. The pulse energy increases from 95 μJ to 1.2 mJ with decreasing pulse repetition rate from 58 kHz down to 1667 Hz. The achieved 1.2 mJ energy was limited by the onset of stimulated Raman scattering. Higher pulse energies of up to 1.5 mJ have also been obtained, however, accompanied with a significant Raman spectral component.

Amplified 10 nm bandwidth pulses were recompressed in a diffraction-grating compressor to as short as ~400 fs. The compressor throughput efficiency is ~52%. However, for this first experimental demonstration a significant pedestal has been observed at the highest pulse energies, which has been identified with the beating-like phase modulation occurring due to SPM induced phase replicated from the modulated shape of the stretched pulses (see discussion earlier). This modulation can be eliminated by a careful design of the system to avoid satellite pulse formation, which can produce significant amplitude ripple on the profile of the stretched chirped pulses.

Volume Bragg Grating Based Compact High-Energy Fiber CPA

As it was briefly discussed earlier, volume Bragg gratings can potentially replace conventional diffraction-grating stretchers and compressors in fiber

CPA systems, thus providing compact CPA circuits for handling the highest energies achievable with a fiber amplifier. Beam size in a fiber grating is essentially not scalable, because it is confined in the fiber as the fundamental mode of the core. A volume chirped-period Bragg grating can be considered as a three-dimensional equivalent of a chirped fiber Bragg grating and therefore has no such limitation; propagating beam size is in principle limited only by transverse dimensions of the volume grating. This permits to significantly increase the energies of recompressed pulses simply by increasing the beam size propagating in such a 3D Bragg compressor. It is useful to note that a beam in a chirped-period volume Bragg grating should be well collimated in order to avoid inducing spatial chirp in the reflected beam, which occurs for the focused beam. This requires the grating length L to be much shorter than the Rayleigh length corresponding to the beam waist radius $\omega : L \ll k\omega^2$. Here $k = 2\pi n/\lambda$ is the beam propagation constant, λ is the wavelength and n is the refractive index of the grating medium. As an illustrative example, one can envision a $\sim 10\,\text{cm}$ long and $\sim 1\,\text{mm}$ diameter volume grating compressor, for which a simple numerical estimate predicts that potentially it could compress nanosecond-long stretched pulses and achieve recompressed pulse energies at about the millijoule level.

In principle, such volume Bragg gratings could be implemented using the same fabrication techniques and exploiting the same UV-induced photosensitivity mechanisms as currently used in producing fiber Bragg gratings [43]. The key technological challenge, however, is the depth-dependent absorption of the UV-light exposure, which can result in transversely inhomogeneous volume grating profile.

The first demonstration of producing chirped volume Bragg grating and its use in a fiber CPA system were reported in Ref. 7. A chirped volume holographic grating ($200\,\mu\text{m} \times 300\,\mu\text{m} \times 5\,\text{mm}$) was written in hydrogen-loaded germano-silicate glass [81]. The chirped grating was formed using an interferometer with a convex mirror in one arm to provide about $4\,\text{nm/cm}$ of linear chirp. The sample was exposed using a pulsed, excimer-pumped, frequency-doubled dye laser at a wavelength 235 nm which produced a fluence at the sample of $0.1\,\text{J/cm}^2/\text{pulse}$. The beam was focused to a 5 mm long horizontal line with a vertical full width of $60\,\mu\text{m}$, and then scanned in five steps along the vertical direction to form a grating approximately $300\,\mu\text{m}$ high. The total exposure at each step was $2.7\,\text{kJ/cm}^2$. The grating depth into the sample was limited by absorption of the UV light, resulting in a roughly exponential profile approximately $200\,\mu\text{m}$ deep. The grating reflectivity exceeded 90% over a bandwidth greater than 2 nm.

The experimental arrangement in which this grating was used for demonstrating a CPA system is shown in Figure 20. The CPA system principally consisted of a femtosecond fiber oscillator, a volume Bragg grating, and

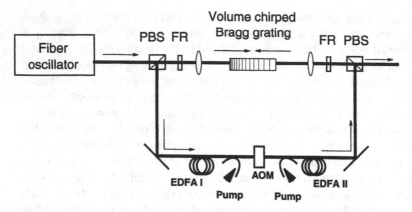

Figure 20 Compact fiber CPA circuit using a volume Bragg grating.

a two-stage Er-doped fiber amplifier with a pulse-selecting acousto-optical gate between them. The seed or amplified pulse was launched into a stretcher or compressor port of the volume grating respectively through a polarizing beam splitter (PBS) and a Faraday rotator (FR). The reflected beam was separated from the incident beam by using the nonreciprocal polarization rotation of the FR. A single grating was used for both pulse stretching and compression in order to take advantage of the phase reciprocity of a chirped Bragg structure [65]. The beam was coupled into the grating by focusing it loosely to an approximately 130 μm spot. The maximum light-collection efficiency achieved with the grating was 30%, which was attributed to the imperfect coupling conditions rather than to the low reflectivity of the grating.

The fiber oscillator was set to generate 2 ps seed pulses at 20 MHz, with a spectral width that did not exceed the ~2.5 nm FWHM reflection spectrum centered at 1542 nm of the grating used in the experiment. Autocorrelation traces of the stretched 25 ps and the recompressed 2.8 ps pulses are shown in Figure 21, indicating that indeed a volume chirped Bragg grating is acting as a pulse stretcher and compressor. The fiber amplifier contains additional dispersion-compensating fiber length, selected to cancel any significant linear-dispersion imbalance. The last amplification stage used a specially designed large-core single-mode fiber for increasing the achievable amplified pulse energies. 108 mW average power, 72 kHz reptition rate, and 1.5 μJ energy stretched pulses were obtained at the output of the amplifier, yielding 500 nJ and 36 mW after recompression in the grating.

Further scaling of available energies, pulse durations and bandwidths should be achievable with the subsequent increases in both the longitudinal and transversal dimensions and the reflection bandwidth of volume gratings.

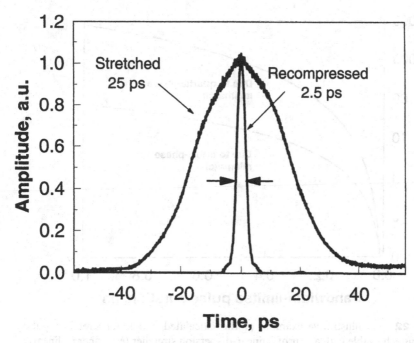

Figure 21 Measured autocorrelation traces of stretched and recompressed pulses in a volume Bragg grating based CPA system shown in Figure 20.

Cubic-Phase Mismatch Compensation Using a QPM Compressor in a High-Energy Fiber CPA

A significant technological step toward improving the compactness and manufacturability of a complete fiber CPA system could be achieved by replacing a diffraction grating stretcher with a fiber or a fiber grating. Indeed, Martinez-type stretching arrangements are much larger and more complicated than Treacy compressors and, considering the necessity to achieve and to maintain fiber-to-fiber coupling through a stretcher in a fiber system, are much more sensitive to alignment tolerances. Furthermore, for broad bandwidth pulses, a Treacy compressor is relatively small, and therefore use of the compressor does not constitute a significant bottleneck with respect to the compactness, robustness, and manufacturability of the overall system.

The main technical challenge in implementing this step is that the dispersion of a bulk diffraction grating compressor has large higher-order terms, typically an order of magnitude larger than those of a fiber-optic stretcher with an equivalent linear dispersion. An illustrative example in Figure 22 shows the calculated maximum stretched pulse durations

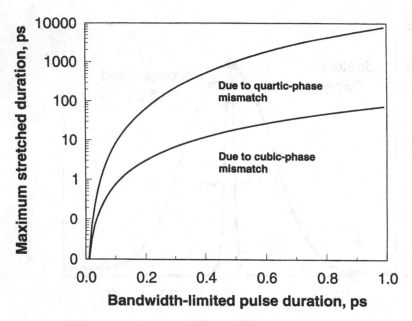

Figure 22 An illustrative example of the calculated maximum stretched pulse durations achievable with a pair of a linear-dispersion stretcher (e.g., fiber or linearly chirped fiber grating) and a diffraction-grating compressor. The maximum stretched duration is limited by the uncompensated higher-order dispersion terms introduced by the diffraction-grating compressor and is shown here as a function of bandwidth-limited incident-pulse duration for the cases of uncompensated second- and third-order GVD terms of the compressor. The specific example is calculated for 1200 l/mm grating at Littrow angle of incidence for 1550 nm light beam. This example shows that due to second-order GVD mismatch stretched pulses can not exceed 100 ps for achieving subpicosecond recompressed pulses.

achievable using a linear-GVD stretcher (a fiber or a linearly chirped FBG) and a diffraction grating compressor as a function of bandwidth-limited pulse duration due to the effect of uncompensated second- and third-order GVD terms. The specific example is for a 1200 lines/mm grating compressor with a near-Littrow angle of incidence (AOI) at $1.55\,\mu m$ wavelength. The relative magnitudes of the higher-order dispersion terms are fixed by the fundamental properties of a Treacy compressor, and therefore compensation of these terms constitutes a very serious problem. In principle, such terms could be compensated using nonlinearly chirped fiber grating stretcher [82], but, due to the prohibitively high accuracy required for controlling the period of this longitudinally-nonlinear chirp during grating fabrication, such a stretcher has not yet been demonstrated.

An alternative solution was demonstrated recently by using a non-linearly chirped QPM grating to compensate for cubic-phase mismatch (second-order GVD) between a linear-GVD fiber or a fiber grating stretcher and a diffraction-grating compressor [55]. An interesting aspect of this solution is that while a QPM grating compressor itself can not provide sufficiently large stretched pulse duration for extracting saturation-fluence limited energies from a fiber amplifier, they can provide a sufficiently large time window for compensating temporal delays associated with the higher order dispersion terms of a diffraction grating compressor.

An example of an Er-doped fiber CPA system using a QPM grating cubic-phase compensator is shown in Figure 23. The basic layout is standard to a fiber CPA in that it consists of a modelocked fiber oscillator (in this particular case providing 300 fs pulses with a center wavelength of 1556 nm), a linear-GVD pulse stretcher, two diode-pumped fiber amplification stages and a pulse compression stage. The principal distinction is that pulse stretching is achieved by using a combination of a linearly chirped fiber

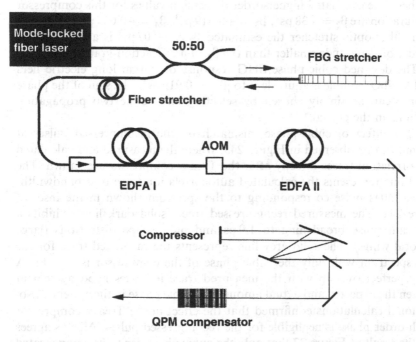

Figure 23 An example of an Er-doped fiber CPA system using a linearly-chirped fiber grating pulse stretcher and a diffraction-grating compressor with an additional QPM compressor. The QPM compressor provides cubic-phase compensation for the second-order GVD of the diffraction-grating compressor.

Bragg grating (6 cm long and 30 nm bandwidth) and a piece (120 m long) of positive dispersion fiber. Such a linear pulse stretcher is entirely monolithic, enabling the assembly of a nearly complete high-energy system (except for the pulse compression stage) using only fiber-optic components. The particular combined fiber-optic stretcher provides a total linear GVD of $+38\,\mathrm{ps}^2$ ($+25\,\mathrm{ps}^2$ in the linearly chirped FBG and $+13\,\mathrm{ps}^2$ in an additional fiber stretcher). For a 10 nm FWHM amplified pulse spectrum, shown in the inset of Figure 24a, this linear GVD produces 300 ps long stretched pulses, which also could be achieved by using a conventional diffraction grating stretcher of approximately 0.5 m length (in folded configuration).

The pulse compressor consists of a conventional diffraction grating arrangement in the Treacy configuration (providing negative linear GVD) and a nonlinear-period poled lithium niobate crystal at the output to compensate the third-order phase mismatch between the linear-GVD stretcher and the Treacy compressor through second harmonic generation. Initially the Treacy compressor had been set at $75.5°$ angle of incidence (AOI) and a grating separation at 38 cm to match the linear GVD of the fiber-optic stretcher. The calculated higher-order dispersion values for this compressor configuration are $\beta_2 = -38\,\mathrm{ps}^2$, $\beta_3 = +0.919\,\mathrm{ps}^3$, $\beta_4 = -0.037\,\mathrm{ps}^4$. Note that for the fiber-optic stretcher the estimated $\beta_3 \approx +0.07\,\mathrm{ps}^3$ is approximately an order of magnitude smaller than the β_3 of a diffraction-grating compressor. The designed cubic-phase SHG response of a 6 cm long electric field poled $LiNbO_3$ sample is equivalent to $|\beta_3| = 0.919\,\mathrm{ps}^3$. The sign of the phase response can be simply chosen by selecting one of the two propagation directions in the crystal.

The effect of cubic-phase mismatch on the recompressed pulses at 1556 nm can be observed in Figure 24a, where the measured and calculated autocorrelation traces directly after the Treacy compressor are shown. The dotted line represents the calculated autocorrelation trace of a bandwidth-limited 350 fs pulse corresponding to the spectrum shown in the inset of Figure 24a. The measured recompressed trace (solid dark line) exhibits a significant pulse broadening to 1.9 ps and pulse-shape distortions (large temporal wings). The solid grey line represents the calculated trace for the same spectrum with only the cubic phase of the compressor is present. A nearly perfect overlap with the measured trace indicates good agreement between the expected and actual amounts of cubic phase in the system. Also, additional calculations confirmed that the effect of the Treacy compressor fourth-order phase is negligible for the recompressed pulses. All this agrees with the result of Figure 22 that only the cubic phase has to be compensated in the present compressor for 300 ps stretched pulses.

The results of compensating this cubic phase through SHG generation in nonlinear-period poled lithium niobate are shown in Figure 24b. The solid

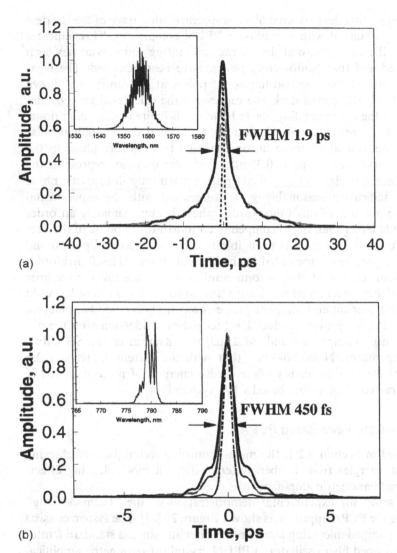

Figure 24 (a) The effect of cubic-phase mismatch on the recompressed pulses at 1556 nm. Calculated autocorrelation trace (dotted line) for bandwidth-limited 350 fs pulse corresponding to the spectrum in the insert. The measured autocorrelation trace (solid dark line) and the calculated trace (solid grey line) with only the cubic phase mismatch from the diffraction-grating compressor left uncompensated. (b) The result of compensating this cubic-phase mismatch using SHG in nonlinearly chirped QPM compressor. The measured autocorrelation trace (solid dark line) of compressed pulses at 780 nm. The calculated trace (dotted dark) for the bandwidth-limit of the generated SH spectrum shown in the insert. Solid grey line is the calculated trace with the residual fourth-order phase from the fiber grating stretcher.

dark line represents the measured 680 fs autocorrelation trace of the shortest 450 fs pulses obtained with a nonlinear PPLN compressor. This indicates that indeed the cubic phase of the diffraction grating compressor has been compensated and that femtosecond pulses have been achieved. The measured spectrum of these second-harmonic pulses at 778 nm is also shown in the inset and the dotted dark line represents the calculated autocorrelation trace of the corresponding 300 fs bandwidth-limited pulse. Additional dispersion measurements and calculations revealed that the residual pulse distortion was caused by some uncompensated fourth-order phase in the system corresponding to $|\beta_4| \approx 0.35\,ps^4$. The solid grey line represents the calculated autocorrelation trace of SH pulses when only this quartic phase is present, indicating reasonably good agreement with the experimental results. The amount of this fourth-order phase is approximately an order of magnitude larger than the uncompensated fourth-order phase of the Treacy compressor (which itself is negligible for the recompressed pulses) and therefore can not be compensated in the current system. This fourth-order phase component is attributed to some nonlinearity in the chirp of the fiber grating stretcher. In a refined nonlinear QPM grating design this order could be taken into account and complete phase compensation could be achieved.

The CPA system has produced $\sim$5 µJ pulses at 1556 nm after the diffraction grating compressor and $\sim$1.25 µJ at 778 nm after the SH cubic-phase compensator. Note, however, that with the current stretched pulse duration of 300 ps significantly higher pulse energies of more than 100 µJ can be extracted from a fiber based CPA system [76].

Multi-Mode Core Fiber Based PCPA

As discussed in Section 4.2.1, the most promising technique for achieving the highest energies from a fiber system relies on mode-cleaning effects in an optical parametric amplifier.

The setup for experimental demonstration of the "beam-cleaning" effect using the PCPA approach is shown Figure 25 [27]. The system consists of a diode-pumped microchip laser, an Yb fiber amplifier, a standard femtosecond Er-droped fiber oscillator, a PPLN crystal for parametric amplification and a diffraction grating pulse compressor. The fiber amplifier is cladding-pumped with a fiber bundle pigtailed broad-stripe laser diode module. At the output of the amplifier, pulses at 1065 nm are separated from the pump beam at 976 nm using a dichroic mirror. Coupling optics are selected such as to provide pump injection into a 300 µm internal cladding (NA = 0.4) of Yb fiber from the pumped end and signal injection into a 25 µm central core (NA = 0.17) form the seeded end. Amplified pump pulses at 1064 nm from a fiber amplifier then are combined with 1556 nm stretched

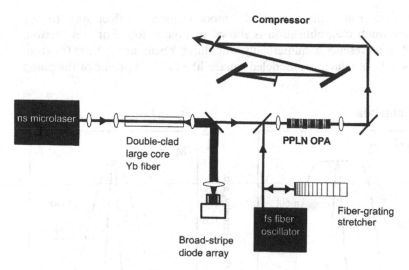

Figure 25 Multimode-core fiber based parametric chirped pulse amplification (PCPA) system for demonstrating mode cleaning though spatial solitary wave formation in PPLN OPA. (From Ref. 27).

femtosecond pulses from a fiber oscillator using another dichroic mirror. The microchip laser is actively Q-switched and synochronized to a free-running femtosecond fiber oscillator. The microchip laser can be operarted from a few hertz to ~10 kHz, producing 750 ps, ~3 µJ pulses. The fiber oscillator produces ~4 mW of 300 fs pulses at 50 MHz, which are temporally stretched in a chirped fiber grating to ~400 ps. The fiber grating is integrated into the fiber oscillator module. Parametric amplification of these pulses is achieved in a 2 cm long and 0.5 mm thick crystal of electric field poled lithium niobate (PPLN) with a poling period of 29.5 µm. Amplified stretched pulses at 1556 nm are recompressed using a standard Treacy-type diffraction grating pair compressor. Use of a fiber grating stretcher allows significant reduction in the size of the system.

The highest amplified pulse energies at the output of the multimode-core Yb fiber amplifier reached up to 250 µJ and were peak-power limited by the onset of stimulated Raman scattering. Up to ~32 % pump-to-signal energy conversion has been achieved in a parametric PPLN amplifier for confocal focusing of both pump and signal input beams. Recompression yielded 50–60 µJ, 1.6 ps pulses at 1556 nm. The pulse duration can be further reduced down to the bandwidth limit by the additional compensation of the cubic-phase mismatch between the fiber grating stretcher and diffraction grating compressor.

The experimentally measured "mode-cleaning" effect due to the optical parametric amplification is shown in Figure 26b. For comparison, Figure 26a represents a numerically calculated result using Eqs. (8) from Section 4.2.1., in which the modeled square-like far-field profile of the pump

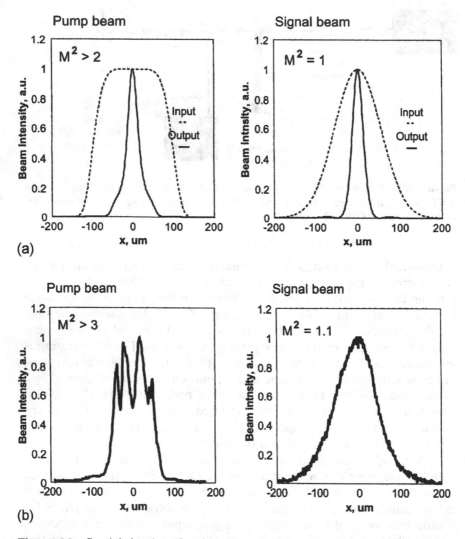

Figure 26 Spatial cleaning of multimode pump beams in a fiber-based PCPA system shown in Figure 25. (a) Numerically simulated and (b) measured far-field profiles of non-diffraction-limited pump beam at the input and of diffraction-limited signal beam at the output of an OPA.

beam at the input (with $M^2 \sim 2.6$) produces a diffraction limited signal beam (with $M^2 = 1$) at the output of an optical parametric amplifier Indeed, as predicted by the theoretical model, a highly multimodal ($M^2 > 3$) pump beam from a large core Yb fiber (shown here in the far-field) produces a diffraction-limited and Gaussian-shaped signal beam with $M^2 = 1.1$ at the output of a 2 cm PPLN crystal. This confirms that spatially multimode output from a large core fiber amplifier can be efficiently converted into a diffraction-limited amplified signal beam by using optical parametric amplification.

4.4 SUMMARY AND FUTURE OUTLOOK

The recent introduction of new fiber-core scaling techniques and of various practical approaches to compact chirped pulse amplification has led to the rapid development of high-energy and high-power femtosecond fiber systems. Currently, femtosecond pulses with energies of up to 1 mJ and average powers of more than 10 W have been achieved, constituting a significant step toward practical turnkey high-power ultrafast technology. In fact, the first commercial fiber CPA systems based on the described approaches have already been introduced.

Further development is anticipated in the directions of making high-energy fiber systems more compact and exploring avenues of achieving even higher powers and energies. The ultimate power and energy limits in fibers are set by a variety of nonlinear effects in the fiber core and is determined by the maximum usable core sizes. In general, the stability of fundamental mode propagation in multimode fibers decreases with the increase of fiber core diameter and the current anticipation is that the maximum usable diameter for diffraction-limited operation should be in the range of about 50–75 μm. At these core sizes nonlinear spatial effects and, possibly, thermal effects are anticipated to become a significant factor. Use of fiber-based PCPA could potentially be less susceptible to such limitations and could provide an alternative technological path for further power and energy scaling.

APPENDIX

Here we show that for strongly stretched pulses in a CPA system phase distortions caused by SPM in a fiber amplifier are determined by the spectral shape of the amplified pulses and are equivalent to introducing nonlinear GVD such that

$$\beta_n^{SPM} = P_0 \gamma z_{eff} \frac{d^n |\tilde{U}_0 \tilde{U}(\omega)|^2}{d\omega^n}\bigg|_{\omega=0} \tag{A1}$$

Here $\beta_n = d^n \varphi(\omega)/d\omega^n$, and $\varphi(\omega)$ is frequency-domain phase with $n = 2,3,4,\ldots$ corresponding to different dispersion orders.

The derivation of Eq. (A1) is based on two key assumptions. First, we assume that the stretcher GVD is linear:

$$\Delta t = \beta_2^{\text{stretcher}} \Delta\omega \tag{A2}$$

In reality this condition is usually not accurately fulfilled, but as long as the time delay produced by higher-order dispersion terms of a pulse stretcher is much smaller than the delay produced by the linear GVD, the effect of these higher order terms can be ignored.

Second, we assume that pulses are strongly chirped, such that $\Delta T_{\text{stretched}} \gg \Delta t_{\text{bandwidth-limited}}$, which is always true for femtosecond pulses stretched to ~ 1 ns duration. For such strongly chirped pulses the temporal shape $U(t)$ and its spectral shape $\tilde{U}(\omega)$ [$\hat{U}(\omega)$ is a Fourier transform of $U(t)$] are similar to each other:

$$U(t) = \tilde{U}_0 \tilde{U}(\omega) = \tilde{U}_0 \tilde{U}(t/\beta_2^{\text{stretcher}}) \tag{A3}$$

which can be shown using the method of stationary phase [83]. For further analysis it is convenient to introdcue normalized pulse amplitude $U_{\text{norm}}(t)$ such that $P(t) - P_0|U_{\text{norm}}(t)|^2$ and $U_{\text{norm}}(0) = 1$. Here $P(t)$ is the pulse instantaneous power and P_0 is peak power at $t = 0$. Then from Eq. (A3) it follows that $\tilde{U}_{\text{norm}}(0) = 1/\tilde{U}_0$ where the normalized spectrum $\tilde{U}_{\text{norm}}(\omega)$ is a Fourier transform of $U_{\text{norm}}(t)$. The value $\tilde{U}_0$ can be found by using Parseval's theorem:

$$E = P_0 \int_{-\infty}^{+\infty} U_{\text{norm}}^2(t)\, dt = P_0 \int_{-\infty}^{+\infty} \tilde{U}_{\text{norm}}^2(\omega)\, d\omega \tag{A4}$$

Indeed, from Eqs. (A5) and (A3) it follows that

$$\int_{-\infty}^{+\infty} U_{\text{norm}}^2(t)\, dt = \int_{-\infty}^{+\infty} \tilde{U}_{\text{norm}}^2(\omega)\, d\omega = \frac{1}{\tilde{U}_2^0 \beta_2^{\text{stretcher}}} \int_{-\infty}^{+\infty} U_{\text{norm}}^2(t)\, dt \tag{A5}$$

which leads to

$$\tilde{U}_0 = \frac{1}{\sqrt{\beta_2^{\text{stretcher}}}} \tag{A6}$$

According to Eq. (10) from the main text of the chapter the solution of non-linear Schrödinger equations (Eq. (9) in the main text) without the dispersion term leads to the expression for the SPM induced instantaneous phase:

$$\phi_{SPM}(t) = P(t)\gamma z_{eff} = P_0 \gamma z_{eff} |U_{norm}(t)|^2 \tag{A7}$$

The Taylor series expansion of the phase is

$$\phi(t) = \alpha_0 + \alpha_1 t + \frac{1}{2}\alpha_2 t^2 + \cdots + \frac{1}{n!}\alpha_n t^n + \cdots + \tag{A8}$$

where

$$\alpha_n = \frac{d^n \phi(t)}{dt^n} \tag{A9}$$

Then from Eq. (A7)

$$\phi_n^{SPM}(t) = \alpha_n^{SPM}\frac{t^n}{n!} = P_0 \gamma z_{eff}\frac{d^n |U_{norm}(t)|^2}{dt^n}\frac{t^n}{n!} \tag{A10}$$

and subsequently

$$\alpha_n^{SPM} = P_0 \gamma z_{eff}\frac{d^n |U_{norm}(t)|^2}{dt^n} \tag{A11}$$

Furthermore,

$$\frac{d^n |U_{norm}(t)|^2}{dt^n} = \frac{d^n \tilde{U}_0^2 |\tilde{U}_{norm}(\omega)|^2}{d\omega^n}\frac{d^n \omega}{dt^n}$$
$$= \tilde{U}_0^2 \frac{d^n |\tilde{U}_{norm}(\omega)|^2}{d\omega^n}\frac{1}{(\beta_2^{stretcher})^n} \tag{A12}$$

Substituting (A2) into (A8) leads to

$$\phi(t) \equiv \varphi(\omega) = \alpha_0 + \alpha_1 \beta_2^{stretcher}\Delta\omega + \cdots + \alpha_n \frac{(\beta_2^{stretcher}\Delta\omega)^n}{n!} + \cdots \tag{A13}$$

Finally, substituting (A11) and (A12) into (A13) gives

$$\frac{d^n \varphi(\omega)}{d\omega^n} = \beta_n^{SPM} = P_0 \gamma z_{eff}\frac{d^n(\tilde{U}_0^2 |\tilde{U}_{norm}(\omega)|^2)}{d\omega^n} \tag{A14}$$

In other words, the action of SPM on the phase of stretched and amplified pulses is equivalent to the action of nonlinear dispersion with nonlinearity terms determined by β_n^{SPM}. To summarize, this action is solely determined by two factors: the maximum phase shift at the peak of the pulse due to SPM, $\phi_{peak}^{SPM} = P_0 \gamma z_{eff}$, and the spectral shape of the amplified pulses, $\tilde{P}(\omega) = \tilde{U}_0^2 |\tilde{U}_{norm}(\omega)|^2$:

$$\beta_n^{SPM} = \phi_{peak}^{SPM} \frac{d^n \tilde{P}(\omega)}{d\omega^n} \tag{A15}$$

REFERENCES

1. O. E. Martinez, J. P. Gordon, and R. L. Fork, "Negative group-velocity dispersion using refraction," J. Opt. Soc. Am. A 1, 1003–1006 (1984).
2. O. E. Martinez, "3000 times grating compressor with positive group velocity dispersion: application to fiber compensation in 1.3–1.6 mm region," IEEE J. Quantum Electron. QE-23, 59–64 (1987).
3. Treacy E. B., "Optical pulse compression with diffraction gratins," IEEE Journal of Quantum Electron., QE-5, 454–458 (1969).
4. D. Strickland and G. Mourou, "Compression of amplified chirped optical pulses," Opt. Commun., Vol. 56, pp. 219–221 (1985).
5. A. Galvanauskas, M. E. Fermann, D. Harter, K. Sugden, I. Bennion, "All-Fiber Femtosecond Pulse Amplification Circuit Using Chirped Bragg Gratings" Appl. Phys. Lett., Vol. 66, pp. 1053–1055 (1995).
6. A. Galvanauskas, M. A. Arbore, M. M. Fejer and D. Harter "Chirped pulse amplification circuits for fiber amplifiers, based on chirped-period quasi-phase-matching gratings," Opt. Lett., Vol. 23, pp. 1695–1697 (1998).
7. A. Galvanauskas, A. Heaney, T. Erdogan, D. Harter, "Use of volume chirped Bragg gratings for compact high-energy chirped pulse amplification circuits," in Conference on Lasers and Electro-Optics, Vol. 6, Optical Society of American, Washington, D.C., (1998), p. 362.
8. P. Maine, D. Strickland, and P. Bado, "Generation of ultrahigh peak power pulses by chirped pulse amplification," IEEE J. Quantum Electron. QE-24, 398–403 (1988).
9. J. D. Love, Proc. Inst. Electr. Eng. Part J, Vol. 13, p. 225 (1989).
10. R. Olshansky, "Mode coupling effects in graded-index optical fibers," Appl. Opt., Vol. 14, pp. 935–945 (1975).
11. D. Gloge, "Optical power flow in multimode fibers," Bell System Techn. J., Vol. 51, pp. 1767–1783 (1972).
12. R. Olshanky, "Distortion losses in cabled optical fibers," Appl. Opt., Vol. 14, pp. 20–21 (1975).
13. W. A. Gambling, D. N. Payne, and H. Matsumura, "Mode conversion coefficients in optical fibers," Appl. Opt., Vol. 14, pp. 1538–1542 (1975).

14. M. E. Fermann, "Single-mode excitation of multimode fibers with ultrashort pulses," Opt. Lett., Vol. 23, pp. 52–54 (1998).

15. J. M. Sousa and O. G. Okhotnikov, "Multimode Er-doped fiber for single-transverse-mode amplification," Appl. Phys. Lett., Vol. 74, pp. 1528–1530 (1999).

16. J. P. Koplow, D. A. V. Kliner, L. Goldberg, "Single-mode operation of a coiled multimode fiber amplifier," Opt. Lett., Vol. 25, pp. 442–444 (2000).

17. L. F. Mollenauer, R. H. Stolen, J. P. Gordon, and W. J. Tomlinson, "Extreme picosecond pulse narrowing by means of soliton effect in single-mode optical fibers," Opt. Lett. 8, 289–291 (1983).

18. W. J. Tomlinson, R. H. Stolen, and C. V. Shank, "Compression of optical pulses chirped by self-phase modulation in fibers," J. Opt. Soc. Am. B 1, 139–149 (1984).

19. G. P. Agrawal, Nonlinear Fiber Optics, Academic Press, San Diego, CA (1995).

20. K. J. Blow, N. J. Doran, and D. Wood, "Generation and stabilization of short soliton pulses in the amplified nonlinear Schrodinger equation," J. Opt. Soc. Am. B, 5, 381–390 (1988).

21. I. Y. Khrushchev, A. B. Grudinin, E. M. Dianov, D. V. Korobkin, V. A. Semenov, and A. M. Prokhorov, Electron. Lett. 26, 456 (1990).

22. D. J. Richardson, V. A. Afanasjev, A. B. Grudinin, D. N. Payne, "Amplification of femtosecond pulses in a passive, all-fiber soliton source," Opt. Lett. 17, 1596–1598 (1992).

23. V. V. Afanasjev, V. A. Vysloukh, V. N. Serkin, "Decay and interaction of femtosecond optical solitons induced by the Raman self-scattering effect," Opt. Lett. 15, 489–491 (1990).

24. K. Tamura and M. Nakazava, "Pulse compression by nonlinear pulse evolution with reduced optical wave breaking in erbium-doped fiber amplifiers," Opt. Lett. 21, 68–70 (1996).

25. A. Dubietis, G. Jonusauskas, and A. Piskarskas, "Powerful femtosecond pulse generation by chirped and stretched pulse parametric amplification in BBO crystal," Opt. Commun., Vol. 88, pp. 437–440 (1992).

26. A. Galvanauskas, A. Hariharan, D. Harter, M. A. Arbore and M. M. Fejer, "High-energy femtosecond pulse amplification in a quasi-phase-matched parametric amplifier," Opt. Lett. 23, 210–212 (1998).

27. A. Galvanauskas, A. Hariharan, F. Raksi, K. K. Wong, D. Harter, G. Imeshev, M. M. Fejer, "Generation of diffraction-limited femtosecond beams using spatially multimode nanosecond pump sources in parametric chirped pulse amplification systems," in Conference on Lasers and Electro-Optics, Optical Society of America, Washington DC, (2000), pp. 394–395.

28. Menyuk, R. Schiek, L. Torner, "Solitary waves due to $\chi^{(2)} : \chi^{(2)}$ cascading," J. Opt. Soc. Am. B, Vol. 11, pp. 2434–2443 (1994).

29. L. Torner, C. B. Clausen, M. M. Fejer, "Adiabatic shaping of quadratic solitions," Opt. Lett., Vol. 23, pp. 903–905 (1998).

30. Y.-H. Chuang, D. D. Meyerhofer, S. Augst, H. Chen, J. Peatrross, and S. Uchida, "Suppression of the pedestal in a chirped-pulse-amplification laser," J. Opt. Soc. Am. B 8, 1226–1235 (1991).

31. M. D. Perry, T. Ditmire, and B. C. Stuart, "Self-phase modulation in chirped-pulse amplification," Opt. Lett. 19, 2149–2151 (1994).

32. A. Braun, S. Kane, and T. Norris, "Compensation of self-phase modulation in chirped-pulse amplification laser systems," Opt. Lett. 22, 615–617 (1997).

33. G. Tempea and T. Brabec, "Theory of self-focusing in a hollow waveguide," Opt. Lett. 23, 762–765 (1998).

34. G. Fibich and A. L. Gaeta, "Critical power for self-focusing in bulk media and in hollow waveguides," Opt. Lett. 25, 335–337 (2000).

35. S. Kane and J. Squier, "Grating compensation of third-order material dispersion in the normal dispersion regime: sub-100-fs chirped-pulse amplification using a fiber stretcher and grating-pair compressor," IEEE J. Quantum Electron. QE-31, 2052–2057 (1995).

36. R. G. Smith, "Optical Power handling capacity of low loss optical fibers as determined by stimulated Raman and Brillouin scattering," Appl. Opt., Vol. 11, pp. 2489–2494 (1972).

37. R. W. Boyd, Nonlinear Optics, Academic Press, Boston, Mass. (1992), Sec. 6.2.

38. B. C. Stuart, M. D. Feit, S. Herman, A. M. Rubenchik, B. W. Shore, and M. D. Perry, "Optical ablation by high-power short-pulse lasers," J. Opt. Soc. Am. B 13, 459–468 (1996).

39. L. Zenteno, "High-power double-clad fiber lasers," J. Lightw. Techn. 11, 1435–1446 (1993).

40. F. W. Dabby and J. R. Whinnery, "Thermal self-focusing of laser beams in lead glasses," Appl. Phys. Lett 13, 284 (1968).

41. M. K. Davis, M. J. F. Digonnet, and R. H. Pantell, "Thermal effects in doped fibers," J. Lightw. Techn. 16, 1013–1023 (1998).

42. D. C. Brown and H. J. Hoffman, "Thermal, stress, and thermo-optic effects in high average power double-clad fiber lasers," IEEE J. Quantum Electron. QE-37, 207–217 (2001).

43. R. Kashyap, Fiber Bragg Gratings, Academic Press, San Diego, CA, 1999.

44. M. A. Arbore, O. Marco, M. M. Fejer, "Pulse compression during second-harmonic generation in aperiodic quasi-phase-matching gratings," Opt. Lett. 22, 865–867 (1997).

45. M. A. Arbore, A. Galvanauskas, D. Harter, M. H. Chou, M. M. Fejer, "Engineerable compression of ultrashort pulses by use of second-harmonic generation in chirped-period-poled lithium niobate," Opt. Lett. 22, 1341–1343 (1997).

46. A. Feisst and P. Koidl, "Current induced periodic ferroelectric domain structures in $LiNbO_3$ applied for efficient nonlinear optical frequency mixing," Appl. Phys. Lett. 47, 1125–1127 (1985).

47. J.-P. Meyn and M. M. Fejer, "Tunable ultraviolet radiation by second-harmonic generation in periodically poled lithium tantalite," Opt. Lett, 22, 1214–1216 (1997).

48. Q. Chen and W. P. Risk, "Periodic poling of $KTiOPO_4$ using an applied electric field," Electron. Lett. 30 (18), 1516–1517 (1994).

49. H. Karlsson, F. Laurell, P. Henriksson, and G. Arvidsson, "Frequency doubling in periodically poled RbTiOAsO$_4$," Electron. Lett. 32 (6), 556–557 (1996).
50. G. Imeshev, M. A. Arbore, M. M. Fejer, A. Galvanauskas, M. Fermann, D. Harter, "Ultrashort-pulse second-harmonic generation with longitudinally nonuniform quasi-phase-matching gratings: pulse compression and shaping," J. Opt. Soc. Am. B 17, 304–318 (2000).
51. G. Imeshev, M. A. Arbore, S. Kasriel, M. M. Fejer, "Pulse shaping and compression by second-harmonic generation with quasi-phase-matching gratings in the presence of arbitrary dispersion," J. Opt. Soc. Am. B 17, 1420–1437 (2000).
52. G. Imeshev, A. Galvanauskas, M. Arbore, and M Fejer, "Ultrafast SHG with chirped QPM gratings in the high conversion efficiency regime," presented at the Annual Meeting '97 for the Center for Nonlinear Optical Materials, Stanford University (1997).
53. D. Marcuse, Theory of Dielectric Optical Waveguides, Academic Press, Boston, MA (1991).
54. J. U. Kang, Y. J. Ding, W. K. Burns, and J. S. Melinger, "Backward second-harmonic generation in periodically poled bulk LiNbO$_3$," Opt. Lett. 22, 862–864 (1997).
55. K. Green, A. Galvanauskas, K. K. Wong, and D. Harter, "Cubic-phase mismatch compensation in femtosecond CPA systems using nonlinear-chirp-period poled LiNbO$_3$," in Nonlinear Optics: Materials, Fundamentals and Applications, OSA Technical Digest (Optical Society of America, Washington DC, 2000), pp. 113–115.
56. G. Meltz, W. W. Morey, and W. H. Glenn, "Formation of Bragg gratings in optical fibers by transverse holographic method," Opt. Lett. 14, 823–825 (1989).
57. K. Sugden, I. Bennion, A. Molony, and N. J. Copner, "Chirped gratings produced in photosensitive optical fibres by fibre deformation during exposure," Electron. Lett. 30, 440–442 (1994).
58. K. O. Hill, B. Malo, F. Bilodeau, D. C. Johnson, and J. Albert, "Bragg grating fabrication in monomode photosensitive optical fiber by UV exposure through a phase mask," Appl. Phys. Lett. 62, 1035–1037 (1993).
59. R. Kashyap, P. F. McKee, R. J. Campbell, and D. L. Williams, "A novel method of writing phot-induced chirped Bragg gratings in optical fibres," Electron. Lett. 12, 996–997 (1994).
60. F. Ouellette, "Dispersion cancellation using linearly chirped Bragg grating filters in optical waveguides," Opt. Lett. 12, 847–849 (1987).
61. J. A. R. Williams, I. Bennion, K. Sugden, N. J. Doran, "Fibre dispersion compensation using a chirped in-fibre Bragg grating," Electr. Lett. 30, 985–987 (1994).
62. R. Kashyap, S. V. Chernikov, P. F. McKee and J. R. Taylor, "30 ps chromatic dispersion compensation of 400 fs pulses at 100 Gbits/s in optical fibres using an all fibre photoinduced chirped reflection grating," Electr. Lett. 30, 1078–1080 (1994).

63. K. O. Hill, F. Bilodeau, B. Malo, T. Kitagawa, S. Theriault, D. C. Johnson, J. Albert, and K. Takiguchi, "Chirped in-fiber Bragg gratings for optical fiber dispersion compensation," Opt. Lett. 19, 1314 (1994).

64. S. Radic, G. P. Agrawal, A. Galvanauskas, "Nonlinear effects in femtosecond pulse amplification using chirped fiber gratings," Optical Fiber Communications Conference, February 25–March 1 (1996) San Jose, California, paper ThF3.

65. A. Galvanauskas, M. E. Fermann, D. Harter, J. D. Minelly, G. G. Vienne, and J. E. Caplen, "Broad-area diode-pumped 1 W femtosecond fiber system," in Conference on Lasers and Electro-Optics, Vol. 9, Optical Society of America, Washington, D. C., (1996), pp 495–496.

66. M. A. Arbore, M. M. Fejer, M. E. Fermann, A. Hariharan, A. Galvanauskas, D. Harter, "Frequency doubling of femtosecond erbium-fiber soliton lasers in periodically poled lithium niobate," Opt. Lett. 22, 13–15 (1997).

67. A. Galvanauskas, M. E. Fermann, M. A. Arbore, M. M. Fejer, J. D. Minelly, J. E. Caplen, K. K. Wong and D. Harter, "Robust high-power and wavelength-tunable femtosecond fiber system based on engineerable PPLN devices," in Nonlinear Optics: Materials, Fundamentals and Applications Topical Meeting, August 10–14 (1998) Princeville, Kauai, Hawaii, paper WB2.

68. M. Hofer, E. Fermann, A. Galvanauskas, D. Harter, and R. S. Windeler, "High-power 100-fs pulse generation by frequency doubling of an erbium/ytterbium-fiber master oscillator power amplifier," Opt. Lett. 23, 1840–1842 (1998).

69. A. Galvanauskas and M. E. Fermann, "13-W femtosecond fiber laser," CLEO 2000, San Francisco, CA, postdeadline paper CDP3.

70. K. Tamura and M. Nakazawa, "Pulse compression by nonlinear pulse evolution with reduced optical wave breaking in erbium-doped fiber amplifier," Opt. Lett. 21, 68–70 (1996).

71. M. E. Fermann, A. Galvanauskas, M. L. Stock, K. K. Wong, D. Harter, and L. Goldberg, "Ultrawide tunable Er soliton fiber laser amplifier in Yb-doped fiber," Opt. Lett. 24, 1428–1426 (1999).

72. J. Limpert, A. Liem, T. Gabler, H. Zellmer, A. Tunnermann, "High average power ultrafast Yb-doped fiber amplifier," in Advanced Solid-State Lasers, Optical Society of America, Washington DC (2001), pp. 215–217.

73. A. Galvanauskas, M. E. Fermann, P. Blixt, J. A. Tellefsen, D. Harter, "Hybrid Diode-Laser Fiber – Amplifier Source of High Energy Ultrashort Pulses," Opt. Lett. 19, 1043–1045 (1994).

74. D. Taverner, D. J. Richardson, L. Dong, J. E. Caplen, K. Williams, and R. V. Penty, "158-μJ pulses from a single-transverse-mode, large-mode-area erbium-doped fiber amplifier," Opt. Lett. 22, 378–380 (1997).

75. D. Taverner, A. Galvanauskas, D. Harter, D. J. Richardson, and L. Dong, "Generation of high-energy pulses using a large-mode-area erbium-doped fiber amplifier," in Conference on Lasers and Electro-Optics, Vol. 9, Optical Society of America, Washington, D. C. (1996), pp. 496–497.

76. J. D. Minelly, A. Galvanauskas, D. Harter, J. E. Caplen, L. Dong, "Cladding-pumped fiber laser/amplifier system generating 100 µJ energy picosecond pulses," in Conference on Lasers and Electro-Optics, Vol. 11, Optical Society of America, Washington, D.C. (1997), pp. 475–476.
77. G. C. Cho, A. Galvanauskas, M. E. Fermann, M. L. Stock, D. Harter, "100 µJ and 5.5 W Yb-fiber femtosecond chirped pulse amplifier system," in Conference on Lasers and Electro-Optics, Optical Society of America, Washington DC (2000), p. 118.
78. A. Galvanauskas, G. C. Cho, A. Hariharan, M. E. Fermann, and D. Harter, "Generation of high-energy femtosecond pulses in multimode-core Yb-fiber chirped-pulse amplification systems," Opt. Lett. 26, 935–937 (2001).
79. J. D. Love and W. M. Henry, "Quantifying loss minimisation in single-mode fibre tapers," Electron. Lett. 22, 912–913 (1986).
80. A. Galvanauskas, Z. Sartania, M. Bischoff, "Millijoule femtosecond fiber CPA system," in Advanced Solid-State Lasers, January 28–31, 2001, Seattle, Washington, postdeadline paper PD3; A. Galvanauskas, Z. Sartania, M. Bischoff, "Millijoule femtosecond all-fiber system," in Conference on Lasers and Electro-Optics (CLEO 2001), Baltimore, MD, May 6–11 (2001) invited paper CMA1.
81. T. Erdogan, A. Partovi, V. Mizrahi, P. J. Lemaire, W. L. Wilson, T. A. Strasser and A. M. Glass, "Volume gratings for holographic storage applications written in high-quality germanosilicate glass," Appl. Opt. 34, 6738 (1995).
82. A. Galvanauskas, A. Hariharan, D. Harter, "Hybrid short-pulse amplifiers with phase-mismatch compensated pulse stretchers and compressors," US Patent No. 5,847,863, issue date December 8 (1998).
83. M. Born and E. Wolf, Principles of Optics, Pergamon Press, New York (1980) Appendix III.

5

Ultrafast Single- and Multiwavelength Modelocked Semiconductor Lasers: Physics and Applications

Peter J. Delfyett

University of Central Florida, Orlando, Florida, U.S.A

5.1 INTRODUCTION AND FUNDAMENTALS

5.1.1 Introduction

The generation of ultrashort optical pulses is an extremely exciting and rapidly growing field. This is primarily due to the fact that these short optical pulses allow the probing of fundamental physics, both in structure and time scale, of light–matter interactions in the femtosecond and attosecond regimes. Additionally, the short temporal duration of these pulses can be exploited for a variety of commercial applications in areas encompassing material processing, optical microscopy, data storage, optical communications, ranging, and novel noninvasive imaging.

To make ultrafast technologies amenable to commercial applications, the modelocked laser source must meet specific criteria so that the use of the technology will enhance the application and also make it more profitable. As a result, much work has gone into developing modelocked lasers that are compact, electrically efficient, and sufficiently robust to survive in a wide range of commercial environments. To meet this need, semiconductor gain media become attractive candidates as modelocked sources owing to their small size (a few hundred micrometers on a side), excellent wall plug efficiency (>50%), and robustness (can be packaged in standard size electronic packages).

This chapter is devoted to the fundamental physics behind the generation of ultrashort, high-power, multiwavelength optical pulses from semiconductor lasers. A key factor in these developments is the ability to understand the underlying gain dynamics and exploit these effects to assist in robust pulse generation.

The chapter is organized as follows. First, a review of the underlying gain dynamics is provided, which includes the associated nonlinear optical effects that influence pulse generation and propagation. From these physics, we provide a simple recipe that will allow for robust modelocked optical pulses. Next, a review is presented of several measurement techniques that are useful for characterizing the generated optical pulses; this review also provides insight into the fundamental physics that control modelocking dynamics. Third, a discussion on intracavity spectral filtering is provided. The techniques discussed here are intended to provide the reader with a background on the incorporation of spectral shaping techniques that can influence the temporal characteristics of the output pulses. These techniques are then applied to the realization of several types of modelocked diode lasers. These sources demonstrate ultrashort pulse width, high average and peak powers, multiwavelength operation, and ultralow noise. To conclude the chapter, we highlight specific application areas, including optical communications and optical signal processing, and show how these modelocked semiconductor lasers can be used to improve performance in these application areas.

5.1.2 Background Physics

The generation of ultrashort optical pulses from semiconductor lasers has been a very active area of research owing to the potentially major impact that these devices can have in a broad range of applications, ranging from medical science to communication/signal processing applications and even entertainment. However, the physics of generating ultrashort pulses is very different from the physics of other solid-state ultrafast lasers. In this section we initially examine the ultrashort gain dynamic response, or the impulse response, of a semiconductor gain medium and show how these physical effects play a role in determining the optical pulse characteristics in a mode-locked semiconductor diode laser. To illustrate the operation of a mode-locked semiconductor, we use a model of an external cavity incorporating a semiconductor gain medium, such as a semiconductor optical amplifier that is pumped by either a constant direct current or a combination of direct and radio-frequency current. To achieve active modelocked operation, we choose a radio-frequency current with a frequency related to the longitudinal mode spacing of the external cavity. In addition, a nonlinear saturable absorber can be included in the laser to initiate passive modelocked

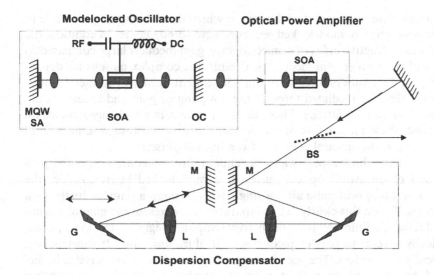

Figure 1 Schematic of an external cavity hybrid modelocked semiconductor diode laser. The oscillator contains a saturable gain and saturable loss medium with dc and RF drive signals. Also shown are an optical power amplifier and a dispersion compensator.

operation. The combined action of both passive and active modelocking is referred to as hybrid modelocking. Figure 1 shows a simple schematic of an external cavity, hybrid modelocked semiconductor diode laser that includes a saturable gain and saturable loss medium. In addition, it should be noted that the presence of the gain, absorber, and any additional optics introduces dispersion into the cavity. Also shown in Figure 1 is a secondary semiconductor optical amplifier used as an optical power amplifier to increase the optical power produced by the modelocked oscillator and a one-to-one telescope between an inverted grating pair to compensate for any pulse spreading that occurs during the pulse generation process.

5.1.3 Concept of Classic Gain Saturation vs. Carrier Heating Induced Gain Saturation

In semiconductor gain media, the gain of the material is depleted by the passage of an ultrashort pulse as it propagates through the medium. One important and distinctive property of the time-varying gain of semiconductor media compared to other gain media is that the effective impulse response of the gain depends on the temporal duration of the optical pulse that is propagating through the gain medium. This property greatly affects the overall gain dynamics and the way in which these dynamics play a role in pulse generation in modelocked semiconductor diode lasers. In addition, the nonlinear refractive index of semiconductor gain media is large compared to those of most other gain media. As a result, the complex pulsewidth-dependent gain dynamics induces large time-dependent nonlinear refractive index changes that are mediated through the coupling of gain and index, via the Kramers–Kronig relations. These time-dependent index changes induce an ultrafast phase modulation of the optical pulse, which results in an overall distortion in the spectral content of the optical pulse.

To show how a pulsewidth-dependent gain dynamic affects the production of ultrashort optical pulses in a modelocked laser, consider the effects on an optical pulse after a single pass through a semiconductor gain medium. To see this clearly, let us separate and represent the gain of a semiconductor medium for two distinctive temporal regimes: (1) long pulses, typically several to tens of picoseconds in duration, and (2) short pulses, typically 1 ps or less. The specific time scales mentioned are physically dictated by mechanisms characteristic of intraband relaxation mechanisms and are explained below.

The reduction of the small signal gain for slow gain dynamics, as a function of time, can be given by the expression [1]

$$G_S(t) = \frac{G_0}{G_0 - (G_0 - 1)\exp\left[-(1/E_{\text{sat}}) \int\limits_{-\infty}^{t} P_{\text{in}}(\tau)\,d\tau\right]} \tag{1}$$

In this expression, $G_S(t)$ is the instantaneous value of the saturated gain for slow gain dynamics, and G_0 is the initial value of the small signal gain before the presence of the optical pulse. E_{sat} is the saturation energy of the gain medium, and $\int_{-\infty}^{t} P_{\text{in}}(\tau)\,d\tau$ is the fractional pulse energy contained in the leading part of the pulse up to time t. This expression is the classic, well-known expression for the gain depletion that an optical pulse would experience in most gain media. It should be noted that a similar expression can be used to describe the physical effects of saturable absorption.

For semiconductor gain media, however, the gain is complicated by the presence of ultrafast physical processes that redistribute the carriers in the conduction and valence bands. This redistribution of carriers has the effect of contributing an additional component of gain reduction that can quickly recover on a subpicosecond time scale. As an example, not only does an ultrashort optical pulse in a semiconductor gain medium experience gain, but also the presence of the optical pulse can induce transitions of carriers to higher energy levels, thus reducing gain. This effect of gain reduction can be identified with two physical effects: (1) free carrier absorption and (2) two-photon absorption.

In addition to the effects of free carrier absorption and two-photon absorption, spectral hole burning also reduces the gain experienced by an optical pulse. Spectral hole burning is an effect that depletes carriers with an energy directly proportional to the laser photon energy without affecting the gain at other energies. Generally, in semiconductor gain media, the gain recovery due to spectral hole burning is on the order of 100 fs and does not play a role for pulses on the order of 1 ps. However, owing to the effects of spectral hole burning, two-photon absorption, and free carrier absorption, the carriers redistribute themselves within 100 fs due to carrier–carrier scattering, resulting in a heated carrier distribution. The effect of carrier heating is to spread the Fermi distribution, with the resulting heated carrier distribution reducing the gain over a wide energy range. An important result of carrier heating is that the hot distribution will quickly cool back to the lattice temperature by phonon emission, generally on a time scale of 1 ps, which results in a small increase in the optical gain.

The effect of carrier heating and cooling can be included in the transient gain dynamics by modeling the additional gain reduction and its recovery by [2]

$$
G_F(t) = -h \int_{-\infty}^{t} \left[1 - \exp\left(\frac{-\tau}{\tau_i}\right) \right] \exp\left(\frac{-\tau}{\tau_2}\right) u(\tau) I(t - \tau) \, d\tau' \tag{2}
$$

where $G_F(t)$ represents the fast transient gain dynamics, τ_i is the intraband thermalization time associated with the recovery from spectral hole burning, τ_2 is the relaxation time of the heated carrier temperature back to the ambient lattice temperature, also referred to as the cooling time, $I(t)$ is the pulse intensity, and h is a constant.

Figure 2 clearly shows how the effects of the fast and slow gain dynamics contribute to shaping an ultrashort optical pulse as it propagates through a semiconductor gain medium. Shown in Figure 2a are plots of the time-varying gain showing both slow gain depletion and gain depletion with the combined effects of both fast and slow dynamics. Also shown is the optical pulse intensity used to compute both slow and fast transient gain saturation effects. In Figure 2b are three optical pulse shapes, showing the input and output pulse shapes from semiconductor gain media whose gain dynamics follow that shown in Figure 2a. The salient feature to note from Figure 2a is that the slow gain dynamic follows the integrated pulse energy, i.e., the gain is depleted according to the integral of the pulse intensity. The fast gain dynamic contains an additional term that increases the gain depletion but undergoes a fast partial gain recovery. The output pulse shape follows [3]

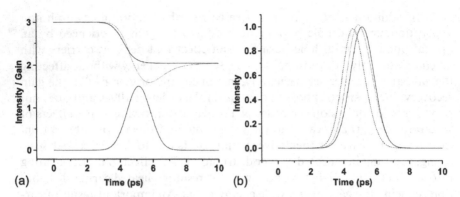

Figure 2 (a) Plots of the both the fast and slow gain dynamics. Also shown is the input optical pulse shape used in calculating the gain dynamic. (b) Input and output pulse shapes for both fast and slow gain dynamics. Note the net effective pulse shortening associated with slow gain dynamics and the net effective pulse broadening due to fast gain dynamics.

$$I_{OUT\ S,F}(t) = G_{S,F}(t)I_{IN}(t) \tag{3}$$

where $I_{OUT\ S,F}$, and $G_{S,F}$ are the output pulse shape and gain for the slow and fast dynamics, and I_{IN} is the input optical pulse.

From Eq. (3) the salient features of Figure 2b show that both gain dynamics shift the position of the optical pulse to earlier times. However, the fast dynamic does not reduce the trailing edge of the amplified pulse compared to the slow gain dynamic. This is to say that the fast gain dynamic provides additional gain for both leading and trailing edges of the pulse, compared to the peak pulse position. The slow dynamic amplifies the leading edge and reduces the gain for the trailing edge of the pulse. The net effect is that the slow gain dynamic provides for an overall pulse width reduction when combined with other modelocking mechanisms, whereas the fast dynamic immediately shows a net pulse width increase. This effect means that the fast gain dynamic imposes a limit on the temporal pulse duration that can be achieved from conventional modelocked semiconductor diode lasers. This is due to the fact that in a modelocked laser the generated optical pulse evolves from intensity noise and a temporal window of gain provided by a modelocking mechanism. Because the temporal duration of the optical pulse decreases as the laser undergoes modelocking, the fast gain dynamic limits the temporal duration by providing a pulse-broadening mechanism that counteracts other pulse-shortening mechanisms due to modelocking.

5.1.4 Linear and Nonlinear Effects Induced by Gain Dynamics—Dispersion and Self-Phase Modulation

Before we can consider how gain dynamics play a role in modelocking, we must consider other linear and nonlinear effects. The primary linear and nonlinear effects associated with pulse propagation in semiconductor gain media are dispersion and self-phase modulation, respectively. Because it was observed above that the fast gain dynamics of carrier heating and cooling serve as a mechanism that limits pulse-shortening mechanisms in modelocked semiconductor diode lasers, it can be assumed that the primary gain dynamic experienced by an optical pulse inside a modelocked semiconductor is due to the slow gain dynamic. Given that the primary gain dynamic is due to conventional slow gain depletion that follows the integrated pulse energy, one can quickly show how this gain depletion influences the nonlinear properties of the semiconductor gain media. For a detailed discussion on fast gain dynamic and the associated nonlinear effects, the reader is directed to Ref. 4.

Self-Phase Modulation

In the limit of slow gain dynamics, the predominant nonlinear effect is due to the time-varying index of refraction that is coupled to the time-varying gain change, whereas the dominant linear effect in pulse propagation is dispersion. It should be recalled that gain depletion also has an effect on distorting the pulse shape; however, this was discussed earlier. Because the main nonlinear effect is associated with the time-varying index induced by the change in gain due to gain depletion, a simple model of self-phase modulation can be employed to obtain insight as to its effect on an ultrashort optical pulse propagating through a semiconductor gain medium.

It should be recalled that the instantaneous frequency of a light beam can be changed, or modulated, if the light beam experiences a time-varying index of refraction. The instantaneous frequency can be derived as [5]

$$\omega_{inst}(t) = \omega_0 - \frac{\omega_0 L}{c} \frac{\partial n(t)}{\partial t} \tag{4}$$

where ω_{inst} is the instantaneous frequency as a function of time, ω_0 is the initial carrier frequency of light, L is the length of the nonlinear medium, c is the speed of light, and $n(t)$ is the time-varying index of refraction. From this equation, it is clear that the instantaneous frequency of a light beam can be modified if it experiences a time-varying refractive index. Specifically, the instantaneous frequency is directly proportional to the temporal derivative of the index. Figure 3 schematically shows the interplay of the gain, index,

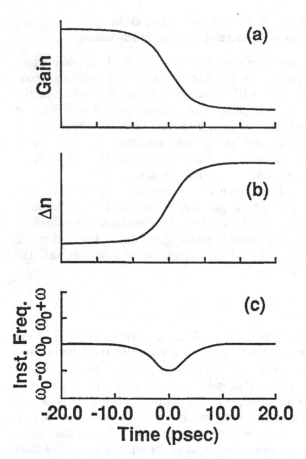

Figure 3 Plots of (a) the slow gain depletion, (b) the associated time-varying refractive index, and (c) the resultant instantaneous frequency chirp impressed on an optical pulse as it undergoes amplification.

and instantaneous frequency for the case of a semiconductor gain medium with slow gain dynamics. As noted above, the gain is depleted according to the integrated pulse energy. The refractive index generally follows the temporal behavior of the time-varying gain, except with opposite sign, i.e., $n(t) \sim -G_S(t)$. Thus, as carriers are removed from the semiconductor gain medium via stimulated emission, the refractive index increases. Because the instantaneous frequency is proportional to the derivative of the index, but the index changes proportionally to the integrated pulse energy, the instantaneous frequency varies in time in direct proportion to the pulse intensity. These three effects are highlighted in Figure 3, which shows the dynamic of slow gain depletion (a), the time-varying index (b), and the resulting instantaneous frequency that is impressed on the optical pulse (c).

Linear Dispersion

We are now ready to describe how the gain dynamics along with the associated gain-induced phase modulation combine with linear dispersion to contribute to pulse shaping and distortion. Using standard linear system theory, it is possible to predict the pulse shape exiting a linearly dispersive medium if the input pulse characteristics and the system transfer function are known. If the input pulse can be represented by a complex electric field $E(t)$ and its spectrum $E(\omega)$ is obtained via the Fourier transform, then the output pulse is simply

$$E_0(t) = F^{-1}\{E_{\text{IN}}(\omega)|H(\omega)| \exp[i\varphi(\omega)]\}$$

where F^{-1} denotes the inverse Fourier transform, $|H(\omega)|$ is the magnitude of the transfer or filter function, and $\varphi(\omega)$ is the spectral phase, which is related to the dispersion of the transfer function.

Saturable Absorption

The predominant effect for saturable absorption is a shortening of the leading edge of an ultrashort optical pulse. Generally speaking, the action of the saturable absorber is to attenuate light of low power and transmit light of high intensity. The standard equation that governs the saturation process is given by Eq. (1) except that the small signal gain G_0 is replaced by the linear loss.

By including the effects of gain saturation and its associated integrating self-phase modulation, saturable absorption, and linear dispersion, one can obtain a clear picture of how a short optical pulse evolves within the cavity of a passively modelocked diode laser. Figure 4 shows curves of the pulse intensity and the instantaneous frequency or chirp impressed on the modelocked pulse within the diode laser cavity. Figure 4a shows an initial Gaussian pulse shape with a parabolic chirp impressed on the pulse owing to the time-varying index associated with gain depletion. Figure 4b shows the optical pulse after propagating through a linearly dispersive medium with positive dispersion. The salient feature to note in this pulse shape is the steepened leading edge, which is caused by a slowing down of the high-frequency components at the leading and trailing edges of the laser pulse. Finally, Figure 4c shows the pulse shape and instantaneous frequency after interaction with the saturable absorber. Note the shortened leading edge due to saturable absorption. More important, it should be noted that the saturable absorber removes the part of the pulse that contains the

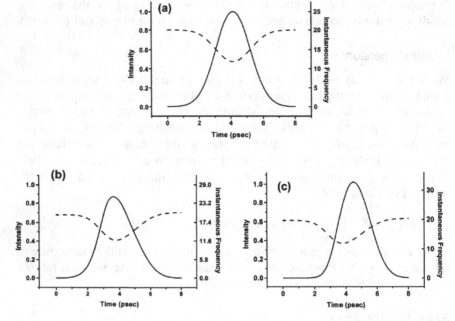

Figure 4 Simulation of the intensity pulse profiles and the corresponding instantaneous frequency of pulses. (a) After propagating through a medium with an integrating nonlinear self-phase modulation. (b) Pulse in (a) after a dispersive medium with
quadratic dispersion. (c) Pulse from (b) including saturable absorption.

frequency chirp that evolves from short wavelengths to long wavelengths,
i.e., the part of the chirp that evolves from blue to red. The resultant pulse
still exhibits an asymmetrical pulse shape with a fast rise time, with a trailing
or long tail. The residual chirp that is left is primarily linear and can be
removed by using standard chirp, or dispersion, compensation techniques.

These simple concepts lead to the suggestion of a recipe that can be
employed to generate high-power, ultrashort optical pulses from semiconductor diode lasers:

1. Allow for broadband, linearly chirped pulses to oscillate within
   the modelocked diode oscillator. This avoids excess gain saturation due to carrier heating effects and also allows for higher output power from the laser oscillator.
2. Obtain high output power by external amplification of the generated linearly chirped pulses. This avoids spectral distortion due to
   extreme gain saturation and its associated self-phase modulation.
   It should also be noted that this approach is similar to a technique

called "chirped pulse amplification" that is employed in high-power solid-state lasers to avoid detrimental nonlinear optical effects that can distort the optical pulse.

3. Compensate for the linear chirp by using dispersion compensation techniques.

By properly incorporating this approach, ultrashort optical pulses of $\sim 200\,f$ can be achieved, with high peak and average power. It should be noted that the roles of external amplification and dispersion compensation should not be reversed, because amplifying a pulse with a duration of less than 1 ps extracts less energy than longer pulses owing to carrier heating effects.

Given these initial basic concepts as a fundamental framework, the remaining portions of this chapter are used to (1) experimentally measure and verify the magnitude of the underlying physical effects, (2) highlight how these concepts can be exploited to generate high-peak-power, ultrashort optical pulses, and (3) demonstrate how one might be able to realize unique operating conditions of a modelocked semiconductor diode laser. Specifically, this chapter delves into the production of multiple-wavelength emission from modelocked semiconductors. In this case, we show that semiconductor lasers can emit multiple modelocked wavelength channels such that the emission is identical to having multiple synchronized, independent modelocked diode lasers, each operating at a different wavelength. Finally, to conclude, we highlight some potential applications using modelocked semiconductor diode lasers.

5.2   MEASUREMENTS OF FUNDAMENTAL EFFECTS

5.2.1   Spectrally Resolved Pump-Probe FROG for Pulse Propagation Experiments (Verification of Linear and Nonlinear Gain Dynamics)

In this section, combined temporally and spectrally resolved pump-probe techniques are used to directly observe the gain and nonlinear behavior of semiconductor optical amplifiers (SOAs). The experimental approach relies on measuring the spectrum of a weak probe pulse in a conventional two-beam time-resolved pump-probe experimental arrangement [6]. By employing this technique in an electrically pumped semiconductor optical amplifier, one can simultaneously measure the transient gain of the device and determine how this transient gain can influence the spectrum of an optical pulse as it propagates through the semiconductor optical amplifier.

The experimental arrangement is schematically represented in Figure 5. In this technique, a 300 MHz hybrid modelocked external cavity

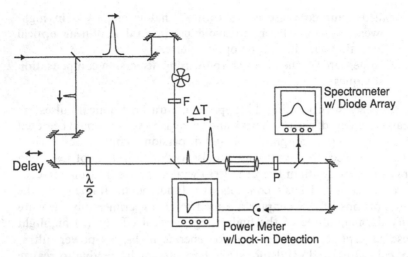

Figure 5 Schematic diagram of a two-beam spectrally resolved pump-probe measurement system.

semiconductor diode laser system is used as the source for generating the pump and probe pulses. The laser system comprises a low-power ($\sim$1 mW) hybrid modelocked diode laser oscillator, generating optical pulses of $\sim$5–10 ps in duration. The optical pulses are passed through a semiconductor optical amplifier to increase the average output power to $\sim$30 mW. The optical pulses are linearly chirped owing to the gain dynamics of the modelocking process and are passed through a dual grating dispersion compensator to reduce the optical pulse duration to $\sim$0.5 ps. The resulting output pulses at 830 nm can be spectrally filtered or nonoptimally compressed to allow for variable pulse duration, if desired. It should be noted that in these experiments the modelocked diode laser is used as a convenience, and similar experiments could be performed with other conventional modelocked laser systems. It should also be noted that the laser used for these experiments is the prototypical modelocked diode laser of Figure 1.

The device under test is an angled stripe semiconductor optical amplifier and is identical to the gain device employed in the external cavity modelocked diode laser system [7]. The pump-probe setup is realized by employing a modified Michelson interferometer. The pump and probe beams are created by directing the output through a pellicle beam splitter. The probe is directed to an adjustable delay half-wave plate and recombined with the pump with a polarization beam splitter. Both beams are then injected into the SOA. It should be noted that the pump polarization is aligned for maximum gain. The probe's beam is selected with a polarizer and detected with a spectrometer and linear diode array readout system. By varying the relative delay between the pump and probe beams, the spectrum of the probe can be directly measured. This provides information on how the gain dynamics induced by the strong pump can affect the instantaneous frequency of an ultrashort optical pulse as it propagates through an SOA under large signal input power conditions.

In Figure 6a, the temporally resolved probe spectrum is shown, with an input average pump power of 5 mW and pulse width of 2 ps. It should be noted that the input powers of the pump are sufficient to induce non-linear optical effects owing to transient changes in the gain and index. It should also be noted that the input pulse was broadened to 2 ps by spectrally filtering the modelocked output. In this case, the dynamic that is being investigated is primarily due to the slow gain depletion. The salient feature of the experimentally measured time-resolved probe spectrum is that the probe spectrum is unaffected at early delays, i.e., the small signal gain experienced by the probe serves only to provide a small gain. As the pump and probe overlap, the transmitted spectral intensity and spectral center wavelength are reduced and shifted to lower frequency, respectively. The reduction of the transmitted probe spectral intensity (~40%) is simply due to the reduction of optical gain, i.e., cross-gain saturation. The spectral shift, however, is due to a cross-phase modulation caused by the time-varying index induced by the passage of the pump. In this case it must be pointed out that the phase modulation caused by the pump is mediated through the time-varying index caused by the large change in

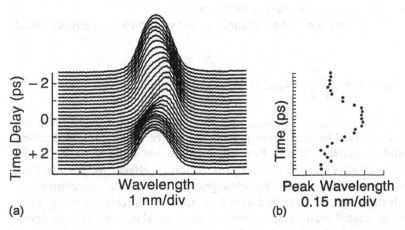

Figure 6 (a) Spectrally resolved probe spectrum plotted versus delay time. (b) Peak wavelength shift of the probe spectrum plotted versus delay.

carrier concentration caused in term by the change in gain. In the present case of an input pulse duration of 2 ps, the temporal variation of the gain depletion can be approximated by the integral of the pulse intensity. It should again be noted that the gain dynamics have a strong dependence on the input pulse duration and as a result can yield substantially different nonlinear behavior, depending on the input pulse parameters.

In the present case, because the primary nonlinearity is the index change caused by the change in carrier concentration, the time-varying index then follows the integral of the pulse intensity. This integrating nonlinearity should be directly contrasted with the more common Kerr-type nonlinearity in which the index follows the pulse intensity directly and plays one of the key roles in Kerr lens modelocked solid-state lasers.

To see the effect of the temporally resolved spectral shift of the probe, Figure 6b plots the peak wavelength shift versus time delay. In this plot it is clearly shown that the peak wavelength initially shifts toward longer wavelengths with a subsequent recovery to the initial central wavelength. This effect is easily understood by considering the induced nonlinearity caused by the change in gain. As the gain is depleted, the index is increased, which results from the reduction of carriers participating in gain. Because the gain depletion follows the integral of the pulse intensity, the time-varying index experiences an identical temporal variation, although opposite in sign. Considering the definition of the instantaneous frequency of an optical pulse propagating through a medium with a time-varying index, e.g., Eq. (4), it becomes clear that if the time-varying index $n(t)$ is an integrating nonlinearity, then the instantaneous frequency will directly follow the intensity. An examination of the peak wavelength shift versus delay displays this feature and can be seen in the simulation of Figure 3.

In these experiments, the measured probe spectrum can be represented by

$$I_{\text{probe}}(\omega, \tau) = \left| \int_{-\infty}^{\infty} dt\, E_{\text{probe}}(t) h(t - \tau) \exp\left(-j\omega t\right) \right|^2 \tag{5}$$

In this expression, I_{probe} and E_{probe} are the spectral intensity and temporal field, respectively. The medium response, $h(t)$, represents the time-dependent linear and nonlinear effects. For the experiments considered here, the response function is due to transient gain saturation effects and the resulting time-varying index induced by changes in the carrier concentration. Although these specific data focus only on transient effects for pulse durations greater than 1 ps, a complete phenomenological model has been developed that agrees with experiments for pulse durations ranging from tens of picoseconds to ~100 fs [4].

It should be noted that the time-marginal of the measured time frequency distribution gives the time-varying small signal gain of the probe as a function of time delay τ, i.e.,

$$\int_{-\infty}^{+\infty} T(\omega, \tau)\, d\omega = G(\tau) \tag{6}$$

because the probe spectrum is experimentally obtained from the time-resolved gain measurement. However, it should be noted that the frequency-marginal does not yield the integrated spectrum of a single probe pulse through the semiconductor optical amplifier, because the phase information is lost for different wavelengths that occur at different times. Nonetheless, it should be observed that the peak wavelength shift versus time delay directly measures the instantaneous frequency, or chirp, that is impressed upon a pulse as it propagates through semiconductor gain media.

From these measurements, the maximum frequency deviation obtained can be estimated from the maximum gain depletion. The single-pass small signal gain $\exp(gL)$ of these devices is ~ 100, including coupling losses. The maximum phase change can be obtained from the maximum change in gain divided by 4, where a factor of 2 is included for the electric field gain and a second factor of 2 is included owing to the relative strengths of the real and imaginary parts of the susceptibility, χ_r and χ_i, according to the Kramers–Kronig relation, i.e., $\chi_{r_{max}} = \chi_{i_{max}}/2$ [8]. Thus, the maximum frequency deviation Δf can be given

$$\Delta f = \frac{1}{2\pi}\left(\frac{gL}{4\tau_p}\right) \tag{7}$$

with τ_p as the pulse width. For pulses 1 ps in duration, this gives a maximum frequency deviation of 180 GHz, corresponding to 0.4 nm, which is in good agreement with the observed frequency shifts and other experimental observations.

Summarizing, these experiments clearly show that large frequency modulation can be impressed on a short optical pulse as it propagates through a semiconductor gain medium. The primary mechanism is self-phase modulation; however, owing to the integrating nature of the nonlinearity, an instantaneous frequency that resembles the optical pulse is produced. This parabolic frequency sweep, or cubic temporal phase, plays an important role in the pulse production dynamics in hybrid modelocked diode lasers.

5.2.2 Intracavity Gain Dynamics Experiments (Demonstration of Linear and Nonlinear Gain Dynamics on Intracavity Modelocked Pulse Propagation)

In this section, we spectrally resolve the second harmonic intensity auto-correlation function (also referred to as SHG-FROG for second harmonic generation—frequency-resolved optical gate) to measure the intracavity optical pulses in a hybrid modelocked laser [9]. The resulting data provide insight into the role of the intracavity nonlinear gain dynamics in the pulse-shaping and chirping dynamics of the modelocking process in a hybrid modelocked diode laser.

The information provided from the spectrally resolved pump-probe measurements discussed in the preceding section will be used as a guide to understand the dynamics of the modelocked diode laser as it undergoes modelocked operation. The results of these experiments show that the integrating nonlinearity associated with gain depletion, coupled with group-velocity dispersion, leads to asymmetrical intensity pulse profiles with predominantly cubic temporal phase, whereas saturable absorption coupled with group-velocity dispersion tends to linearize the chirp.

The preferred technique to be used for these experiments is to measure the spectrally resolved intensity cross-correlation, or short-time intensity spectrogram, where the gating or pump pulse has a pulse duration less than that of the probe pulse to be measured, i.e.,

$$I_{\text{SHG}}(\omega - 2\omega_0, \tau) = \left| \int_{-\infty}^{\infty} dt\, E_{\text{probe}}(t) E_{\text{gate}}(t - \tau) \exp\left(-j\,\omega t\right) \right|^2 \qquad (8)$$

In order to assess the effect of the nonlinear dynamics on the chirping properties of the generated optical pulses, the instantaneous frequency of the optical pulses can be directly measured by spectrally resolving the cross-correlated intensity profiles at the second harmonic frequency. Those were measured at three locations of the cavity, i.e., the laser output, before the saturable absorber, and after the saturable absorber. Experimentally, this is achieved by using a pellicle within the cavity to sample the optical pulse train at the appropriate locations and cross-correlate this signal with the dispersion-compensated output pulses. It should be noted that in these measurements, the intracavity pulses are on the order of 5–10 ps, while the dispersion-compensated optical pulses are $\sim$0.5 ps.

To identify the phase modulation impressed on the intracavity optical pulses, the intracavity gain dynamics need to be measured first, because the impressed chirp is directly related to the gain dynamics. The intracavity gain

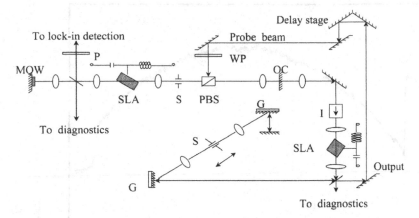

Figure 7 Experimental setup for intracavity gain dynamic measurement. MQW, multiple quantum well saturable absorber; PBS, polarizing beam splitter; SLA, semiconductor laser amplifier, or semiconductor optical amplifier; OC, output coupler; G, diffraction grating; S, slit; I, optical isolator; P, polarizer; and WP, half-wave plate.

and absorption measurements are performed by using the external cavity hybrid modelocked semiconductor as both the laser generating the probing pulse and the modelocked laser under test. The experimental configuration is schematically depicted in Figure 7. The output pulses from the laser system are injected into the laser oscillator using an intracavity polarizing beam splitter. To monitor the gain and absorption dynamics, the transmitted probe beam is partially deflected from the cavity by using a pellicle beam splitter and detected by standard lock-in techniques.

The time-resolved intracavity probe gain measurements are shown in Figure 8. The salient features are the two transient reductions of gain superimposed on a sinusoidal varying gain. The transient gain reductions are due to the intracavity pulse passing through the SOA. The important observation is that the SOA exhibits conventional gain dynamics, which lead to an integrating nonlinearity. Other fast dynamics, such as carrier heating and cooling effects, are not observed. If these effects were observable in the gain dynamics they would substantially modify the chirping effects discussed earlier. In addition, it should be noted that the gain recovered to its initial value within 350 ps, in contrast to the nominal gain recovery time of 1 ns measured in conventional pump-probe measurements. This fast recovery is due to the time-varying pumping rate associated with the RF bias current. This reduced recovery time generates a transient regime of unsaturated gain and is responsible for the production of broad optical spectra from

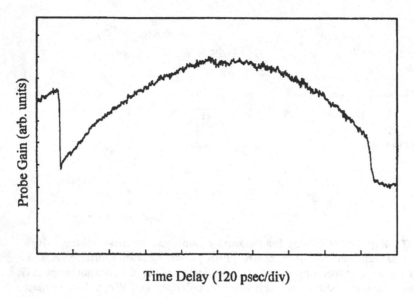

Time Delay (120 psec/div)

Figure 8   Time-resolved intracavity gain dynamic measurement, showing transient gain depletion due to pulse passage through the SOA and long-term sinusoidal gain recovery due to the RF modulation frequency.

modelocked diode lasers. It should also be noted that the pulses do not pass through the SOA device at the point of maximum gain. This is due to the location of the SOA device with respect to the rear reflector. Because the SOA device is not located at the rear reflector, its displacement forces the optical pulses to travel through the SOA at times when the optical gain is approximately the same for each pulse. This manifests itself as the optical pulses passing through the SOA at times symmetrically displaced from the point of maximum gain. In our case, this corresponds to the round-trip time between the SOA and the rear reflector, ~1 ns.

The time-resolved intracavity saturable absorption was also measured. The time-resolved reflectance of the MQW mirror exhibits a relative increase of 20% with a rise time of 10 ps, corresponding to the integrated pulse intensity. The absorber exhibits only a slow recovery, with an exponential absorption recovery time constant of 280 ps. It should be noted that the recovery time measured in these experiments differs from previously measured absorber recovery times [10,11]. These differences may be attributed to differences in the MQW absorber design and differences between the optical transverse mode profiles of the lasers used in this and prior measurements.

To demonstrate the pulse-shaping effects induced by the saturable absorber, input and output pulse intensity profiles were measured. This was achieved by cross-correlating the intracavity pulses with the compressed output optical pulse. Cross-correlation information is then obtained by examining the side lobes of the three-peak correlation signal. Figure 9a shows the intensity pulse profile of the optical pulse before the saturable absorber, and Figure 9b the intensity pulse profile of the optical pulse after the saturable absorber. The salient feature is the reduction of the rising edge of the optical pulse from 6 ps to 3 ps, representing a change of nominally 50% per round-trip. It should be noted that the shoulders in the correlation traces are artifacts created by the two-pulse correlation technique employed.

In Figure 10 are the resulting spectrograms of the output pulse (Fig. 10a) and the intracavity optical pulse before and after the saturable absorber (Fig. 10b and 10c, respectively). It should be noted that the

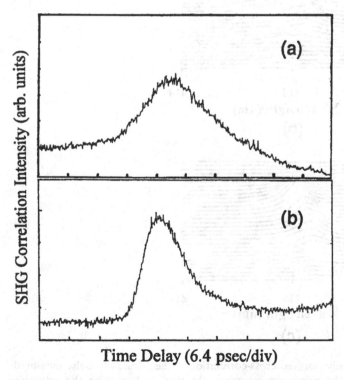

Figure 9 Intracavity temporal pulse profiles (a) before and (b) after the intracavity saturable absorber.

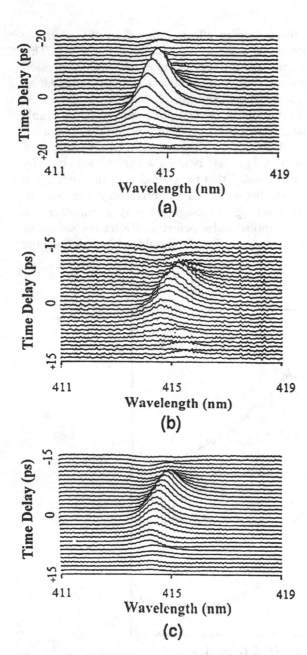

Figure 10 Spectrally resolved cross-correlation of the intracavity pulse measured at (a) the output, (b) before the saturable absorber, and (c) after the saturable absorber.

time-marginal of these SHG spectrograms yields the intensity cross-correlation, i.e.,

$$\int I_{SHG}(\omega - 2\omega_0, \tau) \, d\omega = \int d\omega \left| \int_{-\infty}^{\infty} dt \, E_{probe}(t) \, E_{gate}(t-\tau) \exp(-j\omega t) \right|^2$$

$$= I_{probe} \otimes I_{gate} \tag{9}$$

In each case, the spectrograms show optical pulses with fast rising edges and slower trailing edges. In addition, the spectrograms show that the instantaneous frequencies of the optical pulses are not constant, that is, the center or carrier wavelength varies throughout the pulse duration. In these traces, the center wavelength is upchirped and tends to vary linearly over a major portion of the optical pulse. Quantitatively, the spectrograms show a total wavelength variation of 0.8 nm at the second harmonic wavelength, implying a total wavelength chirp of 1.6 nm at the fundamental wavelength. It should be noted that the chirp exists over the duration of the pulse, implying a nonlinear dispersion of $\sim$5 ps/nm.

The measured linear chirp of the intracavity pulses seems to contradict the observed chirp measured in the single-pass experiments in Figure 6. However, the intracavity pulse shapes and the corresponding spectrograms can be easily explained once the intracavity nonlinear dynamics are considered. It was shown earlier that a pulse propagating through a semiconductor optical amplifier will have a time-dependent frequency impressed upon it, and the instantaneous frequency will exactly follow the optical pulse shape. This occurs through self-phase modulation, where the nonlinearity is an integrating nonlinearity. Under this condition, the instantaneous frequency varies in direct proportion to the optical intensity. The instantaneous frequency, defined as $\omega_{inst} = \omega - \partial\phi/\partial t$, implies that the predominantly parabolic frequency sweep corresponds to the cubic temporal phase. It should be noted that an instantaneous Kerr nonlinearity leads to a primarily linear frequency sweep or quadratic temporal phase. The immediate consequence of the integrating nonlinearity is that GVD will slow down the blue frequency components at the front and rear of the pulse, whereas the center portion of the pulse (red) is allowed to propagate toward the front of the pulse. This effect yields the pulse-steepening effect observed in the spectrograms. It should be noted that this is in direct contrast to conventional modelocked lasers with instantaneous nonlinearities; i.e., GVD symmetrically broadens the optical pulse. When the optical pulse impinges upon the saturable absorber, the front of the optical pulse is removed. In addition to the reduction of the rise time of the optical pulse, the chirp on the rising edge is also removed, resulting in an optical pulse with a predominantly linear chirp impressed over the main portion of the optical pulse.

These effects can be simulated by computing the optical intensity profiles and the instantaneous frequencies, using the nonlinear dynamics described above. It should be noted that the simulation employs a simple model, using normalized nonlinear parameters to demonstrate the salient features of the pulse-shaping process.

For this model, the integrating nonlinearity operates on the temporal phase of the pulse intensity; i.e., the output complex electric field is related to the input field by

$$E_{out}(t) = E_{in}(t) \exp\left[iC \int_{-\infty}^{t} I_{in}(t)\, dt\right] \tag{10}$$

where C is a constant that is related to the change in refractive index.

Linear pulse propagation effects are included as a lumped element that operates linearly on the spectral phase of the optical pulse, i.e.,

$$E_{out}(\omega) = E_{in}(\omega) \exp\left[i\frac{n(\omega)z}{c}\right] \tag{11}$$

In the present case, only the quadratic spectral phase, i.e., GVD, is considered.

Finally, saturable absorption is modeled as a slow saturable absorber, which is appropriate for the multiple quantum well absorber used in the modelocked laser. Equation (1), which describes the output temporal pulse intensity, can also be recast according to

$$I_{out}(t) = I_{in}(t) \frac{G_0 \exp[u(t)]}{1 + G_0\{\exp[u(t)] - 1\}} \tag{12}$$

where $u(t)$ is the integrated pulse energy normalized to the saturation value of the medium [1,3]; i.e.,

$$u(t) = \frac{2\sigma}{h\nu} \int_{-\infty}^{t} I(t)\, dt \tag{13}$$

G_0 is the small signal linear transmittance, and I_{out} and I_{in} are the output and input intensity pulse profiles to the multiple quantum well saturable absorber. The simulation shows the results of an initially transform-limited optical pulse undergoing phase modulation due to gain depletion, followed by linear dispersion, and finally experiencing saturable absorption.

In Figure 11a is the optical pulse shape and its corresponding
instantaneous frequency after a Gaussian transform limited pulse has
undergone phase modulation and encountered a finite amount of dispersion. Note the asymmetrical temporal broadening and the modification of
the instantaneous frequency. In Figure 11b are the results after the saturable absorber. Note the faster rising edge of the pulse and also note that
the chirp is now predominantly linear over the main portion of the optical
pulse. This simple model of SPM, GVD, and saturable absorption supports the experimental results and provides a clearer understanding of
the pulse-shaping and chirping dynamics of hybrid modelocked diode
lasers. These results suggest that by increasing the intracavity gain, power,
and GVD, larger chirping can be obtained, yielding shorter pulses after
dispersion compensation.

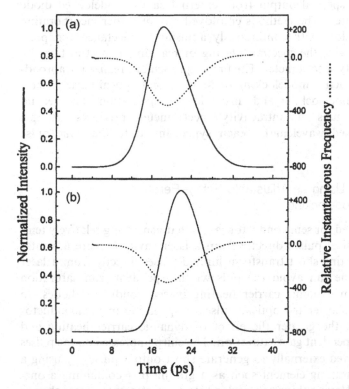

Figure 11 Simulation of the intensity pulse profiles and the corresponding instantaneous frequency of pulses (a) after propagating through a medium with an integrating nonlinear self-phase modulation and a dispersive medium with quadratic
dispersion and (b) after the intracavity saturable absorber.

5.3 PRODUCTION OF ULTRASHORT, HIGH POWER OPTICAL PULSES

5.3.1 Intracavity Spectral Shaping in External Cavity Modelocked Semiconductor Diode Lasers

In this section, we experimentally investigate the potential for artificially tailoring the output modelocked spectrum of an external cavity semiconductor diode laser. The main reason for artificially broadening the output spectrum of the modelocked laser diode is to avoid gain-narrowing effects in the gain medium, which will lead to the generation of temporally broad pulses. It was noted in the foregoing that the gain dynamics of semiconductor lasers are highly nonlinear and prevent the direct generation of ultrashort optical pulses; thus it is reasonable to conclude that alternative methods for achieving an ultrabroad spectral output need to be investigated. Two methods for generating broad spectral output from external cavity modelocked diode lasers are highlighted. The methods employed rely on either incorporating an intracavity etalon, with continuously adjustable reflectance and plate separation, to provide the spectral filtering or realizing a spectral filter by using an intracavity spectrometer. The first approach can realize ultrabroadband emission by active modelocking or femtosecond optical pulse generation by hybrid modelocking and dispersion compensation. The second technique, which uses the intracavity spectrometer, produces multiple tunable modelocked wavelengths, each synchronized to the same pulse repetition rate.

Spectral Filtering Using an Adjustable Fabry–Perot Etalon—Active Modelocking

It was demonstrated for semiconductor lasers that generating relatively temporally broad optical pulses directly from a laser cavity is more advantageous than generating short transform-limited pulses directly from a laser cavity, because one can avoid the pulsewidth-dependent gain saturation effect arising from dynamic carrier heating in semiconductor diodes. In other words, the shorter the optical pulses propagating in semiconductor optical amplifiers, the greater the effect of dynamic carrier heating and the pulsewidth-dependent gain saturation. The chirp impressed on the pulses can then be removed externally to generate short output pulses by using a dispersion-compensating element such as a grating pair containing a one-to-one telescope. Having this as a standard strategy, efforts to create shorter pulses were pursued in two directions. The first step is to generate a broader optical spectrum from a laser cavity, and the second step is to develop means to compensate the nonuniform spectral phase externally to generate the transform-limited short pulses. In the second step, the characterization of the spectral phases of the optical pulses is an important prerequisite for optimum dispersion compensation.

In this section, an intracavity spectral shaping technique is introduced to increase the spectral width of intracavity pulses. Both temporal information via intensity (TIVI) and Gerchberg–Saxton (GS) algorithms are employed to characterize the spectral phase of the output pulses [12]. Then the spectral phase compensation is performed by using a conventional pulse-shaping setup with a liquid crystal spatial light modulator (SLM). It was found that the intracavity spectral shaping technique allows not only the control of spectral intensity but also the control of chirp, to some degree, owing to the nonlinearity of the semiconductor gain media.

Intracavity Spectral Shaping

Because the temporal pulse width of intracavity pulses (10 ps) is much longer than the intraband transition time (~100 fs) of semiconductors, semiconductor optical amplifiers (SOAs) can be regarded simply as homogeneously broadened, two-level gain media. For homogeneously broadened gain media, the gain-narrowing effect due to mode competition is the major obstacle to obtaining a broad optical spectrum. Typically, SOAs have a gain bandwidth of $\sim 10^{13}$ Hz. Modelocked semiconductor laser spectra, however, generally have bandwidths less than $\sim 2.5 \times 10^{12}$ Hz. One way to avoid the gain-narrowing effect is to use an intracavity spectral shaping technique, where artificial loss is introduced in the laser cavity [13]. By controlling the loss profile in the spectral domain, it is possible to control the laser output spectrum to produce an arbitrary spectral shape. In our case, in order to generate a broad optical spectrum, an intracavity spectral shaping element that mimics the inverse spectral gain profile is necessary. In other words, the loss is high where the gain is high and low where it is low, so net gain has a flat spectrum to suppress the gain-narrowing effect.

This gain-flattening filter was realized by employing a Fabry–Perot etalon. The main idea behind the intracavity etalon is immediately understood when one considers the standard formalism for a Fabry–Perot etalon. The salient feature of this approach is that by adjusting the incident angle one adjusts the etalon reflectance from 0 (at Brewster's angle) to nearly 100% (at grazing incidence) and hence controls the finesse. By adjusting the plate separation using a piezoelectric actuator, one controls the free spectral range. This provides complete control of the etalon's transmission function. In our case, the idea is to operate the etalon such that two transmission peaks lie in the wings of the gain spectrum, while the minimum transmission lies at the peak of the gain, as shown in Figure 12a.

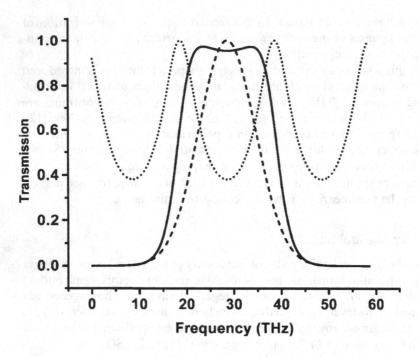

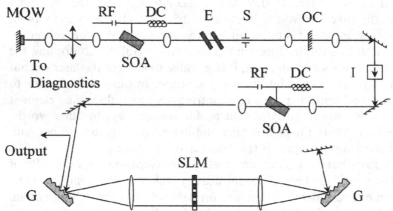

Figure 12 (a) Characteristic of the Fabry–Perot etalon transmission when optimized to flatten the gain spectrum of an optical amplifier. Also shown are the gain profile and the resulting flattened gain spectrum. (b) Schematic diagram of external cavity modelocked semiconductor laser system with an intracavity spectral shaping element. MQW, multiple quantum well saturable absorber; SOA, semiconductor optical amplifier; S, slit; OC, output coupler; I, optical isolator; G, diffraction grating; E, etalon; SLM, spatial light modulator.

A schematic illustration employing intracavity spectral filtering by using a controllable etalon is shown in Figure 12b and is based on the canonical geometry illustrated in Figure 1. The external cavity modelocked semiconductor diode laser is composed of a single angled-stripe AlGaAs/ GaAs/AlGaAs double-heterostructure semiconductor optical amplifier (SOA) [7], which is used as the gain element. The SOA employs a gain-guided double heterostructure. The current channel for gain guiding was realized by the diffusion of zinc through a GaAs cap layer, which plays the role of a current-blocking layer. To suppress Fabry–Perot modes due to imperfect antireflection coating on the facets, the angle of the gain stripe is inclined from the normal to the facets by $\sim 5°$. The key feature of these devices is the thin active region, which is 80 nm. The thin active region and the large optical mode cross section associated with gain-guided devices are the attributes that allow the production of optical pulses of 100 pJ. The typical device length is 350 μm when it is employed as a modelocked oscillator gain element and 500 μm when the device is used as a single-pass power amplifier. The peak emission of the devices is located at ~ 830 nm. The external cavity is composed of a 70% reflective output coupler and a rear reflector that is either a 100% reflector (for active modelocking) or a multiple quantum well (MQW) saturable absorber mirror (for passive and hybrid modelocking). The saturable absorber design uses GaAs wells of 6, 6.5, 7, 7.5, 8, 8.5, and 9 nm separated by 10 nm $Al_{0.3}Ga_{0.7}As$ barriers. This structure is repeated 30 times to provide a sufficient amount of nonlinear absorption. The variation of the well width creates an artificial inhomogeneously broadened absorption region, thus creating the possibility for the support of a wider modelocked spectrum or tunability. The emitted light is collected and collimated on both sides of the SOA using two large numerical aperture lenses. An intracavity adjustable slit is also employed to control the transverse mode profile to force the oscillation of a single transverse mode. The light is focused on both the rear reflector and the output coupler to increase the cavity stability, and re-collected and collimated after the output coupler. An intracavity bandpass filter is employed with the intracavity etalon in an active modelocked regime to assist in the spectral shaping. For intracavity spectral diagnostics, a pellicle is used to deflect a small portion of light to a spectrometer with a linear diode array for real-time readout purposes. Modelocking is achieved by applying a dc bias current to bring the laser close to the continuous wave (CW) threshold combined with an RF signal through a bias tee, where the radio frequency is chosen to match the fundamental mode spacing of the external cavity, which is ~ 300 MHz. The intracavity etalon comprises two thin-film polarizing beam splitters mounted on a rotatable stage to provide continuous tuning of the incidence angle, with the etalon spacing controlled by piezoelectric transducers.

The angle and the gap of the etalon are optimized to obtain the broadest optical spectrum and the most stable operation while the laser is hybridly modelocked. As a result, it was found that the optimum angle, where the reflectivity is ~1.5%, is near Brewster's angle.

A series of output spectra for active modelocking is shown in Figure 13. It should be noted that from the top plot to the bottom, the separation of the peaks in the spectral modulation increases as the gap of the etalon is reduced. This clearly shows the extent of control this spectral shaping technique can provide. The broadest spectrum created with this technique is shown in Figure 14. The resultant spectral width of 16 nm is substantially broadened compared to the typical spectral width of ~3 nm under actively moelocked operation in an identical cavity without the spectral flattening element.

Generally speaking, in conventional actively modelocked diode lasers, the increase of spectral width can lead to the generation of optical pulses with uncorrelated phase, resulting in non-transform-limited optical pulses. The large amount of complex structured spectral phase is difficult to compensate, thus preventing the generation of short pulses. This type of complex

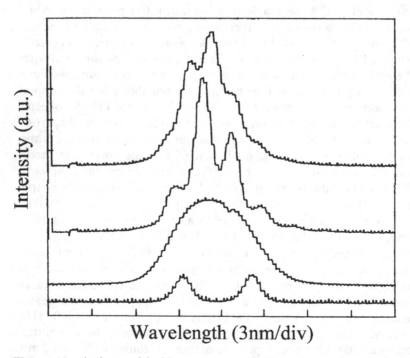

Figure 13 Active modelocking output spectra for various etalon spacings.

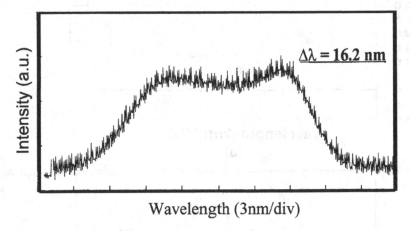

Figure 14 Active modelocking output spectrum (FWHM 16 nm).

phase is due to the fact that the mechanism of side band generation in active modelocking is not sufficiently strong to lock the phases of all the longitudinal modes contained within the modelocked spectrum. In the case at hand, the optical pulse duration as measured by autocorrelation techniques was $\sim 50\,\mathrm{ps}$, much longer than what would be predicted from a time–bandwidth relation. Nonetheless, the important feature demonstrated is the potential of employing an intracavity spectral filter to broaden the output spectra of actively modelocked diode lasers. It should be noted that although the second-order temporal conherence is not optimized for nonlinear optical effects and associated applications, the first-order temporal coherence is maintained. As a result, low-coherence applications, such as optical coherence tomography, that also require good focusability and potentially higher power may benefit from this type of modelocked diode source [14].

Hybrid Modelocking and Spectral Phase Characterization and Compensation. Pulses with organized phase hybrid modelocking are achieved by inserting the saturable absorber into the laser cavity. The resulting optical spectrum and autocorrelation are shown in Figure 15. The spectral width was decreased to less than 8 nm. It is assumed that the reduction of spectral width is due to the nonuniform absorption spectrum of the saturable absorber. Even though the saturable absorber has a broad absorption bandwidth ($\sim 30\,\mathrm{nm}$) due to the band-gap-engineered, inhomogeneously broadened multiple quantum well structure, the absorption spectrum was not perfectly uniform over the bandwidth. owing to the extra complexity in the loss spectrum caused by the saturable absorber, achieving a flat net gain and loss is

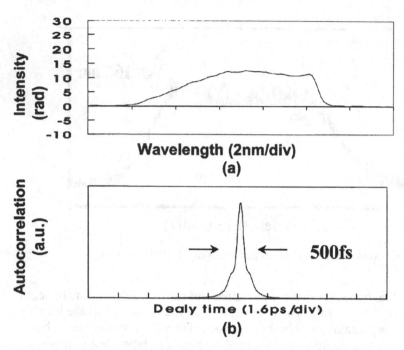

Figure 15 Output characteristics of the hybrid modelocked operation. (a) Output spectrum. (b) Intensity autocorrelation after dispersion compensation.

more difficult for the hybrid modelocked case. Figure 15b is the intensity autocorrelation trace of output pulses after pulse compression by a dual grating dispersion compensator for which the insertion loss is $\sim 60\%$. The width of the autocorrelation shows that the pulses are ~ 1.6 times the transform limit (considering the output spectrum). It should also be noted that the wings of the autocorrelation trace indicate that there is some residual chirp. Even though the dual grating dispersion compensator eliminates the quadratic spectral phase (linear chirp), the higher order phase components remain uncorrected.

It is critical to accurately characterize the phase of pulses for optimum dispersion compensation [15,16]. Phase measurement of semiconductor lasers can be more difficult than for other lasers, owing to their complexity of phase and low output pulse energy. We now show how the spectral phase can be measured using a two-step process: (1) temporal information via intensity (TIVI) and (2) the Gerchberg–Saxton (G-S) algorithm, where the intensity spectrum and intensity autocorrelation are the sources of information. In the TIVI process, a temporal intensity profile of the pulse is obtained form an intensity autocorrelation trace [12]. This resultant temporal intensity profile is used with the measured intensity spectrum to retrieve a spectral or temporal phase by the G-S algorithm. The resultant temporal intensity has a time reversal ambiguity, but it can be easily resolved by comparison with other measurement results. The retrieved temporal intensity and phase of the intensity autocorrelation and spectrum of Figure 15 are shown in Figure 16 It can be seen that the pulses have a predominantly flat spectral phase over the central portion of the spectrum but a more complex phase structure toward the wings of the spectrum. The quadratic spectral phase (linear chirp) is compensated using the dispersion compensator, yielding the flat spectral phase, but the remaining higher order phase components require more sophisticated techniques for complete removal [17–19]. One approach is to place a liquid crystal spatial light modulator (SLM) in the Fourier transform plane of a conventional pulse-shaping setup [20]. The SLM has 128 pixels with a size of 0.1 mm × 2 mm and is also antireflection-coated. A beam width of 5 mm, an incident angle to the gratings of

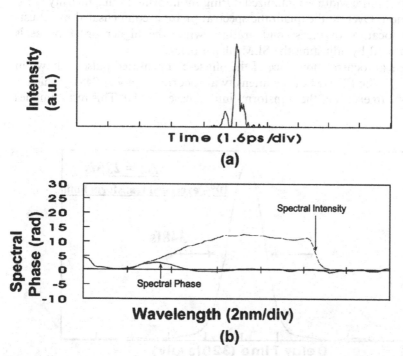

Figure 16 Result of TIVI and G-S algorithm. (a) Retrieved temporal intensity. (b) Retrieved spectral phase (dashed line, spectral intensity; solid line, spectral phase).

44.1°, lenses with a focal length of 300 mm, and gratings with a groove density of 1800 mm^{-1} were used to obtain ~109 μm of spatial resolution at the SLM plane. The appropriate phase distribution to further compress the pulse is superimposed on the spectral phase of the optical pulse by the SLM and is controlled by a computer. The functional form of the phase mask used is

$$\phi(i) = \phi_0 + \phi_2(i - i_0)^2 + \phi_3(i - i_0)^3 + \cdots + \phi_7(i - i_0)^7$$
$$+ \text{ cubic spline fit of } \left[\sum_{k=0}^{16} C_k \delta_{i,(8k)} \right] \tag{14}$$

where i indicates each pixel of the SLM and δ_{ij} is the Kronecker delta symbol. $\phi(i)$ is composed of two parts. The first part is the Taylor series expansion of up to seventh order. The second part is a smoothed form of equally spaced Kronecker delta functions due to the use of cubic spline fitting. This part is designed to provide more degrees of freedom to control the shape of $\phi(i)$. The coefficient ϕ's, C's, and i_0 are manually adjusted so that the output pulse width is minimized during monitoring of the intensity autocorrelation. Overall, the quadratic spectral phase is compensated by adjusting the location of the second grating, while the higher order phase is compensated by adjusting the SLM phase mask.

The autocorrelation trace of the phase-compensated pulse is shown in Figure 17. The FWHM of the intensity autocorrelation was 348 fs, which is only 10% in excess of the transform-limited case (316 fs). This result implies

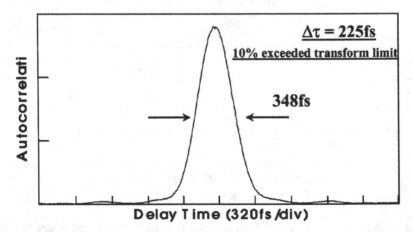

Figure 17   Intensity autocorrelation of the phase-compensated pulse using SLM.

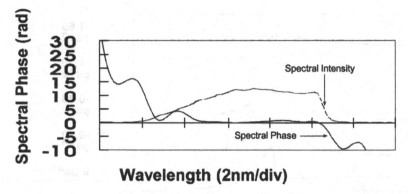

Figure 18 Spectral phase of SLM (dashed line, spectral intensity; solid line, spectral phase).

a temporal pulse duration of 250 fs. The SLM phase mask used is shown in Figure 18. Because this phase mask was used to generate nearly transform-limited pulses, it can be assumed that it represents the actual spectral phase of the optical pulses. The result of the TIVI and the G-S algorithm shown in Figure16 exhibits small structure in the short-wavelength region and flat phase in the center region. However, in the wings of the spectrum, this method does not show an accurate phase value. This is due to the fact that spectral intensity is low in the wings of the spectrum, which leads to a lower signal-to-noise ratio in this region and inaccuracies in the retrieved spectral phase.

Chirp Tailoring. The sign of the chirp for the generated pulses mentioned thus far is positive; i.e., the instantaneous frequency increases in time. However, negative chirp should also be considered because the sign of the nonlinearity can change, depending on the oscillating laser wavelength with respect to the peak gain wavelegth. Figure 19 shows a series of measured intensity autocorrelation traces taken from the output of the laser system shown in Figure 12b after passing through the dual grating dispersion compensator. By varying the grating separation, varying amounts of GVD are added to each pulse. The third plot from the bottom corresponds to zero GVD, the two plots below correspond to positive GVD, and the four plots above correspond to negative GVD. It should be noted that there are sharp peaks that appear to be similar to coherent spikes for both the positive and negative GVD cases. These peaks are not coherent spikes but rather result from the partial compression of the pulses. Similar to a parabolically chirped pulse, part of the pulse is positively chirped and the other part of the pulse is negatively chirped. When negative GVD is

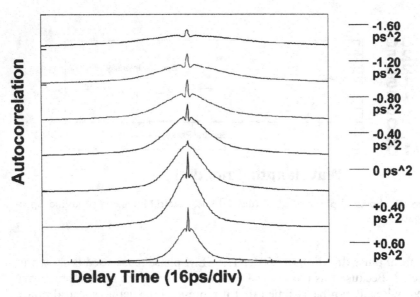

Figure 19 Autocorrelation traces of the predominantly negative chirped pulse for various dispersion compensator setting. The amounts of GVD imposed by the compressor are denoted to the right of the traces.

added to the pulse, the positively chirped part is compressed, creating a sharp peak, while the negatively chirped part is stretched, contributing a pedestal. For the positive GVD cases, the opposite happens. However, the whole envelope of the trace has a minimum width for the second plot from the bottom. This means that the major part of the pulse is negatively chirped but some fraction of the pulse is chirped with a different slope, or positively chirped.

      To explain this effect, the gain dynamics should be reconsidered. For positively chirped pulses, it was assumed that the gain change and the instantaneous index of refraction change due to the gain change are of opposite sign. Because the instantaneous frequency is proportional to the negative time derivative of the index, it was demonstrated that the instantaneous frequency changes from blue to red and then black to blue (temporal cubic phase). If the effect of the saturable absorber is added, pulses become positively chirped when the front edge of the pulses is removed. In order to explain the negative chirp of semiconductor lasers, a small correction is required from this theory. The linewidth enhancement factor plays an important role in this case. The linewidth enhancement factor α is a constant connecting the gain change to the index change in SOAs [8]:

$$\alpha = \frac{\partial \chi_r / \partial N}{\partial \chi_i / \partial N} \tag{15}$$

In this equation, χ_r and χ_i are the real and imaginary parts of the susceptibility and N is the inversion density. The sign of the linewidth enhancement factor depends on whether the laser operates on the low- or high-frequency portion of the gain spectrum. In other words, on the longer wavelength side of the peak of differential gain $\partial g / \partial N$, gain reduction causes an index increase, supporting the observation of positive chirp generation, whereas on the shorter wavelength side, the gain change and index change have the same sign. Thus, on the shorter wavelength side, the index and gain will have similar transient responses and the instantaneous frequency will change from red to blue and then black to red. If the effect of the saturable absorber is added, then pulses will be predominantly negatively chirped. In addition, contrary to the positively chirped case, positive GVD elements in the cavity will oppose the chirp linearization, resulting in a small portion of the pulse having a positive chirp. The location of the peak of differential gain can vary within the gain bandwidth depending on the device structure and material composition. In addition, the peak of differential gain may also be within the absorption region of the device [21,22]. Additional evidence is shown in Figure 20, where laser output spectra are plotted for both positive and negative chirp cases. To emphasize this effect, the laser generates well-defined spectral peaks whose origins are related to the intracavity gain dynamics. Under these conditions, the laser generally operated under slightly different operating conditions than in the short-pulse generation case (compare Figs. 15 and 20). It can be seen that for pulses that are positively chirped, the spectral emissions tend to be located on the long-wavelength side of the gain spectrum. It should also be noted that the two spectral shapes exhibit mirror symmetry. This can be easily understood considering that pulses in the two both cases undergo similar physical processes except for their opposite sign of the chirp. In the bottom traces of Figure 20 are spectra of the amplifier output for both cases. When the laser output pulse is injected into an amplifier SOA, the pulse will experience additional SPM associated with pulse amplification. For the positively chirped pulse, the wavelength will shift to the longer wavelength side owing to SPM, resulting in the enhancement of the right shoulder of the original spectrum. But for the negatively chirped case, there will be a blue shift and an increase in the left shoulder. Again the two spectra are close to being mirror images of each other.

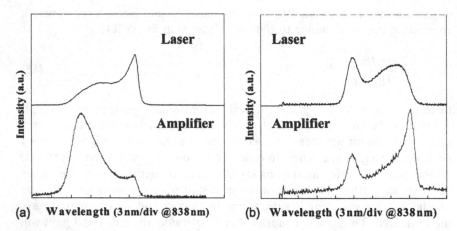

(a)   **Wavelength (3 nm/div @838 nm)**   (b)   **Wavelength (3 nm/div @838 nm)**

Figure 20 Comparison of negatively and positively chirped pulses. (a) Spectra of negatively chirped pulses. Top: Laser output. Bottom: Amplifier output. (b) Spectra of positively chirped pulses. Top: Laser output. Bottom: Amplifier output.

Depending on which side of the differential gain peak the laser spectrum is located on, the sign of the chirp can be positive or negative. Tuning of the laser wavelength was achieved by both the intracavity spectral shaping element and adjustment of the laser cavity alignment. This demonstrates that the intracavity spectral shaping technique can control not only the shape and location of the spectrum but also the chirp of the pulse, which adds an additional degree of freedom to the laser system. In addition, this suggests a fundamental limitation in obtaining a broad optical spectrum from semiconductor lasers. If a spectrum is broad enough to cover the entire gain region, then part of the spectrum will have positive chirp and the other part will have negative chirp. Because this type of chirp (temporal cubic phase) cannot be compensated by using linear optics, even if the spectrum is very broad, it is not recommended for short-pulse generation. Thus, in order to generate a broad spectrum for short pulses from semiconductor lasers, new types of SOAs are required. Recently, a variety of new SOA designs have been developed, such as a multiple quantum well SOA with varying well thickness and single quantum well SOAs that exploit band-filling effects to obtain a broad gain bandwidth [22]. The chirp-tailoring nature of semiconductor lasers also suggests that for small spectral bandwidth laser output pulses can have zero chirp if they are located at the peak of the gain spectrum [23].

5.3.2 High-Power Modelocked External Cavity Semiconductor Lasers Using Inverse Bow-Tie Semiconductor Optical Amplifiers

Thus far we have shown how to generate ultrashort optical pulses with high peak power. However, for some applications, in addition to high peak power, high average power may also be required. Recently, there has been progress in achieving high average power and high pulse energies from semiconductor lasers by employing novel optical amplification techniques and flared or tapered waveguides. The two main techniques employed have been modelocking with an external cavity [24–27] and direct Q-switching in bow-tie structures [28–30]. Whereas Q-switching of bow-tie lasers shows tremendous potential for simple, compact sources of high-power ultrashort optical pulses, modelocking typically generates optical pulses of shorter duration with less temporal jitter and enhanced spectral quality. In this section, we report on both continuous wave (cw) and modelocked operation of an external cavity semiconductor diode laser and amplifier system that employs angle-striped inverse bow-tie semiconductor optical amplifiers. The salient feature of employing inverse bow-tie structures, compared to flared or tapered structures, is that one can achieve the high output powers characteristic of tapered structures while maintaining single-lobed output beams [31] that are characteristic of narrow-stripe devices. The laser oscillator operating in the cw mode produces a maximum output optical power of 700 mW and is tunable over 17 nm. Under modelocked operation, the laser oscillator and amplifier system produces optical pulses 5 ps in duration with an average output power 400 mW at a 1.062 GHz repetition rate. In this case, the optical pulses are chirped, owing to the modelocking process, and the temporal duration was further reduced to 1.3 ps by employing a dual grating dispersion compensator with a 1:1 internal telescope. The output characteristics of this laser system prior to compression imply a pulse energy of 376 pJ with a peak power of 60 W. These characteristics are suitable for exploiting X^3 bulk optical nonlinearities, such as in novel three-dimensional imaging and data storage scenarios and in large-scale commercial printing and marking applications.

Device Structure

The device used as the gain element in the external cavity laser is an inverse bow-tie semiconductor optical amplifier and has been realized by employing a diamond-shaped metal contact as the active stripe of an angled-stripe semiconductor optical amplifier. In this geometry the device contains an adiabatic expander section for the extraction of high output power, followed by an adiabatic contracter to assist in removing astigmatic beam characteristics from conventional flared optical amplifier devices. Figure 21 illustrates the design of a 1 mm long diamond-shaped stripe at an angle of 6° with respect to the facets, linearly tapered from 10 μm width at the facet to about 60 μm width at the center [8]. The device was fabricated using an InGaAs-AlGaAs graded index separate-confinement heterostructure with compressively stained double quantum well layers (GRIN-SCH-MQW). The structure consists of the following layers: 1.2 μm thick graded doping $Al_{0.6}Ga_{0.4}As$ n-clad layer, 0.12 μm thick undoped linearly compositional grade $Al_{0.60-0.35}Ga_{0.40-0.65}As$ confining layer, ~6.07 nm thick undoped $In_{0.08}Ga_{0.92}As$ quantum well, 30 nm thick undopted $Al_{0.35}Ga_{0.65}As$ barrier, ~6.0 nm thick updoped $In_{0.08}Ga_{0.92}As$ quantum well, 0.12 μm thick undoped linearly compositional graded $Al_{0.35-0.60}Ga_{0.65-0.40}As$ confining layer, and a 0.30 μm thick GaAs p^+ contact layer. The quantum wells are under 0.57% compressive strain with a composition of ~$In_{0.06}Ga_{0.92}As$. The structure was grown by low-pressure metal-organic chemical vapor deposition (MOCVD), and the processed devices were antireflection-coated to suppress any residual Fabry–Perot spectral modulation and to increase the overall light output.

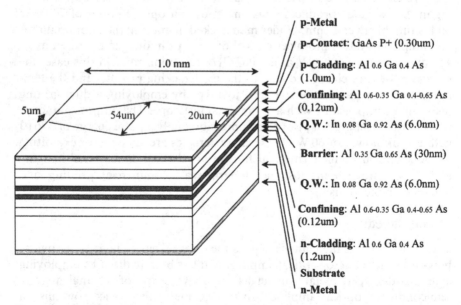

Figure 21 Schematic representation of the angle-striped inverse bow-tie semiconductor optical amplifier.

Device Performance

The output characteristics of the diamond-shaped semiconductor optical amplifier are summarized in Figure 22. In Figure 22a, the spontaneous emission output spectrum is shown, with an emission peak centered at ~ 860 nm, showing a smooth spectral profile without any noticeable Fabry–Perot peaks. In Figure 22b, we show the light output power characteristics versus input driving current. The devices exhibit a smooth superlinear light vs. current curve with the knee located at 400 mA and yielding a maximum output spontaneous emission power of 190 mW per facet with 1.7 A of drive current. It should be noted that in order to maximize the device performance, the thermal issues associated with high-power laser diodes with large driving currents need to addressed. In these experiments, the device is mounted p side down on a gold-coated copper stud. The current supplied enters the device through a gold foil contact in order to withstand the large driving currents. The SOA chip and gold-copper stud are then mounted on a Peltier thermoelectric cooling element, which is controlled by a feedback control circuit to realize automatic temperature stabilization to within 0.1°C. Additional cooling is provided by flowing 65°F water through a modified copper mounting block to assist in removing heat generated from the hot side of the thermoelectric cooler. The water cooling was sufficient to maintain the copper mounting block and hot side of the Peltier element at room temperature. It should be noted that by cascading thermoelectric cooling elements, one can achieve similar heat removal capabilities without the need of a flowing water source.

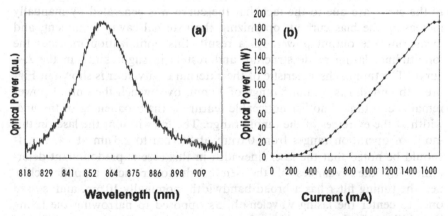

Figure 22   (a) Output spontaneous emission spectrum of the semiconductor optical amplifier. (b) Optical output power of the SOA versus input direct current.

Experimental Setup

The experimental configuration for the external cavity laser system is realized by using the standard configuration of Figure 1. The SOA is placed in an external cavity consisting of a 20% output coupler and either a maximum reflectance rear mirror or a multiple quantum well saturable absorber mirror. The light is collected and collimated from the semiconductor optical amplifier chip by using two large numerical aperture collimating lenses. A telescopic arrangement is also placed around the output coupler to increase the cavity stability. An intracavity vertical slit was employed to control the transverse mode profile, and an intracavity bandpass filter was employed for investigating tuning characteristics. An additional focusing lens was used on the rear reflector to increase the cavity stability and to establish passive modelocking when employing the multiple quantum well saturable absorber mirror. Direct current and an RF sinusoid were applied to the SOA using a bias tee. For modelocked operation, the resultant output pulse train was passed through an optical isolator and subsequently amplified in an identical SOA device to increase the output optical power. Optical power measurements were made by using a calibrated power meter, and temporal and spectral measurements were made by employing a commercially available autocorrelator and a 0.25 m spectrometer combined with a linear diode detector array.

CW Laser Performance

In Figure 23a, we show the $L-I$ curve of the external cavity laser in continous wave operation. The $L-I$ curve shows that the laser has a threshold current of ~ 350 mA and yields a maximum output power of over 700 mW at a dc bias current of 2 A. It should be noted that the slight kinks in the curve are due to the fact that the curve was generated by manually increasing the bias current, optimizing the external cavity alignment, and measuring the output power. As a result, this optimization modifies the longitudinal lasing mode structure and results in slight kinks in the $L-I$ curve. The tuning characteristic of the external cavity laser is shown in Figure 23b and shows a tuning range of 17 nm, over which the output power remains constant. Another noticeable feature is the broadening of the linewidth at the extremes of the tuning range. The linewidth of the laser in this mode of operation ranges from 0.6 nm at 855 nm to 2.3 nm at 843 nm. It should be noted that the laser linewidth in this case is predominant determined by the dynamics of the semiconductor optical amplifier chip; i.e., the tuning filter has a broad bandwidth, nominally 10 nm, and is used only to center the lasing wavelength, as opposed to narrowing the lasing linewidth. A further narrowing of the lasing spectrum can be accomplished by using a grating loaded external cavity configuration. With a 1200 g/mm

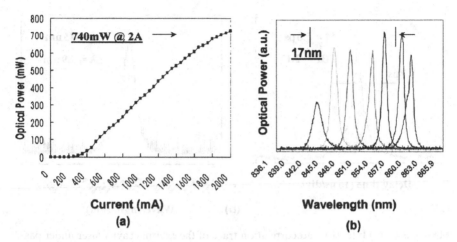

Figure 23 (a) Continuous wave *L–I* curve, showing a threshold at ~350 mA and a maximum output power greater than 700 mW at a bias of 2 A. (b) Tuning characteristic of the external cavity laser, showing a tuning bandwidth of 17 nm.

grating employed as the rear reflector, a cw lasing linewidth of 0.44 nm was measured, which is the resolution limit of the grating spectrometer and linear diode detector array.

Modelocked Operation

By replacing the rear highly reflecting mirror with a semiconductor multiple quantum well saturable absorber, passive modelocking can be achieved. It should be noted that the design of the multiple quantum well saturable absorber must provide nonlinear absorption within the gain bandwidth of the semiconductor optical amplifier. For the present experiments, the intra-cavity filter is used to force the laser to operate within the excitonic absorption band. To achieve passive modelocking, the laser was biased with 556 mA of direct current and produced an average output power of 36 mW. The laser operates at a repetition rate of 1.062 GHz, which corresponds to the fourth harmonic of the fundamental longitudinal mode spacing of 265.5 MHz. This is due to the interplay between the gain recovery time of the SOA chip, which was measured to be 0.3 ns, and the saturable absorber recovery time, which can vary from 150 to 280 ps, depending on the device design, processing, and cavity configuration. Figure 24 shows the autocorrelation and optical spectrum of the passively modelocked laser. From the autocorrelation trace, it is observed that the laser produces optical pulses with an autocorrelation FWHM of 8.7 ps. The optical spectrum

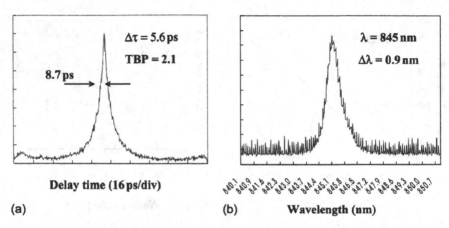

Figure 24 (a) Intensity autocorrelation trace of the external cavity laser under passive modelocking conditions, showing an autocorrelation FWHM of 8.7 ps. (b) Output optical spectrum of the external cavity laser under passive modelocking conditions, showing an FWHM spectral width of 0.9 nm.

obtained had an FWHM spectral width of 0.9 nm and was centered at ~ 845 nm because of the location of the excitonic absorption band of the semiconductor multiple quantum well saturable absorber mirror. If a hyperbolic secant squared optical pulse is assumed, the laser generates optical pulses of 5.6 ps, with a time bandwidth product of 2.1 implying that the pulse is 6.4 times the Fourier transform limit.

Hybrid modelocking can be achieved by applying an RF current to the laser oscillator. In this case, the RF frequency is chosen to be equal to the fourth harmonic of the cavity fundamental frequency in order to enhance the pulse production determined by the saturable absorber and gain dynamics. Figure 25 presents the autocorrelation trace and optical spectrum for the laser oscillator operating under hybrid modelocking conditions. For these results, the dc bias to the oscillator was 600 mA, with ~ 50 mW of RF power coupled into the diode. It should be noted that in this case the radio frequency primarily serves to stabilize the pulse production, because the relative RF and direct currents imply a small depth of modulation of the cavity round-trip losses. A double stub tuner was employed to increase the amount of RF current coupled into the device. By systematically adjusting the tuner, one can maximize the coupled Rf power. From the autocorrelation trace and optical spectrum, it is observed that the main influence of the RF currents is to temporally broaden the autocorrelation pulsewidth to 9.5 ps and broaden the optical spectrum to 1.3 nm. If hyperbolic secant squared pulses are assumed, the optical pulsewidth is 6.2 ps,

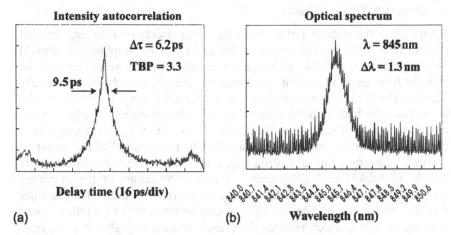

Figure 25 (a) Intensity autocorrelation trace of the external cavity laser under hybrid modelocking conditions, showing an autocorrelation FWHM of 95 ps. (b) Output optical spectrum of the external cavity laser under hybrid modelocking conditions, showing an FWHM spectral width of 1.3 nm.

implying a time–bandwidth product of 3.3, which is 10 times the transform limit. This slight temporal and spectral broadening effect is due to the increased chirp impressed on the optical pulse owing to the RF driving signal. The output power and central wavelength remain unchanged at 36 mW and 845 nm, respectively. Small satellite peaks are observable in the autocorrelation trace. These are due to the reduced effective reflectivity of the diode facet caused by the inverse bow-tie geometry. In this case, the inverse bow-tie geometry has flared ends, and any residual reflected light can more easily couple back into the gain medium, leading to satellite pulses, in contrast to narrower single-stripe devices, which produces satellite-free optical pulses. This effect can be eliminated by a modification of the device design, similar to concepts employed for conventional tapered devices. It should be noted that the large time–bandwidth products obtained in both passive and hybrid modelocked operation are due to the large nonlinearities with gain depletion in the semiconductor optical amplifier and saturable absorption in the MQW mirror. The SOA dynamic imparts a large cubic temporal phase, which becomes linearized due to the combined effects to group velocity dispersion and saturable absorption. The resulting optical pulse is non-transform-limited with a predominantly linear chirp, which can be removed with dispersion compensation techniques.

Amplification Characteristics

By injecting the optical pulses from the laser oscillator into an identical inverse bow-tie SOA, the output pulse train can be amplified. It should be noted that for good amplification, the oscillator and amplifier must be isolated from each other. Without isolation, the primary effect is to initiate lasing of the SOA external amplifier caused by light being from reflected the output coupler back into the external semiconductor optical amplifier stage, as compared to effecting the modelocked operation. Under these conditions, with a direct current of 1 A applied to the external SOA amplifier, 400 mW of average output is obtained, after subtracting the approximately 40 mW of background spontaneous emission contained in the amplified output. It should be noted that the bias current to the SOA amplifier was limited to 1 A in order to avoid excessive amounts of amplified spontaneous emission, which reduces the pulse contrast, at higher bias current. This effect can be minimized by employing a pulse bias to the SOA in order to provide maximum gain during the passage of the pulse through the SOA and reducing the spontaneous emission between pulses. The average power and pulse width generated from the laser and obtained after amplification imply a pulse energy of 376 pJ and a peak power of 60 W, which is sufficient for exploiting a variety of χ^3 optical nonlinearities, for nonlinear photonic applications. It should be noted that by optimizing the multiple quantum well saturable absorber, one can achieve modelocked operation at the gain maximum. This will allow for greater amplified power due to larger injection powers and a better matching of the injected laser wavelength with the gain peak of the external optical amplifier.

By lowering the dc bias current of the oscillator to 423 mA and employing a narrower transverse mode filter, the satellite pulses can be removed. Further optimization of the laser cavity provided for an additional spectral broadening to 1.5 nm. Under these conditions, the laser produced a measured optical autocorrelation FWHM of 4.2 ps, implying a pulse width of 2.7 ps, with a time–bandwidth product of 1.7 (five times the transform limit) and a average output power of 17 mW (see Fig. 26). Subsequent amplification of the optical pulses produced an average output power 218 mW while maintaining the optical pulse width. Owing to the chirp impressed on the optical pulse during the modelocked process, optical dispersion compensation techniques can be employed to further reduce the pulse width. Injecting the pulse train into a standard dual grating compressor employing an internal 1:1 telescope operating in a double-pass geometry, the pulse width was further reduced to a deconvolved width of 1.3 ps, yielding a time–bandwidth product of 0.8 (see Fig. 27).

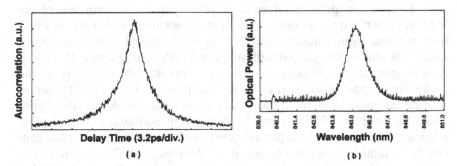

Figure 26 (a) Intensity autocorrelation trace of the external cavity laser under optimized hybrid modelocking conditions, showing an autocorrelation FWHM of 4.2 ps. (b) Output optical spectrum of the external cavity laser under optimized hybrid modelocked conditions, showing an FWHM spectral width of 1.5 nm.

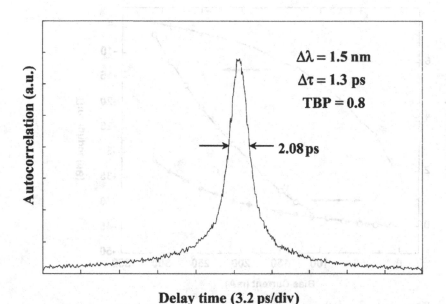

$\Delta\lambda = 1.5$ nm

$\Delta\tau = 1.3$ ps

TBP = 0.8

2.08 ps

Delay time (3.2 ps/div)

Figure 27 Intensity autocorrelation trace of the optical pulses after a double pass through the dual grating dispersion compensator, showing a deconvolved pulse width of 1.3 ps, assuming a hyperbolic secant pulse shape.

To obtain additional information regarding the performance of the
SOA amplifier stage, the output power was measured with varying injection currents while maintaining the injected optical power constant.
Figure 28 shows the output optical power, after subtracting the background spontaneous emission, versus bias current. For comparison, the
background spontaneous emission is also plotted. The resulting curves
provide information of the transparency point of the SOA amplifier, from
which the coupling efficiency can be inferred. In addition, if the coupling
efficiency and the input–output relations are known, the single-pass gain
can be obtained. From the figure, it is observed that the output power
curve crosses the spontaneous emission curve near 70 mA, indicating
the current bias point for transparency. The injected optical power in this
case was measured to be 17 mW, with an output amplified power of
180 μW, implying a coupling efficiency of −19.9 dB. From the amplified
output powers obtained in the modelocking experiments described above,
the included coupling loss implies single-pass gains in excess of 30 dB.
The large coupling loss in this case is due to a nonoptimized transverse
mode matching between the laser oscillator and the SOA amplifier. The

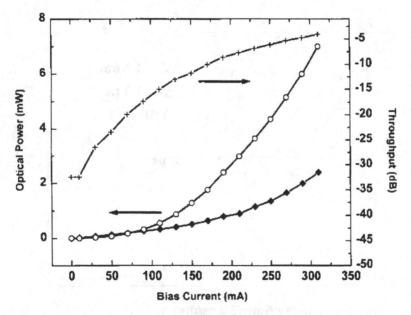

Figure 28 Input–output characteristics of the SOA amplifier. The injected optical
power is constant at 17 mW. (◯) Stimulated emission output power; (◆) spontaneous emission output. Right-hand axis denotes the throughput (P_{out}/P_{in}) in decibels
(+).

improper mode match is due to the 2:1 telescope employed to allow the incorporation of an optical isolator, with a small clear aperture, between the oscillator and the SOA amplifier. The input and output transverse mode profiles of the SOA amplifier are shown in Figure 29. It should be noted that by employing a large clear aperture optical isolator, better mode matching can be achieved, which would lead to lower coupling losses and high amplified output powers. Incorporating an optimized optical injection scheme, in addition to achieving higher output powers, allows the injected light to dominate the stimulated emission process and reduces the competition with amplified spontaneous emission. In addition, this would then allow the dc bias level of the SOA amplifier to be increased. In this configuration, full exploitation of the single-pass gain characteristics can be achieved, leading to average modelocked output powers of $\sim 700\,\text{mW}$, which is comparable to that observed in the cw lasing regime.

From these results we have shown that novel inverse bow-tie semiconductor optical amplifiers can be used in external cavity semiconductor lasers for high-power operation. Continuous wave tunable operation was demonstrated with a tuning bandwidth of $\sim 17\,\text{nm}$ and cw average output power of 700 mW. In addition, modelocked operation in both passive and hybrid modelocked regimes was investigated. In these experiments, typical pulse widths of 5 ps were generated from the oscillator, and with subsequent amplification a maximum average output power of 400 mW was

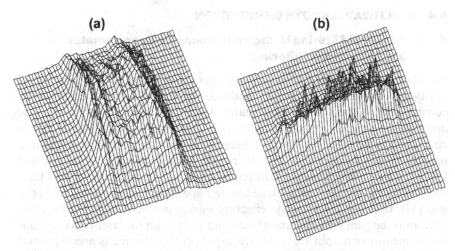

Figure 29   (a) Input and (b) output transverse optical mode profiles of the SOA amplifier.

obtained, at a repetition rate of 1.062 GHz, implying a pulse energy of 367 pJ and a peak power of 60 W. Further optimization of the laser oscillator combined with dispersion compensation techniques allowed the production of optical pulses of 1.3 ps. Shorter optical pulses and higher output powers may be accomplished by redesigning the multiple quantum well saturable absorber mirror so that lasing can occur at the gain peak of the semiconductor optical amplifier device. In addition, this will allow for higher injection power and also provide for a better spectral match to the external optical amplifier. By exploiting other semiconductor material systems, SOA devices operating at 810 or 980 nm may allow for even higher output power. In this case, however, appropriate multiple quantum well saturable absorber mirrors need to be designed. Further optimization can be achieved by employing low output impedance RF amplifiers or by using a well-designed impedance-matching circuit to better couple the RF power into the oscillator chip. This will provide for a larger depth of modulation and enhance the modelocking stability. Finally, because of the fast gain recovery time, a shorter cavity can be employed to make the fundamental longitudinal mode spacing compatible with the modelocking frequency of ∼1 GHz. Incorporating these modifications, we expect that modelocked optical pulse trains with an average power approaching 2 W can be achieved from an all-semiconductor-diode laser system.

5.4 MULTIWAVELENGTH GENERATION

5.4.1 Spectral Filtering Using an Intracavity Spectrometer and Amplitude Filtering

An alternative method for intracavity spectral filtering employs an intracavity spectrometer [32]. In this method, spectral filtering is achieved by constructing a spectral filtering plane within the optical cavity. This is done by replacing the output coupler with a diffraction grating–lens–mirror combination (see Fig. 30). In this optical geometry, the grating is placed in a near Littrow configuration. The diffracted light is collected by a lens one focal distance away from the diffraction grating. The collected light is then focused to a spectral filtering plane located one focal distance away from the lens [33]. In this plane a highly reflecting mirror is placed to reflect the light back into the gain medium, and the output light from the laser is taken from the zeroth-order output from the grating. Spectral filtering is accomplished by placing a suitable transmission mask directly in front of mirror $M1$. For example, if four equally spaced slits are placed at the filtering plane, the laser

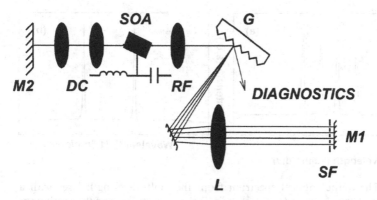

Figure 30 External cavity semiconductor laser incorporating an intracavity spectrometer and amplitude filtering. SOA, semiconductor optical amplifier; G, grating (1800 line/mm); SF, spatial filter; L, 150 mm achromatic lens; M, end mirrors.

can be made to oscillate with four equally spaced wavelengths. The linearity of the frequency spread can easily be calculated by examining the linear dispersion in the filtering plane. Starting with the grating equation, $a(\sin\theta - \sin\varphi) = m\,\lambda$, one can differentiate the grating equation to obtain the angular dispersion $d\theta/d\lambda$ and then multiply by the focal length f to give the linear dispersion. The linear dispersion, L, is then represented by $L = fm/a\cos\theta$ (mm/nm). In these equations θ and φ are the output and input angles, respectively, m is the order, λ is the wavelength, and a is the grating spacing. The experimental configuration employed a gold-coated 1800 g/mm diffraction grating and an achromatic lens with a focal length of 160 mm. For modelocked operation, a suitable mask design for the generation of four independent wavelengths separated by 2 nm, each generating nearly transform-limited pulses of 10 ps duration, implies a mask design with a slit spacing of 750 μm, with each slit having a width of 38 μm.

Under cw operating conditions, gain competition between independent wavelength components typically prevents continuous generation of multiple wavelengths from a single-stripe semicondictor laser. Although dual wavelength operation from a semiconductor laser has been demonstrated in the past, the device employed uses either a wide-stripe or multistripe device structure [34]. However, by exploiting the transient unsaturated gain within the gain medium of a modelocked laser, it becomes possible to sustain robust, stable multiwavelength generation from a single-stripe semiconductor diode laser. Figure 31a illustrates the output modelocked optical spectrum from the multiwavelength laser with a four-slit amplitude mask, demonstrating the robustness of multiwavelength generation from a diode

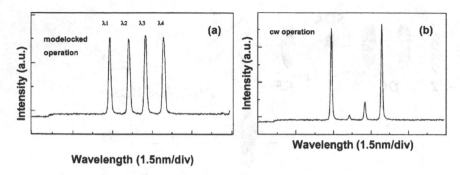

Figure 31 The output optical spectrum from the multiwavelength laser with a four-slit amplitude mask under (a) active modelocking operation and (b) continuous wave operation.

laser in modelocked operation. As a comparison, if the laser is operated in a continuous fashion, the gain competition prevents the four wavelengths from simultaneously lasing (Fig 31b). In these results, the bias current is 165 mA, with 0.5 W of RF current for modelocked operation, showing the emitted spectra centered at 835 nm. The output power from the laser oscillator is measured to be 1.5 mW and can be increased to 25 mW by employing a second SOA device ($l = 5007\,\mu\text{m}$) as a single-pass power amplifier.

To demonstrate the flexibility of the spectral filtering technique, a four-slit mask was fabricated in a fanned shape; i.e., the four slits originated from a single location at the bottom of the mask, with the distance between slits increasing toward the top of the mask. By translating the mask horizontally and vertically with respect to the filtering plane, the placement of the four wavelength modelocked spectrum can be tuned within the gain bandwidth, and the channel separation can be varied. Figure 32a shows the tuning characteristic of the four-wavelength packet, demonstrating tunability over 18 nm, and Figure 32b shows the channel variation from 1 nm to 2.5 nm. These data suggest the possibility of accurately generating a multiplicity of modelocked wavelengths at well-defined frequencies, which is important for wavelength division multiplexing (WDM) data links, and also the possibility of generating wavelength packets with over 20 independent wavelengths.

Figure 33 illustrates the temporal output characteristics of the four-wavelength modelocked diode laser. In Figure 33a are autocorrelation traces of the composite four-wavelength output (top) and the autocorrelation traces of the individual modelocked channels. From this figure it is observed that the autocorrelation traces measure optical pulse widths of

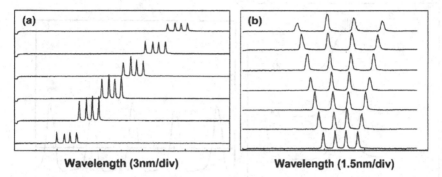

Figure 32 Wavelength-tuning curves. (a) Tuning of the center wavelength over 18 nm with a constant wavelength separation and (b) detuning of the wavelength channel spacing from 1 nm to 2.5 nm with a constant center wavelength.

~ 12 ps (assuming a Gaussian pulse shape) for each wavelength channel and for the composite output. In this case, the laser is harmonically modelocked at a 2.57 GHz repetition rate, for an overall aggregate pulse rate of 10 GHz (Fig. 33b). An important attribute of the composite autocorrelation trace (Fig. 33a, top) is that the pulse width obtained is identical to that of the individual autocorrelation traces. This implies that the wavelength channels are temporally synchronized, i.e., all channels are simultaneously produced in time. In the limit of uncorrelated temporal jitter and static temporal skew between wavelength channels, the resulting composite autocorrelation trace would necessarily be broadened by an amount equal to the temporal jitter over the measurement plus the fixed overall temporal delay between the leading and trailing optical pulses. In Figure 34 is a single-shot, time-resolved spectrum measured with a spectrometer and a streak camera. In this setup, the time-resolved spectrum was generated by passing the composite output pulse train through a grating–lens combination in order to image the spectrum on the input slit of a streak camera. Triggering of the streak camera was accomplished by synchronization with the active modelocking signal generated with a synthesizer. From this spectrogram, it is observed that the four wavelengths are temporally synchronized.

Another important attribute of the composite autocorrelation trace is the temporal modulation superimposed on the intensity envelope. This modulation is inversely proportional to the frequency difference between adjacent wavelength channels. It should be noted that in early work on actively modelocked laser diodes, periodic coherent spikes were observed on the autocorrelation trace. In these results the temporal spikes are due to the residual Fabry–Perot reflectance of the cleaved diode facets. This

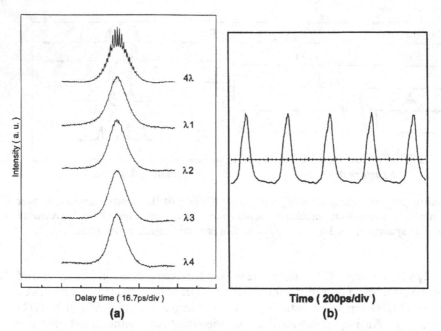

Figure 33 The temporal output characteristics of the four-wavelength modelocked diode laser. (a) Autocorrelation traces of the composite four-wavelength output (top) and the individual modelocked channels. (b). Oscilloscopic trace of output optical pulses, showing a channel rate of ~2.5 GHz.

effect, if not completely eliminated, produces temporal coherent spikes spaced at the round-trip time of the isolated diode chip. The Fabry–Perot effect can be viewed as a form of spectral filtering that incorporates both periodic amplitude and phase filtering. It is the periodic phase filtering component that produces the multiple reflections and causes the periodic spikes in the autocorrelation trace. For the present four-wavelength laser, it must be stressed that the Fabry–Perot modes of the isolated diode chip have been competely eliminated by employing both angle-striping techniques and suitable dielectric antireflection coatings. This implies that the spectral filtering achieved by the intracavity spectrometer and mask provides only pure amplitude filtering, without any periodic phase component. In the limit of completely random phase between wavelength components, the resulting beat frequency will randomly move within the envelope of the composite optical pulse, thus eliminating the observation of temporal beats across the entire autocorrelation trace. As a result, the observation of clearly defined temporal beats in the composite autocorrelation suggests that a

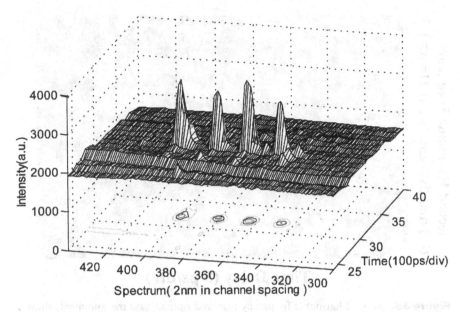

Figure 34 The single-shot time-resolved spectrum as measured with a streak camera, showing temporally synchronized four-wavelength channels.

more detailed investigation needs to be performed to completely understand the nature of the observed temporal beat signals.

To obtain a better understanding of the temporal beat observed in the composite autocorrelation trace, a second harmonic frequency-resolved optical gate measurement (SHG-FROG) was performed, with the resulting FROG trace shown in Figure 35. The trace shows seven clearly defined spectral bands, corresponding to the convolution of the four-channel input spectra. In addition, temporal modulation is clearly observed in the FROG trace throughout the temporal delay. These data show that the temporal modulation observed in the autocorrelation results from the temporal modulation measured in the FROG trace, because the autocorrelation function is the marginal of the FROG. More important, the large temporal modulation observed for all delays in the FROG trace clearly shows that well-defined spectral phase relations exist between all spectral components, thus making the four generated modelocked wavelength channels phase-coherent with each other.

For this phase relationship to occur, a temporal modulation of the refractive index must occur at a rate equal to the frequency difference. In the present case, because the modulation frequency of the cavity is only

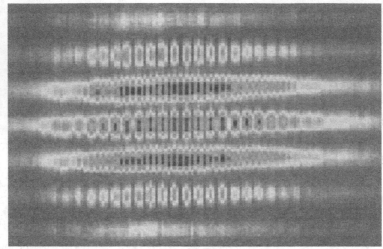

Time Delay (8 ps/div)

Figure 35 Second harmonic frequency-resolved optical gate measurement, showing seven spectral bands corresponding to the four-wavelength input spectrum. Additional temporal modulation is observed for all delay times, showing the phase-locked nature of the multiwavelength modelocked spectra.

2.5 GHz, the ultrafast modulation must occur through the nonlinear gain dynamics of the semiconductor optical amplifier. The temporal modulation on the autocorrelation trace has a period of 1.6 ps, which is sufficiently fast for the dynamic carrier heating and cooling effects to modulate the gain and hence the index [35,36]. It is precisely the same mechanism that is exploited in nonlinear four-wave mixing (FWM) and wavelength conversion in SOAs [37].

To verify the possible coupling through four-wave mixing, optical spectra were measured on a logarithmic scale in order to provide the dynamic range to observe FWM side bands. In Figure 36a is a spectrum of the four-wavelength modelocked laser, showing the existence of FWM side bands for wavelength channels with a frequency difference of Δv, 2 Δv, and 3 Δv. This suggests that there is energy exchange between wavelength channels. It should be noted that the variation of intensity of the FWM side bands versus frequency detuning is due to the frequency dependence of the nonlinear susceptibility and the fact that each side band can be composed of multiple side bands generated by different combinations of the four wavelengths. To verify the energy exchange between channels, the laser was operated with only two wavelengths with a frequency difference of Δv,

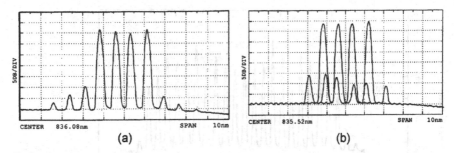

CENTER 836.08nm SPAN 10nm (a)   CENTER 835.52nm SPAN 10nm (b)

Figure 36 Intracavity four-wave mixing spectra. (a). The spectrum of the four-wavelength modelocked diode laser, showing the existence of a multiplicity of FWM sidebands. (b) Three overlaid spectra highlighting the generation of the FWM side bands with respect to the four-wavelength channel location.

the FWM side bands were measured, and their frequency positions were noted. In Figure 36b is a plot of three different spectra overlaid to highlight the generation of the FWM side bands with respect to the four modelocked wavelength channels. From this figure it is seen that all of the FWM side bands lie within the bandwidth of the four modelocked channels. These side bands serve as seeding mechanisms to assist in establishing a correlated spectral phase relation between wavelength channels.

From the spectra in Figure 32a, the tuning characteristic suggests the possibility of supporting many more independent wavelength channels. To fully exploit the gain bandwidth of the diode laser, a mask was fabricated with over 20 evenly spaced slits to accommodate as many wavelength components as possible. In Figure 37a is a spectrum of the modelocked laser employing the new mask design. In this figure it is seen that over 25 wavelength channels are generated, with 21 components above the -3 dB line. By temporally interleaving the output pulse train, a 5 GHz modelocked signal is generated, implying a total aggregate pulse rate of 105 GHz. This is displayed in the oscilloscope trace, which employs a 20 GHz sampling oscilloscope and a 10 GHz photodetector. The temporal interleaving was accomplished by employing three beam splitters with three optical delay stages, providing a 2^3 increase in repetition rate. The the input power to the interleaving delay stages was 25 mW with an overall insertion loss of 10 dB. From the autocorrelation traces, assuming a deconvolved pulsewidth of 12 ps, an additional eightfold interleaving can be achieved, for a total pulse rate of 840 GHz. As shown earlier, high average modelocked powers can be achieved by employing SOA devices (400 mW) so that large insertion losses associated with multiple interleaved steps can be tolerated if required.

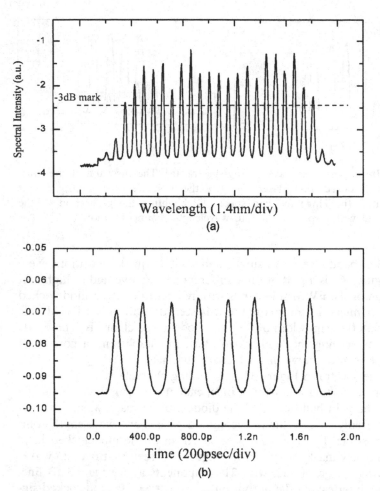

Figure 37 (a) The spectrum of a modelocked multiwavelength laser employing the new mask with over 20 slits to accommodate more than 20 wavelengths. (b) Temporally interleaved output pulse train, showing a channel rate of ~5 GHz.

5.4.2 Coherent Coupling in Multiwavelength Modelocked Lasers

The intracavity gain dynamics discussed earlier has shown a transient regime of unsaturated gain. This effect allows the gain to completely recover well within a single cavity round-trip. Under normal continuous wave operating conditions of semiconductor diode lasers, the gain primarily behaves in accordance with homogeneously broadened gain media, which clamps the gain across the optical spectrum and generally allows for the production of narrow optical spectra. This is due to the time scale involved in the gain dynamics. However, under modelocked operation, the transient regime of unsaturated gain allows the gain to recover completely across the entire gain bandwidth and allows for the production of a spectrally broad output. This effect, as shown above, allows the production of multiwavelength modelocked emission from an actively modelocked external cavity diode laser. It was also noted above that intracavity four-wave mixing occurs within the laser cavity during the modelocking process and may contribute to establishing a correlation between the wavelength components.

In this section we use the spectrally resolved second harmonic intensity autocorrelation FROG, or SHG-FROG, technique to understand the fundamental mechanisms responsible for phase correlation between the four modelocked wavelengths. This is done by directing the multiwavelength modelocked optical pulse train into an autocorrelator and measuring the second harmonic spectrum as a function of delay. By modeling the electric field spectrum of the input pulse and comparing theoretical and experimental SHG-FROG traces, fundamental information on the nature of the phase coupling can be obtained.

In this experiment, the SHG-FROG is represented as

$$
I_{\text{SHG}}(\omega - 2\omega_0, \tau) = \left| \int_{-\infty}^{\infty} dt E_{\text{sig}}(t) E_{\text{sig}}(t - \tau) \exp(-j\omega t) \right|^2 \tag{16}
$$

This distribution also has the property that the time-marginal is the intensity autocorrelation function.

5.4.3 Experimental Results

Recall Figure 35, which displays the four-wavelength SHG-FROG spectrogram. This FROG trace was obtained by measuring 128 spectra, with ~ 666.67 fs delay between spectra. The FROG trace shows seven spectral bands and significant modulation along the temporal delay. The modulation periods are measured to be 2.67 ps on the second, fourth, and sixth spectral bands and 1.34 ps on the third and fifth bands.

Owing to the four sampling or delay steps in one 2.67 ps period, the fine structure within this short temporal regime cannot be completely resolved. In Figure 38, a higher resolution SHG-FROG spectrogram is

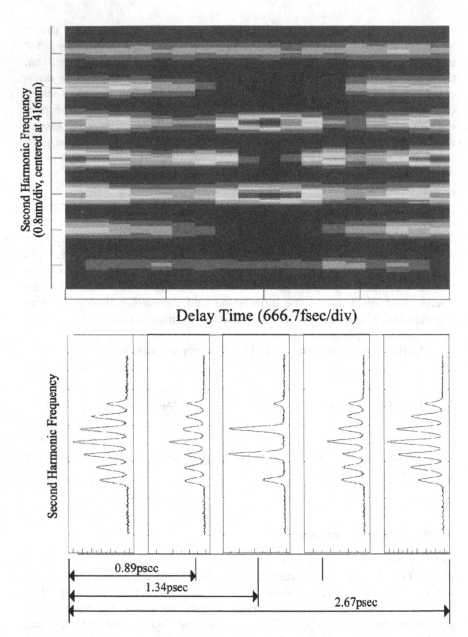

Figure 38 SHG-FROG spectrogram, showing the fine structure associated with high beat frequencies. The temporal delay spans one period of the lowest beat frequency.

displayed that represents the spectrogram within a single 2.67 ps period, revealing the fine structure. Also shown in Figure 38 is a series of second harmonic spectra along the temporal delay highlighting the spectral shape at different temporal delays. In this experiment, fivefold increase in the sampling rate is employed. The salient feature in Figure 38 is the additional modulation resolved on the central band with a modulation period equal to 0.89 ps.

To explain the experimental results and further extract information about phase correlation, an analytic expression may be derived for signals that contain multiple spectral components.

In the multiwavelength laser, let each of the wavelength bands be represented as $e_n(v) \exp(j\phi_n)$, where ϕ_n is the spectral phase of the nth wavelength band. The N-wavelength spectrum with a frequency channel spacing of Δ is then given by

$$e(v) = \sum_{n=0}^{N-1} e_n(v - n\Delta) \exp[j\phi_n(v - n\Delta)] \tag{17}$$

With $e(v)$ defined as the complex envelope of the optical spectrum, the ambiguity function may then be defined as

$$|X(f,\tau)| = \left| \int_{-\infty}^{\infty} dv\, e(v)\, e^*(v - f)\, \exp(j2\pi v\tau) \right| \tag{18}$$

By changing the order of summation and integration when calculating the ambiguity function, one obtains

$$X(f,\tau) = \sum_{n=0}^{N-1} \sum_{m=0}^{N-1} \int_{-\infty}^{\infty} e_n(v - n\Delta) \exp[j\phi_n(v - n\Delta)]\, e_m^*(v - m\Delta - f)$$

$$\times \exp[-j\phi_m(v - m\Delta - f)]\, \exp(j2\pi v\tau)\, dv \tag{19}$$

Letting $U_n = e_n(v) \exp(j\phi_n)$, $U_m = e_m(v) \exp(j\phi_m)$, and $v_1 = v - n\Delta$, we get

$$X(f,\tau) = \sum_{n=0}^{N-1} \exp[j2\pi n\Delta\tau] \sum_{m=0}^{N-1} \int_{-\infty}^{\infty} U_n(v_1) U_m^*\{v_1 - [f - (n - m)\Delta]\}$$

$$\times \exp(j2\pi v_1\tau)\, dv$$

$$= \sum_{n=0}^{N-1} \exp[j2\pi n\Delta\tau] \sum_{m=0}^{N-1} X_{n,m}[f - (n - m)\Delta, \tau] \tag{20}$$

The integral in Eq. (20) resembles the ambiguity function except for the two envelope functions, which may not necessarily be the same. Following Refs. 38 and 39, we can rewrite the double sum in a form that collects terms involving the same delay. Using $p = n - m$, we note that

$$\sum_{n=0}^{N-1}\sum_{m=0}^{N-1} = \sum_{p=-(N-1)}^{0}\left.\sum_{n=0}^{N-1-|p|}\right|_{m=n-p} + \sum_{p=1}^{N-1}\left.\sum_{m=0}^{N-1-|p|}\right|_{n=m+p} \tag{21}$$

Using Eqs. (20) and (21) we get

$$X(f,\tau) = \sum_{p=-(N-1)}^{0}\left[X_{n,m}(f-p\Delta,\tau)\sum_{n=0}^{N-1-|p|}\exp(j2\pi n\Delta\tau)\right]$$

$$+ \sum_{p=1}^{N-1}\left[\exp(j2\pi p\Delta\tau)X_{n,m}(f-p\Delta,\tau)\sum_{m=0}^{N-1-|p|}\exp(j2\pi m\Delta\tau)\right] \tag{22}$$

The following relation is used to replace the sum of exponents:

$$\sum_{n=0}^{N-1-|p|}\exp(j2\pi n\Delta\tau) = \exp[j\pi\tau(N-1-|p|)\Delta]\frac{\sin[\pi\tau(N-|p|)\Delta]}{\sin(\pi\tau\Delta)} \tag{23}$$

Using Eqs. (22) and (23) yields two complementary sums for positive and negative values of p, which can be combined into one sum, resulting in

$$|X(f,\tau)| = \left|\sum_{p=-(N-1)}^{N-1}\exp[j\pi\tau(N-1+p)\Delta]\right.$$

$$\left.X_{n,m}(f-p\Delta,\tau)\frac{\sin[\pi\tau(N-|p|)\Delta]}{\sin(\pi\tau\Delta)}\right| \tag{24}$$

In cases where the separation between wavelength bands is larger than the width of the individual wavelength channel, the absolute value can be taken inside the sum, yielding

$$|X(f,\tau)| = \sum_{p=-(N-1)}^{N-1}|X_{n,m}(f-p\Delta,\tau)|\left|\frac{\sin[\pi\tau(N-|p|)\Delta]}{\sin(\pi\tau\Delta)}\right| \tag{25}$$

Equation (25) is the ambiguity function of the N-wavelength spectrum, where $X_{n,m}(f-p\Delta,\tau)$ represents the cross-ambiguity function between any pair of the N wavelengths. This expression also represents an analytical SHG-FROG spectrogram of an N-wavelength spectrum including all spectral phase information, by squaring the expression and translating the output frequency to the second harmonic frequency. One can obtain the SHG-FROG spectrogram for any given equally spaced multiwavelength optical spectrum based on this expression. It should be noted that the cross-ambiguity functions physically correspond to the sum-frequency terms in the measured FROG trace, whereas the direct harmonic frequencies are generated when $n = m$.

As an example, consider the multiwavelength spectrum that is composed of identical-intensity spectral envelopes and a constant spectral phase. For simplicity, assume that the constant phase is zero, and define $e_n(v) = e_m(v) = e_c(v)$, Then

$$|X(f, \tau)| = \sum_{p=-(N-1)}^{N-1} |X_c(f - p\Delta, \tau)| \left| \frac{\sin[\pi\tau(N - |p|)\Delta]}{\sin(\pi\tau\Delta)} \right| \tag{26}$$

where $X_c(f - p\Delta, \tau)$ is the ambiguity function of an individual wavelength. If the spectral intensity envelope of the individual wavelength is further assumed to be a rectangular shape with a spectral width of δ, we can replace $X_c(f - p\Delta, \tau)$ with the ambiguity function of the rectangular signal obtained from [38,39]

$$|X(f, \tau)| = \sum_{p=-(N-1)}^{N-1} \left| \left(1 - \frac{|f - p\Delta|}{\delta}\right) \frac{\sin[\pi\delta(1 - |f - p\Delta|/\delta)\tau]}{\pi\delta(1 - |f - p\Delta|/\delta)\tau} \right|$$
$$\times \left| \frac{\sin[\pi\tau(N - |p|)\Delta]}{\sin(\pi\tau\Delta)} \right| \tag{27}$$

From Eq. (27) one can note that the cut of the ambiguity function $|X(f, \tau)|$ along frequency f (set $\tau = 0$) is a periodic reproduction of the marginal obtained in the case of a single wavelength, but with linearly decreasing amplitudes with decreasing distance from the origin. The spacing between peaks is equal to the N-wavelength channel separation Δ. On the other hand, the marginal of $|X(f, \tau)|$ along delay τ (setting $f = 0$) is more complicated. Nonetheless, this marginal generally shows a modulated sinc function. In Figure 39, a contour plot of $|X(f, \tau)|^2$ for the flat-phase four-wavelength optical signal is shown. The salient feature is the similarity of the modulation fringes on each of the seven stripes between this plot and the experimental SHG-FROG spectrogram.

To assist in understanding the experimental SHG-FROG result, a simpler schematic can be provided after careful consideration of Eqs. (25)–(27).

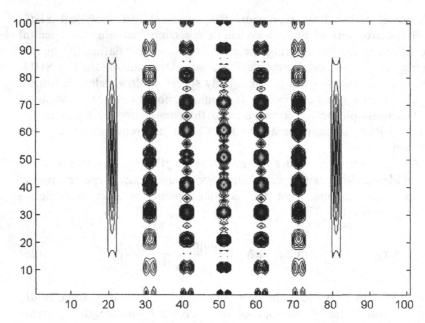

Figure 39 Contour plot of the derived analytic expression for a coherent, four-wavelength SHG-FROG spectrogram.

In Figure 40, the output spectra of a four-wavelength second harmonic generation process is plotted. The top four bars represent the four fundamental wavelength bands with center frequency ω, channel separation Δ, unit amplitude, and rectangular spectral envelope. The lower seven bars represent the second harmonic electric field spectrum. Note that the rectangular envelope in the fundamental spectrum has been transformed into a triangular shape at the second harmonic, because the product of the temporal fields corresponds to the convolution of their spectra. The seven-peak spectrum includes both the direct doubled fundamental frequencies and all combinations of sum frequencies. In addition, it must be noted that each doubled or sum frequency component is modulated at a rate proportional to the difference in photon energy involved in the SHG process, e.g., 0, Δ, 2Δ, and 3Δ. These difference terms are composed of contributions from all possible combinations of fundamental terms on each SHG spectral position.

By using this general theory, the SHG-FROG spectrogram is easily explained. The fundamental four-wavelengths are converted into seven second harmonic wavelength bands. The input rectangular four-wavelength spectral field envelope in Figure 31a is converted into a quasi-triangular

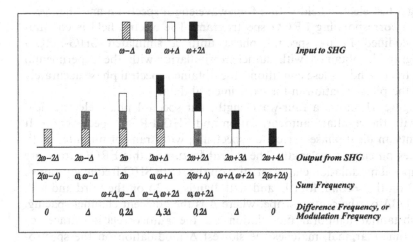

Figure 40 Schematic diagram showing the possible harmonic and sum frequencies generated by a second harmonic crystal, with four uniform input wavelength bands.

second harmonic field spectrum, as shown in Figure 39. The variation of the SHG spectral shape in Figure 39 is the modulation effect resulting from the difference frequency terms and is composed of three modulation rates, 1/2.67, 1/1.34, and 1/0.89 THz, which correspond to the difference frequencies between wavelength bands. The term 0 merely represents a dc component, i.e., constant intensity, on the first, third, fifth, and seventh stripes. These data show that the temporal modulation observed in the autocorrelation trace results from the temporal modulation measured in the FROG trace, because the autocorrelation function is the marginal of the FROG. In addition, it must be stressed that the modulation is observed throughout the entire temporal delay, suggesting that each wavelength band is correlated with each of the others, i.e., the phase difference between wavelength bands is constant throughout the temporal delay. In the limiting case of a completely uncorrelated phase, the temporal modulation would "washout" and would be observable only at times near zero delay.

The analytic expression for the two-dimensional frequency–time plot is not easy to implement for an arbitrary spectral phase, compared to direct numerical simulation. Therefore, to determine the nature of the phase-correlated multiwavelength emission, computer-simulated SHG-FROG spectrograms are used to analyze the phase relation within the four-wavelength spectrum. The phase information is obtained by first creating a test electric field with a four-wavelength spectrum and then generating a corresponding FROG spectrogram. The electric field is continuously modified in its spectral phase until a simulated SHG-FROG spectrogram is obtained with sufficient similarity with the experimental FROG trace. Under these conditions, the obtained spectral phase accurately reflects the phase relation in the experimental data.

Figure 41 shows a four-wavelength, flat spectral phase electric field along with the resultant autocorrelation and SHG-FROG spectrogram. It represents an ideal phase-correlated spectrum, with transform-limited optical pulses on each wavelength. The salient feature in this FROG trace are the temporal modulation characteristics on the seven SHG frequency bands, i.e., $1/\Delta$ on the second, fourth, and sixth bands; $1/2\Delta$ on the third and fifth bands; $1/3\Delta$ on the central stripe, where Δ is the adjacent channel spacing. The substantial temporal modulation on the autocorrelation trace, or FROG time-marginal, matches the slowest Δ modulation on the spectrogram. Again, the modulation exits throughout the entire temporal delay and is consistent with what is observed in the experimental SHG-FROG spectrograms.

To distinguish the experimentally obtained SHG-FROG from a completely random phase field, the second simulation is made by modeling a four-wavelength fundamental spectrum with random spectral phase between channels. Physically, this represents the case of multiple synchronized modelocked lasers operating with different wavelengths with noise burst optical pulses being combining to form the multiwavelength spectrum. The result from this simulation is plotted in Figure 42. The FROG spectrogram shows a disorderly and unsystematic temporal modulation pattern, which is distinctly different from our experimental FROG results. The marginal FROG trace confirms the random-phase nature by the strong central coherent spiked and the absence of significant temporal modulation across its intensity envelope. In the limit of completely random phase between wavelength components, the resulting beat frequency can be considered to randomly shift within the envelope of the composite optical pulse, thus eliminating the observation of temporal beats across the autocorrelation trace. This simulation result demonstrates that the experimental data are not a result of a phase-uncorrelated four-wavelength electric field.

To accurately determine the phase correlation between wavelength bands, intracavity effects such as the laser gain dynamics and intracavity group-velocity dispersion need to be considered, because they substantially modify the spectral phase of the output. These effects are included by using multiple quadratic phase terms associated with each wavelength channel to represent the linear chirp on each of the four-wavelength pulses

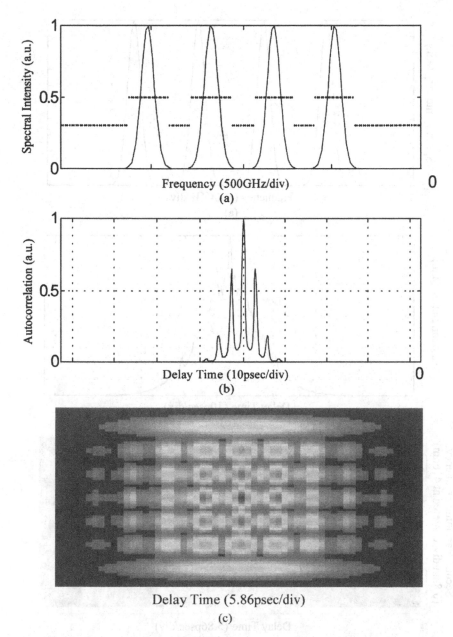

Figure 41 Numerical simulation of the output from an SHG-FROG, with four coherent modelocked wavelength bands.

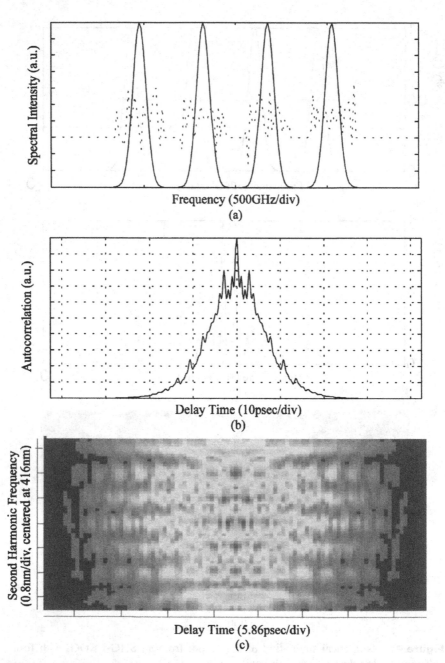

Figure 42 Numerical simulation of the output from an SHG-FROG, with four incoherent (noise burst) modelocked wavelength bands.

due to the gain dynamics. An additional overall quadratic phase is employed to represent the GVD introduced by the intracavity bulk grating. Finally, a small random phase component is introduced to represent the coupling fluctuation between wavelength channels due to four-wave mixing. As a result, the field spectrum is modeled with four identical quadratic phase terms, with a peak-to-peak magnitude of $\sim 1.6\pi$ on each wavelength. These spectral phase terms are modulated by a broad quadratic phase term of $\sim 0.4\pi$ over the spectral range, and each phase component has a small random fluctuation of less than 10% of the maximum phase magnitude.

In Figure 43, the spectral intensity and phase (a), the resultant marginal FROG trace (b), and SHG-FROG spectrogram (c) are shown. Compared to the flat phase electric field, it can be seen that the optical pulse is broadened to the same width as that in the experimental by the added linear chirp. In addition, the SHG-FROG spectrogram has modulation features nearly identical to those observed in the flat phase field and the experimental spectrogram. The small phase perturbation add a slight dc background to the third and fifth bands. In addition, the modulation on this spectrogram is very distinct for all delays, as opposed to the random phase case.

As an additional check, recall that the time-marginal of the FROG trace is equal to the SHG autocorrelation. Figure 44 shows two overlaid traces, with the solid curve showing the theoretical marginal FROG trace obtained from the above simulation, and the dashed curve showing the experimental SHG autocorrelation trace independently measured from the output of the four-wavelength laser. Good agreement between the two traces is achieved. In both traces, the temporal modulation clearly shows the presence of the 3Δ term, whereas the 2Δ term simply becomes the intensity background and is absent from the autocorrelation trace. The good agreement achieved between the simulated and measured FROG traces and their marginals shows the correlated spectral phase relation between modelocked wavelength bands.

5.5 APPLICATIONS OF MODELOCKED DIODES

5.5.1 Hybrid Optical Time and Wavelength Division Multiplexing (HOTWDM) Using Multiwavelength Modelocked Semiconductor Diode Lasers

Ultrahigh-speed data links will become widespread with the deployment of broadband switched digital networks and services such as teleconferencing, video telephony, and computer services. The key hurdle in the commercial

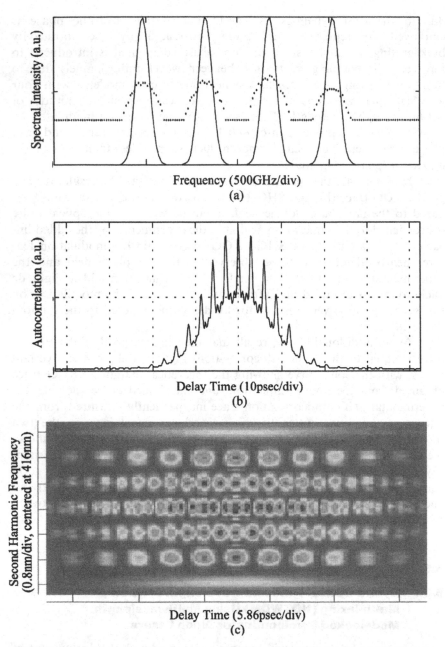

Frequency (500GHz/div)
(a)

Delay Time (10psec/div)
(b)

Delay Time (5.86psec/div)
(c)

Figure 43 Numerical simulation of the experimental SHG-FROG, which includes intracavity gain dynamics and dispersive effects.

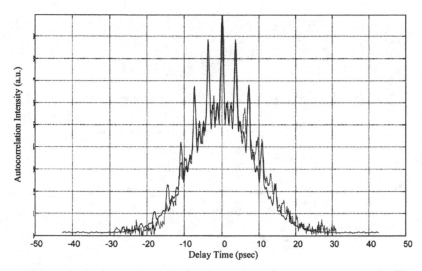

Figure 44 Overlaid plots of the theoretical marginal from the SHG-FROG simulation, with the experimentally measured composite intensity autocorrelation.

development of these networks is the availability of cost-effective photonic technologies that will enable the generation, transmission, and processing of vast amounts of data. The present state-of-the-art optical communications and signal processing typically rely on a wavelength division multiplexed (WDM) or a timed division multiplexed (TDM) hardware platform. In either case, to achieve the maximum data transmission and processing rates, the user is required to operate at the limits of either technology. An alternative approach to achieving ultrahigh-speed signal processing and transmission rates is to combine the advantages of the two technologies, i.e., to develop a hybrid technology that relies on both WDM and TDM technologies. In this case, this hybridized data format is allowed to exploit the advantages of both technologies to provide the user with the required state-of-the-art processing and transmission rates without pushing an individual technology to its limits. To demonstrate the potential of this approach, an ultrahigh-speed optical link that employs these unique technologies is schematically represented in Figure 45. To facilitate this approach, we use a multiwavelength modelocked semiconductor laser that generates over 20 independent wavelengths from a single device. Data are transmitted by modulating each wavelength at a data rate of 5 Gbit/s. Owing to the modelocked nature of the semiconductor laser, the optical data are transmitted in the return-to-zero (R-Z) data format. At the receiver, the data are demultiplexed by employing an all-optical clock recovery

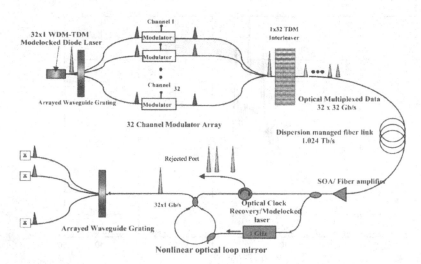

Figure 45 Schematic diagram of a hybrid optical time–wavelength division multi-plexed optical link, using a multiwavelength modelocked laser, all-optical clock recovery, and a nonlinear optical loop mirror for temporal demultiplexing.

oscillator in conjunction with an ultrafast nonlinear optical loop that uses a semiconductor optical amplifier as the nonlinear element. Owing to the temporal duration of the data bits ($\sim$10 ps), aggressive temporal interleaving of the data should allow for single-wavelength channel data rates of 40 Gbit/s, implying a total aggregate data transmission of 800 Gbit/s from a single semiconductor laser diode.

5.5.2 Ultrahigh-Speed Optical Sampling for Analog-to-Digital Applications

Another scenario that can take advantage of the unique output from the multiwavelength modelocked semiconductor laser is to employ the multi-wavelength modelocked laser as a novel sampling stream that can be used to sample ultrawideband RF and microwave signals. As an example, we use the multiwavelength modelocked diode laser as an optical sampling tool for applications relating to wideband analog-to-digital conversion and sig-nal processing. A simple architectural scheme that can exploit our multiwa-velength technology is illustrated in Figure 46. The salient feature of this architecture exploits the steplike dispersion characteristic that can easily be realized by employing WDM demultiplexers and fixed optical tap delay lines, identified as the wavelength dispersion in Figure 46. In the simplest

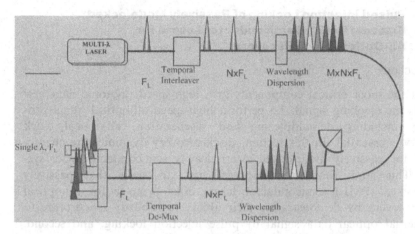

Figure 46 Schematic diagram using a multiwavelength modelocked diode laser in a high-speed optical sampling system. De-Mux, demultiplexer.

scheme, a multiwavelength laser operating at a pulse repetition rate of F with M wavelengths can be converted to an optical sampling stream with a sampling rate equal to $F \times M$, by converting the multiple wavelengths that are generated synchronously (in parallel) to a chirped serial temporal stream of pulses by using dispersion to force each separate wavelength component to experience a different optical delay. In this case, the overall effect is to convert a low bit rate sampling stream into a high bit rate sampling stream. At the receiver end, WDM demultiplexers can then be employed to separate out each individual wavelength component (which occurs at the low bit rate) and conventional off-the-shelf components can be employed for signal processing. This allows for ultrahigh-speed signal processing that only relies on low-speed electronics. It should be noted that this general concept can be extended to include an additional sampling rate multiplication factor by including temporal interleaving and subsequent optical time domain demultiplexing.

To achieve each of the above networking and signal processing capabilities, three important signal processing functions need to be demonstrated: (1) optical clock recovery, (2) all-optical switching of multiple wavelength packets, and (3) conversion of temporally parallel wavelength channels to a temporally serial wavelength sequence, e.g., optical parallel-to-serial conversion. In the next section, we show how modelocked semiconductor lasers can be used to realize these three important signal processing function.

5.5.3   Pulsed Injection Locking of Passively Modelocked External Cavity Semiconductor Lasers for All-Optical Clock Recovery

Introduction

One of the most critical components in a high-speed photonic data network is the clocking signal. To perform high-speed all-optical signal processing, including demultiplexing and regeneration, all-optical clock recovery is essential. In this section, clock recovery dynamics are characterized for passively modelocked externalcavity semiconductor laser systems. This study has two purposes: first, to show that passively modelocked (PML) semiconductor lasers offer promise for all-optical clock recovery by demonstrating their ability to robustly synchronize to an external optical data signal by pulse injection locking, and second, to characterize the critical performance factors for all-optical clock recovery systems and to provide insight into the experimental techniques for measuring their characteristics.

System Description

Figure 47 illustrates an experimental schematic for all-optical clock recovery by pulse injection locking of a passively modelocked semiconductor laser. The data signal, generated by a hybridly modelocked semiconductor laser system, is injected into the clock oscillator by using a pellicle beam splitter with low reflectance (~4%) to minimize cavity losses in the clock oscillator and to minimize insertion loss for the data signal (~0.2 dB). The clock recovery oscillator consists of a semiconductor optical amplifier (SOA)

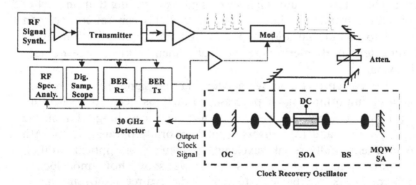

Figure 47   Experimental scheme for all-optical clock recovery.

inside an external optical cavity, with a multiple-quantum-well (MQW) saturable absorbed (SA) located at the back reflector.

To characterize the optical clock signal, the clock oscillator's output is received by a high-speed photodetector (30 GHz). The output of the photodetector is connected to one of several diagnostic systems, including an RF spectrum analyzer for spectral analysis, a digital sampling oscilloscope (DSO) for temporal characterization, and a bit error rate (BER) receiver for signal error analysis. For some experimental measurements on the DSO, the signal is amplified by a low-bandwidth amplifier (500 MHz) to minimize aliasing effects due to the high-frequency components (from the short pulse width) of the clock signal.

Theory of Operation

The mechanisms that establish passive modelocking in the clock recovery oscillator are also responsible for phase locking to an injected data signal for clock recovery. The theory of passive modelocking with a slow saturable absorber, where the pulse width is much shorter than the absorption recovery time, is well understood [40,41], and models have been presented for pulse injection locking [42,43] of PML lasers.

This interplay of gain saturation and saturable absorption in the PML laser is the predominant factor in the robust clocking dynamics of the system. The presence of additional optical flux, through pulse injection of a data signal, alters the dynamics of the saturation processes, causing the clock oscillator to synchronize to the injected data stream. It should be noted that the additional optical flux alters the modelocking process more easily if the gain/absorption is unsaturated. This implies that more injected power is required to achieve synchronization when the injected data signal arrives at the saturating medium immediately after the inherent cavity pulse has saturated the medium. Less injected power is required when the medium has had sufficient time to recover.

When an injected data pulse passes through the SOA before the inherent cavity pulse, it starts to saturate the gain as it is amplified. Then, when the inherent cavity pulse passes through the SOA, the gain saturates sooner, reducing the trailing edge of the inherent pulse more quickly. Likewise, as the injected data pulse passes through the saturable absorber, it starts to bleach the absorber, reducing the absorption of the leading edge of the inherent cavity pulse. The net result of the altered pulse shaping is to shift the inherent clock pulse toward the injected data signal. The experimental results presented in this paper show that only a very small data signal is required to achieve clock synchronization.

Error-Free Clock Recovery

Regenerated clock signals from the injection-locked PML modelocked laser are illustrated in Figure 48. Figure 48a shows the synchronized clock signal when the PML laser was fundamentally modelocked. Harmonic passive modelocking and subsequently harmonic clock recovery were demonstrated by modifying the gain and absorption saturation energies of the SOA and SA, respectively. The gain saturation energy was changed by adjusting the dc current bias of the SOA, which alters the carrier injection rate and, subsequently, the carrier concentration. The absorption saturation energy was changed by adjusting the location of the focal spot on the SA, due to non-uniformities in the absorber caused by the lift-off process, proton implantation, and the fabricating growth process. Figures 48b–48d show clock signals at 622 MHz, 933 MHz, and 1.244 GHz, respectively, when a 311 MHz all-1's data stream was used for clock synchronization. Harmonic clock recovery from the same PML laser, as demonstrated in Figure 48, shows the potential capability of a reconfigurable clocking rate from a single

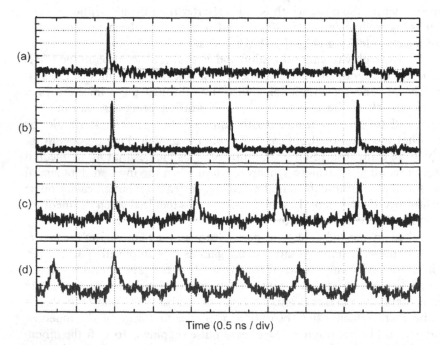

Time (0.5 ns / div)

Figure 48 Regenerated clock signal from a passively modelocked laser at the fundamental (a) 311 MHz and at the harmonics (b) 622 MHz, (c) 933 MHz, and (d) 1.244 GHz.

clock oscillator. The experimental results show that the clock pulses were degraded with higher harmonic operation as a result of cross-gain modulation between the multiple intracavity pulses due to the long recovery time (~ 1.1 ns) of the SOA. It should be noted that fundamental clock recovery has been demonstrated at 10 Gbit/s and higher, typically using monolithic devices [44,45].

A bit error rate (BER) measurement was performed to verify error-free operation of the clock oscillator. The BER measurement was performed using a bit test pattern of [10101010]. With 140 μW of average injected data power, the clock oscillator remained synchronized to the data signal, with no missed clock bits, for over 10 min (error rate $<5 \times 10^{-12}$). Injecting data with average power as low as 50 μW resulted in at least 1 min of error-free operation.

In general, short bursts of errors occurred with increasing frequency as the amount of injected power was decreased, because the clock slipped momentarily out of synchronization. This continued until the injected power was reduced below a minimum at which the clock completely lost synchronization with the incoming data stream. Figure 49 illustrates the synchronization characteristics of the clock oscillator in the RF power spectrum of

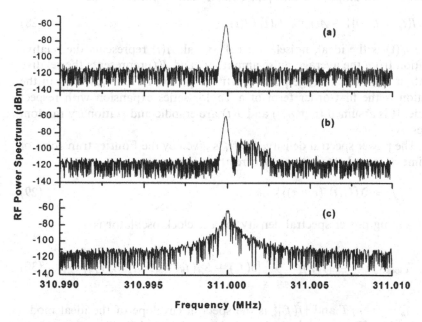

Figure 49 Radio-frequency power spectrum of clock signal when injected power is (a) above threshold, (b) at threshold, and (c) below threshold.

the clock signal as the injected data power was decreased. The injection threshold for this system was measured to be 3–6 μW average power (10–20 fJ pulse energy), which is less than 1% of the clock intracavity power.

Phase Noise (Timing Jitter) and Amplitude Noise

Phase noise and amplitude noise are important parameters to characterize for clocking systems, because they adversely affect the clocking performance. High timing stability is a fundamental requirement for clocking systems, which implies that the phase noise should be as small as possible. Amplitude noise is especially significant when a clock signal is to be used for data regeneration or optical logic that relies on the clock intensity (e.g., nonlinear and interferometric switching systems) where it will be directly mapped onto the processed data signal.

Standard RF power spectrum techniques were used to measure the phase noise and amplitude noise of the clock oscillator [46,47]. It should be noted that by performing AM and PM noise measurements in the frequency domain, additional information regarding the physical mechanism of these noise sources can be identified, e.g., the roll-off of the noise band, the frequency band of the noise, etc. The modelocked signal from the clock oscillator, with amplitude and phase noise, can be expressed as

$$I(t) = I_0(t)[1 + A(t)] + I_0(t)TJ(t) \tag{28}$$

where $I_0(t)$ is the ideal, noiseless, clock signal; $A(t)$ represents the relative deviation from the average pulse amplitude; and $J(t)$ represents the relative deviation from the average pulse repetition period T. The last term in the equation is the first-order term of a Taylor series expansion with respect to time. It is assumed that $A(t)$ and $J(t)$ are ergodic and stationary random processes.

The power spectral density, $S_p(f)$, is given by the Fourier transform of the time-averaged intensity autocorrelation function:

$$S_p(f) = \Im\{\langle I(t)I(t+\tau)\rangle\} \tag{29}$$

The resulting power spectral density for the clock oscillator is

$$S_{\text{Clock}}(f) = \frac{2\pi}{T^2}|\tilde{f}(f)|^2 \sum_{n=-\infty}^{+\infty} [\delta(f_n) + S_A(f_n) + (2\pi n)^2 S_J(f_n) \tag{30}$$

where $f_n = f - n/T$ and $|\tilde{f}(f)|^2$ is the spectral envelope of the ideal modelocked pulses. The term n is the harmonic number, and S_A, S_J are the amplitude and phase noise power densities, respectively.

Figure 50 shows the noise power spectra of the clock recovery oscillator on the fundamental and tenth harmonics over a frequency offset range from 100 Hz to 10 MHz. Clock performance for both free-running and injection-locked operation are displayed.

The fractional pulse-to-pulse rms timing jitter is calculated using

$$\frac{\Delta t}{T} = \frac{1}{2\pi} \left(\frac{P_2 - P_1}{P_c(n_2^2 - n_1^2)} \right)^{1/2} \tag{31}$$

where n_2 and n_1 are the harmonic numbers, P_1 and P_2 are the total integrated powers contained in the noise side bands at the harmonics, and P_c is the carrier power. Using Eq. (31) for the power spectra in Figure 50, the clock oscillator's fractional timing jitter is measured to be 2×10^{-3}.

The amplitude noise can also be determined from the noise power spectra. The relative pulse energy fluctuation is calculated using

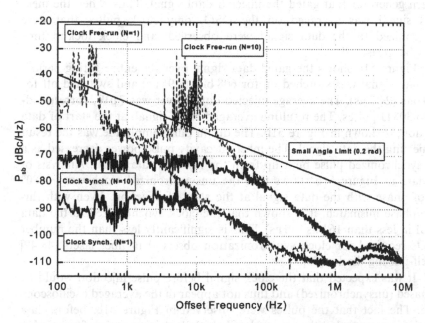

Figure 50 Noise power spectrum at the fundamental and tenth harmonic of the clock oscillator when synchronized to data (solid) and free-running (dashed), measured from 100 Hz to 10 MHZ offset frequencies.

$$\frac{\Delta E}{E} = \left(\frac{n_2^2 P_2 - n_1^2 P_1}{P_c(n_2^2 - n_1^2)}\right)^{1/2} \tag{32}$$

Using Eq. (32) for the power spectra in Figure 50, the clock oscillator's relative pulse energy fluctuation is measured to be 0.05.

Lockup Time

It is important to measure the time required to achieve synchronization of the clock oscillator to an injected data stream, because clock lockup time will affect system latency. To measure the lockup time of the clock recovery oscillator, the injected data signal was gated using an SOA modulator or a Mach–Zehnder lithium niobate intensity modulator. The data signal was terminated for a long temporal period to ensure the loss of clock synchronization. When the data signal was re-established, time was required for the clock laser to resynchronize to the data signal.

The clock output trace was detected with the high-speed photodetector, amplified by the low-bandwidth (500 MHz) amplifier to reduce aliasing, and measured by the DSO. The DSO was triggered by the same pattern generator that gated the injected data signal. Thus, when the measured signal was averaged on the DSO, only clock pulses that were synchronized to the data signal were observed, and clock lockup time could be measured.

Figure 51a shows the gated data signal over the entire gating period. The data signal was switched on for 608 bits (1.95 μs) and switched off for 7584 bits (24.3 μs). The average injected data power was 7 μW, corresponding to 303 fJ pulses. The resulting averaged clock signal, at the start of data injection, is shown in Figure 51b. The dip in pulse amplitude was the result of injecting the data signal. The inherent cavity pulse power decreased as a new synchronized pulse built up from the injected data signal. Analysis of the data in Figure 51b shows that the unsynchronized clock was ~180° out of phase with the data signal at the start of injection. Even with this large phase mismatch, it is shown that the clock synchronized to the data signal in less than 16 bits (50 ns). This is significantly less than the number of bits required for clock synchronization observed in fiber lasers [48,49] or self-pulsing diodes [50].

It was expected that the clock signal before pulse injection would be dephased (unsynchronized) and thus not appear in the averaged oscilloscope trace. The fact that the pulses were observed in Figure 51b, before data injection, suggests that the accumulated clock dephasing arising from timing jitter was quite small. Loss of synchronization was, rather, a result of a slight mismatch between the free-running clocking rate and the data rate

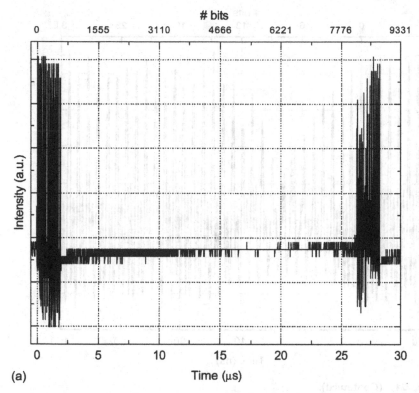

Figure 51   (a) Gated data signal, (b,c) Clock signal at the start of data injection (b) with and (c) without slight frequency offset.

and is deterministic in nature. It should be noted that in this case the clock and data signals will dephase at the beat frequency. By carefully adjusting the clock's natural repetition rate, it was possible to accurately match the data rate, in which case no dephasing occurred and the clock remained synchronized, even with no data signal present, for all 7584 bits (24.3 µs). This is illustrated in Figure 51c. It should also be noted that the clock signal did not actually dephase in 24.3 µs, but rather the experimental measurement was limited by the experimental techniques, which required that the high-speed DSO be triggered slowly to be able to average over the entire gated bit pattern. It is likely that the clock dephasing time is even longer than what was demonstrated.

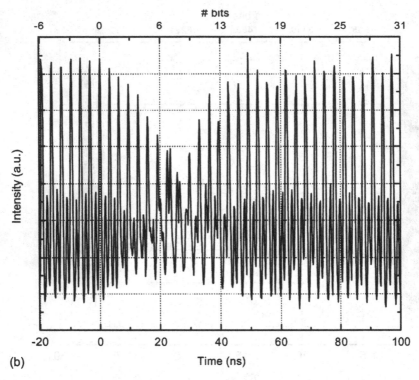

Figure 51   (Continued).

Frequency-Locking Bandwidth

The nominal frequency of the clock oscillator is controlled by its cavity length. This frequency will vary slightly as environmental parameters (temperature, dc bias, etc.) slightly alter the effective cavity length. Deviations in the data signal frequency can also exist, especially if the clocking system is used in an optical network, where data streams may come from a number of independent sources. Thus, to ensure proper clock operation, the clocking system should have a large frequency-locking bandwidth.

To demonstrate the frequency-locking bandwidth of the clock recovery oscillator, the clock's performance was observed while varying the input data signal's carrier frequency. The data rate was varied by adjusting the frequency of the signal generator that provided the RF bias modulation for the actively modelocked data laser. Figure 52a shows the maximum upper and lower frequency offsets achieved for a given average injected data power. The inset graph shows the total frequency-locking bandwidth of the

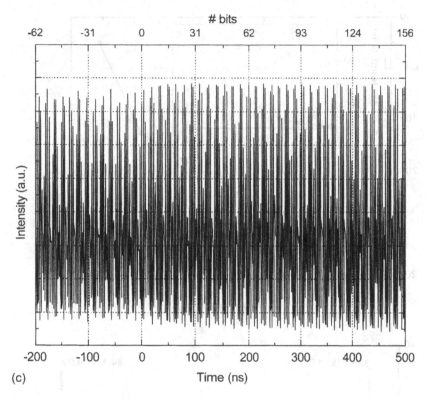

(c)

Figure 51 (Continued).

clock oscillator as a function of injected data power. At the clock recovery threshold power, there was almost no locking bandwidth. The clocking rate must match the data repetition rate exactly in order to achieve clock synchronization. As the injected data power was increased, the locking bandwidth increased until very large powers, where the increase rolled off. The maximum locking bandwidth observed was 833 kHz, which corresponds to a fractional bandwidth of 2.9×10^{-3}. Larger locking bandwidths may be possible with greater injected data power. It should be noted that these results of locking bandwidth are in good agreement with the locking bandwidth in clock recovery experiments at 10 GHz [51], showing that significant insight into the clock recovery process can be obtained by using lower data rates and extrapolating the results to higher data rates.

At first, one might suspect that the shift in clock frequency is caused by a change in the index of refraction in the SOA due to data injection. However, the magnitude and sign of the resulting change in index of refraction

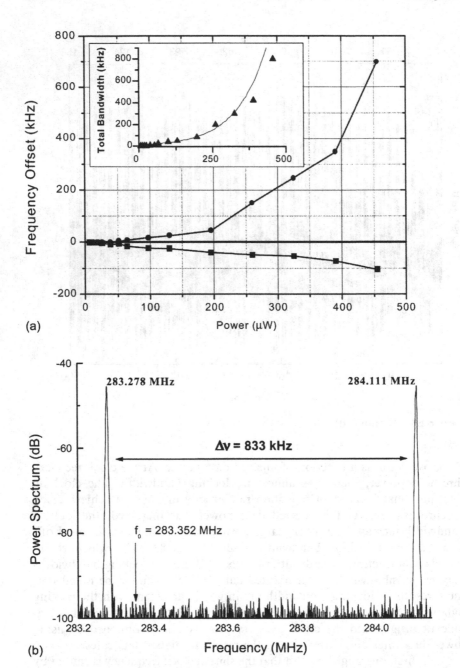

Figure 52 (a) Repetition rate detuning bandwidth vs. injected power. (b) RF power spectrum of clock signal as injected data rate is detuned.

do not support this conclusion. Injecting the data signal reduces the carrier concentration through gain depletion, which increases the index of refraction and the effective cavity length. This implies that injecting a data signal *decreases* the clock frequency, whereas the experimental results show that the clock frequency is more easily increased. Ignoring the sign and considering only the magnitude of the frequency shift, to achieve the positive frequency shift observed in Figure. 52b (752 kHz), the index of refraction of the SOA would need to change by a factor of 2.83, which is physically unrealistic. It should be noted that by performing this experiment at a low data rate, we can conclusively determine that the frequency shift is not due to changes in the refractive index, which would not be clearly evident if the clock were operating at 10 GHz, where small changes in the index could make a significant impact.

Immunity to Bit Pattern Effects

The clock oscillator should operate properly independently of the bit pattern that is encoded onto the data signal. To show immunity to bit pattern effects, two bit patterning experiments were performed: one to observe the clocking behavior for a data signal with few 1's (because 1's are the data pulses that cause the synchronization effects in the clock) and one to observe the clocking behavior for a data signal with a long series of 0's.

First, a subharmonic data signal was used to show locking to a sparse signal with few 1's. Stable clock regeneration was achieved using a data signal at one-fiftieth of the clock repetition rate. Figure 53 shows the output pulse stream from the clocking system, operating at 283 MHz, when a 5.66 MHz data signal was injected. The bold trace indicates the injected data signal. It should be noted that the sinusoidal modulation on the recovered optical clock is due to the aliasing incurred by displaying the entire clock stream and two consecutive logical 1's within the data packet on the DSO. This modulation is not actually present on the clock signal.

This experiment also demonstrates the capability of subharmonic injection locking of a clock oscillator. This becomes especially important when the clock oscillator operates at ultrafast speeds, where it may be necessary to achieve synchronization or phase stabilization with a slower optical signal. Synchronization of an ultrafast clock has been demonstrated in this way, through subharmonic pulse injection, at up to 40 GHz [52].

The second experiment, previously discussed in the section on lockup time, concerns locking the clock oscillator to a data signal that contains a long string of 0's. The data signal consisted of 76 words of data 1's (608 bits) followed by 948 words of all 0's (7584 bits). Figure 51c showed the clock signal at the end of the long string of zeros, showing that the clock remained

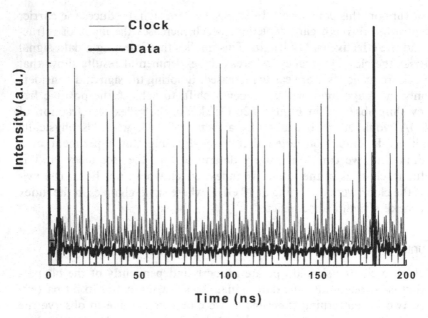

Figure 53   Clock recovery for the 1/50 subharmonic data input.

synchronized. This shows that the clock remains synchronized for over 24.3 µs in the absence of a data signal.

The bit pattern immunity is a result of two conditions. First, that the pulse-to-pulse timing jitter, which is the predominant mechanism that limits locking stability, is very small. Second, that the pulse energy required to achieve clock synchronization is small, due to the high gain and low saturation energy of the SOA.

Summarizing injection locking of modelocked diodes, we see that the robust clocking dynamics of injection-locked PML semiconductor lasers make them ideal candidates for systems that require all-optical clock recovery. In addition, the experimental characterization of the clock recovery dynamics provides insight into the characteristics of modelocking and parameters that drive clocking performance for many all-optical clock recovery techniques.

5.5.4   Multiwavelength All-Optical TDM Switching Using a Semiconductor Optical Amplifier in a Loop Mirror

Many high performance optical communication and signal processing systems employ architectures based on wavelength division multiplexed (WDM) formats or optical time division multiplexed (OTDM) formats.

As noted above, a hybrid approach may be able to exploit the parallelism of WDM architectures and the speed of OTDM. As an example, in an optical fiber communication link using a conventional WDM approach, individual users are assigned a specific time slot and wavelength channel. At the receiver, the aggregate data channel is initially wavelength-filtered and subsequently time-demultiplexed to select a single data channel. Increasing the bandwidth of the user generally requires faster signal processing at the receiver site. By employing a hybrid approach, one can increase the user's data rate without increasing the TDM channel rate by encoding data on multiple wavelengths and transferring wavelength bits in parallel [53,54]. However, in order for the user to receive parallel data across multiple wavelength channels in a single time slot, a receiver component capable of multiwavelength TDM demultiplexing is necessary.

In this section, an all-optical demultiplexer capable of simultaneously switching multiple wavelength channels is demonstrated. The demultiplexer uses a semiconductor optical amplifier asymmetrically placed inside a loop mirror to achieve multiwavelength switching with high-speed capability. This multiwavelength demultiplexer, combined with the optical clock recovery unit discussed above, can be capable of demultiplexing parallel WDM coded data streams.

Experimental Setup

The experimental system is illustrated in Figure 54. The data laser transmitter is a multiwavelength modelocked diode laser operating with four modelocked wavelength channels with 1 nm nominal channel spacing at 2.5 Gbit/s. The control laser is a standard hybrid modelocked external cavity diode laser with single-wavelength output. Each laser is amplified by a single pass through a semiconductor optical amplifier (SOA) to provide 15–20 mW of optical power.

The active RF drive for both lasers is provided by RF signal synthesizers with locked time bases, to produce synchronized output data pulses at 311 Mbit/s. It should be noted that optical clock recovery techniques could be employed to establish synchronized data and control pulses. The data laser is optically multiplexed for a 2.5 × 4 Gbit/s data rate.

The optical switch is a free-space terahertz optical asymmetrical demultiplexer (TOAD) [55]. It consists of a nonlinear loop mirror and uses a semiconductor optical amplifier as the nonlinear optical element. The maximum switching speed for this type of loop mirror is primarily determined by the response of the nonlinear element. In our case, the nonlinear optical response induced by the control signal is due to the change in refractive index in the

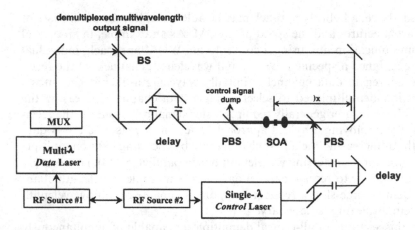

Figure 54 Schematic of the multiwavelength-switching experiment setup. SOA, semiconductor optical amplifier; BS, beam splitter; PBS, polarization beam splitter;)x, distance from the loop midpoint; MUX, optical multiplexer.

SOA caused by gain depletion. It has been shown that demultiplexing can be achieved for single-wavelength data rates up to 250 Gbits/s [56].

The nonlinear element in the optical loop is a 500 μm long angle-striped GaAs-AlGaAs SOA. The switching window is controlled by the distance of the nonlinear element from the midpoint of the loop. For this experiment, the switching window was set to 250 ps by adjusting the delay leg inside the loop.

The control and data signals are launched into the optical switch with orthogonal polarizations. The gain of the SOA is polarization-dependent; therefore, the polarizations of the input signals are selected for maximum gain for the control signal and minimum gain for the data signal. This minimizes cross-talk between the control and data signals.

The control signal delay is adjusted such that the control pulse arrives at the SOA between the counterpropagating data pulse components, to enable switching. In this arrangement, only one of the counterpropagating data pulse components experiences a phase shift from gain depletion in the SOA, which causes constructive interference at the output port and destructive interference at the rejected port for all wavelength channels. Thus, for a single time slot, all wavelength channels are switched simultaneously.

To measure the performance of all-optical switching of multiple wavelength channels, the switched data output was measured in both temporal and spectral domains. To accurately measure switching contrast in the spectral domain, the optical multiplexer was bypassed to allow the data laser to operate at the control laser rate, i.e., 311 Mbit/s. This allows each four-wavelength data pulse packet to be switched and shows contrast without accumulating leakage from multiple rejected data pulse packets.

Switching was achieved by using control pulses with an energy of 1.6 pJ/pulse. Figure 55a shows the switch output in the spectral domain, measured with a spectrometer and a linear diode array readout system. It can be seen from this figure that all wavelength components are switched with high contrast and minimal distortion. It should be noted that variation in the peak intensity for each wavelength channel is mapped from the input data laser and is not a factor of the switching response. Figure 55b shows the output of the switch in the time domain, measured with a 30 GHz photodetector. This figure shows high switching contrast in the time domain. For both plots in Figure 55, the bold trace shows the output of the switch in the presence of a control signal. The dotted trace shows the output of the switch without a control signal. The signal contrast appears to be at least 10 dB in the time domain and even greater in the spectral domain. By integrating the energy in the wavelength channels with and without a control signal, we observed contrasts of at least 13 dB, with some channels showing contrast as great as 23 dB.

Across the optical spectrum, the switching contrast is shown to be as good as or better than that shown in the time domain. The enhanced contrast in the spectral domain can be attributed to better spatial filtering of the switch output signal. Because the switch employs a free-space interferometer, the nonoptimized mode patterns contribute to the switching contrast. For the time domain measurement, the output beam is spatially filtered through an aperture. However, the low sensitivity of the photodetec-

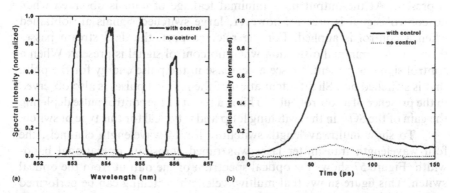

Figure 55 Output intensity from multiwavelength switch in (a) spectral domain and (b) time domain.

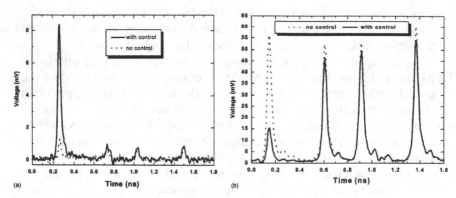

Figure 56 (a) Switch output and (b) rejected port output of the optical switch showing demultiplexing of 2.5 Gbit/s data.

tor requires a large aperture in order to have sufficient power to observe. The spectrometer used for the spectral measurements, however, has much better sensitivity, which allows a decrease in the aperture size and improves the observation of the contrast. Additional filtering was also provided by the spectrometer's input slit.

To show the demultiplexing capability of the switch, the data laser was multiplexed up to 2.5 Gbit/s with an encoded bit pattern of 11110000 using an optical multiplexer. Figure 56a shows the output of the optical demultiplexer with 2.5 Gbit/s data input. Figure 56b shows the output from the rejected port of the optical switch with the same 23.5 Gbit/s data input. In both cases, the bold trace shows the demultiplexed data in the presence of the control signal. The broken trace shows the data when no control signal is present. At the output port, minimal leakage of data is observed when no control signal is present; however, large switched signals are obtained when a control is applied. For the rejected port, the data stream passes through with minimal distortion when no control signal is present. When a control signal is present, we see a decrease in the pulse energy for the pulse that is switched out. Slight attenuation of the rejected pulses is also observed in the presence of a control pulse. This is a result of the control pulse depleting the gain of the SOA in the switching loop and is typical for this type of switch.

To show multiwavelength switching for 14 wavelength channels, the four-wavelength transmitter laser was tuned across its operational bandwidth. Figure 57 shows the optical spectrum of the output from the optical switch. This figure shows that multiwavelength switching can be performed across a wide spectral bandwidth with high contrast. The alternating bold and light traces show the separate data sets for each tuning of the transmitter

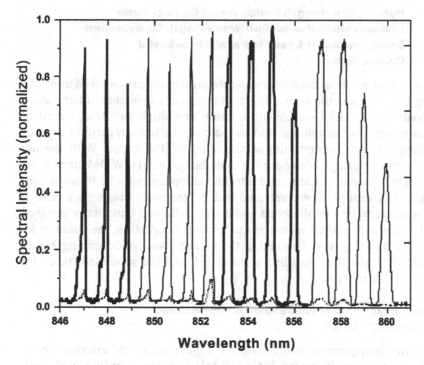

Figure 57 Multiwavelength switching across 14 nm bandwidth.

laser. The broken line shows the leakage from the switch when no control is present.

We expect that this type of optical switching can be performed over a wider spectral bandwidth than demonstrated. A mismatch between the gain bandwidths of the data laser and the SOA in the switch ultimately limited the switching bandwidth in our experiment. Only three wavelengths were used in the shortest wavelength measurements because the data laser was operating near the edge of its tuning bandwidth. On the long-wavelength side, the measurement was limited by the gain bandwidth of the SOA in the optical switch.

These experimental results show that hybrid WDM/TDM architectures incorporating multiwavelength TDM switching may be useful for providing high data rates to the end user by utilizing parallel bit transmission on multiple wavelength channels.

5.5.5   Hybrid Wavelength Division and Optical Time Division Multiplexed Multiwavelength Modelocked Semiconductor Laser in Parallel-to-Serial Conversion

In this section we show the performance of an actively modelocked multiwavelength semiconductor laser using a commercially available, off-the-shelf, fiberized grating based wavelength division demultiplexer. Also, we introduce a unique and simple method to increase the pulse repetition rate, or sampling rate, by incorporating a secondary WDM device. With the use of the multiwavelength laser source and the secondary WDM device, we can realize a simple, efficient, and ultrafast all-optical parallel-to-serial conversion. The results shown are analyzed by temporal diagnostics from a high-speed sampling oscilloscope and a spectrally and temporally resolved streak camera diagnostic measurement system. In addition, the characteristic of the filter function containing the arrayed waveguide grating router (AWG) filter and the delay line is simulated, and the simulated trace of a streak camera is shown.

Experimental Setup—Multiwavelength Modelocked Semiconductor Laser

The cavity configuration is illustrated in Figure 58. In the external cavity, a 2.3 mm, long index-guided InGaAsP-InP multiple quantum well semiconductor optical amplifier is used as the gain medium. The diode device exhibits a full width at half-maximum (FWHM) spontaneous emission spectral bandwidth of 28 nm with a central emission wavelength of 1556 nm. To produce multiple wavelengths, a commercially available WDM filter (insertion loss ~4 dB) is incorporated inside the cavity as a

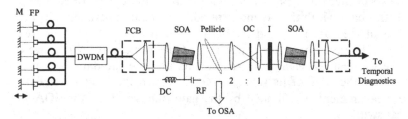

Figure 58   Schematic of the multiwavelength modelocked semiconductor diode laser system. M-mirror; FP-fiberport; FCB-fiber coupling bench; DWDM-dense wavelength division multiplexer; SOA-semiconductor optical amplifier; OC-output coupler; I-isolator; OSA-optical spectrum analyzer.

wavelength-selective element. The spontaneous emission spectrum of the diode and WDM spectrum from each channel are shown in Figure 59. It should be noted that the intensity distributions of the spontaneous emission and WDM channels are normalized for clarity. The 16-channel WDM filter [57] is designed to have the transmission channels ranging between 1533 to 1557 nm, with 200 GHz (1.6 nm) channel spacing. Because the gain peak of the SOA device is near 1556 nm, the WDM channels closest to this wavelength are employed, i.e., 1550.6, 1552.2, 1553.8, 1555.4, and 1557 nm. The output fibers are connectorized and connected to commercially available fiber collimators. The resultant light from the WDM collimated channels travels to individual mirrors and is reflected back into the fiber. The position of each mirror and the mirror-to-fiber coupling efficiency can be adjusted independently to optimize the cavity length and to obtain a uniform spectral intensity for each wavelength, respectively. It should be noted that the stability of the laser was good, with minor adjustment of the end mirrors required after a period of about 1 h.

The cavity is actively modelocked by modulating the injection current applied to the SOA at a repetition rate of 169.5 MHz, which is the seventh harmonic of the fundamental cavity frequency and corresponds to a cavity with an optical path length of approximately 6.2 m. Although the optical

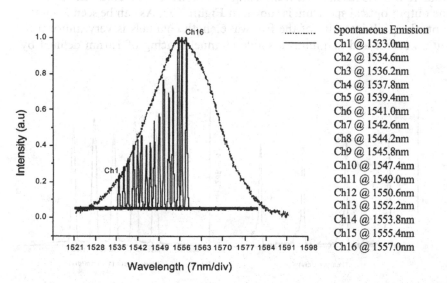

Figure 59 Output spectral characteristics of spontaneous emission and the 16-channel DWDM filter.

gain in semiconductor laser diodes is homogeneously broadened and normally suffers from gain competition effects under pure continuous wave operation, the regime of transient unsaturated gain allows the oscillation of multiple wavelengths under modelocked operation.

The laser employs a 30% output coupler placed at the focus of a 2:1 telescope. This configuration improves the cavity stability and reduces the transverse mode field diameter to allow transmission through an optical isolator. The average output power from the laser is approximately 0.6 mW and is subsequently amplified by an additional semiconductor optical amplifier with 80 mA bias current. The output power after amplification is approximately 6 mW, providing an overall single-pass gain of 10, including coupling losses. Without any injection to the amplifier, the spontaneous emission is 0.25 mW at 80 mA bias current, providing relatively high contrast optical pulses.

Experimental Results—Multiwavelength Semiconductor Laser

The performance of the multiwavelength modelocked semiconductor laser is characterized in both the temporal and spectral domains. The threshold current of this laser cavity is measured to be approximately 58 mA. The harmonically modelocked pulse train output is shown in Figure 60a, operating at a repetition rate of 169.5 MHz. Each pulse contains five wavelengths and thus leads to a 847.5 MHz pulse train using ~ 170 MHz electronics. In addition, the output optical spectrum is shown in Figure 60b. As can be seen from the plot, the peak intensity of the five wavelength channels is very uniform to within 0.4 dB and maintains a stable channel spacing of 1.6 nm defined by

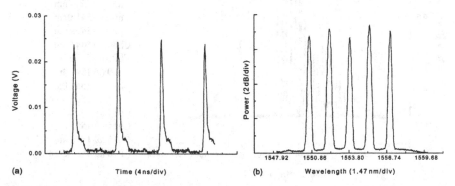

Figure 60 Output characteristics of the multiwavelength modelocked diode laser. (a) Time domain; (b) spectral domain.

the WDM filter used in the cavity. Output pulse duration and spectral width of 350 ps and 0.13 nm, respectively, are obtained. It should be noted that the spectral width is almost the limit of the resolution of the spectrometer (0.1 nm). The corresponding time–bandwidth product is calculated to be about 35.

Experimental Setup—Parallel-to-Serial Wavelength Conversion

The parallel-to-serial conversion configuration, or wavelength-to-time mapping, is shown in Figure 61. The five-wavelength modelocked semiconductor laser from the cavity is used as a source in our experiments. The output light is incident on a Hitachi eight-channel AWG [58] to separate all the wavelengths into their respective individual channels. To temporally distribute five wavelengths uniformly within one modelocked period, the optical path of one wavelength channel is fixed and serves as a reference, and the other wavelengths are delayed with respect to the fixed wavelength. The amount of delay necessary to achieve this task is first calculated and then realized by preparing individual fibers to provide the required optical path difference between channels. This approach provides a rough but cost-effective way to achieve the desired temporal distribution of wavelengths. The precise adjustment of delay can be achieved by using commercially available hybrid fiber/free-space optical delay lines. Once each wavelength channel is properly delayed, they are combined with a standard 8×1 coupler. Owing to the insertion loss associated with both the AWG and the coupler, the output serial pulse train is amplified using an erbium-doped fiber amplifier (EDFA), and the resultant amplified signal is directed to the temporal and spectral diagnostics.

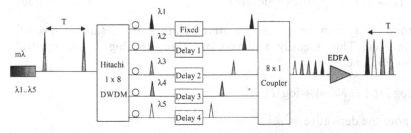

Figure 61 Schematic of the parallel-to-serial wavelength conversion mλ-multi-wavelength laser; T-modelocking period; DWDM-dense wavelength division multiplexer; EDFA-erbium-doped fiber amplifier.

Theory and Simulated Results—Parallel-to-Serial Wavelength Conversion

The output pulse from the oscillator contains five wavelengths simultaneously and can be expressed as

$$E_{in}(t) = A \sum_{n=1}^{5} \exp\left[-\pi\left(\frac{t - t_0}{a}\right)^2\right] \exp(-it\omega_n) \tag{33}$$

where A is the constant amplitude of each of the individual electric fields, a is the pulse duration of each pulse, t_0 is an arbitrary delay to denote the time of peak value of the electric field, and ω_n are the n individual carrier frequencies. The composite multiwavelength pulse is subsequently passed through the AWG filter and the fiber delay line. The resulting output pulse can be expressed as

$$E_{out}(t) = B \sum_{n=1}^{5} \exp\left[-\pi\left(\frac{t - t_n}{b}\right)^2\right] \exp(-it\omega_n) \tag{34}$$

where B is the amplitude of each of the individual electric fields, b corresponds to the output pulse duration, and t_n represents the temporal delay associated with the peak value of the n individual pulses. The filter function can be derived by dividing the Fourier transform of the output pulse by the Fourier transform of the input pulse:

$$F(\omega) = \frac{\Im[E_{out}(t)]}{\Im[E_{in}(t)]} \tag{35}$$

Assuming that the Fourier transform of $f(t)$ is $\tilde{F}(\omega)$, i.e.,

$$f(t) \rightarrow \tilde{F}(\omega) = |f(\omega)| \exp[i\Phi(\omega)] \tag{36}$$

the group delay can be found by manipulating the resulting spectral function $F(\omega)$. This is easily achieved by taking the log of the spectral function $F(\omega)$, i.e.,

$$\log \tilde{F}(\omega) = i\Phi(\omega) + \log|F(\omega)| \tag{37}$$

Next, note the derivative

$$\frac{d}{d\omega} \log \tilde{F}(\omega) = \frac{\tilde{F}(\omega)}{\tilde{F}(\omega)} = i\frac{d\Phi(\omega)}{d\omega} + \frac{d}{d\omega} \log|F(\omega)| \tag{38}$$

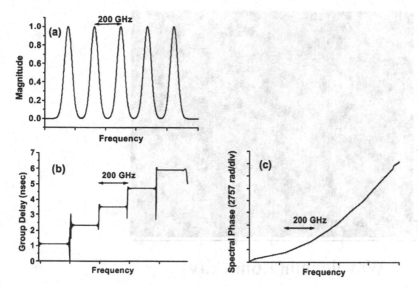

Figure 62 Filter function of the parallel-to-serial converter (a) magnitude; (b) ground delay; and (c) spectral phase.

where $\tilde{F}'$ is the derivative of $\tilde{F}$, and recognize that $\tilde{F}$ can be represented in terms of its real and imaginary parts, $\tilde{F} = F_R + iF_I$. By rationalizing the denominator, the derivative of the log function can be simplified as

$$\frac{d}{d(\omega)} \log \tilde{F}(\omega) = \frac{F_R F'_R + F_I F'_I + i(F_R F'_I - F_I F'_R)}{|\tilde{F}(\omega)|^2} \qquad (39)$$

By equating Eqs.(6) and (7), the group delay can be found as

$$\text{Group delay} \equiv \frac{d}{d\omega} \Phi(\omega) = \frac{F_R F'_I - F_I F'_R}{|F(\omega)|^2} \qquad (40)$$

Finally, if desired, the spectral phase can be found by integrating the group delay. Figure 62 shows the magnitude, the group delay, and the phase of the filter function. As shown in Figure 62b, the group delay can be expressed as a summation of unit step functions, i.e.,

$$\frac{d}{d\omega} \Phi(\omega) = \sum_{n=0}^{4} u(f - nf_0) \qquad (41)$$

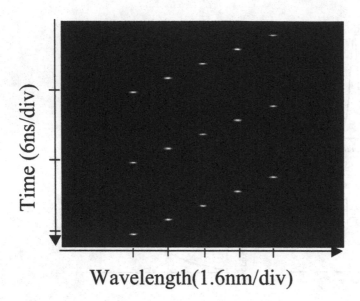

Figure 63 Simulation of a spectrally resolved streak camera plot, showing the evolution of wavelength versus time.

It should be noted that the spikes or noise on the group delay are due to dividing by values close to zero between wavelengths. As a final note on the relationship between the magnitude and phase of the filter function, one might assume that because the filter is linear and time-invariant, the magnitude and phase should be related by a Hilbert transform, or the Kramers–Kronig relations. However, this relation only occurs for filters that are minimum phase filters. Other linear time-invariant filters exist, which are classified as maximum phase filters, and for filters of this type the amplitude and phase response do not obey the Kramers–Kronig relations [59].

The output signal enters the streak camera. A gate function in the streak camera scans the incoming signal. Therefore, the output pulse can be expressed as

$$\text{STFT}_{E(t)}(\omega, \tau) = \int\limits_{-\infty}^{\infty} dt[x(t) \cdot g(t - \tau)] \ \exp(-i\omega t) \qquad (42)$$

where $\text{STFT}_{E(t)}$ represents the "short time Fourier transform" (STFT) of the signal $E(t)$ and $g(t)$ is the gating function [60]. In the present simulation, $g(t)$ is chosen as

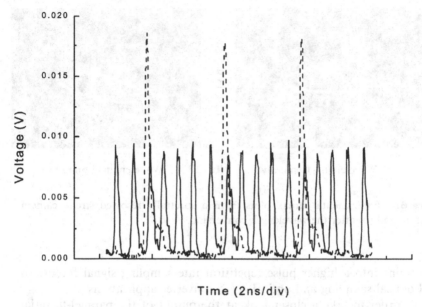

Time (2ns/div)

Figure 64 Modelocked pulse train of multiwavelength modelocked diode laser. Dashed trace: before the parallel-to-serial wavelength conversion. Solid trace: after the parallel-to-serial wavelength conversion.

$$g(t) = \exp\left[-\pi\left(\frac{t}{d}\right)^2\right] \tag{43}$$

The simulated $STFT_{E(t)}$ output is shown in Figure 63, which clearly shows that the Five wavelengths are now evenly separated in time.

Experimental Results—Parallel-to-Serial Wavelength Conversion

Figure 64 illustrates the time domain measurements before and after the wavelength conversion, as measured with a sampling oscilloscope and high-speed photodetector. The instrumentation bandwidth for these measurements was 20 GHz. The dashed line is the pulse train measured before the conversion. Again, each modelocked pulse contains five wavelengths simultaneously. The solid line indicates the pulse train measured after the parallel-to-serial conversion. As can be seen from the plot, the amplitude of the converted pulse train is uniform owing to the ability to independently control the intracavity gain to each channel separately. As a result, this technique can be suitable for converting lower bit rate multiwavelength optical

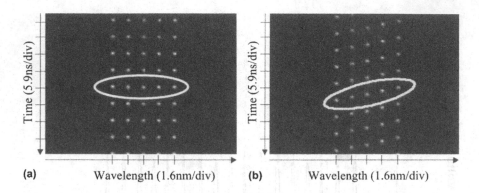

Figure 65 Experimental streak plots from a spectrally resolved streak camera (a) before and (b) after the wavelength conversion.

pulse trains into a higher pulse repetition rate sampling signal for use in novel optical sampling and optical A/D converter applications.

In order to take a closer look at the output of the parallel-to-serial conversion process, the STFT of the output laser pulse train is measured by temporally resolving the output spectrum. This is achieved by using an optical spectrometer in conjunction with a streak camera. In this configuration, the two-dimensional image intensifier detector array of the streak camera can be employed so that the temporal and spectral characteristics can be observed simultaneously. The output from the streak camera displays the resulting temporally resolved spectral output of the laser and is displayed in Figure 65. For the case of a parallel wavelength output shown in Figure 65a, i.e., before the conversion process, the image clearly shows that five wavelengths exist simultaneously. It should be noted that the streak camera is triggered with a repetition rate that is one-hundredth of the modelocked repetition rate. In this figure, the total temporal span is 50 ns, which contains approximately nine parallel wavelength packets. Figure 65b shows the experimental results after the parallel-to-serial conversion process. In this experiment, λ_1 is fixed, while λ_2 is delayed with respect to λ_1 by one-fifth of the modelocked frequency, λ_3 is delayed with respect to λ_1 by two-fifths, and so on. Because of the arrangement of the delay line of wavelengths with respect to one another in this experiment, the set of five wavelengths is now tilted upward toward the longer wavelength side. The overall result is that each wavelength now exists at a different time and the overall pulse repetition frequency of the laser is increased. Alternatively, the parallel-to-serial conversion process can be envisaged as a wavelength-to-time mapping. This simple process of employing multiple wavelengths that can be independently controlled leads to the potential for future applications in ultrafast parallel optical signal processing and novel communication architectures using this mixed time–frequency optical output.

In this section we have shown how to realize a five-wavelength harmonically modelocked semiconductor laser using a commercially available WDM filter. We showed that by employing the WDM filter inside the cavity, multiple wavelengths are generated with a well-defined channel spacing of 1.6 nm. It should be recognized that the spectral match between the WDM filter and the SOA gain spectrum is critical to obtaining the maximum number of wavelengths to be modelocked. The relative amplitude of the output pulses and the cavity length are controlled independently for each wavelength. The utility of this unique source was demonstrated by increasing the nominal pulse repetition frequency of the modelocked laser by performing a parallel-to-serial wavelength conversion using an arrayed waveguide grating (AWG) router.

In these experiments, a fivefold increase in the original pulse repetition rate was achieved. The key benefit of this approach is that low-speed electronics can be used in conjunction with WDM technology to generate high-speed sampling pulse trains, without the need for high-frequency electronics.

5.6 SUMMARY AND FUTURE OUTLOOK

This chapter has focused on the current state of the art in external cavity modelocked semiconductor diode lasers. We first reviewed the important fundamental physics underlying optical gain saturation and the related optical nonlinear effects associated with pulse propagation in semiconductor gain media. Next we identified intracavity spectral shaping as a way to avoid detrimental gain narrowing in modelocked semiconductor lasers, as a way to generate broad modelocked output spectra, the idea being that we can compensate or compress the optical pulses outside of the modelocked oscillator. This resulted in the generation of ultrashort optical pulses of ~ 0.2 ps duration. In addition, we identified a unique approach to generating high average output power by adiabatically tapering the gain stripe of a semiconductor optical amplifier.

Related to the spectral gain flattening that was achieved by the use of a Fabry–Perot intracavity etalon operating exactly off-resonance, we showed that the idea of intracavity spectral filtering could be extended by using a spectrometer within the external cavity. This resulted in the realization of generating multiple modelocked locked-wavelength channels that can be individually controlled in terms of the absolute wavelength and channel separation. Four-wave mixing was also identified as a potential physical mechanism for mediating wide-range wavelength correlation.

Finally, to show that the technologies are amenable to commercial activities related to high-speed networking and signal processing, we showed how to realize a multiwavelength modelocked semiconductor diode laser using a commercially available wavelength division multiplexed filter. We also highlighted two key applications that employ the multiwavelength modelocked diode laser: (1) hybrid optical time–wavelength division multiplexed optical links and (2) a photonic analog-to-digital converter. Each of these applications provides state-of-the-art optical networking and signal processing capabilities by exploiting the unique nature of multiwavelength modelocked diode lasers.

In the future, it is anticipated that researchers will exploit new regimes of modelocking with semiconductor gain media, not just in terms of the ultrashort pulses that the technique can provide but also in terms of multi-wavelength generation, for a host of applications ranging from imaging to telecommunications and signal processing to remote sensing. It is hoped that this chapter has been able to provide the reader with a glimpse of what is to come and to inspire new ideas in the area of ultrafast photonics.

ACKNOWLEDGMENTS

Wish to acknowledge the efforts of S. Gee, B. Mathason, H. Shi, and I. Nitta. In addition, the work presented in this chapter has been supported by the National Science Foundation, AFOSR, and DARPA. I also thank collaborators at Sarnoff Corp., G. Alphonse, J. Connolly, J. Abeles, and the OIDA-JOP Program.

REFERENCES

1. J Verdeyen. Laser Electronics. 3rd ed. Englewood Cliffs, NJ: Prentice-Hall, 1995.
2. K Hall, G Lenz, AM Darwish, EP Ippen. Subpicosecond gain and index non-linearities in InGaAsP diode lasers. Opt Commun 111: 589–612 (1994).
3. JC Dieles, W Rudolph. Ultrashort Laser Pulse Phenomena. San Diego, CA: Academic Press, 1996.
4. MY Hong, YH Chang, A Dienes, JP Heritage, PJ Delfyett, S Dijaili, FG Patterson. Femtosecond self- and cross-phase modulation in semiconductor laser amplifiers. IEEE J Selected Topics Quantum Electron 2:523–539 (1996).
5. RR Alfano, ed. The Supercontinuum Laser Source. NewYork: Springer Verlag, 1989.

6. PJ Delfyett, LT Florez, N Stoffel, T Gmitter, NC Andreadakis, Y Silberberg, JP Heritage, GA Alphonse. High-power ultrafast laser diodes. IEEE J Quantum Electron 28(10):2203–2219 (1992).
7. GA Alphonse, DB Gilbert, MG Harvey, M Ettenberg. High power superluminescent laser diodes. IEEE J Quantum Electron 24:2454 (1988).
8. A Yariv. Quantum Electronics. 3rd ed. New York: Wiley, 1989.
9. R Trebino, DJ Kane. Using phase retrieval to measure the intensity and phase of ultrashort pulses: frequency-resolved optical gating. J Opt Soc Am A 11:2429 (1993).
10. PW Smith, Y Silberberg, DAB Miller, J Opt Soc Am B 2:1228 (1985).
11. Y Silberberg, PW Smith, DAB Miller, B Tell, AC Gossard, W Wiegmann. Appl Phys Lett 16:701 (1985).
12. J Peatross, A Rundquist. Temporal decorrelation of short laser pulses. J Opt Soc Am B 15(1):216–222 (1998).
13. CPJ Barty, G Korn, F Raksi, C Rose-Retruck, J Squier, AC Tien, KR Wilson, VV Yakovlev, K Yamakawa. Regenerative pulse shaping and amplification of ultrabroadband optical pulses. Opt Lett 21:219–221 (1996).
14. JG Fujimoto, B Bouma, G Tearney, S Boppart, C Pitris, E Swanson, J Southern, M Brezinski. High speed high resolution optical coherence tomography with use of femtosecond lasers. Conference on Lasers and Electro-Optics, Vol 11. 1997 OSA Tech Dig Series, Washington, DC: Opt Soc Am, 1997, p 163.
15. DJ Kane, R Trebino. Characterization of arbitrary femtosecond pulses using frequency-resolved optical gating. IEEE J Quantum Electron 29(2):571–579 (1993).
16. C Iaconis, IA Walmsley. Spectral phase interferometry for direct electric-field reconstruction of ultrashort optical pulses. Opt Lett 23(10):792–794 (1998).
17. A Sullivan, J Bonlie, DF Price, WE White. 1.1-J, 120-fs laser system based on Nd:glass-pumped Ti:sapphire. Opt Lett 21(8):603–605 (1996).
18. BE Lemoff, CPJ Barty. Quintic-phase-limited, spatially uniform expansion and recompression of ultrashort optical pulses. Opt Lett 18(19):1651–1653 (1993).
19. D Yelin, D Meshulach, Y Silberberg. Adaptive femtosecond pulse compression. Opt Lett 22(23):1793–1795 (1997).
20. AM Weiner, JP Heritage, EM Kirschner. High resolution femtosecond pulse shaping. J Opt Soc Am B5:1563–1572 (1988).
21. M Osinski, J Buus. Linewidth broadening factor in semiconductor lasers—an overview. IEEE J Quantum Electron 23:9–29 (1987).
22. CF Lin, BL Lee. Extremely broadband AlGaAs/GaAs superluminescent diodes. Appl Phys Lett 71:1598–1600 (1997).
23. T Yamanaka, Y Yoshikuni, K Yokoyama, W Lui, S Seki. Theoretical study on enhanced differential gain and extremely reduced linewidth enhancement factor in quantum well lasers. IEEE J Quantum Electron 29:1609–1616 (1993).
24. A Mar, R Helkey, J Bowers, D Mehuys, D Welch. Modelocked operation of a master oscillator power amplifier. IEEE Photon Tech Lett 6:1067–1069 (1994).
25. L Goldberg, D Mehuys, D Welch. High power modelocked compound laser using a tapered semiconductor amplifier. IEEE Photon Tech Lett 6:1070–1072 (1994).

26. L Goldberg, D Mehuys. Blue light generation using a high power tapered amplifier modelocked laser. Appl Phys Lett 65:522–524 (1994).
27. A Mar, R Helkey, W Zou, D Bruce, J Bowers. High power modelocked semiconductor laser using flared waveguides. Appl Phys Lett 66:3558–3560 (1995).
28. KA Williams, J Sarma, IH White, RV Penty, I Middlemast, T Ryan, FR Laughton, JS Roberts. Q-switched bow-tie lasers for high energy picosecond pulse generation. Electron Lett 30(4):320–321 (1994).
29. B Zhu, IH White, KA Williams, FR Laughton, RV Penty. High peak power picosecond optical pulse generation from Q-switched bow-tie laser with a tapered traveling wave amplifier. IEEE Photon Tech Lett 8(4):503–505 (1996).
30. KA Williams, IH White, RV Penty, FR Laughton. Gain switched dynamics of tapered waveguide bow-tie lasers: experiment and theory. IEEE Photon Tech Lett 9(2):167–169 (1997).
31. GA Alphonse, N Morris, MG Harvey, DB Gilbert, JC Connolly. New high power single mode superluminescent diode with low spectral modulation. Conference on Lasers and Electro-optics, Vol 11. OSA Tech Dig Ser. Washington, DC:Opt Soc Am, 1997, pp 107–108.
32. MB Danilov, IP Christov. Amplification of spatially dispersed ultrabroadband laser pulses. Opt Commun 77:397–401 (1990).
33. AM Weiner. Femtosecond optical pulse shaping and processing. Prog Quantum Electron 19(3):161–237 (1995).
34. K Zhu, O Nyairo, IH White. Dual wavelength picosecond optica pulse generation using an actively modelocked multichannel grating cavity laser. IEEE Photon Tech Lett 6:348–351 (1994).
35. PJ Delfyett, Y Silberberg, GA Alphonse. Hot-carrier thermalization induced self-phase modulation in semiconductor traveling wave amplifiers. Appl Phys Lett 59:10–12 (1991).
36. CT Hultgren, EP Ippen. Ultrafast refractive index dynamics in AlGaAs diode laser amplifiers. Appl Phys Lett 59:635–637 (1991).
37. J Zhou, N Park, JW Dawson, KJ Vahala. Highly nondegenerate four-wave mixing and gain nonlinearity in a strained multiple-quantum-well optical amplifier. Appl Phys Lett 62:2301–2303 (1993).
38. AW Rihaczek. Principles of High Resolution Radar. New York: McGraw-Hill, 1969.
39. N Levanon. Radar Principles. New York: Wiley, 1988.
40. GHC New. Pulse evolution in mode-locked quasi-continuous lasers. IEEE J Quantum Electron 10:115–124 (1974).
41. HA Haus. Theory of mode mocking with a slow saturable absorber. IEEE J Quantum Electron 11(Sept):736–746 (1975).
42. M Margalit, M Orenstein, HA Haus. Injection locking of a passively modelocked laser. IEEE J Quantum Electron 32(Jan):155–160 (1996).
43. M Margalit, M Orenstein, HA Haus. Noise in pulsed injection locking of a passively modelocked laser. IEEE J Quantum Electron 32(May):796–801 (1996).

44. H Yokoyama, T Shimizu, T Ono, Y Yano. Synchronous injection locking operation of monolithic mode-locked diode lasers. Opt Rev 2:85–88 (1995).
45. R Ludwig, A Ehrhardt, W Pieper, E Jahn, N Agrawal, H-J Ehrke, L Kuller, HG Weber. 40 Gbit/s demultiplexing experiment with 10 GHz all-optical clock recovery using a modelocked semiconductor laser. Electron Lett 32:327–329 (1996).
46. D von der Linde. Characterization of the noise in continuously operating modelocked lasers. Appl Phys B 39:201–217 (1986).
47. U Keller, KD Li, M Rodwell, DM Bloom. Noise characterization of femtosecond fiber Raman solution lasers. IEEE J Quantum Electron 25(Mar):280–287 (1989).
48. LE Adams, ES Kintzer, JG Fujimoto. Performance and scalability of an all-optical clock recovery figure eight laser. IEEE Photon Technol Lett 8(Jan):55–57 (1996).
49. M Margalit, M Orenstein, G Eisenstein. Multimode injection locking buildup and dephasing in passively mode locked fiber laser. Opt Commun 129(Sept):373–378 (1996).
50. P Barnsley. All-optical clock extraction using two-contact devices. IEE Proc J 140(Oct):325–336 (1993).
51. A Ehrhardt, R Ludwig, W Pieper, E Jahn, HG Weber. Characterization of an all optical clock recovery operating in excess of 40 Gb/s. Tech Dig ECOC 196, ThB1.5.
52. H Kurita, T Shimizu, H Yokoyama. Experimental investigations of harmonic synchronization conditions and mechanisms of mode-locked laser diodes induced by optical-pulse injection. IEEE J Selected Topics Quantum Electron 2(Sept):508–513 (1996).
53. ML Loeb, GR Stilwell Jr. High-speed data transmission on an optical fiber using a byte-wide WDM system. J Lightwave Technol 6:1306–1311 (1988).
54. S-K Shao. WDM coding for high capacity lightwave systems. J Lightwave Technol 12:137–148 (1994).
55. JP Sokoloff, PR Prucnal, I Glesk, M Kane. A terahertz optical asymmetric demultiplexer (TOAD). IEEE Photon Technol Lett 5:787–790 (1993).
56. I Glesk, JP Sokoloff, PR Prucnal. Demonstration of all-optical demultiplexing of TDM data at 250 Gb/s. Electron Lett 30:339–341 (1994).
57. JP Laude, JM Lermer. Wavelength division multiplexing/demultiplexing (WDM) using diffraction gratings. In: JM Lerner, ed. Application, Theory, and Fabrication of Periodic Structures, Diffraction Gratings and Moire Phenomena II. Proc SPIE 503: 22–28 (1984).
58. MK Smit, C van Dan. PHASER-based WDM-devices: principle, design, and applications. IEEE J Selected Topics Quantum Electron 2(2):236–250 (1996).
59. G Lenz, BJ Eggelton, CR Giles, CK Madsen, RE Slusher. Dispersive properties of optical filters for WDM systems. IEEE J Quantum Electron 34(8):1390–1402 (1998).
60. GF Boudreaux-Bartels. Mixed time frequency signal transformations. In: AD Poularikas, ed. Transforms and Applications Handbook. Boca Raton, FL: CRC Press, 1996, Chap 12.

6

Overview of Industrial and Medical Applications of Ultrafast Lasers*

Gregg Sucha

IMRA America, Inc., Ann Arbor, Michigan, U.S.A.

6.1 INTRODUCTION

Timing is everything, as the old adage goes, and this is a pivotal time for ultrafast laser technology. Ultrafast lasers are on the verge of opening up a large variety of engineering and industrial applications. These "real-world" applications range from high-tech, such as high-speed circuit testing and biological imaging, to more everyday applications such as inspection of packaged food. The markets that will be influenced include industries such as telecommunications, automotive, electronics, medical (device manufacturing, diagnostics, therapy), dental, and ophthalmic and the manufacture and inspection of consumer goods. Although many of the concepts and techniques that make these ultrafast applications possible have proven in the laboratory, there are still barriers to their becoming mainstream. The primary obstacle arises from the complexity, size, and cost of conventional ultrafast laser systems. However, as can be seen from the first half of this book, there have also been significant advances in developing compact, rugged, turnkey ultrafast lasers for use in many of these applications. These new generations of lasers will need to operate very robustly under sometimes harsh conditions. There has been significant progress to date, and progress continues on several fronts, including solid-state, fiber, and diode lasers.

---

*The author is an employee of IMRA America, Inc., a commercial developer of ultrafast laser technology. Any opinions, findings, and conclusions or recommendations expressed in this material are those of the author and do not necessarily reflect the views of IMRA America, Inc.

Ultrafast laser technology is quite mature from a scientific standpoint, having existed as a discipline for over 20 years.[†] Chemists and physicists developed ultrafast lasers for the purpose of measuring extremely fast physical processes such as molecular vibrations, chemical reactions, charge transfer processes, and molecular conformational changes. All these processes take place on time scales of femtoseconds to picoseconds.[‡] In the realm of condensed matter physics, carrier relaxation and thermalization, wavepacket evolution, electron-hole scattering, and countless other processes also occur on these incredibly fast time scales. This then, can be considered to be the first application of ultrashort laser pulses, according to the current definitions of ultrafast. The evolution of ultrafast laser technology and applications can be traced to its current status by referring to Ref. 1; the progress toward commercial applications has been published more recently in conference proceedings [2]. Perhaps nothing better illustrates the remarkable capabilities of very short light pulses than the ability to capture on film the flight of a bullet as it tears through a playing card, as shown in the famous pictures produced by Harold Edgerton, who pioneered fast strobe photography. The high speed and short duration of light pulses can "freeze" the action of just about any processes that one can think of. But the speed—or more accurately, the *brevity* of pulse duration—of light sources has increased dramatically since the days of high-speed strobe photography, which used light flashes on the order of microseconds. Today's ultrafast lasers produce pulses that are millions of times faster than the strobe flashes used for strobe photography.

For the most part, the technical and performance requirements for so-called real-world applications of ultrafast laser technology are already satisfied by current state-of-the-art ultrafast laser systems. In spite of this, however, the list of real commercial successes is still quite small. Many of the applications that are on the verge of being commercialized are based on techniques that were explored over 15 years ago. These seminal works, technologies, and techniques are *enabling* technologies that have had the potential to far outstrip existing technologies, especially in terms of speed. These new ultrafast technologies matured as far as they could in the research phase but in many cases have been put on hold due to the lack of a market and/or the lack of a practical ultrafast laser sources. But this situation is changing. Well-packaged, turnkey ultrafast lasers are being

---

[†]One can trace the evolution of ultrafast laser technology and its applications by referring to volumes of the proceedings on Ultrafast Phenomena [1].

[‡]The prefixes for these extremely small quantities are pico-and femto-, defined as 10^{-12} and 10^{-15}, respectively.

rolled out by various vendors, targeted not only toward the scientific market (which has been the mainstay of the ultrafast laser industry thus far) but also toward specific industrial applications. However, one thing has not changed yet: Ultrafast lasers are still substantially more expensive than conventional pulsed lasers, such as Q-switched lasers, which are now commonplace in so many sectors of industry. Nevertheless, the number of applications being developed reflects the growing confidence in various sectors of industry that ultrafast laser technology is emerging from the laboratory environment and will someday occupy a place on the factory floor, in the hospital or clinic, or in deployable systems such as telecom networks. In each case, the marketplace will ultimately decide whether or not the extra cost of ultrafast lasers is offset by the value added to the users' applications.

A comprehensive treatment of all potential ultrafast applications would be almost impossible. And to limit such a treatment to a single book of this size would be genuinely impossible. The emphasis of this second section of the book is thus on ultrafast applications that have been—or are being—developed commercially or that are undergoing clinical tests in medical, dental, or ophthalmic environments or that appear to be particularly promising for future markets. In this spirit, the editors invited a number of experts in various subfields to contribute chapters for this applications section of the book. These chapters are

7. Micromachining (*S. Nolte*)
8. Structural changes induced in transparent materials with ultrashort laser pulses (*C.B. Schaffer, A. O. Jamison, J. F. García, and E. Mazur*)
9. Rapid scanning time delays for ultrafast measurement systems (*G. Sucha*)
10. Electro-optic sampling and field mapping (*J. F. Whitaker and K. Yang*)
11. Terahertz wave imaging and its applications (*Q. Chen and X.-C. Zhang*)
12. Phase-controlled few-cycle light (*G. Tempea, R. Holzwarth, A. Apolonski, T. W. Hänsch, and F. Krausz*)
13. Ultrahigh bit rate communication systems (*M. Nakazawa*)
14. Nonlinear microscopy with ultrashort pulse lasers (*M. Müller and J. Squier*)
15. Optical coherence tomography (*J. G. Fujimoto, M. Brezinski, W. Drexler, I. Hartl, F. Kärtner, X. Li, and U. Morgner*)
16. Ultrafast lasers in ophthalmology (*R. M. Kurtz, M. A. Sarayba, and T. Juhasz*)

This list leaves out several other applications that are being commercially developed and a great many that are not yet being commercially developed but are nonetheless very promising. This introductory chapter to the applications section of the book is designed to give a somewhat fuller view of the entire range of ultrafast laser applications beyond those that are presented in the subsequent chapters. We harmonize these chapters with each other and also with other ultrafast applications within a larger context. To accomplish this, we create a sort of hierarchy—an "ultrafast applications matrix"— bounded on one side by characteristics of ultrafast laser technology and on the other side by market-specific applications. In this chapter we discuss several aspects of ultrafast lasers, starting with the unique and useful properties of ultrashort laser pulses as well as the physical effects and techniques that are exploited for their use in applications. Brief comments about the chapters are folded in among more detailed discussions of other applications that are not covered elsewhere. Toward the end of the chapter, examples are given of ultrafast-laser-based instrumentation systems that are available commercially at this time.

One will notice, upon reading these chapters, that the various authors have quite different perspectives and that the applications are at somewhat differing levels of maturity in the technological sense. Some are in the research phase, whereas others are being commercialized. Another way of saying this is that the applications fall at various levels in the matrix, either toward the laser technology side or toward the market-specific side. One clear example of the latter is the chapter on ophthalmic surgical applications, some of which have FDA approval and are undergoing serious commercialization efforts including advanced clinical trials.

6.2   ULTRASHORT PULSE LASER TECHNOLOGY, APPLICATIONS, AND MARKETS

In discussing ultrafast laser applications, several natural questions arise: Why use ultrashort laser pulses? Which lasers can be used for medical applications? For a given ultrafast laser, how many different applications can it be used for, and which markets will these applications fall into? There are any number of ways in which one could consider and arrange applications of ultrafast lasers. In our attempt to understand the complex relationships among laser properties, specific techniques, applications, and markets, we can construct a matrix such as the one in Figure 1. The matrix, according to this construction scheme, has three main layers. The first layer is the simplest, listing the special properties of ultrashort pulses that make them uniquely suited to their applications. The second layer, labeled

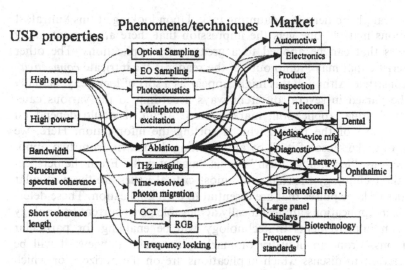

Figure 1 of laser applications, techniques, and ultrashort laser properties. Heavy connecting lines emphasize links to micromachining applications (automotive, pharmaceutical, etc.) and the laser properties (high power and high speed) that are critical to micromachining. Lighter connecting lines show links for other applications.

"Phenomena/techniques" is a listing of the physical phenomena and the techniques that are exploited by using ultrashort pulses; e.g., multiphoton excitation, terahertz imaging, optical coherence tomography (OCT). In some sense, these can be considered applications in themselves. However, this depends on one's perspective—an ultrafast laser researcher would call OCT an application, whereas a physician who was using OCT for medical diagnosis would call it a tool. The third and last layer, labeled "Market," lists various examples of major business sectors and industries, such as medical, automotive, and microelectronics, as well as some small application areas that are not exactly market segments at this time. Note here that the medical sector is subdivided into various subcategories including diagnosis, therapy, and device manufacturing. Each of the other business sectors could be similarly divided into application areas such as manufacturing and process control, but this has not been done for the sake of clarity because the matrix is already quite crowded. The elements of these three layers are linked by connecting lines going from left to right, connecting a useful laser property (e.g., high speed) to a technique (e.g., terahertz imaging) and finally terminating on a specific application (e.g., product inspection).

    Organizing and making sense of the complex network that connects applications, markets, laser properties, laser technologies, laser techniques, and physical phenomena is a daunting task. Upon looking at this "ultrafast applications matrix" one gets the impression that there are a great many techniques that can be applied to a great many applications. The other characteristic that may "jump out" at the reader is the intricate *connectivity* of the construct. Although this might appear as a sort of hierarchy, actually it can be parsed in several different ways according to the various categories (technique, application, market, etc.). Again, it would depend on the perspective of the person who is sorting the information. Here, we attempt to make sense of it all by approaching the matrix from two opposite directions: "technology push" versus "market pull." From the perspective of the technology providers, the most critical issues are the particular properties of laser pulses that are essential to each application. These determine which applications are technically achievable given current technology and which improvements in technology will be enabling for particular applications. From an industry or marketplace point of view, it will be more relevant to discuss which applications are on the horizon, or which are already being established in a particular area of industry, based on the particular needs of that industry. From this point, we can then define the system or instrument specifications and then look backward to the laser technology that drives it to see, first, if there are any conventional (*i.e.*, non-ultrafast) lasers that can fulfill the requirements. If this is not the case, we then determine whether there is an ultrafast laser technology that can enable that application. From the point of view of the systems integrator, the technique (e.g., multiphoton fluorescence) is of primary importance. From this standpoint, one must then work in both directions—looking forward to understand the specific needs of the various applications markets for that technology, and looking backward to the laser technology that will be able to drive the technology.

6.2.1 Properties of Ultrashort Pulses

We start from the perspective of the laser technology provider. Rather than talk about the laser technology itself, which has received a thorough exposition in the first half of this book, it is useful to discuss the special properties and capabilities of ultrashort laser pulses that are unique and that make them especially useful for a number of applications. These properties are

Short duration
High peak power
Broad spectral bandwidth
Short coherence length
Structured spectral/coherence properties

Some of these properties go "against the grain" in terms of the traditionally espoused advantages of lasers over other types of light sources. For example, since the laser was invented, we have been taught that high spectral purity and extremely narrow spectral linewidths (achieved simultaneously with high brightness and directionality) are the hallmarks of laser technology and are the very properties that enable lasers to do so many new things that cannot be done with more conventional light sources. Yet we will discuss in this chapter, and in Chapters 12 and 15, applications in which the broad spectral bandwidth of ultrashort laser pulses is considered to be a valuable asset.

The applications covered by the chapters in this section can be roughly divided into categories that are based on the five properties listed above; i.e., those that use the high peak power of ultrashort pulses and those that use the very high speed, the broad bandwidth, the short coherence length, or the structured spectral coherence. These distinctions are reasonably good, but in some cases the boundaries are a bit blurred (i.e., the properties are interconnected) or else a particular application may require a combination of properties. For example, femtosecond micromachining relies on the short duration of the pulse to help provide the extremely high peak powers needed to drive the ablation process. However, the very clean and deterministic ablation that characterizes femtosecond micromachining also depends critically on the very short duration of the pulse; high power alone is not sufficient for clean ablation. It is also necessary that the pulses not exceed a certain duration (<1 ps). The reasons for this time limit are related to the complex physics involved in the ablation process, as is discussed in some detail in Chapter 7.

High Speed

The most obvious property of ultrashort pulses is high speed. Ultrashort light pulses are critical for time-resolving fast processes. It is helpful here to define what we mean by "ultrashort" laser pulses. How short is ultrashort? There are no hard and fast rules for this definition, but for the purposes of this book, and in accordance with the rough "conventions" of the field, we will say that laser pulses that are shorter than approximately 10 ps fall in the ultrashort regime. The majority of applications use laser pulses with durations falling in the range of 100 fs to 1 ps.

Ultrashort laser pulses are far shorter and their ability to time-resolve fast processes is far greater than anything that can be accomplished by means of conventional or state-of-the-art electronics. However, the speed of electronics-based technology continues to increase. Chip sets capable of 80 Gbit/s are being contemplated and tested for the field of optical telecommunications, and digital optoelectronic integrated circuits (ICs) operating at

80 Gbit/s have been demonstrated [3]. In addition, automotive radar and other applications will require the development of ICs that can operate at frequencies in the range of 60 to 80 GHz [4]. As the speed of electronic circuits continues to increase, current testing technologies, which are based on electronics are being outstripped. There is an increasing need to resort to new diagnostic techniques that are capable of measuring these high-speed signals. These new techniques, based on electro-optic sampling using ultrafast lasers, are treated in Chapter 10, and a time delay scanning technique applicable to this and other high-speed instrumentation is covered in Chapter 9. Optical telecommunications at high bit rates requires optical pulses that are properly tailored (including the pulse width) for the particular format (RZ or NRZ) and bit rate being used. Chapter 13 discusses the key ultrafast technologies required for optical time division multiplexed (OTDM) communication systems operating at bit rates of 40 Gbit/s, 80 Gbit/s, and 1.28 Tbit/s (terabits per second). These key technologies include not only the ultrafast laser but also other components such as nonlinear optical loop mirrors (NOLMs) for optical clock recovery and demultiplexing. For the highest bit rates of 1.28 Tbit/s, pulse widths of 200–400 fs are used.

Unlike the previous examples, it is not immediately obvious that the generation and detection of terahertz radiation requires ultrashort laser pulses. It might seem that this could be accomplished by pushing the state of the art in microwave and millimeter-wave technology. Indeed, terahertz technologies using electronics and even cw lasers are being pursued. Nevertheless, the field of terahertz optics has been greatly advanced over the last several years, made possible primarily by ultrafast laser technology. This terahertz technology is described in Chapter 11, including demonstrations of a number of applications including medical imaging and surface metrology.

High Peak Power

For laser pulses of various shapes, the pulse energy E_P, peak power P_P, and pulse duration τ_P are approximately related by the very simple equation

$$P_P \approx \frac{E_P}{\tau_P} \tag{1}$$

Thus, the extremely short duration of femtosecond laser pulses implies that very high peak powers can be generated even for moderate pulse energies. For example, a typical small fiber laser produces optical pulses with energies of only ~ 1 nJ. However, because the pulses have durations of only 100 fs $(10^{-13}$ s) this gives a peak power level of over 10 kW. Intensities on the order of $1\,MW/cm^2$ to $10\,GW/cm^2$ commonly occur when such pulses are focused through a microscope objective—but with average powers on the order of only tens to hundreds of milliwatts so that samples don't burn—making practical the technique of multiphoton excitation that can be used in myriad applications such as nonlinear confocal microscopy [5], two-photon polymerization for microfabrication, and multiphoton writing of multilayer optical memories. Nonlinear microscopy using ultrafast lasers is discussed in Chapter 14.

For pulse energies at the microjoule and millijoule level (typically achieved only in amplified laser systems), the peak output powers exceed $10\,MW$ and $10\,GW$, respectively, for pulse widths of $\sim 100\,fs$. When focused down to small spot sizes on the order of $30\,\mu m^2$, these pulses give on-target intensities of 1–$1000\,TW/cm^2$. At such high intensities, the electric field is sufficiently high to induce field ionization and/or multiphoton ionization, tearing electrons out of their orbits and accelerating them sufficiently to induce avalanche ionization. For ultrashort laser pulses, these processes occur so quickly that ablation occurs before thermal diffusion and transport can occur to the surrounding regions of material. In this way, the heat-affected zone (HAZ) and shock-affected zone (SAZ) are greatly reduced, resulting in very clean surface cutting of all types of materials. The physical mechanisms, results, and applications are described in detail in Chapter 7. This technique (also called femtomachining) has recently become one of the most active and exciting areas in ultrafast research and development, with the potential for huge economic rewards, because it has implications for ultrafine, nanoscale manufacturing in such diverse markets as electronics, medicine, dentistry, the automotive industry, pharmaceuticals, and many others. This type of physics carries over also to laser surgical applications, where it is desired to make very clean, precise cuts in delicate tissue with mininal collateral damage. Applications of these unique capabilities in ophthalmology are described in Chapter 16.

At somewhat more modest pulse energies (5–50 nJ) and peak powers (100 kW to 1 MW), and with tight focusing to micrometer spot sizes creating very high intensities, similar multiphoton ionization interactions can induce microexplosions that change material properties, most notably the local density, and thereby also the refractive index. This approach has potential applications in areas of optical data storage and waveguide writing, as detailed in Chapter 8.

Broad Bandwidth

Broad spectral bandwidth is an essential property for very short pulses, whether they be optical, microwave, or electrical. As expressed by the

Heisenberg uncertainty principle, the time–bandwidth product of any pulse cannot fall below a limit that is written approximately as

$$\Delta t \times \Delta v \geqslant 0.4 \tag{2}$$

where Δt is the temporal width of the pulse in seconds and Δv is the spectral width of the pulse in hertz, both expressed as full width at half-maximum (FWHM). The factor 0.4 is not exact and actually depends on the details of the pulse shape. These details give slight variations, but the general rule is inviolable: Ultrashort pulses must necessarily have broad spectral bandwidth.

Not many ultrafast laser applications rely strictly on bandwidth. Usually the requirement for broad bandwidth occurs in combination with some other desired property.* There have been, however, in the last several years, demonstrations of the use of the broad bandwidth of femtosecond lasers for optical communications. For example, ultrafast lasers may well provide an economical solution as transmitters for broadband WDM optical access systems [6].

The need for a broad spectral bandwidth—as opposed to the usual narrowband output of conventional lasers—begs the question: Why not use a lightbulb or other similar light source if you want broad bandwidth? The answer is that conventional broadband sources have low brightness, whereas broadband ultrafast lasers have higher brightness by many orders of magnitude. The brightness is absolutely critical to applications that require tight focusing or high spatial resolution along with good signal-to-noise (S/N) ratio, so that data or image acquisition times remain reasonable.

Short Coherence Length

As with narrow spectral bandwidth, long coherence length is one of those properties of laser light that is traditionally thought of as being beneficial. This is especially true for such applications as holography, were the depth of field of the image demands a rather long coherence length. However, *short* coherence length is an essential property for high resolution optical coherence tomography (OCT). Because the longitudinal resolution of OCT depends directly on the coherence length of the light source, it is desirable to have a source with as short a coherence length as possible. The coherence length, L_C, of a light source is given approximately by the relation

---

*Indeed, scientists (especially spectroscopists) very often express a desire for a laser that could violate the uncertainty principle—a laser that could, say, generate ultrashort pulses of 100 fs or shorter while somehow simultaneously retaining the very narrow spectral bandwidths that make lasers so useful for spectroscopic investigations.

$$L_C = \lambda^2/\Delta\lambda \tag{3}$$

where λ is the wavelength of the light and $\Delta\lambda$ is the spectral bandwidth. More precise definitions of coherence length can be found in Chapter 15. For nearly transform-limited pulses, the coherence length and the pulse-width are directly related by

$$L_C \sim c\tau \tag{4}$$

To give a more physical idea of the relation between pulsewidth and coherence length, a 100 fs laser pulse has a coherence length of about 30 µm.

Once again, we can ask why we should use an ultrafast laser instead of another light source that possesses short coherence length, such as a thermal source (lightbulb) or an LED. In OCT applications, superluminescent diodes (SLDs) are currently used in commercial systems, and ultrafast lasers are not currently competitive owing to their expense and complexity. However, as described in Chapter 15, ultrafast lasers can generate even broader bandwidth than an SLD, thus giving more than five fold improvement in longitudinal resolution. Ultrafast lasers also have brightness values that are orders of magnitude higher than is possible for SLDs, which can be essential in some circumstances.

Another application that depends upon short coherence length is video projection. Laser-based red–green–blue (RGB) illumination systems for video projection suffer from certain drawbacks. One of the most serious is speckle. A laser source with long (>10 mm) coherence length produces considerable speckle upon hitting a target such as a projection screen, thereby degrading the image quality. The severity of the speckle effect depends on the relationship between the coherence length and the statistics of the surface roughness of the projection screen. Ultrafast lasers produce greatly reduced speckle and therefore could potentially serve as the best sources for laser projection systems [7]. The high peak power of picosecond pulses also is of important in that it allows efficient frequency conversion of infrared (IR) laser pulses into the various visible wavelengths.

Structured Spectral Coherence Properties

One other property of ultrashort pulse lasers that is exploited in certain applications is the uniquely structured spectral coherence properties. Whereas an individual ultrashort pulse possesses broad spectral bandwidth and short coherence length, a repetitive train of ultrashort optical pulses possesses a "comb" structure in both the temporal and spectral domains (see Fig. 8 of Chap 12). Consider the illustrative example of a modelocked laser oscillator producing optical pulses with 200 fs duration and having a repetition frequency of 50 MHz. If one examines the optical spectrum with sufficient spectral resolution, one will find a very large number (>100,000) of narrow optical frequency components that are very nearly evenly spaced by 50 MHz. What is more, each of these frequency components bears a strongly ordered coherence relationship to the others in the comb.

Frequency metrology is one area in which the coherence properties of this frequency comb are of primary important. Using nonlinear techniques described in Chapter 12, the coherence is carefully manipulated, providing a comb of individual frequencies that have extraordinary stability and that have the potential to be used as highly accurate frequency standards. Here, the relative coherence between the various spectral components and the broad spectrum are of primary importance. Other properties such as short pulse width and high peak power are of secondary or tertiary concern except that they are exploited for nonlinear processes such as spectral broadening (supercontinuum generation) that are essential to the phase-locking process described in Chapter 12.

Other Laser Parameters and Technical Issues

In addition to the above four properties of ultrashort pulses, there are other operational parameters of USPL lasers (as well as other types of lasers) that must be defined and controlled for the particular application at hand. These parameters include wavelength, tunability, pulse shape, and repetition frequency. Operating wavelength, for example, is determined by issues such as tissue absorption in medical applications, and fiber loss and infrastructure considerations as in telecommunications, where wavelengths in the range of 1500–1600 nm are of key interest. Repetition frequencies commonly range from kilohertz (for micromachining) to megahertz (for instrumentation and multiphoton applications) to several gigahertz (for telecom applications). A modelocked laser with 10 GHz repetition rate is described in Chapter 13. Additionally, users will often want to specify other parameters such as beam mode output quality (quantified by the M^2 parameter), noise properties (rms amplitude, timing jitter, and phase noise), and pointing stability. It is almost always desirable to keep these last parameters at a minimum, because they represent deviations of the laser behavior from the ideal.

One other key technology issue related to laser applications is *beam delivery*. In many laser instrumentation or laser tool systems, fiber-optic beam delivery is the method of choice, because it offers a significant reduction in system complexity and greater flexibility in system design and use. For ultrashort lasers, however, fiber delivery is particularly problematic for a number of reasons. First, there is dispersion, which will broaden ultrashort pulses to unacceptably long pulse widths. Second, because of the high peak power of ultrashort laser pulses, nonlinear effects (e.g., Raman generation, self-phase modulation, and spectral broadening) are often very strong in fibers, leading to pulse broadening or even a total breakup of the pulse or possibly even burning of the fiber itself. This presents a conundrum to the systems integrator: The very properties that make ultrashort laser pulses so useful also make fiber-optic delivery difficult. Although there are some technological solutions, such as compensating the fiber dispersion using pre- or postcompensators, these solutions present an unwelcome increase in complexity to the overall system, especially when they require the use of bulk optics. This is an area that needs further advances.

6.2.2 Applications (Phenomena/Techniques)

Parsing via the second level of the ultrafast applications matrix, we find the listing "*Phenomena/techniques*," and this is the layer to which most of the applications chapters refer. Upon more careful examination of the applications, we find that there are layers within layers in this matrix. The subsequent, detailed chapters also extend into the third level, discussing various *specific* applications of the techniques described.

Applications at the phemona/techniques level can be categorized into a few main groups:

High-speed testing and instrumentation techniques—Optical sampling, electro-optic sampling, ultrafast A/D converters, femtosecond holography

Terahertz applications—Terahertz generation, detection, spectroscopy, and imaging

Multiphoton applications—optical beam induced current (OBIC), photochemical alteration for writing of optical memories, photopolymerization for nanofabrication, laser scanning fluorescence microscopy (LSM), fluorescence-based analytical applications for genomics and drug discovery, as well as medical diagnostics and therapeutic applications such as fluorescence lifetime imaging (FLIM) and photodynamic therapy

Materials processing applications—Ablation, micromachining, surface modification, index modification of materials for waveguide writing, cutting, and photodisruption for ophthalmic procedures

Metrology applications—Frequency metrology, OCT, picosecond ultrasonics, precision ranging and LIDAR

Imaging applications—Ballistic photon imaging, optical Kerr gating, FLIM

Signal processing applications—Femtosecond space–time holography, frequency domain pulse shaping

Each of these general techniques is applicable to a variety of applications in various markets. For the purpose of illustration, we take the example of multiphoton applications and expand upon it. The ultrafast applications matrix is redrawn in Figure 2 to emphasize this category. More details are brought out by revealing some of the intervening layers between the techniques and the market-specific applications.

Multiphoton Applications

Multiphoton excitation is made possible by the high peak power of focused ultrafast laser pulses and is used in a variety of ways including photochemical alteration, polymerization, and fluorescence. There is one significant omission from this list: ablation. Strictly speaking, ablation, as it is used for femtosecond micromachining and materials processing, is a multiphoton process (this is indicated in Figure 2 by dashed lines). However, according to the nomenclature used in the laser field, these applications do not usually fall under the rubric of "multiphoton applications." The best established use for multiphoton excitation (MPE) is in laser scanning microscopy (LSM). This

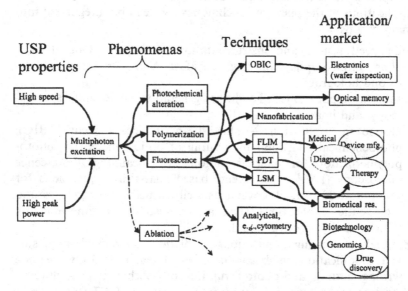

Figure 2 Ultrafast laser applications matrix emphasizing the phenomenon of multiphoton excitation. The intervening layers between the second and third layers are revealed, distinguishing phenomena, techniques, and applications. Specific types of multiphoton techniques include OBIC, photochemical alteration for optical memories, photopolymerization for nanofabrication, laser scanning microscopy, and fluorescence-based analytical applications for genomics and drug discovery.

technique uses the nonlinear absorption characteristic of two-photon absorption (TPA) to strongly localize the optical excitation, especially along the axial direction, thereby giving superior sectioning capabilities to conventional confocal microscopy [5]. This technique also reduces the photodamage to other areas of the specimen. Superior localization of excitation and sectioning ability, illustrated in Figure 3, is the *key feature* of most multiphoton applications. The other key advantage of two-photon excitation (TPE) is that it uses longer wavelengths than single-photon excitation (SPE), and because longer wavelengths are more weakly scattered in scattering media, TPE results in superior penetration into scattering media.

Other multiphoton applications that use these same advantages include photochemical alteration for writing of optical memories, photopolymerization for nanofabrication, fluorescence-based analytical applications for genomics and drug discovery, and medical diagnostics and therapeutic applications such as fluorescence lifetime imaging and photodynamic therapy. Here, we describe some of these applications in more detail.

Optical Beam Induced Current. Two-photon excitation has been demonstrated as an attractive mechanism for optical beam induced current (OBIC) for integrated circuit testing [8]. OBIC is an established technique for circuit diagnostics whereby an optical beam is used to induce photocurrents at selected locations in an integrated circuit. The characteristics of these induced photocurrents give information about the conditions existing in the circuit, including the logic states. Usually this is done using single-

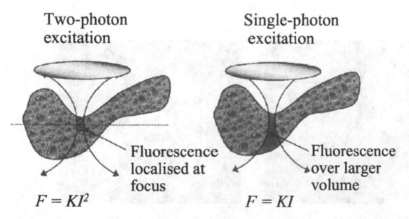

Two-photon excitation

Single-photon excitation

Fluorescence localised at focus

$F = KI^2$

Fluorescence over larger volume

$F = KI$

Figure 3 Comparison of single-photon excitation (right) with two-photon excitation. Note the tighter localization of two-photon excitation due to the nonlinear (quadratic) dependence of excitation strength on the local beam intensity.

photon excitation at wavelengths that are strongly absorbed by the substrate material. Because of this, excitation must be done from the top side, with the limitations that many underlying features cannot be accessed by the excitation beam. However, by using two-photon excitation at long wavelengths that lie in the transparency region of the substrate, excitation can be done from the *back side* of the IC. Additionally, the tight focusing and nonlinear effects provide very carefully localized excitation at selected depths into the substrate.

Two-Photon Polymerization. Fabrication of three-dimensional objects by laser-induced polymerization (by UV lasers) is already a commercial reality. These types of systems use UV-curing polymer resins plus a computer-controlled laser scanning system to selectively write patterns and thereby polymerize a 3D solid, layer by layer, from a bath of the resin. The depth resolution (in the Z dimension) is more or less determined by the absorption depth of the resin at the curing wavelength. However, two-photon excitation of the process, as with two-photon microscopy, confines the region of excitation to the focal depth of the Gaussian beam geometry at the focal point. For tight focusing, the confocal parameter, and therefore the excitation region, can be on the order of 1 μm. This has enabled researchers to demonstrate unprecedented resolution and the finer features in the fabrication of small structures using two-photon photopolymerization [9]. In a recent demonstration, researchers fabricated a 3D statue of a bull (shown in Fig. 4) with dimensions of 10 μm long by 7 μm tall, using a tightly focused femtosecond laser to induce two-photon photopolymerization [10]. By careful control of the beam and of the polymerization exposure process, they

 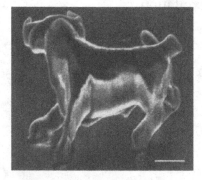

Figure 4 Miniature statue of a bull fabricated by two-photon excited photopolymerization. Dimensions of the bull are 10 μm long by 7 μm high. Scale bars are 2 μm in length. (Courtesy of H. Sun and S. Kawata.)

were able to obtain spatial feature resolution of 150 nm. The most significant commercial potential of this technique is, of course, the fabrication not of statues but of other types of structures. The same group has also fabricated three-dimensional photonic structures and memory structures based on the same technique. Additionally, such tiny structures could be used for micro-machines, MEMS, or for medical applications such as drug delivery, and sensing. The potential for medical applications is illustrated nicely by the fact that the fabricated bull was approximately the same size as a red blood cell, suggesting that such tiny microsensors or micromachines could deliv-ered anywhere in the body.

Other Applications

Metrology. Interferometry provides very accurate readings of surface contours and small mechanical motions of objects. Subwavelength precision is routinely achieved in many types of measurement situations. However, the standard optical metrological techniques such as interferometry, moiré fringes, and triangulation are not adequate in many cases. One of the most difficult situations is one in which one needs accurate information about a remote target. Although interferometry gives outstanding performance and accuracy with respect to surface features and small motions, the dynamic range is problematic. One might also want information on the absolute range to the object. Distance metrology using ultrashort laser pulses can extract this type of information. A method using precision beat-frequency measurements between the various laser modes has been demonstrated that can measure the distance to a target placed 240 m distant from the laser system. The range to the target was measured to a precision of 50 μm, even at this large range [11]. In addition, using a two-wavelength method (using the fundamental and second harmonic of the laser pulses) the index of refraction of air can be measured to an accuracy of 6 parts per million (ppm).

Photonic Analog-to-Digital Conversion. A demand for A/D conver-ters operating at multigigahertz sampling rates with better than 12 bit reso-lution has emerged in the areas of radar and microwave communications. Optical sampling of high-speed electrical signals offers a solution whose tim-ing precision and linearity of the sampling function are superior to those of conventional electronic A/D converters. The low jitter and high-speed advantages of photonics (especially modelocked lasers) can be combined, via time demultiplexing, with numerous electrical A/D converters to form *photonic* A/D conversion systems of various types. Some of these systems rely on ultrafast photoconductive switching technology based on LT grown GaAs MSM switches having short recombination lifetimes for very high time resolution. The use of phase-encoded optical sampling enhances the linearity of the sampling function, a crucial parameter for accurate A/D conversion [12].

Femtosecond Pulse Shaping and Processing. Femtosecond pulse shaping systems based on diffraction gratings can be used to Fourier synthesize ultrashort laser pulses with almost arbitrary shapes, via phase and amplitude modulation. This is implemented by employing masks and/or programmable spatial light modulators in the Fourier plane of a pulse-shaping system [13]. These types of pulse-shaping operations have also been combined with holographic techniques, creating a new techniques called femtosecond space–time holography [14]. The capabilities of such systems, when combined with nonlinear optical processes, include a number of signal processing applications such as time reversal, encoding and decoding of bit streams for code division multiple access (CDMA) formats, optical thresholding, and dispersion compensation. These types of operations show great promise for application to optical communication networks.

6.2.3 Applications (Market-Specific)

Parsing the last level of the applications matrix, several techniques or ultrafast laser-driven technologies may apply to a particular market area. Here we consider a number of very specific examples of ultrafast applications applied to particular market areas. We use as examples the sectors of medical (including dental and ophthalmic) applications, information technology, and telecommunications.

Medical and Dental Applications

In the medical field, several different techniques are being researched for various specific applications. Multiphoton excitation is applicable to photodynamic therapy for the cure of many diseases as well as to multiphoton confocal imaging diagnostics of, for example, various types of skin cancer. Moderately powered ultrashort pulses can be used for soft tissue ablation, which has been shown to be especially effective as a laser microkeratome for LASIK procedures. Hard tissue ablation for dental procedures is also under investigation. Micromachining can be used for manufacturing medical devices for drug delivery (requiring micrometer-sized orifices) and for the fabrication of stents, for example. OCT uses the short coherence length of ultrashort pulses to perform precision imaging of features of tissue on even a subcellular level. Terahertz imaging, driven by femtosecond lasers, has been demonstrated for imaging abnormalities with submillimeter resolution in both hard and soft tissues. A special two-wavelength picosecond laser is being used for two-color differential absorption imaging of infant brain oxygenation levels [15]. Fluorescence lifetime imaging of both fluorophores and autofluorescence provides better sensitivity and accuracy than intensity-based fluorescence measurements and can distinguish local perturbations in the heterogeneous environment of tissues [16]. FLIM has been shown to distinguish between healthy skin and malignant lesions [17]. Subpico-second time-gated transillumination imaging uses the early-arriving photons (or ballistic photons) to view objects in scattering media such as bones inside tissue [18]. A summary of medical applications for ultrashort pulse lasers (USPLs) is given in Ref. 19.

Laser Dentistry. Lasers are used in dentistry for a number of proce-dures including the treatment of oral malignancies, periodontal disease, den-tal caries prevention and control, and removal of hard tissue for cavity restoration, preparation, and sterilization. Patients' fear of the dental drill is pervasive and is so acute in many cases that the patients wait until more severe and painful complications develop before they seek treatment. This has provided the driving force behind the ongoing effort to develop new laser-based modalities that will replace the dentist's handpiece. However, as pointed out by Neev and Squier [20], the success of various types of con-ventional lasers in surgical and dental procedures is limited by two interac-tion characteristics: low removal rates and high collateral damage. Conventional (non-ultrafast) laser technology seems inevitably to present a trade-off between these two factors. Collateral damage includes effects such as cracking due to the laser-generated shock waves and also the effects of non-local heating. This is especially critical in dental procedures, because temperature increases as small as several degrees Celsius can permanently damage the pulp of the tooth. These effects have been seen with a number of laser systems including ruby, CO_2 Er:YAG, and ArF eximer lasers [21].

Owing to these types of drawbacks, researchers are striving to exploit the advantages of operation in the "cold" ablation regime. The known advantages of ultrashort laser pulse interactions with materials thus appear very promising for dental applications. Research directly comparing USPLs (50 fs and 350 fs) with conventional (1 ns) lasers for cutting and drilling of enamel and dentin show clear advantages for the USPLs. Ablation thresh-olds are several fold smaller for USPLs (~ 0.5 J/cm^2) while simultaneously providing cleaner cuts with less collateral damage. Smaller temperature rises ($<1°C$) are incurred when using ultrashort pulses. Niemz [21] also reported that the quality of the cavities cut by raster scanning a beam of ultrashort pulses over the surface of extracted teeth is superior to those cut with con-ventional lasers. When using femtosecond laser pulses (100 fs, 780 nm, 50 µJ, 1 kHz) raster scanned while focused down to 30 µm spot size, material removal rates were found to be $0.18\,\text{mm}^3\,\text{min}^{-1}\,\text{mJ}^{-1}$. Figure 5(a) shows a cavity $1\,\text{mm} \times 1\,\text{mm}$ in area by approximately $300\,\mu\text{m}$ deep drilled by raster scanning 150 fs laser pulses over the surface of the tooth. Figure 5b is a greatly magnified view of the machined surface showing open pulpal channels. The very straight sides and square corners of the laser-drilled cavity are quite striking in appearance. The difference between nanosecond and femtosecond ablation is illustrated very clearly by comparing Figures 5c and 5d. Femtosecond pulses clearly produce much more precise ablation than the nanosecond pulses from the Er:YAG laser.

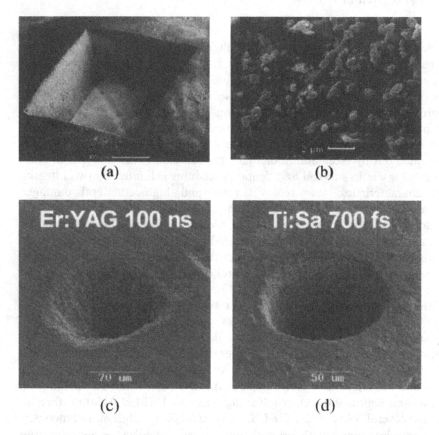

Figure 5 (a) Cavity $1\,\text{mm} \times 1\,\text{mm}$ using 150 fs laser pulses; (b) detail of cavity showing open pulpal channels; (c), (d) holes drilled with Er:YAG laser (ns pulses) and Ti:Sa laser (fs pulses) respectively, showing the superior ablation quality obtained with femtosecond pulses. (Courtesy of Dr. A. Kasenbacher, Traunstein, Germany and Dipl. Phys. J. Serbin, Laser Zentrum Hannover e. V., Hannover, Germany.)

Other issues, such as the selectivity of ablation rates for healthy versus carious tissues, come into play. For example, using 30 ps pulses from an Nd:YLF system, material removal rates are found to be $0.18 \, mm^3 \, min^{-1} \, mJ^{-1}$ for healthy enamel and up to $1.5 \, mm^3 \, min^{-1} \, mJ^{-1}$ for carious enamel, showing one order of magnitude difference in selectivity. The use of plasma-mediated ablation has an additional advantage in the area of diagnostics: The plasma discharge can be monitored and spectrally analyzed to give an on-line health diagnosis of the irradiated area. Whereas the plasma spark induced on healthy tooth tissue shows distinct and strong calcium emission lines, the demineralized composition of decayed tooth tissue shows a strong reduction in the calcium emission lines [21]. This in situ diagnostic capability will aid the dentist in selectively removing diseased tissue.

Photodynamic Therapy. Photodynamic therapy (PDT) has been developed as a minimally invasive treatment of diseased tissue. In PDT, a photosensitive agent is introduced into the afflicted area of tissue, which is then locally illuminated by a light source. This illumination photoactivates the agent, which in turn selectively destroys the diseased tissue, thereby preventing the further spread of the disease. The photosensitive agents are either exogenous (administered) or endogenous (naturally occurring photosensitive materials in the patient). PDT is used for treatment of melanomas, for example. Typically, the illumination is carried out at wavelengths in the visible or near-infrared (NIR) region of the spectrum. Because tissue absorption and scattering are very strong at wavelengths of less than 600 nm, PDT is typically limited to surface treatments. For this reason, much recent work has focused on the development of PDT agents that can be photoactivated by light at wavelengths greater than 600 nm. Because the tissue absorption decreases significantly at IR wavelengths, it should then be possible to treat subsurface disease sites.

Two-photon excitation (TPE) or multiphoton excitation (MPE) provide a means of exciting the PDT agents at nonresonant wavelengths greater than 700 nm. As with multiphoton microscopy, the TPA cross sections are several orders of magnitude smaller than the SPE cross sections, thus requiring very high peak powers to activate the PDT agent. Femtosecond laser pulses provide the necessary high peak power (kilowatts to megawatts) while still maintaining low average powers to prevent photothermal damage to the healthy tissue. Again, as with multiphoton microscopy, because TPE and MPE are nonlinear excitation processes, it is possible to spatially localize the excitation in two or even three dimensions by focusing the beam. Experimental characterization of photosensitive compounds under TPE and MPE has been carried out on both exogenous agents (PHOTOFRIN®) and endogenous photosensitizers related to melanomas using femtosecond lasers at excitation wavelengths in the range of 730–1045 nm [22]. Initial research in this area appears encouraging, in the sense that the results of photosensitization of PDT agents with MPE appear to be roughly equivalent to those using conventional SPE. All other things being equal, the potential for improved localization of PDT treatment points to the possibility of real advantages in using ultrafast lasers for PDT.

Functional Diagnostic Imaging of Oxygenation Levels. The large variety of medical imaging modalities is constantly being expanded for various diagnostic situations. Chapter 15 not only describes the imaging technique of OCT but also presents many examples of specific diagnostic applications. Another type of *functional* diagnostic imaging has been developed for a specific application related to conditions such as oxygenation levels in the brains of premature infants. These levels must be carefully monitored to help ensure proper development of the brain. An imaging system to perform this function—not just monitoring oxygen levels, but imaging them—has been developed recently that uses picosecond laser pulses. This system uses a combination of two techniques: photon time-of-flight measurement (in which the arrival times of photons at different detectors is measured) and differential absorption measurement at two different wavelengths. This particular type of time-resolved imaging uses multiple sources and detectors, relying on the differences in the time of flight of photons in various types of tissues under various conditions [15].

The imaging device, called MONSTIR (multichannel opto-electronic near-infrared system for time-resolved image reconstruction), is illustrated schematically in Figure 6. Picosecond pulses at two specially chosen wavelengths alternately illuminate the infant's head. The two wavelengths are chosen on the basis of the absorption spectra of oxyhemoglobin and deoxyhemoglobin. The wavelengths are chosen with the criterion being to maximize the difference in absorption signal between the two wavelengths, depending upon the oxygenation of the tissue. To perform imaging, the system requires multiple sources and multiple detectors. The procedure is implemented using a 32-channel time-resolved system based on state-of-the-art time-correlated single-photon counting (TCSPC) instrumentation.

Pulses of light from a dual-wavelength fiber laser are coupled successively into a series of optical fibers so that the point of illumination on the tissue surface is varied *sequentially*. Each pulse has a duration of around 1 ps. Meanwhile the transmitted light is collected by 32 detector fiber bundles *simultaneously*. The source and detector fibers are arranged over the surface of the neonatal head. Each of the 32 detector fiber bundles is coupled to the cathode of a microchannel-plate photomultiplier tube (MCP-PMT) via a variable optical attenuator. By measuring the delay between these pulses

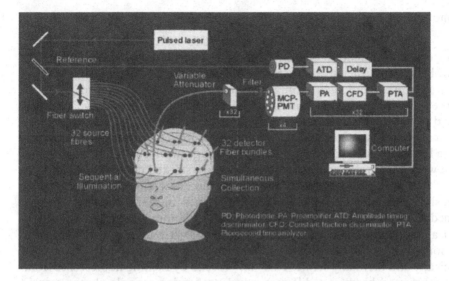

Figure 6 MONSTIR system for imaging oxygenation levels in the brain of neo-nates. A picosecond light pulse at a selected wavelength is incident on the surface of the head. Photon migration times are accurately measured by an array of sensors positioned around the head, and the timing information is processed to give a mapping of oxygen levels. (Courtesy of J. Hebden, University College of London, UK.)

and a reference signal received directly via the laser, histograms of photon flight times (temporal point-spread functions, or TPSFs) are gradually built up within the storage memory of the device. Each represents the TPSF for a distinct line of sight across the head. For 32 separate source positions, the instrument can acquire a total of 1024 TPSFs. By processing this information via a reconstruction algorithm called temporal optical absorption and scattering tomography (TOAST), three-dimensional images of absorption and scattering can be constructed.

Terahertz Imaging for Diagnosis of Melanomas. Terahertz pulse imaging (TPI) is being considered as a diagnostic imaging technique for the diagnosis of basal cell carcinomas. Usually, the diagnosis is done visually, followed by biopsy for any suspicious lesions. Imaging through tissue of any substantial thickness using terahertz radiation is problematic due to the strong water absorption of the radiation. However, initial experiments indicate that TPI used in reflection mode shows a contrast between cancerous tissue and healthy tissue [23]. This technique appears promising as a diagnostic that could circumvent unnecessary biopsies by indicating the type and depth of tumors before surgery.

Information Technology

Three-Dimensional Multilayer Optical Memory. The public's appetite for digital information is growing rapidly, and there is increasing demand for better and faster access to this information. This is made possible via faster networks, and data storage systems need to be developed that can meet the ever more demanding requirements for higher storage densities and faster data transfer rates and random access capabilities. Due to limitations on areal storage densities of optical recording technology, researchers are looking to volumetric or multilayer storage in which a single disk would be written with data on many layers that are stacked closely together.

Optical recording based on two-photon absorption seems a very promising technology for future WORM (write once read many) storage media. [24–26]. The process uses two write photons to excite a transition of a suitable molecule to a high-energy state, whereupon it combines with another type of dye molecule and becomes a new written "bit" that is chemically and thermally stable. This "bit" is read by exciting fluorescence from another read beam at a different wavelength using a single-photon transition. For the writing process, the two photons can be of the same wavelength or of different wavelengths, the main criterion being that the two photon energies combined must exceed the energy needed to excite the molecule to the upper states.

The advantage of using two-photon excitation for the writing process is, once again, very similar to the advantages of using two-photon excitation is confocal microscopy; i.e., the greater excitation at the tight focus ensures that only the focal volume is excited by the beam, giving very localized writing in a very thin plane (several micrometers thick). In this way, the layers of information can be spaced by about 30 µm. It is estimated that this technology can achieve 200 Gbyte storage capacity with a 120 mm diameter disk that is 10 mm thick. By using parallel readout with multiple readout heads, data transfer rates of gigabits per second should be possible.

Telecommunications

Applications in telecommunications can be divided into two main categories: deployable systems and instrumentation (or test and measurement) systems. Ultrafast lasers are finding application in both of these areas.

Telecom-Deployable Systems. The potential role of ultrafast technology in deployable telecommunication systems has been summarized by Knox [27]. Telecommunications applications and their relation to ultrafast laser properties can be represented differently from the applications matrix used in this chapter, as shown in Figure 7. Here it is seen that ultrafast technology has the potential to extend the performance of OTDM, WDM, and CDMA systems as well as chirped pulse (CPWDM) systems. Clearly, the high speed enables high bit rates [for sources and for demultiplexing (demux) functions], while the broad bandwidth allows for large numbers of WDM channels. Because of the fundamental limits of the time–bandwidth product, there is a direct trade-off between the number of channels and the per-channel bit rates. At one extreme, the record for single-wavelength OTDM system speed (1.28 Tbit/s) is discussed in Chapter 13. At the other extreme, broadband pulses from ultrafast lasers have been spectrally sliced to provide up to 1021 WDM channels at lower bit rates of 74 Mbit/s and spaced by 9.5 GHz [6], and it appears than it should be possible to obtain more than 10,000 channels by spectral slicing methods. Ultrafast lasers, although considerably more expensive than the current WDM sources (consisting of a DFB laser and a high-speed modulator) may well provide the most economical broadband source for WDM systems with hundreds of channels, because WDM systems currently require many narrowband sources (one source for each channel) whereas a single ultrafast

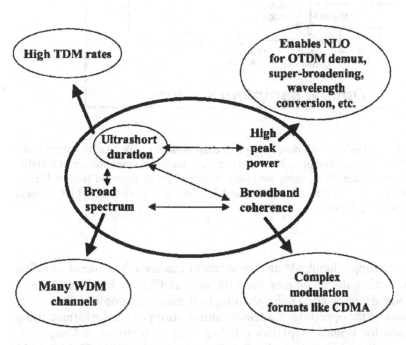

Figure 7 Deployable telecom systems applications that utilize specific properties of ultrafast lasers. (After Ref. 27.)

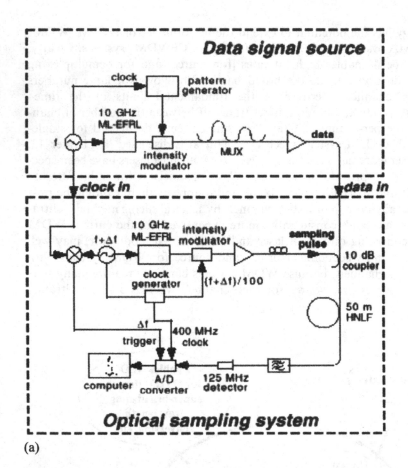

Figure 8 (a) High-speed optical sampling diagnostic for high bit rate optical communications systems. Sampling is carried out via parametric amplification in a highly by nonlinear optical fiber, using sampling pulses of 1.6 ps duration from a USPL. (b) Eye diagrams a bit rates of 160, 200, and 300 Gbit/s. (Courtesy of J. Li, Chalmers University of Technology.)

laser could supply hundreds or thousands of channels. The use of ultrafast lasers for other more complex formats such as CDMA has been demonstrated using femtosecond pulse-shaping techniques and nonlinear processes [28]. Many of the operations for telecommunications can be performed using semiconductor optical amplifiers (SOAs). This is discussed in Chapter 5. A very condensed review of ultrafast laser applications in OTDM/WDM systems can be found in Ref. 29.

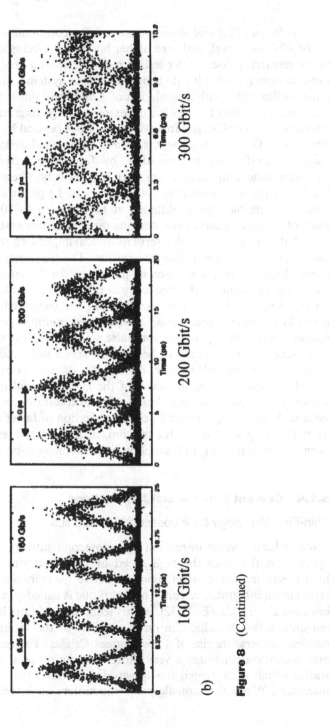

Figure 8 (Continued)

Telecom (Test and Measurement). Telecommunications is advancing to the 40 Gbit/s level, and even higher bit rates are being developed, at least in the research phase. Faster testing methods are required for evaluation of these extremely high bit rate communications systems, and electronic sampling oscilloscopes with a resolution of ~10 ps will not suffice. Optical sampling using ultrashort pulses has the extremely high time resolution to enable characterization of bit streams even at bit rates beyond hundreds of gigabits per second. Optical sampling has been demonstrated using nonlinear effects such as sum-frequency generation, by four-wave mixing in SOAs, and by parametric amplification in high-nonlinearity fibers [30]. Using this last technique with optical sampling pulses of 1.5 ps duration, eye diagram measurements have been obtained at bit rates up to 300 Gbit/s. Refresh rates of 5 frames/s have been demonstrated with this system (see Fig. 8).

Ultrafast lasers can also serve as optical impulse sources for more standard types of high-speed characterization. The frequency response of high-speed photoreceivers is an essential parameter for determining performance. Instead of operating in the frequency domain, it is possible to obtain the frequency response by Fourier transforming the measured impulse response taken in the time domain using a high-speed sampling oscilloscope, with a femtosecond laser as an optical impulse source [31]. One would not call this an ultrafast application per se, because the speed and bandwidth of the measurement are limited by the bandwidth of the sampling oscilloscope (50 GHz). Nevertheless, the speed of the laser is useful in the sense that it does not contribute to degradation of time resolution in the measurement system. At much higher speeds, characterization of faster components, such as 80 Gbit/s photonic ICs for telecom, at least in the research phase, has been carried out using EO sampling with subpicosecond pulses [3].

6.2.4 Current Commercial Applications

Thin-Film Metrology by Picosecond Ultrasonics

Semiconductor wafer inspection by picosecond ultrasonics is a metrology application that uses the high speed of ultrashort optical pulses to give high-precision, non-contact, nondestructive measurements of the various layers in an integrated circuit. The technique is called picosecond ultrasonic laser sonar, or PULSE[TM] [32]. In this technique, a short laser pulse is directed onto a thin metallic film on the surface of the wafer, inducing a very sudden temperature rise of a few degrees Celsius. The resulting rapid thermal expansion generates a very short acoustic pulse at the surface of the wafer, which then propagates downward through the various layers and interfaces. When the acoustic pulse encounters an interface, it is partially reflected back toward the surface of the wafer, whose reflectivity is probed by a second optical pulse in a classic "pump-probe" type of setup. A mapping of the time dependence of the surface reflectivity gives information on the reflected acoustic signal (timing, strength, and width of the echos) that can be analyzed to determine film thickness, the number of layers, interlayer roughness and adhesion, interlayer contamination, and the reactions between layers. Very high precision is possible, depending upon the velocity of sound in the various materials. Typical metallization processes require <5% process control repeatability, which in turn requires ±0.5% measurement repeatability. These processes also call for layers as thin as 50 Å, which demands measurement precision or repeatability of ±0.25 Å. For a "typical" sound velocity of 5 km/s (5 nm/ps), a film thickness accuracy of 1 Å requires that the return pulse be located to an accuracy of 20 fs. Ultrashort lasers and measurement systems are indeed capable of this type of accuracy.

Commercial wafer inspection systems based on this technology have been developed (available from Rudolph Technologies Inc.) for the microelectronics industry. This system, called the *Meta*PULSE^{TM} (see Fig. 9) can simultaneously determine the individual thicknesses of five or more layers in a multilayer metal (MLM) film stack. Film thickness accurations 1–3% and repeatability of <0.5% are routinely achieved. The *Meta*-PULSE^{TM} system (with its integrated femtosecond laser source) is unique in that, unlinke other methods, it can operate even on opaque material layers, allowing it to perform metrology on the three-to-five-layer metal film stacks used to interconnect transistors in semiconductor chips. The PULSE technique is also nondestructive, allowing it to be used in production on product wafers.

Terahertz Spectroscopy and Imaging

Technology for terahertz generation and detection is opening a frequency region of the electromagnetic spectrum for which there is very little existing technology. Although we cannot predict the full impact of this technology at this time, it is probably safe to predict that it will have a large impact in many fields, similar to the discovery of X-rays, which revolutionized medical imaging and many other fields. This is discussed in some detail in Chapter 11. One of the properties that makes terahertz radiation (also called T-rays in the literature) so useful is the peculiar combination of optical properties of various materials at these frequencies. Paper, plastics, and other materials that are commonly used for packaging are transparent to terahertz radiation, whereas water is highly absorbing and metals are highly reflective. This combination of properties makes T-rays useful for many applications such as

Figure 9 Rudolph Technologies *Meta*PULSE 200 wafer inspection system uses picosecond ultrasonics to accurately measure multilayer structures for integrated circuit fabrication. (Courtesy of Rudolph Technologies, Inc.)

spectroscopy (with features complementary to those of FTIR spectroscopy), package inspection, and moisture sensing, to name just a few.

Although any work involving terahertz radiation used to be strictly confined to the research laboratory, progress has been made in demonstrating and constructing compact, portable terahertz systems that employ small femtosecond fiber lasers [33]. In fact, commercial T-ray systems are now being offered that are rugged, modular, and easy to use. The T-RAY™ 2000 analytical system (available from Picometrix) uses state-of-the-art terahertz transceivers that are connected to the ultrafast laser source via fiber-optic cables, enabling easy positioning of the transmitter, receiver, and target specimen without the difficult alignment that is usually so much a part of terahertz measurements [34]. The system shown in Figure 10 also

Figure 10 Picometrix T-Ray 2000 THz imaging system features terahertz transmitter and receiver pair with fiber-optic delivery of the ultrafast laser pulses. (Courtesy of Picometrix, Inc.)

has a control box that contains a high-speed time-delay scanning system. A grating dispersion compensator module makes it possible to use fiber delivery between the various modules. The bandwidth of the system spans the range of 20 GHz to 2 THz. When used for T-ray imaging, submillimeter spatial resolution is possible. Analysis modes in the software include time-domain, FFT, and various imaging modes.

High-Speed Optical Sampling

In the previous section, optical sampling was mentioned as a means to measure high-speed optical signals. Optical sampling via sum-frequency generation has been introduced into commercially available instrumentation. The Agilent 86119A Optical Sampling System employs a femtosecond fiber laser (IMRA Femtolite) and sum frequency generation in a nonlinear crystal to time-resolve high bit-rate optical signals from a remote source. The optical sampling system is completely compatible with Infiniium digital communication analyzer (DCA) mainframes and associated plug-in modules, to provide a complete solution to high bit-rate diagnostics. The system has a transition time of <600 fs, and enough optical bandwidth range

Figure 11   IntraLase Pulsion FS laser system performs femtosecond laser microkeratome procedures for vision correction surgery. (Courtesy of IntraLase Corporation.)

(1530–1610 nm) to cover the C- and L-optical communications bands. A proprietary, bit rate agile timebase gives <200 fs jitter, so that signals exceeding 160 Gbit/s can be measured with this system.

Ophthalmic Laser Procedures

In Chapter 16, Kurtz et al. describe a series of ophthalmic surgical laser techniques that can be performed only with *femtosecond* lasers. The

Pulsion™ FS laser (available from IntraLase Corporation) is a commercial system for performing the femtosecond laser microkeratome procedure for vision correction surgery described in Chapter 16. The Pulsion™ FS (see Fig. 11) is the only all-laser alternative to the mechanical microkeratome. The entire system contains subsystems for applanation (flattening of the cornea), alignment, and beam steering/scanning, as well as an ultrafast laser source. Procedures are controlled by integrated software.

6.3 CONCLUSION

This chapter and this section of the book are not intended to be comprehensive in their coverage. Those who are experienced in the ultrafast laser field will undoubtedly note many omissions in the area of applications. However, the hope here is that we present a reasonably accurate "snapshot" of the current state of ultrafast laser applications and their potential place in real-world markets and provide a glimpse of the future as well. If the reader draws the conclusion that there are indeed a large number of interesting and commercially useful applications for ultrafast lasers, then we have been successful.

The time has come for ultrafast lasers to make their mark in the commercial realm, and it is clear that they are being considered for numerous practical, commercial applications beyond the laboratory. Many who are involved in the ultrafast field have known for several years that this technology will have an impact in a commercial sense. Yet the road has been long. Our adeptness at understanding and harnessing the powers of nature has not been matched by our ability to understand and influence the powers of commerce. Nevertheless, many needs have been identified that can take advantage of the unique capabilities of ultrafast lasers. The convergence of emerging markets with the availability of reliable turnkey, inexpensive ultrafast lasers will result in a revolution not only in the laser industry but hopefully in *many* industries, including health care.

ACKNOWLEDGMENTS

I am grateful to S. R. Bolton for a critical reading of this chapter. Thanks go also to H. Sun, S. Kawata, A. Kasenbacher, J. Serbin, J. Hebden, and J. Li as well as IntraLase Corp., Picometrix Inc., and Rudolph Technologies, Inc. for contributed figures and technical input.

REFERENCES

1. Proc. Ultrafast Phenomena Conf., Vols 4, 14, 23, 38, 46, 48, 53, 55, 60, 62, and 63. Springer-Verlag Ser Chem Phys New York: Springer-Verlag.
2. See SPIE conference proceedings referred to throughout this chapter.
3. K Sano, K Murata, T Akeyoshi, H Kitabayashi, E Sano. Monolitic digital optoelectronic IC's toward 100 Gbit/s. In: Y-K Chen, W Knox, M Rodwell, eds. OSA Trends Opt Photonics, Vol 49. Ultrafast Electronics and Optoelectronics, Washington, DC: Opt Soc Am, 2001, pp 26–33.
4. T Löffler, T Pfeifer, HG Roskos, H Kurtz. Detection of free-running electric signals up to 75 GHz using a femtosecond-pulse laser. IEEE Photon Technol Lett 7:1189–1191 (1995).
5. W Denk. Two-photon excitation in functional biological imaging. J Biomed Opt 1 (3):296–304 (1996).
6. BC Collings, ML Mitchell, L Boivin, WH Knox. A 1021 channel WDM system. IEEE Photon Technol Lett 12:906–908 (2000).
7. A Nebel, B Ruffing, R Wallenstein. A high power diode-pumped all-solid-state RGB laser source. Conference on Lasers and Electro-Optics, Vol 6. 1997 OSA Tech Dig Ser. Washington, DC: Opt Soc Am, 1998, Postdeadline paper CPD3.
8. C Xu, W Denk. Two-photon beam induced current imaging through the backside of integrated circuits. Appl Phys Lett 71:2578–2580 (1997).
9. H-B Sun, T Kawakami, Y-X Ye, S Matuso, H Misawa, M Miwa, R Kaneko. Real three-dimensional microstructures fabricated by photopolymerization of resins through two-photon absorption. Opt lett 25:1110–1112 (2000).
10. S Kawata, H-B Sun, T Tanaka, K Takada. Finer features for functional microdevices. Nature 412:697 (2001).
11. K Minoshima, H Matsumoto. High-accuracy measurement of 240-m distance in an optical tunnel by use of a compact femtosecond laser. Appl Opt 39: 5512–5517 (2000).
12. JC Twichell, PW Juodawlkis, JL Wasserman, RC Williamson, GE Betts. Extending the performance of optically sampled time demultiplexed analog-to-digital converters. Conference on Lasers and Electro-Optics. OSA Tech Dig. Washington, DC: Opt Soc Am, 2000, pp 624–625.
13. AM Weiner, DE Leaird, JS Patel, JR Wullert. Programmable femtosecond pulse shaping by use of a multielement liquid-crystal modulator. Opt Lett 15:356–368 (1990).
14. Y Ding, RM Brubaker, DD Nolte, MR Melloch, AM Weiner. Femtosecond pulse shaping by dynamic helograms in photorefractive multiple quantum wells. Opt Lett 22:718–720 (1997).
15. FEW Schmidt, ME Fry, EMC Hillman, JC Hebden, DT Delpy. A 32-channel time-resolved instrument for medical optical tomography. Rev Sci Instr 71:256–265 (2000).
16. R Jones, K Dowling, MJ Cole, D Parsons-Karavassilis, PMW French, MJ Lever, JD Hares, AKL Dymoke-Bradshow. Fluorescence lifetime imaging for

biomedicine using all-solid-state ultrafast laser technology. In: KR Murray, J Neev eds. Commercial and Biomedical Applications of Ultrafast Lasers. Proc SPIE 3616:86–91 (1999).

17. R Cubeddu, A Pifferi, P Taroni, A Torricelli, G Valentini, F Rinaldi, E Sorbellini. Fluorescence lifetime imaging: an application to the detection of skin tumors. IEEE J Selected Topics Quantum Electron 5:923–927 (1999).

18. ME Zevallos, SK Gayen, BB Das, M Alrubaiee, RR Alfano. Picosecond electronic time-gated imaging of bones in tissues. IEEE J Selected Topics Quantum Electron 5:916–922 (1999).

19. JE Marion II, B-M Kim. Medical applications of ultra-short pulse lasers, In: KR Murray, J Neev, eds. Commercial and Biomedical Applications of Ultrafast Laser. Proc SPIE 3616:42–50 (1999).

20. J Neev, J Squier. Preliminary characterization of hard dental tissue ablation with femtosecond lasers. In J Neev, ed. Applications of Ultrashort-Pulse Lasers in Medicine and Biology. Proc SPIE 3255:105 (1998).

21. MH Niemz. Ultrashort laser pulses in dentistry: advantages and limitations. In: J Neev, ed. Applications of Ultrashort-Pulse Lasers in Medicine and Biology. Proc SPIE 3255:84 (1998).

22. EA Wachter, MG Petersen, HC Dees. Photodynamic therapy with ultrafast lasers. In: Murray, Neev eds. Commercial and Biomedical Applications of Ultrafast Lasers. Proc SPIE 3616:66 (1999).

23. RM Woodward, B Cole, VP Wallace, DD Arnone, R Pye, EH Linfield, M Pepper, AG Davies. Terahertz pulse imaging in in-vitro basal cell carcinoma samples. OSA Trends in Optics and Photonics (TOPS) 56, Conference on Lasers and Electro-Optics (CLEO 2001). Tech Dig, Postconf ed. Washington, DC: Opt Soc Am, 2001, pp 329–330.

24. MM Wang, SC Esener, FB McCormick, I Cokgör, AS Dvornikov, PM Rentzepis. Experimental characterization of a two-photon memory. Opt Lett 22:558–560 (1997).

25. H Zhang, EP Walker, W Feng, Y Zhang, JM Costa, AS Dornikov, S Esener, P Rentzepis. Multi-layer optical data storage based on two-photon recordable fluorescent disk media. Proc 18th IEEE Symp Mass Storage Systems, San Diego, CA, Apr 17–20, 2001.

26. http://www.call-recall.com/ie/index.htm

27. WH Knox. Ultrafast technology in telecommunications. IEEE J Selected Topics Quantum Electron 6:1273–1278 (2000).

28. C-C Chang, HP Sardesai, AM Weiner. Code-division multiple-access encoding and decoding of femtosecond optical pulses in a 2.5 km fiber link. IEEE Photon Tech Lett 10:171 (1998).

29. T Morioka. Ultrafast optical technologies for large-capacity OTDM/WDM transmission. In: Y-K Chen, W Knox, M Rodwell, eds. OSA Trends in Optics and Photonics, Vol 49, Ultrafast Electronics and Optoelectronics. Washington DC: Opt Soc Am, 2001, pp 34–36.

30. J Li, J Hansryd, PO Hedekvis, PA Andrekson, SN Knudsen. 300 Gbit/s eye-diagram measurement by optical sampling using fiber based parametric ampli-

fication. OSA Trends in Optics and Photonics (TOPS), Vol 54, Optical Fiber
Communication Conference. Tech Dig Postconf ed. Washington, DC: Opt
Soc Am, 2001, paper PD31.

31. TS Clement, PD Hale, KC Coakley, CM Wang. Time-domain measurement of
the frequency response of high-speed photoreceivers to 50 GHz. Tech Dig,
Symp Opti Fiber Meas, NIST SP 953. September 2000, pp 121–124.

32. C Thomsen, HT Grahn, HJ Maris, J Tauc. Surface generation and detection of
phonons by picosecond light pulses. Phys Rev B 34:4129 (1986).

33. M Li, X-C Zhang, G Sucha, DJ Harter. Portable THz system and its applications. In: KR Murray, J Neev eds. Commercial and Biomedical Applications of
Ultrafast Lasers. Proc SPIE 3616:126–135 (1999).

34. JV Rudd, D Zimdars, MW Warmuth. Compact, fiber-pigtailed, terahertz imaging system. In: J Neev, KR Murray, eds. Commercial and Biomedical Applications of Ultrafast Lasers II. Proc SPIE 3934:27–35 (2000).

7
Micromachining

Stefan Nolte
Friedrich Schiller University, Jena, Germany

7.1 INTRODUCTION

Material processing with pulsed lasers has been an intensive research topic
since the invention of the laser in 1960. Nowadays, the laser is used as an
efficient and qualified tool in many industrial processes such as heavy industrial cutting, hardening, and welding. In the microfabrication industry, however, the laser has not yet become a universal instrument. In general, one
needs a special laser for a particular microstructuring application. Excimer
lasers are used, for example, for the micromachining of ceramics and polymers, and Nd:YAG-lasers are used for microdrilling and marking. Moreover, for several applications, particularly for the precise microstructuring
of metallic materials, the use of lasers with pulse durations in the range of
nanoseconds to microseconds is limited due to thermal or mechanical
damage (melting, formation of burr and cracks, changes in the morphology,
etc.).These limitations have stimulated widespread research activities to
minimize collateral damage and thermal diffusion out of the irradiated area
by using ultrashort laser pulses, including investigations on the ablation of
dielectrics (e.g., Küper and Stuke, 1989; Du et al., 1994; Pronko et al.,
1995a; Stuart et al., 1995, 1996; Varel et al., 1996; Ashkenasi et al., 1998;
Lenzner et al., 1998; von der Linde and Sokolowski-Tinten, 2000) and
metals (e.g., Preuss et al., 1995; Krüger and Kautek, 1995; Momma et al.,
1996; Nolte et al., 1997; Feuerhake et al., 1998; Wellershoff et al., 1999)
as well as attempts to produce submicrometer structures (Pronko et al.,
1995b; Simon and Ihlemann, 1996).

To summarize the results of this research: The femtosecond laser
seems to be an excellent and universal tool for microfabrication. Metals,

semiconductors, dielectrics, polymers, etc., transparent and opaque materials (hard and fragile), can be microstructured with femtosecond pulses so that no postprocessing is required. Furthermore, the surrounding areas are not affected or damaged, which opens the possibility for new fields of application.

The advantages of ultrashort pulse laser ablation have been demonstrated in a spectacular way by cutting explosives (Fig. 1) (Perry et al., 1999). Explosives cannot be cut with long laser pulses because they are ignited due to the thermal load, even for 600 ps pulses (see Fig. 1b). In contrast, the irradiation with femtosecond pulses results in clean cuts (see Fig. 1a). Thermal transfer and shock waves are substantially smaller than necessary for ignition. Furthermore, no chemical reaction products are observable in this case, and the laser-cut surface is chemically identical to the original material. Such applications clearly demonstrate the advantages of ultrashort pulse laser ablation.

These properties have spurred interest not only from the industry but also from biology and medicine. However, other criteria such as process throughput, yield, reliability, and costs also have to be taken into account. Here, ultrashort pulse lasers still have some drawbacks. But with the recent developments of reliable, compact, and easy-to-operate laser sources this technology is becoming a serious competitor to conventional microprocessing techniques.

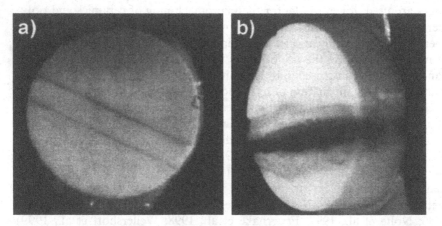

Figure 1 Cuts in an explosive pellet produced by a Ti:sapphire laser operating at 120 fs (a) and 600 ps (b). Although no evidence of thermal influence of the cutting surfaces is observable in the case of femtosecond pulses (a), thermal load in the long-pulse case caused the pellet to ignite and burn (b). (From Perry et al., 1999.)

The theoretical and experimental basics of ultrashort-pulse laser micromachining are summarized in the following section. In Section 7.3, several techniques for the fabrication of microstructures as well as concrete application examples are highlighted. The possibilities for submicrometer machining are then discussed in Section 7.4.

7.2 FEMTOSECOND LASER ABLATION: BACKGROUND

Because of the much faster energy deposition, ablation with femtosecond laser pulses is different from long-pulse (nanosecond) laser ablation. The differences are schematically shown in Figure 2. In the case of long-pulse laser ablation, the electrons and the lattice remain in thermal equilibrium. During the laser pulse, heat diffuses out of the irradiated area and the material expands.

In contrast, heating with ultrashort laser pulses is a strongly nonequilibrium process. At first, the laser radiation is absorbed inside a surface layer by bound and free electrons. This is accompanied by the excitation and ionization of the material and heating of free electrons by inverse Bremsstrahlung, followed by fast energy relaxation within the electron subsystem. Later, energy transfer from the electrons to the lattice (atomic subsystem), bond breaking, and material expansion take place. For laser fluences close

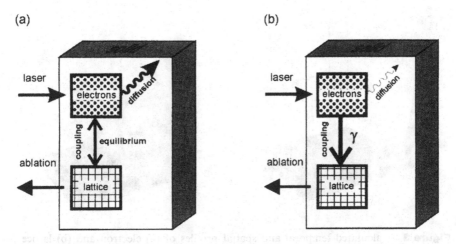

Figure 2 Schematic of the ablation of solid targets with (a) long laser pulses and (b) ultrashort laser pulses.

to the ablation threshold the electron–ion energy transfer occurs on a pico-second time scale.

Figure 3 shows a calculated temporal and spatial temperature distribution of electrons and lattice for a Cu target irradiated by a 100 fs pulse at an intensity of 10^{13} W/cm^2. The heating of the electron subsystem occurs very fast, whereas the lattice remains relatively cold during the laser pulse. The energy transfer to the lattice takes place on a longer time scale. Therefore, the electrons are heated up to very high transient temperatures.

For laser fluences slightly above the ablation threshold, heat diffusion out of the irradiated area is minimal for ultrashort laser pulses. At the focus, one can speak of the formation of a high density plasma as is schematically shown in Figure 4. In the case of dielectrics, the electron heat transport into the target is strongly suppressed. Electrons are not able to escape due to the charge separation force keeping the plasma neutral. In metals electron heat transport is allowed, because the hot electrons moving into the target can be replaced by cold electrons from the adjacent region (return current). Despite this, there exist additional physical reasons (see Sec.7.2.2) for the reduction of the electron heat transport in metals for ultrashort pulses.

The complex physics associated with ultrashort-pulse laser ablation makes a detailed theoretical analysis difficult. In principle, the dynamic behavior of the ablation process can be simulated by molecular dynamics (MD) calculations (e.g., Hermann et al., 1998). In these calculations the interaction of single molecules with their neighbors is considered. MD simulations permit several qualitative predictions of ultrashort-pulse laser ablation,

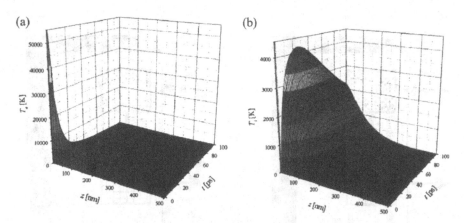

Figure 3 Calculated temporal and spatial profiles of (a) electron and (b) lattice temperatures in copper after irradiation with a 100 fs pulse at an intensity of 10^{13} W/cm^2. Ablation has not been taken into account.

(a) (b)

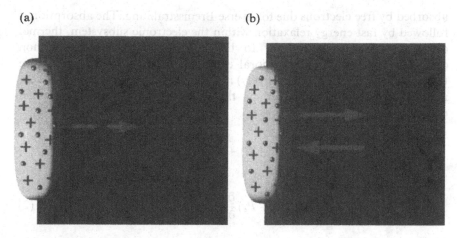

Figure 4 Schematic of femtosecond laser ablation and electron heat transport in dielectric (left) and metal (right) targets.

such as the increase in the ablation rate and the decrease in the ablation threshold with decreasing pulse duration. MD simulations also show that the main material removal occurs on a time scale of a few picoseconds, even for femtosecond laser pulses.

However, MD simulations require enormous computational power. In practice, only very small interaction volumes with dimensions of a few nanometers (containing only several tens of thousands of atoms) can be modeled, which is far from experimental conditions. To compare simulations with experimental results, the calculations have to be rescaled, making quantitative data questionable.

An alternative theoretical description of the ablation process is based on consideration of the average behavior of many atoms. Such a description is possible for ultrashort laser pulses on the basis of a two-temperature model (TTM) in which the electrons and the lattice are characterized by their temperatures.

A full treatment of the ablation process including the plasma hydrodynamic is possible only numerically. Although this yields good agreement with experimental results (e.g., Komashko et al., 1999), a simplified analytical solution, which is more instructive for the reader, is discussed here.

7.2.1 Ablation of Metals

This section presents a theoretical description of femtosecond laser ablation of metal targets on the basis of a TTM. In the case of metals, laser energy is absorbed by free electrons due to inverse Bremsstrahlung. The absorption is followed by fast energy relaxation within the electronic subsystem, thermal diffusion, and energy transfer to the lattice due to electron–phonon coupling. The spatial and temporal evolution of the electron and lattice temperatures (T_e and T_i, respectively) in a thin surface layer with subsequent material expansion is described by the following set of one-dimensional equations.

$$C_e \frac{dT_e}{dt} = -\frac{\partial Q(x)}{\partial x} - \gamma(T_e - T_i) + S - P_e \frac{\partial u}{\partial x} \tag{1}$$

$$C_i \frac{dT_i}{dt} = \gamma(T_e - T_i) - (P_i + P_c) \frac{\partial u}{\partial x} \tag{2}$$

$$\rho \frac{du}{dt} = -\frac{\partial}{\partial x}(P_c + P_e + P_i) \tag{3}$$

$$\frac{\partial \rho}{\partial t} + \frac{\partial \rho u}{\partial x} = 0 \tag{4}$$

where x is the direction perpendicular to the target surface, $d/dt = \partial/\partial t + \partial u/\partial x$, and C_e and C_i are the heat capacities (per unit volume) of the electron and lattice subsystems, respectively. The parameter γ characterizes the electron–lattice coupling, ρ and u are the density and velocity of the evaporated material, P_e and P_i are the thermal electron and ion pressures, P_c is the elastic (or "cold") pressure, which is positive for compression and negative for expansion, $Q(x) = -k_e(T_e) \, \partial T_e/\partial x$ is the heat flux, and $S = I(t)A\alpha \exp(-\alpha x)$ is the laser heating source term. Here, k_e is the electron thermal conductivity, A and α are the surface absorptivity and the material absorption coefficient, and $I(t)$ is the laser intensity. Equations (1) and (2) are energy conservation equations for the electron and ion subsystems. Equation (3) expresses Newton's law, and Eq. (4)—the continuity equation—describes the conservation of mass.

In spite of the obvious simplicity of the above equations, their application for modeling of femtosecond laser ablation is problematic. This is because of a lack of reliable information on several parameters that enter into these hydrodynamic equations. For example, there is insufficient information on the equations of state that could describe the evolution of the electron, ion, and cold pressures.

A simple model for ultrashort-pulse laser ablation can be obtained by neglecting the material expansion completely and declaring that one needs a

certain amount of energy to initiate ablation. In this case the last terms in
Eqs. (1) and (2) (containing $\partial u/\partial x$) can be omitted and the hydrodynamic
equations reduce to a one-dimensional two-temperature diffusion model
proposed by Anisimov et al. (1974).

$$C_e \frac{\partial T_e}{\partial t} = -\frac{\partial Q(x)}{\partial x} - \gamma(T_e - T_i) + S \tag{5}$$

$$C_i \frac{\partial T_i}{\partial t} = \gamma(T_e - T_i) \tag{6}$$

A discussion of the laser–metal interaction for long (nanosecond) and ultrashort (femtosecond) laser pulses follows.

Long (Nanosecond) Pulses

If the laser pulse duration is much longer than the lattice heating time
($\tau_i = C_i/\gamma$, which is of the order of $\tau_i \approx 0.01$–1 ns), then thermalization
between the electron subsystem and the lattice takes place during the laser
pulse. In this case the electrons and the lattice can be characterized by a
common temperature $T = T_e = T_i$ and Eqs. (5) and (6) reduce to the well-
known one-dimensional heat diffusion equation

$$C_i \frac{\partial T}{\partial t} = \frac{\partial}{\partial x} k_e \frac{\partial T}{\partial x} + I(t)A\alpha \exp(-\alpha x) \tag{7}$$

which describes long-pulse laser heating. Solutions of Eq. (7) can be found,
for example, in Prokhorov et al. (1990).

Ultrashort (Femtosecond) Pulses

For femtosecond laser pulses, the energy transfer to the lattice during the
laser pulse and the heat conduction can be neglected in a first approximation. In this simplified case, the ablation rate and the ablation threshold
depend only on the optical penetration depth (see, e.g., Preuss et al.,
1995; Krüger and Kautek, 1995; Chichkov et al., 1996). However, as will
be shown below, the interplay of energy exchange between electrons and lattice and the heat diffusion have important consequences, even for ultrashort-pulse ablation. When the heat capacity, the thermal conductivity,
and therefore the electron thermal diffusivity are treated as constants, one
can find the following equilibrium temperature distribution for the electrons
and the lattice after thermal relaxation (Nolte et al., 1997):

$$T_i \approx \frac{F_a}{C_i}\left(\frac{1}{l^2 - \delta^2}\right)\left[l\exp\left(\frac{-x}{l}\right) - \delta\exp\left(\frac{-x}{\delta}\right)\right] \tag{8}$$

Here the optical penetration depth is obtained by $\delta = 1/\alpha$ and the electron thermal diffusion length is given by $l = \sqrt{D\tau_a}$. The duration of the ablation process, τ_a, is determined by the time necessary for the energy transfer from the electrons to the lattice. F_a is the absorbed laser fluence.

Two cases, illustrated in Figure 5, can be distinguished: the optical penetration depth exceeding the thermal diffusion length, $\delta > l$, and vice versa, $l > \delta$. For these cases the following formulas for the equilibrium temperatures can be obtained from Eq. (8):

$$T_i \approx \frac{F_a}{C_i \delta} \exp\left(\frac{-x}{\delta}\right) \qquad (\delta > l) \tag{9}$$

$$T_i \approx \frac{F_a}{C_i l} \exp\left(\frac{-x}{l}\right) \qquad (\delta < l) \tag{10}$$

In this simplified model, significant ablation takes place when the energy of the lattice $C_i T_i$ (per unit volume) exceeds a certain threshold value. This threshold value can be estimated in a first approximation as the heat of evaporation $\rho\Omega$, where ρ is the density and Ω is the specific heat of evaporation per unit mass. The condition for significant ablation, $C_i T_i \geq \rho\Omega$, can be written as

$$F_a \geq F_{th}^{\delta} \exp\left(\frac{x}{\delta}\right), \qquad F_{th}^{\delta} \approx \rho\Omega\delta \qquad (\delta > l) \tag{11}$$

$$F_a \geq F_{th}^{l} \exp\left(\frac{x}{l}\right), \qquad F_{th}^{l} \approx \rho\Omega l \qquad (\delta < l) \tag{12}$$

where F_{th}^{δ} and F_{th}^{l} are the corresponding thresholds for the absorbed laser fluence.

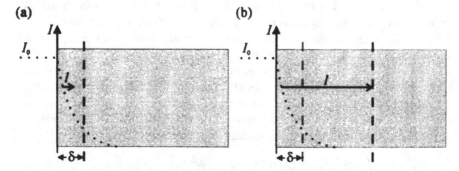

Figure 5 Schematic of the two ultrashort pulse heating regimes: (a) $\delta > l$; (b) $\delta < l$.

The ablation depths are easily derived from Eqs. (11) and (12):

$$L \approx \delta \ln\left(\frac{F_a}{F_{th}^{\delta}}\right) \qquad (\delta > l) \tag{13}$$

$$L \approx l \ln\left(\frac{F_a}{F_{th}^{l}}\right) \qquad (\delta < l) \tag{14}$$

This means that two logarithmic scaling laws are obtained from the TTM. These two ablation regimes have also been observed experimentally for sub-picosecond pulse ablation of Cu (Nolte et al., 1997) (see Fig. 6) and for Au and Ag (Furusawa et al., 1999). With respect to Eqs. (13) and (14), these two logarithmic dependencies can be attributed to the optical penetration depth and the electronic heat conduction, respectively.

In Figure 6 the ablation depth per pulse is shown as a function of laser fluence for 150 fs laser pulses (wavelength 780 nm) irradiating a Cu target. The two ablation regimes, that means the two different logarithmic dependencies, are clearly visible.

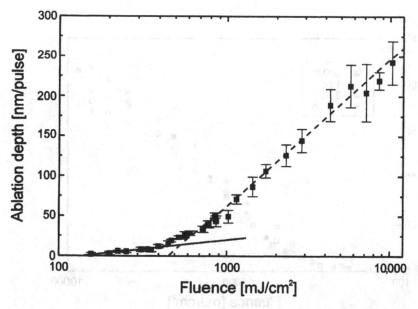

Figure 6 Ablation depth per pulse for copper as a function of incident laser fluence for 150 fs Ti:sapphire-laser radiation ($\lambda = 780$ nm). (——)$L = 10\ln(F/F_{th}^{\delta})$nm; (– – –) $L = 80\ln(F/F_{th}^{l})$nm.

The fact that the thermal diffusion length can be smaller than the optical penetration depth for fluences F smaller than $0.5 \, \text{J/cm}^2$ is related to changes in the heat conduction due to the strong nonequilibrium between electrons and lattice. This is discussed in more detail in Section 7.2.2.

Figure 7 shows a comparison of ablation rates for Cu obtained with different pulse durations between 0.5 and 4.8 ps. The ablation regime dominated by the optical penetration depth is accessible only for subpicosecond pulses. For pulses $\tau_L \geq 1 \, \text{ps}$ there is already significant heat diffusion during the laser pulse. Therefore, the energy cannot be deposited in an area determined by the optical penetration depth.

In the case of Cu, the ablation rates are only slightly dependent on pulse duration for pulses up to $\sim 5 \, \text{ps}$. When the pulse duration is increased further, the ablation rates decrease because of the increasing energy losses caused by heat diffusion.

7.2.2 Heat Conduction in Metals

Because the electronic heat capacity is much smaller than the lattice heat capacity, electrons can be heated to very high transient temperatures. When $k_B T_e$ remains lower than the Fermi energy $E_F = m v_F^2 / 2$, the electron heat

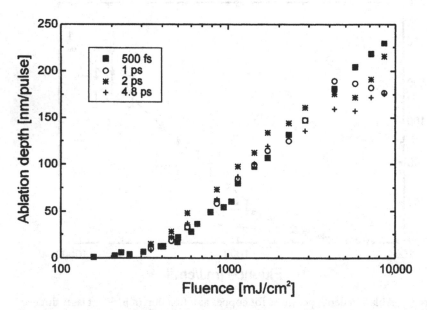

Figure 7 Ablation depth per pulse for copper and different pulse durations from 500 fs to 4.8 ps ($\lambda = 780 \, \text{nm}$).

capacity and the electron thermal conductivity are given by (e.g., Ashcroft and Mermin, 1976)

$$C_e \approx \pi^2 \frac{k_B T_e}{E_F} \left(\frac{N_e k_B}{2} \right) \quad \text{and} \quad k_e = \frac{C_e v_F^2 \tau}{3}$$

where N_e is the electron density and τ is the electron relaxation time, which is determined by electron–phonon and electron–electron collisions $1/\tau = 1/\tau_{e-ph} + 1/\tau_{e-e}$. When the lattice temperature is higher than the Debye temperature, $T_i \geq \Theta = \hbar \omega_D / k_B$, where ω_D is the Debye frequency (the maximum phonon frequency), all oscillation modes of the lattice are excited, and a good approximation for the electron–phonon collision frequency is given by (e.g., Abrikosov, 1988) $1/\tau_{e-ph} \sim k_B T_i / \hbar$. The electron–electron collision frequency can be estimated by (e.g., Abrikosov, 1988) $1/\tau_{e-e} \sim (k_B T_e)^2 / \hbar E_F$.

In the case of material processing with long laser pulses, the electronic and atomic subsystems are in equilibrium, $T_e = T_i$, and the condition $\tau_{e-ph} < \tau_{e-e}$ or $\tau \approx \tau_{e-ph}$ is fulfilled. With ultrashort laser pulses a strong overheating of the electronic subsystem occurs $(T_e \gg T_i)$, and $\tau_{e-ph} > \tau_{e-e}$ becomes possible. In this regime the electron relaxation time is determined by the electron–electron collisions $\tau \approx \tau_{e-e} \sim T_e^{-2}$, and the dependencies of the thermal conductivity and thermal diffusivity, $k_e \sim T_e^{-1}$ and $D = k_e/C_e \sim T_e^{-2}$, are valid. This corresponds to a rapid decrease in the thermal diffusivity and in the electron thermal losses with increasing electron temperature (and thus laser fluence) (Kanavin et al., 1998). This effect is responsible for the reduced influence of heat diffusion when processing metals with ultrashort laser pulses.

7.2.3 Ablation of Dielectrics and Semiconductors

The physics of ultrashort-pulse laser ablation of metals (as described earlier) and other solids (semiconductors, dielectrics, polymers, etc.) is essentially the same. The main difference is that in metals the electron density can be considered constant during the interaction with an ultrashort laser pulse. In other materials, the electron density changes due to interband transitions and multiphoton and electron impact (avalanche) ionization followed by optical breakdown. Thus, an additional equation is necessary for the description of the temporal evolution of the free electron density for describing nonmetallic samples. This can be satisfied by the following rate equation for multiphoton and avalanche ionization (Stuart et al., 1996):

$$\frac{\partial N_e}{\partial t} = \sigma^{(n)} N_0 I(t)^n + \beta I(t) N_e(t) \tag{15}$$

Here N_e is the (time-dependent) electron density, N_0 is the initial atomic density, $\sigma^{(n)}$ is the n-photon ionization cross section, $I(t)$ is the laser intensity, and β is the avalanche coefficient [$\beta \approx 0.011 \, \text{cm}^2 \, \text{ps}^{-1} \, \text{GW}^{-1}$ for fused silica (Perry et al., 1999)]. The first term in Eq. (15) describes the n-photon ionization, where n is the smallest value satisfying $n\hbar\omega \geq E_g$ ($\hbar\omega$ is the photon energy and E_g is the band gap of the material). The second term in Eq. (15) corresponds to avalanche ionization and is therefore proportional to the number of free electrons.

The material becomes opaque to the laser radiation when the density of free electrons reaches a critical value

$$N_e^{\text{cr}} = \frac{\varepsilon_0 m \omega^2}{e^2} = \frac{1.11 \times 10^{21}}{\lambda^2} \quad [\text{cm}^{-3}] \quad (16)$$

Here m and e denote the mass and charge of the electron, ω the laser light frequency, and λ the laser wavelength (measured in micrometers).

Figure 8 illustrates the change of the electron density with time in a fused silica sample irradiated with a 100 fs laser pulse at an intensity of $I = 1.17 \times 10^{13} \, \text{W/cm}^2$. As can be seen, first free electrons are generated by multiphoton ionization. These electrons then serve as seed electrons for the avalanche process.

This behavior is in contrast to ablation with long laser pulses. In this case, the intensity is not high enough to generate a significant number of free electrons by multiphoton ionization. Thus, ablation of dielectrics with long laser pulses relies on the presence of some free electrons (due to defects) for the start of the avalanche process. As a consequence, the threshold for long-pulse ablation is strongly dependent on the quality of the target material (number of defects) and is therefore not well defined. It scales with the pulse duration as $F_{\text{th}} \sim \tau_L^{1/2}$ (which holds for dielectrics as well as for metals), because heat diffusion out of the irradiated area is the determining process.

In contrast, ultrashort-pulse laser ablation of dielectrics does not rely on the presence of free electrons due to defects, resulting in a very well defined ablation threshold. Although there is no significant dependence of the threshold fluence on pulse duration for metals (Corkum et al., 1988; Stuart et al., 1996) for laser pulses $\tau_L < \sim 100 \, \text{ps}$ (because heat diffusion is strongly reduced), for dielectrics and semiconductors the threshold fluence is still decreasing with decreasing pulse duration. This is due to the fact that the absorption process is strongly nonlinear and the absorption increases with increasing intensity, i.e., decreasing pulse duration. However, the threshold fluence decreases more slowly than $F_{\text{th}} \sim \tau_L^{1/2}$ and approaches asymptotically the limit where all free electrons are created solely by multiphoton ionization (Stuart et al., 1996) (see Fig. 9).

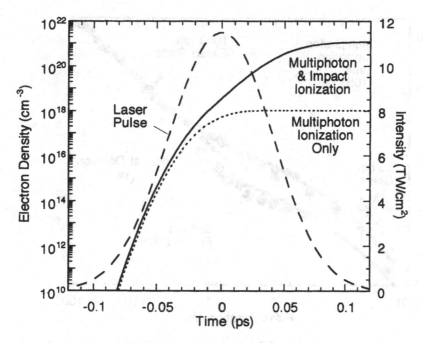

Figure 8 Calculated temporal evolution of the electron density for a 100 fs, 1053 nm pulse (dashed curve) of peak intensity 11.7 TW/cm^2 in fused silica. Multiphoton ionization (dotted curve) starts the avalanche; the solid curve is the total electron density including electrons generated by impact ionization. (From Stuart et al., 1996.)

7.3 MICROABLATION

When an ultrashort laser pulse interacts with a solid target, the hydrodynamic motion of the ablated material during the laser pulse is negligible, and absorption occurs directly at the target surface. The absence (or unimportance) of the hydrodynamic motion during the laser pulse is one of the advantages of ultrashort laser pulses that allow precise and controllable energy deposition inside the solid target. In the following section, two different regimes of ultrashort-pulse laser ablation will be discussed: ablation at relatively high and low intensities.

7.3.1 High-Intensity Ablation

At high laser intensities, the absorption of laser radiation by a solid target results in the creation of a dense, high-temperature plasma. In this case,

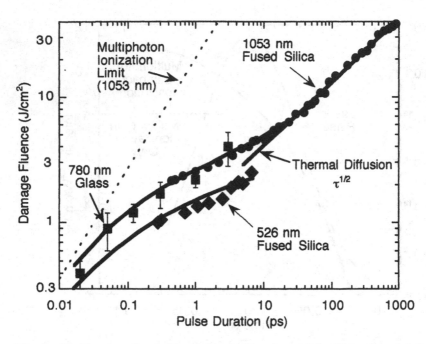

Figure 9 Damage thresholds of fused silica for visible (526 nm) and infrared (1053 nm) laser radiation. Below ~10 ps there is a deviation from the $\tau_L^{1/2}$ scaling, indicating the transition to an ionization-dominated damage mechanism. (From Perry et al., 1999.)

the ablation process has the following scenario. The absorbed laser energy is first deposited as kinetic (thermal) energy of the electrons. This initiates an electron thermal wave propagating inside the target (see, e.g., Hughes, 1975). Collective electron motion and energy transfer from the electrons to the ion subsystem (which occurs on a longer time scale owing to the large difference in electron and ion masses) initiate ablation of the target material and an ablation (or rarefaction) wave propagating behind the thermal wave. The energy transfer to the ions (which means an energy loss for the electron subsystem) slows down the thermal wave. At $t = \tau_a$ (which is approximately the duration of the ablation process), the ablation wave overtakes the front of the thermal wave.

During the ablation process a shock wave is created owing to the recoil pressure of the ablated material. At $t = \tau_a$ the formation of this shock wave is completed and the so-called hydrodynamic separation (separation of the shock wave from the heat wave) occurs (see, e.g., Zeldovich and Raizer, 1966). For times $t > \tau_a$, ablation due to the shock wave can occur, which

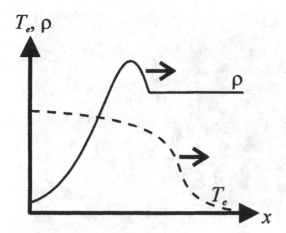

Figure 10 Schematic illustration of the high-intensity ablation process. The thermal wave propagating into the solid is followed by an ablation wave. The formation of a shock front is indicated.

corresponds to the hydrodynamic regime of ablation. The formation of the thermal and the shock wave are schematically illustrated in Figure 10.

At the laser fluences considered here, the target material is removed in plasma, liquid, and vapor phases. In this case the heat penetrates deep into the target. Between the front of the thermal wave and the ablation front a considerable part of the material is molten. Due to the plasma expansion and the vaporization process, a strong recoil pressure is created that expels the molten material, which results in the formation of droplets and debris on the target surface. Moreover, the expelled liquid material resolidifies at the edges of the hole, resulting in the formation of burr. This can be seen in Figure 11, which shows a scanning electron microscopic (SEM) image of a hole drilled in a stainless steel plate with 100 pulses at an energy of $E = 2\,\mathrm{mJ}$ ($F \approx 200\,\mathrm{J/cm^2}$) and a pulse duration of $\tau_L = 200\,\mathrm{fs}$.

At higher fluences, the recoil pressure is increased, and the ablation has an explosive character. The burr decreases, because the particles are ejected at higher velocities and do not resolidify at the edges of the hole. Figure 12 shows two examples of holes drilled in a Cu plate. The laser pulse energies were $E = 5\,\mathrm{mJ}$ ($F \approx 250\,\mathrm{J/cm^2}$) and $E = 70\,\mathrm{mJ}$ ($F \approx 3600\,\mathrm{J/cm^2}$), respectively. In both cases the pulse duration was $\tau_L = 180\,\mathrm{fs}$. Although the burr decreases with increasing intensity, the high ablation pressure causes strong deformations of the surrounding material (Luft et al., 1996).

It is important to note that relatively high ablation rates (e.g., the ablation rate is $>50\,\mathrm{\mu m/pulse}$ in Fig. 12b) can be achieved with ultrashort laser

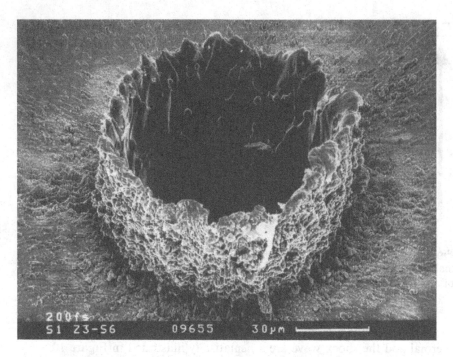

Figure 11 Scanning electron microscopic image of a hole drilled in stainless steel
($E = 2\,\text{mJ}$, $F \approx 200\,\text{J/cm}^2$, $\tau_L = 200\,\text{fs}$, 100 pulses, hole depth 140 µm).

pulses. In the case of long-pulse ablation, the achievable ablation rate is limited by the so-called plasma shielding (see, e.g., Bäuerle, 1996). Above a certain intensity the laser light is strongly absorbed within the expanding plasma and does not reach the target. Although this may result in a higher overall absorption of laser radiation (Komashko et al., 1999), the part of the energy that is transferred to the target and contributes to the ablation process remains small. For ultrashort pulses, plasma shielding does not occur, because there exists a temporal separation between the energy deposition and the ablation process (the plasma expansion occurs mainly after the laser pulse).

Regarding the achievable machining precision, ultrashort pulses do not provide considerable advantages in this high-intensity regime compared to long laser pulses, because the ablation process is accompanied by strong thermal and mechanical effects on the target material. However, precision can be substantially increased when working in the low-fluence regime, which is discussed in the next section.

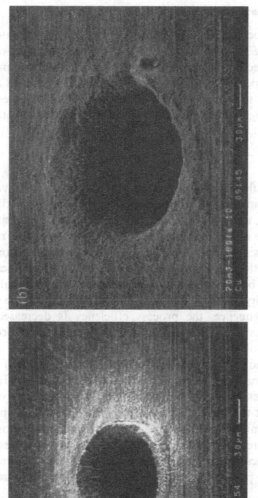

Figure 12 Scanning electron microscopic images of holes drilled in copper. (a) $E = 5$ mJ, $F \approx 250$ J/cm^2, $\tau_L = 180$ fs, 100 pulses, depth 180 μm; (b) $E = 70$ mJ, $F \approx 3600$ J cm^2, $\tau_L = 180$ fs, 10 pulses, depth >500 μm.

7.3.2   Laser Ablation in a Low-Fluence Regime

High-precision processing can be achieved with ultrashort laser pulses at low laser fluences $F < 10\,\mathrm{J/cm^2}$. Even the delicate structuring of explosives or metallic workpieces with negligible heat-affected zones and practically no burr formation is possible in this regime, for which a detailed analysis has already been given in Section 7.2. In this section some examples of possible applications are highlighted, including the drilling and microstructuring of different materials.

Drilling

Ultrashort laser pulses allow the precise and minimally invasive processing of practically all materials. This includes transparent materials, such as glasses or crystals, as well as metallic workpieces, the processing of which is difficult because of the high thermal conductivity and low melting temperatures. When using conventional long-pulse lasers (nanosecond to microsecond pulse duration), the melting and subsequent resolidification of the material leads to burr generation. Furthermore, for materials with high thermal conductivity, there is significant energy diffusion to the surrounding areas. As a consequence, the process efficiency is decreased and large heat-affected zones are generated in which the material properties are altered.

Figure 13 shows a comparison of holes drilled with (a) nanosecond and (b) femtosecond laser pulses in a 100 μm thick steel foil. In both cases the fluence has been chosen to be just above the threshold to drill through the foil. Note that the fluence necessary to drill the hole is approximately one order of magnitude smaller for ultrashort than for nanosecond laser pulses. This is a direct consequence of the reduced energy losses due to heat diffusion and indicates the higher processing efficiency. Moreover, in the case of ultrashort pulses, the heat-affected zones are minimized ($\ll 1$ μm), (Luft et al., 1996).

The presence of burr and large droplets on the target surface is also noticeable in the case of long laser pulses (Fig. 13a), whereas the use of ultrashort pulses results in clean ablation with sharp and burr-free edges and minimal debris (Fig. 13b). The ablation with long laser pulses takes place in the liquid and vapor phases, whereas ultrashort pulse ablation is dominated by the formation of vapor and plasma phases.

Another important aspect for industrial use is the reproducibility of the ablation process. In the case of long laser pulses, the presence of the liquid phase leads to unstable drilling conditions, making additional techniques necessary to improve the hole quality (for example, the use of a gas jet to blow out the molten material). In contrast, ablation with ultrashort pulses

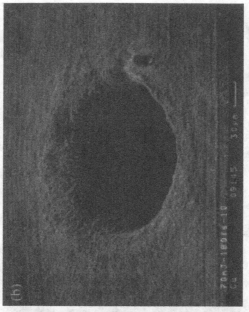

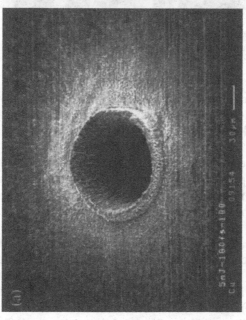

Figure 13 Holes drilled in a 100 μm thick stainless steel foil with (a) 3.3 ns and (b) 200 fs Ti:sapphire laser pulses. The pulse energies and fluences used were $E = 1\,mJ$, $F = 4.2\,J/cm^2$ and $E = 120\,\mu J$, $F = 0.5\,J/cm^2$, respectively. In both cases 10,000 pulses were used.

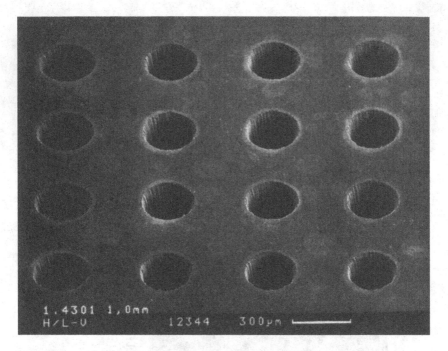

Figure 14 Array of holes drilled with 150 fs Ti:sapphire laser pulses at an energy of 1.3 mJ in 1 mm thick stainless steel.

is practically free of molten material, resulting in an intrinsically stable and reproducible process. Figure 14 shows an example of an array of holes drilled in a 1 mm thick steel plate with 150 fs pulses, demonstrating the high reproducibility.

Apart from these advantages, the high intensities associated with ultra-short pulses make a special beam-forming and -guiding system necessary to generate well-defined microstructures with high quality and reproducibility. Although the exact hole diameter depends on pulse energy when simply focusing, the use of the imaging (mask projection) technique, where an aperture is geometrically imaged by a lens onto the target surface, is one possibility for producing holes with well-defined diameters. However, the imaging technique in combination with high-energy ultrashort laser pulses suffers from the inevitable intermediate focal point in front of the image plane (target surface). The high intensities in this area result in plasma breakdown and nonlinear effects (beam deformation and filamentation, self-focusing, etc.) in the ambient air. To avoid these detrimental effects, the ambient pressure has to be reduced, making the use of vacuum equipment necessary (see Fig. 15). Clearly, such a requirement represents a serious drawback for the widespread acceptance of the ultrashort pulse technology.

As an alternative, the use of diffractive optical elements (DOEs) has turned out to be an appropriate and elegant method for the beam shaping of ultrashort laser pulses (Momma et al., 1998). The use of DOEs in general makes it possible to produce practically any desired intensity distribution with high efficiency and without an intermediate focal point. As a consequence, nonlinear effects in the air are minimal, and high-precision processing becomes possible at atmospheric pressure. Moreover, the beam path can be shortened significantly compared with the imaging technique.

Another aspect that affects the hole quality when drilling high aspect ratio holes with ultrashort laser pulses is the polarization state of the laser radiation. Because of the polarization-dependent reflectivity at the walls inside the hole, a nonuniform intensity distribution can develop, leading to deformation of the hole shape at the rear surface (Nolte et al., 1999b).

To maintain the original hole shape, the reflection anisotropy due to the linear polarized laser radiation has to be avoided. This can be accomplished either by converting the linear polarization to circular polarization, e.g., by using a quarter-wave plate, or by rotating the direction of the linear polarization during the drilling process. The latter can be achieved by using a rotating half-wave plate that turns the polarization direction from shot to shot. Note that similar influences of the polarization state on the processing quality are well known from conventional laser cutting.

Figure 16 shows examples of high aspect ratio holes drilled in quartz by focusing 120 fs pulses onto the target surface (Varel et al., 1997). The increase in hole depth with increasing number of pulses can be seen. Moreover, a characteristic narrowing of the holes to a diameter of $\sim 21\,\mu m$, smaller than the focus diameter ($\sim 30\,\mu m$), is observable over the first 200–300 μm. The hole diameter then remains constant until the bottom of the hole is reached. Although Figure 16 still shows some evidence of damage to the surrounding material (microcracks and a darkened region — probably caused by mechanical stress), holes produced with a reduced fluence show much less evidence of damage.

Promising applications of femtosecond laser drilling can be found in the field of generating high quality small holes with defined diameters less than 100 μm, which is rather difficult for conventional techniques. Potential applications include the drilling of spinnerets, the production of hydraulic or pneumatic components, the structuring of biosensors (von Woedtke et al., 1997), and the drilling of fuel injection nozzles for the automotive industry.

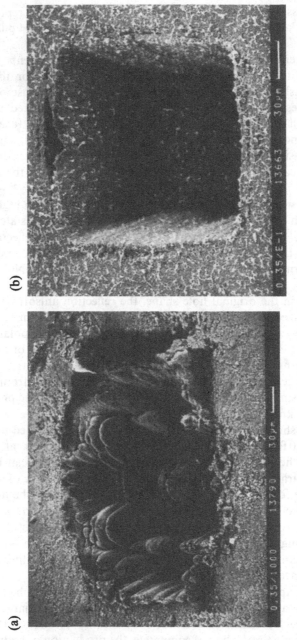

Figure 15 Influence of atmospheric pressures of (a) 1000 mbar and (b) 0.1 mbar when using ultrashort laser pulses in combination with the imaging technique to ablate stainless steel ($E = 350\,\mu J$, $F = 1.35\,J/cm^2$, $\tau_L = 150\,fs$).

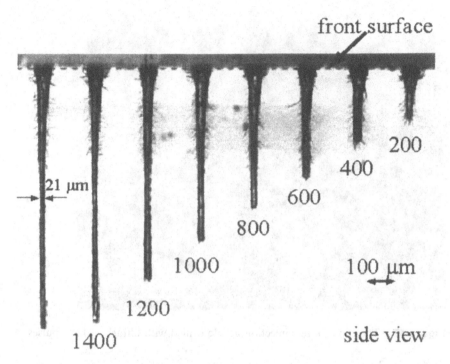

Figure 16 High aspect ratio holes produced in quartz with 790 nm laser pulses under vacuum conditions ($< 10^{-4}$ mbar). The image shows the increase in hole depth with the number of pulses for a pulse duration of 120 fs and a fluence of 42 J/cm^2. (From Varel et al., 1997.)

Although the current requirements for fuel injection nozzles can be fulfilled with conventional, non-laser-based technology (electrical discharge machining), future nozzle designs may require smaller holes that can be produced with ultrashort laser pulses. This is demonstrated in Figure 17, which shows the cap of a fuel injection nozzle drilled with femtosecond pulses.

Microstructuring

Apart from the drilling with ultrashort laser pulses discussed in the former section, many applications can be found within the field of microstructuring. Here, too, one of the most important aspects is the minimal damage of the surrounding material, which applies to not only industrial but also medical applications.

Flexible microstructuring is possible by focusing the laser radiation onto the workpiece and scanning the beam or moving the sample.

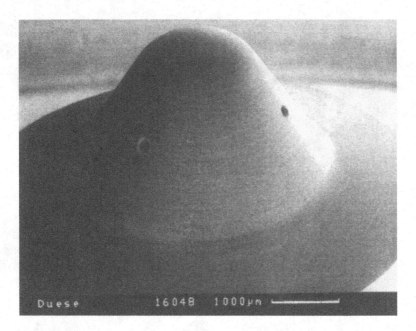

Figure 17   The cap of a fuel injection nozzle drilled with ultrashort laser pulses.

Figure 18 shows as an example a detail of an electron acceleration grid structure for a streak camera with 20 μm periodicity and 7 μm strut size produced in a 7 μm thick Ni foil using ultrashort pulses. The free-standing single struts are >300 μm long and are not deformed due to thermal or mechanical (ablation pressure) influences.

Another promising application is the structuring of medical implants such as stents. The stenting procedure is used as an alternative to bypass surgery when coronary blood vessels are blocked by plaque. For this purpose the coronary stent, a precisely slotted tube with a diameter of 1–2 mm, is inserted into the blocked vessel by means of a balloon catheter and expanded by inflation of the balloon. It stays inside the vessel and ensures sufficient blood flow. To achieve the desired plastic deformation during the inflation, the stent has to be structured appropriately. The ultrashort-pulse laser opens the possibility of using materials that are favorable from the medical point of view but that cannot be structured with the necessary precision by conventional (laser) techniques (Fig. 19a). Moreover, thinner struts can also be manufactured because of the smaller heat-affected zones, making stents for smaller vessels possible.

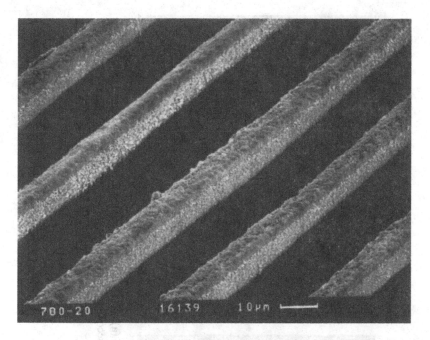

Figure 18 Detail of an electron acceleration grid for a streak camera with 20 μm periodicity and 7 μm strut size in a 7 μm thick nickel foil, produced by translating the workpiece with respect to the femtosecond laser beam focus.

In addition, the production of bioresorbable stents—stents that dissolve after a certain time, when the vessel has stabilized—becomes possible. For this purpose, special alloys and organic polymers are under investigation. Apart from a burr-free structuring of these materials, it is especially important not to change the specific material properties to ensure dissolution in the predefined manner. These requirements can be met with ultrashort laser pulses (Fig. 19b).

Apart from the applications highlighted above, there exist many other examples where microstructuring with femtosecond laser pulses can produce superior results compared to conventional (laser) machining techniques. Often, these improvements are based on the minimally invasive character of the ultrashort-pulse ablation process. Figure 20 shows a survey of some materials that can be structured by femtosecond laser pulses without the formation of large heat or mechanically affected zones, which means that they are microcrack-free. This, of course, is also of special importance for direct

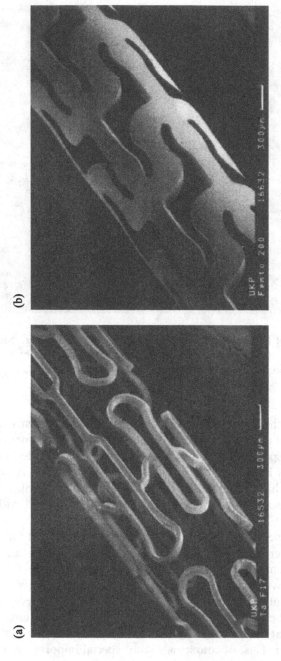

Figure 19 Prototypes of medical implants (stents) made of (a) metal (titanium) and (b) a bioresorbable material (organic polymer). In both cases the stents were structured with femtosecond laser pulses. No postprocessing techniques were applied.

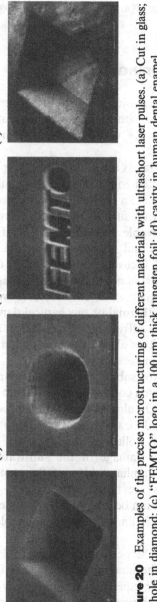

Figure 20 Examples of the precise microstructuring of different materials with ultrashort laser pulses. (a) Cut in glass; (b) hole in diamond; (c) "FEMTO" logo in a 100 μm thick tungsten foil; (d) cavity in human dental enamel.

medical applications, for example in ophthalmology (see Chap. 16) and dentistry (Fig. 20d).

7.4 NANOSTRUCTURING

Several applications, e.g., in the field of photolithography, require the generation of structures with submicrometric features. Various techniques to produce such structures with femtosecond lasers are discussed in this section.

When long-pulse lasers are considered, their use is often limited by the formation of heat-affected zones, which can be substantially larger than the desired structures. This is an especially severe problem for materials with a high thermal conductivity, such as metals. However, even for other materials the minimal achievable structure size is often limited by thermally induced damage (stress and microcracks) of the surrounding area. Therefore, the possibility of focusing the laser beam on a spot as small as the structure to be produced is not sufficient to generate these fine structures.

In contrast, the generation of structures with ultrashort laser pulses is not accompanied by thermal or mechanical damage. Thus, structure sizes are limited only by the diffraction of the optical system, i.e., the minimum focal spot achievable. This allows the production of structures with a resolution in the submicrometer range.

One application for submicrometric structuring is the repair of photolithographic masks, which consist of an $\sim$100 nm thick chrome layer on top of a quartz substrate. On almost every mask produced, defects in the form of excess chrome in undesirable places can be found. Such defects must be removed with minimal collateral damage (i.e., complete removal of the excess chrome and no damage of the substrate), which would result in reduced transmission or phase shifts of the transmitted light. When non-laser-based techniques are taken into account, the defect repair can be achieved using focused ion beams. Although these systems have a resolution below 25 nm, their use is limited due to inevitable damage of the substrate (Haight et al., 1999).

Mask repair can be performed by using a femtosecond laser system in combination with the imaging geometry (Haight et al., 1999). For this purpose, a rectangular aperture irradiated by a femtosecond pulse is imaged with a high numerical aperture microscope objective (demagnifying the aperture by a factor of 100). Different spot sizes are obtained by varying the aperture. Typical transmissions of repaired sites in excess of 95% relative to similar areas requiring no repair have been achieved. Moreover, phase distortions caused by the repaired site are hardly measurable.

As semiconductor industry demands on photolithographic masks increase, repair tools with further improved resolution are also required. The reduction of the laser wavelength (Simon and Ihlemann, 1996) is, of course, one possible way to achieve a higher resolution. Another option is to make use of the well-defined ablation threshold associated with ultra-short pulse ablation. The principal idea is based on the fact that a spatial Gaussian beam profile is maintained when the beam is focused. The focal diameter is defined as the beam radius, where the intensity is $1/e^2$ of the maximum value (regardless of the absolute value), but the pulse energy can always be adjusted in such a way that only the peak of the Gaussian distribution is above the ablation threshold (see Fig. 21). By using this approach, sub-diffraction-size holes in a 600 nm thick silver film on a glass slide substrate have been produced (Pronko et al., 1995b). For this purpose, the pulses of a 200 fs Ti:sapphire amplifier system ($\lambda = 800$ nm) have been focused with a $10 \times$ microscope objective, resulting in a spot size

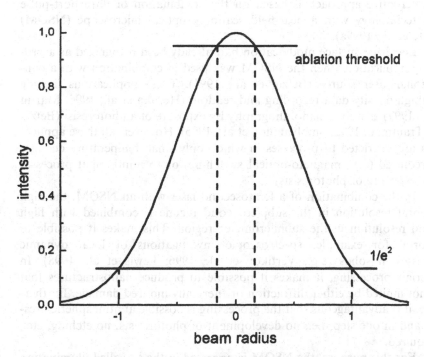

Figure 21 Schematic of the intensity distribution in the focus of a Gaussian laser beam. Due to careful adjustment of the pulse energy, only the fluence in the central part of the beam is above the ablation threshold, making sub-diffraction-limited structuring possible.

of ~3.0 μm. By careful adjustment of the pulse energy to achieve a fluence only slightly larger than the ablation threshold in the central part of the beam, holes just 300 nm in diameter and with a depth of 52 nm have been produced. Note that the hole diameter is approximately one order of magnitude smaller than the beam diameter.

This approach is based on two important aspects. One is the well-defined ablation threshold associated with ultrashort-pulse ablation; the other is the minimal heat diffusion out of the irradiated area. This allows an ablation process determined only by the optical properties of the material and the laser radiation and not limited by thermal or mechanical damage. When the same setup was used with 7 ns pulses, the minimum hole size achieved was 2 μm, and the holes showed strong thermal influence (melting of the chrome layer).

Despite the promising results, the success of this approach for achieving sub-diffraction-limited resolution depends on the stability of the laser system. With just small energy fluctuations, the hole size changes drastically. An alternative approach is based on the combination of ultrashort-pulse laser technology with a near-field scanning optical microscope (NSOM) (Nolte et al., 1999a).

Localized surface modification had already been recognized as a promising application when the NSOM was used in combination with a conventional laser source (Betzig et al., 1991). These applications include ultrahigh-density data recording and readout (Hosaka et al., 1996; Martin et al., 1997) as well as nanolithography by exposure of a photoresist (Betzig and Trautman, 1992; Smolyaninov et al., 1995). However, all these applications are restricted to processes in which only small temperature changes are required (e.g., magneto-optical switching) or to nonthermal processes (e.g., exposure of photoresists).

In the combination of a femtosecond laser with an NSOM, the high temporal resolution in the subpicosecond regime is combined with high spatial resolution in the submicrometer region. This makes it possible to perform, for example, spectroscopic investigations of local dynamic processes in solids (e.g., Vertikov et al., 1996; Lewis et al., 1998). In materials processing, it makes it possible to produce nanostructures that are not limited by either diffraction or thermally induced damage. Furthermore, it is advantageous that the processing is possible at atmospheric pressure and in one step; i.e., no development of photoresists, no etching, etc., is required.

For this purpose, the NSOM is operated in the so-called illumination mode, where the femtosecond pulses are coupled into a hollow fiber. A hollow fiber is used to minimize the temporal broadening of the pulse when propagating in the fiber. The fiber tip, which is tapered down to a diameter

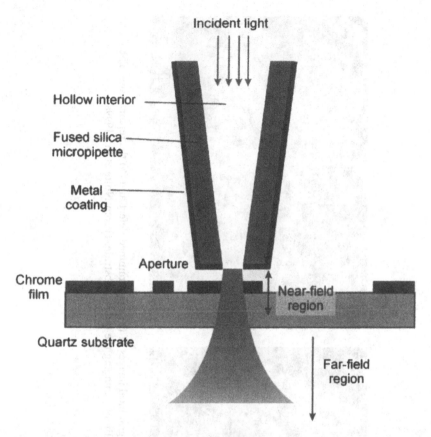

Figure 22 Schematic diagram of the setup for near-field structuring. The femto-second pulses are coupled into a tapered hollow fiber with an output aperture of only a few hundred nanometers.

on the order of a few hundred nanometers, is brought into near-field contact with the photomask surface (distance ~50–100 nm). The aperture of the fiber tip determines the minimum feature size that can be structured. A schematic setup is shown in Figure 22. The repair of photolithographic masks can be performed with this setup when the fiber tip is scanned over the defect. This allows a defined removal of defects with high spatial resolution. An example of such a repair is shown in Figure 23. Figure 23a shows an atomic force microscopic (AFM) image of a programmed defect on a photolithographic mask. This defect (excess chrome) was removed completely by scanning the fiber tip in parallel traces with a separation of 100 nm (Fig. 23b). For these experiments the second harmonic radiation of a

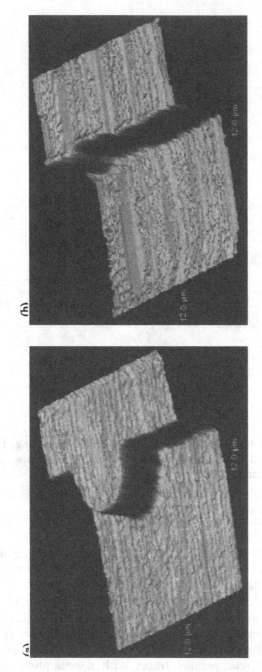

Figure 23 Atomic force microscopic topographical images of (a) a programmed defect and (b) the repaired site. The defect (excess chrome) was removed completely without damaging the quartz substrate. (From Nolte et al., 1999a.)

Ti:sapphire femtosecond laser ($\lambda = 390$ nm) was coupled into a fiber with a rather large tip diameter of 690 nm. The quartz substrate was not damaged by the laser radiation.

In addition to the high-resolution removal capabilities of this approach, the same setup can be used as a conventional NSOM, first to detect the exact position of the repair site, then to inspect the site. This makes the combination of an NSOM with a femtosecond laser an interesting possibility for high-resolution mask repair.

7.5 CONCLUSION

Ultrashort laser pulses show great potential for precise micromachining. The laser energy can be deposited efficiently, rapidly, and localized in practically any material (metal, semiconductor, dielectric, polymer, etc., whether transparent or opaque). The ablation thresholds are well defined and much lower than those for long laser pulses, because heat diffusion is minimized. This is also the reason for the minimal thermally and mechanically influenced zones when processing with ultrashort laser pulses.

Because of these advantages, ultrashort-pulse lasers can be viewed as universal tools for precise microstructuring. They will become serious competitors to conventional techniques and, in addition, open up completely new areas in laser-based micromachining that cannot be accessed with today's techniques.

At present, ultrashort-pulse lasers have to prove that they can meet industrial expectations with respect to reliability, throughput, and costs. But with the growth of the electronics, medical, and environmental sensing markets and the general push to smaller structures, the prospects for this new technique are very promising.

ACKNOWLEDGMENTS

I thank B. C. Stuart (Lawrence Livermore National Laboratory), D. Ashkenasi (Laser- und Medizin-Technologie GmbH, Berlin), C. Momma (Cortronik GmbH), B. N. Chichkov Laser Zentrum, and the short-pulse laser group of Hannover e.V., LZH for their assistance. The results shown in Figures 1, 8, and 9 were obtained under the auspices of the University of California, Lawrence Livermore National Laboratory, and the U.S. Department of Energy.

REFERENCES

Abrikosov AA. (1988). Fundamentals of the Theory of Metals. Amsterdam: North-Holland.

Anisimov SI, Kapeliovich BL, Perel'man TL. (1974). Electron emission from metal surfaces exposed to ultrashort laser pulses. Sov Phys—JETP 39:375–377.

Ashcroft NW, Mermin ND. (1976). Solid State Physics. Philadelphia: Saunders.

Ashkenasi D, Varel H, Rosenfeld A, Henz S, Herrmann J, Campbell EEB. (1998). Application of self-focusing of ps laser pulses for three-dimensional micro-structuring of transparent materials. Appl Phys Lett 72:1442–1444.

Bäuerle D. (1996). Laser Processing and Chemistry. Berlin: Springer-Verlag.

Betzig E, Trautman JK. (1992). Near-field optics:microscopy, spectroscopy, and sur-face modification beyond the diffraction limit. Science 257:189–195.

Betzig E, Trautman JK, Harris TD, Weiner JS, Kostelak RL. (1991). Breaking the diffraction barrier: optical microscopy on a nanometric scale. Science 251:1468–1470.

Chichkov BN, Momma C, Nolte S, von Alvensleben F, Tünnermann A. (1996). Femtosecond, picosecond and nanosecond laser ablation of solids. Appl Phys A 63:109–115.

Corkum PB, Brunel F, Sherman NK, Srinivasan-Rao T. (1988). Thermal response of metals to ultrashort-pulse laser excitation. Phys Rev Lett 61:2886–2889.

Du D, Liu X, Korn G, Squier J, Mourou G. (1994). Laser-induced breakdown by impact ionization in SiO_2 with pulse widths from 7 ns to 150 fs. Appl Phys Lett 64:3071–3073.

Feuerhake M, Klein-Wiele J-H, Marowsky G, Simon P. (1998). Dynamic ablation studies of sub-micron gratings on metals with sub-picosecond time resolution. Appl Phys A 67:603–606.

Furusawa K, Takahashi K, Kumagai H, Midorikawa K, Obara M. (1999). Ablation characteristics of Au, Ag, and Cu metals using a femtosecond Ti:sapphire laser. Appl Phys A 69:S359–S366.

Haight R, Hayden D, Longo P, Neary T, Wagner A. (1999). MARS: femtosecond laser mask advanced repair system in manufacturing. J Vac Sci Technol B 17:3137–3143.

Herrmann RFW, Gerlach J, Campbell EEB. (1998). Ultrashort pulse laser ablation of silicon: an MD simulation study. Appl Phys A 66:35–42.

Hosaka S, Shintani T, Miyamoto M, Hirotsune A, Terao M, Yoshida M, Fujita K, Kämmer S. (1996). Nanometer-sized phase-change recording using a scanning near-field optical microscope with a laser diode. Jpn J Appl Phys 35:443–447.

Hughes TP. (1975). Plasmas and Laser Light. Bristol, UK: Adam Hilger.

Kanavin AP, Smetanin IV, Isakov VA, Afanasiev YuV, Chichkov BN, Wellegehau-sen B, Nolte S, Momma C, Tünnermann A. (1998). Heat transport in metals irradiated by ultrashort laser pulses. Phys Rev B 57:14698–14703.

Komashko AM, Feit MD, Rubenchik AM, Perry MD, Banks PS. (1999). Simulation of material removal efficiency with ultrashort laser pulses. Appl Phys A 69:S95–S98.

Krüger J, Kautek W. (1995). Femtosecond-pulse laser processing of metallic and semiconducting thin films. Proc Laser-Induced Thin Film Processing. SPIE 2403:436–447.

Küper S, Stuke M. (1989). Ablation of polytetrafluorethylene (Teflon) with femtosecond UV excimer laser pulses. Appl Phys Lett 54:4–6.

Lenzner M, Krüger J, Sartania S, Cheng Z, Spielmann C, Mourou G, Kautek W, Krausz F. (1998). Femtosecond optical breakdown in dielectrics. Phys Rev Lett 80:4076–4079.

Lewis MK, Wolanin P, Gafni A, Steel DG. (1998). Near-field scanning optical microscopy of single molecules by femtosecond two-photon excitation. Opt Lett 23:1111–1113.

Luft A, Franz U, Emsermann A, Kaspar J. (1996). A study of thermal and mechanical effects on materials induced by pulsed laser drilling. Appl Phys A 63:93–101.

Martin Y, Rishton S, Wickramasinghe HK. (1997). Optical data storage read out at 256 Gbits/in.^{2}. Appl Phys Lett 71:1–3.

Momma C, Chichkov BN, Nolte S, von Alvensleben F, Tünnermann A, Welling H, Wellegehausen B. (1996). Short-pulse laser ablation of solid targets. Opt Commun 129:134–142.

Momma C, Nolte S, Kamlage G, von Alvensleben F, Tünnermann A. (1998). Beam delivery of femtosecond laser radiation by diffractive optical elements. Appl Phys A 67:517–520.

Nolte S, Momma C, Jacobs H, Tünnermann A, Chichkov BN, Wellegehausen B, Welling H. (1997). Ablation of metals by ultrashort laser pulses. J Opt Soc Am B 14:2716–2722

Nolte S, Chichkov BN, Welling H, Shani Y, Lieberman K, Terkel H. (1999a). Nanostructuring with spatially localized femtosecond laser pulses. Opt Lett 24:914–916.

Nolte S, Momma C, Kamlage G, Ostendorf A, Fallnich C, von Alvensleben F, Welling H. (1999b). Polarization effects in ultrashort-pulse laser drilling. Appl Phys A 68:563–567.

Perry MD, Stuart BC, Banks PS, Feit MD, Yanovsky V, Rubenchik AM. (1999). Ultrashort-pulse laser machining of dielectric materials. J Appl Phys 85:6803–6810.

Preuss S, Demchuk A, Stuke M. (1995). Sub-picosecond UV laser ablation of metals. Appl Phys A 61:33–37.

Prokhorov AM, Konov VI, Ursu I, Mihailescu IN. (1990). Laser Heating of Metals. Bristol, UK: Adam Hilger.

Pronko PP, Dutta SK, Du D, Singh RK.(1995a). Thermophysical effects in laser processing of materials with picosecond and femtosecond pulses. J Appl Phys 78:6233–6240.

Pronko PP, Dutta SK, Squier J, Rudd JV, Du D, Mourou G. (1995b). Machining of submicron holes using a femtosecond laser at 800 nm. Opt Commun 114:106–110.

Simon P, Ihlemann J. (1996). Machining of submicron structures on metals and semiconductors by ultrashort UV-laser pulses. Appl Phys A 63:505–508.

Smolyaninov II, Mazzoni DL, Davis CC. (1995). Near-field direct-write ultraviolet lithography and shear force microscopic studies of the lithographic process. Appl Phys Lett 67:3859–3361.

Stuart BC, Feit MD, Rubenchik AM, Shore BW, Perry MD. (1995). Laser-induced damage in dielectrics with nanosecond to subpicosecond pulses. Phys Rev Lett 74:2248–2251.

Stuart BC, Feit MD, Herman S, Rubenchik AM, Shore BW, Perry MD. (1996). Optical ablation by high-power short-pulse lasers. J Opt Soc Am B 13: 459–468.

Varel H, Ashkenasi D, Rosenfeld A, Herrmann R, Noack F, Campbell EEB. (1996). Laser-induced damage in SiO_2 and CaF_2 with picosecond and femtosecond laser pulses. Appl Phys A 62:293–294.

Varel H, Ashkenasi D, Rosenfeld A, Wähmer M, Campbell EEB. (1997). Micromachining of quartz with ultrashort laser pulses. Appl Phys A 65:367–373.

Vertikov A, Kuball M, Nurmikko AV, Maris HJ. (1996). Time-resolved pump-probe experiments with subwavelength lateral resolution. Appl Phys Lett 69:2465–2467.

von der Linde D, Sokolowski-Tinten K. (2000). The physical mechanisms of short-pulse laser ablation. Appl Surf Sci 154–155:1–10.

von Woedtke T, Abel P, Krüger J, Kautek W. (1997). Subpicosecond-pulse laser microstructuring for enhanced reproducibility of biosensors. Sensors Actuators B 42:151–156.

Wellershoff S-S, Hohlfeld J, Güdde J, Matthias E. (1999), The role of electron-phonon coupling in femtosecond laser damage of metals. Appl Phys A 69:S99–S107.

Zeldovich YaB, Raizer YuP. (1966). Physics of Shock Waves and High-Temperature Hydrodynamic Phenomena. New York: Academic Press.

8

Structural Changes Induced in Transparent Materials with Ultrashort Laser Pulses

Chris B. Schaffer, Alan O. Jamison, José F. Garc®a, and Eric Mazur

Harvard University, Cambridge, Massachusetts, U.S.A.

8.1 INTRODUCTION

In recent years, femtosecond lasers have been used for a multitude of micro-machining tasks [1]. Several groups have shown that femtosecond laser pulses cleanly ablate virtually any material with a precision that consistently meets or exceeds that of other laser-based techniques, making the femto-second laser an extremely versatile surface micromachining tool [2–7]. For large-band-gap materials, where laser machining relies on nonlinear absorption of high-intensity pulses for energy deposition, femtosecond lasers offer even greater benefit. Because the absorption in a transparent material is non-linear, it can be confined to a very small volume by tight focusing, and the absorbing volume can be located in the bulk of the material, allowing three-dimensional micromachining [8,9]. The extent of the structural change produced by femtosecond laser pulses can be as small as or even smaller than the focal volume. Recent demonstrations of three-dimensional microma-chining of glass using femtosecond lasers include three-dimensional binary data storage [8,10] and the direct writing of optical waveguides [9–12] and waveguide splitters [13]. The growing interest in femtosecond laser microma-chining of bulk transparent materials makes it more important than ever to uncover the mechanisms responsible for producing permanent structural changes.

To micromachine a transparent material in three dimensions, a femto-second laser pulse is tightly focused into the material. The laser intensity in the focal volume can become high enough to ionize electrons by multipho-ton ionization, tunneling ionization, and avalanche ionization [14–18]. This nonlinear ionization leads to optical breakdown and plasma formation, but only in the focal volume where the laser intensity is high. For a transparent material, the laser focus can be located in the bulk of the sample, so the energy deposition occurs inside the material. Initially, the energy deposited by the laser pulse resides only in the electrons, with the ions still cold. On a 10 ps time scale, the electrons collisionally heat the ions, and the two systems reach thermal equilibrium at a temperature of about 10^5 K. On a 10 ns time scale, the electrons recombine with the ions, producing the spark character-istic of optical breakdown. After the plasma recombines, the remaining energy in the material is in the form of thermal energy. This thermal energy diffuses out of the focal volume on a microsecond time scale.

For laser pulse energies above a certain material-dependent energy threshold, permanent structural changes are produced in the material. Sev-eral different mechanisms can lead to these changes, each producing a dif-ferent morphology—from small density variations to color centers to voids. For example, if the melting temperature of the material is exceeded in some small volume, the material can resolidifiy nonuniformly, leading to density and refractive index variations. The laser pulse can induce the formation of color centers in some materials, leading to refractive index changes in the focal volume [12,19]. At higher laser energy, hot electrons and ions may explosively expand out of the focal volume into the surround-ing material [20]. This explosive expansion leaves a void or a less dense central region surrounded by a denser halo. Densification of glass by femtosecond laser pulses in a manner analogous to ultraviolet-induced densification has also been suggested [13]. Which of these mechanisms causes structural change depends on laser, focusing, and material para-meters.

For very tight focusing, the minimum size of a structurally altered region is under 1 μm [2,8]. These microscopic material changes are the build-ing blocks from which more complex three-dimensional devices can be micromachined. By scanning the laser focus through the material, a three-dimensional object can be built up out of the changes made by each laser pulse. In addition, as described above, different laser and material para-meters can lead to different structural change morphologies, allowing another degree of freedom in device fabrication.

In this chapter we present a systematic study of the morphology of structural changes induced in bulk glass by tightly focused femtosecond laser pulses. Using high-resolution optical microscopy we investigate the dependence of the morphology of the structural change on laser energy and the number of laser shots as well as on the focusing angle. We use electron microscopy to obtain a more detailed look at the morphology. Our observations show a transition from a structural change mechanism dominated by localized melting or densification to one dominated by an explosive expansion. We also present a new method for bulk micromachining with high-repetition-rate femtosecond lasers that causes material changes by a cumulative heating of the material around the focus by multiple laser pulses. Finally, we discuss applications of femtosecond laser micromachining in three-dimensional binary data storage and in direct writing of single-mode optical waveguides in bulk glass.

8.2 THE ROLE OF PULSE DURATION

Previous work has shown that for pulses longer than a few hundred femtoseconds, the structural changes produced by the laser pulse become larger than the focal volume, degrading the machining precision [8,15,21,22]. In addition, such pulses often cause cracks that radiate from the focal region. The greater extent of structural change and the cracks are due to the greater energy that is required to reach the intensity threshold for nonlinear ionization with these longer pulses.

In transparent materials, the threshold for producing permanent structural change coincides with the threshold for optical breakdown and plasma formation [15]. If the laser intensity exceeds the threshold intensity, a critical density plasma is formed as the temporal peak of the pulse passes through the laser focus. This plasma strongly absorbs energy from the second half of the laser pulse, depositing enough energy to produce permanent structural change. This plasma formation and absorption explains the larger damage observed for longer laser pulses. For long laser pulses the energy required to reach the threshold intensity for plasma formation is higher, and consequently more energy is deposited in the material by the strong absorption of the second half of the pulse, leading to more extensive material changes. For femtosecond laser pulses, breakdown occurs with less energetic laser pulses, leading to the deposition of less energy and smaller structural changes.

For high-precision micromachining, damage outside the focal volume must be minimized. Experiments in a variety of transparent materials suggest that pulse durations of a few hundred femtoseconds or shorter are necessary for the structural change produced near the threshold to be confined to the focal volume. All the work described in this chapter was carried out with laser pulses with a duration of 110 fs or less.

8.3 EXPERIMENTAL TECHNIQUES

To achieve micrometric machining precision, not only must the laser pulses be of femtosecond duration and near threshold energy, they must also be focused with high numerical aperture (NA) microscope objectives. Furthermore, care must be taken to ensure that the laser is focused to a diffraction-limited spot size. For this tight focusing, the energy threshold for producing material changes is low enough that an unamplified laser system can be used for micromachining. In this section, we discuss the proper use of microscope objectives and describe a modified femtosecond laser oscillator that is used for unamplified micromachining.

8.3.1 Focusing with Microscope Objectives

A diagram of the focusing setup used in the experiments discussed in this chapter is shown in Figure 1. A 110 fs, 800 nm laser pulse from a regeneratively amplified Ti:sapphire laser is focused inside the bulk of a transparent material by a microscope objective with a numerical aperture in the range of 0.25–1.4. The sample is moved by a computer-controlled stage, and single or multiple laser pulses are fired into the sample to produce arrays of structures. We then examine these arrays using differential interference contrast (DIC) optical microscopy and scanning electron microscopy (SEM).

To achieve the highest laser intensity for a given pulse energy, the laser pulse must be delivered to the sample without temporal distortion and must

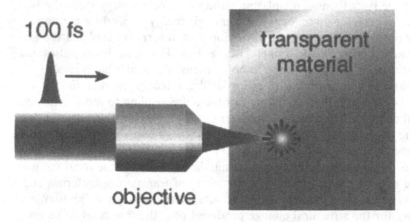

Figure 1 Diagram of the focusing setup used to produce structures in bulk glass using femtosecond laser pulses.

be well focused inside the sample. It is difficult to directly measure the laser pulse duration at the focus of the high-NA objectives, so, instead, we minimize the observed threshold for producing structural change by adjusting the grating-based pulse compressor on the laser system. We generally find that for all but the 1.4 NA oil immersion objective there is very little change in threshold due to compressor adjustment for transform-limited 110 fs pulses. This lack of change reflects the moderate dispersion of the objectives and the moderate bandwidth of our regeneratively amplified laser system.

To ensure diffraction-limited focusing, the back aperture of the microscope objective is illuminated as uniformly as possible and is completely filled. The sample, Corning 0211, is the glass used to make cover slips, and the laser is focused 170 µm beneath the surface, where the microscope objective is designed to have minimal spherical aberration. For microscope objectives with NAs greater than about 0.5, the threshold for structural change increases dramatically when the laser is not focused at the optimal depth in the material. This increase is caused by degradation of the laser focus due to spherical aberration. For machining applications, this problem can be overcome by using an objective with a collar that compensates for varying amounts of spherical aberration.* For the experiments described here, the laser is always focused at the depth where the threshold for producing structural change is the lowest.

For objectives with NA less than about 0.45, the power required to reach the threshold intensity for structural change approaches the critical power for self-focusing for the material. As a result, self-focusing and other nonlinear propagation effects must be taken into account [23–25]. Self-focusing can also play a role when above-threshold energy is used to produce structural changes.

8.3.2 Long-Cavity Oscillator

Using a light-scattering technique, we measured the threshold for permanent structural change in Corning 0211 to be only 5 nJ for 110 fs pulses focused by a 1.4 NA oil immersion microscope objective [9]. This low threshold opens the door to using unamplified laser systems for micromachining. In Section 8.5.2 we discuss how micromachining with an oscillator can differ from micromachining with amplified lasers, and in Section 8.6 we show some of the applications of this oscillator-only micromachining. Here we describe the long-cavity laser oscillator used for these experiments.

Because of losses in the prism compressor, beam delivery optics, and the microscope objective, it is necessary to start with about 15 nJ of laser

---

*These objectives are available from several companies, including Zeiss, Leica, and Nikon.

energy to deliver 5 nJ to the sample. This energy is slightly higher than the output of standard Ti:sapphire laser oscillators. To achieve this higher output energy, we constructed a long-cavity Ti:sapphire laser oscillator [26,27]. A one-to-one imaging telescope, based on a mirror with a 2 m radius of curvature, is inserted into the output coupler end of a Ti:sapphire laser. The telescope images the old position of the output coupler to its new position, keeping the spatial modes of the laser the same while increasing the length of the cavity. The longer cavity allows more gain to build up in the laser crystal before the pulse depletes it. The result is higher energy pulses at a lower repetition rate. Our laser operates at 25 MHz and routinely produces 20 nJ, 30 fs pulses.

The 40 nm wide bandwidth of the 25 MHz oscillator makes dispersion compensation essential, especially when using a 1.4 NA objective. We use a prism compressor to minimize the energy threshold for producing a structural change in the sample. Because optical breakdown is an intensity-dependent process, this minimizes the pulse width. We do not measure the pulse duration at the laser focus. Other groups have shown by direct measurement that, using only a simple prism pair for dispersion compensation, pulses as short as 15 fs can be achieved at the focus of high-NA objectives [28].

8.4 SINGLE-SHOT STRUCTURAL CHANGE MORPHOLOGY

In order to reliably machine a structure, we must know the nature and extent of the structural change produced by a single laser pulse. To this end, we studied the morphology of the structural change produced for various laser and focusing parameters in bulk borosilicate glass (Corning 0211) by tightly focused femtosecond laser pulses.

8.4.1 Role of Laser Pulse Energy and Focusing Angle—Optical Microscopy

We first look at differences in the structural changes produced by different laser pulse energies focused by different microscope objectives. Figure 2 shows a DIC micrograph of an array of structures produced in Corning 0211 using single 40 nJ, 110 fs laser pulses focused by a 0.65 NA microscope objective. The diameter of the structures in the image is about 0.5 μm, the resolution limit of the optical microscope used to take the image. The structures shown in Figure 2 are difficult to visualize without a contrast enhancing microscopic technique such as DIC, because the refractive index change is small. The inset in Figure 2 shows a side view taken in reflection of a structure produced using 110 fs, 40 nJ pulses focused by a 0.65 NA

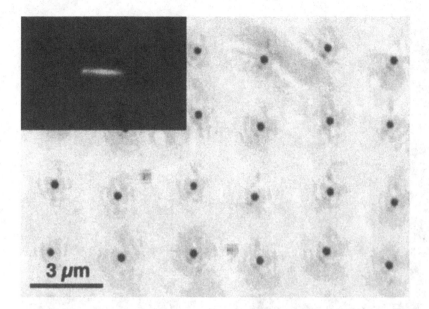

Figure 2 Differential interference contrast optical micrograph of an array of structures produced with single 40 nJ, 110 fs laser pulses focused by a 0.65 NA objective. The laser pulse is incident perpendicular to the plane of the image. The inset shows, at the same scale, a side-view optical micrograph of a structure taken in reflection. The laser pulse is incident from the left of the inset.

objective. The inset is on the same scale as the rest of the image. The elongation of the structure reflects the oblong focal volume of the 0.65 NA objective. The structures shown in Figure 2 are produced with an energy that is about twice the observed threshold for permanent structural change for this objective.

Figure 3 shows side-view DIC optical micrographs of structures produced in Corning 0211 using 110 fs laser pulses with different laser energies and focused by different microscope objectives. Figure 3a shows structures produced by 50 nJ pulses focused by a 0.45 NA microscope objective. The cylindrical structures are similar to the structures shown in Figure 2, except that the cylinders are about 4 μm long, reflecting the longer confocal parameter of the 0.45 NA objective compared to that of the 0.65 NA objective. Figure 3b shows structures produced by 15 nJ pulses focused by a 1.4 NA oil immersion microscope objective. The structures are only slightly elongated along the beam propagation direction and have a length of approximately 1.5 μm. The shape and extent of the structural change are again determined by the focal volume.

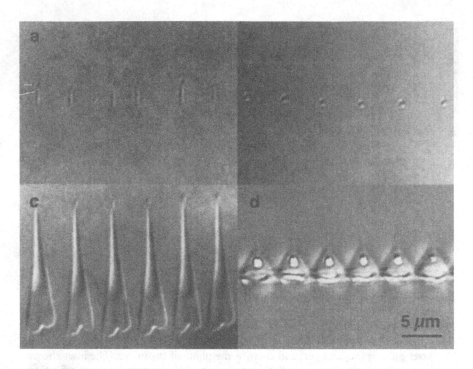

Figure 3 Side-view DIC optical images of structures produced in bulk glass using single 110 fs laser pulses with various laser energies and focusing conditions: (a) 500 nJ, 0.45 NA; (b) 500 nJ, 1.4 NA; (c) 50 nJ, 0.45 NA; (d) 15 nJ, 1.4 NA. The laser pulse is incident from the bottom of the figure.

For laser energies that exceed the threshold energy by about a factor of 10 or more, the structures show a more complicated morphology than the cylindrical structures shown in Figures 2, 3a, and 3b. Figure 3c shows structures produced with 500 nJ pulses focused by a 0.45 NA objective. The structure has a conical shape with the base of the cone oriented toward the incident direction of the laser. The extent of the structural modification is much larger than the focal volume of the objective. The structures are also formed closer to the laser than the focal plane of the objective. This upstream shift in position is due to self-focusing of the laser pulses in the material. Using third harmonic generation microscopy, Squier and Müller [29] took three-dimensional images of structures similar to those shown in Figure 3c.

Figure 3d shows structures produced with 500 nJ pulses focused by a 1.4 NA oil immersion microscope objective. The conical structures have a steeper cone angle than the structures shown in Figure 3c, and the base of the cone shows indications of cracking. Also visible in Figure 3d is a high-contrast, spherically shaped region in the center of each cone. These spherical regions are easily visible under standard white light transmission microscopy, indicating that the refractive index change is larger in these regions than for the rest of the structure. Electron microscopy reveals that the spherical regions are voids inside the glass.

8.4.2 Role of Laser Pulse Energy and Focusing Angle— Electron Microscopy

Although the optical micrographs in Section 8.4.1 show the rough features of the structural change morphology, it is difficult to determine details of the morphology from these optical images because the structures are at the resolution limit of an optical microscope. Because these details provide clues to the mechanisms responsible for producing the structural changes, we turn to scanning electron microscopy (SEM).

Because SEM is a surface-imaging tool, it is necessary to expose the bulk structural modifications before imaging. The samples can be prepared by polishing down to the level of the structures [8]. This method allows the diameter of structures that are too small to be resolved optically to be accurately determined. Polishing, however, smooths out many small-scale features that may provide important information about the mechanism for structural change.

For this study, we prepared SEM samples using the steps outlined in Figure 4. A thin sample is scribed on one side using a diamond glass cutter. The femtosecond laser beam is focused into the sample through the unscribed surface, and the sample is translated at about 5 mm/s during the exposure to a 1 kHz pulse train, producing a row of structures spaced by about 5 μm. Multiple 5 μm spaced rows of structures are written 170 μm beneath the surface of the sample in the region just above the diamond scribe line. After irradiation, the samples are fractured along the scribe line. The fracture plane goes through some structures, bringing them to the surface so they can be imaged in the microscope. Before imaging, the sample is coated with 5–10 nm of graphite to make it conductive. In addition to preserving small-scale features that are smoothed out by polishing, this technique provides side-view images, which are hard to obtain by polishing because of the small diameter of the structures.

Figure 5 shows SEM images of structures produced by 110 fs laser pulses focused by a 0.45 NA objective. The laser pulse energies are 140 nJ in Figure 5a, 250 nJ in Figure 5b, and 500 nJ in Figure 5c. The SEM image

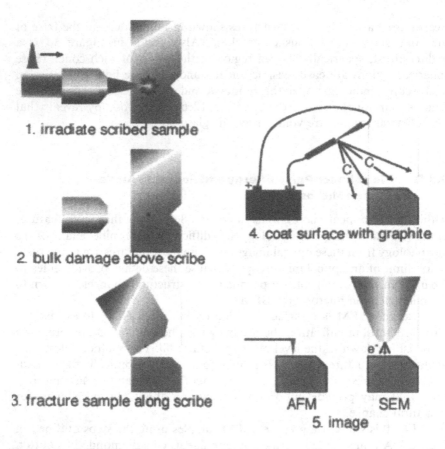

1. irradiate scribed sample

2. bulk damage above scribe

3. fracture sample along scribe

4. coat surface with graphite

AFM   SEM

5. image

Figure 4 Schematic diagram showing the procedure used to prepare samples for imaging in the scanning electron microscope.

of the structures looks different depending on what part of the structure is bisected by the fracture plane. For the structures shown in Figure 5, the fracture plane went through the center of each structure. There is a transition in morphology between the structures produced with 140 nJ and those produced with 500 nJ. The structures produced with 500 nJ show clear evidence of a void at the center of the structure, whereas the structures produced with 140 nJ show only very small surface relief. The structures made with 250 nJ represent a transition between these two morphologies. We previously observed voids similar to those in Figure 5c in SEM studies of samples that were polished to bring the structures to the surface [8].

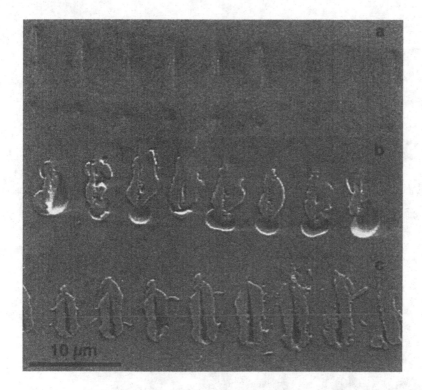

Figure 5 Side-view SEM images of structures produced in bulk glass using single 110 fs laser pulses focused by a 0.45 NA microscope objective. The laser energies are (a) 140 nJ, (b) 250 nJ, and (c) 500 nJ. The laser pulse is incident from the bottom of the figure.

We performed atomic force microscopy on structures similar to those shown in Figure 5. We found that the lobes on either side of the void in Figure 5c rise a few hundred nanometers from the surface. The outside edges are sharp, suggesting that a fracture occurred. The structures shown in Figure 5a have a very smooth surface relief of about 100 nm.

Figure 6 shows a similar series of SEM images of structures produced by 110 fs pulses focused by a 1.4 NA oil immersion objective. The laser energies are 36 nJ in Figure 6a, 140 nJ in Figure 6b, and 500 nJ in Figure 6c. Note the difference in scale relative to Figure 5. There is again a transition in morphology, from small surface relief in Figure 6a, to a void in Figure 6b, to extensive cracking of the material in Figure 6c. Voids like those in Figure 6b produce the high contrast spherical structures in Figure 3d.

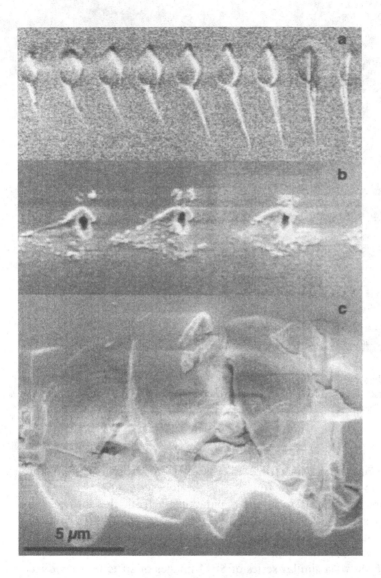

Figure 6   Side-view SEM images of structures produced in bulk glass using single 110 fs laser pulses focused by a 1.4 NA oil immersion microscope objective. The laser energies are (a) 36 nJ, (b) 140 nJ, and (c) 500 nJ. The laser pulse is incident from the bottom of the figure.

8.4.3 Connecting Morphology to Mechanism

The production of the cone-shaped structures in Figures 3c and 3d can be explained as follows. Because the pulse energy is significantly above the threshold for permanent change, a critical density plasma is formed at the focus by the leading edge of the pulse. This leading edge produces the structural change at the tip of the cone. Because the remainder of the pulse contains more energy than is necessary to produce a critical density plasma at the focus, it produces a plasma upstream from the laser focus, where the laser size is larger. Successive time slices of the pulse produce successively larger diameter structural changes that are located farther upstream from the focus, building up the cones seen in Figures 3c and 3d. At the peak of the pulse, structural change is produced the farthest from the laser focus, forming the base of the cone. The second half of the laser pulse is strongly absorbed by the plasma formed by the peak of the pulse, leading to the extensive structural changes at the base of the cone seen in Figures 3d and 6c. This mechanism for producing structural change leads to a structure that extends over a region much larger than the focal volume and contains a wide range of material modification, from small density and refractive index changes near the tip of the cone, to voids inside the cone, to cracking at the base of the cone. From a micromachining perspective, it is clear that the highest precision is achieved when the laser energy is close to the threshold for structural change, and the extent of the structural change is determined by the focal volume of the objective as in Figures 2, 3a, and 3b.

The voids evident in the SEM images in Figures 5c and 6b suggest an explosive mechanism. After laser excitation, hot electrons and ions explosively expand out of the focal region into the surrounding material, leaving a void or a less dense central region surrounded by a denser halo [20]. The lobes on either side of the void in Figure 5c and the ring-shaped surface relief in Figure 6c are due to trapped stress in the material resulting from the expansion that was released when the sample was fractured. As is clear from Figure 3d, the void is not necessarily formed at the laser focus, but rather where high enough energy density is achieved to drive the expansion.

Closer to threshold the changes are more subtle. The surface relief shown in Figure 5a and 6a is caused by fracturing either just above or just below material with a different density. A density change also explains the small refractive index change of the structures shown in Figures 2, 3a, and 3b.

How, then, does femtosecond laser irradiation lead to density changes in the material? One possibility is that the material heated by the laser pulse melts and resolidifies nonuniformly. For laser energies close to the threshold, the electrons and ions are not hot enough to drive the explosive expansion described above. Instead, they recombine, leaving molten material that cools and then resolidifies. Because of strong gradients in temperature, the resolidification is nonuniform, leading to density changes in the material. Another possibility is that the laser pulse drives a structural transition by directly (i.e., nonthermally) breaking bonds in the material [13]. Femtosecond laser pulses are known to induce such nonthermal structural changes in semiconductors [30]. Furthermore, silica glasses are known to undergo densification when exposed to ultraviolet light, presumably due to the ionization in the glass [31].

The different structural change morphologies and mechanisms described above can be used to tailor the structure produced by a single laser pulse to a particular machining application. The small density and refractive index changes produced near threshold are suitable for direct writing of optical waveguides and other photonic devices, and the voids formed at higher energy are ideal for binary data storage because of their high optical contrast.

8.5 MULTIPLE-SHOT STRUCTURAL CHANGE MORPHOLOGY

In the experiments discussed so far, structural changes are produced by single laser pulses. We now examine the morphology of the structures produced when multiple laser pulses irradiate the same spot in a sample. For multiple-shot structural change, the repetition rate of the laser is an important factor. If the time between successive laser pulses is shorter than the roughly 1 µs it takes for energy deposited by one pulse to diffuse out of the focal volume, then a train of pulses irradiating one spot in the sample raises the temperature of the material around the focal volume [9,32]. In effect, the laser serves as a point source of heat that is localized inside the bulk of the material. Before considering this cumulative heating effect, we look at the morphology of structures produced with multiple pulses arriving at a low repetition rate, where cumulative heating does not occur.

8.5.1 Low Repetition Rate

At repetition rates below about 1 MHz, the material cools between successive laser pulses. Multiple pulses amplify the structural change produced by a single pulse, leading to higher constrast structures. Figure 7 shows a DIC optical micrograph of an array of structures in Corning 0211 produced using 110 fs laser pulses from a 1 kHz laser focused by a 1.4 NA oil immersion objective. Each structure is made with a different combination of laser

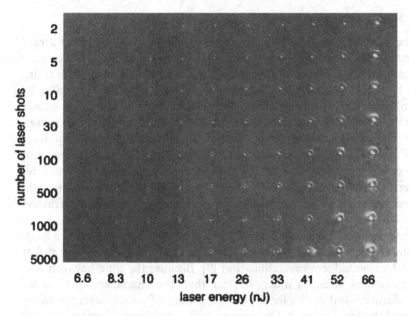

Figure 7 Differential interference contrast optical micrograph of structures produced with various energies and various numbers of 110 fs laser pulses focused by a 1.4 NA microscope objective. The laser pulses are incident perpendicular to the plane of the image.

energy and number of incident shots. The energy ranges from 6.6 to 66 nJ, while the number of shots incident on each spot ranges from 2 to 5000. The diameter of the structures increases with increasing laser energy, and the optical contrast increases with increasing number of incident laser pulses. The desired optical contrast and size scale of the structure can be set, almost independently, by choosing an appropriate laser energy and number of laser shots on a single spot. For the range of energies shown in Figure 7, the structures retain the slightly elongated morphology shown in Figure 3b. It is only at higher energies that the conical structures shown in Figure 3d appear.

For this 1.4 NA focusing, no observable changes are produced by irradiating one spot in the sample with many pulses that have too little energy to induce a structural change with one laser pulse. In contrast, we do observe a decrease in the threshold for optically observable structural changes for an increasing number of incident laser pulses when focusing with a 0.25 NA objective. Other researchers have observed a decrease in the surface ablation threshold of transparent materials with multiple-pulse irradiation [33,34].

8.5.2   High Repetition Rate

Using the 25 MHz laser oscillator described in Section 8.3.2, we make arrays of structures and analyze them using DIC optical microscopy. Figure 8 shows an optical micrograph of structures made with 30 fs, 5 nJ pulse trains focused by a 1.4 NA objective. The number of incident pulses increases by factors of 10, from 10^2 on the left to 10^5 on the right. The structures shown in Figure 8 are much larger than the structures produced with single pulses, and the size of the structures increases with increasing number of laser pulses. Figure 9 shows the radius of the structures as a function of the number of incident laser pulses. Side-view microscopy reveals that the structures are spherical. There is a series of rings evident in the structures on the right in Figure 8, suggestive of regions in the structure with different refractive indices.

The structures shown in Figure 8 are produced by a cumulative heating of the material around the laser focus by many successive laser pulses followed by nonuniform resolidification [9]. Because the time between successive pulses, at 40 ns, is much shorter than the characteristic time for thermal diffusion out of the focal volume, a train of pulses heats the material around the focal volume. Over many pulses a volume of material much larger than the focal volume is heated above the melting temperature for the glass. The larger the number of incident laser pulses, the larger the radius out to which the glass melts. After the pulse train is turned off, the material

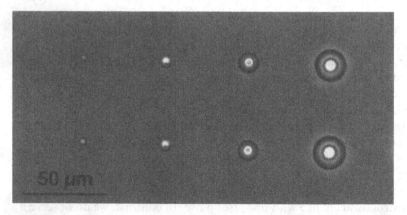

Figure 8   Optical microscopic image of structures produced with multiple 5 nJ, 30 fs laser pulses from a 25 MHz oscillator focused by a 1.4 NA objective. The number of pulses used increases by factors of 10 from 10^2 on the left to 10^5 on the right. The laser pulses are incident perpendicular to the plane of the image.

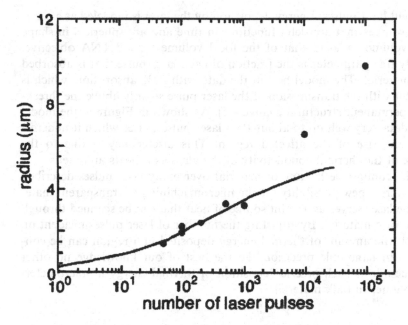

Figure 9 Radius of the structure produced with 30 fs, 5 nJ laser pulses incident at 25 MHz and focused by a 1.4 NA objective as a function of the number of incident laser pulses. The line represents the calculated maximum radius out to which the melting temperature of the glass is exceeded for a given number of incident laser pulses.

cools and, because of the temperature gradients, resolidifies nonuniformly, leading to the optical contrast shown in Figure 8. The structure stops growing after about 10^7 pulses. At that point, the structure itself disturbs the laser focus, causing the intensity to drop below the threshold intensity and therefore preventing further energy deposition.

A nonthermal mechanism cannot account for the structures shown in Figure 8. First, the laser pulse train never directly irradiates most of the material where refractive index changes are observed. Only the submicrometer-sized focal volume at the center is directly irradiated, whereas the structures shown in Figure 8 extend up to 10 µm from the focal spot. Second, the size of the structure increases with the number of incident laser pulses, indicating that as more pulses deposit more energy, a larger region melts.

We modeled this cumulative heating effect using a thermal diffusion equation. The maximum radius out to which the temperature exceeds the melting temperature for the glass was calculated for different numbers of incident laser pulses. Energy deposition by the laser is modeled as a series of heat sources that are delta functions in time and are spherical in shape with a volume equal to that of the focal volume for a 1.4 NA objective. The only free parameter is the fraction of each laser pulse that is absorbed by the material. The model best fit the data with 30% absorption, which is consistent with the transmission of the laser pulse slightly above the threshold for permanent structural change [35]. As shown in Figure 9, the model fits the data very well up to about 1000 laser pulses, after which it underestimates the size of the affected region. This discrepancy is due to the decrease of the thermal conductivity of the glass as it heats and melts.

The cumulative heating of material over many laser pulses described above offers a new possibility for the micromachining of transparent materials. The laser serves as a point source of heat that can be scanned through the bulk of a material. By adjusting the number of laser pulses incident on one spot, the amount of thermal energy deposited in a region can be controlled with nanojoule precision. To the best of our knowledge no other technique allows such precise deposition of thermal energy in micrometer-sized volumes in bulk material.

8.6 APPLICATIONS

Femtosecond lasers have been used to micromachine several different devices in a wide variety of materials [1]. Here we concentrate on two applications that have been explored by our group: three-dimensional binary data storage and direct writing of optical waveguides. Both of them require bulk micromachining of a transparent material.

8.6.1 Three-Dimensional Binary Data Storage

The structures shown in Figure 2 can be used as bits for three-dimensional binary data storage [8,10]. The presence of a structure represents a 1 and the absence of a structure, a 0. Multiple planes of data are written inside the same sample, with a minimum spacing of about 1 μm within a plane and about 5–10 μm between planes, depending on the microscope objective used to write the data. Readout is accomplished with either transmission, reflection, or scattered light microscopy using an objective whose confocal parameter is smaller than the spacing of data planes.

The oscillator-only machining discussed in the previous section has important implications for the three-dimensional data storage application. In previous demonstrations, an amplified laser operating at a kilohertz repetition rate was used to write the data. The structures shown in Figure 3b, however, can be produced with pulses from an unamplified megahertz repetition rate laser. Using an oscillator provides two major benefits. First, the cost of an unamplifed laser system is much less than that of an amplified laser. Second, the higher repetition rate of the oscillator allows writing speeds on the order of 50 Mbit/s (for the 25 MHz oscillator described in Sec. 8.3.2).

8.6.2   Direct Writing of Single-Mode Optical Waveguides

The refractive index changes discussed in Sections 8.4.1 and 8.5.2 can be used to produce waveguides in bulk glass. By slowly translating the sample along the propagation direction of the laser beam, other groups have demonstrated that the cylindrical structures produced by single laser pulses can be smoothly connected, forming a waveguide [11–13]. Recently, we showed that the cumulative thermal effect discussed in Section 8.5.2 can be used to produce optical waveguides [9].

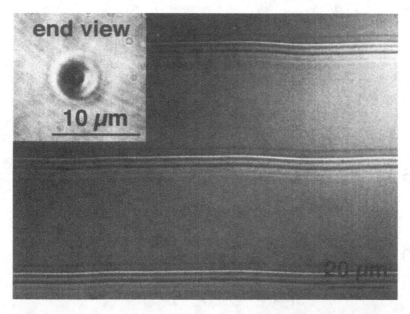

Figure 10   Differential interference contrast optical microscopic image of waveguides written inside bulk glass using a 25 MHz train of 5 nJ, 30 fs pulses focused by a 1.4 NA microscope objective. The sample was translated at 20 mm/s perpendicular to the incident direction of the laser beam. The inset shows the end face of a waveguide.

Figure 10 shows a DIC image of cylindrical structures written inside bulk Corning 0211 glass by translating the sample at 20 mm/s perpendicular to the incident direction of a 25 MHz train of 5 nJ, 30 fs laser pulses that is focused about 110 μm beneath the surface of the sample with a 1.4 NA objective. The cylinders form single-mode optical waveguides [9]. The inset in Figure 10 shows an end view of one of these waveguides after the glass was cleaved to produce a clean face. The same ring structure as in Figure 8 is evident in the end view.

Figure 11 shows the near-field output profile of one of these waveguides for 633 nm laser light. Light from a He:Ne laser was coupled into a 10 mm long waveguide using a 0.25 NA microscope objective. The output end of the waveguide was imaged onto a CCD camera using a 0.65 NA objective. As the data show, the waveguide has a single-mode, near-Gaussian output profile for visible wavelengths. Direct writing of optical waveguides and other photonic devices in three dimensions in bulk material may become important for the telecommunications industry. This task can be accomplished using just a femtosecond laser oscillator.

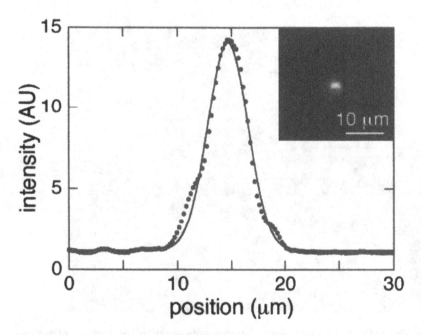

Figure 11 Waveguide output profile at 633 nm. The points correspond to a cross section of the near-field mode, and the curve represents a best-fit Gaussian. The inset shows a raw CCD image of the near-field mode.

8.7   CONCLUSIONS AND FUTURE QUESTIONS

Using optical and electron microscopy, we characterized the morphology of the structural change produced in bulk glass by tightly focused femtosecond laser pulses with various laser energies and repetition rates and under various focusing conditions. We find that near the threshold for permanent structural change, the structures produced by single pulses consist of small density and refractive index changes with a shape that reflects the focal volume of the focusing objective. Under very tight focusing conditions, the threshold for structural change is only about 5 nJ, allowing micromachining with unamplified lasers. At higher laser energies, the shape of the structure is conical, with the tip of the cone oriented along the laser propagation direction. This conical shape is produced by different time slices of the laser pulse producing breakdown at different positions upstream from the laser focus. Voids are also often formed in the material at high laser energies, suggesting an explosive mechanism. At high repetition rate, energy is deposited faster than it can diffuse out of the focal volume, resulting in a cumulative heating effect over many pulses. The laser serves as a point source of heat inside the material, melting a region of glass whose volume depends on the number of incident laser pulses. As an application of this cumulative heating effect, we discussed direct writing of single-mode optical waveguides inside bulk glass.

We are currently further investigating the mechanisms responsible for producing permanent structural change, in particular at laser energies near the threshold, and developing applications for the cumulative thermal effect described in Section 8.5.2. Recently we also found that femtosecond laser pulses can induce chemical, rather than structural, changes in bulk transparent materials. For example, one can locally carbonize polystyrene by exposing it to intense femtosecond laser pulses, opening the door to micrometer-scale nonlinear photochemistry. This technique shows great promise for the fabrication of micromechanical devices.

ACKNOWLEDGMENTS

We wish to thank Dr. André Brodeur, Jonathan B. Ashcom, and Willie Lieght for assistance with the experiments and Professor Nicholas Bloembergen for useful discussions. This work is funded by the National Science Foundation, by a Materials Research Science and Engineering Center grant, and by a grant from the Harvard Office for Technology Transfer.

REFERENCES

1. S Nolte, G Kamlage, F Korte, T Bauer, T Wagner, A Ostendorf, C Fallnich, H Welling. Microstructuring with femtosecond lasers. Adv Eng Mater 2:23 (2000).
2. X Liu, D Du, G Mourou. IEEE J Quantum Electron 33:1706 (1997).
3. X Zhu, AYu Naumov, DM Villeneuve, PB Corkum. Appl Phys A 69(suppl):S367 (1999).
4. S Nolte, C Momma, H Jacobs, A Tünnermann, BN Chichkov, B Wellegenhausen, H Welling. J Opt Soc Am B 14:2716 (1997).
5. M Lenzner, J Krüger, W Kautek, F Krausz. Precision laser ablation of dielectrics in the 10-fs regime. Appl Phys A 68:369 (1999).
6. W Kautek, J Krüger, M Lenzner, S Sartania, Ch Speilmann, F Krausz. Appl Phys Lett 69:3146 (1996).
7. MD Perry, BC Stuart, PS Banks, MD Feit, V Yanovsky, AM Rubenchik. J Appl Phys 85:6803 (1999).
8. EN Glezer, M Milosavljevic, L Huang, RJ Finlay, T-H Her, JP Callan, E Mazur. Opt Lett 21:2023 (1996).
9. CB Schaffer, A Brodeur, JF Garcia, E Mazur. Micromachining bulk glass by use of femtosecond laser pulses with nanojoule energy. Opt Lett 26:93 (2001).
10. M Watanabe, S Juodkazis, H-B Sun, S Matsuo, H Misawa, M Miwa, R Kaneko. Appl Phys Lett 74:3957 (1999).
11. KM Davis, K Miura, N Sugimoto, K Hirao. Opt Lett 21:1729 (1996).
12. K Miura, J Qui, H Inouye, T Mitsuyu, K Hirao. Appl Phys Lett 71:3329 (1997).
13. D Homoelle, S Wielandy, AL Gaeta, NF Borrelli, C Smith. Opt Lett 24:1311 (1999).
14. D Du, X Liu, G Korn, J Squier, G Mourou. Appl Phys Lett 64:3071 (1994).
15. BC Stuart, MD Feit, S Herman, AM Rubenchik, BW Shore, MD Perry. J Opt Soc Am B 13:459 (1996).
16. BC Stuart, MD Feit, S Herman, AM Rubenchik, BW Shore, MD Perry. Phys Rev B 53:1749 (1996).
17. M Lenzner, J Kruger, S Sartania, Z Cheng, Ch Spielmann, G Mourou, W Kautek, F Krausz. Femtosecond optical breakdown in dielectrics. Phys Rev Lett 80:4076 (1998).
18. C Schaffer, A Brodeur, E Mazur. Laser-induced breakdown and damage in bulk transparent materials induced by tightly focused femtosecond laser pulses, Meas Sci Technol 12:1784 (2001).
19. J Qiu, K Miura, K Hirao. Jpn J Appl Phys 37:2263 (1998).
20. EN Glezer, E Mazur. Appl Phys Lett 71:882 (1997).
21. H Varel, D Ashkenasi, A Rosenfeld, M Wahmer, EEB Campbell. Appl Phys A 65:367 (1997).
22. D Ashkenasi, H Varel, A Rosenfeld, F Noack, EEB Campbell. Appl Phys A 63:103 (1996).

23. MJ Soileau, WE Williams, M Mansour, EW Van Stryland. Opt Eng 28:1133 (1989).
24. JH Marburger. Prog Quantum Electron 4:35 (1975).
25. AL Gaeta. Catastrophic collapse of ultrashort pulses. Phys Rev Lett 84:3582 (2000).
26. SH Cho, U Morgner, FX Kartner, EP Ippen, JG Fujimoto, JE Cunningham, WH Knox. OSA Tech Dig: Conf Lasers and Electro Opt 99:470 (1999).
27. AR Libertun, R Shelton, HC Kapteyn, MM Murnane. OSA Tech Dig: Conf Lasers Electro Opt 99:469 (1999).
28. DN Fittinghoff, JA Squier, CPJ Barty, JN Sweetser, R Trebino, M Müller. Opt Lett 23:1046 (1998).
29. JA Squier, M Müller. Third-harmonic generation imaging of laser-induced breakdown in glass. Appl Opt 38:5789 (1999).
30. JP Callan, AM-T Kim, L Huang, E Mazur. Ultrafast electron and lattice dynamics in semiconductors at high excited carrier densities. Chem Phys 251:167 (2000), and references therein.
31. M Douay, WX Xie, T Taunay, P Bernage, P Niay, P Cordier, B Poumellee, L Dong, JF Bayon, H Poignang, E Delevaque. Densification involved in UV-based photosensitivity of silica glasses and optical fibers. J Lightwave Technol 15:1329 (1997).
32. HS Carslaw, JC Jaeger. Conduction of Heat in Solids. London: (Oxford, 1959), p 256.
33. M Lenzner, J Krüger, W Kautek, F Krausz. Incubation of laser ablation in fused silica with 5-fs pulses. Appl Phys A 69:465 (1999).
34. A Rosenfeld, M Lorenz, R Stoian, D Ashkenasi. Ultrashort-laser-pulse damage threshold of transparent materials and the role of incubation. Appl Phys A 69(suppl):S373 (1999).
35. CB Schaffer, A Brodeur, N Nishimura, E Mazur. Laser-induced microexplosions in transparent materials: microstructuring with nanojoules. Proc SPIE 3616:143 (1999).

9

Rapid Scanning Time Delays for Ultrafast Measurement Systems

Gregg Sucha
IMRA America, Inc., Ann Arbor, Michigan, U.S.A.

9.1 INTRODUCTION

As ultrafast lasers find their way into new applications areas, there remain some technological barriers to realizing their full potential. One of the key issues for practical applications of ultrafast lasers is the implementation of an adjustable time delay between optical pulses or between an optical pulse and a signal to be measured. The manifold applications of ultrafast lasers described in this book exploit various properties of the ultrashort pulses, such as high peak power, short coherence length, broad bandwidth, or, of course, the short pulse duration. This chapter is most relevant to applications that utilize the high speed of ultrashort pulses for various types of time-resolved measurements. Generally speaking, these applications can be classified into two main categories: (1) pump-probe measurements and (2) external signal measurements. High-speed sampling of optical waveforms for telecommunications is an example of the latter type. An incoming signal from a device under test (DUT) is sampled using an ultrashort laser pulse from a local laser. This requires various kinds of asynchronous sampling techniques. Here, however, we concentrate on laser systems that can be used for the first type of measurements system, i.e., a pump-probe system.

   A large number of applications involving ultrafast lasers require an adjustable time delay for the optical pulses in order to time-resolve the fast processes being measured. Adjustable timing delays between optical pulses are an indispensable part of ultrashort pulse applications: time-

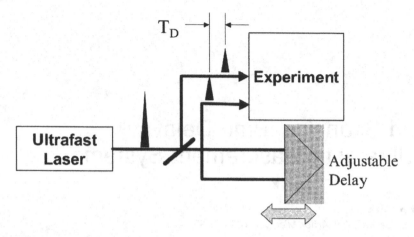

Figure 1   Schematic of typical measurement system using a single ultrafast laser. The time delay between the excitation and probe pulses is usually implemented by using a mechanical scanning delay line.

resolved spectroscopy, biological and medical imaging [1], fast photo-detection and optical sampling [2], terahertz generation [3], detection and imaging [4], and optical time domain reflectometry (OTDR). Figure 1 shows a schematic diagram of a "typical" pump-probe type of measurement system using an ultrafast laser and a scanning delay unit. Usually the timing delay is implemented in the form of moving mirrors or rotating glass blocks, for example. Although this is adequate for many types of experiments, it does impose severe limitations on the attainable scan speed and range and the long-term scan accuracy, making ultrafast laser methods impractical for many other types of measurements and applications. Thus, it can be fairly stated that the systems for implementation of adjustable or scanning time delays can be the "showtoppers" for some applications. Section 9.1.1 describes and compares various methods for time delay scanning.

9.1.1   Comparison of Scanning Techniques

Applications of short optical pulses require that one set of optical pulses be delayed with aspect to another set of optical pulses, the reason being that most measurements are made by time gating (sampling) the measured signal using short optical pulses. Very often the temporal delay must be known to very high accuracy (on the order of 10 fs). These methods can be classified broadly as optomechanical methods and as equivalent time

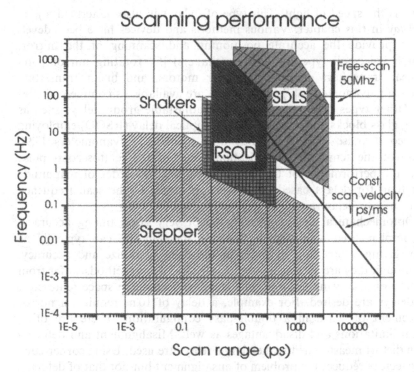

Figure 2 Performance chart of scanning delay systems employing various opto-mechanical methods (steppers and shakers) or equivalent time scanning (free-running) systems as well as an SDLS system. The straight line depicts a contour representing a constant slew rate of 1 ps/ms.

sampling methods. The key performance parameters of the scanning methods are the scan range and the scan frequency. Temporal dynamic range and timing accuracy are also important factors. The performance of these various scanning methods is charted in Figure 2, and is described below.

Optomechanical Methods

A number of optomechanical delay methods have been developed over the past two decades. The conventional method for delaying and scanning optical pulses is to reflect the pulses from a mirror and physically move the mirror (using some mechanical means) through some distance D. The resulting time delay, ΔT, is given by

$$\Delta T = 2D/c$$

where c is the speed of light. This type of delay will also be termed a *physical delay* in this chapter. Various methods and devices have been developed to provide the accurate positioning and scanning of the mirror, including voice-coil types of devices (shakers) [5], rotating mirror pairs [6], linear translators employing stepper motors, and linear translators employing galvanometers [7], all of which are available as commercial products. Other types of physical delays use adjustable group delays such as rotating glass blocks and a rapid scanning optical delay (RSOD) employing femtosecond pulse shapers (FPSS) and scanning galvanometers [8,9]. Because of the temporal magnification effect of the femtosecond pulse shaper, the performance of the RSOD is at least an order of magnitude higher than for delays based on shakers, in terms of the scan frequency at a given scan delay range.

Optomechanical delays are capable of very good timing accuracy. When implemented with stepping motors, the step size can typically be $0.1\,\mu m$, giving a precision of $\sim 1\,fs$ in the delay step size and accuracy. But although they are very accurate, the physical delay methods suffer from a number of disadvantages, the chief one being the large space required if large delays are desired. For example, a delay of 10 ns requires a mirror displacement of 1.5 m in a double-pass configuration. There are other physical limitations and disadvantages as well. Misalignment and defocusing can distort measurements when large delays are used. Using corner-cube retroreflectors reduces the problem of misalignment but not that of defocusing. The defocusing effect can occur when the scan length is an appreciable fraction of the confocal parameter of the beam. For example, a time delay of 10 ns entails a change in free-space propagation of 3 m. Thus, to minimize the effects of defocusing, the confocal parameter of the beam being delayed should be considerably larger than 3 m. This can be very critical in applications where the on-target pulse intensity (governed by spot size) and the alignment must not vary as the time delay is scanned.* Working through some numbers would be illustrative: If we assume that the confocal parameter needs to be approximately 10 times as large as the scan range, then we require that $Z_R = 30\,m$. At a wavelength of 1550 nm, this requires a beam radius of $w_0 = 12\,mm$. This is impractically large for many situations. The need for large mirror displacement can be reduced by multipassing the delay line (double-passing the delay line cuts the required mirror displacement in half), but this does not alleviate the defocusing problem. Multi-

---

*In photoconductive sampling measurements, both the pump and probe must be focused onto $5\,\mu m$ gaps in a transmission line. The measured photoconductive response is sensitive to both the spot size and the alignment of both beams.

passing causes its own set of problems in that alignment procedures are more complicated and the optical losses are increased.

Yet another limitation is imposed by the scanning rates and scanning frequencies that can be achieved simultaneously. It is often desirable to signal-average while scanning rapidly (>30 Hz) in order to provide "real-time" displays of the measurement in progress. The advantages of rapid scanning were pointed out by Edelstein et al. [7], who showed that rapid scanning of the time delay moves the data acquisition frequency range out of the baseband noise of the lasers. This has the beneficial effect of reducing the intrascan laser fluctuations that plague measurements that use slow scan mechanisms such as stepper motors. It was also pointed out that one of the key advantages of using a rapid scanning system is that is gives immediate visual feedback, enabling the user to optimize experimental parameters in real time. Rapid scanning measurement techniques are capable of high sensitivity as demonstrated in time-resolved experiments on coherent optical phonons [10]. Kütt et al. [10] measured reflectivity changes of $\Delta R \sim 10^{-6}$ with a femtosecond laser and a mechanical delay scanning at 100 Hz. In using the mechanical rapid scanning systems, however, the scanning range is limited at such high scan frequencies. The galvanometer-driven translator gives 300 ps of delay at 30 Hz, whereas the RSOD gives 10 ps of delay at 400 Hz and 100 ps at lower frequencies [8]. These figures represent approximate practical upper limits on simultaneously achievable scan ranges and frequencies using mechanical means. Any further increase of scanning range and/or frequency with such reciprocating devices would cause high levels of vibration, which can be disruptive to laser operation. Rotating glass blocks avoid the vibration problem and are capable of higher scan speed but lack adjustability of the scan range.

Equivalent Time Scanning Methods

Equivalent time sampling (also known as asynchronous sampling) methods provide temporal scanning without the need for any mechanical motion. Pulse timing delays can be produced by two free-scanning lasers [11–14] in which two modelocked lasers are set to different repetition frequencies. This produces a time delay sweeping effect, with the sweep repetition frequency at the offset frequency between the two lasers. Time delay sweeping can also be effected by slewing the RF phase between two electronically synchronized modelocked lasers [15]. Another novel approach uses stepped mirror delay lines employing acousto-optic deflectors as the dispersive elements [16].

The nonmechanical methods, in particular, are capable of very high scanning speeds, both in terms of slew rate (delay/time) and in terms of scanning repetition frequency. For example, in the subpicosecond optical sampling (SOS) technique [12], two independent modelocked Ti:sapphire users (each with a nominal repetition rate of 80 MHz) can be set to have slightly different cavity lengths, causing the lasers to scan through each other at an offset frequency of $\Delta v \sim 80$ Hz. This offset frequency can be stabilized to a local oscillator to minimize the timing jitter and give good timing accuracy. In this particular case, because the laser repetition rate was near 80 MHz, the total scan range was about 13 ns. Thus, temporal scanning can be achieved without any moving mechanical delay lines. Accurate timing calibration near $\Delta T = 0$ is achieved by cross-correlating the two lasers in a nonlinear crystal, the resulting signal being used to trigger the data acquisition electronics.

Although this method generates a reasonably high scan frequency and slew rate, its chief drawback is that it can be highly wasteful of data acquisition time. For example, if one is interested in only a 100 ps or 10 ps scan range instead of the full 13 ns scan range, then only 1% or 0.1% respectively, of the 12.5 ms scan time is useful, and the remaining 99% (or 99.9%) is "dead time." Thus, useful data are acquired at a duty cycle of only 1% or less. This increases data acquisition time by a factor of 100 or 1000. This can be partially circumvented by using lasers with higher repetition rates (e.g., $v_1 = 1$ GHz). However, this solution is unacceptable for many applications that require flexibility of adjustment of scan ranges. This issue—temporal dynamic range—is critical for the development of powerful, flexible instrumentation and for particular applications. For example, pump-probe measurements of semiconductors are frequently conducted over a wide dynamic range of time intervals. Carrier lifetimes on the order of several nanoseconds make a 1 GHz laser unacceptable, because residual carriers from the previous laser pulse would still be present when the next pulse arrived. Yet at the same time it is often desirable to zoom in on a much narrower time scale—say, 10 ps—to look at extremely fast dynamics on the order of 100 fs. Thus, the free-scanning laser technique lacks the versatility of scan range selection that is required in many applications. This can be likened to using an oscilloscope that has only one sweep range.

9.1.2 Synchronization of Modelocked Lasers

Timing stabilization and synchronization of modelocked lasers are technologies that are very closely related to time delay scanning. Often it is necessary to stabilize the repetition frequency and the timing of a laser to a master RF clock or to synchronize two modelocked lasers. Several methods have been used to stabilize the timing between two modelocked lasers in cases where the lasers were actively modelocked [17], passively modelocked [18–22], and regeneratively modelocked [15], or with combinations of passively of passively and actively modelocked lasers [23]. The methods used for synchronization can be divided into two main types: (1) passive optical methods and (2) electronic stabilization.

The highest synchronization accuracy is achieved by passive optical methods in which the two lasers interact via optical effects [18–21]. These optical effects (e.g., cross-phase modulation) cause a rigid locking effect between the two lasers, which become synchronized to within less than one pulse width (<100 fs). Although these give the most accurate synchronization, the time delay between the lasers is rigidly fixed, so that in order to scan the time delay between them one must use the conventional physical scanning delay methods. Electronic stabilization using simple RF phase detection gives the most flexibility in terms of adjusting the relative time delay. These systems can maintain timing accuracy on the order of a few picoseconds (~3 ps) over long periods of time. Such systems are commercially available for stabilizing a Ti:sapphire laser to an external frequency reference or for synchronizing two modelocked Ti:sapphire lasers (See, e.g., Ref. 24). Timing stabilization of better than 100 fs has been achieved by use of a pulsed optical phase-locked loop (POPLL). This is a hybrid optoelectronic method [23] in which the electronic stabilizer circuit derives the timing error signal from an optical cross-correlator. But this suffers from the same lack of timing adjustability as the passive optical methods.

9.2 SCANNING DUAL LASER SYSTEM

9.2.1 Description of SDLS Method

The SDLS method is another type of equivalent time sampling method that solves many of the problems and shortcomings of the other methods. The SDLS combines the best features of asynchronous sampling (high slew rates and large time delays) and rapid optomechanical scanning (adjustability of scan range and frequency). In the SDLS method the time delay between two modelocked lasers is repetitively swept out in a controllable manner at a given sweep speed. The basic scanning dual laser system is illustrated in Figure 3. This system, like the free-scanning laser system, consists of two lasers (master and slave lasers) that have nearly identical cavity lengths and repetition rates [25]. The cavity length of the slave laser is controlled by a PZT-mounted cavity mirror. However, unlike the free-scanning lasers, the master and slave lasers are not allowed to totally scan through each other. Rather, the two lasers are electronically synchronized via a phase-locked loop (PLL) circuit. While the master laser is held at a constant repetition frequency (v_1) or allowed to drift of its own accord, the repetition frequency of the slave laser (v_2) is dithered about v_1. This dithering of the

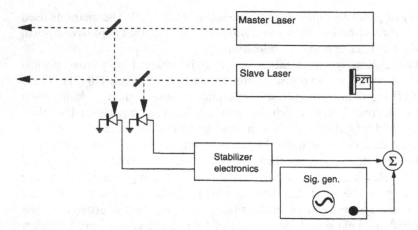

Figure 3 Schematic of scanning laser system showing master and slave lasers and stabilizer electronics (consisting of a phase-locked loop circuit). The signal generator applies the dithering signal to the PZT-mounted end mirror of the slave laser cavity to effect time delay scanning.

repetition frequencies is accomplished by changing the cavity length (L_2) of the slave laser at a "high" frequency (in the range of 30 Hz to 1 kHz) while its *average* repetition rate is slaved to that of the master laser by a "slow" PLL circuit whose bandwidth is less than the scan frequency.

A quantitative analysis of the SDLS technique is very simple. We consider the example in which a square wave is applied to the slave laser PZT at a scanning frequency of f_s. Then the cavity length mismatch is a rapidly varying function of time,

$$\Delta L(t) = \Delta L_0 \cdot \mathrm{Sq}(f_s t) \tag{1}$$

where ΔL_0 is the amplitude of the square wave displacement, and $\mathrm{Sq}(x)$ is the square wave function of unity amplitude ($-1 \leq \mathrm{Sq}(x) \leq 1$). This gives linear scanning of the time delay in both the positive and negative directions for one-half of the scan cycle (i.e., a triangle wave). In the time stationary case, a constant cavity length mismatch of ΔL produces an offset frequency of

$$\Delta \nu = -\frac{c \, \Delta L}{2L^2} \tag{2a}$$

which can be rewritten as

$$\frac{\Delta \nu}{\nu} = -\frac{\Delta L}{L} \tag{2b}$$

In the SDLS method, however, the cavity length is dithered at a frequency that is higher than the offset frequency so that the pulses never have a chance to completely walk through each other. Thus, the scan frequency satisfies the condition

$$f_s \gg \Delta v \tag{3}$$

The time-varying pulse delay, $T_D(t)$, defined as the relative time delay between the pulses from the two lasers, is proportional to the time integral of the cavity length mismatch:

$$T_D(t) = \frac{1}{L} \int_0^t \Delta L(t') \, dt' \tag{4}$$

Figure 4a shows a scheme in which a square wave is applied to the PZT in laser 2. The instantaneous repetition rate, v_2, dithers around v_1, and the relative time delay, $T_D(t)$, scans back and forth linearly in time. The scan range, ΔT_{max} (defined as the largest change in relative time delay), is dependent on the base repetition rate, v_1, the scan frequency, f_s, and the scan distance, ΔL, according to the equation

$$\Delta T_{max} = \frac{\Delta L_0}{2L} \cdot \left(\frac{1}{f_s}\right) \tag{5a}$$

or, in terms of the offset frequency,

$$\Delta T_{max} = \frac{1}{2} \left(\frac{\Delta v}{v}\right) \left(\frac{1}{f_s}\right) \tag{5b}$$

At this point we will also define a few quantities that are useful as figures of merit for comparison with mechanical scanning methods. The scan rate

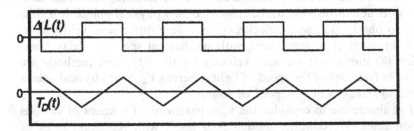

Figure 4 Time-dependent cavity length difference, $\Delta L(t)$, and the predicted scan trajectory, $T_D(t)$, for square-wave dithering, assuming the integral relationship between $T_D(t)$ and $\Delta L(t)$.

(or sweep rate), R_{scan}, is the rate at which $\Delta T_D(t)$ changes per unit time (in the laboratory time frame). This is given simply by

$$R_{\text{scan}} \equiv \frac{\partial T_D(t)}{\partial t} = \frac{\Delta L}{L} \tag{6a}$$

or, in more convenient laboratory units,

$$R_{\text{scan}} \equiv 10^9 \frac{\Delta L}{L} \quad (\text{ps/ms}) \tag{6b}$$

The temporal sampling grid is given by the relative pulse advance between the lasers per round trip of the laser pulse in the cavity,

$$\delta t_g = 2 \, \Delta L / c \tag{7}$$

This represents the ultimate temporal resolution limit for a given cavity length mismatch. Simply put, the pulse advances by the difference in cavity length in one round-trip.

Because the pulse timing advance keeps accumulating with each round-trip, the scanning laser system can be thought of as a "regenerative delay line." One can also define a scanning velocity, which is the amount of spatial pulse advance per unit time. This is given by

$$V_{\text{scan}}(t) = \frac{2\Delta L(t)}{T_R} = \frac{\Delta L(t)}{L} c \tag{8}$$

where $TR = \frac{2L}{c}$ is the cavity round-trip time. Note that these figures of merit apply equally well to the SDLS method and to asynchronous, free-scanning lasers. It is useful at this point to make another comparison with conventional mechanical scanning. When using physical delays (e.g., a moving mirror), the scan velocity is generally twice the speed of the moving mirror. However, with both the SDLS method and the free-scanning lasers, the scan velocity does not correspond to the velocity of any physical object. It is thus possible to obtain high scan velocities—even supersonic scan velocities—that are impractical to implement with mechanical scanning delay lines. Equation (8) shows that the scan velocity for the RF-phase methods are measured in fractions of the speed of light, whereas V_{scan} for physical delays is limited to fractions of the speed of sound.

It is illustrative to consider the scan parameters for lasers of various repetition rates. We consider modelocked lasers with repetition rates of 10 MHz, 100 MHz, and 1 GHz, and we use Eqs. (5)–(8) to calculate the scan parameters (δt_g, ΔT_{max}, R_{scan}, and V_{scan}) given the cavity length dithering parameters f_s and ΔL_0 for the case of square-wave dithering. The results

Table 1 Scan Parameters for 10 MHz Laser System ($L = 15$ m)

Scan freq. f_s (Hz)	ΔL_0 (μm)	Offset freq. Δv (Hz)	Scan rate R_{scan} (ps/ms)	Scan vel. V_s (m/s)	Grid δt_g (fs)	Scan range ΔT_{max} (ps)
10^1	0.15	0.1	1	3	1	500
10^2	0.15	0.1	1	3	1	50
10^3	0.15	0.1	1	3	1	5
10^1	1.50	1	10	30	10	5000
10^2	1.50	1	10	30	10	500
10^3	1.50	1	10	30	10	50
10^1	15.00	10	100	300	100	50,000
10^2	15.00	10	100	300	100	5000
10^3	15.00	10	100	300	100	500

are given in Tables 1, 2, and 3 for a number of practical dithering parameters. These data show, for example, that the attainable scan velocities
can be quite high. For a 1 GHz laser system, a cavity length displacement
of 1.5 μm produces a scan velocity of 3 km/s (approximately Mach 10). This
would be difficult to achieve reproducibly with a physical delay such as a
moving mirror. The tables also show that it is possible to obtain scan ranges
of nanoseconds at high scan frequencies (>100 Hz) that can give real-time
data acquisition. An example of this is shown later in this chapter. This type
of scanning performance is not feasible with physical delays.

These tables give a quantitative idea of the achievable scan rates and
ranges, and they are helpful in determining the dithering parameters, f_s
and ΔL_0 (given in the two leftmost columns) that are required for producing

Table 2 Scan Parameters for 100 MHz Laser System ($L = 15$ m)

Scan freq. f_s (Hz)	ΔL_0 (μm)	Offset freq. Δv (Hz)	Scan rate R_{scan} (ps/ms)	Scan vel. V_s (m/s)	Grid δt_g (fs)	Scan range ΔT_{max} (ps)
10^2	0.15	10	10^2	30	1	500
10^3	0.15	10	10^2	30	1	50
10^4	0.15	10	10^2	30	1	5
10^2	1.50	100	10^3	300	10	5000
10^3	1.50	100	10^3	300	10	500
10^4	1.50	100	10^3	300	10	50
10^3	15.00	1000	10^4	3000	100	5000
10^4	15.00	1000	10^4	3000	100	500

Table 3 Scan Parameters for 1 GHz Laser System ($L = 0.15\,\text{m}$)

Scan freq. f_s (Hz)	ΔL_0 (μm)	Offset freq. Δv (Hz)	Scan rate R_{scan} (ps/ms)	Scan vel. V_s (m/s)	Grid δt_g (fs)	Scan range ΔT_{max} (ps)
10^2	0.15	10^3	10^3	300	1	—[a]
10^3	0.15	10^3	10^3	300	1	500
10^4	0.15	10^3	10^3	300	1	50
10^2	1.50	10^4	10^4	3,000	10	—[a]
10^3	1.50	10^4	10^4	3,000	10	—[a]
10^4	1.50	10^4	10^4	3,000	10	500
10^3	15.00	10^5	10^5	30,000	100	—[a]
10^4	15.00	10^5	10^5	30,000	100	—[a]

[a]Total walkoff (1 ns maximum).

the desired set of scan parameters (δt_g, ΔT_{max}, R_{scan}, and V_s). We take as an example a pair of 100 MHz lasers with a nominal cavity length of $L = 1.5\,\text{m}$. Usually, when setting up a measurement, the user wants to specify two things: the scan range, ΔT_{max}, and the sampling grid δt_g (given in the two rightmost columns). From these quantities we can use Table 2 to look up the required scanning amplitudes and frequencies. Suppose, for example, that we wish to scan over a 500 ps range. If we choose $\delta t_g = 10\,\text{fs}$, then the cavity length displacement is $\Delta L_0 = 1.5\,\mu\text{m}$. The requirement that $\Delta T_{\text{max}} = 500\,\text{ps}$ then dictates a scan frequency of $f_s = 100\,\text{Hz}$, using either the table or Eq. (5a). In reality, for delay ranges larger than 500 ps, it is a simple matter to set the scan range by observing the pulses scanning on an oscilloscope. This is discussed in a later section.

For other types of applications, different parameters will be critical and will determine the scan parameters. For example, in experiments involving photoacoustic or photoelastic effects, it may be crucial to match V_s with an acoustic velocity in the sample. This strictly determines ΔL_0, depending upon the repetition frequency of the lasers. For a pair of 1 GHz lasers and a desired scan velocity of $V_s = 3\,\text{km/s}$, Table 3 shows that the required cavity length mismatch is $\Delta L_0 = 1.5\,\mu\text{m}$. Then, if a 500 ps scan range is desired, this dictates that $f_s = 10\,\text{kHz}$.

9.2.2 Synchronized Dual Fiber Laser

An SDLS constructed from a pair of passively modelocked Er:fiber lasers is shown in Figure 5. These are electronically synchronized using a simple PLL circuit [26]. The construction and performance of two different SDLS systems will be discussed, the first based on a pair of 4.6 MHz lasers [27]

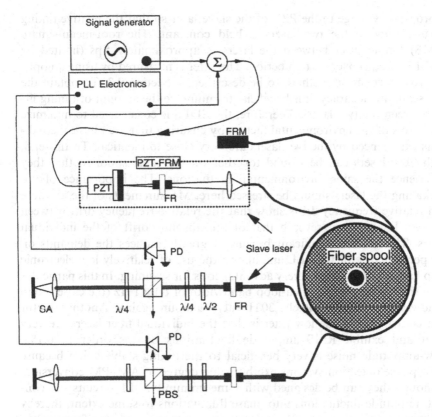

Figure 5 Dual fiber laser system with co-wrapped fibers. Both master and slave lasers are wrapped on the same spool. The master laser is terminated by a Faraday rotator mirror (FRM); the slave laser is terminated by a Faraday rotator mirror assembly (PZT-FRM) containing a PZT-mounted mirror. Other components are half-wave plate ($\lambda/2$), quarter-wave plate ($\lambda/4$), polarizing beam splitter (PBS), Faraday rotator (FR), and saturable absorber (SA).

($T_0 = 216$ ns) and the second based on a pair of 50 MHz lasers ($T_0 = 20$ ns). For the first SDLS, the two lasers have essentially identical cavity lengths and have repetition rates near $\nu = 4.626$ MHz However, the slave laser has one cavity mirror mounted on a PZT stage, allowing a 45 μm adjustment range of the cavity length in order to implement scanning and synchronization. Both lasers produce pulses of approximately 600 fs duration at a wavelength of 1550 nm.

The two lasers are synchronized using a PLL circuit that contains simple phase detection and low-pass filters and amplifiers. The PLL applies an appropriate voltage to the PZT of the slave laser so that the relative timing (phase) between the two lasers is held constant. The root-mean-square (RMS) timing jitter between the lasers is approximately 5 ps (limited by the PLL circuit) over a 3 min period. The jitter is measured by using an optical cross-correlation method, to be described in Section 9.3. To obtain the best scanning accuracy, it is beneficial to minimize the amount of timing jitter between the two lasers. Therefore, the SDLS is constructed to minimize the effects of the environmental factors by ensuring that the physical conditions experienced by the two lasers are very close to identical. To this end, both fiber lasers can be wound together on the same spool so that they experience the same environmental fluctuations. The importance of co-packaging the lasers should be stressed here. Measurements of the absolute and relative frequency drifts show that the relative frequency drift between the two lasers is one-seventh that of the absolute drift of the individual lasers. This is very significant because it greatly reduces the demands on the performance of the PLL and allows the use of relatively low electronic loop bandwidths, which is very advantageous for scanning. In this particular case, the circuit had an open-loop bandwidth of only 1 Hz (the closed-loop bandwidth is approximately 30 Hz at maximum gain). Another factor that contributes to the low jitter is that the individual fiber lasers are very quiet and exhibit RMS amplitude fluctuations on the order of 0.1%. Low-amplitude noise is very beneficial to the timing stabilization because most phase detection systems exhibit some degree of AM–PM conversion. Although they can be designed with some immunity, PLL circuits still convert amplitude fluctuations into phase fluctuations to some extent, thereby generating timing errors. Even small timing errors are significant when applications demand subpicosecond accuracy.

9.2.3 Scanning Performance

Once the stabilizer circuit is activated and the lasers are synchronized, the timing of the slave laser can be scanned by adding a dithering voltage to the feedback voltage at a summing point and applying the combined signal to the PZT. The scanning effect is readily observed by viewing the pulses from both lasers on an oscilloscope that is trigged by pulses from only the master laser. Figure 6 shows scope traces of scanning pulses from a 4.6 MHz dual laser system and from a 50 MHz dual laser system. For the 4.6 MHz system, Figure 6a shows a 10 ns scan range at 30 Hz. Fifteen consecutive scans are overlapped in this 0.5 s exposure. For the sake of clarity, the scan range of the slave laser pulse has been displaced from coincidence with the master laser pulses in order to show the 10 ns scan "envelope." For the calibration and pump-probe measurements discussed in later sections,

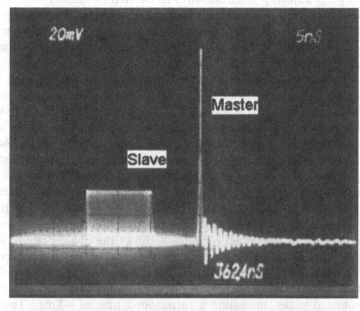

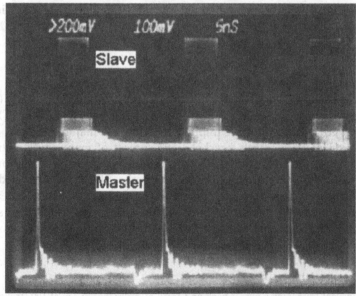

Figure 6 Oscilloscope photos of laser pulses during scanning. (a) Pulses from 5 MHz lasers scanning with $T_{max} = 10$ ns, $f_s = 30$ Hz; (b) pulses from 50 MHz lasers scanning with $T_{max} = 5$ ns, $f_s = 100$ Hz (0.5 s exposure).

the scan range is adjusted (electronically) to overlap the master laser pulse. Performance for the 50 MHz SDLS is illustrated in Figure 6b, which shows a 5 ns scan range at $f_s = 100$ Hz. Note that this type of scanning performance would be highly impractical (and perhaps even *dangerous*) to implement using a mechanical method such as shakers.*

As we shall see in a later section, the linearity of the scanning is a key consideration for ease of use. Although a perfectly linear scan trajectory (constant scan velocity) is most desirable, it is a virtual certainly that the stabilizer circuit will induce a nonlinearity in the scan trajectory as it tries to counteract the timing error induced by the scanning. Also, owing to mechanical inertia and resonances, the PZT does not respond instantaneously to the applied voltages and may even overshoot its position if a square wave voltage is applied. This would also cause the scan trajectory to deviate from linearity. For reasonably large scan ranges (>1 ns) these effects can be easily monitored in real time by observing the output of the phase detector in the circuit.[†] Figure 7 shows the phase detector output and the applied dithering voltage for the case of an applied square wave at 100 Hz for a scan range of 1 ns. As expected, square wave dithering produces nearly linear scanning. The triangular wave in Figure 7 exhibits rounded peaks, indicating a turnaround time of ~1 ms.[‡] The linearity of the scan trajectory is affected by the PLL gain and bandwidth.

9.3 TIMING CALIBRATION

The implementation of SDLS is reasonably straightforward, but the real challenge is encountered when trying to obtain high timing accuracy (sub picosecond) during the scanning procedures. Time base distortion (TBD) and timing jitter in waveform sampling systems are major concerns, and the correction of these effects has been the subject of many investigations and much development work [28,29]. The SDLS system is no exception to this. The sampled waveforms obtained using the SDLS are distorted by the nonlinear scan trajectory mentioned above (a type of TBD) and also by random timing jitter occurring both within a single scan and from scan to scan. The waveform data measured using the SDLS require signal processing to remove distortions and the effects of noise.

---

*This is equivalent to shaking a mirror back and forth by a distance of 1 m at 100 strokes per second.
[†]The phase detector output has a responsivity of ~11mV/ns for the 5 MHz system and ~100 mV/ns for the 50 MHz system.
[‡]The PZT has an unloaded small-signal bandwidth of 8 kHz.

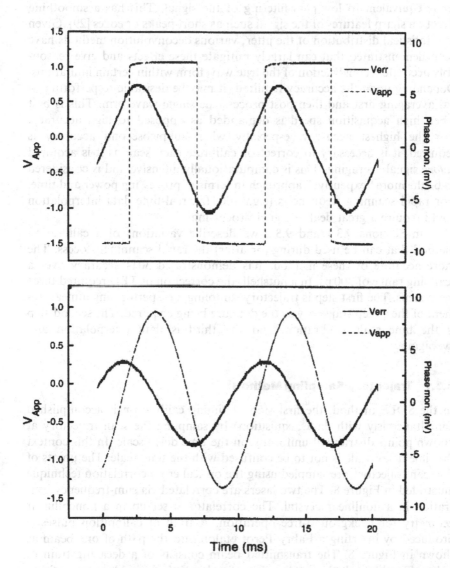

Figure 7 Output of PLL phase detector (——) and applied voltage waveforms (- - -) for linear scanning (square wave dithering) in the upper figure and for sinusoidal scanning in the lower figure. The phase detector output gives visual confirmation of the time integral relationship between $T_D(t)$ and $\Delta L(t)$ in the scanning technique.

It is known that additive signal averaging in the presence of timing jitter is equivalent to low-pass filtering of the signal. This has a smoothing effect on sharp features of the signal such as short peaks or edges [29]. Given the statistical distribution of the jitter, various deconvolution methods have been demonstrated that can largely mitigate these effects and give reasonably accurate reconstruction of the true waveform within certain limitations. Depending upon the accuracy required, it may be desirable to perform signal averaging first and then post process the single waveform. This is best when high acquisition speed is demanded as opposed to high accuracy. For the highest accuracy, especially when subpicosecond accuracy is required, it is necessary to correct or calibrate each scan as it is acquired *before* signal averaging. This is computationally intensive and is considered to be the more "expensive" approach in terms of processing power and time. For rapid scanning frequencies (e.g., 100 Hz), real-time data interpolation could require a great deal of signal processing.

In Sections 9.3.1 and 9.3.2 we describe variations of a calibration method that can be used during (or after) the rapid scanning process. The more accurate of these methods has demonstrated 30 fs accuracy over a scanning range of 300 ps. In a nutshell, the correction of TBD required three main steps. The first step is trajectory sampling, i.e., performing a measurement of the sweep trajectory as the data are being acquired. The second step is the time scale calibration, and the third is data interpolation and averaging.

9.3.1 Trajectory Sampling Method

In the SDLS method, the first step of timing calibration is accomplished (simultaneously with data acquisition) by sampling the scan trajectory at known points distributed uniformly on the time delay scale. In this context the time delay scale is not to be confused with the time scale. The points of the scan trajectory are sampled using the optical cross-correlation technique illustrated in Figure 8. The two lasers are correlated via sum-frequency generation in a nonlinear crystal. The correlator is set up in a noncollinear geometry for background-free operation. A train of calibration pulses is produced by inserting a Fabry–Perot etalon into the path of one beam as shown in Figure 8. The transmitted beam consists of a decaying train of pulses. To obtain the largest number of pulses, it is desirable to use a fairly high surface reflectivity $(R > 90\%)$. Passing pulses through a 5 mm thick etalon with surface reflectivities of $R = 90\%$ produces a decaying train of pulses spaced in delay by 50 ps. In this way, the SDLS system acquires experimental data (via the Y channel) while simultaneously acquiring the necessary timing data (via the X channel) as shown in Figure 9. By trig-

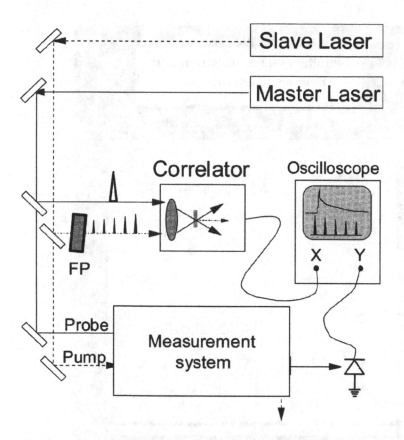

Figure 8 Generic ultrafast sampling system consisting of SDLS, timing calibration unit (comprising an optical cross-correlator with Fabry–Perot etalon), and an unspecified measurement system (e.g., electro-optic sampling).

gering the data acquisition from the first peak of each calibration scan, the first peaks are automatically aligned.

Figure 10 shows the detail of a single calibration sweep with both the real-time scale (a few milliseconds) and the delay scale (a few hundred picoseconds). With the lasers scanning an interval of $\Delta T_{max} = 1$ ns, up to seven calibration pulses can be clearly seen, giving accurate timing calibration over a 300 ps time interval. The inset shows an individual cross-correlation pulse and its width ($\tau_{CC} = 1.2$ ps) as calibrated by the pulse spacing. Each cross-correlation peak takes about 7 μs to sweep out. The sweep rate is $R_{scan} = 165$ ps/ms, which for $\nu = 4.6$ MHz gives a temporal sampling grid of $\delta_{t_g} = 36$ fs.

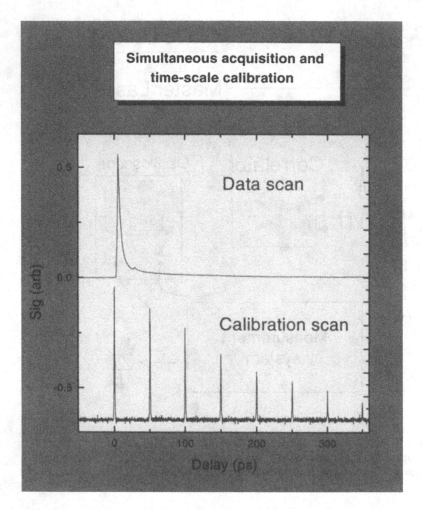

Figure 9 Recorded data traces of experimental data (upper trace) and timing calibration data (lower trace) that were acquired during a pump-probe measurement of a semiconductor thin film. This illustrates the simultaneous acquisition of the desired experimental data, along with the timing calibration waveform.

9.3.2 Time Scale Calibration

The experimental data and the time scale data shown in Figure 9 can be either from a single scan or averaged over many scans. Depending upon how the data were acquired, it is possible, by analyzing the timing

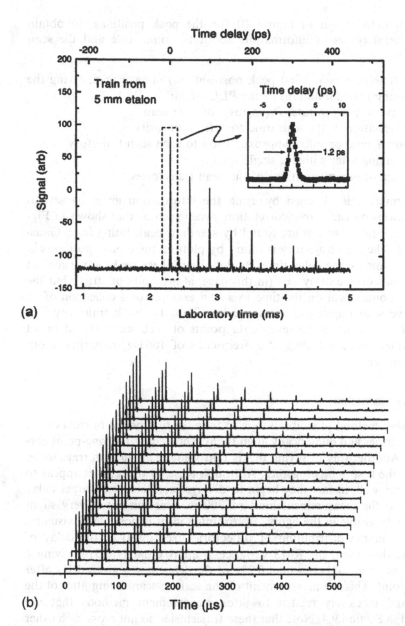

Figure 10 (a) Cross-correlation signal (single sweep) of pulse train with 50 ps pulse spacing (created by 5 mm etalon). Inset shows close-up of the first cross-correlation peak (τ_{CC} = 1.2 ps). (b) Cross-correlation signal (40 consecutive sweeps) taken at 100 Hz scan frequency.

calibration data shown in Figure 10 for the peak positions, to obtain several crucial pieces of information about the time scale and the scan trajectories:

> The time-zero point (first peak position, T_0) of each scan, giving the timing jitter limitations of the PLL circuit
> The average sweep rate, $R_s(\text{ps/ms})$, of each scan
> The deviation of the scan trajectory from linearity
> A polynomial or spline approximation to each scan trajectory
> The timing jitter within a single scan
> The scan-to-scan variations of the scan trajectories.

This information is obtained by using the information given by several peak positions on each cross-correlation sweep such as that shown in Figure 10. The peak positions are found by standard peak fitting (e.g., Gaussian peak). The scan trajectory is found by plotting these peak positions in laboratory time vs. the delay time. By definition, these peaks are spaced at even intervals on the delay axis (in this case, at multiples of 50 ps), but the spacing is nonuniform on the time axis. An example of a collection of 40 consecutive scan trajectories is shown in Figure 11. Each trajectory is a cubic polynomial fit to the seven data points of each scan. This data set was obtained while scanning at a frequency of 100 Hz, requiring about 0.5 s to acquire.

First-Point Correction

The simplest method of calibration is to take the first points of each of the scan trajectories and numerically align them. This is called a "one-point correction." Another way of saying this is that we take all 40 scan trajectories and "tie" the first points together. Although these 40 trajectories appear to overlap very well, in fact they begin to diverge increasingly at larger delay times up to the measurement limit of 300 ps. This fan-out is very slight and is barely visible in the figure. By zooming in, it is possible to visualize the spread in the distribution of trajectories (see inset) at the delay of 250 ps. At this point, the RMS variance is approximately 0.5 ps, giving a timescale accuracy of better than 0.2% of the delay at any given delay after the first point. This is a measurement of the scan-to-scan timing jitter of the system and is closely related to jitter measurement methods that are described in Section 9.4. Note that these trajectories do not cross each other and appear to have a common functional form, suggesting that the scan trajectories jitter in a similar, regular fashion. The RMS jitter of the trajectories is plotted in Figure 11b as a function of time in microseconds. If one plots the RMS jitter versus delay (in picoseconds), then up to time delays of

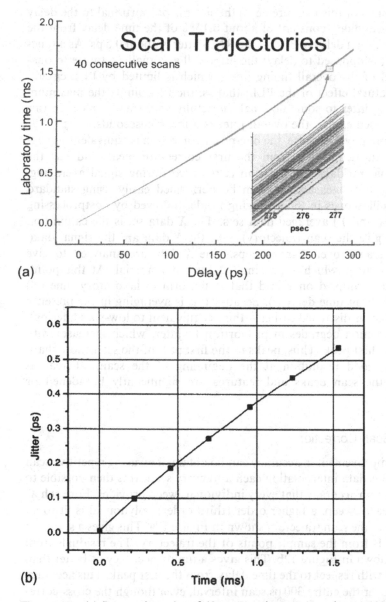

Figure 11 (a) Scan trajectories of 40 consecutive sweeps, taken at 100 Hz scan frequency, spanning 300 ps of time delay. Each trajectory has seven data points and is fitted with a cubic polynomial. A slight deviation from linearity (time base distortion) is apparent. Inset: Magnified view of nonoverlapping scan trajectories. (b) Jitter vs. time for the same 40 scans.

350 ps (the limits of this measurement) the jitter is proportional to the delay with a proportionality constant of about 0.15% of the time delay from the first pulse; i.e., for a delay of 300 ps, the RMS jitter is about 0.5 ps. As a function of time (as opposed to delay) the jitter will eventually saturate to one-half the level of the overall timing jitter, which is limited by PLL circuit, which is a natural effect of the PLL; that is, the PLL limits the maximum relative timing jitter to some nominal (hopefully very small) value. In this case, as discussed earlier, the overall jitter is a few picoseconds.

In a more physical sense, the one-point correction is equivalent to triggering the data acquisition from the first correlation peak and has the advantage that no data interpolation is required during signal averaging. Thus, this type of measurement can be performed easily using standard digitizing oscilloscopes in the averaging mode, followed by postprocessing of the two (X and Y) averaged data sets. The X data set is the calibration data (which give the scan trajectory), and the Y data are the signal data. Both are averaged over many sweeps. The X data are analyzed to give the scan trajectory, which approximated by a polynomial. At this point, the Y data are sampled on a grid that is uniform in laboratory time (T) but nonuniform in time delay (t). Because this is averaging in the presence of timing jitter (as discussed earlier—this is equivalent to low-pass filtering), the time resolution degrades in proportion to jitter, which increases with time after the first peak. Thus, peaks in the first part of the scan show sharp features and good resolution at the beginning of the scan, whereas at the end of the scan peaks and features are significantly broadened or smeared out.

Repetitive Scan Correction

Significant improvement in accuracy can be obtained by using repetitive scan correction plus data interpolation each and every scan. It is then possible to preserve the high accuracy that every individual sweep provides. For each X-dataset for each sweep, a higher order (third-order) polynomial is fit to all seven points of the scan trajectory shown in Figure 12a. This gives a standard deviation 30 fs from the sample points of the trajectory. The residuals from this fit are shown in Figure 12b. This gives a timing accuracy of better than 1 part in 10^4 with respect to the time delay from the first peak. This accuracy is obtained over the entire 300 ps scan interval, even though the cross-correlation width is on the order of 1 ps. The high single-scan accuracy is due to the large sweep rate of the lasers, which puts the scanning rate far above the baseband timing jitter noise. This is discussed in more detail later.

This higher level of timing accuracy could not be obtained by simple averaging followed by postprocessing. As we saw in the previous section

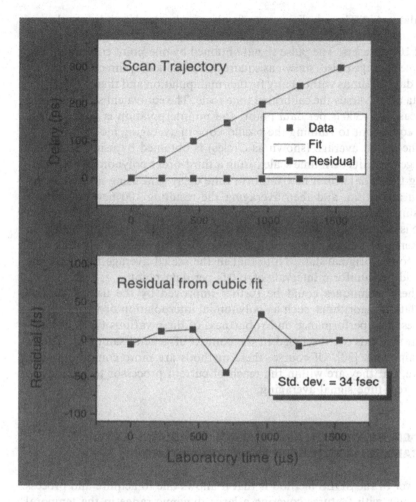

Figure 12 Scan trajectory with cubic fit (a) and residual from cubic fit (b). RMS deviation from cubic fit is 30 fs for the 300 ps scan range.

on first-point correction, simple averaging gives about 0.5 ps of accuracy at a delay of 300 ps. The features of the data will be correspondingly smoothed, as if a low-pass filter had been applied to the data. The high accuracy (30 fs) offered by polynomial fitting to the scan trajectory can be obtained only by fitting and correction *each scan* and then using the time scale information to interpolate the data on a uniform time delay grid. These effects are illustrated in Figure 13, which shows the results obtained with a dataset of 100 scans. For this test, the timing calibration data serve as both the X and Y datasets.* The misalignment of pulses is clearly shown by overlaying 10 plots in Figure 13a. The lower frame (Fig. 13b) shows the signals averaged over all 100 datasets. The pulse signal obtained by one-point correction, i.e., the simple average pulse (shown as squares), is obtained by directly averaging the 100 data columns without any further manipulation and then plotting the new data array versus the calibrated time scale. The equivalent sampling grid in this case is ~500 fs per data point, and no interpolation is done. This is exactly equivalent to running the oscilloscope in averaging mode.

The scaled average (shown as circles) is obtained by using each data scan to generate its own time scale (using a third-order polynomial fit), interpolating the Y data on a new uniform time delay grid using a linear interpolation algorithm, and then averaging the resulting 100 new waveforms into a single waveform. The improvement in signal quality using the scaled average is clearly shown in comparison with the simple averaging method in the presence of timing jitter, which shows significant pulse broadening with respect to the original data. Note that in the scaled average the new time delay grid had uniform intervals of 100 fs per data point.

These techniques could be further improved by the use of various interpolation algorithms, such as polynomial interpolation or various types of splines. The performance and robustness of these various types of interpolation methods are discussed in the context of general sampling systems by Rolain et al. [30]. Of course, these methods are more computationally intensive, but they are within the reach of current processor technology to provide real-time signal averaging.

9.4 DEMONSTRATION APPLICATION: MEASUREMENT OF CARRIER LIFETIMES IN SEMICONDUCTORS

The power of the SDLS method is that it allows one to acquire and process time signals with features covering a large dynamic range in the temporal sense. Condensed matter experiments provide an excellent venue for demonstrating the versatility and dynamic range of the SDLS method. Various phenomena, such as carrier lifetime in semiconductors, can have widely varying time response ranging from femtoseconds to nanoseconds, depending upon conditions, and the SDLS technique has proven itself useful for measuring carrier lifetimes in InGaAs via a simple pump-probe technique.

---

*This is valid for a test of the algorithm. Additionally, in order to best illustrate the effects of timing jitter on various signal-averaging algorithms, a dataset with very sharp features (short pulses) is required. The timing calibration data are therefore well suited to this task.

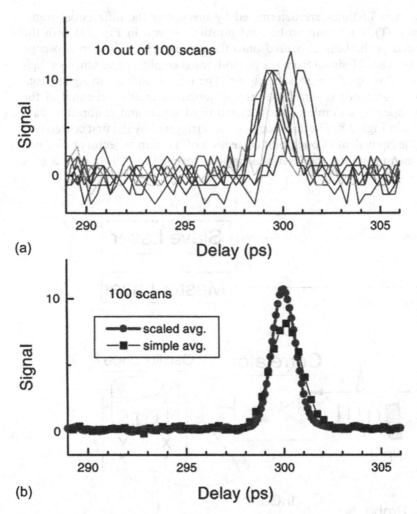

Figure 13 (a) Overlaid plots of 10 scans out of a 100-scan dataset, showing closeup of the seventh peak near a delay of 300 ps. Scan-to-scan timing jitter induces temporal misalignment of pulses. (b) Average of 100 scans using simple averaging (squares) and using repetitive scan calibration with interpolation, denoted as scaled averaging (circles).

For illustrate purposes, data from two different InGaAs samples are shown. One InGaAs sample is radiation-damaged to reduce the carrier lifetime to a few picoseconds, whereas the other is undamaged, having a carrier lifetime of several nanoseconds.

Carrier lifetimes are determined by measuring the differential transmission (DT) in a pump-probe configuration (shown in Fig. 14) with the pump and probe beams focused onto the sample with a 10 × microscope objective. The DT signal from the photodiode is amplified and sent to a digitizing oscilloscope for signal averaging. The timing calibration signal from the cross-correlator is simultaneously acquired on another channel of the oscilloscope. This simultaneous acquisition of signal and calibration data is shown in Figure 8. The data acquisition is triggered by the first correlation peak (the equivalent of one-point correction of the scan trajectory). For one set of measurements, the laser was linearly scanned at 100 Hz over a scan

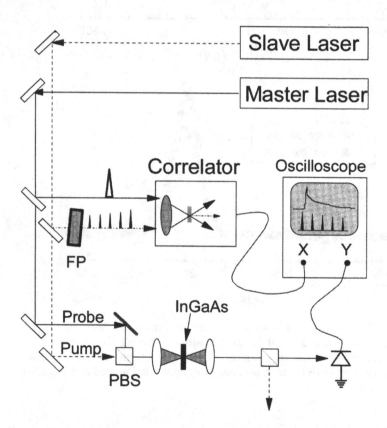

Figure 14 Collinear pump-probe setup for measurement of carrier lifetimes in InGaAs, using SDLS for time scanning and using cross-correlator with etalon for time scale calibration. Pump and probe beams are combined (before the sample) and split (after the sample) by using polarizing beam splitters (PBS).

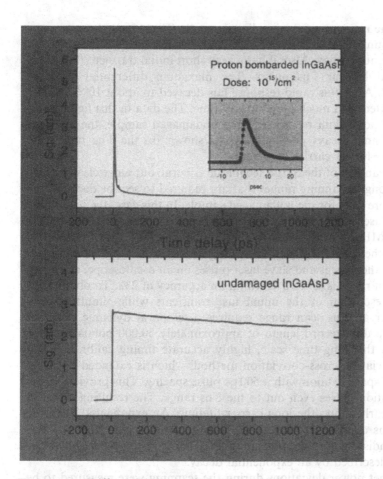

Figure 15 Differential transmission signal from pump-probe measurement of InGaAs samples scanning over a range of $\Delta T_{\max} = 1200$ ps for (a) proton-bombarded sample and (b) undamaged sample. Insets show initial fast transients of DT signal.

range of 1.2 ns. The signal was averaged over 19,000 scans, which requires about 10 min,* resulting in the excellent signal-to-noise (S/N) ratio shown in the time-resolved DT signal (Fig. 15) for both samples. The S/N ratio is limited by the detector noise and the 8-bit analog-to-digital (A/D) input of the oscilloscope.

---

*Although the scan frequency is 100 Hz, the scope is able to average only 30 scans/s with a 5000-point record, which increase the acquisition time from 3 min to 10 min.

The time response of the irradiated sample (Fig. 15(a)) shows a number of interesting features. First of all, the carrier relaxation is clearly not a simple exponential decay. Also, it has a very short initial transient that is not fully resolved by our 1 ps pulses. The maximum differential signal is $\Delta T/T = 10^{-1}$. After this rapid response has decayed to about 10^{-2}, the signal is dominated by a much longer decay time. The data in this figure comprise at 5000-point data record. For the undamaged sample, the response rises rapidly and decays only slightly, as shown on the 1 ns time scale because of the longer carrier lifetime.

The advantage of the SDLS technique is borne out more clearly as we employ the longer scanning ranges that are required to see the decay of the DT signal more fully for the undamaged sample. In this case, the scan range must be increased to about 10 ns. This scan range is not attainable at 100 Hz with the 4.6 MHz system owing to the limit of the PZT travel. However, by lowering the scan frequency to 30 Hz, an 8 ns scan range is obtained. By observing the master and slave laser pulses on an oscilloscope, as in Figure 6, the scan range is set to 8 ns within an accuracy of 2%. To obtain sub-picosecond resolution of the initial fast transients while simultaneously scanning over an 8 ns scan range requires a temporal dynamic range of $>10^4$. Thus, a data record length of approximately 50,000 points is called for. Even on this long time scale, highly accurate timing calibration can be obtained via the cross-correlation method, which is extended by using a thicker, air-spaced etalon with a 500 ps pulse spacing. This provides a series of calibration pulses even out to the 8 ns range. The resulting signal in Figure 16 clearly shows the long carrier lifetime. An exponential fit to the decay curve is overlaid on the data. The overlap is so close that the two curves are indistinguishable, indicating that the carrier population decay is very well described by an exponential decay.

On-target power deviations during the scanning were measured to be less than 0.1%. We should note that if one were to set up a similar pump-probe measurement with a moving mirror on a large stepper motor stage, it would be quite difficult to align the delay line with sufficient accuracy to give a 10^{-3} deviation of the power density on target over the whole 8 ns scan interval. This is a key advantage of SDLS when scanning large time delays.

9.5 CHARACTERIZATION OF LASER JITTER

9.5.1 Fundamental Limitations and Effects on Measurements

There is ongoing work to develop lasers with lower timing jitter to serve as sources for measurement applications. Modelocked lasers with ultralow

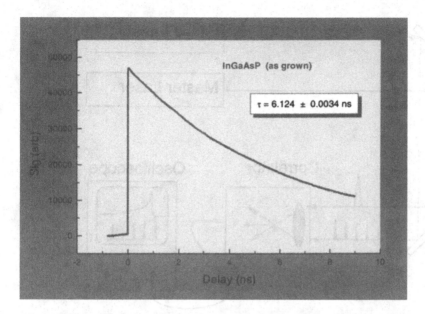

Figure 16 Pump-probe signal from undamaged sample over a scan range of $\Delta T_{max} = 10$ ns. Exponential fit to the data is overlaid. Note the almost perfect fit.

timing jitter will become the essential building blocks of many new kinds of ultrafast measurement systems. The recent development of ultralow-noise modelocked fiber lasers with nearly quantum-limited timing jitter [31] promises great advancements in the realm of high-precision measurement systems employing femtosecond lasers. In measurement applications using temporally scanned dual laser systems, timing jitter is a determining factor in system performance [32], so it is becoming increasingly important to be able to measure jitter more conveniently and with more accuracy than is possible with current methods.

The timing calibration (TBD correction) methods described above are quite effective, giving time resolution well below 1 ps. However, there are still limitations. One of the more obvious limitations is that the sampled jitter trajectory works best near the calibration points, i.e., near the peak of the cross-correlation scans. But we have not yet addressed the question of what happens to the trajectory *between* the calibration points. To what extent have we *undersampled* the trajectory? (There are only seven sampling points in the examples above.) Of course, the experimental data are still affected by the jitter that occurs between the correlation peaks, and the effects of this "hidden" jitter must be addressed. It was mentioned earlier than random

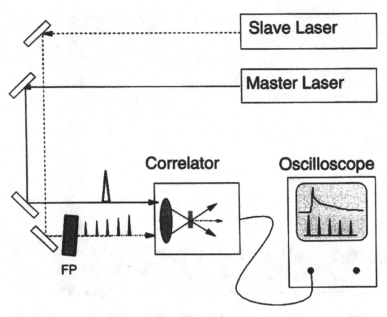

Figure 17 System schematic for the measurement of relative timing jitter between two lasers. The correlator consists simply of a nonlinear mixing crystal (such as PPLN) and a detector (a PMT). An etalon placed in one arm of the correlator serves as a rattle plate to give timing markers at uniform intervals of time delay.

timing jitter has the effect of low-pass filtering and smooths the sharp features of the signal. The effects of this are most severe in the case of simple averaging in real time, where the effects of scan-to-scan timing jitter can be quite strong. But it also occurs in the case of scan-to-scan calibration before averaging, where the pseudo-filtering effects of jitter certainly occur between the calibration points. The question is how to quantify these effects.

Timing jitter within the SDLS has a few different origins. One is related to the limitations of the PLL circuit. Current commercial laser systems on the market give good synchronization with long-term timing jitter or drift of only a few picoseconds. However, this result does not necessarily hold in the context of rapid scanning. The most fundamental effect is the intrinsic timing jitter of the lasers. Under the conditions we have described in the SDLS system—that is, in the case where the electronics noise to the laser PZT is not a dominating factor and where the PLL bandwidth is much less than the scan frequency—then it is the intrinsic timing jitter of the *laser itself* that is the limiting factor in obtaining timing accuracy. It is therefore critical to measure the relative and absolute timing jitter of the lasers in order to understand and predict the ultimate limits of timing accuracy obtainable with SDLS.

In Sections 9.5.2–9.5.4 we describe a method for the measurement and analysis of the relative timing jitter between two lasers. This method gives very high sensitivity to small amounts of jitter ($\sim$4 fs when measuring 1 ps pulses), and the analysis methods give conceptually simpler interpretations of the jitter data [33]. The method and analysis are particularly well suited to scanning dual laser systems, because they yield graphic representations that directly show the growth of jitter as a function of time and also the expected jitter values between calibration points. Ultimately this method should have subfemtosecond sensitivity when measuring laser pulses of 100 fs or less. Also, correlations among power, pulse width, and jitter can be extracted from the data in a straightforward manner. The method presented here is also more accurate than previous time domain jitter measurements, which require deconvolution to determine jitter and discard much of the interesting time-dependent information.

9.5.2 Measurement of System Jitter

As a first step toward characterization, the overall timing jitter (i.e., the locking accuracy) between the two lasers must be measured. This is essentially a test of the PLL circuit's ability to maintain synchronization on the slow time scale ($>$1 s) and of the intrinsic laser timing fluctuations on a faster time scale ($<$100 ms). These measurements are performed to high accuracy using optical cross-correlation in a similar fashion to the timing calibration method described earlier. With the feedback circuit enabled and the lasers locked, the relative time delay between the lasers is scanned repetitively over a predetermined time delay (in this case, approximately 100 ps at a scan rate of 102 Hz). Timing calibration is obtained by inserting an etalon into one arm of the correlator to give a sequence of pulses with a known temporal spacing. In each successive scan, the cross-correlation peak (and its replicas from the etalon) appears at a slightly different time delay due to the jitter between the lasers. The jitter is measured as the fluctuation in time delay between the first cross-correlation pulse of each scan and the rising edge of the scan clock (i.e., the TTL pulses from the function generator that drives the scanning delay). The peak delays of the correlation are sampled to form a time series of the jitter data. This is different from previous cross-correlation methods in which the data are averaged, [15] or are acquired by using very slow scanning, which itself gives an averaging effect. In the rapid scanning method, statistics (primarily standard deviation) are generated from this time series of peak arrival times. These measurements have revealed an RMS timing jitter of 5.4 ps over a measurement period of 10 min. For the 4.629 MHz repetition rate (216 ns repetition period), this corresponds to a phase accuracy of one part in 40,000, or 30 arc-seconds [34]. Because this method gives a time series of the jitter, one can also easily Fourier transform the data to five spectral characteristics of the jitter noise. This is useful in revealing, for example, resonant frequencies in the entire feedback loop, so that the circuit characteristics can be tailored properly. In the present case, the measured jitter represents the limitations of the particular PLL circuit that was used for stabilization. Commercial versions of modelocked laser synchronization systems are capable of reducing the relative timing jitter to ~1–2 ps.

This rapid scanning cross-correlation method has distinct advantages and is the most accurate method for measuring low-frequency (<1000 Hz) timing jitter. Unlike the established RF spectrum analyzer techniques [35], no assumptions need to be made about whether the timing noise is random or has some specific, systematic time dependence. Any combination of random and/or systematic timing noise can be measured in this way as long as the data are sampled at a sufficiently high rate. Also, although the RF techniques are commonly used to measure the jitter of a single laser, they cannot easily be adapted to measure the relative timing jitter between two separate lasers. Additionally, the time series of the jitter gives valuable information about the sources of the timing jitter. This information is lost in the slow scanning and averaging methods.

9.5.3 Method for Measuring Laser Jitter

As mentioned at the beginning of this section, the PLL jitter does not represent the limit of timing accuracy attainable with the SDLS method. Rather, it is the intrinsic timing jitter of the lasers themselves, and there is a method for measuring this critical parameter. The timing jitter measurement method is a simplified variant of the SDLS method. It uses the same timing calibration apparatus as that shown in Figure 8 but without the application experiment. Additionally, the PLL is disabled and the lasers are allowed to drift (in timing) with respect to each other. A small cavity length mismatch is intentionally introduced, which causes the lasers to scan through each other at a given offset frequency (typically $\Delta f \sim$ 5–50 Hz). As the lasers sweep through each other, time-delay calibration is obtained from the same cross-correlator with the Fabry–Perot etalon inserted into one arm. Again, in this case, a 5 mm etalon gives pulses every 50 ps. This is essentially asynchronous sampling of the scan trajectory between two free-running lasers. Figure 18a shows a waterfall plot of 20 consecutive cross-correlation sweeps (5000 points each) at a rate of 5 scans/s, for a total acquisition time of 4 s. The data acquisition is triggered by the first peak of each sweep. In Figure

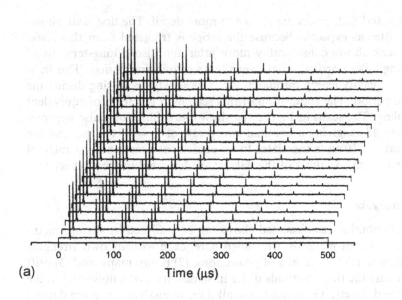

(a)

Time (μs)

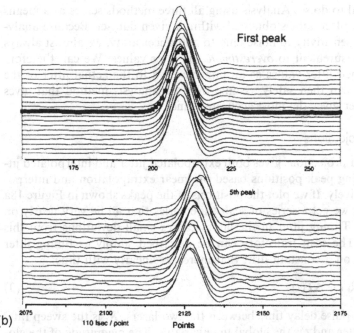

First peak

5th peak

110 fsec / point

Points

(b)

Figure 18 (a) Cascade plot of 20 consecutive cross-correlation scans acquired in 4 s during free-running operation (PLL disabled) between the two lasers set to have an offset frequency of 5 Hz. (b) Detail of first and fifth peaks for 20 scans. Drift and jitter are apparent for the set of fifth peaks.

18b the first and fifth peaks are shown in more detail. The first peak shows very little jitter as expected because the scope is triggered from this peak. The fifth peak shows considerably more jitter and also a long-term trend of increasingly later arrival times over the 4 s acquisition period. This indicates that the cavity length mismatch, ΔL, was slowly decreasing during the acquisition period. The acquisition of this scan data is a form of equivalent time sampling. The scans in Figure 18a show one calibration pulse approximately every 48 µs, giving a nominal sweep rate of $R_s = 1.04$ ps/µs, and the sampling rate of 10 ms/s translates to an equivalent time-sampling interval of 104 fs per point. Analysis of the datasets is discussed in the next section.

9.5.4 Analysis

The datasets obtained as discussed above can be analyzed by various methods. Three different methods are shown here, each with its own strengths and weaknesses: (1) the method of projections; (2) return maps; and (3) drift fitting. Because the three methods differ in sensitivity to the noise and to the nature of the datasets, it is useful to use all three to analyze any given dataset if it is practical to do so. Analysis using all three methods serves as a means to cross-check jitter values obtained within a given dataset. Because analytical errors or sensitivity to noise tend to add uncertainty, we almost always expect the measurement to *overestimate* the jitter values. We can therefore use the obtained jitter values as an upper bound to the actual jitter of the laser, and we can then safely assume that the analysis method that gives the lowest jitter gives the most accurate value.

Method of Projections

The method of projections gives both extrapolated jitter and interpolated jitter by predicting peak positions based on linear extrapolation and interpolation, respectively. If we plot the positions of the peaks shown in Figure 18a for each scan, we get a series of scan trajectories that all lie very nearly on straight lines. Timing jitter in the lasers causes small deviations from this straight line. The scan trajectory can be described simply as timing jitter superimposed on a constant drift that increases linearly with time:

$$T(t) = R_s t + \Delta T(t) \tag{9}$$

where T is the pulse delay time between the two lasers, R_s is the sweep rate, $\Delta T(t)$ is the jitter and t is the global time variable. The magnitude of the global time scale varies from microseconds to seconds depending upon the sweep rate, which is set to any desired value by adjusting the cavity length difference. Although we tend to treat the jitter as a small quantity for short times, it should be kept in mind that for passively modelocked lasers, $\Delta T(t)$ increases without bound for large times. The jitter term can be found by subtracting the constant drift term, which in turn is estimated by plotting a line through any two points of the trajectory and extending it through the time range of the remaining points. This projected (extrapolated or interpolated) trajectory provides an estimate of where the subsequent or intervening peaks should occur in the absence of timing jitter. The deviations from this projection comprise a "jitter trajectory."

A good estimate of the jitter requires a reasonably large number of scans so that statistics can be accumulated. Figure 19a shows the jitter trajectories for the 20 consecutive scans, where we have used projections between the zeroth and second points (a two-point projection) and between the zeroth and third points (a three-point projection). This collection of jitter trajectories (referred to here as a "jitter tree") is analyzed statistically to give the time dependence of the jitter. At each point in time (i.e., every 48 μs)

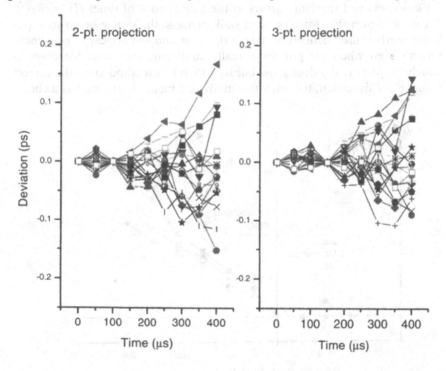

Figure 19 Jitter trees generated from two- and three-point projections. A two-point projection has one data point between the two "squeezing points," whereas a three-point projection has two intervening data points.

we calculate the spread (i.e., standard deviation) of the points of the jitter tree at that time.

The amounts of jitter calculated from the standard deviations of the jitter trees are given in Figure 20. For the dataset shown, the jitter growth is found to be linear in time for all the degrees of projections ($m = 1, \ldots, 4$). One might expect that the projected jitter would be less as the order m increases. The jitter from the two-point projection is slightly lower than that from the one-point projection, as expected. However, the three- and four-point jitter is actually slightly greater than the two-point, although the difference is quite small. The smallest jitter growth rate is for the two-point projection, which gives a jitter growth rate of $R_J = 240\,\text{fs/ms}$ or 0.24 ppb.

For all orders $m = 1, \ldots, 4$, the jitter grows linearly with time on this time scale. This is significant because it differs from the theoretically expected time dependence, which predicts that jitter growth is a random walk process and therefore grows as the square root of time, $J(t) = R_J\sqrt{t}$. It can be shown that for a random walk process, the simple linear extrapolation method used here *overestimates* the jitter and will appear to give linear growth even when the growth is really sublinear. However, for cases in which the jitter really does grow linearly, then this method gives the correct value. For this reason, the other two analytical methods are used as a check.

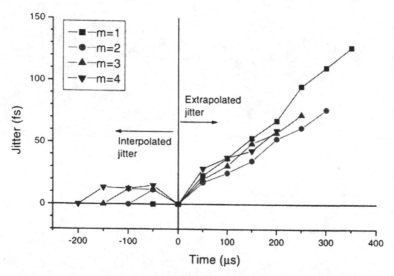

Figure 20 Timing jitter vs. time calculated from projections for $m = 1, 2, 3, 4$. The data plotted at negative times (i.e., between the two squeezing points) is used to obtain estimates for interpolated jitter between two timing markers.

Additional information can be obtained from the jitter trees, namely the *interpolated* jitter values. We wish to estimate the timing jitter that occurs *between* the timing markers. This is estimated by the standard deviation of the data points between the "squeezed" points (at times T_1 and T_N) of the jitter trees, as illustrated in Figure 20 for negative times. It can be seen that the jitter is zero near the two squeezed points (by definition) and is largest at the midpoint of the interval. In reality, of course, near the markers the delay is known to within the fitting accuracy of that peak, which is better than ±5 fs for the first five peaks. For any time between the markers the time delay is known less accurately, with the greatest uncertainty occurring at the midpoint between the markers. The maximum jitter at this midpoint is approximately one-fourth of the timing jitter expected at T_N when using N-point projections. For example, consider the data plotted for the four-point projection in the interval $-200\,\mu s < T < 0$. The maximum interpolated jitter between the squeezed points occurs at the midpoint between them and is about 15 fs, which is about one-fourth of the jitter measured at $T = 200\,\mu s$ (as plotted for $T > 0$). This provides guidelines for using a scanning dual laser system or other type of asynchronous measurement system. If a given accuracy is required for a time-resolved measurement, then the expected jitter interpolated between the timing markers helps to determine the required density of timing markers. In this case, a marker spacing of 200 µs gives timing accuracy of 15 fs or better at this scan speed. Extrapolating a bit, a marker spacing of 1 ms gives timing of ±60 fs or better.

Return Maps

The second method of analysis employs return maps. Return maps are an important class of embedding techniques that find much use in time series analysis and studies of nonlinear dynamics and chaos. To construct a return map from a time series $X = x_1, x_2, \ldots, x_n, \ldots$, we formulate a new series $Q(X) = q_1, q_2, \ldots, q_n$ from the series X. The quantity q_n can be any function of the elements x_n. There is an almost endless variety of return maps that can be formed from these time series data. However, we try to use the simplest ones possible. In the simplest case, $q_n = x_n$. The next level of complexity would be to define, for example, $q_n = x_n - x_{n-1}$, and so on. Then we plot the elements of the series as a function of the previous element, i.e., we plot q_n vs. q_{n-1}. The resulting scatter plot gives useful graphical information about the original time series, and it can then be further analyzed mathematically to yield quantities of interest.

In our particular case, return maps provide a method of analysis that avoids the overestimate errors of the projection method under the special condition that we use *consecutive intervals of equal duration*. This method also works well for datasets that have very few scans, in which case it is necessary to extract jitter information from individual scans. As an example, in a subsequent measurement of the jitter over longer time intervals, only four scans were acquired due to memory limitations. This would give only four jitter trajectories in a jitter tree, giving unreliable statistics. However, the timing information available *within a single scan* can be used by making return maps of the time intervals, Δ_K^M, between the peaks and then measuring the spread of the data. Using the data shown in Figure 18a, we form return maps of Δ_{k+1}^1 vs. Δ_k^1 and Δ_{k+2}^2 vs. $\Delta_k^2, \ldots, \Delta_{k+m}^m$ vs. Δ_k^m, etc., where the intervals Δ_k^m are defined as illustrated in Figure 21a. The index m indicates the order of the interval, i.e., the number of peaks skipped between the endpoints of the interval. The subscript k is the index of the pulse at the start of the interval in question. The return maps shown in Figure 21b are essentially scatter plots of these interval values. The scatter plots are then fitted with a straight line through the origin, and the standard deviation of the point from this line gives the timing jitter. The same analysis that shows that the projection method overestimates the jitter also shows that we can obtain the correct jitter values provided that the intervals plotted on the ordinate and abscissa of the return maps are consecutive end-to-end intervals.

Drift Fitting

The previous two methods treat each scan independently, followed by statistical analysis of the group. However, it is also possible to obtain estimates of the jitter by processing the peaks *collectively* without considering the dynamics occurring within individual scans. The key to this method is that one must be able to properly account for long-term trends in peak arrival times. Equation (9) assumes that the scan rate, R_S, is constant over time, which is a reasonably accurate assumption within the time frame of an individual scan. However, there are long-term drifts in the offset frequency (and thus the sweep rate) that must be considered. The previous two methods are not degraded by long-term variations, because each scan is treated independently. However, if the long-term variations occur on a slower time scale than the offset frequency, then they can also be calculated and nulled out, and the jitter can be extracted. Figure 18b shows the zeroth and fourth peaks of 20 consecutive scans accumulated over a 4 s time period. The zeroth peak shows very little (± 30 fs), as expected because the oscilloscope is triggered from this peak. (The 30 fs jitter is attributable to the combination of simple threshold triggering and the equivalent sampling interval of 104 fs/point.) The fourth peak, however, shows considerably more jitter and also reveals a long-term trend of increasingly later arrival times, indicating that

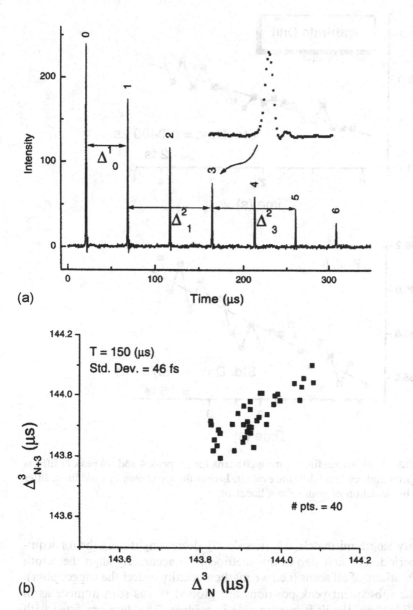

Figure 21 (a) Definition of time delay intervals Δ_k^m for cross-correlation peak data. For accurate results for a random walk process, return maps must be constructed from adjacent equal-duration time intervals such as Δ_1^2 and Δ_3^2 as shown. (b) Return maps of peak arrival times for time intervals of 150 μs.

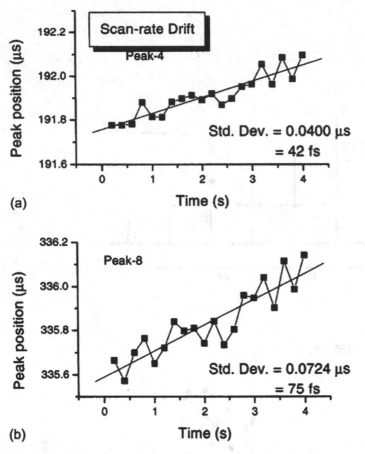

(a)

(b)

Figure 22 Peak arrival times during 20 scans for (a) peak 4 and (b) peak 8. Effects of both jitter and scan rate drift are evident. Linear fits are shown by solid lines. Jitter is given by deviation of points from linear fit.

the cavity length mismatch. ΔL, was slowly decreasing during the 4 s acquisition period. The first step in this method is to accurately align the zeroth timing markers of all scans (i.e., we mathematically cancel the trigger jitter). Then the subsequent peak positions are plotted versus scan number as in Figure 22. Here, the drift in scan rate is evident. The data are fitted with a straight line, and the standard deviation of the data points from the linear fit gives another estimate of the timing jitter.

Under some conditions, a linear fit to the thermal drift is not sufficient. If the dataset exhibits a long-term drift that is very nonlinear during the time of acquisition (say, over approximately 5 s), then fitting these data with a line will obviously result in an overestimate of the jitter. Fitting using higher order polynomials gives the correct estimate of jitter in these cases, but only with a judicious choice of the polynomial order. There is a point where the fit goes beyond getting rid of slow trends and begins to cancel out relevant jitter information. If the polynomial order is set too high, the system will overfit the data, thus artificially removing some of the jitter and resulting in an *underestimate*. A seventh-order fit is appropriate for typical datasets, although this is dependent on the measurement conditions.

The jitter estimates from all three analytical methods (jitter trees, return maps, and drift fitting) are plotted in Figure 23. Note that the three methods give roughly equivalent estimates. Drift fitting gives the lowest jitter estimate although it agrees closely with the lowest value obtained by projections. A more detailed comparison follows in a subsequent section.

9.5.5 Factors Affecting Accuracy

The timing accuracy and sensitivity of the jitter measurement (as well as timing calibration) method are determined by a number of factors including the signal-to-noise ratio, the cross-correlation width (~1 ps), the equivalent sampling interval (10 Mega samples/s, or approximately

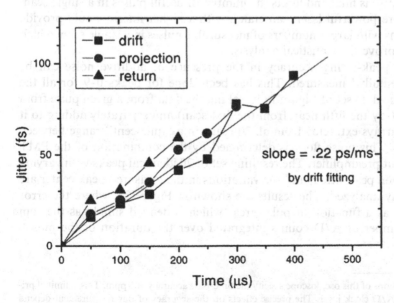

Figure 23 Comparison of jitter vs. time obtained by three different analytical methods (projections, return maps, and drift fitting) using the same dataset.

104 fs/point), the signal digitization (8 bits), the time base accuracy of the oscilloscope, and, to a lesser extent, the timing jitter itself.

As one category of potential sources of error, we consider possible distortions to the time scale of the sampled waveform due to analog-to-digital (A/D) clock jitter in the oscilloscope, which is itself a form of time base distortion (TBD). Typical digitizing oscilloscopes have a specified time base accuracy of 10 ppm of the full range. Because the full time base range of 500 μs is equivalent to 500 ps, the time base error is less than 5 fs across the entire scan. This error does not apply to the zeroth peak, but presumably it does apply to the other peaks. It represents a significant error for the early pulses (1– 3) in this case. This error can be minimized by using a thinner etalon and adjusting the time base to a faster sweep speed. Alternatively, one can use an oscilloscope with a more accurate time base.*

Another source of error occurs in the peak fitting. The relevant factors for this type of data are the additive noise and the dynamic range of the A/D conversion. Because of the pulse ringdown in the calibration etalon, the ninth pulse intensity is about one-tenth as strong as that of the zeroth pulse. With the 8-bit A/D converter of the oscilloscope providing 256 counts at full scale, the first peak is set to about 200 counts to avoid saturation of the input. Under these conditions, the tenth peak is only about 20 counts maximum. This can potentially degrade the fitting accuracy even if the S/N ratio is high and limits the number of useful pulses in a single scan. Using more powerful lasers, an etalon with a higher finesse would provide pulse trains with larger numbers of measurable pulses in a single scan, which would improve the statistical analysis.

The peak-fitting accuracy in the presence of additive noise can be "experimentally" measured. This has been done for peaks 0–9 for all the scans in the data set of Figure 18 by taking the data from a given pulse from one scan (say the fifth peak from the first scan) and separately adding to it 20 noise arrays extracted from all 20 scans in a "quiescent" range between the peaks. This noise floor is determined by the combination of the PMT and current preamplifier. The resulting set of 20 identical peaks with varying noise is then peak-fitted, and the variations in the measured peak center are statistically analyzed. The results are shown in Figure 24, where the error is plotted as a function of pulse area (which is defined simply as the sum of the number of A/D counts integrated over the duration of the pulse).

---

*Newer versions of this oscilloscope specify the time-base accuracy at 3 ppm. This is limited primarily by A/D clock jitter. The precise effects on the accuracy of our measurement depend upon the frequency-dependent statistics of the clock jitter, which are not available. The 5 fs error contribution is a worst-case estimate based on the 10 ppm specification.

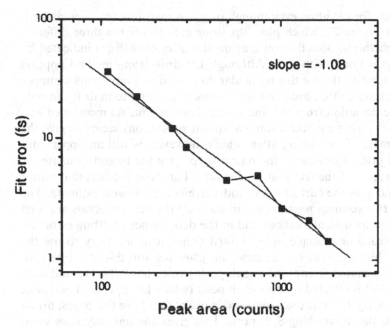

Figure 24   Peak-fitting error versus peak area (in A/D counts) for cross-correlation peaks in the presence of additive noise.

As expected, the fit error becomes larger as the peak height and are a get smaller. The location error of the first five peaks is less than 5 fs. Clearly, the pulse area and S/N ratio are important factors determining this accuracy. Other, lesser factors include the 8-bit digitization and the timing jitter of the laser itself. However, these effects give errors of only about 2 fs or less.

The information on peak-fitting accuracy provides a check on the jitter results and helps in selecting the proper peaks and intervals for measuring the jitter without compromising the accuracy of the measurement. This is discussed in the next section.

9.5.6   Comparison of Analytical Methods

The three analytical methods have different strengths and weaknesses based on how they treat the data and on how sensitive they are to peak-fitting errors. It has already been pointed out that return maps and drift fitting do not suffer from the overestimate effect that occurs with the projection method. On the other hand, the drift-fitting method is more sensitive to long-term drift effects that are not a problem for the projection or return map method. Not surprisingly, the three methods do not give exactly the same answers for the jitter, even though they are analyzing the same dataset. Looking at Figure 23, which plots the jitter growth for the three different analytical methods, note that once again the jitter growth (as indicated by drift fitting) is linear with time. Although the drift-fitting method appears to give the lowest jitter for this particular dataset, it still represents an upper limit to the timing jitter, because fluctuations in the long-term drift can only increase the deviation from the linear or polynomial fit. As mentioned earlier, the drift-fitting method (using two-point projection) seems to give the lowest estimate of the timing jitter, which, however, is still an upper limit on the real jitter. Conversely, the return maps give the largest estimate for the jitter in spite of the fact that mathematical analysis predicts that return maps should give the correct value and therefore the lowest estimate. The reason for this seeming paradox lies in the way the peak intervals are used in the various analytical methods and in the dependence of fitting errors on the peak area. For example, using return maps, it is necessary to use the eighth and higher peaks to calculate the jitter for fourth-order intervals, so that the estimated jitter at the fourth peak (which has $S/N = 63$ and peak fit error of 4 fs) is affected by the eighth peak (which has $S/N = 14$ and peak fit error of 36 fs). But it is these high-order peaks that have the largest fitting error due to the diminishing S/N ratio. This gives anomalously large error contributions (and therefore a larger estimate for jitter) for the low-order peaks.

The projection method suffers from the overestimate deficiencies discussed earlier in Section 9.5.4. The fact that it gives a lower estimate than the return maps (in spite of the mathematical deficiencies) demonstrates the importance of the peak-fitting errors. These can be reduced by increasing the S/N ratio of all the peaks and by making the peak heights more uniform. This may be done, for example, by using an etalon with higher finesse. The drift-fitting method suffers the least from peak-fitting errors. Returning to our example, the fourth peaks effectively form their own dataset, and therefore the calculated jitter value does not depend quite as strongly upon the fitting or errors associated with any higher order peaks such as the sixth- or eighth-order peaks. For this reason, the drift fitting gives the lowest jitter value and lowest error. Even if the drift in scan rate has some short-term ($\sim$5 Hz) fluctuations, they can be mathematically cancelled by high-order polynomial fitting. However, as can be seen in the example above, the peak timing is relatively well behaved over the acquisition period, resulting in an estimate that agrees well with the other methods. The key to ensuring the accuracy of this method is to sweep the lasers at a high enough offset frequency that long-term fluctuations in the sweep rate are small from one scan to the next. This dataset was acquired at a rate of 5 scans/s, so even 1 Hz fluctuations might affect the accuracy.

Unlike the method of projections, the accuracy of the drift-fitting method should not depend upon the functional dependence of the jitter growth, i.e., whether it grows linearly or sublinearly. It should give the true functional dependence of $J(T)$ in any case as long as the drift is slow compared with the offset frequency as in the present case. Likewise, the return map method theoretically gives the correct jitter estimate for a random walk where $J(T) \propto T^{1/2}$. The data for all three methods, however, still show a largely linear dependence on time, indicating that the jitter really does grow linearly over this time range.

9.6 DISCUSSION

9.6.1 Issues Affecting System Time Resolution

So far we have concentrated primarily on laser and system timing jitter as being the main limiting factors to the performance of scanning laser systems, and this is justified to a large extent. However, there are a number of other effects and considerations that play a part in the overall system performance. It is important to remember that the metrics for system performance include not only timing accuracy but also scan frequency, scan range, slew rate, temporal dynamic range, and dynamic adjustability range. These other factors include such effects as linearity (or nonlinearity) of the scan trajectory, slew rate vs. jitter growth, electronic noise in the PLL circuit, and laser pulse width, among others. These come into play in sometimes subtle ways depending upon the mode of operation, whether it be first-point correction, repetitive scan calibration, or asynchronous (free-running). In the following, we discuss how these various phenomena affect the system performance in its various modes of operation.

In any of the three modes of operation, higher accuracy is obtained by scanning as rapidly as possible. As pointed out in the introductory sections, rapid scanning takes the data acquisition frequency range out of the baseband noise of the laser. For rapid mechanical scanning (using a single laser) this reduces the effects of the baseband amplitude noise, while timing jitter of the laser is inconsequential. With electronically synchronized laser systems, however, laser timing jitter has a significant effect, and rapid scanning is necessary to avoid degradation of time resolution. Ultimately, if the scanning is slowed down to a very low slew rate, then the timing accuracy one could expect is simply equal to the overall system jitter, which is determined by the PLL circuit. In most cases, this is on the order of a few picoseconds.

9.6.2   Performance Optimization

Temporal Dynamic Adjustability Range

The dynamic adjustability range is basically the number of decades over which one can adjust the temporal scanning range. It is defined as the maximum/minimum ratio of the useful scan ranges. For either of the first two operating modes of the SDLS (first-point or repetitive calibration), the temporal dynamic range is limited primarily by overall system jitter. This is because the system jitter determines the minimum useful scan range. For example, with a typical PLL jitter of 2 ps, the minimum useful scan range can be no less than 10 ps. So with a maximum scan range of 10 ns, the dynamic adjustability range is about 10^3.

The SDLS scanning method provides great timing flexibility, not only in the scan range and frequency but also in the delay range. Just as some oscilloscopes have delaying time bases, the SDLS system can, for example, scan any 1 ns subinterval of the $T_0 = 217$ ns pulse repetition period. This is possible because the PLL circuit uses phase detection at the fundamental repetition frequency $\nu_1 = 4.6$ MHz rather than a higher harmonic. In order to provide the smallest possible timing jitter between two lasers, previous stabilization techniques have employed phase detection at high harmonics of the laser repetition frequency [36]. However, the calibration technique described in Section 9.3.2 successfully compensates for the somewhat larger overall timing jitter, as demonstrated by the fact that $\sim$30 fs accuracy could be obtained (with a third-order fit) even though the overall timing jitter between the two lasers was several picoseconds from scan to scan. The high-harmonic stabilization method also has the drawback that the adjustability of T_D is limited to a small subinterval of the laser repetition period, T_0, and in addition it has multiple locking points within the repetition period, which is also undesirable. In the present system, using phase detection at the fundamental, ν_1, it is possible in principle to slew the relative timing delay electronically over the full 216 ns timing interval if desired.

Timing Accuracy and Temporal Dynamic Range

By temporal dynamic range, we mean the ratio between the scan range and the timing accuracy. This value can be quite high. For example, when using repetitive scan correction, the timing accuracy is found to be 30 fs for a 300 ps scan range, giving a temporal dynamic range of 10^4. Also, for free-scanning lasers, values of up to 10^5 have been experimentally measured. The difference between these results is related to the various contributions alluded to in the previous section—in this case, electronic noise from the PLL circuit.

When using first-point correction (the simplest and fastest method for averaging), the overall jitter of the PLL is effectively canceled by the DAQ triggering scheme. However, this works very well only if the scan trajectory is linear. If it is nonlinear, then the overall jitter is canceled only at the beginning of the scan in the vicinity of the triggering point (a series of such nonlinear scan trajectories was shown in Figure 11). But in the middle to latter part of the scan, the nonlinearity of the trajectory works in combination with overall jitter to give a substantial degree of apparent scan-to-scan jitter, thus degrading the time resolution in this area of the scan. Thus, scan linearity is a key factor in optimizing the accuracy when using first-point correction. Using the repetitive scan calibration method described in Section 9.3.2, it is possible to obtain a relative timing jitter of less than one part in 10^4 by using a high-order polynomial correction on each scan. Overall timing jitter, as limited by the PLL, is effectively canceled even though the scan trajectory is nonlinear. Thus, scan nonlinearity becomes much less of an issue when using repetitive high-order scan correction.

During first-point correction mode, assuming that scan linearity holds, there is still the issue of intrinsic laser timing jitter, which again degrades the latter part of the scan. There is a very simple and useful way of thinking about these timing and jitter noise effects. As can be seen in 17c, the RMS jitter with respect to the first correlation peak (i.e., with one-point correction) grows linearly with time (at some point, of course, it saturates, but we consider only the short time range shown). We define an average "jitter growth rate," R_J, given by the slope of this line with respect to the "laboratory" time axis. In the case of 17c, the jitter growth rate is $R_J = 300\,\text{fs}/300\,\mu\text{s} = 10^{-9}$. Alternatively, we can also define the "fractional jitter-to-scan ratio," $\rho_{JS} \equiv R_J/R_{\text{scan}}$, which is given by the slope of the line with respect to the time delay axis in Figure 17c, where we find $\rho_{JS} = 6 \times 10^{-3}$. This figure of merit gives the expected fractional timing accuracy that can be obtained for a given R_J and R_S. This information can be used in a very simple way: Once R_J has been measured for a pair of lasers, one can then specify the minimum scan rate that is required to give a certain degree of timing accuracy. For example, if we desire a fractional timing accuracy of $\rho_{JS} = 10^{-3}$, then the scan rate must be 1000 times the jitter growth rate, requiring a minimum scan rate of $R_S = 10^{-6}$, which is six times as large as the value of $R_S = 17 \times 10^{-9}$ that was used in the measurement. Note that the jitter growth has contributions not only from the laser itself but also from such external factors as spurious electrical noise, vibrations, and air turbulence. In fact, the timing jitter is particularly sensitive to these factors—much more so than is amplitude noise. It is therefore imperative to isolate the lasers from the external environment as much as possible. Intrinsic laser jitter is ultimately limited by quantum-mechanical effects [37].

Another factor not yet mentioned is the laser pulse width. At first glance it would appear that the shorter the laser pulse, the higher the accuracy. But this is not necessarily true for the SDLS or asynchronous systems. This has to do with the calibration method being used. To illustrate, if the cavity length mismatch of the two lasers is set to be $\Delta L = 15\mu m$ then the temporal grid (or equivalently, the pulse timing advance per round-trip) is $\delta_{t_G} = 100\,fs$. If the laser pulse width is $100\,fs$, then only a few laser shots will contribute to the measured cross-correlation pulse. A Gaussian fit to these few points cannot be expected to give highly accurate results. However, if the pulse width is $1\,ps$, then 20–30 pulses will make up the correlation signal, giving much better statistics to give a more accurate fit.

The dynamic range of the SDLS can be pushed beyond the 10^4 level stated earlier. A comparison of the jitter during scanning and free running shows that there is increased jitter during scanning compared with two free-running lasers. This is probably due to electronic noise from the PLL that gets applied to the PZT. A test of the "ultimate" timing accuracy can be done by isolating the laser jitter from other external effects and fitting the scan trajectories of two free-running lasers without any voltage applied to the PZT (note that this is somewhat different from the three jitter analysis methods described in Sec. 9.5.4). Under these conditions, the only deviation from linearity is due to the intrinsic laser timing jitter [38]. This measurement (when using a $500\,ps$ etalon spacing) gives an RMS residual from the scan trajectory of $50\,fs$ over a $5\,ns$ scan range, giving a temporal dynamic range of 10^5.

9.7 SUMMARY

In this chapter we have described a new scanning technique (the scanning dual laser system, or SDLS) for ultrafast lasers. This method makes it possible to achieve simultaneously both high-frequency scanning and adjustable scan range of up to several nanoseconds. Although the SDLS method was demonstrated using a pair of passively modelocked fiber lasers, it is a more general method and is applicable to many types of modelocked lasers including solid-state and external cavity diode lasers. To increase the applicability of the SDLS method to many more applications, it is feasible to implement it using femtosecond fiber lasers with average powers of hundreds of milliwatts [39]. It should also be possible (although somewhat more difficult) to implement SDLS with actively modelocked lasers as well. The utility of SDLS as a measurement system has been demonstrated by measuring the carrier relaxation times of InGaAs samples. The temporal dynamic range of this particular measurement spanned four orders of magnitude

(1 ps to 10 ns). With better time resolution, this could be increased to six orders of magnitude.

SDLS has a number of features that give it advantages over mechanical scanning or free-scanning laser systems: high scan frequency, long scan range, selectivity of scan parameters, high accuracy (presently 30 fs), large temporal dynamic range, high scan velocities, large time delays in a small package (up to 200 ns), no misalignment of beam or defocusing effects for large delays, and little mechanical vibration. With these advantages and features, the SDLS technique should give ultrafast laser systems the kind of timing flexibility and controllability that is expected of oscilloscopes and other practical instruments. This type of system could be the "engine" in a variety of ultrafast measurement systems, and many applications stand to benefit. Such applications include high-speed electronic measurement (such as electro-optic or photoconductive sampling), terahertz generation and detection, terahertz spectroscopy, ultrafast analog-to-digital converters, metrology, fluorescence lifetime measurement applications, wafer characterization, and telecommunications applications such as component testing or optical waveform measurement. Many other applications are sure to become important, most of which we cannot foresee at this time. But SDLS based on ultrafast fiber lasers stands to play an important role in bringing these applications out of the laboratory and into the commercial arena.

REFERENCES

1. JG Fujimoto, S De Silvestri, EP Ippen, CA Puliafito, R Margolis, A Oseroff. Femtosecond optical ranging in biological systems. Opt Lett 11:150–152 (Mar 1986).
2. Y Chen, S Williamson, T Brock. 1.9 Picosecond high-sensitivity sampling optical temporal analyzer. Appl Phys Lett 64(5):551–553 (1994).
3. Q Wu, X-C Zhang Ultrafast electro-optic field sensors. Appl Phys Lett 68(21):1604–1606 (1996).
4. BB Hu, MC Nuss. Imaging with terahertz waves. Opt Lett 20:1716–1718 (1995).
5. RF Fork, FA Beisser. Real-time intensity autocorrelation interferometer. Appl Opt 17:3534–3535 (1978).
6. ZA Yasa, NM Amer. A rapid scanning autocorrelation scheme for continuous monitoring of picosecond laser pulses. Opt Commun 36:406–408 (1981).
7. DC Edelstein, RB Romney, M Scheuermann. Rapid programmable 300 ps optical delay scanner and signal-averaging system for ultrafast measurements. Rev Sci Instrum 62:579–583 (1990).
8. KF Kwong, D Yankelevich, KC Chu, JP Heritage, A Dienes. 400-Hz mechanical scanning optical delay line. Opt Lett 18(7):558–560 (1993).

9. KC Chu, K Liu, JP Heritage, A Dienes. Scanning femtosecond optical delay with 1000 × pulse width excursion. Conference on Lasers and Electro-Optics. OSA Tech Dig Ser, Vol 8. Washington, DC: Opt Sco Am, 1994, Paper CThI23.

10. W Kütt, W Albrecht, H Kurz. Generation of coherent phonons in condensed media. IEEE J Quantum Electron QE-28:2434–2444 (1992).

11. KS Giboney, ST Allen, MJW Rodwell, JE Bowers. Picosecond measurements by free-running electro-optic sampling. IEEE Photon Tech Lett 6:1353–1355 (Nov 1994).

12. JD Kafka, JW Pieterse, ML Watts. Two-color subpicosecond optical sampling technique. Opt Lett 17:1286–1288 (Sept 15, 1992).

13. MH Ober, G Sucha, ME Fermann. Controllable dual-wavelength operation of a femtosecond neodymium fiber laser. Opt Lett 20:195–197 (Jan 15, 1995).

14. A Black, RB Apte, DM Bloom. High-speed signal averaging system for periodic signals. Rev Sci Instrum 63:3191–3195 (1992).

15. DE Spence, WE Sleat, JM Evans, W Sibbett, JD Kafka. Time synchronization measurements between two self-modelocked Ti:sapphire lasers. Opt Commun 101:286–296 (Aug 15, 1993).

16. R Payaket, S Hunter, JE Ford, S Esener. Programmable ultrashort optical pulse delay using an acousto-optic deflector. Appl Opt 34(8):1445–1453, (1995).

17. MJW Rodwell, DM Bloom, KJ Weingarten. Subpicosecond laser timing stabilization. IEEE J Quantum Electron QE-25:817–827 (Apr 1989).

18. JM Evans, DE Spence, D Burns, W Sibbett. Dual-wavelength self-modelocked Ti:sapphire lasers. Opt Lett 13:1074–1077 (Jul 1, 1993).

19. MRX de Barros, PC Becker. Two-color synchronously mode-locked femtosecond Ti:sapphire laser. Opt Lett 18:631–633 (Apr 15, 1993).

20. DR Dykaar, SB Darak. Sticky pulses: two-color cross-mode-locked femtosecond operation of a single Ti:sapphire laser. Opt Lett. 18:634–637 (Apr 15, 1993).

21. Z Zhang, T Yagi. Dual-wavelength synchronous operation of a mode-locked Ti:sapphire laser based on self-spectrum splitting. Opt Lett 18:2126–2128 (Dec. 15, 1993).

22. DA Pattison, PN Kean, JWD Gray, I Bennion, NJ Doran. Actively mode-locked dual-wavelength fiber laser with ultra-low inter-pulse-stream timing jitter. IEEE Photon Tech Lett 7:1415–1417 (Dec. 1995).

23. SP Dijaili, JS Smith, A Dienes. Timing synchronization of a passively mode-locked dye laser using a pulsed optical phase locked loop. Appl Phys Lett 55:418–420 (July 1989).

24. Spectra Physics Lok-to-Clock™ system, Spectra-Physics Lasers, Inc., Mountain View, CA.

25. G Sucha, ME Fermann, DJ Harter, M Hofer. A new method for rapid temporal scanning of ultrafast lasers. IEEE-JSTQE 2:605–621 (1996).

26. G Sucha, M Hofer, ME Fermann, F Haberl, D Harter. Synchronization of environmentally coupled, passively modelocked fiber lasers. Opt Lett 21:1570–1572 (1996).

27. ME Fermann, LM Yang, ML Stock, MJ Andrejco. Environmentally stable Kerr-type mode-locked erbium fiber laser producing 360-fs pulses. Opt Lett 19:43–45 (1994).
28. CM Wang, PD Hale, KJ Coakley, Least-squares estimation of time-base distortion of sampling oscilloscopes. IEEE Trans Instrum Measure 48:1324–1332 (1999).
29. WL Gans. The measurement and deconvolution of time jitter in equivalent-time waveform samplers. IEEE Trans Instrum Measure IM-32:126–133 (1983).
30. Y Rolain, J Schoukens, G Vandersteen, Signal reconstruction for non equidistant finite length sample sets: a "KIS" approach. IEEE Trans Instrum Measure 47:1046–1052 (1998).
31. S Namiki, CX Yu, HA Haus. Observation of nearly quantum-limited timing jitter in an all-fiber ring laser. J Opt Soc Am B 13:2817 (1996).
32. G Sucha, ME Fermann, D Harter, M Hofer. A new method for rapid temporal scanning of ultrafast lasers. IEEE J-STQE 2:605 (1996).
33. G Sucha, ME Fermann, DJ Harter. Time-domain jitter measurement of mode-locked fiber lasers. In: T Elsaessar, JG Fujimoto, DA Wiersma, W Zinth, eds. Ultrafast Phenomena XI, Vol 63. Berlin: Springer-Verlag, 1998.
34. G Sucha, M Hofer, ME Fermann, D Harter. Synchronization of environmentally coupled, passively modelocked fiber lasers. Opt Lett 21:1570–1572 (1996).
35. D von der Linde. Characterization of the noise in continuously operating mode-locked lasers. Appl Phys B 39:201–217 (1986).
36. DE Spence, JM Dudley, K Lamb, WE Sleat, W Sibbett. Opt Lett 19:481 (1994).
37. HA Haus, A Mecozzi. Noise of mode-locked lasers. IEEE J Quantum Electron QE-29:983 (1993).
38. G Sucha, Noise considerations for ultrafast measurements. Proc LEOS Annual Meeting, Orlando, FL 1998, Paper WJ5.
39. JD Minelly, A Galvanauskas, ME Fermann, D Harter, JE Caplen, ZJ Chen, DN Payne, Femtosecond pulse amplification in cladding-pumped fibers. Opt Lett 20:1797 (1995).

10

Electro-Optic Sampling and Field Mapping

J. F. Whitaker and K. Yang

University of Michigan, Ann Arbor, Michigan, U.S.A.

10.1 INTRODUCTION

The interaction of optical signals with electrical elements has been an important part of the fields of solid-state and microwave electronics for many
years. This association, especially through the areas of fiber-optic communications and optical computing, is ultimately leading to the situation whereby
integrated optical and electronic systems are becoming just as important as
either one of them on its own (Knox, 2000). Visible and infrared (especially
the latter) photoconductors, photodiodes, phototransistors, laser diodes,
electro-optic modulators, and a variety of other components are commonly
used in optoelectronic telecommunications systems to convert optical signals to electric signals and to manipulate optical radiation. Furthermore,
light can be used to control microwaves through the use of *p-i-n* and
IMPATT diodes as well as other two- and three-terminal devices (e.g., see
Mourou et al., 1985). Extending beyond traditional microwaves, optics is
also found to be useful in the generation of submillimeter-wave radio
frequencies (RF), commonly referred to as the terahertz (THz) regime
(see, e.g., Coleman, 2000, and references therein). The utility of the optical–
electronic interactions in nearly all of these instances is further enhanced—
or in some cases allowed only by—the employment of ultrashort-duration
laser pulses from ultrafast lasers, compared to continuous-wave laser light
or incoherent light. In these applications the short pulses are used as extremely fast trigger signals or for packing as much information into as small a
time period as possible in ultrahigh bit rate communications systems.

This chapter deals with a complementary aspect of the relationship between optics and electronics: the application of short pulses of laser light to the detection and measurement of electric signals. That is, essentially because their very short duration output bursts of energy can be exploited as short-duration sampling gates, ultrafast lasers have been combined with techniques that make them an effective tool in the field of diagnostic testing and the determination of the electrical characteristics of a component. The two principal optically based measurement techniques for ultrafast electrical testing that have been developed over the past 20 years are electro-optic sampling (Valdmanis et al., 1983; Kolner et al., 1983) and photoconductive sampling (Auston, 1984). Although each has been continually developed and extensively profiled throughout its history, their use has been limited almost entirely to the research laboratory. However, with the importance of the ultrafast time domain in optical telecommunications as well as the emergence of new applications such as microwave electric field mapping in wireless and radar environments, the interest in electro-optic sampling in particular has been enhanced. It is in this context that this chapter is presented. After a short background section, the fundamentals of electro-optic sampling are considered, followed by descriptions of some of the principal applications of the technique.

10.2 BACKGROUND

10.2.1 Motivation

Testing can be considered to be as integral a part of electronics as design, simulation, and fabrication, although the development of new characterization technology has lagged somewhat behind the efforts that were necessary to produce large-scale integrated (LSI) circuits, fast-switching digital circuits, millimeter-wave monolithic microwave integrated circuits (MMICs), and high-frequency discrete devices. Conventional, purely electronic time domain measurement instruments, such as the sampling oscilloscope, have measurement bandwidths that, for the most advanced instruments, are now approaching the 100 GHz barrier. The extremely useful RF network analyzer has exhibited a frequency domain measurement capability in the 100 GHz range for many years and has well-established calibration standards and sophisticated on-wafer probe adapters for contacting a device under test (DUT). In addition, less conventional but nonetheless extensively researched measurement systems that have also made their way into the marketplace include ones such as the electron beam (e-beam) tester (see, e.g., Plows, 1990).

So what, then, with so many alternatives available, is the true role of the ultrafast laser in electronics testing? First of all, despite the increasing bandwidth of new instrumentation, the frequency of operation of electronic devices can still exceed that of the instruments. For instance, it may be necessary to accurately characterize harmonics of a waveform that have frequency content in excess of an instrument's bandwidth. In other cases, a novel device may exploit a new phenomenon and have a simulated cutoff frequency that is extraordinarily high. The goal of a measurement may also be to determine the intrinsic response of an unpackaged component in order to gather more information on fundamental device physics. A sampling gate that had subpicosecond resolution, such as the fastest optoelectronic gates that are triggered by ultrashort laser pulses, could then be used to reproduce such waveforms and signals.

Second, and introduced by the suggestion that one may wish to measure unpackaged devices, conventional instrumentation cannot access many regions of interest within a circuit or an array of elements because of inherent limitations on the physical size, impedance, and other aspects of electronic probes. For instance, a metal probe or antenna cannot be inserted into the near field of a broadcasting antenna because of the high risk of disturbing the radiating fields. However, a small, high-impedance, low-invasive optoelectronic gate that can be easily reached by a train of ultrafast laser pulses either through free-space propagation or via optical fiber would serve as a probe that has the ability to extract signals from intermediate locations between the inputs and outputs of a circuit or within an antenna. The use of such a probe alleviates the need to de-embed interconnections and cables that may place limitations on the bandwidth of signals transmitted to a conventional instrument, thus allowing potentially greater accuracy in measurements or the opportunity to acquire the true response of the DUT only. This benefit of high-impedance, internal node probing is not exclusively an advantage of ultrafast optics, but rather of optics alone (Quang et al., 1995). However, if one wishes to obtain high-frequency information from a circuit, it will undoubtedly be necessary to utilize a short-pulse laser.

Finally, ultrafast lasers play an important role in electrical measurements because electric signals generated optoelectronically via short-pulse laser excitation, such as the transients that appear at the output of a telecommunications photodetector having a high bit rate, time domain multiplexed (TDM) optical input, may be too brief to accurately detect by means of ordinary instruments. That is, it is necessary to use a short-pulse laser to detect and reproduce an ultrafast electric signal photoexcited by a short-duration optical pulse. This fact, combined again with the ability of optical probes to access locations that conventional probes cannot, may create a significant role for optically based testing in the field of high bit rate optical telecommunications, as evidenced by investigations at NIST and other laboratories into accuracy and calibration issues (Williams et al., 2001).

10.2.2 Time Domain Investigations

The need for a sampling gate and a repetitive waveform occurs when the time an electron spends traveling between the deflection plates of an ordinary cathode ray tube oscilloscope is not short compared to the temporal characteristics of the waveform. If an electron experiences a changing field as it passes through the deflection plates, then a distorted version of the actual waveform will be displayed. A sampling oscilloscope will also exhibit distortion if the interaction time of the input signal with the sampling gate pulse is not smaller than the temporal characteristics of the input. In this case the voltage to be measured while the sampling gate is open will change if the gate pulse is not short enough.

Electro-optic (EO) modulation uses the birefringence induced in a noncentrosymmetric crystal by an electric field via the Pockels effect in order to transfer information from an RF signal to an optical beam (Yariv and Yeh, 1984). If that optical beam is a train of ultrafast pulses from a mode-locked laser and the pulses are variably delayed with respect to the signal being measured, then the modulation of the beam can take on an added role and be referred to as electro-optic sampling. On account of the short duration of the laser pulses, which then behave as sampling gates, the electro-optic sampling technique has the highest temporal resolution of any method used to characterize electric signals. For example, using one variety of EO sampling known as external EO sampling, along with a 100 fs ($1\,\text{fs} = 1 \times 10^{-15}\,\text{s}$) full width at half-maximum (FWHM) amplitude laser pulse, a guided electric transient with a rise time of less than 300 fs has been measured (Valdmanis, 1990). Free-space-radiated THz beam transients that have been detected via electro-optic sampling and a 12 fs duration laser pulse have been observed to have periods of 30 fs (Han and Zhang, 1998).

The impulse response of an electronic or optoelectronic DUT is in general an electric transient that exists as an aperiodic energy signal in time. Systems that are capable of measuring electric events that exhibit rapidly changing features rely on sequential sampling to capture the nature of the signal. Through necessity, then, the energy signal to be measured must be repeated many times so that a fast gate can allow the acquisition over successive small time intervals of the amplitude of the waveform. According to sampling theory, if the values of the time waveform are taken close enough together, then these values can be used to accurately recover intervening values and reconstruct the waveform. The transient is thus made to be periodic for the purpose of extracting information about its amplitude as it varies with time. If the transient to be measured is created by part of the same laser pulse that is ultimately modulated during EO sampling, then there is virtually perfect synchronization between the excitation and sampling pulses. However, if some other pulse generator or a continuous wave (cw) RF source creates the signal to be measured, then it is necessary to have some synchronization between the electric source and the modelocked laser.

One advantage of measuring in the time domain with electro-optic sampling is that regardless of whether a cw microwave signal, a guided electric impulse, or a free-space THz waveform is being extracted, the detection is coherent, and the signal is proportional to the true electric field. The amplitude and the phase information of the signal are thus preserved. This not only allows the straightforward determination of the rise, fall, and delay times within a microwave digital circuit, for instance, but it also yields information that could be used to perform vector network analysis on an analog component. There has historically been resistance to time domain measurements from the analog microwave community, however, because small-signal analysis needs to consider only a single frequency at a time and there is no particular advantage to measuring the time characteristics of a cw signal of one frequency. On the other hand, the measurement of distorted, harmonic-rich sinusoidal waveforms is one area in which time-resolving microwave signals can benefit from electro-optic sampling.

10.2.3 Frequency Domain Investigation

Conventional characterization of microwave components is performed in the frequency domain using a vector network analyzer and 50 ohm impedance-matched on-wafer probes to contact the DUT. Broadband measurements are obtained by stepping the input cw microwave signal through many different frequencies. In comparing alternative optoelectronic time domain measurements with frequency domain measurements, a number of trade-offs become evident. The time domain measurements typically use a short-pulse input signal that contains all of the frequencies of interest, and because ultrafast optics can be used with fast photoconductivity to generate extremely short duration electric inputs (Whitaker, 1993), this technique also allows testing at much higher frequencies than conventional network analysis. The other principal advantage to using transient measurements performed in the time domain is that broadband information can be obtained with a single measurement (Frankel et al., 1992). This is accomplished through the use of computer algorithms employing a fast Fourier transform (FFT) analysis of transients that possess features such as rising or falling edges that are of short duration (Nicolson, 1968). The pulse shape and the spectrum of the electric signal generated using optically based techniques can even be tailored to desired shapes, which may be advantageous in certain applications.

As mentioned earlier, electro-optic sampling in the time domain approach can be used to characterize device and circuit nonlinear effects, and both the amplitude and the phase information of the fields measured are also preserved. Furthermore, electro-optic sampling characterization is well suited for the multiport measurements that are necessary for acquiring a complete set of device attributes such as the well-known scattering parameters, or S parameters. One drawback, however, is the concern that the peak energy of an electric transient input may actually drive a DUT into an unintended nonlinear regime of operation when it is the small-signal response that is desired. This, coupled with other factors such as the emergence of network analyzers operating out to frequencies well in excess of 100 GHz and the low level of interest for direct device characterization of, for instance, transistors possessing cutoff frequencies modeled to be many hundreds of gigahertz, has greatly restricted the use of electro-optic sampling for S-parameter measurements.

One field where the electro-optic time domain measurement and conversion to the frequency domain via FFT has remained popular is in terahertz spectroscopy, where electro-optic probe tips or plates have been employed to capture the free-space radiating THz beams after their propagation through a material of unknown dielectric or conductive properties (Wu et al., 1996).

10.2.4 Electric Field Mapping

The area with perhaps the greatest potential for utilizing ultrafast electro-optic sampling may actually be the spatial determination of the amplitude and phase of the electric field of cw microwave signals (Yang et al., 1998). Here the temporal nature of the signal is not necessarily displayed, nor is there information conveyed regarding the frequency of the signal. Instead, the amplitude and phase of the electric field of a single frequency are extracted at many points near a device under test by using a harmonic-mixing-based electro-optic sampling technique, and then the points are sorted to allow the observation of the spatial distribution of a microwave or millimeter-wave signal. This has led to the description of this technique as electric field mapping. Although this can also be accomplished without an ultrafast laser, by directly modulating a cw optical beam with the microwave signal to be measured in an electro-optic crystal, the bandwidth of the measurement will be limited by the bandwidth of the photodetector used to sense the modulated beam. An ultrafast source, on the other hand, will still act as a sampling gate pulse in the electro-optic crystal, and thus a very high bandwidth can be achieved. This application of electro-optic sampling for the spatial mapping of electric fields is described in detail in Section 10.4.2.

10.2.5 Electro-Optic Sampling Embodiments

Internal Electro-Optic Sampling

Electro-optic sampling can be essentially categorized into three generations with respect to the way in which the EO probe is applied. The so-called internal EO sampling embodiment represents the first generation of EO sampling (Fig. 1a). Internal probing uses the substrate of a DUT as the EO sensor crystal (Kolner and Bloom, 1986; David et al., 1996), and since several relevant semiconductors such as GaAs and InP are electro-optic, this method can be used to extract an electric field from within the structure of a circuit. Although it is not necessary for the substrates that are used for internal EO sampling to be semiconductors, such as in the very first EO sampling experiments that sensed fields from microstrips fabricated on lithium tantalate (LiTaO$_3$) (Valdmanis and Mourou, 1986), it is especially useful to be able to probe the signals and fields from within single-level, semiconductor high-speed integrated circuits using this technique (Heutmaker et al., 1988). As shown in Figure 1a, internal EO sampling can be implemented in either a transmission or reflection geometry.

The internal EO method uses a relatively simple measurement scheme compared to other EO field-mapping techniques, because it does not require a probe crystal that is separate from the DUT. Also, because an external electro-optic crystal is not required, a sophisticated fabrication process for the electro-optic material, which is crucial for other EO embodiments, can be avoided. As in Figure 1a, the optical beam is directly focused into the substrate of the DUT, and the reflected or transmitted beam is detected to measure the electric field inside the substrate (see Sec. 10.3 for details). However, the advantage of having an electro-optic in situ probe for an IC is also a significant drawback to this method. That is, it can be applied only to circuits fabricated on substrates that exhibit the electro-optic effect, which thus excludes silicon and its derivatives (SiGe and silicon-on-insulator, known as SOI), and the technique cannot be applied to observe radiated fields. To overcome the limitations of the internal EO sampling method, a quasi-internal EO sampling technique using a hybrid geometry of a semiconductor DUT connected to an adjacent EO crystal with a coplanar electrode structure was employed to characterize a high-speed transistor (Meyer et al., 1985). The output signal from the transistor propagated to the transmission line on a LiTaO$_3$ substrate, where the transient was extracted with internal EO sampling. This method eventually led to

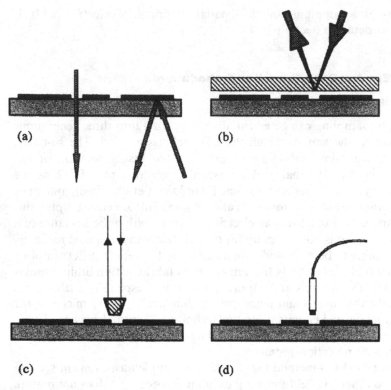

Figure 1 The principal electro-optic sampling geometries, as they would be used to sense electric fields on a coplanar waveguide transmission line. A cross section of the coplanar waveguide is drawn, where the center conductor is the signal line and the flanking, larger conductors are the ground lines. (a) Internal electro-optic sampling in transmission (left) and reflection (right) modes; (b) external electro-optic sampling with a plate of electro-optic material covering the waveguide; (c) external electro-optic sampling with a small, positionable, crystal probe tip; (d) an optical-fiber-mounted electro-optic probe. (Adapted from Yang, 2001.)

the development of external EO sampling probes and the free-space EO field-mapping technique.

External Electro-Optic Sampling

External EO sampling, the second generation of EO field-sensing methods, uses a crystal that is external to the DUT and is the technique employed for field mapping and terahertz detection, the two most common applications of EO sampling currently in use. Figure 1b depicts the external EO probing technique in a geometry whereby a large EO crystal plate is suspended over or resting on the top of the DUT (as in e.g., Meyer and Mourou, 1985). The incident train of ultrafast laser pulses is focused into the EO plate at the location where the measurement is desired, and then the reflected or transmitted beam from the EO crystal is detected to make the electric field measurement. To maximize the reflection of the probe beam and enhance the sensitivity of the system, a high-reflectivity coating is often applied to the EO crystal. However, if one wishes to detect the electric field only above metallized areas of the DUT, then it is possible to use the laser reflection from the metal under the probe.

The larger electro-optic plate can also be replaced by a much smaller version that is diced and polished to a very fine tip from the same material as the larger slab (Nees and Mourou, 1990). Electro-optic sampling in this fashion, as depicted in Figure 1c, takes on the characteristics of a true probe, as the position of the crystal or the DUT can be freely changed, allowing fields in a wide variety of locations to be sensed without presenting a large piece of dielectric material to an extensive surface area of a potentially high-frequency circuit. The ultrafast laser beam holding the electrical information returns from the probe via total internal reflection (shown), reflection from a coating deposited onto the bottom of the probe, or from reflection off the DUT metallization. With a free-space-propagating optical beam being focused to the bottom of the small EO probe tip, which then has a separate alignment with respect to both the laser beam and the DUT, one of the drawbacks of this system is that it requires a rather sophisticated alignment procedure to optimize the sensitivity of the detection. Because of this, it is extremely difficult to scan the probe over the DUT to map out an electric field distribution, and thus either the DUT must be scanned if one wishes to map fields or else this style of probe is used primarily for time-resolved measurements (rather than spatially resolved measurements) at a specific position.

It should be evident that when an external EO crystal is used, the electric field can be measured regardless of the DUT substrate. Furthermore, by selecting an appropriate EO crystal, as described in Section 10.3, three orthogonal electric field components can be successfully measured (see, e.g., Yang et al., 2000a). The external probing method can be used not only for guided wave structures but also for radiating structures, because it uses a probe crystal positioned where the field would normally exist in free space. In addition to the two-dimensional movement of the DUT in the plane of its surface for field mapping, the probe also can be positioned at many different heights from the DUT. Because of the three-dimensional freedom of movement of the probe, the near- to far-field transition of an antenna can also be measured (Yang, 2001). This three-dimensional measurement capability provides substantial information on radiating structures and, in particular, can be used to reveal information leading to the improvement of complicated radiating structures such as grid oscillators or amplifier arrays.

Fiber-Mounted Electro-Optic Probe

The third generation of EO sampling embodiments further improves upon the concept of the freely positionable probe but eliminates the inconvenience associated with free-space optical propagation. This is accomplished through the use of an optical fiber coupling to a small mechanically or chemically micromachined electro-optic probe (Sasaki and Nagatsuma, 2000; Wakana et al., 2000; Yang et al., 2001a). Owing to the flexibility and small diameter of the optical fiber, the fiber-based EO field-mapping system demonstrates exceptional measurement versatility. When attached to a computer-controlled translation stage, for instance, it can be easily guided to a desired measurement location without any need to disturb the position of the DUT, and the probe is also then especially amenable to scanning over a DUT in order to map out the electric field. Without any requirement to guide the laser beam to the probe with free-space optics, the fiber-based system can be easily used to detect electric fields within enclosed microwave and millimeter-wave structures such as horn antennas, waveguides, or cavities, where it is only necessary to have access through a hole much smaller than the microwave wavelength to be detected. Furthermore, the probe crystal and its support structure can be made so small that the interference of the fiber-based EO system on the DUT can be expected to be very low (Reano et al., 2001).

10.3 PRINCIPLES OF ELECTRO-OPTIC SAMPLING

The technique of electro-optic sampling is a novel union of the fields of optics and electronics. As stated in the introduction, it allows the short time durations easily obtained for optical fields to be used in the measurement of electric fields whose existence is too brief to otherwise be measured electronically. In addition, because it is a technique that can take advantage of small, freely positionable probes, EO sampling measurements can be used to separate out the response of a DUT from external cables, in most cases from external connections altogether.

10.3.1 Pockels Effect

As the electric field of a microwave or other relatively low frequency signal interacts with an electro-optic crystal, the optical dielectric properties of the crystal are changed. This leads to the well-known phenomenon of the Pockels effect. Crystals that have different indices of refraction along different crystal axes are known as birefringent crystals. When the index in one direction varies by a different amount than that in another direction when under the influence of an external stimulus, then the birefringence is said to be induced. When the birefringence is induced by an RF electric field, then the change in polarization of an optical beam propagating through the crystal is described by the Pockels effect. In EO sampling, one can think of the incident laser beam as probing the birefringence induced by a low-frequency electric field. The laser may emit a continuous beam (to measure dc or low microwave frequency electric fields) or a train of short-duration pulses (to measure any fields, including transient or microwave fields that vary over extremely short time durations).

As shown in Figure 2, one can consider that the sampling laser beam is split into two orthogonal components within the birefringent crystal, for an incident beam linearly polarized at 45° to the optic axis. In this case the orthogonal polarization components would be parallel to the x and z directions of the EO crystal, with the optic axis in the z direction. The polarization components traveling along the major axes experience different indices of refraction and retardation in the birefringent material, and thus a phase shift results between the two. In general, this difference in phase, δ, is given in radians by

$$\delta = \frac{2\pi}{\lambda} l \, \Delta n \tag{1}$$

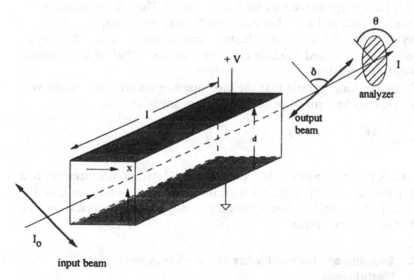

Figure 2 An electro-optic crystal between voltage-biased electrodes is used to demonstrate the Pockels effect and polarization rotation that occurs as a result of electric-field-induced birefringence.

where λ is the wavelength of the laser, l is the interaction distance of the laser within the crystal, and n is the index of refraction (Δn is the birefringence).

The shaded surface areas contacted by the voltage source on the top and bottom of the crystal represent electrodes, such as those on a traveling wave electro-optic modulator, that may be used to apply an electric field to the crystal. Without this field, any inherent birefringence (sometimes called static birefringence) due to the material causes a fixed phase shift between the polarizations, as given by Eq. (1). However, birefringence in the crystal is further induced, and the phase shift of the optical beam polarization components are further influenced, by the applied electric field through the index of refraction and the electro-optic relation

$$\Delta n = -\frac{1}{2}n^3 rE \tag{2}$$

where r is the electro-optic coefficient, a tensor, and E is the applied electric field. Combining Eqs. (1) and (2), the field-induced phase change can thus be given as

$$\delta_f = -\frac{\pi}{\lambda}n^3 rl\frac{V}{d} \tag{3}$$

where V is the voltage applied across the distance d. The voltage necessary to induce a phase shift of 180° is known as the half-wave voltage, V_π, and this is typically around 2700 V for a lithium tantalate modulator. This device is in one of the standard Pockels cell configurations, that of a transverse electro-optic modulator.

It is interesting to note that the birefringence induced by a given voltage change may be written in terms of this quantity as

$$\delta_f = \pi\frac{\Delta V}{V_\pi} \tag{4}$$

and that this phase change would typically be $10^{-4}\,\pi$ for a measurement of a 250 mV amplitude signal. Unless δ is an integral number of 2π radians, the polarization of an optical beam passing through the EO crystal will be altered from its input value.

10.3.2 Sensing of Electric Fields Through Intensity Modulation

The polarization axis of the optical beam is modulated during the EO sampling process by an amount that is quantified by using the angle δ, indicated in Figure 2, and determined by the length of the beam path in the EO crystal, the strength of the electric field in the crystal, and the EO coefficient of the crystal. Thus, for a given EO crystal that has fixed physical size with a certain EO coefficient, the angle δ reveals the strength of the electric field in the crystal (or bias voltage V across a distance). From this principle, if one places an optical polarizer in the output beam path with a certain orientation, the output signal can be obtained as a form of intensity modulation. For example, once the polarizer is aligned to the polarization direction of the input beam ($\theta = 0$ in Fig. 2), the output optical signal should be maximized if the external bias voltage equals zero (so that there is no electric field in the crystal), given that there is no static birefringence. If the polarizer is oriented to obtain maximum output power with a finite bias voltage (or finite electric field in the crystal, $\theta = \delta$ in Fig. 2), then the output power is reduced as the bias voltage (or electric field) decreases. The detailed mathematical presentation of this is better left to a textbook (e.g., Yariv and Yeh, 1984), although here it is important to introduce the connection and interaction between the unknown RF electric field and the intensity of the ultrafast laser pulse train.

In most applications, a waveplate and a polarizer are inserted in the output beam path of an electro-optic probe crystal as an analyzer in order to convert the polarization rotation (or retardation signal) of the laser into a change in intensity. This intensity modulation introduced by the electro-optic effect can be expressed as

$$I = I_0 \sin^2\left(\frac{\Gamma_0 + \Delta\Gamma}{2}\right) \tag{5}$$

where I_0 is the input intensity, Γ_0 is the static phase retardation due to the external analyzer wave plate and any static birefringence, and $\Delta\Gamma$ is the retardation induced by the electro-optic effect and the unknown electric field in the crystal (where $\Delta\Gamma$ is now the same as δ_f). In electro-optic sampling, the analyzer is typically adjusted so that with no applied electric field ($\Delta\Gamma = 0$) the retardance term is $\Gamma_0 = \pi/2$ (analogous to an optical biasing). This is the quarter-wave bias condition, or 50% transmission point, in contrast to the 0% transmission found if $\Gamma_0 = 0$ and 100% transmission if $\Gamma_0 = \pi$. With no static phase retardation, 100% modulation can also be achieved by applying the electric field that corresponds to the half-wave voltage introduced in the previous section. It should be recognized from Eq. (5) that optically biasing the electro-optic modulator at the quarter-wave point allows small changes in the induced birefringence to alter the argument of the sine squared term so that the output intensity changes in virtually a linear fashion with applied electric field (see Fig. 3). Electro-optic sampling probes

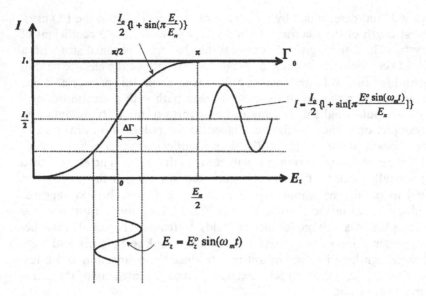

Figure 3 Output intensity, or transmission function, of an electro-optic modulator. By monitoring the change in intensity of a laser beam passing through the device, the electric field applied to an electo-optic sampling probe is revealed. (From Yang, 2001; reprinted with permission.)

are fabricated from dielectrics and semiconductors that, when combined with the typical geometries employed, yield a device half-wave voltage in the kilovolt range. Thus, when using sensitive instrumentation that also allows typical intensity changes of the order of 10^{-6} to be detected, the large half-wave voltage helps to increase the dynamic range of EO sampling for larger fields so that the total dynamic range can be extended to a level approaching 100 dB. It should also be noted that EO sampling of small signals is still possible even if the quarter-wave optical bias condition is not utilized, and in one instance, for poor quality EO crystals such as ZnTe that are commonly observed to scatter some of the incident laser probe beam, it is preferable to operate far away from this quarter-wave bias (Jiang et al., 1999).

Figure 3 also depicts how a sinusoidal input waveform, polarized in a direction E_z along the optic axis of the electro-optic probe, would induce a change in output intensity I when the analyzer had been used to initially bias the modulator to the quarter-wave condition. The expression describing the modulation function at the top of the figure is the same as the sine squared modulation function of Eq. (5) under the optical bias condition of $\Gamma_0 = \pi/2$.

$$I = I_0 \sin^2 \left(\frac{\pi}{4} + \frac{\Delta\Gamma}{2} \right) \tag{6}$$

This can be transformed using trigonometric identities to yield

$$I = \frac{I_0}{2} [1 + \sin(\Delta\Gamma)] = \frac{I_0}{2} \left[1 + \sin\left(\pi \frac{E_z}{E_\pi} \right) \right] \tag{7}$$

where the induced birefringence is written in terms of electric field instead of voltage. The variable E_π is considered to be the half-wave electric field, where the maximum transmitted intensity in Figure 3 is reached when $E_z = E_\pi/2$ because we consider there to be no electric field applied ($E_z = 0$) at the quarter-wave optical bias point. Because for EO sampling we typically expect to be measuring microwave, millimeter-wave, and sub-millimeter-wave electric fields, we can also express E_z in Eq. (7) as a time-varying function. Thus the optical intensity change that we expect from the sensing of an unknown electric field is

$$I = \frac{I_0}{2} \left[1 + \sin\left(\pi \frac{E_z^0 \sin(\omega_m t)}{E_\pi} \right) \right] \tag{8}$$

where E_z^0 and ω_m are the unknown electric field amplitude and frequency.

10.3.3 Free-Space Electro-Optic Probing Using Gallium Arsenide

As shown in the preceding sections, the electro-optic sampling technique is a practical application of crystal structures the optical properties of which change in the presence of electric fields. Thus, the degree to which an electro-optic probe is sensitive to a given electric field vector strongly relies on the crystal structure of the material. For the measurement of guided electrical signals, it has historically been common when employing pulsed visible or near-infrared Ti:sapphire lasers to use external sampling with crystals of $LiTaO_3$ and bismuth silicate (BSO) to sense the tangential and normal orientations, respectively, of the electric field. These crystals have been used to acquire a great deal of information on many microwave and millimeter-wave devices and structures, and the first EO field-mapping experiments to extract three orthogonal components of an electric field pattern were conducted using these crystals. Similarly, a majority of the groups performing electro-optic probing of free-space radiating terahertz beams use the visibly transparent ZnTe crystal.

However, despite the great number of successful EO sampling experiments using these crystals, they have a number of drawbacks when applied to microwave-component measurements. These shortcomings primarily result from the desire to produce a probe that can be inserted into the near field of a DUT while perturbing it as little as possible. To minimize the invasiveness, the size of the probes is minimized. Thus, crystals are polished so that they have footprint areas as small as a 20 μm × 10 μm, although, owing to the nature of the materials and the mechanical processes necessary to polish them, this is a somewhat difficult proposition. Mechanical polishing also has a low yield and places serious limitations on just how small a probe can be made. Furthermore, the dielectric constants of conventional electro-optic crystals are usually very large (e.g., $LiTaO_3$ has $\varepsilon_r \approx 44$; BSO has $\varepsilon_r \approx 50$), and these permittivities can also lead to distortion of the local electric field distribution.

Ease of microfabrication and the desire to reduce the permittivity of the electro-optic probe has encouraged the use of GaAs as an electro-optic probe material. First, standard solid-state fabrication processes such as photolithography and wet and dry etching techniques can be used for GaAs (Williams, 1984), making it more convenient to miniaturize GaAs than BSO or $LiTaO_3$. In addition, GaAs has a relatively small dielectric constant ($\varepsilon_r \approx 13$). Initially, however, because of the typical crystal orientation employed for GaAs, external GaAs EO probe tips were used only for sensing fields normal to the DUT (Heutmaker et al., 1990; Shinagawa and Nagatsuma, 1990). Because an EO probe that could measure only normal field components would not have wide utility for characterizing microwave devices and thus hold no advantage over crystals such as $LiTaO_3$, miniaturized versions of GaAs probes that can sense tangential electric fields emanating from a DUT were developed using modified etching techniques and are now in use (Yang et al., 2000b). It is also important to note that to avoid the photogeneration of carriers in semiconductor EO crystals, laser wavelengths corresponding to photon energies less than the bandgap of the semiconductor must be used. For instance, GaAs must be probed by photon energies less than its 1.42 eV bandgap energy. Thus, pulse-compressed Nd:YAG sources operating at 1.06 μm wavelength and modelocked Ti:sapphire lasers tuned to >880 nm have been employed.

Gallium arsenide is known to have a zinc blende crystal lattice, so that it possesses identical structure in each of the (100), (010), and (001) planes (Miller indices, as in Sze, 1985). All three of these orientations have electro-optic sensitivity to electric fields that are polarized normal with respect to their crystal plane. In Figure 4 we indicate graphically that if a probe laserbeam is incident normal to one of the three directions listed above, then EO sensitivity will occur when a low-frequency electric field is applied in a

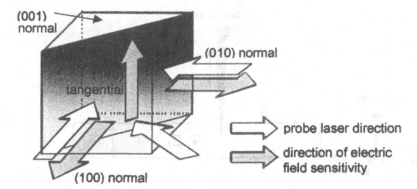

Figure 4 Depiction of the electro-optic sensitivity of several crystal planes of GaAs to an applied electric field. The shaded arrows indicate the polarization of the applied electric field, and the open arrows indicate the propagation direction of the incident probe laser pulse train. (Adapted from Yang, 2001.)

direction parallel to the laser propagation direction. Thus, if the crystal of Figure 4 were fashioned into a probe so that the laser light was directed through the crystal normal to the (100) or (010) planes (as shown), then the resulting electro-optic modulator/probe would be sensitive to a normal electric field guided by a transmission line or radiated by an antenna. On the other hand, when the probe laser beam is aligned normal to the (110) plane, the GaAs crystal exhibits sensitivity to the low-frequency electric field component that is parallel to the plane. Thus, through the use of a (110)-oriented GaAs wafer (this plane is also depicted in Fig. 4), a probe beam incident normal to the GaAs surface will acquire information on the low-frequency applied electric field that is polarized in the same plane. This is a very important characteristic of the GaAs crystal for the detection of tangential field components, and as a result the same type of electro-optic material can be used to measure both normal and transverse electric fields from a DUT. This is advantageous because then full-field measurements can be obtained using probes that would each have the same degree of invasiveness. In practice, a (100)-oriented probe would be used to sense a normal electric field, and a (110)-oriented probe would measure one tangential field vector and then be rotated by 90° at the test point in order to capture the second, orthogonal component of the transverse electric field.

Figure 5 shows a typical measurement configuration for a GaAs (110)-oriented probe tip. This is a cross-sectional view of a (110) GaAs crystal and also a top view of Figure 4, where the x' and y' axes are defined in the crystal directions as shown. The probe laser beam is aligned in the direction of the y' axis to detect the electric field (E_z) that is parallel to z', which is the same as

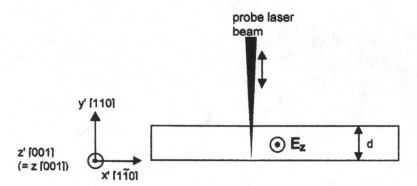

Figure 5 Cross-sectional view of a GaAs electro-optic probe that would be sensitive to a tangential electric field. This view represents the top view of Figure 4, and y' and x' are shown with their corresponding crystal directions. The probe laser direction is normal to the (110) plane, and the applied electric field, E_z, is parallel to the (110) plane and along the z axis. (From Yang, 2001; reprinted with permission.)

the z axis. A high-reflection coating can be deposited on one side of the EO probe in order to easily retrieve the optical beam containing the electric field information, and thus the probe beam passes through the EO crystal twice. This also enhances the interaction length of the probe beam in the EO crystal and leads to an increase in electric field sensitivity. Furthermore, this is reflected in the expression for the phase retardation due to the applied electric field E_z for the situation in Figure 5, which is written based on Eqs. (1) and (2).

$$\Gamma = \frac{2\pi}{\lambda} n_0^3 r_{41} E_z \times 2d \qquad (9)$$

where d is the thickness of the GaAs, n_0 is the GaAs refractive index, and r_{41} is the relevant electro-optic coefficient from the electro-optic tensor for GaAs (Yariv and Yeh, 1984).

An example of the coupling of a transverse electric field into a GaAs EO probe will help to illustrate the retardation induced on a laser by an RF field to be measured. First, one must consider the fact that the electric field passes through a transition between, typically, air and the GaAs EO probe. This is depicted in Figure 6, which shows the difference in the electrical properties between air (medium I) and GaAs (medium II). Because we consider the z component (or z' component) of the electric field as the polarization being sensed, this E_z should be parallel to the air/GaAs interface, and the propagation direction of the field can be assumed to be perpendicular to the air/GaAs interface. The E_z generated from a DUT propagates

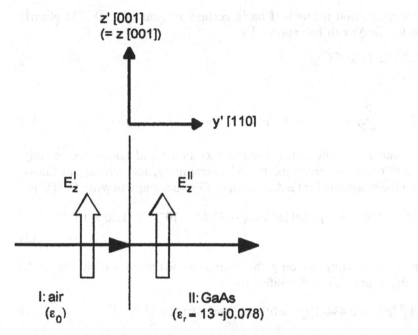

Figure 6 Electric field propagation between air (I) and GaAs (II). The propagation direction is assumed to be normal to the boundary and is displayed with open arrows. The incident electric field is assumed to be polarized parallel to the z' axis, as indicated by the open arrows. (From Yang, 2001; reprinted with permission.)

via air along the y' direction to the interface. Because the distance from the GaAs field sensor to a microwave or millimeter-wave DUT is often extremely small with respect to the wavelength (typically in a range from several tens to hundreds of micrometers), the electric field in the air just before the interface can be considered the electric field on the DUT. This actually provides EO probes with the unique capability to sense the near-field pattern of an antenna or to resolve the contributions of individual unit cells to the near-field distribution of an array.

The electric field propagating along the y' direction in medium I (air) from Figure 6 can be expressed as

$$\mathbf{E}^I = \hat{z}' E_z^I e^{-jk_0 y'} \tag{10}$$

where E_z^I is the amplitude of the electric field in the air and k_0 is the propagation constant of air.

The electric field in medium II (GaAs) is determined by the transmission constant T and complex propagation constant γ (based on the assumption that medium II has a certain amount of loss). The electric field in the GaAs can be expressed as

$$\mathbf{E}^{II} = \hat{z}' T E_z^I e^{-j\gamma y'}$$
(11)

where

$$T = \frac{2Z}{Z + Z_0}, \qquad \gamma = j\omega\sqrt{\mu\varepsilon}$$
(11)

where Z and Z_0 are the intrinsic impedance of air and GaAs, respectively, and μ and ε are the permeability and permittivity, respectively, of GaAs. By using the constants for GaAs shown in Figure 6, one can write Eq. (11) as

$$\mathbf{E}^{II} = \hat{z}' (0.434 + j0.001) E_z^I \exp[-\omega(3.61 \times 10^{-11} + j1.20 \times 10^{-11})y']$$
(12)

and by concentrating on only the amplitude variation within the GaAs probe, this expression is simplified to

$$|\mathbf{E}^{II}(y')| = 0.434[E_z^I(y' = 0)] \exp[-\omega 3.61 \times 10^{-11} y']$$
(13)

One also should take into account that there is often a nonuniform electric field distribution in the GaAs along the y' direction, because the signal becomes attenuated or the field changes characteristics as it propagates farther away from its source. Thus Eq. (9) becomes

$$\Gamma = \frac{2\pi}{\lambda} n_0^3 r_{41} \times 2 \times \int_0^d E_z(y') \, dy'$$
(14a)

$$\Gamma = \frac{2\pi}{\lambda} n_0^3 r_{41} \times [E_z^I(y' = 0)] \times \frac{0.868}{3.6 \times 10^{-11}} (1 - e^{3.6 \times 10^{-11} \omega d})$$
(14b)

This is the $\Delta\Gamma$ retardation experienced by an optical beam in a GaAs EO probe induced by an applied RF or transient electric field from, for instance, a microwave device under test. This retardation value would be plugged back into Eq.(5) in order to determine the optical intensity modulation arising from the interaction of the laser with the RF signal in the EO crystal.

10.4 ELECTRO-OPTIC SAMPLING SYSTEM

As discussed in Section 10.2, electro-optic sampling has most often been implemented in one of two ways. The first uses a variable time delay between excitation and probe laser pulse trains in a standard "pump-probe" type of configuration, resulting in the acquisition of time domain transient waveforms. The second involves synchronizing a single laser pulse train with a cw RF signal and then performing harmonic mixing to extract either a time domain reproduction of the sinusoid or the amplitude and phase of the single-frequency signal. When the amplitude and phase approach is taken, then a system can be developed to map out the characteristics of an electric field pattern in space. The EO sampling systems are described in the following subsections.

10.4.1   Time Domain Electro-Optic Sampling System

The schematic for an EO sampling system used to time-resolve very short duration transients generated using ultrafast pulses and a photodetector or photoconductive switch is shown in Figure 7. The laser pulse train is first divided into two beams using an appropriate beam splitter. The probe, or sampling beam, is linearly polarized, if necessary, and then directed by a dichroic beam splitter into the EO sampling crystal, in this case an external probe tip. While the dichroic mirror is reflecting the red and infrared wavelengths typically used in EO sampling, shorter wavelengths from an illuminator are transmitted to the probe and DUT so that a microscope eyepiece can be used to align the probe tip relative to the circuit elements and the probe beam relative to the tip and circuit. The returning laser light from the probe tip is reflected from a slightly different part of the dichroic beam splitter so that it is directed to another mirror that is used to guide the beam into the compensator, which is the quarter-wave plate of the analyzer described previously. The polarizer in the analyzer is often a Rochan or Wollaston polarizer, because they allow easy access to two orthogonal linear polarizations (the transmitted and rejected beams).

The trigger or excitation beam passes through a variable delay so that the electrical signal generated by the photoexcited switch or DUT can sweep through the optical sampling gate pulses (i.e., the phase is shifted mechanically rather than electrically). The delay rail, often a stepper motor, stops at each of many positions to allow the probe beam to dwell at constant intervals in time along the electric waveform. The signals from the photodetectors at the output of the analyzer (usually p-i-n photodiodes) can be integrated by the detection electronics at each of the delay-rail positions to improve the signal-to-noise ratio. The outputs of the diodes, $\{50\% + \Delta_{intensity} + noise\}$ and $\{50\% - \Delta_{intensity} + noise\}$, are subtracted in the differential amplifier to give $2\Delta_{intensity}$, further improving the signal-to-noise ratio by $3\,dB$. Note that $\Delta_{intensity}$ is the change in intensity resulting from the $\Delta\Gamma$ birefringence change of the EO crystal induced by the electric

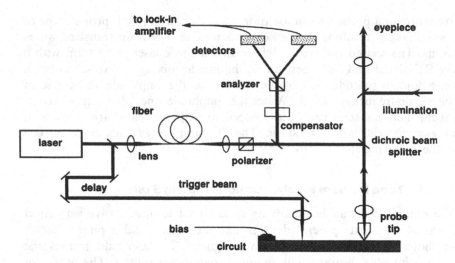

Figure 7 Schematic of a time domain electro-optic sampling system. The fiber is included only as a flexible guide to couple the laser pulses to the optical components, which are mounted on a single optical breadboard attached to an *x-y-z* translation stage. The illumination and eyepiece allow one to observe in detail the bottom of the probe tip, the circuit features under the tip, and the probe beam location within the tip. (From Valdmanis, 1990; reprinted with permission.)

field excited by the trigger laser pulses. The photodetectors do not have to have a high bandwidth in this pulsed laser EO sampling application, because the laser pulse repetition rate is usually under 100 MHz. If a cw laser were employed instead, then this beam would need to be modulated by the fast waveform to be measured, and a detector would then have to have a large enough bandwidth that it could faithfully convert the modulated optical beam into an electronic replica.

A mechanical chopper or an acousto-optic modulator can be placed in the trigger beam path, or a periodic voltage bias may be applied to the photoexcited signal source, in order for the probe beam leaving the EO probe to become modulated. A lock-in amplifier, which detects signals only at this modulation frequency, is used to gain further immunity from the $1/f$ noise present in the system and to integrate the signal from the slow detectors at each interval the chopper is open. Often enough noise can be removed that a detector shot-noise-limited sensitivity of $10 \, \mu V/Hz^{1/2}$ can be attained. Finally, the output from the lock-in is sent to a computer, where multiple traces of the sampled waveform can be averaged together on a display, the signal-to-noise ratio increasing as the square root of the number of averages.

Electro-Optic Sampling Time Resolution

The primary reason for the application of short optical pulses to the measurement of electrical signals is the unprecedented temporal resolution that is attainable only through these means. Subpicosecond electrical measurements are well within the capability of the external EO sampling technique, as is evident in the waveform of Figure 8. The 900 fs FWHM and the 500 fs rise and fall times of the waveform were measured using a $LiTaO_3$ external probe positioned at the coplanar-stripline output line from a photoconductive switch excited by $\sim$100 fs laser pulses from a Ti:sapphire oscillator. The switch–detector gap was fabricated on a low-temperature-grown GaAs (LT-GaAs) epitaxial layer (growth temperature of 190°C and annealing conditions $T_A = 600$°C for 10 min), a material known for its ultrafast trapping time (Gupta et al., 1992). Guided signals of much shorter duration have also been captured, essentially at their point of origin (Keil and Dykaar, 1992), and radiating terahertz beams generated with $\sim$10 fs duration laser pulses have been characterized using ZnTe EO probes to have features much shorter than 100 fs (Han and Zhang, 1998). (Examples of THz waveforms measured via EO sampling can also be seen in Chapter 15.) The bandwidth of signals that can be measured through EO sampling is thus found to

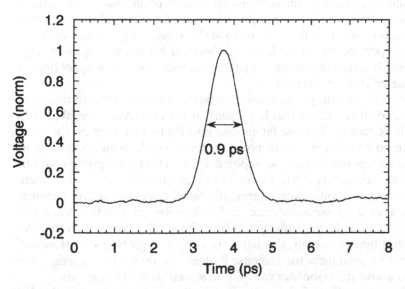

Figure 8 Subpicosecond electric waveform at the output of a low-temperature-grown GaAs photoconductive detector, as measured by a $LiTaO_3$ external electro-optic probe.

extend easily into the terahertz regime. Most striking, the magnitude of the Fourier transform of subpicosecond electric pulses that have been *guided* over significant distances indicates that it is possible to achieve usable signal strength in circuits even at a frequency well past 1 THz (Cheng et al., 1994).

Ideally, the EO system employed to modulate the light intensity using an electric field would have a bandwidth that is infinitely large and a temporal resolution that is infinitely small. In this case the profile of an electrical signal could always be translated onto an optical beam with a perfect correspondence between temporal points. Practically, however, the temporal resolution of optically based measurement systems is limited by a number of factors, the most important of which for EO sampling will now be briefly discussed.

The electro-optic material that serves as the medium for interaction between the optical and electric signals provides one fundamental limitation. That is, the response can be affected by how fast the field-dependent birefringence of the crystal reacts. This is generally determined by the absorption that results from a lattice resonance in the crystal. Typically centered around a frequency in the terahertz regime, this absorption will cause the crystal to lose its sensitivity to the changing electric field. Starting at low frequencies and moving up, EO crystals will usually exhibit acoustic, lattice, and electronic resonances, with dielectric dispersion becoming negligible as one moves away from the corresponding absorption peaks. The acoustic resonance, which varies according to the size of the crystal and how it is mounted, is typically on the order of 10 MHz, a frequency that is usually less than the repetition rate of the laser and therefore below the frequency range of interest. It is thus the lattice resonance that will provide the upper limit as far as the crystal is concerned.

For the commonly used lithium tantalate, it has been shown that there is a vibrational resonance that has a peak at 6.3 THz (Auston et al., 1985), so that if frequencies increase far enough into the terahertz regime there will be enhanced absorption and the probe response will drop out. A reasonable limit on the temporal response imposed by a $LiTaO_3$ EO probe could be given as approximately 300 fs. Compared to the other limitations to be mentioned, this time response would generally be adequate to perform measurements on existing semiconductor and millimeter-wave RF devices and components. For the measurement of broadband THz beams with frequencies several times this limit, this lattice resonance is problematic. However, because the crystal dielectric response flattens out above the resonance, frequencies above the resonance can still be sensed with EO sampling.

The interaction of the optical and electric signals within the EO modulator can also serve as a limitation. Assuming the modulator can respond linearly to a given applied electric field, then its temporal resolution is determined by the convolution time of an optical probe pulse with a portion of an electric wave as they copropagate in the birefringent EO probe material. For example, if a voltage wave traveling along a planar transmission line is probed in such a way that the optical beam path is normal to the plane, then the resolution would be the time it takes an optical pulse to traverse the region of the crystal that contains the applied electric field lines, convolved with the transit time of a portion of the electric signal across the diameter of the optical beam. This convolution time can be minimized either by using temporally shorter probe laser pulses or by decreasing the duration of the electric signal that must spend time in the beam waist of the probe pulses. The latter can be accomplished through the use of a more tightly focused optical beam or through velocity matching of the optical and electric signals (so that the same brief portion of electric transient modulates the optical pulse through the entire interaction length).

The final constraint to temporal response to be considered, and one that pertains to guided signals, results from the imperfect performance of simple circuit elements such as transmission lines. Unless observing radiation that has been launched into free space, all measurement techniques must use some electrode configuration to transmit a signal from the point where it is generated to the point where it is sampled. Dispersive and attenuative effects due to metal lines, dielectric substrates, and the geometries in which they are used have the potential to eliminate any chance of attaining the desired resolution in an experiment. However, if care is taken to limit the propagation distance of the electric signal, or if the truly in situ capabilities of EO probing are exploited, then extraordinary resolution is possible.

Electro-Optic Sampling Sensitivity

In order to attain the high level of sensitivity possible with EO sampling, special care must be taken in numerous parts of the experiment. One principal area involves $1/f$ noise alluded to earlier. This pink noise arises from temperature, power line, laser intensity, and other fluctuations, and it is a particular nuisance at measurement frequencies below 1 kHz. Because the noise power per unit of bandwidth increases toward DC for pink noise, it is thus useful if the measurement bandwidth is kept small and the center frequency of the passband is kept away from DC. The lock-in amplifier is well suited to these tasks, and its time constant (TC) can be used to control the bandwidth of its final filtering stage, along with a low-noise mixer. The mixer allows a low-noise audio lock-in amplifier, which can detect signals modulated at low frequency (100 kHz maximum frequency), to measure signals modulated at a center frequency of several megahertz (Chwalek and

Dykaar, 1990). Alternatively, a low-noise RF lock-in amplifier can be employed. For a lock-in filter with a 12 dB/octave roll-off, the bandwidth of the measurement is determined by the time constant as $B = 1/(8 \times TC)$, and in order to increase the ratio of the signal voltage to the noise by the factor n, the bandwidth must be decreased by n^2 (for white noise). A typical lock-in time constant used is 100 ms, yielding a bandwidth of 1.25 Hz.

Of course, the time constant cannot be increased indefinitely to decrease the bandwidth and enhance the signal-to-noise ratio (SNR), because the data acquisition time will increase to an unreasonable length. In fact, for the best SNR to be obtained from measurements using laser pulse activated input transients, one of two potentially time-consuming methods is necessary: A narrow bandwidth (larger TC) on the lock-in must be used and the signals acquired after several very slow sweeps of the optical excitation beam delay averaged together; or a larger bandwidth (smaller TC) must be used and the waveforms after many faster sweeps of the delay must be averaged together. Either process may take as long as several minutes, although for typical device and circuit measurements, much less than a minute is usually necessary to acquire any one waveform.

10.4.2 Electro-Optic Field-Mapping System

If the pulse repetition rate from the laser can be stabilized (i.e., if the laser has low phase noise) and it can be phase-locked to a sinusoidal electric signal (i.e., a cw signal from a synthesizer), then the EO sampling technique can be applied to measure the amplitude and phase of the cw electric field. This is particularly useful for characterizing microwave and millimeter-wave devices, integrated circuits, and radiating structures. If the EO probe is repositioned in a regular pattern over a device under test, then the electric field pattern that is present can also be mapped.

Harmonic Mixing for Single-Frequency Sampling

A number of techniques can be used to stabilize the laser cavity, although most involve the generation of an error signal that is fed back to a piezoelectric driver on one mirror of the laser cavity when the laser repetition rate deviates from an expected value (see, e.g., Pfeifer et al., 1998 and references therein). The laser repetition frequency is then also divided down to create a 10 MHz signal that can typically be used as a reference by an RF synthesizer in order to synchronize the phases of the optical and electric signals. This synchronization is represented schematically in Figure 9, where a common 10 MHz reference is derived from the laser 80 MHz repetition rate and input

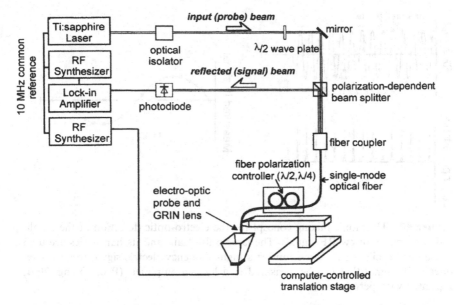

Figure 9 Electro-optic field-mapping system schematic. (Adapted from Yang, 2001.)

to two RF synthesizers. One synthesizer provides an input microwave signal to a device under test, while the other supplies a constant-phase 3 MHz reference signal to a lock-in amplifier that will receive the photodiode output signal corresponding to the EO modulation.

The 3 MHz input to the lock-in amplifier results from a harmonic mixing technique that uses the laser repetition frequency and the components of its Fourier series spectrum as a local oscillator (LO). During the measurement, the RF input to the DUT is set to be the sum of an integer multiple n of the 80 MHz laser pulse frequency and a small offset intermediate frequency (IF). Figure 10 displays the laser pulse train, the RF signal, and the measured IF signal as a function of time. Because the pulse-repetition frequency of the laser is much smaller than the operating frequency of the microwave or millimeter-wave DUT, there is virtually always some integer multiple of the laser repetition frequency that is very close to the actual operating frequency of the DUT. (The maximum difference between the actual operating frequency and the integer multiple of the laser repetition frequency can be 80 MHz.) If one uses the radio frequency without any intermediate frequency, the laser pulses in the EO probe always sample the exact same position on the RF signal. In this case, the measured signal would show a uniform value with respect to time. However, by adding

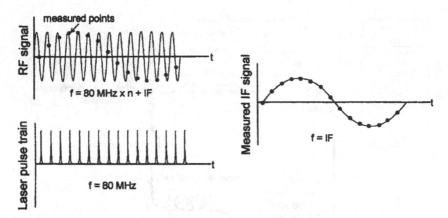

Figure 10 Harmonic mixing concept for the electro-optic detection of the ampli-tude and phase of cw RF signals. The laser pulse train and its harmonics are used like a local oscillator to down-convert a high-frequency electric signal to a low-fre-quency IF signal that can be measured in a lock-in amplifier. (From Yang, 2001; reprinted with permission.)

the IF to the RF, each laser pulse detects a slightly different position on the RF signal. As a result, the measured EO signal has a very low intermediate frequency as shown in Figure 10. Thus, the electrical network after the photodiode (Fig. 9) does not need to handle high-frequency signals. The intermediate frequency can be selected independently of the operating fre-quency of the DUT, but an odd-number IF is selected intentionally, because most electronic equipment is operated by an even-number frequency power source (60 HzAC), and thus the possibility of noise from the AC line voltage can be minimized. With the cw RF signal synchronized to the laser pulses, the lock-in amplifier captures both the amplitude and the phase of the RF signal from the DUT. The 3 MHz reference signal not only allows the 3 MHz mixed-down EO signal to be detected by the lock-in amplifier, but as the EO probe moves to different spatial locations and senses electric fields with potentially different phases, the lock-in is also able to compare these phases with the fixed phase of the reference signal.

Optical System and Polarization Control

The optical system used for EO field scanning and mapping will have much greater positioning flexibility if optical fiber laser delivery is employed, as depicted in Figure 9. All of the optical components introduced for "free-space" EO probing in Figure 7 are still present in the fiber-based system, although they are configured somewhat differently, and several new components are added to accommodate the optical fiber. For instance, the laser input (probe) beam is passed through an optical isolator to suppress any backward reflection from the fiber coupler that could disturb the phase stability of the laser system. The half-wave plate is used to rotate the polarization of the input beam so that it is transmitted through the polarization-dependent beam splitter. This probe beam is focused into the single-mode optical fiber through a microscope objective and then propagates through the waveguide until it reaches the EO probe, which is mounted on a GRIN lens.

As mentioned earlier, from a size standpoint, it is most convenient to use a semiconductor such as GaAs for the EO probe for such a fiber-based system, because this material can be easily micromachined to a small tip. Figure 11 indicates how a 500 µm × 500 µm × 200 µm GaAs EO probe tip can be situated so that the 4 µm diameter output from the single-mode fiber is imaged by a GRIN lens onto the bottom face of the GaAs. In the work described in Yang et al. (2001a), the laser wavelength is tuned to 900 nm and the power input to the fiber is attenuated to 15 mW to avoid band gap and two-photon absorption, respectively, in the GaAs. A high-reflection thin-film dielectric coating, consisting of alternating layers of MgF_2 and ZnSe, was deposited on the face of the GaAs after the wet etching step that forms the probe tips. The signal beam modulated by the presence of the applied electric field inside the GaAs was then reflected back through the GRIN lens and the optical fiber to the polarization dependent beam splitter, which directed the signal beam to the photodiode. To achieve two-dimensional mapping, the probe end of the fiber is enclosed within a length of quartz tube that is attached by a connecting arm to a computer-controlled X–Y translation stage.

The resultant EO signal incident on the photodiode is obtained from the polarization difference between the input and reflected (signal) beams. Because both the input and signal beams travel along exactly the same path, a special polarization manipulation scheme is required to separate one beam from the other. The polarization-dependent beam splitter shown in Figure 12 (labeled BS) transmits the horizontally polarized beam. The figure also indicates the polarization conditions at a number of critical points in the system. The input beam polarizations are displayed using solid circles and lines, and the polarizations for the reflected signal beam are shown as dotted lines. The horizontally polarized laser output (position IN1, Fig. 12), rotated by the optical isolator (IN2) and the half-wave plate (IN3), propagated through the beam splitter (IN4) and was focused into the single-mode optical fiber by the fiber coupler (IN5). From this point, it is necessary to manipulate the polarization so that it will be aligned correctly with respect to the principal axes of the GaAs EO probe. At the same time, the signal beam returning from the probe should have a certain

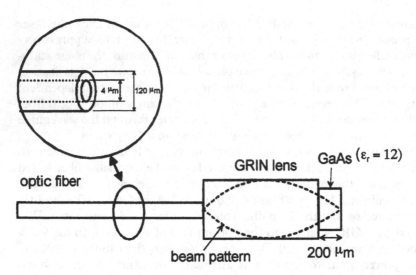

Figure 11   GaAs EO probe assembly using a GRIN lens. The core and cladding of the single mode optical fiber is magnified in the circle. (From Yang, 2001; reprinted with permission.)

amount of polarization rotation so that it can be separated from the detection beam when it arrives back at the beam splitter. To accomplish this, two in situ polarization control loops from a "butterfly wing" polarization controller are used. The first loop serves the function of a quarter-wave retarder, and the other acts as a half-wave retarder. In both cases, the planes of the loops are equivalent to the fast axes of the wave plates. Thus, one could control the polarization by setting the loop planes at a certain angle from the horizontal (or vertical) axis.

One way to manipulate the polarization for EO sampling in the fiber-based system is to first set the fast plane of the quarter-wave loop to be 22.5° off the horizontal axis. Thus, a linearly horizontally polarized beam is changed into an elliptically polarized beam that is 22.5° off from the horizontal axis (IN6, Fig. 12). The half-wave loop provides arbitrary control of the polarization (θ in IN7) of the input beam to the GaAs probe. Thus the polarization axis can be aligned to the optic axis of the GaAs probe tip. The probe beam that travels through the GaAs is reflected back to the GRIN lens, and in the course of travel through the GaAs the polarization of the beam is slightly altered by the presence of the electric fields in the EO crystal. The angle δ shown in OUT1 in Figure 12 represents the electro-optically modulated polarization angle (the same as $\Delta\Gamma$). The reflected (signal) beam passes through the half-wave loop again, so that the total

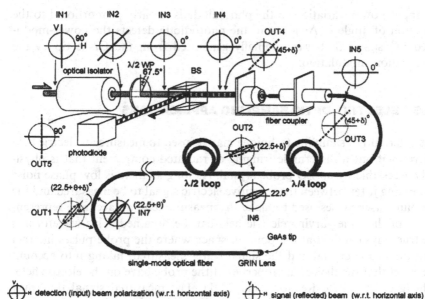

Figure 12 Diagram of the polarizations of the input and signal beams in a fiber-based electro-optic field mapping system. The input beam is drawn as a thick gray line, and dark gray bands are used to indicate the modulated signal beam. The polarization conditions of the input beam are displayed with solid lines (or ellipses) in the solid circular axes (axes labeled IN). The dashed circular axes and dotted lines (or ellipses) are used to indicate polarization angle for the signal beam (axes labeled OUT). The horizontal axis is used as a reference for the polarization angles. (From Yang, 2001; reprinted with permission.)

retardation becomes one full wavelength, and thus the initial phase retardation denoted as θ is canceled. As a result, the signal beam after the half-wave loop (OUT2) has polarization identical to that of the detection beam at position IN6, except that the signal beam includes an additional EO signal that is expressed as the polarization angle δ. Because the signal beam now passes the quarter-wave loop a second time, the total polarization retardation is equal to a half-wavelength with a 22.5° offset. Thus the signal beam at OUT3 is linearly polarized at (45 + δ)° off the horizontal axis.

The returning signal beam into the polarization-dependent beam splitter (OUT4) will then be divided so that the vertical component is dumped into the isolator, while the orthogonal component is directed onto the photodiode. With the signal beam from the EO probe having a polarization angle of (45 + δ)°, approximately half of this is detected and sent to the lock-in amplifier. Because the angle δ is much smaller than the fixed 45° angle, the input power variation to the photodiode is linearly proportional to the variation of angle δ. As a result, the photodiode detects the power-modulated EO signal that was originally generated from the GaAs tip by the polarization modulation.

10.5 ELECTRO-OPTIC SAMPLING APPLICATIONS

The external EO sampling technique may be used to measure an electric field wherever it has a discernible fringing or radiated component that is physically accessible by an EO probe crystal. As long as there is low phase noise and timing jitter between the repetitive electric signal to be measured and the sampling laser pulses used to make the measurement, an accurate representation of the time-varying electric field can be obtained. This waveform is the true physical field at the point in space where the probe pulses interact with the probe crystal and the signal being measured, taking into account the effect that the dielectric properties of the probe have on the electric field. (Cheng et al., 1995; Reano et al., 2001). The measured signal thus may include contributions from reflections at discontinuities and imperfect terminations, radiated fields, and cross-talk signals from neighboring lines. For optimum characterization of guided signals at the input or output of devices and circuits, measurements are therefore typically made far from reflection sources using very thin probe crystals so as to eliminate these problems (Frankel et al. 1991a). To distinguish between input and reflected waveforms, it may also be necessary to move an EO probe some distance down a transmission line from the DUT, either using a very high bandwidth transmission line or extracting the amplitude and phase factors for the line and then correcting any measured signals for the distortion (Frankel et al., 1992). For characterization of radiated signals from antennas using the fiber-based EO probe technique, the probe may simply be inserted into the near or far field of the DUT. In both cases, it is still possible to get essentially all the electric field information present at a given location.

10.5.1 Electro-Optic Sampling of Guided Electric Waveforms

As the temporal properties of signals in digital circuits, optical telecommunications photoreceivers, and other components become shorter and the bandwidth capabilities of modern electronic and photonic devices increase, it is necessary to extract and resolve the characteristics of short-duration electric signals. Similarly, it is also necessary to understand how transmission lines behave when carrying short-duration electric waveforms, in order to develop improved packages and interconnects. Electro-optic sampling has been shown to be well suited to the study of these structures due to both its temporal resolution and its nature as a noncontact measurement technique. Not only can observations of the actual signals propagating on transmission lines be made, but quantitative information on the propagation factor of the line between dc and the submillimeter-wave regime can also be determined.

As an example, the result of a series of measurements conducted on several coplanar stripline (CPS) transmission lines are shown in Figure 13 (Cheng et al., 1994). In one case the substrate was GaAs with a 1 μm layer of ultrafast-relaxation, epitaxial LT-GaAs grown on the surface, whereas in a second case the LT-GaAs was removed from its GaAs substrate by epitaxial lift-off (Yablonovitch et al., 1989) and bonded on a fused silica substrate of significantly lower permittivity. In both cases the LT-GaAs filled a photoconductive gap fabricated in the CPS, allowing the generation of the short input pulse labeled as the 0.2 mm propagation distance in Figure 13. This reference point was defined by placing the electro-optic tip and optical probe beam at a position 200 μm from the photoconductive switch pulse-generation site, and this input waveform is approximately 1 ps at its FWHM point and has a rise time of 0.5 ps. The probe tip and probe beam were then moved to positions 1.2 mm, 2.2 mm, and 3.2 mm from the reference point, and the measurements shown in Figure 13 were made.

The waveforms in this figure demonstrate the distortion effects on a guided electric signal that has frequency content extending from direct current to approximately 1 THz. On a conventional interconnect from a high-speed photodetector, the measurements might be used to determine if an interconnect will be the factor that limits the bandwidth of the detector. The waveforms also indicate that if one wishes to determine the intrinsic response of a device, for instance, it may be necessary to extract a waveform from a location very near the active device. The fact that the transmission structure is an open-boundary one, so that the field lines experience both the substrate dielectric and air, causes significant pulse dispersion that can be clearly observed. Qualitatively, the modal dispersion resulting from the dielectric inhomogeneity causes the higher frequencies to travel at a lower velocity than the low frequencies. This leads to a reduction of the peak amplitude of the pulse as part of its energy is delayed in time with respect to the rest of the pulse. This high-frequency energy appears as an oscillatory, ringing behavior after the main pulse of the waveform has passed. The amplitude of the propagating signal also decreases due to frequency-dependent skin effect and radiative losses. Whereas the former depends on the quality (conductivity) of the transmission line electrodes, the latter occurs because the guided mode travels at a higher velocity than

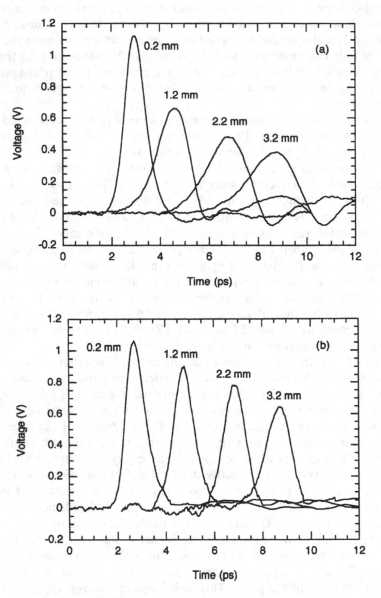

Figure 13 External electro-optic sampling measurements of the propagation of a 1 ps FWHM, 0.5 ps rise time electric pulse on coplanar waveguide transmission lines fabricated on (a) GaAs and (b) quartz substrates.

that of the electromagnetic signal in the substrate (Rutledge et al., 1983). This results in a wave being launched into the substrate, with the corresponding attenuation of the guided waveform increasing with frequency as f^2–f^3, rather than the $f^{1/2}$ due to the skin effect (Grischkowsky et al., 1987; Frankel et al., 1991b).

This loss is quantified in Figure 14, where the attenuation from 20 GHz to 1.0 THz is plotted after performing a fast Fourier transform of each temporal waveform and comparing the propagation factors for the output to those of the input. Here it becomes clear how powerful the high-resolution, time domain EO sampling technique can be. The functional form of the attenuation increase can easily be determined, and, correspondingly, it is also possible to observe the decrease with frequency of the phase velocity, which leads to the increase of rise time, pulse width, and ringing (Cheng et al., 1994). The attenuation response of a CPS fabricated on a 1.4 μm thick, silicon dioxide/silicon nitride membrane is also included in Figure 14 as an inset. The data for the CPS on GaAs and fused silica are limited because the attenuation is large at high frequencies, and thus the signal-to-noise ratio is not high enough to extract the attenuation data. There is virtually no radiation loss for the membrane line, because it has a uniform dielectric of air both above and below the CPS electrodes. In contrast, because of the dielectric mismatch of the CPS lines on GaAs and fused silica, there is a dramatic increase in loss with frequency, above 200 GHz for the former and 350 GHz for the latter. This high radiation attenuation will make it impossible to operate such coplanar lines at very high frequencies.

It should be noted that it is only necessary to compare the Fourier transforms of two of the time domain waveforms in Figure 13 to obtain the information in Figure 14. However, because the high frequencies in the spectrum attenuate rapidly with propagation distance, it is best to compare the input signal with a waveform a short distance away so that the highest frequencies are not lost in the noise. Because the measurements are made in the time domain with 100 fs resolution, all of the amplitude and phase information for a broad range of frequencies is obtained, and a passive element such as a section of transmission line can be characterized after using EO sampling to acquire only two waveforms.

10.5.2 Electro-Optic Field Mapping

Electric field distributions can be resolved for single-frequency RF, microwave, and millimeter-wave signals in one of two ways: in one dimension by scanning the EO probe over a line through the field of interest, and in two dimensions by forming a plane through a procedure such as a compu-

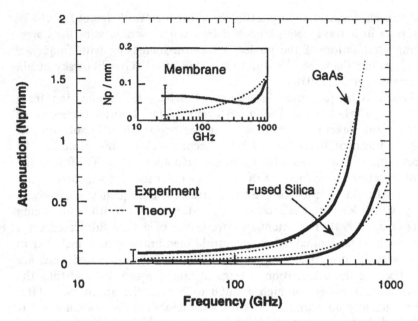

Figure 14 Attenuation, in nepers per millimeter (NP/mm) of the coplanar striplines of Figure 13, as reduced from the Fourier transforms of the propagating experimental time domain waveforms captured via electro-optic sampling (solid lines). The attenuation from a CPS fabricated on a silicon-based membrane, which exhibits none of the radiation loss plaguing the other CPS lines, is included in the inset. The theoretical development (dotted lines) of the waveform propagation is described by Cheng et al. (1994). (From Cheng et al., 1994; reprinted with permission.)

ter-controlled raster scan. Three-dimensional spatial scans can also be achieved by changing the distance between the probe and DUT and then combining multiple planes, although the total scanning time would prohibit the same spatial resolution in the separation of the planes as can be obtained between measurement points within the plane. Such field-mapping scans can be performed to capture normal or tangential electric fields on either guided wave structures (Yang et al., 1998) or radiating elements (Kamogawa et al., 1994; Pfeifer et al., 1998; Yang et al., 2000a), and they can be used to obtain both near-field and far-field distributions. In almost all cases the experimental maps provide a different way to look at the fields compared with the results from standard electromagnetic models, and they often reveal some attributes of the field that are unexpected. The measurements can also be used to validate complicated computer models, especially at high frequencies where models are relatively untested.

One-Dimensional Electro-Optic Field Mapping

One-dimensional line scans of an electro-optic probe over a DUT are superior for acquiring quantitative information on both the amplitude and phase of the electric field, and they allow detailed comparisons of the field from different parts of the DUT. These scans also take appreciably less time to complete than the two-dimensional scans, often on the order of 1 min. These measurements have been used both to verify the operation of microwave components and to study the physical behavior of complex circuits that have multiple parallel paths with potentially complex phase relationships.

In a simple example, the normal and tangential electric field components for an even-mode coplanar waveguide driven with a 15 GHz sine wave from a microwave synthesizer were scanned (Yang et al., 1998). Figure 15 illustrates the cross section of the CPW and the measured one-dimensional amplitude and phase maps. These field distributions are intuitively what one would anticipate from the structure. For example, the line scan in Figure 15a indicates the presence of a high EO signal on the center conductor, a minimum EO signal in between the metal lines, and a relatively low EO signal on the ground conductor, corresponding to the strength of the normal electric field component. The results thus indicate a high electric field concentration on the center conductor and a relatively weak electric field on the ground planes, with the normal fields on the center conductor and on the ground planes having vectors that point in opposite directions. This is a result of the observed phase change of 180° between the center conductor and the two ground planes.

Corresponding to the strength of the tangential field component in the transverse direction (i.e., the horizontal field component), the EO signal in Figure 15b is a maximum above the CPW gaps, with a phase change of 180° taking place in the middle of the center conductor. The EO signal for the transverse field drops off to a small value over the top of the metal electrodes, following the "perfect electric conductor" boundary condition. From the combined results on the normal and tangential electric fields obtained by the EO field mapping, one can conclude that the electric field is initially generated at the center conductor in the normal direction, and as the field travels from the center conductor into air it changes its direction from normal to two opposite tangential directions. Finally, the field changes direction toward the ground planes. Because the width of the ground plane is much bigger than that of the center conductor, the electric field density is expected to be lower on the ground plane than on the center conductor, as reflected in the EO signal strength. These one-dimensional field maps demonstrate the excellent discrimination in electric field components available using external EO sampling.

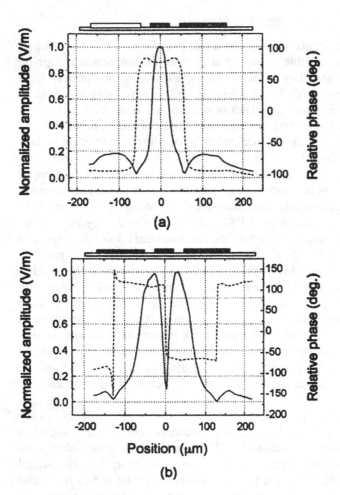

Figure 15 One-dimensional electro-optic mapping of the (a) normal field and (b) tangential field components of the even mode on a coplanar waveguide (transverse direction). The frequency of operation is 15 GHz. The solid line is the normalized amplitude, and the dashed line is the phase (the center conductor has a width of 40 μm, with spacing of 24 μm). (From Yang et al., 1998; reprinted with permission.)

Two-Dimensional Electro-Optic Field Mapping

Electro-optic scans of a full plane of an electric field are useful to give an overview of the electrical behavior of a large area of a circuit or antenna. One can immediately notice, for instance, if a signal is a propagating wave or a standing wave, indicating whether there could be an impedance mismatch or even a short or open circuit present in the circuit. The ways in which such faults affect the internal operation of a circuit can also be visualized. Perhaps even more significant, the expanding use of microwaves in areas such as satellite communications, wireless and mobile communications, navigation, and remote sensing has fueled an increasing demand for accurate characterization of microwave antennas. In particular, antenna arrays, such as those found in quasi-optical power-combining systems and phased array radars, can benefit from a diagnostic technique that could determine the amplitude and phase of the electric field radiated from the individual elements (Yang et al., 2001b).

One simple 2-D field-mapping example can be taken from measurements on a microstrip-fed patch antenna. The DUT, outlined in white in the panels of Figure 16, was designed for a resonance frequency of 4 GHz. The antenna has 15 μm thick Cu metallizations on both sides of a 2.55 mm thick Duroid substrate ($\varepsilon_r = 10.3$). The antenna is 8 mm × 11.18 mm, and impedance-matching insets, which have 0.5 mm width and 6.1 mm length, are fabricated along the feed line. The center conductor of the feeding coaxial cable is connected to the antenna through the 2.2 mm wide and 34 mm long microstrip line. During the two-dimensional field mapping, the unit step size of the scanning was 200 μm × 250 μm along the x and y axes, and 80 × 80 points were taken. The total elapsed time for the scanning was ∼ 30 min.

Figure 16 shows the amplitude and phase of the three orthogonal electric field components for the patch antenna. Each amplitude component is normalized to the maximum value within its measurement window. The probe is situated so that it scans at a distance of approximately 100 μm above the plane of the patch antenna. Because the length of the antenna (along the y direction) is designed to be a half-wavelength at the resonance frequency, the measured potential reaches its maximum value at the edges of the patch close to the feed line and at the farthest extent from the feed. Following the potential distribution, the maximum electric field in both the x and y tangential directions, which has been measured using both a LiTaO$_3$ probe and a (110)-oriented GaAs probe, can be observed around the edges of the patch and near the ends of the long dimension. The probe is rotated 90° in order to distinguish between the x and y directions.

For the measurement of the normal electric field component, either the BSO crystal or (100)-GaAs-type probe can be used, as well as any other EO probe with sensitivity to the normal field. The normal field measurement result clearly displays the peak electric field around both the near and far edges of the antenna (relative to the feed end) and concentrated over the metal of the antenna as expected. The x and y components (tangential fields), on the other hand, have their peak amplitudes outside the antenna

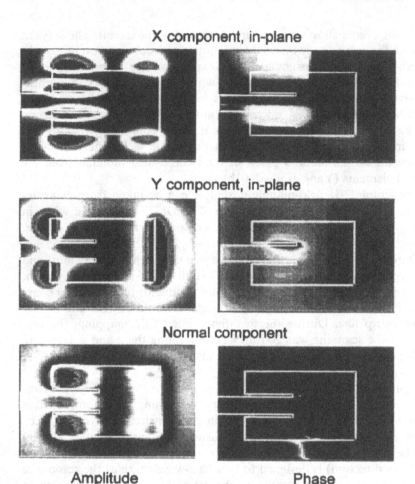

Figure 16 Electro-optic field-mapping measurements of three orthogonal components of the amplitude and phase electric field patterns from a patch antenna at 4.003 GHz. The perimeter of the antenna is indicated by the solid white line, and the probe was scanned at a distance of ~100 μm above the antenna. (Adapted from Yang et al., 2000a.)

pattern around the edges, where the probe captures the electric field as it becomes parallel to the plane of the patch. For the electric field x component, the peak amplitudes are observed along the corners of the long side edges of the antenna and between the patch and the matching-section insets. The four peaks around the corners of the antenna are explained by the potential distribution on the antenna. Due to the phase difference of the voltage between the feed line and the near-side edge of the antenna, which happens to be 180° as indicated by the light and dark shades in the phase part of Figure 16, two peak amplitudes also occur at the impedance-matching insets. For the y component, the electro-optic field map measurement shows peak amplitudes on both the near- and far-side edges of the patch. These two y-component peak amplitudes also correspond well to the expected potential distribution on the antenna.

It should also be noted that the phase of the y component is essentially uniform across the patch, indicating that the high-amplitude field components at the ends of the patch will add together in phase as they propagate into the far field. However, for the normal and x components, half of the strong portions of the field patterns are 180° out of phase with the other half. This comparison points out one of the most important distinctions of radiating structures. In general, most of a radiating structure is designed to deliver information to a point spatially remote from it, i.e., in the far field. Under far-field conditions, the dimensions of the radiating structures are very small compared to the distance between the observation point and the structure. As a result, the fields with opposite polarity (or 180° out of phase) that are also in close proximity to each other cancel out. Thus, the total electric field in both the normal and x components will approach zero because of the cancelation of the out-of-phase terms present in the near-field patterns.

Figure 17 displays the radiation pattern from a more complicated network, that of a pair of monolithic patch antennas. The amplitude and phase have also been measured simultaneously for this network, but they are shown instead as a "composite" electric field distribution. Because the surface plot includes not only the amplitude but also the phase, the value of the electric field varies from positive to negative numbers along the vertical axis. The positive electric field represents a positive y component, whereas the electric field in the negative y direction has a negative value. Theoretically, the patch antenna has two main radiation edges, and thus it should have two identical peak radiating patterns on each edge. However, a small section of the feed edge is sacrificed for the feed microstrip line. As a result, the patch shows a long radiation peak on the front-side edge whereas the feed-side pattern is divided in two. Figure 17 clearly demonstrates the difference between the two microstrip patches and microstrip transmission lines. Basically, the electric field generated from the microstrip has characteristics almost identical to those of the field from the patch. However, Figure 17 shows three positive field peaks at the edge of the patch, whereas the field around the microstrip displays a negative value on the patch side and a positive value on the other side. Therefore, in this case one can easily see that the

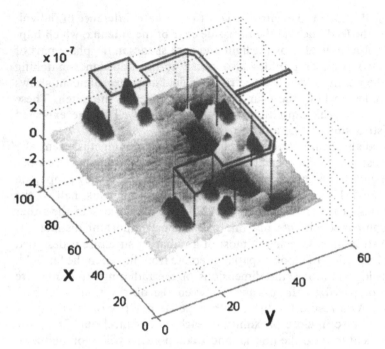

Figure 17 Electro-optic field-mapping measurement of the y component of the electric field from a simple network of two patch antennas. The amplitude and phase information are combined, so that the vertical axis, representing normalized values of electric field strength, has both positive and negative values. (From Yang, 2001; reprinted with permission.)

patches with the uniform-phase electric field are the dominant radiating structures of the array. This highlights one reason that accurate phase measurement is so crucial for a complete near-field analysis and emphasizes one of the most outstanding features of the EO near-field mapping technique.

In order to demonstrate the application of EO field mapping to a complex system, a quasi-optic (QO) power-combining amplifier has also been probed. The array, produced at Lockheed-Martin and also using microstrip patch structures for the input and output antennas, has five columns of active cells and was fed with a plane wave from a horn antenna (Hubert et al., 1999; Yang et al., 2001f). The white circuit outline in Figure 18 shows the location of the patch antennas, the microwave amplifiers (MMIC), and the coupling network that brings the received plane wave from the back of the planar array up to the top surface that is displayed.

Figure 18 shows one of the measurement results obtained by electrooptical mapping at a distance of $< 100\,\mu m$ above the array surface.

The field maps displayed are the normalized amplitude (Fig. 18a) and phase, in degrees (Fig. 18b), of the y component, which was the dominant field component for this array. Because of the high spatial resolution of the EO measurement technique, even though a large area with more than a dozen devices was scanned, the two individual radiating edges of each patch antenna are easily distinguished in Figure 18a. It is quite noticeable that one element at the bottom of the middle column was malfunctioning. Also, it was found that this malfunctioning element disturbed the phase distribution for the array significantly. Despite the electrical failure of this element, most of the patch antennas otherwise show a quite uniform phase distribution.

As the measurement distance from the array increased, it was noticed that the output electric field of the surrounding elements around the malfunctioning element expanded to the area over the malfunctioning element. As a result, the absence of electric field above the malfunctioning element became less significant as the distance increased. Thus, it can be concluded that it is impossible to detect malfunctioning elements above a certain

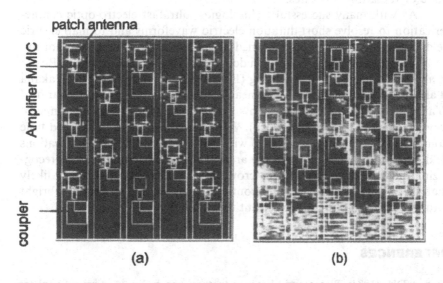

(a)　　　　　　　　　　(b)

Figure 18 Electro-optic field-mapping measurement of the radiating electric field component for a quasi-optic power-combining array of microstrip patch antennas operating at $\sim 35\,\text{GHz}$. (a) Normalized amplitude and (b) phase (in degrees) from the array at a measurement height of less than $100\,\mu\text{m}$. The light areas of the amplitude plot indicate the high-field regions, and the dark areas indicate the presence of essentially no electric field. The phase map displays a high degree of nonuniformity in the bottom half of the plot. (Adapted from Yang et al., 2001b.)

distance from the array with any certainty. This phenomenon emphasizes the importance of accurate near-field measurements—not possible with conventional far-field measurements or low-resolution near-field measurement techniques—for validating the integrity of systems such as QO power-combining arrays.

1.6 CONCLUSION

Throughout the past two decades, the field of optical electronics has grown to the point where it is clearly understood that many future electronic systems will rely on ultrafast optics for enhanced speed and bandwidth capability. It appears that advanced testing and probing at the highest frequencies and shortest time durations will also follow the same trend of exploiting the many benefits ultrafast lasers and optically based tests have to offer. This chapter has introduced electro-optic sampling as the highest bandwidth and most flexible technique for the measurement of electric fields around microwave and millimeter-wave components, whether based on guided-wave or radiating structures.

As with many successful technologies, ultrafast electro-optic characterization to resolve short-duration electric waveforms and to map electric field distributions have both spawned novel applications that were not initially envisioned. These have included temperature measurement (Reano et al., 2001), magnetic field sensing (Elezzabi and Freeman, 1996; Wakana et al., 2000), the transformation of near-field antenna patterns to the far field (Yang, 2001), and EO measurements within the interior of enclosed microwave packages (Yang et al., 2001a). With the demands for high-speed time domain measurements expanding within the optical telecommunications field, and with the engineers in the area of wireless communications recognizing the benefits of near-field microwave characterization, it is very likely that these applications will only continue to grow and maintain a bright future for electro-optic measurement techniques.

REFERENCES

Auston DH. (1984). Picosecond photoconductors: physical properties and applications. In: CH Lee, ed. Picosecond Electronics. Orlando, FL: Academic Press, pp 73–116.

Auston DH, Cheung KP, Valdmanis JA, Smith PR. (1985). Ultrafast optical electronics: from femtoseconds to terahertz. In: GA Mourou, DM Bloom, C-H Lee, eds. Picosecond Electronics and Optoelectronics. Berlin: Springer-Verlag, pp 2–8.

Cheng H, Whitaker JF, Weller TM, Katehi LPB. (1994). Terahertz-bandwidth characteristics of coplanar transmission lines on low permittivity substrates. IEEE Trans Microwave Theory Tech 42:2399–2406.

Cheng H-J, Whitaker JF, Herrick KJ, Dib N, Katehi, LPB, Coutaz JL. (1995). Electro-optic probes: high-permittivity crystals vs. low-permittivity polymers. In: Ultrafast Electronics and Optoelectronics. 1995 OSA Tech Dig Ser. Washington, DC: Opt Soc Am, pp 128–130.

Chwalek JM, Dykaar DR. (1990). A mixer-based electro-optic sampling system for submillivolt signal detection. Rev Sci Instrum 61:1273–1276.

Coleman PD. (2000). Reminiscences on selected millennium highlights in the quest for tunable terahertz-submillimeter wave oscillators. IEEE J Selected Topics Quantum Electron 6:1000–1007.

David G, Tempel R, Wolff L, Jäger D. (1996). Analysis of microwave propagation effects using 2D electro-optic field mapping techniques. Opt Quantum Electron 28:919–931.

Elezzabi AY, Freeman MR. (1996). Ultrafast magneto-optic sampling of picosecond current pulses. Appl Phys Lett 68:3546–3548.

Frankel MY, Whitaker JF, Mourou GA, Valdmanis JA. (1991a). Experimental characterization of external electrooptic probes. IEEE Microwave Guided Wave Lett 1:60–62.

Frankel MY, Gupta S, Valdmanis JA, Mourou GM. (1991b). Terahertz attenuation and dispersion characteristics of coplanar transmission lines. IEEE Trans Microwave Theory Tech 39:910–916.

Frankel MY, Whitaker JF, Mourou GA. (1992). Optoelectronic transient characterization of ultrafast devices, IEEE J Quantum Electron 28:2313–2324.

Grischkowsky D, Duling IN III, Chen JN, Chi C-C. (1987). Electromagnetic shock waves from transmission lines. Phys Rev Lett 59:1663–1666.

Gupta S, Whitaker JF, Mourou G. (1992). Ultrafast carrier dynamics in III-V semiconductors grown by molecular-beam epitaxy at very low substrate temperatures. IEEE J Quantum Electron 28:2464–2472.

Han PY, Zhang X-C. (1998). Coherent, broadband midinfrared terahertz beam sensors. Appl Phys Lett 73:3049–3051.

Heutmaker MS, Cook TB, Bosacchi B, Wisenfeld JM, Tucker RS. (1988). Electro-optic sampling of a high-speed GaAs integrated circuit. IEEE J Quantum Electron 24:226–233.

Heutmaker MS, Harvey GT, Cruickshank DG, Bechtold PE. (1990). Electrooptic sampling of silicon integrated circuits using a GaAs probe tip. Dig 17th Int Conf on Quantum Electron IQEC 90:50–53.

Hubert J, Mirth L, Ortiz S, Mortazawi A. (1999). A 4 watt Ka-band quasi-optical amplifier. IEEE MTT-S Int Microwave Symp Dig. New York: IEEE, 551–554.

Jiang Z, Sun FG, Chen Q, Zhang X-C. (1999). Electro-optic sampling near zero optical transmission point. Appl Phys Lett 74:1191–1193.

Kamogawa K, Toyoda I, Nishikawa K, Tokumitsu T. (1994). Characterization of a monolithic slot antenna using an electro-optic sampling technique. IEEE Microwave Guided Wave Lett 4:414–416.

Keil UD, Dykaar DR. (1992). Electro-optic sampling and carrier dynamics at zero propagation distance. Appl Phys Lett 61: 1504–1506.

Knox WH. (2000). Ultrafast technology in telecommunications. IEEE J Selected Topics Quantum Electron 6: 1273–1278.

Kolner BH, Bloom DM. (1986). Electrooptic sampling in GaAs integrated circuits. IEEE J Quantum Electron QE-22:79–93.

Kolner BH, Bloom DM, Cross PS. (1983). Electro-optic sampling with picosecond resolution. Electron Lett 19:574–575.

Meyer KE, Mourou GA. (1985). Two dimensional E-field mapping with subpicosecond resolution. In: GA Mourou, DM Bloom, C-H Lee, eds. Picosecond Electronics and Optoelectronics. Berling: Springer-Verlag, pp 46–49.

Meyer KE, Dykaar DR, Mourou GA. (1985). Characterization of TEGFETs and MESFETs using the electrooptic sampling technique. In: GA Mourou, DM Bloom, C-H Lee, eds. Picosecond Electronics and Optoelectronics. Berlin: Springer-Verlag, pp 54–57.

Mourou GA, Bloom DM, Lee C-H, eds. (1985). Picosecond Electronics and Optoelectronics. Berlin: Springer-Verlag.

Nees J, Mourou G. (1990). Noncontact electro-optic sampling with a GaAs injection laser. Electron Lett 22:918–919.

Nicolson AM. (1968). Broad-band microwave transmission characteristics from a single measurement of the transient response. IEEE Trans Instrum Measure 17:395–402.

Pfeifer T, Loeffler T, Roskos HG, Kurz H, Singer M, Biebl EM. (1998). Electrooptic near-field mapping of planar resonators. IEEE Trans Antenna Propagat 46: 284–291.

Plows G. (1990). Electron-beam probing. In: RB Marcus, ed. Measurement of High-Speed Signals in Solid State Devices. San Diego: Academic Press, pp 336–382.

Quang D Le, Erasme D, Huyart B. (1995). MMIC-calibrated probing by CW electrooptic modulation. IEEE Trans Microwave Theory Tech 43:1031–1036.

Reano RM, Yang K, Whitaker JF, Katehi LPB. (2001). Simultaneous measurements of electric and thermal fields utilizing an electro-optic semiconductor probe. IEEE Trans Microwave Theory Tech 49:2523–2531.

Rutledge DB, Neikirk DP, Kasilingham DP. (1983). Integrated-circuit antennas. In: KJ Button, ed. Infrared and Millimeter Waves. New York: Academic Press, pp 1–90.

Sasaki A, Nagatsuma T. (2000). Millimeter-wave imaging using an electrooptic detector as a harmonic mixer. IEEE J Selected Topics Quantum Electron 6:735–740.

Shinagawa M, Nagatsuma T. (1990). Electro-optic sampling using an external GaAs probe tips. Electron Lett 26:1341–1343.

Sze SM. (1985). Semiconductor Devices: Physics and Technology. New York: Wiley, pp 8–12.

Valdmanis JA. (1990). Electro-optic measurement techniques for picosecond materials, devices, and integrated circuits. In: RB Marcus, ed. Measurement of High-Speed Signals in Solid State Devices. San Diego: Academic Press, pp 136–219.

Valdmanis JA, Mourou G. (1986). Subpicosecond electro-optic sampling: principles and applications. IEEE J Quantum Electron 22:69–78.

Valdmanis JA, Mourou G, Gabel CW. (1983). Subpicosecond electrical sampling. IEEE J Quantum Electron QE-19:664–667.

Wakana S, Ohara T, Abe M, Yamazaki E, Kishi M, Tsuchiya M. (2000). Fiber-edge electrooptic/magnetooptic probe for spectral-domain analysis of electromagnetic field. IEEE Trans Microwave Theory Tech 48:2611–2616.

Whitaker JF. (1993). Optoelectronic applications of LTMBE III-V materials, Mater Sci Eng B22:61–67.

Williams DF, Hale PD, Clement TS, Morgan JM. (2001). Calibrating electro-optic sampling systems. IEEE MTT-S Int Microwave Symp Dig. New York: IEEE, pp 1527–1530.

Williams RE. (1984). Gallium Arsenide Processing Techniques. Dedham, MA: Artech House.

Wu Q, Litz M, Zhang X-C. (1996). Broadband detection capability of ZnTe electro-optic field detectors. Appl Phys Lett 68:2924–2926.

Yablonovitch E, Hwang DD, Gmitter TJ, Flores LT, Harbison JP. (1989). Van der Waals bonding of GaAs epitaxial liftoff films onto arbitrary substrates. Appl Phys Lett 56:2419–2421.

Yang K. (2001). Application of ultrafast optical techniques to the characterization of mm-wave integrated circuits and radiating structures. PhD Thesis, Univ Michigan, pp 42–45.

Yang K, David G, Robertson S, Whitaker JF, Katehi LPB. (1998). Electro-optic mapping of near-field distributions in integrated microwave circuits. IEEE Trans Microwave Theory Tech 46:2338–2343.

Yang K, David G, Yook J-G, Papapolymerou I, Katehi LPB, Whitaker JF. (2000a). Electrooptic mapping and finite-element modeling of the near-field pattern of a microstrip patch antenna. IEEE Trans Microwave Theory Tech 48:288–294.

Yang K, Katehi LPB, Whitaker JF. (2000b). Electro-optic field mapping system utilizing external gallium arsenide probes. Appl Phys Lett 77:486–488.

Yang K, Katehi LPB, Whitaker JF. (2001a). Electric-field mapping system using an optical-fiber-based electro-optic probe. IEEE Microwave Wireless Comp Lett 11:164–166.

Yang K, Marshall T, Forman M, Hubert J, Mirth L, Popovic Z, Katehi LPB, Whitaker JF. (2001b). Active-amplifier-array diagnostics using high-resolution electro-optic field mapping. IEEE Trans Microwave Theory Tech 49:849–857.

Yariv A, Yeh P. (1984). Optical Waves in Crystals. New York: Wiley, pp 220–287.

11

Terahertz Wave Imaging and Its Applications

Qin Chen and X.-C. Zhang
Rensselaer Polytechnic Institute, Troy, New York, U.S.A.

11.1 INTRODUCTION

Imaging is generally understood as the measurement and replication of the intensity distribution of an active source emitting an electromagnetic wave or the backscattering profile of a passive object or scene [1]. The functionality of imaging can be greatly extended by incorporating spectroscopic techniques in the imaging system. For example, organic functional groups in the specimen can be identified and imaged by their select pattern of absorption wavelength [2]. Depending on what electromagnetic radiation is employed, imaging has been historically categorized as microwave imaging or optical imaging. The application of microwave and optical imaging to basic research and everyday life situations has been extremely successful. Compared with the long history of microwave and optical imaging, terahertz (THz) wave imaging based on optoelectronic THz time domain spectroscopy is in its infancy, emerging only in 1995 [3].

This chapter discusses THz wave imaging and its promising applications. It is organized as follows. In Section 11.2, the introduction of the scanning two-dimensional THz wave imaging system is followed by the description of THz time domain spectroscopy. In Section 11.3 an all-optical scanning THz wave imaging system that employs electro-optic crystals as the THz source and sensor is discussed. An electro-optic THz transceiver and its application to THz time-of-flight imaging are described in Section 11.4. A dynamic aperture technique used to improve the spatial resolution of THz wave imaging beyond the diffraction limit is introduced in Section

11.5. The unique features and applications of THz wave imaging are demonstrated by various THz wave images. The chapter is summarized Section 11.6.

11.2 TERAHERTZ WAVE IMAGING AND TIME DOMAIN SPECTROSCOPY

The term "terahertz" applies to a spectral region that occupies a very large portion of the electromagnetic spectrum between the microwave and infrared optical bands. It is of great importance owing to the rich variety of physical and chemical processes that occur in the terahertz region [4,5]. Although reliable optical and microwave sources have been available for a number of years, it has been relatively difficult to efficiently generate and detect THz radiation. Most terahertz sources have been either low-brightness emitters such as thermal sources or cumbersome molecular vapor lasers, whose far-infrared radiation (with selectable wavelength) is narrowband. Conventionally, *incoherent* detectors such as bolometers and pyroelectric detectors have been used to detect the intensity of THz radiation, but these suffer from low detectivity, cryogenic requirements, and the noise caused by the thermal background radiation. Coherent heterodyne detection of far infrared radiation is limited, by the frequency of the local oscillator and mixer, to less than 1 THz [6]. Since the advent of terahertz time domain spectroscopy during the mid-1980s, these difficulties have been overcome [7–13].

11.2.1 Terahertz Time Domain Spectroscopy

Terahertz time domain spectroscopy (THz-TDS) is based on electromagnetic transients that are generated and detected optoelectrically by femtrosecond laser pulses. These terahertz transients are single-cycle bursts of electromagnetic radiation of typically less than 1 ps duration. Their spectral density spans the range from below 0.1 THz to more than 3 THz. The brightness of these THz transients greatly exceeds that of conventional thermal sources because of the high spatial coherence. The temporally gated detection technique allows direct measurement of the THz electric field in the time domain with a time resolution of a fraction of a picosecond. The detection is thus *coherent*; i.e., both the amplitude and the phase of the THz spectrum can be extracted from the Fourier transform of the detected THz time domain waveform. This is very useful for applications that require the measurement of the real and imaginary parts of the dielectric function. The sensitivity of the gated detection technique is orders of magnitude higher than that of traditional incoherent detection. In addition to this benefit, time-gated coherent detection is immune to incoherent far-IR radiation, making it possible to perform spectroscopy of high temperature materials even in the presence of a strong blackbody radiation background [14].

Two main mechanisms are employed for the generation of THz radiation in a typical THz-TDS system: photoconduction and optical rectification [15,16]. In the first, photoconductors switched by an ultrafast laser pulse function as a radiating antenna. Based on their structure, the antennas can be classified as elementary Hertzian dipole antennas, resonant dipole antennas, tapered antennas, or transmission line or large-aperture antennas [17,18]. For THz generation via optical rectification, electro-optic crystals are used as the THz source [19–23]. With the incidence of an ultrafast pulse on the electro-optic crystals, the different frequency components within the bandwidth of the fundamental optical beam form a polarization that oscillates at the beat frequency between these frequency components. This time-varying dielectric polarization produces a transient dipole that radiates broadband electromagnetic waves. In comparison with the THz radiation from the photoconductive antenna, THz optical rectification radiation has less power but shorter pulse duration and larger bandwidth. The average power level of THz optical rectification radiation can reach several microwatts, depending on the pump power of the ultrafast laser sources.

The photoconductive dipole antenna (PDA) has been the workhorse for coherent detection of THz radiation [10–13] since its first demonstration by Grischkowsky et al. PDAs can generally measure THz radiation from below 100 GHz to over 4 THz. PDAs consist of a coplanar transmission line and a dipole antenna fabricated on a photoconductive material that has an ultrashort carrier lifetime. Subpicosecond response times are readily achievable in PDA chips made of regular GaAs wafers, low-temperature-grown GaAs (LT-GaAs), radiation-damaged silicon on sapphire, and polysilicon composite materials. However, the detection bandwidth of PDAs is limited by the intrinsic resonant behavior of the dipole structure, whose resonant frequency is determined by the dipole length.

Free-space electro-optic sampling (FS-EOS), an alternative coherent detection scheme for THz radiation, was first demonstrated by Wu and Zhang and coworkers [24–29] and is becoming increasingly popular. FS-EOS detects the polarization change of the optical probe beam induced by the THz electric field via the electro-optic Pockels effect in an electro-optic crystal. FS-EOS gives a signal directly proportional to the THz electric field. Because the EO effect is almost instantaneous on the THz time scale, the detection bandwidth is much higher that that of a PDA. A detection bandwidth of up to 70 THz has been reported [30]. In FS-EOS, the choice of sensor crystals is determined by the matching between the phase velocity of the THz wave and the group velocity of the ultrafast probe pulse. For a THz-TDS system using a common Ti:sapphire ultrafast laser, ZnTe is the best sensor crystal for EO sampling, because the velocity-matching condition is well satisfied in ZnTe at an optical wavelength of 822 nm, which also makes ZnTe the best electro-optic crystal for THz optical rectification generation.

Figure 1 shows a schematic of the experimental setup for coherent THz time domain spectroscopy. It consists of a femtosecond laser, a computer-controlled optical scanning delay line, an optically gated terahertz photoconducting dipole emitter, a set of off-axis paraboloidal mirrors for collimating and focusing the terahertz beam, and a ZnTe sensor for terahertz EO sampling in a balanced detection geometry. The output of a femtosecond laser is split into pump and probe pulses by a beam splitter. The pump pulses generate single-cycle bursts of THz radiation with ~1 ps time duration from the photoconductive emitter. To realize EO sampling, the THz pulses and probe optical pulses should arrive at the ZnTe sensor at the same time and travel through the crystal collinearly. The temporal measurement of the THz electric field is achieved by varying the time delay between the pump and probe pulses.

Figure 2a shows a typical time domain THz waveform measured by using the above THz-TDS system with a large-aperture THz emitter. Figure 2b plots the corresponding spectrum obtained by a numerical fast Fourier transform (FFT), showing that the broadband radiation has a peak frequency around 0.3 THz. Using a phase-sensitive lock-in amplifier and

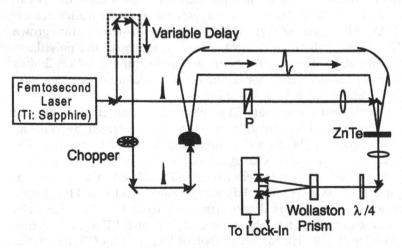

Figure 1 Schematic of the experimental setup of THz time domain spectroscopy.

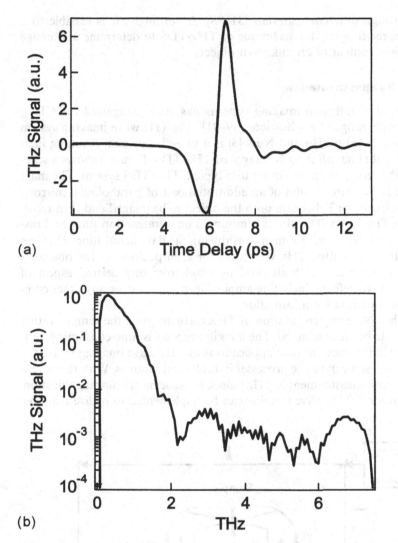

Figure 2 (a) Typical time domain THz waveform and (b) its spectrum from a large-aperture antenna.

time-gated detection, the signal-to-noise ratio of the detected THz electric field can readily exceed 10,000:1, achieving the shot-noise limit [27].

In general, most chemical compounds have strong, highly specific frequency dependent absorption and dispersion in the THz frequency range, which leads to characteristic time domain waveforms when THz radiation passes through different materials [31–38]. Accordingly, it is feasible to a certain degree to apply the technique of THz-TDS to determine and image the chemical content of an unknown object.

11.2.2 Terahertz Imaging

The value of far-infrared imaging systems has been recognized for a long time in a wide range of applications [39–41]. The THz wave imaging system first demonstrated by Hu and Nuss [3] is a two-dimensional scanning imaging system that has all the advantages of THz-TDS. Figure 3 shows a schematic of the setup, which is similar to a typical THz-TDS system. The main difference is the introduction of an additional pair of paraboloidal mirrors, which can focus the THz beam onto the object to be imaged and then recollimate the THz beam. The object is mounted on a translation stage and raster-scanned in the X and Y dimensions during the data acquisition. One can measure the transmitted THz waveform for each position of the object. A THz wave image can be built pixel by pixel from any desired aspect of the measured waveform, including amplitude, phase, and/or any other combination of quantitative information.

In the above implementation of THz wave imaging, the sample, rather than the THz beam, is scanned. The imaging process is time-consuming, putting severe limitations on some applications of THz wave imaging such as in situ studies of some dynamic processes in biological samples. With the advent of electro-optic measurement of THz pulses, the scheme of unit-magnification, two-dimensional THz wave imaging can be implemented to realize real-time

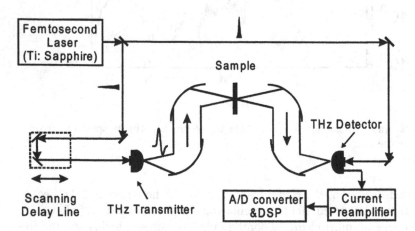

Figure 3 Schematic of experimental setup of 2D scanning THz wave transmission imaging.

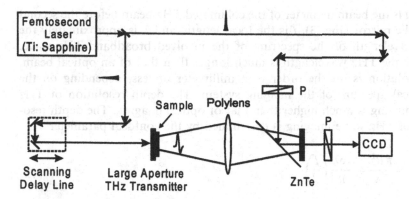

Figure 4 Schematic of the experimental setup for 2D THz wave imaging with a CCD camera.

imaging [42,43]. The schematic of this imaging setup is shown in Figure 4. It consists of a femtosecond laser, an optically gated THz large- aperture antenna, the object to be imaged, a high-density polyethylene lens, and a high-speed CCD camera. In this setup, the spatial distribution of the THz electric field after the sample is mapped onto the EO crystal by the polyethylene lens in a 2f-2f, unit magnification imaging scheme and is up-converted to the optical region by an expanded probe beam via the Pockels effect. The corresponding optical image is then recorded by the CCD camera. The data acquisition rate for this imaging technique is limited only by the speed of the CCD camera, which can be as high as 70 frames/s [44]. This imaging system has the unique capability of making single-shot THz wave images [45].

In this chapter, we concentrate on scanning THz wave imaging.

11.3 AN ALL-OPTICAL SCANNING TERAHERTZ WAVE IMAGING SYSTEM

The spatial resolution of THz wave imaging in the scanning imaging setup is determined by the focal spot size of the THz beam at the sample position. Because the THz electric field is measured with the THz wave imaging system (rather than intensity), a factor of $\sqrt{2}$ should be observed in the different spatial distributions of the electric field versus the intensity distribution of the electromagnetic wave, assuming a Gaussian beam. The spatial resolution of THz wave imaging can be expressed by

$$R = \sqrt{2}\,\frac{4\lambda}{\pi}\left(\frac{f}{d}\right)$$

where d is the beam diameter of the collimated THz beam before the second parabolic mirror (Fig. 3), f is the focal length, and λ is approximately the peak wavelength of the spectrum of the involved broadband THz wave. Because the THz wavelength is much longer than that of an optical beam, the resolution is on the order of a millimeter or less, depending on the numerical aperture of the imaging system. The depth resolution of THz wave imaging is much higher than that of optical imaging. The depth resolution of THz wave imaging is determined by the confocal parameter

$$b = \frac{\pi R^2}{2\lambda} = \frac{16\lambda}{\pi} \left(\frac{f}{d}\right)^2$$

In the conventional scanning THz wave imaging system, photoconducting antennas are used as the sources and detectors of THz radiation. The typical peak frequency of the measured THz radiation lies at ~ 0.3 THz, which is associated with the lifetime of photocarriers and the antenna structure. The spatial resolution and measurable THz bandwidth of the above THz wave imaging system are limited. Because the focal size of a THz beam is proportional to the wavelength, a desirable imaging system should have a THz source that radiates THz pulses with a higher peak frequency and a THz sensor with a faster response. Electro-optic crystals are good candidates because of the nearly instantaneous optical response and THz optical rectification, giving generation and detection with higher peak frequencies.

An all-optical scanning THz wave imaging system was implemented by using electro-optic crystals as a THz source and a THz sensor. This THz wave imaging system can improve diffraction-limited spatial resolution through the use of THz radiation with higher peak frequencies. Its application can extend to the mid-infrared range. With a 12 fs laser and thin EO crystals (on the order of 20 pm) as the THz source and sensor, a system spatial resolution of ~ 50 μm was demonstrated [47].

11.3.1 Terahertz Generation and Detection by Zinc Blende Crystals

When an intense pulsed optical beam with a broad frequency spectrum is incident upon an electro-optic nonlinear crystal, the nonlinear interaction between any two frequency components within the bandwidth will induce a dielectric polarization and radiate electromagnetic waves at their beat frequency. The spectrum of this radiation has a broader bandwidth and a higher peak frequency, than that from photoconducting THz antennas. By the dipole radiation approximation in the far field, the amplitude of the radiated THz electric field is proportional to the second time derivative of the optically induced dielectric polarization. The magnitude and direction of the THz electric field depend on the crystal orientation, owing to the orientational dependence of the dielectric polarization induced in electro-optic crystals.

The coherent detection of the free-space THz pulse can also be effectively realized with high signal-to-noise ratio (SNR) in the electro-optic nonlinear crystal once the phase-matching condition is satisfied. The detection is based on the Pockels effect, where the THz transient electric field provides a bias field to induce a birefringence in the crystal. By collinearly propagating through the electro-optic crystal with a THz pulse, an optical pulse can synchronously probe the transient index change induced by the THz electric field. The detection scheme for the optical probe beam can be either the balanced detection method, where a quarter-wave plate and a pair of balanced photodetectors are used, or the "cross" method, where two crossed polarizers are used to null out the optical transmission when no THz wave is present [48]. In both schemes, the phase change of the probe optical beam induced by the Pockels effect is converted to a variation of light intensity to realize highly sensitive detection by a lock-in amplifier. Because the phase change from the induced birefringence is proportional to the transient THz electric field for the case of balanced detection, the THz temporal waveform can be traced out by sampling the light intensity of the optical probe beam while the temporal delay between the THz pulse and the optical probe beam is varied continuously. As in the case of EO THz generation via optical rectification, the signal level of EO sampling *detection* also shows a dependence on the crystal orientation relative to the electric fields of the THz and optical probe beams. The optimum orientations of the zinc blende crystal used as the THz wave emitter and sensor are listed in Table 1, where L is the thickness of the crystal in the propagation direction of the THz beam and λ is the wavelength of the optical probe beam [49]. For the sensor, the results are applicable for both the balanced and crossed detection schemes. The maximum amplitude of the generated THz electric field and the maximum THz induced phase retardation from the (110) zinc blende crystal are $\sqrt{4/3}$ times and $\sqrt{3/2}$ times as large as those from the (111) zinc blende crystal, respectively. Therefore, in practice, a (110) zinc blende crystal is preferred in order to obtain a better signal-to-noise ratio (SNR).

11.3.2 Characterization of an All-Optical Terahertz Imaging System

Figure 5 schematically illustrates the all-optical scanning THz wave imaging system. The laser source is a regeneratively amplified Ti:sapphire

Table 1 Optimum Orientations of Zinc Blende Crystals as THz Wave Emitters[a,c] and Sensors[b,c]

	THz emitter			THz sensor	
	(110) Crystal	(111) Crystal		(110) Crystal	(111) Crystal
Polarization direction of pump beam	$\langle \pm 1, \mp 1, 1\rangle$	Arbitrary	Polarization direction of THz beam	$\langle \pm \mu 1, 0\rangle$	$\langle \pm 1, \mp 1, 0\rangle$
Polarization direction of generated THz beam	$\langle \mp 1, \pm 1 -1\rangle$	3θ from the polarization direction of the optical pump beam[d]	Polarization direction of probe beam	$\langle \pm 1, \mp 1, 0\rangle$ or $\langle 0, 0, \pm 1\rangle$	$\langle \pm 1, \mp 1, 0\rangle$ or $\langle \pm 1, \pm 1, \mp 2\rangle$
Magnitude of THz electric field	E_{THzmax} $\propto \sqrt{\dfrac{4}{3}}\, d_{14}E_{pump}^2$	$E_{THz} = \text{constant}$ $\propto d_{14}E_{pump}^2$	Magnitude of induced phase retardation	$\Gamma_{max} = \dfrac{2\pi L}{\lambda} n_0^3\gamma_{41} E_{THz}$	$\Gamma_{max} = \dfrac{\sqrt{8/3}\pi L}{\lambda} n_0^3\gamma_{41} E_{THz}$

[a]There is no THz generation from (100) crystals along the propagation direction of the pump beam.
[b]There is no induced phase retardation from the (100) crystals.
[c]For the case of normal incidence.
[d]θ is the angle between the polarization of the optical pump beam and the $(-1, -1, 2)$ direction of the (111) crystal.

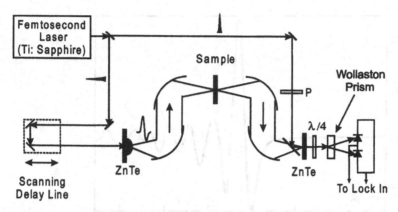

Figure 5 Schematic of experimental setup of all-optical scanning THz wave imaging system with balanced detection geometry.

laser (Coherent RegA 9000). Pulses of 830 nm wavelength and 250 fs are produced at a repetition rate of 250 kHz, providing a pulse energy of 4 μJ. The peak optical power is 16 MW. The THz beam is generated by the pump pulse via optical rectification in a 2 mm thick (110) ZnTe crystal. The sensor is a 4 mm thick (110) ZnTe crystal. An aplanatic hyper-hemispherical silicon lens is attached to the back of the ZnTe emitter to increase the coupling efficiency of THz radiation from ZnTe to free space and to improve the quality of the THz beam for the optics downstream. The first two parabolic mirrors with ∼10 cm off-axis focal length are used to focus the THz beam to the sample to be imaged. The last two are used to collect and focus the transmitted THz beam to the ZnTe sensor. Either balanced or crossed detection geometry can be used for the EO sampling. The sample to be imaged is mounted on a 2D translation stage with a minimum resolution of 1 μm in both the horizontal and vertical directions. The computer-controlled scanning delay line can also be replaced by a galvanometer, which allows faster scanning in order to record the whole THz waveform at each image pixel. The typical temporal THz waveform detected by this system is shown in Figure 6a. The system dynamic range is over 50,000. A Fourier transformation of this waveform reveals its spectrum with a peak at ∼0.9 THz, as shown in Figure 6b.

It has been experimentally verified that the size of the collimated THz beam between the first two parabolic mirrors is frequency-independent, and its diameter is measured to be ∼2.8 cm. At the focal point of the THz beam, its size is frequency-dependent with diameter ∼1.1 mm for the peak frequency of the THz spectrum. To measure the focal size of the THz beam

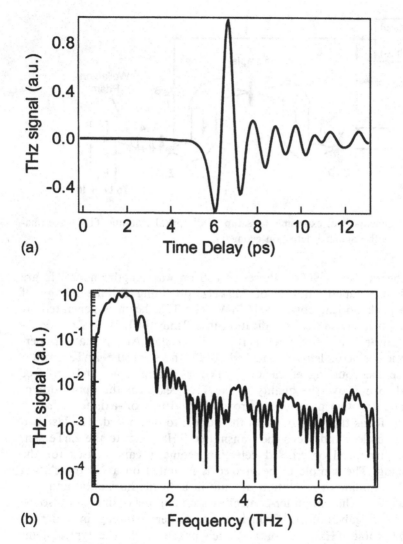

Figure 6   (a) Typical time domain THz waveform and (b) its spectrum from the all-optical THz wave imaging system.

experimentally, one can systematically scan a razor blade across the THz focal point (using a step size of 100 μm) and record the corresponding THz waveforms. The FFT of this set of waveforms yields information on how the amplitude of each THz frequency component changes with the position of the razor blade. From the distance between 90% and 10% of

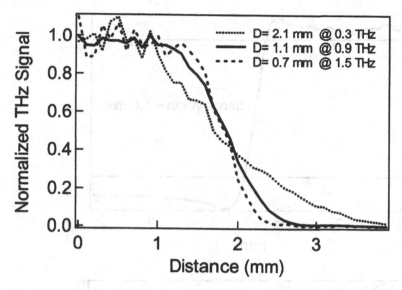

Figure 7 Terahertz focal spot size for different frequencies.

the maximum signal level, the beam size of each THz frequency component can be estimated. Figure 7 shows the results for three different frequencies: 0.3 THz, 0.9 THz, and 1.5 THz.

The spatial resolution of the scanning THz wave imaging system is determined by the THz beam focal size, which is wavelength-dependent. Compared with the conventional THz wave imaging system with photoconductive emitter and detector, the THz beam focal size in an all-optical THz wave imaging system should be smaller, because the generated THz radiation in this system has a higher peak frequency. The diffraction-limited spatial resolution is improved, as can be seen from the following experimental results.

Figure 8 shows the razor blade scanning measurement of the THz focal size. Figure 8a is the result when a (110) ZnTe crystal was used as the THz source. Figure 8b is the result when a GaAs photoconductive antenna is used. From the distance between 90% and 10% of the maximum signal levels, the spatial resolution was estimated to be 1.0 mm and 3.0 mm, respectively. It is seen that the spatial resolution is improved by a factor of 3, because the peak frequency of the THz radiation from the GaAs larger aperture antenna is only 0.3 THz, which is one-third of that from a ZnTe THz optical rectification source.

With a 12 fs ultrafast laser and two 20 μm thick (110) ZnTe crystals as the THz source and sensor, mid-infrared THz radiation was generated and

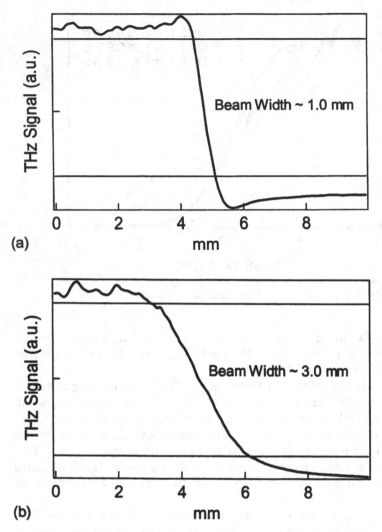

Figure 8 Data traces obtained from scans of a razor blade through the THz beam focal point. (a) THz radiation is from a ZnTe optical rectification source; (b) THz radiation is from a GaAs large-aperture antenna.

detected. The spatial resolution of the all-optical THz wave imaging system was greatly improved. It has been reported that a spatial resolution of ~50 μm has been achieved, allowing the realization of mid-infrared imaging of onion cells [47].

11.3.3   Terahertz Wave Imaging Modes and Applications

Several different imaging modes can be achieved with THz imaging systems, providing different types of information about the specimen or sample. The performance and applicability of the THz imaging system is enhanced by reducing the acquisition times. By introducing a galvanometer-driven time delay into the path of the THz pump beam, it is possible to obtain fast recording of the THz waveform at each image pixel. The acquisition time is sufficiently reduced that THz wave images of a fly on a common house-plant leaf were obtained using this all-optical, rapid-scanning THz wave imaging system.

Figure 9 displays the images in three different modes, showing some unique features of THz wave imaging. In Figure 9a the image is displayed in terms of the peak amplitude of the THz time domain waveform. Note that most of the leaf could not be quite distinguished from the background (free space). The reason is that the peak amplitude of the THz waveform does not change very much when the THz beam goes through free space as opposed to the dry leaf, as can be seen in 9d.

In Figure 9b, the image is displayed in terms of the amplitude of the THz peak frequency. The leaf can now be easily distinguished from the background. This is because absorption of THz radiation by water is quite severe, and the absorption coefficient increases with frequency in the tera-hertz frequency range. This result naturally indicates the application of THz wave imaging in the chemical content mapping of biological objects.

In Figure 9c, the image is displayed in terms of the amplitude at the zero timing of the free-space THz waveform. It is seen that the dry leaf can be distinguished from the background also, because phase information is being taken advantage of. Terahertz pulses experience a phase shift when they go through the sample (leaf). In the time domain, the phase change expresses itself by a small timing shift of the whole THz waveform, as shown in Figure 9d. When the timing between the THz pump and probe beams is fixed at a certain timing position, the image contrast is determined by the difference of THz signals at points *a* and *b* and is thus a great improvement in comparison with 9a. Therefore, the THz wave image in Figure 9c is actually a phase image.

Experimentally, the phase image can be easily realized in an imaging setup without the galvanometer-driven time delay. Terahertz phase images of the watermarks on several common paper currencies have been obtained [50], as shown in Figure 10. Figure 11 illustrates two temporal THz wave-forms obtained from a 100 deutsche mark note. A vertical shift is made to clarify the data. The top one is the THz waveform when the THz beam is transmitted through the watermark. The middle one is the THz waveform

(a)

(b)

Figure 9 Terahertz wave images of a fly on a leaf of a common houseplant (a) displayed in terms of the peak amplitude of THz waveform, (b) displayed in terms of the THz amplitude at peak frequency, (c) displayed in terms of the THz signal intensity at certain fixed timing. (d) Terahertz waveforms transmitted through free space (solid line) and through a leaf (dashed line).

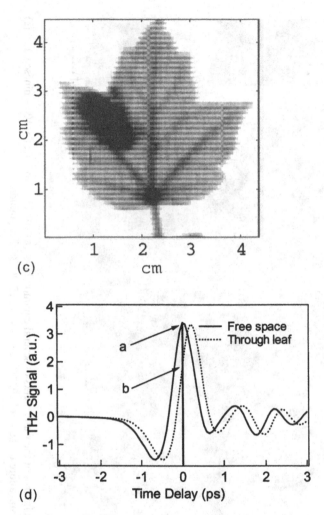

Figure 9 (Continued)

when the THz beam transmits through the normal park of the bill but not the watermark. The waveforms show a timing shift as small as 0.067 ps, which is due to the phase variation of the THz pulses and induced by the different refractive indices of the watermark and normal part of the bill. The difference between the top and middle waveforms is shown on the bottom. During the process of 2D image scanning, the timing of the optical delay line was set at zero, which is the timing position of the peak amplitude of the subtracted waveform. These results demonstrate that the image system

Figure 10 Terahertz wave images of the watermarks of several common bills: (a) 100 U.S. dollars; (b) 100 Chinese yuan; (c) 100 German marks; (d) 50 Singapore dollars; (e) 10,000 Korea won; (f) 20 British pounds.

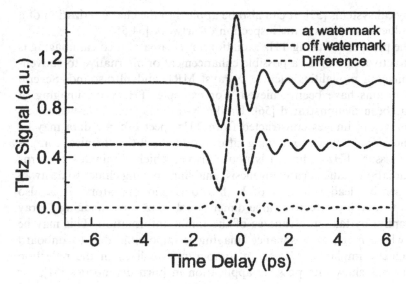

Figure 11 Temporal THz waveforms when the THz beam is focused on and off the watermark. The difference between the two waveforms is shown at the bottom.

is sensitive to a temporal shift as small as 67 fs. Such good sensitivity stems from the phase sensitivity of the gated detection of THz pulses. The method used to set the timing gives the highest image contrast.

The results indicate that THz wave images showing a tiny difference in either refractive index or thickness can be easily obtained in the all-optical THz wave imaging system. If the minimum discernible phase variation is defined as one sampling point's shift of the whole THz time domain waveform then the minimum measurable refractive index difference (Δn) is given by $d \times \Delta n = 1\,\mu m$, for a translation stage with resolution of $1\,\mu m$. $n \times \Delta d = 1\,\mu m$ can be used to calculated the minimum measurable difference of the sample thickness (Δd).

The applications of THz wave imaging are quite diverse. Many molecules have characteristic "fingerprint" absorption spectra in the terahertz region. Terahertz wave imaging systems can detect and map out chemical compositions within an object. The THz wave images of a packaged semiconductor integrated circuit [3] and the voids inside solid plastic or plastic composite parts [51] have been used to illustrate its possible applications in industrial inspection and quality control. Its potential application in the food industry can be seen by the imaging of bacon and also by locating an alien object inside a container of powder or flour [52]. The fast and nonintrusive density mapping of wood shows its application in the study of biological systems [53]. It can also be applied to the characterization of a semiconductor wafer and the inspection of artwork [54,55].

The prospect of using THz radiation in the area of medical imaging is very attractive, serving as a possible enhancement or alternative to conventional imaging modalities such as X-rays, MRI, and ultrasound. Several demonstrations have been achieved. For example, THz wave imaging of teeth has been demonstrated [56]. Unlike X-ray imaging, THz waves can be focused, and images constructed from THz spectroscopic data may be two-dimensional images containing information about the chemical composition of tissue. THz radiation is non-ionizing, which eliminates concerns about radiation-induced carcinogenesis, enabling imaging clinics to do away with expensive lead shielding of both rooms and operators. It is also possible to develop effective three-dimensional THz wave imaging (T-ray tomography) by taking advantage of the timing information. This may be of use where magnetic resonance imaging is impossible due to onboard ferromagnetic implants. THz wave tomography realized in the reflection geometry has shown its possible application in burn diagnostics [51], for example.

In the all-optical THz wave imaging system, THz wave images of a mammographic phantom have been obtained, as shown in Figure 12. The mammographic phantom was originally designed to test the performance of a mammographic system by a quantitative evaluation of the system's ability to image small structures similar to those found clinically. Objects within the phantom simulate calcifications, fibrous calcifications in ducts, and tumors or masses. The detection of these small structures is important in the early detection of breast cancer. The original diameters of the nylon fiber and Al_2O_3 specks are 0.54, 0.4, 0.32, and 0.24 mm, respectively. The thickness of the masses are 0.5 and 0.25 mm, respectively. The results clearly show the ability of the THz wave imaging system in terms of clinical imaging standards. Figure 13 shows the spectral dependence of the experimentally measured absorption coefficient (of the field amplitude) of freshly excised breast tissue. The THz wave image of an alien fiber (diameter = 0.6 mm) hidden inside a freshly excised breast tissue sample is illustrated in Figure 14.

The most promising application of THz wave imaging is the detection of concealed weapons and contraband at airports and other source locations. Conventional systems at high-security checkpoints include metal detectors for personnel and X-ray systems for hand-carried items. These systems have been very effective but have a number of shortcomings. Metal detectors can detect only metal targets, such as ordinary handguns or knives. The effectiveness of these detectors can vary depending on the quantity, orientation, and type of metal. Furthermore, no discrimination is possible between simple innocuous items such as glasses, belt buckles, and keys,

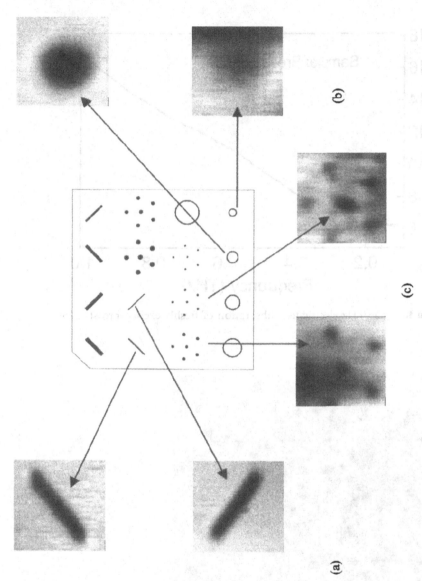

Figure 12 Mammographic phantom and its THz wave images. The central diagram shows the phantom structure, and the surrounding pictures are obtained with the all-optical THz wave imaging system. (a) Nylon fiber; (b) mass; and (c) speck.

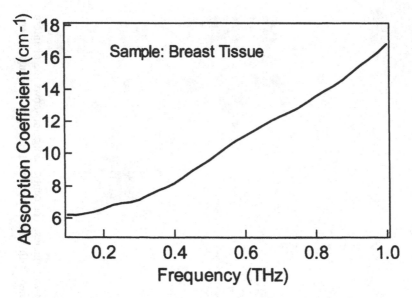

Figure 13 The THz electric field absorption of freshly excised breast tissue.

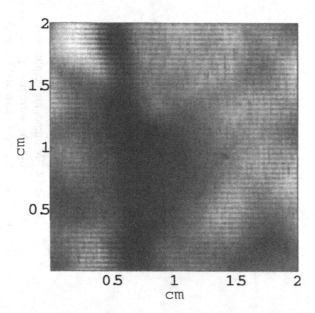

Figure 14 Terahertz wave image of an alien fiber hidden within freshly excised breast tissue.

and actual threats. This leads to a rather high number of nuisance alarms. Modern threats include plastic or ceramic handguns and knives as well as extremely dangerous items such as plastic and liquid explosives. These items cannot be detected with metal detectors. THz wave imaging systems can be used to detect these modern threats. The suitability of THz detection for such security applications has been indicated by the fact that plastic explosives look different under THz radiation and can be distinguished from the molecular structure of suitcases, clothing, and other household materials, as demonstrated by Agrawal et al. [57]. As can be seen from Figure 15. THz waves can readily penetrate common clothing materials. THz wave imaging systems are capable of forming an image of a person as well as many concealed items.

There are some technical and practical considerations that affect the applications of THz wave imaging. For example, the characteristic water absorption spectrum in the THz frequency region is known to have very strong absorption lines, so that even atmospheric moisture is easily detectable. This can be considered an advantage or disadvantage, depending upon the application. For the case where transmission through a significant distance in air is required, the water absorption could be detrimental. However, there are significant intervals in the THz spectrum in which the water absorption is weak, if one needs to avoid this effect. This definitely limits some practical applications of THz wave imaging. On the practical side, one further exciting aspect of THz wave imaging is that of portability and cost. Currently, the ultrafast lasers employed are very large and expensive, but rapid advancements in solid-state laser technology make the arrival of small, inexpensive THz sources a very real possibility.

11.4 TERAHERTZ WAVE TIME-OF-FLIGHT IMAGING WITH AN ELECTRO-OPTIC THz TRANSCEIVER

As discussed in the preceding section, in any traditional experimental setup of THz time domain spectroscopy and THz wave imaging systems, a separate THz wave emitter and sensor are used for the generation and detection of the THz pulses. Electro-optic crystals have been used in THz time domain spectroscopy [58,59], specifically in the all-optical THz wave imaging system. Because the electro-optic detection of the THz beam is basically the reverse process of the rectified generation of THz radiation, it should be possible to implement the THz wave emitter and sensor in the same crystal. Therefore, it is feasible to construct an electro-optic THz transceiver that alternately radiates THz pulses through the effect of optical rectification and detects them via the electro-optic effect [60].

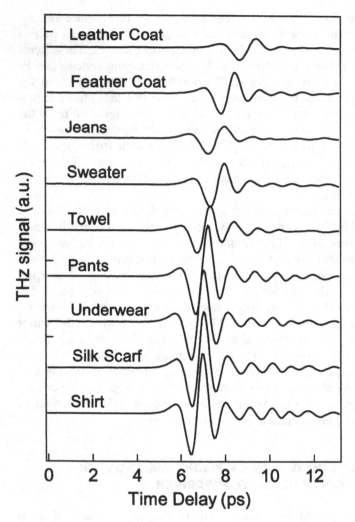

Figure 15 Terahertz wave transmission through 10 different types of cloth. The THz signal is normalized in terms of the peak amplitude of the free-space THz waveform.

Unique applications of such a THz transceiver can be found in THz wave time-of-flight imaging. THz wave ranging, and spectroscopy [61]. Figure 16 shows two typical experimental setups of the electro-optic THz transceiver. The setup in Figure 16a can be used for THz time domain spectroscopy, in which the THz beam going through the sample is collimated.

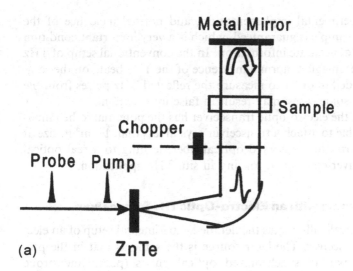

(a)

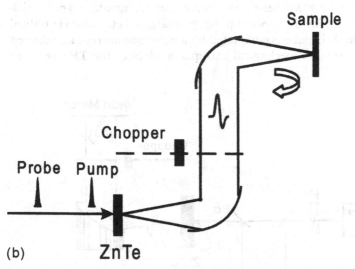

(b)

Figure 16 Schematic of the experimental setup of the electro-optic THz transceiver in two applications. The THz signal is generated and detected in the same ZnTe crystal (a) for spectroscopy and (b) for THz wave time-of-flight imaging.

The setup for THz wave time-of-flight imaging is illustrated in Figure 16b; the THz beam is focused by a second parabolic mirror to reach high imaging resolution. In comparison with the conventional setup of THz wave time-of-flight imaging [62], one immediate advantage of the THz transceiver is that it facilitates the experimental implementation, and normal incidence of the THz beam on the sample is guaranteed, which is a very important condition for the extraction of accurate information. In the conventional setup of THz wave time-of-flight imaging, normal incidence of the THz beam on the sample has to be avoided in order to measure the reflected THz pulses from the sample, which, in some cases, may result in false information.

In principle, the electro-optic transceiver has the potential to be miniaturized. It is possible to attach a transceiver crystal less than $1\,mm^3$ in size at the end of a polarization-preserved optical fiber, leading to a real optical fiber THz transceiver convenient for any in situ THz application.

11.4.1   Experiments with an Electro-Optic THz Transceiver

Figure 17 schematically illustrates the detailed experimental setup of an electro-optic THz transceiver. The laser source is the same as that in the previous section. A pair of synchronized optical pulses (pump and probe pulses) generated by a Michelson interferometer illuminate a 4 mm thick (110) ZnTe crystal. The first optical pulse generates a THz pulse via optical rectification. The THz pulse is collimated by a parabolic mirror and reflected by a metallic mirror. A mechanical chopper modulates the THz beam at

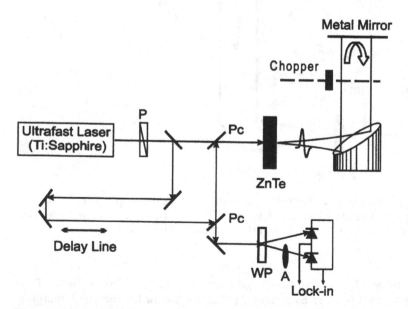

Figure 17   Schematic of the experimental setup of the electro-optic THz transceiver. P, Polarizer; WP, Wollaston prism; Pc, pellicle; A, attenuator.

450 Hz. The relatively low modulation frequency is due to the use of a wide-slot chopper blade to match the large size of the THz beam. The second optical pulse samples the returned THz signal via the electro-optic effect in the same crystal with a lock-in amplifier. It has been theoretically predicted and experimentally verified that the optimum direction of the pump beam polarization is 26° counterclockwise from the Z axis of a (110) ZnTe crystal [49].

Figure 18 shows a set of waveforms measured by moving the metallic mirror along the THz propagation direction with a 1 mm step (round-trip time = 6.6 ps). The first small THz signal (inverted) is a THz pulse reflected from the metallic chopper blade, which was set to be roughly perpendicular to the propagation direction of the THz beam. The position of the chopper was fixed. The second, larger THz signal is a THz pulse reflected from the metallic mirror, and its timing position is shifted with the location of the mirror. The time delay between two THz signals is determined by the round-trip time of the THz pulse traveling between the chopper and the metallic mirror. The reflection from the chopper blade automatically serves as a reference for the system timing calibration. There is an apparent difference of π between the phases of the reflected THz signals from the chopper

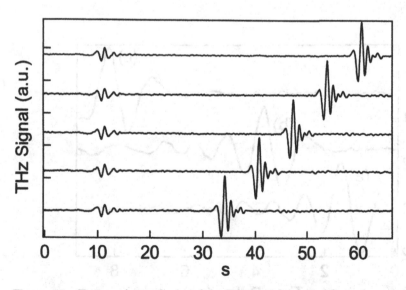

Figure 18 Temporal waveforms of the THz signal reflected from the metallic mirror with 1 mm displacement in each step along the THz propagation direction. The first inverted signal is the reflection from the metallic chopper blade, and the second peak is the signal returned from the metallic mirror.

and those from the metallic mirror, due to the fact that these two signals (measured with the lock-in amplifier) occur during alternating time intervals. Because of the reflected pump beam from the ZnTe, the noise floor is relatively high, leading to a degraded signal-to-noise ratio compared with the previously discussed THz systems. The dynamic range can be improved by increasing the modulation frequency, which should decrease the $1/f$ noise.

11.4.2 Time-of-Flight Imaging with an Electro-Optic THz Transceiver

The delayed THz signal in Figure 18 clearly indicates the possibility of using the THz transceiver for time-of-flight imaging applications. Using the experimental setup shown in Figure 16b, THz wave time-of-flight imaging with an electro-optic transceiver is demonstrated by imaging the surface of a razor blade pasted on a metal mirror. There are three different reflection metal layers in this sample: the metal handle of the razor, the razor surface, and the metal mirror. Figure 19 shows the THz waveforms reflected from the three different metal layers. The timing difference of the peak amplitude indicates the spatial separation of these layers and can be used to construct a time-of-flight image, as shown in Figure 20a. The 2D spatial distribution of

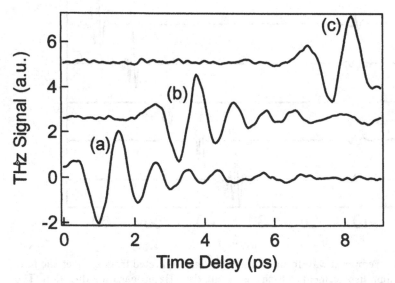

Figure 19 Terahertz waveform reflected from (a) the metal handle of the razor, (b) the razor surface, and (c) the metal mirror.

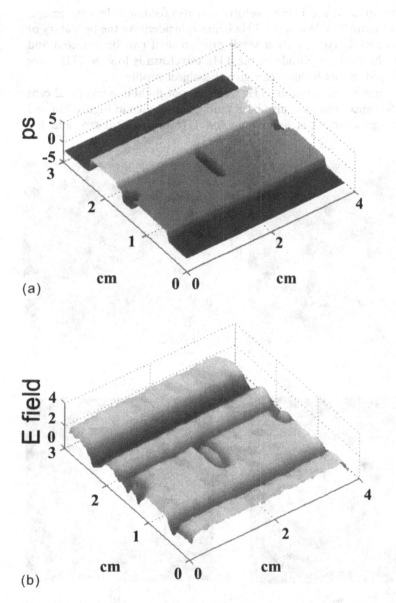

(a)

(b)

Figure 20 (a) Terahertz time-of-flight image of a razor blade. The gray levels represent the timing of the peak amplitude. (b) THz wave image of a razor blade. The gray levels represent the peak amplitudes.

the peak amplitude of the THz waveform can also form a THz wave image, as shown in Figure 20b. When the THz beam is incident on the boundary of the different metal layers, only a small portion of it can be reflected and detected, so the peak amplitude of the THz waveform is low. A THz wave image displayed in this fashion can give the object profile.

In the same imaging system, THz wave time-of-flight images of a 25 cent coin and a 50 pence coin have also been realized, as shown in Figure 21a and 21b. The image contrast is limited by the THz beam focal size and the

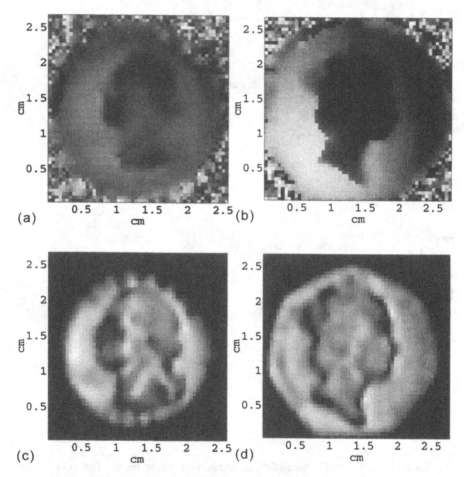

Figure 21 Terahertz wave time-of-flight images of (a) a U.S. quarter and (b) UK 50 pence coin. The gray levels represent the timing of the peak amplitude. THz wave images of (c) U.S. quarter and (d) a UK 50 pence coin. The gray levels represent the peak amplitudes within a certain timing window.

flatness of the metal surface. In the case of a slightly tilted surface, the image can be displayed in terms of the peak amplitude within a certain short-time window to get some additional information on the object, as can be seen in Figures 21c and 21d. The width of the short-time window is decided by the degree of tilt. If two imaged regions are on two different reflection layers and their spatial separation is large enough, the image can be displayed in the above fashion at two suitably chosen central timing positions. Terahertz wave time-of-flight imaging can still be implemented without displaying the image in terms of the timing of the peak amplitude.

Terahertz wave time-of-flight imaging is also applicable to situations where the sample under study consists of a number of spatially separated dielectric layers. Another possible use is in the area of biomedical diagnostics, where the samples may consist of only reflecting surface but with far more complex morphology. An example of such an applications is the study of surface or near-surface skin properties, such as in the diagnosis of burn depth and severity [51]. A reliable noninvasive THz probe of burn depth would be of great value to clinicians, who currently have no such technology.

11.5 NEAR-FIELD THz WAVE IMAGING WITH A DYNAMIC APERTURE

According to Abbe's law, the spatial resolution that can be achieved when imaging with electromagnetic waves is limited by the wavelength of the radiation employed. However, the diffraction limit to spatial resolution is not fundamental but rather arises from the assumption that the light source is typically many wavelengths away from the sample of interest. With the lateral scanning of a light source in close proximity to a sample, one can generate an image at a resolution that is functionally dependent on only the source size and the source-to-sample separation, each of which can, in principle, be made much smaller than the wavelength of the employed radiation. Conventionally, in near-field microscopy, the light incident upon one side of an optically opaque screen is transmitted through a subwavelength-diameter aperture to realize a tiny source. Early demonstrations of near-field microwave scanning microscopes were reported by Soohoo in 1962 [63] and Ash and Nicholls in 1972 [64]. In a recent development, a spatial resolution of $\lambda/10^6$ was achieved [65]. In the optical region, a technological breakthrough in near-field optical microscopy was developed in 1991 by Betzig and Trautmann [66], who used a tapered optical fiber as a near-field probe. The improvement made single-molecule imaging (with resolution of $\sim\lambda/40$–$\lambda/120$) possible.

The concept of near-field microscopy has been adopted to improve upon the diffraction-limited spatial resolution of scanning THz wave imaging system [67,68], in which the peak frequency of THz radiation is generally 0.5 THz. The tiny THz sources necessary for near-field imaging can be realized by a tapered metal tube with a nearly circular aperture of less than 100 μm diameter. This type of source has achieved sub wavelength spatial resolution (better than λ/4). The main disadvantage of this kind of near-field imaging is the high-pass filtering of the THz signal due to the waveguide effect of the tapered metal tube, which not only decreases the THz signal but also seriously limits the transmitted THz bandwidth.

The near-field THz wave imaging with a dynamic aperture [69] discussed in this section does not require the introduction of a physical aperture. In this technique, the near-field THz aperture is realized by a transient photocarrier layer induced by an optical gating beam. The size of the photocarrier layer is determined by the focal size of the optical beam, which can be as small as several micrometers and can be easily adjusted by moving the focusing lens. The thickness of the photocarrier layer is determined by the absorption depth of the optical beam on the semiconductor material, which is generally on the order of several micrometers. Therefore, the high-pass filtering of the THz signal due to the waveguide effect can be avoided, and the transmitted THz bandwidth can be maintained, which is important for THz-TDS on a micrometer scale.

11.5.1 Characterization of a Dynamic Aperture in a THz Wave Imaging System

Figure 22 shows the schematic of the near-field THz wave imaging setup with a dynamic aperture. The laser source is the same as in the previous sections. A THz beam with a peak at 0.9 THz is generated by an optical pump beam through optical rectification in a 2 mm thick (110) ZnTe crystal. The main difference between this setup and the all-optical THz wave imaging system is the introduction of an optical gating beam. This gating beam, which is split off from the same laser output, is focused by a lens with a focal length of 7.6 cm and illuminates a semiconductor wafer (silicon, GaAs, or LT GaAs) along with the THz beam. The average power of the gating optical pulses is ~12 mW, and their arrivial times at the semiconductor samples can be independently controlled by a mechanical delay stage. The experimental setup is actually an optical pump and THz probe experimental setup.

For a typical optical pump and THz probe experiment, when the THz beam is chopped, the transmission of the THz beam through the sample will show a step-function-like variation with the time delay between the THz and

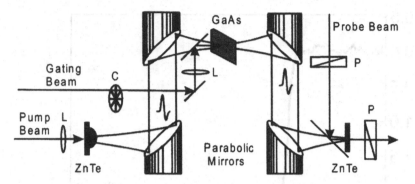

Figure 22 Schematic illustration of near-field THz wave imaging setup with a dynamic aperture. Sample: Silicon, GaAs, or LT GaAs wafers.

optical pump beams. Such variation of the THz wave transmission is not quite discernible when the optical beam is tightly focused on a semiconductor wafer, because the effective modulation area is too small to cause significant modulation of the THz wave transmission. However, by chopping the optical gating beam instead of the THz beam, one can obtain a step-function-like variation of the THz wave transmission with greatly improved sensitivity.

Figure 23 shows the variation of the THz signal transmitted through a GaAs wafer with the time delay between the THz pulse and the gating pulse. The gating beam focal size at the GaAs wafer was estimated as 22 μm by a scanning measurement of a razor blade. The measured THz signal is proportional to the fluctuation of the THz wave transmission during the "on" and "off" states of the optical gating beam chopped by the mechanical chopper. When the optical gating pulse arrives at the GaAs wafer earlier than the THz pulse, the generation of photocarriers will increase the local conductivity of the GaAs wafer and decrease the THz wave transmission in an area determined by the focal size of the optical gating beam. The decrease of the THz wave transmission due to the increase of local conductivity can last (for a time scale of loops) until the electrons and holes diffuse into the sample and eventually recombine [70].

Experiments on a Dynamic Aperture Created on a GaAs Wafer

To investigate how the gating beam size affects the THz signal level, the gating beam size was changed by moving the gating beam focusing lens along the beam propagation direction at a step size of 1 mm. The measured optical gating beam diameters were 22, 38, 65, 88, 126, 160, 184, 212, and 232 μm,

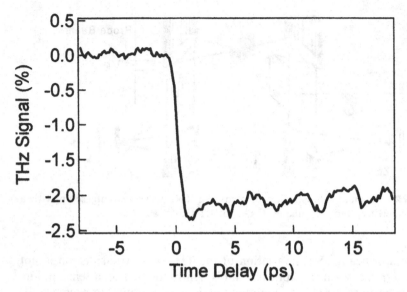

Figure 23 Variation of the THz signal with a time delay between the THz beam and optical gating beam. Negative timing means that the optical gating pulse arrives later than the THz pulse. Sample: GaAs.

respectively. The gating beam power was 12 mW, and a semi-insulating GaAs wafer with resistivity of 2×10^8 ohm·cm was used as the gating material.

By chopping the optical gating beam and moving the time delay between the THz pump and probe beams, a set of THz waveforms were measured, as shown in Figure 24a. A vertical shift is made to clarify the data. To illustrate the modulation effect of photocarriers on THz pulses, a corresponding set of data was also recorded by changing the time delay between the THz and optical gating beams, as shown in Figure 24b. The absolute THz signal in Figures 24a and 24b is calibrated in terms of the peak amplitude of the original THz waveform, which consists of the THz pulses transmitted through the GaAs wafer and was measured by chopping the THz pump beam. It can be seen that when the gating beam size is 22 μm, the peak amplitude of the THz waveform decreases to ~2%. The THz signals after the zero timing in Figure 24b indicate the modulation of photocarriers on THz wave transmission. The FFTs of the set of THz waveforms are normalized and shown in Figure 25. The result clearly shows that the dynamic aperture technique is free of the spatial filter effect, because the thickness of the photocarrier layer on a GaAs wafer is on the order of 1 μm. The THz bandwidth can still be maintained. This is one of the main advantages of this unique technique.

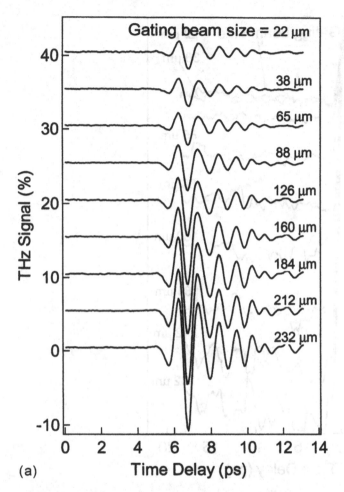

Figure 24 Variation of the THz signal with the gating beam size. The gating beam is chopped. Negative timing means that the optical gating pulse arrives later than the THz pulse. (a) With a time delay between the THz pump and probe beams (a vertical shift is made to clarify data); (b) with a time delay between the THz and optical gating beams.

Figure 26 illustrates how the amplitude of the peak frequency (0.9 THz) varies with the nine different gating beam sizes. It is understandable that not only the size of the optical gating beam but also the conductivity of the local photocarrier layer affects the magnitude of the THz signal. By considering the two factors just discussed, a simulation based on classical aperture

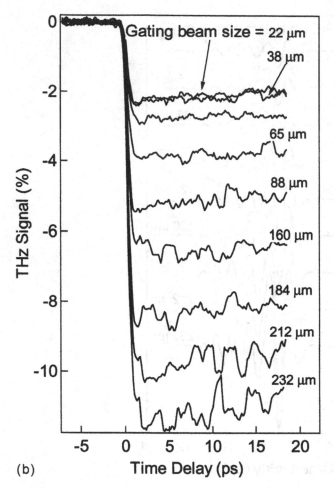

Figure 24 (Continued).

diffraction theory [71] and the Drude model [72] was performed. As shown in Figure 26, the simulation fits and experimental data well.

Experiments on a Dynamic Aperture Created on a Silicon Wafer

Similarly, on an n-doped silicon wafer with resistivity of $\sim$4.5 ohm·cm, a set of THz waveforms similar to those in Figure 24a were measured; then are shown in Figure 27. It can be seen that the decrease of the THz signal with the gating beam size does not follow the same trend as that shown in

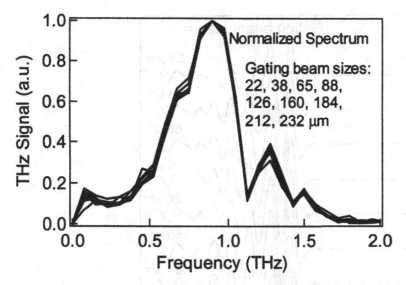

Figure 25 The normalized THz spectrum for nine different gating beam sizes, showing that the dynamic aperture technique is free of the spatial filter effect. Sample: GaAs.

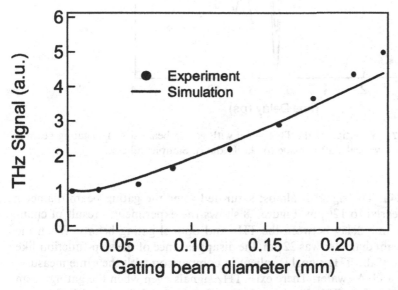

Figure 26 Experimental and simulation results on the variation of THz signal with the gating beam size. Sample: GaAs.

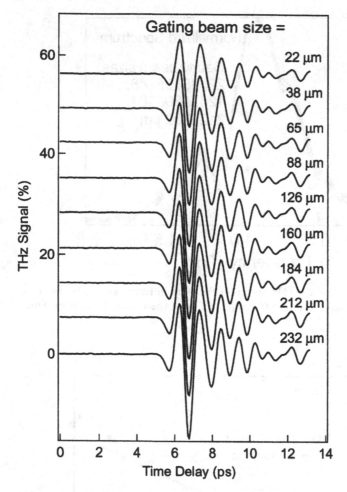

Figure 27 Variation of the THz signal with gating beam size. The gating beam is chopped (a vertical shift is made to clarify data). Sample: Silicon.

Figure 24a. The signal is almost saturated after the gating beam diameter has decreased to 126 μm. Figure 28 shows the experimental result of changing the time delay between the THz and optical gating beams. When the gating beam diameter was 22 μm, the disappearance of the step-function-like variation of the THz signal is salient in comparison with the same measurement on a GaAs wafer. There exist THz signals even when the gating beam arrives later than THz beam, and the magnitude of the signal is approximately four times that of GaAs.

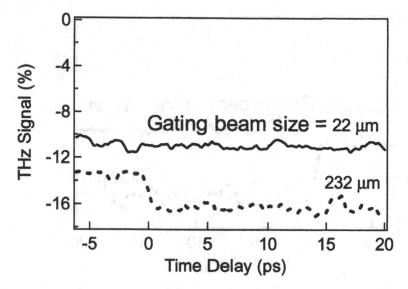

Figure 28 Variation of THz signals with the time delay between the THz beam and chopped optical gating beam. The laser repetition rate was 250 kHz, and the gating power was 13 mW. Negative timing means that the optical gating pulse arrives later than the THz pulse.

The above experimental phenomena are attributed to the local temperature increase induced by the previous gating pulses. The local temperature increase changes the complex refractive index of silicon, especially the imaginary part. Such a temperature effect can last as long as several milliseconds in silicon [73]. Because the repetition rate of our laser is just 250 kHz, the effect can still influence the THz wave transmission even when the gating beam arrives later than the THz beam on the silicon wafer. When the gating beam arrives at the sample earlier than the THz beam, the absence of the photocarrier modulation on the THz beam is due to the small area of the local photocarrier layer, which can modulate only a small portion of the transmitted THz radiation. The modulation effect of the photocarrier is actually buried in the fluctuation of the THz signal caused by the temperature effect. By increasing the gating beam diameter on the sample to 232 μm, as shown in Figure 28, the modulation of the THz beam by the photocarriers can be observed, because the larger gating beam diameter means less light intensity and less temperature increase but greater THz modulation by photocarriers.

The local temperature increase of a silicon wafer caused by the incidence of an optical beam is proportional to the light intensity. Therefore

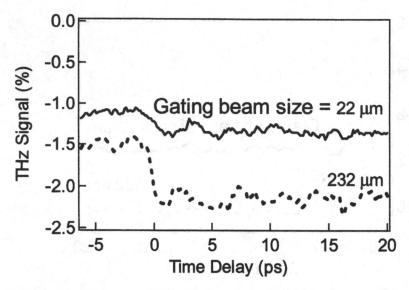

Figure 29 Variation of THz signals with the time delay between the THz beam and chopped optical gating beam. The laser repetition rate was 250 kHz, and the gating power was 1.8 mW. Negative timing means that the optical gating pulse arrives later than the THz pulse.

the THz signal modulated by the temperature effect should decrease when the gating beam power decreases. This phenomenon is shown in Figure 29. The gating beam power was reduced to 1.8 mW, and the THz signal modulated by the photocarriers can be observed when the gating beam size is 22 μm (Fig. 29). By decreasing the laser repetition rate to 10 kHz while maintaining the gating power, the experimental result shown in Figure 30 shows that the measured THz signal was mainly due to modulation of the photocarrier layer.

The above results indicate that the spatial resolution of a THz wave imaging system with a dynamic aperture created on a silicon wafer is limited. In principle, if the modulation of the THz signal is due only to the photocarrier layer, the spatial resolution should be uniquely determined by the diameter of the photocarrier layer or the gating beam size on a semiconductor wafer. However, in the case of a silicon wafer, the existence of the temperature effect actually damages the spatial resolution, because the detected THz signal at a small focal size of the gating beam is mainly due to the temperature modulation and is proportional to the heated area. The heated area can be much larger than the photocarrier layer, owing to the heat diffusion and the relatively long lifetime of the temperature effect.

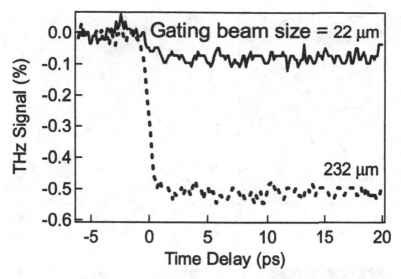

Figure 30 Variation of THz signals with the time delay between the THz beam and chopped optical gating beam. The laser repetition rate was 10 kHz, and the gating power was 1.6 mW. Negative timing means that the optical gating pulse arrives later than the THz pulse.

11.5.2 Terahertz Wave Images

Near-Field THz Wave Images Based on a Dynamic Aperture Created on a GaAs Wafer

When the optical gating beam is incident on a metal line deposited on a GaAs wafer, the THz signal detected in the experimental setup of THz wave imaging with a dynamic aperture will not exist, because there is no THz wave transmission through the metal. Figure 31a shows a THz wave image of a simple metal circuit deposited on a semi-insulating GaAs wafer, obtained with the dynamic aperture technique. The optical gating pulse was focused by a lens with a focal length of 15 cm. Its focused spot size was ~50 μm, and the pulse's timing was set so that it arrived at the sample several picoseconds earlier than the THz pulse. For comparison, the same circuit was also imaged by the conventional THz wave imaging technique in which the optical gating beam was blocked and the THz optical pump beam was modulated by a chopper. The image is shown in Figure 31b. From Figures 31a and 31b it is clear that the spatial resolution and the image contrast are greatly improved by the introduction of a dynamic aperture. The transient photocarrier layer, excited by the optical gating beam, serves as

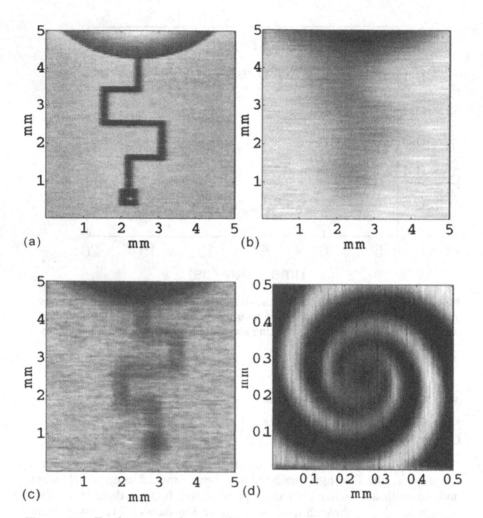

Figure 31 Terahertz wave images of some metal circuits deposited on a GaAs wafer, imaged by (a) near-field technique with a dynamic aperture, (b) the conventional THz wave imaging technique, (c) near-field technique with a dynamic aperture (the sample was flipped), and (d) near-field technique with a dynamic aperture. (Image of a THz antenna, image size 0.5 mm × 0.5 mm.)

a near-field aperture that partially blocks the THz wave transmission in a region much smaller than the focal spot of the THz beam. Because the dynamic aperture and the metal circuit are directly on the sample's front surface, THz propagation is not involved, and the spatial resolution of the THz wave image is thus determined solely by the focal size of the optical gating beam. To test the effect of THz propagation, the circuit sample was flipped and a THz wave image of the circuit on the back side of the wafer was obtained, as shown in Figure 31c. The wafer thickness was ~0.4 mm, so the circuit pattern was 0.4 mm from the dynamic aperture. The THz beam propagated such a distance to reach the pattern, and the spatial resolution and contrast were relatively reduced, clearly showing the diffraction effect. As shown in Figure 31d, the metal pattern of a photoconductive THz antenna whose size is 0.5 mm × 0.5 mm, was also imaged. The result again shows the great improvement of the spatial resolution.

To quantitatively estimate the spatial resolution of the THz wave imaging system with a dynamic aperture, a simple metal pattern deposited on an LT-GaAs wafer was imaged, as shown in Figure 32a. The separation of the two metal films was 50 μm. The optical gating pulse was focused by a lens with f = 7.6 cm, and its focused spot size was 22 μm. As can be in Figure 32b, from the distance between 90% and 10% of the maximum THz signal level, the spatial resolution was estimated to be 36 μm, which is $\sim \lambda_0/10$, where λ_0 is the peak THz wavelength (at 0.9 THz). The difference between the achieved spatial resolution and the gating beam focal size can be attributed to the difference between the coherent and incoherent measurements. The estimated optical gating beam size of 22 μm was based on the measurement of a razor blade scanning, in which the optical light intensity was detected. In the THz wave imaging setup, the THz electric field was measured as a result of the coherent measurement. Theoretically, the spatial resolution of incoherent imaging is $\sqrt{2}$ times better than that of coherent imaging. By considering the above factor, it can be concluded that the spatial resolution of near-field THz wave imaging with a dynamic aperture created on a GaAs wafer is determined by the focal size of the optical gating beam.

The other factor limiting the spatial resolution is the system signal-to-noise ratio (SNR). Decreasing the spot size of the gating beam on the sample can increase the spatial resolution, but it can also decrease the modulation of the THz beam, thereby degrading the SNR. In addition, the laser fluence of the optical gating beam on the GaAs wafer has to be smaller than ~175 mJ/cm² which is the ablation threshold of GaAs [74]. According to classical diffraction theory, the THz signal detected in the imaging system with a chopped optical gating beam decreases in terms of the square law of the gating beam diameter, and a system SNR of 1×10^6 is required to achieve a spatial resolution of 1 μm. The ultimate limit for the achievable spatial resolution is the system signal-to-noise ratio.

It should be emphasized that near-field THz wave imaging with a dynamic aperture and those reported in Refs. 67 and 68 are all accomplished in an illumination mode in which the sample is illuminated by a spatially confined THz wave. The THz wave experiences the diffraction effect before

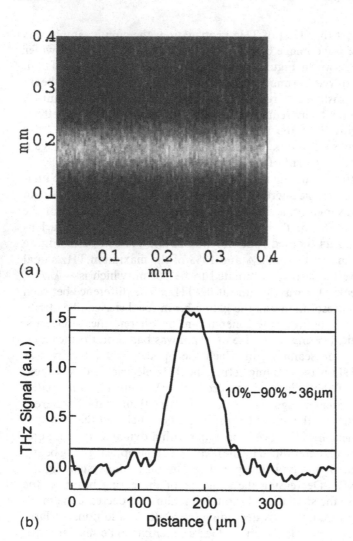

Figure 32 (a) Terahertz wave images of a metal pattern deposited on a LT-GaAs wafer; gap width = 50 μm. (b) The one-dimensional spatial distribution of the THz signal. The distance between 90% and 10% of the maximum THz signal level is 36 μm.

it is detected in the far field. Recently, near-field THz wave imaging in the collection mode was reported [75,76]. A low-temperature (LT) GaAs photo-conducting antenna with a protruding GaAs taper is used as the THz wave detector and is placed close to the sample. The THz wave in this mode does not experience the diffraction before it is detected, and higher spatial resolution is expected to be achieved in this mode. In the collection mode, the $\sqrt{2}$-

fold difference of the spatial resolution between the coherent and incoherent measurements can be avoided. Unfortunately, the waveguide effect that limits the bandwidth of the detected THz pulse still exists and fabrication of such a special antenna detector requires a clean room and is complicated.

To explore some unique imaging capabilities of the imaging system with a dynamic aperture, "THz" was carved on the surface of a GaAs wafer by optically damaging it. The total area affected was 0.3 mm × 0.5 mm. Figure 33 shows the THz wave image of the incised "THz" obtained by using the dynamic aperture technique. It should be noted that this kind of image cannot be obtained by the conventional THz wave imaging technique. To illustrate this, a larger area ($\sim$0.5 cm × 0.5 cm) on the same GaAs wafer was damaged and imaged. The image obtained by the dynamic aperture technique is shown in Figure 34a, and the conventional THz wave imaging technique could not show any difference between the damaged and undamaged areas, as shown in Figure 34b. Conventional THz wave imaging is not sensitive to such a tiny change in the surface quality. However, the depth of THz wave modulation from the photocarriers in those two areas is different, owing to the differences in light absorption and photocarrier lifetime. These imaging results indicate that the dynamic aperture technique has the potential to be used to characterize aspects of semiconductor surfaces such as the doping level and type with a spatial resolution limited by the near-infrared wavelength and the near-field effect.

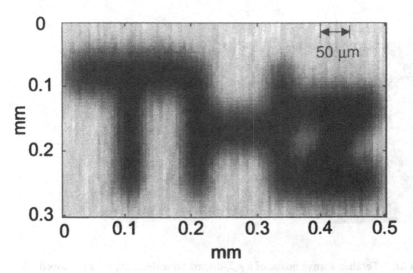

Figure 33 Terahertz wave imaging of the word THz by the near-field technique with a dynamic aperture. The spatial resolution is better than 50 μm.

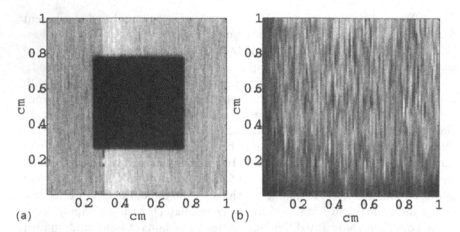

Figure 34 Terahertz wave imaging of an optically damaged area ($\sim 0.5\,\mathrm{cm} \times 0.5\,\mathrm{cm}$) (a) by the dynamic aperture technique and (b) by conventional THz wave imaging.

Near-Field THz Wave Images Based on a Dynamic Aperture Created on a Silicon Wafer

Figure 35 shows THz wave images of a p^+-doped ($5 \times 10^{15}\,\mathrm{cm}^{-3}$) annealed area on an n-doped silicon wafer. Because the wafer conductivity is high in

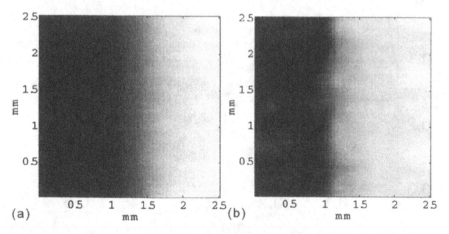

Figure 35 Terahertz wave image of a p^+-doped annealed area on an n-doped silicon wafer (a) by conventional THz wave imaging and (b) by the dynamic aperture technique.

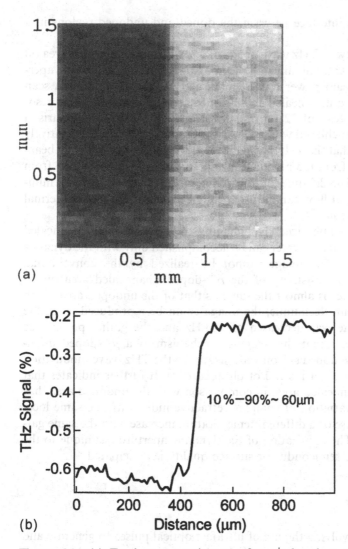

Figure 36 (a) Terahertz wave images of a p^+-doped unannealed area on an n-doped silicon wafer. (b) The one-dimensional spatial distribution of the THz signal. The distance between 90% and 10% of the maximum THz signal level is 60 μm.

the doped and annealed area, the image from the conventional method can, of course, distinguish the doped and undoped regions on the same silicon wafer, as shown in Figure 35a. But the image obtained by the dynamic technique (Fig. 35b) shows a spatial resolution that is much improved, which helps to locate the interface between the doped and undoped regions in a more accurate way.

Figure 36 shows a THz wave image of a p^+-doped unannealed area on an n-doped silicon wafer obtained in the imaging setup with a dynamic aperture. The gating beam power was 1.8 mW, and its size was 22 µm. A scan across the p^+-doped unannealed and n-doped areas indicates a spatial resolution of 60 µm instead of 22 µm, as shown in Figure 36b. In comparison with the resolution achieved with a GaAs wafer, the reason for the relatively poor resolution is that the THz signal detected with the 1.8 mW gating beam power was mainly from the modulation due to the temperature effect from previous pulses. The lifetime of the temperature effect is several milliseconds, and the heat flow can diffuse locally because of the higher thermal conductivity of silicon.

It should be emphasized that imaging of the p^+-doped unannealed area on an n-doped silicon wafer can be accomplished only with the dynamic aperture technique. The image cannot be realized by the conventional method, because the resistivity of the p^+-doped unannealed area on an n-doped silicon wafer is almost the same as that of the undoped area. With the dynamic aperture technique, the image could be reproduced when the laser repetition rate was decreased of 10 kHz and the gating power was 1.6 mW. This reveals that the imaging mechanism of a p^+-doped unannealed area on an n-doped silicon wafer is due to the THz wave modulation of the temperature effect instead of photocarriers. It further indicates that the p^+-doped unannealed area, in comparison with the undoped area, has either a different variation of complex refractive index with the same local temperature increase or a different temeperature increase with the same gating beam power. The application of the dynamic aperture technique to the characterization of semiconductor surface quality is anticipated.

11.6 SUMMARY

The technology involving the use of ultrafast optical pulses to generate and detect terahertz waves is now well developed. The time-gated coherent detection of THz waves allows the direct measurement of THz electric fields in the time domain. Terahertz wave imaging techniques based on THz time domain spectroscopy system are capable of imaging various objects in the THz frequency region. Based on the desired information, THz wave images can be accomplished by displaying the time domain electric field amplitude, spectrum amplitude, phase, or any other combination. The potential applications of THz wave imaging can be found in many fields, including, for example, physics, biology, the semiconductor industry, law enforcement, environmental monitoring, and medical diagnostics. With the development of compact, wall-plug-efficient, and cost-effective femtosecond fiber lasers, THz wave imaging systems are becoming not only practical but also commercially feasible.

REFERENCES

1. D Steinberg. Microwave Imaging with Large Antenna Arrays. New York: Wiley, 1983.
2. D Wetzel, S Levine. Imaging molecular chemistry with infrared microscopy. Science 285:1224 (1999).
3. BB Hu, MC Nuss. Imaging with terahertz waves. Opt Lett 20:1716 (1995).
4. A Finch, PN Gates, K Radcliffe, FF Bentley. Chemical Application of Far Infrared Spectroscopy. New York: Academic Press, 1970.
5. W Chantry. Submillimeter Spectroscopy. New York: Academic Press, 1971.
6. TW Crowe, TC Grein, R Zimmermann, P Zimmermann. Progress toward solid-state local oscillators at 1 THz. IEEE Microwave Guided Wave Lett 6:207 (1996).
7. G Mourou, CV Stancampiano, D Blumenthal. Picosecond microwave pulse generation. Appl Phys Lett 38:470 (1981).
8. DH Auston, KP Cheung, PR Smith. Picosecond photoconducting Hertzian dipoles. Appl Phys Lett 45:284 (1984).
9. PR Smith, DH Auston, MC Nuss. Subpicosecond photoconducting dipole antennas. IEEE J Quantum Electron 24:255 (1988).
10. Ch Fattinger, D Grischkowsky. Point source terahertz optics. Appl Phys Lett 53:1480 (1988).
11. AP DeFonzo, M Jarwala, CR Lutz. Optoelectronic transmission and reception of ultrashort electric pulses. Appl Phys Lett 50:1155 (1987); Far-field characteristic of optically pulsed millimeter wave antennas, Appl Phys Lett 54:2186 (1989).
12. M van Exter, D Grischkowsky. Characterization of an optoelectronic terahertz beam system. IEEE Trans Microwave Theory Tech 38:1684 (1990).
13. MC Nuss, J Orenstein. Terahertz time-domain spectroscopy (THz-TDS). In: G Gruener, ed. Millimeter-Wave Spectroscopy of Solids. Heidelberg, Germany: Springer-Verlag, 1997, and references therein.
14. RA Cheville, D Grischkowsky. Far-infrared terahertz time-domain spectroscopy of flames. Opt Lett 20:1646 (1995).
15. X-C Zhang. Generation and detection of terahertz electromagnetic pulses from semiconductors with femtosecond optics. J Luminescence 66/67:488 (1996).
16. B Greene, P Sateta, D Dykaar, S Schmitt-Rink, SL Chuang. Far-infrared light generation at semiconductor surface and its spectroscopic applications. IEEE J Quantum Electron 28:2302 (1992).
17. I Brener, D Dykaar, LN Pfeiffer, MC Nuss. Terahertz emission from electric field singularities in biased semiconductors. Opt Lett 23:1924 (1996).

18. L Xu, X-C Zhang, DH Auston, B Jalali. Terahertz radiation from larger aperture Si p-i-n diodes. Appl Phys Lett 59:3357 (1991).

19. BB Hu, X-C Zhang, DH Auston, PR Smith. Free-space radiation from electro-optic crystals. Appl Phys Lett 56:506 (1990).

20. YH Jin, X-C Zhang. Terahertz optical rectification. J Nonlinear Opt Phys Mater 4:459 (1995).

21. Y Rice, X-F Ma, X-C Zhang, D Bliss, J Perkin, M Alexander. Terahertz optical rectification from ⟨110⟩ zinc-blende crystals. Appl Phys Lett 64:1324 (1993).

22. A Bonvalet, M Joffre, JL Martin, A Migus. Generation of ultrabroadband femtosecond pulses in the mid-infrared by optical rectification of 15 fs light pulses at 100 MHz repetition rate. Appl Phys Lett 67:2907 (1995).

23. SL Chuang, S Schmitt-Rink, B Greene, P Sateta, A Levi. Optical rectification at semiconductor surface. Phys Rev Lett 68:102 (1992).

24. Q Wu, X-C Zhang. Free-space electro-optic sampling of terahertz beam. Appl Phys Lett 67:3523 (1995).

25. Q Wu, X-C Zhang. Ultrafast electro-optic field sensors. Appl Phys Lett 68:1604 (1996).

26. Q WU, M Litz, X-C Zhang. Broadband detection capability of electro-optic field probes. Appl Phys Lett 68:2924 (1996).

27. Q Wu, X-C Zhang. Design and characterization of traveling-wave electro-optic THz sensors. IEEE-JSTQE 3:693 (1996).

28. A Nahata, DH Auston, TF Heinz, C Wu. Coherent detection of freely propagating terahertz radiation by electro-optic sampling. Appl Phys Lett 68:150 (1996).

29. P Uhd Jepsen, C Winnewisser, M Schall, V Schya, SR Keiding, H Helm. Detection of THz pulses by phase retardation in lithium tantalate. Phys Rev E 53:3052 (1996).

30. A Leitenstorfer, S Hunsche, J Shah, M Nuss. Detectors and sources for ultrabroadband electro-optic sampling: experiment and theory. Appl Phys Lett 74:1516 (1999).

31. M van Exter, C Fattinger, D Grischkowsky. Terahertz time-domain spectroscopy of water vapor. Opt Lett 14:1128 (1989).

32. D Grischkowsky, S Keiding, M van Exter, C Fattinger, Far-infrared time-domain spectroscopy with terahertz beams of dielectrics and semiconductors. J Opt Soc Am B 7:2006 (1990).

33. C Ronne, PO Astrand, SR Keiding. THz spectroscopy of liquid H_2O and D_2O Phys Rev Lett 82:2888 (1999).

34. L Thrane, RH Jacobsen, P Jepsen, RS Keiding. THz reflection spectroscopy of liquid water. Chem Phys Lett 240:330 (1995).

35. JT Kindt, CA Schmuttenmaer. Far-infrared dielectric properties of polar liquids probed by femtosecond terahertz pulse spectroscopy. J Phys Chem 100:10373 (1996).

36. DM Mittleman, RH Jacobsen, R Neelamani, RG Baraniuk, MC Nuss. Gas sensing using terahertz time-domain spectroscopy. Appl Phys B 67:379 (1998).

37. RH Jacobsen, DM Mittleman, MC Nuss. Chemical recognition of gases and gas mixtures using terahertz waveforms. Opt Lett 21:2011 (1996).
38. MC Nuss. Chemistry is right for T-ray imaging. IEEE Circuits Devices 12:25 (1996).
39. TS Hartwick, DT Hodges, DH Barker, FB Foote. Far infrared imagery. Appl Opt 15:1919 (1976).
40. AJ Cantor, PK Cheo, MC Foster, LA Newman. Application of submillimeter wave lasers to high voltage cable inspection. IEEE J Quantum Electron QE-17:477 (1981).
41. NC Currie, FJ Demma, DD Ferris, MC Wicks. Infrared and millimeter-wave sensors for military special operations and law enforcement applications. Int J Infrared Millimeter Waves 17:1117 (1996).
42. Q Wu, T Hewitt, X-C Zhang. Two-dimensional electro-optic imaging of THz beams. Appl Phys Lett 69:1026 (1996).
43. Q Wu, FG Sun, P Campbell, X-C Zhang. Dynamic range of an electro-optic field sensor and its imaging applications. Appl Phys Lett 68:3224 (1996).
44. Z Jiang, XG Xu, X-C Zhang. Improvement of terahertz imaging with a dynamic subtraction technique. Appl Opt 39:2982 (2000).
45. Z Jiang, X-C Zhang. Single-shot spatial-temporal THz field imaging. Opt Lett 23:1114 (1998).
46. Q Chen, Z Jiang, X-C Zhang. All-optical THz image. Proc SPIE 3617:98 (1999).
47. P Han, G Cho, X-C Zhang. Time-domain transillumination of biological tissues with terahertz pulses. Opt Lett 25:242 (2000).
48. Z Jiang, FG Sun, Q Chen, X-C Zhang. Electro-optic sampling near zero optical transmission point. Appl Phys Lett 74:1191 (1999).
49. Q Chen, Z Jiang, M Tani, X-C Zhang. Electro-optic transceivers for terahertz wave applications. J Opt Soc Am B 18:823 (2001).
50. Q Chen, X-C Zhang. Polarization modulation in optoelectronic generation and detection of terahertz beams. Appl Phys Lett 74:3435 (1999).
51. D Mittleman, R Jacobsen, MC Nuss. T-ray imaging. IEEE J Selected Topics Quantum Electron 2:679 (1996).
52. M Herrmann, K Sakai. Objects in powders detected and imaged with THz radiation. Conference on Lasers and Electro-Optics. Washington, DC: Opt Soc Am, 2000, p 479.
53. M Koch. THz imaging: fundamentals and biological applications. Proc SPIE—Int Soc Opt Eng 3828:202 (1999).
54. DM Mittleman, J Cunningham, MC Nuss. Noncontact semiconductor wafer characterization with the terahertz Hall effect. Appl Phys Lett 71:16 (1997).
55. DM Mittleman, M Gupta, R Neelamani, R Baraniuk, J Rudd, M Koch. Recent advances in terahertz imaging. Appl Phys B 68:1085 (1999).
56. DD Arnone, CM Ciesla, N Khammo. Applications of terahertz (THz) technology to medical imaging. Proc SPIE—Int Soc Opt Eng 3828:209 (1999).

57. V Agrawal, T Bork, DW Van Der Weide. Electronic THz reflection spectroscopy for detecting energetic materials. 1998 IEEE Sixth Int Conf Terahertz Electron Proc, p 34.

58. C Winnewisser, P Uhd Jepsen, M Schall, V Schyja, H Helm. Electro-optic detection of THz radiation in LiTaO₃ LiNbO₃ and ZnTe. Appl Phys Lett 70:3069 (1997).

59. A Nahata, A Weling, T Heinz, A wideband coherent terahertz spectroscopy system using optical rectification and electro-optic sampling. Appl Phys Lett 69:2321 (1996).

60. Q Chen, Z Jiang, M Tani, X-C Zhang. Electro-optic terahertz transceiver. Electron Lett 36:1298 (2000).

61. Q Chen, X-C Zhang. Terahertz imaging with an electro-optic transceiver. LEOS Annual Meeting, Rio Grande, Puerto Rico, 2000.

62. D. M. Mittleman, S Hunsche, L Boivin, MC Nuss. T-ray tomography. Opt Lett 22:904 (1997).

63. R Soohoo. A microwave magnetic microscope. J Appl Phys 33:1276 (1962).

64. E Ash, G Nicholls. Super-resolution aperture scanning microscope. Nature 237:510 (1972).

65. B Knoll, F Keilmann, R Guckenberger. Contrast of microwave near-field microscopy. Appl Phys Lett 70:2667 (1997).

66. E Betzig, J Trautmann. Near-field optics: microscopy, spectroscopy, and surface modification beyond the diffraction limit. Science 257:189 (1992), and references therein.

67. S Hunsche, M Koch, I Brener, MC Nuss. THz near-field imaging. Opt Commun 150:22 (1998).

68. K Wynne, D Jaroszynski. Superluminal terahertz pulses. Opt Lett 24:25 (1999).

69. Q Chen, Z Jiang, GX Xu, X-C Zhang. Near-field terahertz imaging with a dynamic aperture. Opt Lett 25:1122 (2000).

70. B Greene, P Sateta, D Dykaar, S Schmitt-Rink, SL Chuang. Far-infrared light generation at semiconductor surface and its spectroscopic applications. IEEE J Quantum Electron 28:2302 (1992).

71. JD Jackson. Classical Electrodynamics. New York: Wiley, 1999, Chap 10.

72. N Katzenellenbogen, D Grischkowsky. Electrical characterization of 4 THz of n- and p-type GaAs using THz time-domain spectroscopy. Appl Phys Lett 61:840 (1992).

73. J England. Time-resolved reflectivity measurement of temperature distributions during swept-line electron-beam heating of silicon. J Appl Phys 70:389 (1991).

74. A Cavalleri, K Sokolowski-Tinten, J Bialkowski, D Von Der Linde. Femtosecond melting and ablation of semiconductors studied with time of flight mass spectroscopy. J Appl Phys 85:3301 (1999).

75. O Mitrofanov, I Brener, MC Wanke, J Federick. Near-field microscope probe for far infrared time domain measurements. Appl Phys Lett 75:591 (2000).

76. O Mitrofanov, I Brener, MC Wanke, J Federici. Terahertz near-field microscopy based on a collection mode detector. Appl Phys Lett 77:3496 (2000).

12

Phase-Controlled Few-Cycle Light

G. Tempea, A. Apolonski, and F. Krausz
Technische Universität Wien, Vienna, Austria

R. Holzwarth and T. W. Hänsch
Max-Planck-Institut für Quantenoptik, Garching, Germany

12.1 INTRODUCTION

12.1.1 Recent Advances in Ultrafast Optics

The 1990s brought about dramatic advances in controlling ultrashort-pulse optical radiation. The quest for ever shorter laser pulses led to pulse durations as short as approximately twice the oscillation period of the carrier field ($T_0 \approx 2.6\,\text{fs}$ at $\lambda_0 = 0.8\,\mu\text{m}$, the center wavelength of the titanium-doped sapphire laser), approaching the limit set by the laser cycle (Cheng et al., 1998; Baltuska et al., 1997; Morgner et al., 1999; Sutter et al., 1999). Furthermore, the frequency sweep (chirp) and amplitude envelope of femtosecond pulses can now be tailored (e.g., Weiner, 2000) for specific applications, such as coherent control of molecular dynamics, often referred to as femtochemistry (Judson and Rabitz, 1992). Recently it was demonstrated that techniques similar to those used for controlling chemical reactions can also be efficiently employed for influencing products of strong-field interactions, such as high-harmonic radiation (Bartels et al., 2000). This process, if driven by few-cycle pulses (Spielmann et al., 1997; Schnürer et al., 1998), is capable of delivering X-ray pulses shorter than the oscillation period of the driving laser (Drescher et al., 2001) and has the potential to produce pulses shorter than 1 fs in duration (e.g., Brabec and Krausz, 2000). The parameters of pulses near 1 fs and possibly attosecond pulses emerging from this process have been predicted to sensitively depend on how the oscillations of the electric field

$E(t) = A(t) \exp[-i(\omega_0 t + \varphi)] + \text{c.c.}$ fit within the amplitude envelope (de Bohan et al., 1998; Tempea et al., 1999). This is determined by φ, which has been referred to as the absolute or carrier envelope phase of light pulses (Xu et al., 1996). This parameter, which could not be accessed experimentally until recently, is the focus of this chapter. Its control and measurement in femtosecond pulses will have dramatic impacts on frequency domain and time domain metrology alike.

Early investigations (Xu et al., 1996) reviewed in Section 12.2 revealed that laser modelocking is able to produce pulses with reproducible envelope and chirp but fails to deliver pulses with constant carrier envelope phase φ, i.e., with reproducible field evolution (Poppe et al., 1998). Recent breakthroughs in stabilization and control of φ (Jones et al., 2000; Apolonski et al., 2000) with subcycle precision (Poppe et al., 2001) in modelocked lasers now pave the way to controlling atomic processes on sub-light-period time scales. Simultaneously, controlling φ in the output of a modelocked laser provides an optical frequency ruler of unprecedented precision for high-resolution spectroscopy and frequency metrology (Holzwarth et al., 2000; Diddams et al., 2000). In Section 12.3 the implementations of phase-controlled modelocking in Boulder (Jones et al., 2000), Garching (Holzwarth et al., 2000), and Vienna (Apolonski et al., 2000) are reviewed. The former two were designed for frequency metrology, and the latter, for time domain measurements. Section 12.4 discusses the different demands that time domain and frequency domain applications make on phase-controlled modelocked laser oscillators. The former calls for a carrier envelope phase jitter much smaller than π over extended periods of time and reliable techniques for measuring this jitter (Poppe et al., 2001), but the cavity length need not be stabilized. By contrast, the latter tolerates a large jitter exceeding π but requires a cavity length stabilized to a reference radio-frequency (RF) or optical oscillator.

The techniques developed so far (Sec. 12.3) for stabilizing φ in the pulse train emitted by a modelocked laser have made it possible to fix the pulse-to-pulse change in φ but failed to control the *value* of the carrier envelope phase. Measurement and control of φ is of central importance for the generation of few-cycle light with controlled evolution of the fields, which will permit the control of atomic processes on a sub-light-period time scale. Perhaps the most striking manifestation of this capability will be the *reproducible* generation of sub-femtosecond XUV/X-ray pulses. Strong-field processes will not only benefit from phase control but may also help to achieve it, as will be shown in Section 12.5. Intense few-cycle pulses with a φ of stabilized and known value together with precisely synchronized sub-femtosecond X-ray pulses will open up a new chapter in the control and measurement of ultrafast microscopic processes.

12.1.2 The Role of the Carrier Envelope Phase in Strong-Field Interactions

The recognition of the fact that not only the full width at half-maximum (FWHM) duration of the pulses but also their complex amplitude envelope, $A(t) = |A(t)| \exp[-i\beta(t)]$, play significant roles in interactions of femtosecond pulses with matter has triggered the development of diagnostic techniques capable of determining these parameters. Frequency-resolved optical gating (FROG) (Trebino et al., 1997) and spectral phase interferometry for direct electric field reconstruction (SPIDER) (Iaconis and Walmsley, 1998) have been the most widely used techniques providing access to both the amplitude envelope $|A(t)|$ and the chirp $\beta(t)$ of ultrashort light pulses. These methods, however, do not allow reconstruction of the full electric field $E(t) = A(t) \exp[-i(\omega_0 t + \varphi)] + \text{c.c.}$ because they fail to determine the carrier envelope or absolute phase φ.

This appears to become a severe deficiency in the few-cycle regime, where the pulse duration approaches the field oscillation cycle T_0 and hence the evolution of the electric field $E(t)$ becomes substantially dependent on the carrier envelope phase. This dependence on φ becomes particularly important in the strong-field regime, where dynamics of processes such as optical field ionization are extremely sensitive to minor changes in the strength of the electric field. Figure 1 depicts the maximum field strength of a pulse with a sech^2-shaped intensity envelope carried at a wavelength of 800 nm (Ti:sapphire laser) as a function of the carrier envelope phase φ for different pulse durations τ. As expected, the phase dependence rapidly decreases with increasing pulse duration. For $\tau \approx T_0$, the variation of the maximum field strength $E_{\max}$ reaches almost 10% and remains appreciable up to $\tau \approx 4T_0$ (of the order of 1%). Changes in $E_{\max}$ of this order of magnitude may substantially change the optical field ionization rate, as we shall see in Section 12.5, and thereby influence strong-field processes triggered by ionization, such as high-order harmonic generation (Brabec and Krausz, 2000).

12.1.3 Impact of Carrier Envelope Phase Control in the Frequency Domain

In the frequency domain the periodic pulse train of a modelocked laser can be described as a comb of equidistant modes at $f_n = f_0 + nf_r$, where f_r is the repetition rate of the laser pulses. It can be readily shown that a finite pulse-to-pulse carrier envelope phase shift $\Delta\varphi$ results in $f_0 = (\Delta\varphi/2\pi)f_r$. Stabilizing φ in addition to $A(t)$ in the laser output implies that the full electric field $E(t)$ should exhibit the same periodicity as the envelope $A(t)$ in the time domain. This corresponds to a harmonic spectrum in

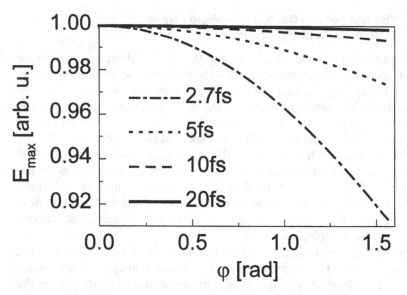

Figure 1 The maximum field strength of a hyperbolic secant pulse with a carrier wavelength of 800 nm, as a function of the absolute phase, for different pulse durations.

the frequency domain, implying $f_0 = 0$, i.e., that phase-locked laser modes are harmonics of the laser pulse repetition rate. The intimate relationship between the temporal evolution of φ and the value of f_0 has two important implications: (1) Phase control in the time domain gives rise to precisely known frequencies of the oscillating modes and (2) control of φ may possibly be achieved by frequency domain techniques (Telle et al., 1999; Reichert et al., 1999).

Locking f_0 and f_r to microwave frequency standards gives rise to an ultrabroadband frequency comb that can be used as a "ruler" in the frequency domain to measure differences between optical frequencies. This was recognized more than 20 years ago when the frequency comb of a mode-locked picosecond dye laser was first used as an optical ruler to measure frequency gaps in sodium (Eckstein et al., 1978), followed by a series of related investigations (Ferguson et al., 1979; Kane et al., 1986; Wineland et al., 1989). The gaps bridged in this manner remained small until the recent availability of broadband femtosecond Ti:sapphire lasers, which offer direct access to terahertz frequencies (Udem et al., 1999a; Holzwarth et al., 2000). Along with extracavity frequency broadening the heterodyne stabilization techniques, Kerr lens modelocked lasers are now capable of synthesizing a stabilized optical frequency comb spanning a full octave (Jones et al., 2000; Holzwarth et al., 2000). With suitable diagnostics these techniques have been shown to be capable of delivering few-cycle pulses with a carrier envelope phase "frozen" to within a tiny fraction of π (Apolonski et al., 2000; Poppe et al., 2001), which is prerequisite for time domain applications.

12.2 CARRIER ENVELOPE PHASE JITTER IN MODELOCKED OSCILLATORS

12.2.1 Origin of Carrier Envelope Dephasing

The carrier envelope phase in a modelocked oscillator was first experimentally addressed several years after sub-10 fs pulses in the few-cycle regime had been generated in modelocked laser oscillators (Xu et al., 1996). The carrier envelope phase φ of consecutive pulses in a train $E_n = A_n(t) \times \exp[-i(\omega_0 t + \varphi_n)] + \text{c.c.}$ (where ω_0 is the carrier frequency) emitted from a modelocked laser is expected to change by $\Delta\varphi_n = \varphi_{n+1} - \varphi_n = \Delta\varphi_0 + \delta_n$. The predictable part, $\Delta\varphi_0$, of this phase change originates from the difference between the effective group velocity v_g and the phase velocity v_p at the carrier frequency in the laser cavity and represents the mean value of $\Delta\varphi_n$ averaged over many pulses, $\Delta\varphi_0 = \langle\Delta\varphi_n\rangle$. The carrier envelope phase shift experienced by a pulse upon propagation through a transparent material of length L and refractive index n can be expressed as

$$\Delta\varphi = 2\pi f_c L \left(\frac{1}{v_p} - \frac{1}{v_g}\right) = 2\pi L \left(\frac{dn}{d\lambda}\right)\Big|_{\lambda_0} = \pi \frac{L}{L_d} \tag{1}$$

where L_d is the propagation length over which φ gets shifted by π. This dephasing length is $L_d \approx 20\,\text{cm}$ in air and $\approx 19\,\mu\text{m}$ in sapphire. Comparing these values with those of the propagation lengths in the respective media in a Ti:sapphire oscillator, we may conclude that the carrier envelope dephasing experienced by a laser pulse during a resonator round-trip amounts to a large integer multiple of 2π plus a rational fraction of 2π. We denote this physically relevant part with $\Delta\varphi_n$ and refer to it throughout this chapter as pulse-to-pulse or round-trip carrier envelope phase shift. The length of the laser cavity can, in principle, be tuned such that the round-trip phase change would be equal to an integer multiple of 2π, and all the pulses in the emitted train would have a constant absolute phase, affected only by small random changes δ_n. However, even small values of δ_n rapidly accumulate to a large ($\gg 2\pi$) jitter of φ.

12.2.2 Measurement of the Pulse-to-Pulse Carrier Envelope Phase Shift

The time-averaged pulse-to-pulse carrier envelope shift $\Delta\varphi_0$ induced by cavity dispersion was first measured by a dispersion-balanced cross-correlator, as depicted in Figure 2 (Xu et al., 1996). For accurate balance of the dispersion in the interferometer arms, a delay section equal to the resonator round-trip was evacuated in the long arm, while pulse splitting and recombination were performed with identical broadband dielectric coatings on opposite sides of the beam splitter. The second harmonic signal produced by the combined pulses in a β-barium borate (BBO) crystal was detected with a photomultiplier (PMT). The recorded interferometric cross-correlation function,

$$G(\tau) = \int \{[E_n(t) + E_{n+1}(t - \tau)]^2\}^2 \, dt \tag{2}$$

is sensitive to the phase difference $\Delta\varphi_n$ between the fields $E_n(t)$ and $E_{n+1}(t)$ of the successive pulses from the laser. The correlator shown in Figure 2 can record the cross-correlation trace only by using a large number of pulses emitted by the laser over an extended period of time. As a consequence, the recorded cross-correlation provided information about the time-averaged value $\Delta\varphi_0 = \langle\Delta\varphi_n\rangle$ only. $\Delta\varphi_0$ can be inferred (up to an integer multiple of 2π) from the delay $\Delta\tau$ of the central fringe with respect to the peak of the envelope of the cross-correlation function as shown in Figure 3 by using the simple relationship

$$\Delta\tau = \Delta\varphi_0/\omega_0 \tag{3}$$

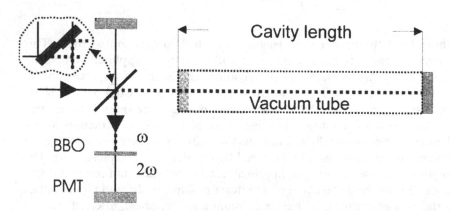

Figure 2 Dispersion-balanced cross-correlator used for the measurement of the pulse-to-pulse change of the carrier envelope phase.

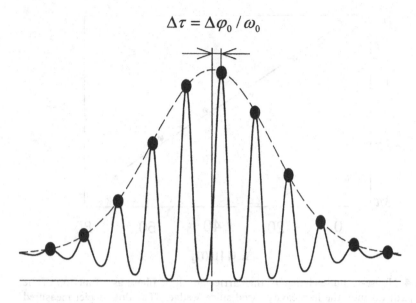

$$\Delta \tau = \Delta \varphi_0 / \omega_0$$

Figure 3 Calculated cross-correlation function of two pulses with absolute phases φ and $\varphi + \Delta \varphi$. The difference $\Delta \varphi$ can be determined from the position of the fringe peaks on the envelope.

In order to control $\Delta \varphi_0$, a pair of wedged fused silica plates were introduced into the laser resonator, contributing to $\Delta \varphi_0$ by $\pi L_g / L_d$ according to Eq. (1). Varying the path length L_g in the thin fused silica wedges allows an investigation of the dependence of $\Delta \varphi_0$ on the resonator dispersion. Owing to the short dephasing length in fused silica ($L_d \approx 29 \, \mu$m), $\Delta \varphi_0$ can be tuned over the full $(0, 2\pi)$ range without notably affecting the pulse shape. Figure 4 compares the measured changes of $\Delta \varphi_0$ (dots) with the calculated values obtained from the wavelength dependence of the refractive index $n(\lambda)$ by using Eq. (1) (solid line).

Although it provides insight into the mechanisms determining the carrier envelope slip in modelocked oscillators and demonstrates that the cavity dispersion (i.e., the difference between phase delay and group delay) provides a means of control on the carrier envelope phase, the above-described experiment does not permit measuring its jitter, for two reasons. First, reliance on many laser pulses for recording $G(\tau)$ prevents fluctuations δ_n of the pulse-to-pulse phase shift around its mean value $\Delta \varphi_0$ from being measured. Second, the accuracy of this technique is very limited ($\approx \pi/10$). Hence even if a single-shot measurement of $G(\tau)$ allowed δ_n to be determined, the upper limit such a measurement could set upon the carrier envelope phase jitter

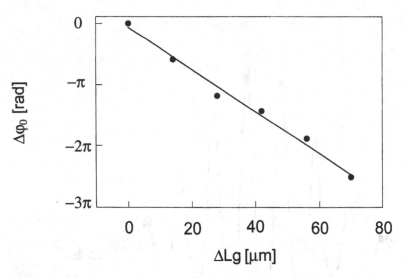

Figure 4 Pulse-to-pulse change of the carrier envelope phase as a function of the optical path through the intracavity fused silica wedges. The dots depict measured data, whereas the line was calculated by means of Eq. (1).

would guarantee phase stability only for a few successive pulses in the output pulse train (over a period of a fraction of a microsecond). Carrier envelope phase stability for extended periods of time calls for techniques capable of determining $\Delta\varphi_n$ with a precision of better than $10^{-6}\pi$, which is far beyond the capability of this cross-correlation technique.

12.2.3 Energy-Noise-Induced Carrier Envelope Phase Jitter

In spite of its shortcomings, this simple interferometric cross-correlation of successive few-cycle laser pulses allows us to assess the influence of pulse energy variations on the pulse-to-pulse phase shift and estimate the resultant jitter of φ. The root mean square (rms) of the random shifts in φ accumulated over a time interval T_m due to fluctuations of the pulse energy W_n (which are assumed to be the main source of phase fluctuations in a wide frequency range) is given by

$$\sigma_{\varphi(W)}(T_m) = \frac{W_0}{\sqrt{2\pi}}\left|\frac{\partial\Delta\varphi}{\partial W}\right|_{W_0}\left(\int\limits_{1/T_m}^{f_r/2} S_W(f)\frac{f^2}{f^2}\,df\right)^{1/2} \tag{4}$$

where W_0 is the mean value of the intracavity pulse energy and $S_W(f)$ is the power spectral density of the pulse energy fluctuations around this mean value (Xu et al., 1996). Both can be routinely measured. The dispersion-balanced cross-correlator enables us to determine the dependence of $\Delta\varphi_0$ on the pulse energy (which is varied by changing the pump power). The measured data shown in Figure 5 (Xu et al., 1996) allows calculation of the slope $W_0|\partial\Delta\varphi/\partial W|_{W_0}$ for Eq. (4). With this slope known, the contribution of the energy-fluctuation-induced phase variations to the overall phase jitter can be inferred from the measured power spectral density $S_W(f)$ of pulse energy variations. The energy-noise-induced jitter $\sigma_{\varphi(W)}(T_m)$ of the carrier envelope phase as obtained from Eq.(4) and the power spectral density $S_W(f)$ are depicted in Figure 6 for a sub-10 fs Ti:sapphire ring oscillator. $\sigma_{\varphi(W)}(T_m)$ drops to $\pi/10$ only for measurement times shorter than 10 μs. This is the time scale over which the sub-10 fs Ti:sapphire oscillator used in these early experiments (Xu et al., 1996) was capable of delivering pulses with a "frozen" carrier envelope phase.

The central objective that initially triggered the quest for phase-stable pulses was coherent control of strong-field interactions. However, this domain of nonlinear optics cannot be accessed with pulses emitted directly from megahertz repetition rate oscillators. The pulse energy delivered by modelocked oscillators needs to be boosted by several orders of magnitude

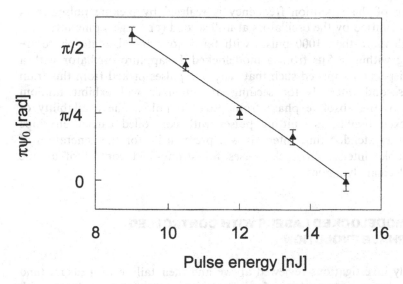

Figure 5 Dependence of the pulse-to pulse change of the carrier envelope phase on the pulse energy, which was varied by means of the pump power.

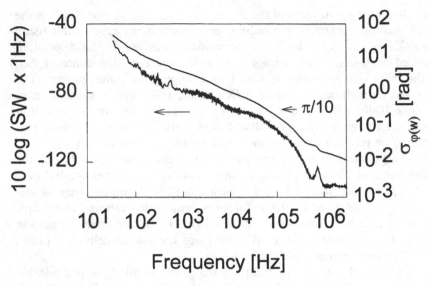

Figure 6   Measured power spectral density $S_w(f)$ of energy variations (normalized to a 1 Hz bandwidth and the calculated rms jitter $\sigma_{\varphi(w)}(f)$.

for the exploration of a wide range of strong-field applications. This can be accomplished only at strongly reduced kilohertz repetition rates. The reduction of the repetition frequency is realized by picking pulses from the train emitted by the oscillators at millisecond (or longer) time intervals. Although more than 1000 pulses with fairly constant absolute phase are emerging within $\approx 5\,\mu s$ from a modelocked Ti:sapphire oscillator with a cavity dispersion adjusted such that $\Delta\varphi_0 = 0$, pulses picked from this train at millisecond intervals for seeding an amplifier will exhibit random changes of the absolute phase from pulse to pulse. The availability of modelocked oscillators emitting pulses with controlled carrier envelope phase over extended time intervals is a prerequisite for the generation of phase-stable, intense few-cycle pulses for strong-field control of atomic and molecular dynamics.

12.3   MODELOCKED LASERS WITH CONTROLLED PHASE EVOLUTION

The early investigations reviewed above and their failure to measure (and control) the carrier envelope phase evolution in femtosecond lasers with satisfactory precision in the time domain triggered proposals for drawing on frequency domain (heterodyne) techniques in achieving this goal (Hänsch, 1997; Telle et al., 1999; Reichert et al., 1999). The Boulder, Vienna, and Garching experiments performed in 2000 (Jones et al., 2000; Apolonski et al., 2000; Holzwarth et al., 2000) put these proposals into practice and created an intriguing symbiosis between frequency domain spectroscopy and ultrafast optics. These experiments made broadband femtosecond oscillators capable of (1) performing direct and accurate measurements of frequency differences on the multi-terahertz scale and, with the development of new diagnostic tools (Poppe et al., 2001), (2) generating few-cycle pulses with a carrier envelope phase controlled to within a tiny fraction of π for extended periods of time ($\gg 1$ s).

12.3.1 Concepts

To gain insight into the basic concept that gave rise to these advances, consider a pulse circulating in a laser cavity with length L with an amplitude envelope $A(t)$ carried at a frequency f_c ($= \omega_0/2\pi$), as illustrated in Figure 7. The reproduction of this amplitude envelope at some fixed position in the activity (e.g., at the output coupler) defines the pulse repetition

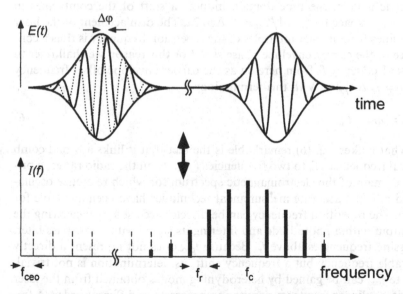

Figure 7 Two consecutive pulses from the train emitted by a modelocked laser and the corresponding spectrum. The pulse-to-pulse phase shift $\Delta\varphi$ results in an offset frequency $f_0 = \Delta\varphi/2\pi T$ because the carrier wave at f_c moves with the phase velocity v_p while the envelope moves with the group velocity v_g.

time $T_r = f_r^{-1}$ by demanding $A(t) = A(t - T_r)$, where T_r is determined by the cavity group velocity at ω_0, $T_r = 2L/v_g$. Because of the periodicity of the envelope function, the electric field can be written as

$$E(t) = \mathrm{Re}\left(A(t)e^{-2\pi i f_c t}\right) = \mathrm{Re}\left(\sum_n A_n \exp[-2\pi i(f_c + nf_r)t]\right)$$

$$= \mathrm{Re}\left(\sum_n A_n e^{-2\pi i f_n t}\right) \tag{5}$$

where A_n are Fourier components of $A(t)$. This equation shows that the resulting spectrum consists of a comb of laser modes that are separated by the pulse repetition frequency. Because f_c is not necessarily an integer multiple of f_r, the modes are generally shifted with respect to harmonics of the pulse repetition frequency by an offset $f_0 < f_r$, $f_n = nf_r + f_0$. Although the envelope $A(t)$ precisely reproduces itself with the period T_r, if the absolute phase φ changes by $\Delta\varphi \neq 0$ from pulse to pulse for the reasons discussed in the preceding section, the full electric field $E(t) = A(t) \times \cos(2\pi f_c + \varphi)$ will fulfill the condition $E(t) = E(t - T_\varphi)$ for a different period that satisfies $T_\varphi = T_r(2\pi/\Delta\varphi)$. This additional periodic modulation of the electric field in the time domain induces a shift of the comb lines in the frequency space by $f_0 = 1/T_\varphi = f_r\,\Delta\varphi/2\pi$. The displacement of the laser spectral lines from integer multiples of the repetition frequency is thus a consequence of the carrier envelope phase slip. For this reason we shall refer to the offset frequency f_0 from here on as the carrier envelope offset frequency $f_{ceo} \equiv f_0 = f_r\,\Delta\varphi/2\pi$. With this terminology,

$$f_n = nf_r + f_{ceo} \tag{6}$$

What makes Eq. (6) remarkable is the fact that it links a broad comb of optical frequencies f_n to two frequencies (f_{ceo}, f_r) in the radio range, a privileged domain of the electromagnetic spectrum for which reference oscillators and reliable, accurate measurement techniques have been available for decades. The repetition frequency can be directly accessed by measuring the laser output with a photodiode and filtering its signal with a low-pass filter, suppressing frequencies above f_r. Because f_{ceo} does not represent a directly measurable frequency but a frequency shift, its determination is not trivial. Access to f_{ceo} can be gained by heterodyning modes obtained from the laser comb via nonlinear frequency conversion processes of different order. A frequency closed to a given mode $f_k = kf_r$ can be generated either from the mode f_n via a qth-order nonlinear process or from the mode f_m via a pth-order nonlinear process ($k, m,$ and n are large integers, such that $nq = mp$):

$$f_{qn} = qf_n = qnf_r + qf_{ceo} \tag{7}$$

$$f_{pm} = pf_m = pmf_r + pf_{ceo} \tag{8}$$

Heterodyning f_{qn} with f_{pm} will give rise to a beat note at

$$\Delta f = qnf_r + qf_{ceo} - pmf_r - pf_{ceo} = (q - p)f_{ceo} \tag{9}$$

If the frequency comb were narrow, the realization of two different nonlinear frequency conversion paths leading to the same spectral line might call for the use of one or more additional phase-locked transfer oscillators (Telle et al., 1999). Fortunately, the advent of photonic crystal fibers (PCFs) allows extracavity broadening to more than one optical octave (Jones et al., 2000; Holzwarth et al., 2000), just as specially designed oscillators with more than 1 MW peak power did in conjunction with standard single-mode fibers (Apolonski et al., 2000). These advances opened the way to the simplest possible implementation of the above concept, characterized by $p = 1$ and $q = 2$ in the above terminology. Measuring f_{ceo} in this case relies on the heterodyne detection of the short-wavelength modes of the comb with the frequency-doubled long-wavelength modes, which can be accomplished if the frequency comb spans a full optical octave (Fig. 8). This requirement can be relaxed by beating the second harmonic of the high-frequency spectral

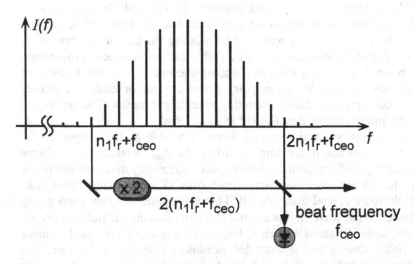

Figure 8 To gain access to the offset frequency f_{ceo}, the infrared part of the femtosecond spectrum is frequency-doubled and a beat signal with the green part is observed.

wing with the third-order harmonic of the low-frequency modes [$q = 3, p = 2$ in Eqs. (7–9)] as proposed by Mehendale et al. (2000) and recently demonstrated by Kärtner et al. (2000). In what follows, we focus on implementations relying on the lowest order nonlinear conversion ($p = 1, q = 2$).

12.3.2   Experimental Implementations

In this subsection we review the experiments performed in Boulder, Vienna, and Garching in 2000, all of which drew on the heterodyne detection of the fundamental and second harmonics of the laser spectrum. Yet the detailed implementations are quite different, and it is instructive to compare the advantages and disadvantages of the schemes employed so far.

Frequency Domain Metrology: Boulder, Colorado, USA

The system demonstrated at the University of Colorado in Boulder (Jones et al., 2000) uses a Kerr lens modelocked (KLM) Ti:sapphire laser pumped with a single-frequency, frequency-doubled Nd:YVO$_4$ laser generating pulses in the range of 10–15 fs at 90 MHz (see Fig. 9). The output spectrum is centered at 830 nm and has a bandwidth of 70 nm. Prisms were employed for intracavity dispersion control. Because the system was conceived mainly for frequency domain applications, stabilizing the repetition rate f_r (in addition to the stabilization of f_{ceo}) was essential. To this end, the retroreflector behind the prism was mounted on a piezoelectric transducer that allowed both tilt and translation. By means of translating the mirror, the repetition frequency f_r can be locked to a highly stable radio-frequency sysnthesizer. As the beam is spectrally dispersed across the end mirror (as a result of the single pass through the prism pair), tilting the mirror leads to a change of the round-trip group delay, providing an efficient means of adjusting the round-trip phase change $\Delta\varphi_0$ (Udem et al., 1999a).

Because the spectral width of 70 nm available at the output of the oscillator is insufficient for implementing the f_{ceo} measurement scheme depicted in Figure 8, spectral broadening was realized by means of nonlinear propagation in air–silica microstructured fiber (Knight et al., 1996) [also termed photonic crystal fiber (PCF)]. This waveguide has near-zero group delay dispersion at wavelengths around 800 nm, allowing the pulse to propagate over large distances before it becomes temporally stretched (Ranka et al., 2000). Owing to the extended nonlinear interaction distance, this novel fiber can induce substantial spectral broadening even at low pulse energy levels. In the experiment of Jones et al. (2000), pulses with an average power as low as 25 mW were coupled into the fiber, generating a continuum

To cross-correlator

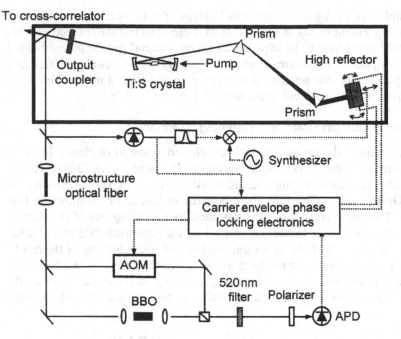

Figure 9 Layout of the phase-stabilized laser system realized at the University of Colorado (Johnes et al., 2000).

extending from 510 nm to 1125 nm (at −20 dB). A dichroic beam splitter spectrally separates the continuum emerging from the fiber: The range between 500 and 900 nm is directed through the arm of the interferometer containing an acousto-optic modulator (AOM). The near-infrared components between 900 and 1100 nm propagate through the other arm, which contains a 4 mm thick β-barium borate (BBO) crystal angle-tuned for efficient second harmonic generation at 1040 nm.

The interferometric setup allows matching the fundamental and frequency-doubled signals both spatially and temporally. Proper synchronization of the two 520 nm pulses is essential, because it enables a large number of modes to contribute to the beating signal, dramatically enhancing the signal-to-noise ratio. A 10 nm spectral range around 520 nm is selected from the recombined beam with a multilayer interference filter. The radio-frequency beats are detected with an avalanche photodiode at $\pm(f_{ceo} - f_{AOM})$, where f_{AOM} is the frequency of the AOM operated at $(7/8)f_r$. The signal-to noise ratio of the beating signal is enhanced by means of a tracking oscillator; the obtained output is further used to generate an error signal programmable to be an integer multiple of $f_r/16$, thus making it possible to lock $\Delta\varphi$ from 0 to 2π in 16 steps of $\pi/8$. As already mentioned, the active element of the loop consists of a piezoelectric mount that permits the round-trip phase to be adjusted by tilting the end mirror placed behind the prism pair. A cross-correlator (Xu et al., 1996) was used to investigate the capability of this scheme to control φ in the time domain. This issue is addressed in more detail in Section 12.4.

Frequency Domain Metrology: Garching, Germany

The f_{ceo}-stabilized femtosecond laser developed at the Max-Planck-Institute for Quantum Optics in Garching, Germany (Holzwarth et al., 2000) is based on a 25 fs Ti:sapphire ring oscillator with a near 1 GHz repetition rate (GigaOptics, model GigaJet). The schematic of this source is shown in Figure 10. The commercial laser was modified by mounting one of the mirrors on a translation stage for coarse control of the repetition rate and another mirror on a piezo transducer for fine-tuning and phase locking of the repetition rate. An electro-optic modulator inserted in the pump beam allowed for fine adjustment and phase locking of f_{ceo} owing to the dependence of the round-trip carrier envelope phase shift on the intracavity pulse energy

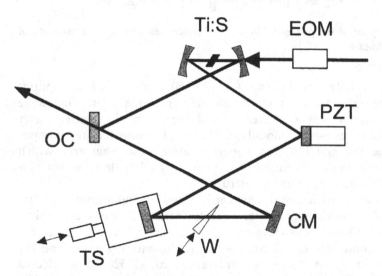

Figure 10 Layout of the high repetition rate, phase-stabilized laser demonstrated in Garching, Germany. The piezo transducer (PZT) and translation stage (TS) are used for coarse adjustment and locking of the repetition rate f_r. The fused silica wedge (W) and an electro-optic intensity modulator (EOM) are used for coarse adjustment and locking of the offset frequency f_{ceo}. All reflectors except the output coupler (OC) are dispersive or "chirped" mirrors. Ti:s-Ti:sapphire laser.

(Xu et al., 1996) (see also preceding section). Controlling f_{ceo} by (weak) modulation of the pump power is superior to f_{ceo} control by mechanical adjustment because (1) misalignment of the laser cavity can be avoided and, more important, (2) the bandwidth of the servo loop, which determines the maximum speed of (random) phase excursions that can be regulated, is substantially increased. Therefore a much tighter lock (with significantly reduced phase excursions) can be realized with such a system.

The pump power can be changed by only about 10% without affecting modelocked operation. The corresponding change of the slipping frequency beat signal f_{ceo} is about 60 MHz. This range is broad enough for phase locking, but it does not permit placement of f_{ceo} at any desired frequency within $f_{rep}/2$ of a few hundred megahertz: For this purpose, a fused silica wedge at Brewster's angle was included in the setup (Xu et al., 1996). This also gives access to the offset frequency f_{ceo} (by tuning the cavity dispersion) and can be used to preset f_{ceo} to the desired position (e.g., 64 MHz as in most experiments) and phase lock it via modulation of the pump power. With 7 W of pump power, more than 600 mW average power is emitted from the femtosecond laser. To generate an octave-spanning comb, pulses with an average power of ≈ 150 mW were coupled in a 35 cm long PCF. The ring design makes the laser almost immune to feedback from the fiber, whereas the high repetition rate increases the available power per mode. Fine adjustment of the spectrum can be achieved by rotating a $\lambda/2$ wave plate in front of the fiber in-coupling. The laser can be operated at repetition frequencies of 1 GHz (single-folded), 624 MHz, and 750 MHz (double-folded). The beat signal at f_{ceo} is generated with a dichroic interferometric setup similar to that described above.

Time-Domain Metrology: Vienna, Austria

In contrast to the two systems described above, the phase-stabilized Vienna laser (Apolonski et al., 2000; Poppe et al., 2001) was developed for time domain applications (Fig. 11). Apart from setting more stringent requirements on phase stabilization (see Sec. 12.4), phase-sensitive strong-field experiments call for *intense few-cycle* pulses. The pulse energy of modelocked lasers cannot be enhanced by merely increasing the pump power, because strong nonlinearities arising in the crystal tend to adversely affect Kerr lens modelocking. However, the pulse energy can be increased at constant power by reducing the repetition rate. This was achieved by incorporating a 1:1 imaging telescope in the cavity, which leaves the radial intensity distribution in the laser medium unchanged. In this manner the resonator length was extended to ≈ 6 m, which corresponds to $f_r = 24$ MHz. Pumped with a single-frequency, frequency-doubled Nd:YVO$_4$ laser ($P_{pump} \approx 4.5$ W),

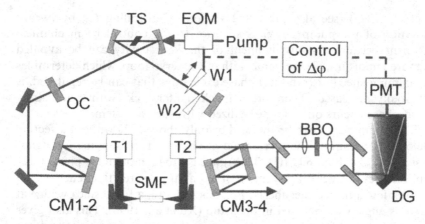

Figure 11 Experimental setup used in Vienna for the generation of phase-controlled 6 fs pulses at a repetition rate of 24 MHz.

this mirror dispersion controlled laser generates a highly stable train of 9 fs, 20 nJ pulses. These pulses exhibit a peak power well above 1 MW, high enough to induce substantial spectral broadening by self-phase modulation in a *standard* single-mode optical fiber (FS-SC-3314, 3M). The broadened spectrum spans a full octave from less than 500 nm to more than 1100 nm. The spectrally broadened pulses are sent through a chirped mirror compressor, yielding sub-6 fs, 6 nJ pulses in a diffraction-limited beam. A dichroic chirped mirror CM4 in the compressor transmits the spectral components near 1080 nm and 540 nm used for stabilizing φ, whereas most of the pulse energy remains available for experiments.

A nonlinear interferometer similar to the one employed by Jones et al. (2000) has been used to beat the second harmonic of the 1080 nm wave packet with the 540 nm wave packet selected from the fundamental spectrum. The resultant beat note arising at f_{ceo} exhibits a satisfactory signal-to-noise ratio (>30 dB in 100 kHz bandwidth) for phase locking. The carrier-slip beat note was phase locked to $f_{local} = 1$ MHz (synthesized by a stabilized local oscillator) by adjusting the pump power with an electro-optic modulator (EOM) (Holzwarth et al., 2000). Once f_{ceo} is manually adjusted by translating a thin fused silica wedge (W2 in Fig. 11) to be within $f_{local} \pm 100$ kHz, the servo loop pulls f_{ceo} to f_{local} and phase locks the carrier-slip beat note to the signal of the local oscillator. It should be noted that in a mirror dispersion controlled (MDC) oscillator, the spatial alignment has no influence on the intracavity dispersion, whereas in a laser containing prism pairs they are coupled. As a direct consequence, f_{ceo} proves to be fairly stable in the MDC laser; in the absence of a feedback loop, f_{ceo} fluctuates in a $\pm 100\,\mathrm{kHz}$ range on a milisecond to second time scale. Under the same conditions, f_ceo drifts over several megahertz in a prism-controlled oscillator (Morgner et al., 2001).

The few-cycle waveforms generated by compression of the fiber output with broadband chirped dielectric mirrors can be characterized by means of the SPIDER technique (Iaconis and Walmsley, 1998). This technique provides access to the amplitude envelope and frequency chirp. Figure 12 shows the electric field and its envelope of a pulse emitted from the carrier envelope offset stabilized oscillator-compressor system described by Poppe et al. (2001). The still unknown value of the absolute phase was assumed to be $\varphi = 1.25\,\mathrm{rad}$ when reconstructing the electric field (thick line in Fig. 12) from the SPIDER measurement. Even though we do not know at which phase of the carrier-slip beat signal any particular value of φ can be realized, Figure 12 provides a representation of the range of waveforms that can be generated by picking pulses from the $f_r = 24\,\mathrm{MHz}$ train at a repetition rate of $f_\mathrm{ceo} = 1\,\mathrm{MHz}$ (or its submultiples) at various phases of the carrier-slip beat signal oscillating at f_ceo. The reproducibility of the carrier envelope phase (and thus electric and magnetic field evolution) in the pulses selected in this manner is illustrated by the thickness of the line in Figure 12, indicating a phase stability of better than $\pm 0.3\,\mathrm{rad}$. The technique employed for

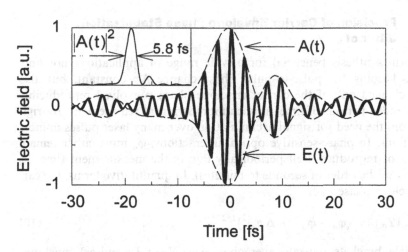

Figure 12 Electric field of a waveform (full line) emitted by the carrier envelope phase locked few-cycle pulse generator in Vienna at a particular setting of the trigger phase from the local oscillator yielding $\varphi = 1.25\,\mathrm{rad}$, as obtained from a SPIDER measurement. The intensity envelope in the inset has a half-width of 5.8 fs. The waveform is reproduced within an accuracy represented by its thickness at the repetition rate of 1 MHz of the local reference oscillator.

measuring the carrier envelope phase jitter with subcycle precision, which is indispensable for time domain applications, is discussed in the next section.

12.4 OPTICAL WAVEFORM SYNTHESIS VERSUS FREQUENCY SYNTHESIS

The synthesis of electromagnetic waveforms, routinely accomplished by radio-frequency signal generators, would provide an exceptional means of controlling strong-field, ultrafast microscopic processes if it were extended to optical frequencies. The recently developed optical frequency synthesizers (Holzwarth et al., 2000; Jones et al., 2000) hold, in principle, the potential of enabling optical waveform synthesis, by which we mean the repeated generation of reproducible waveforms, i.e., phase-stable light pulses, over extended periods of time. In this section we discuss the differences in the demands that time domain and frequency domain applications make on f_{ceo}-locked femtosecond oscillators. The vastly different demands require different diagnostic tools for assessing whether such a system qualifies as a precision (few-cycle) optical waveform synthesizer or optical frequency synthesizer for time and frequency domain metrology, respectively.

12.4.1 Precision of Carrier Envelope Phase Stabilization, Jitter of φ

Waveform synthesis beneficial for a wide range of applications not only requires keeping the pulse-to-pulse phase change $\Delta\varphi$ constant, but also can be claimed only if the carrier envelope phase φ evolved reproducibly over extended periods of time, a much more stringent requirement. It originates from the need for signal accumulation over many laser pulses in most applications. In phase-sensitive optical interactions, φ_n must either remain constant or reproduce itself periodically during the measurement time T_m (typically of the order of seconds to minutes). In quantitative terms, the carrier envelope phase jitter

$$\sigma_\varphi(T_m) = \langle(\varphi_n - \varphi_0 - n\,\Delta\varphi_0)^2\rangle^{1/2} \tag{10}$$

(the angle brackets indicate averaging over $N = f_r T_m$ pulses) must not exceed a tiny fraction of π, preferably for $T_m \gg 1$ s. Although phase locking of the carrier envelope offset frequency $f_{ceo} = (\Delta\varphi_0/2\pi)f_r$ to a stabilized RF local oscillator can substantially reduce this random deviation, the availability of a frequency comb stable enough for accurate frequency metrology does not automatically imply fulfillment of the condition $\sigma_\varphi \ll \pi$. In fact, random phase excursion of phase-locked signals may exceed 2π without substantially compromising the utility of the signal for frequency metrology as long as these large-amplitude excursions are detected and subsequently corrected for by the feedback loop (cycle-slip-free operation). By sharp contrast, random excursions of φ of the order of π or larger may completely prevent phase-sensitive nonlinear effects from being observed in the time domain.

As shown in Section 12.2, phase jitter arises due to a random contribution δ_n to the pulse-to-pulse change $\Delta\varphi_n$, primarily caused by pulse energy fluctuations. Even a relatively small phase noise δ_n superimposed on the mean value of the pulse-to-pulse phase shift $\Delta\varphi_0$ can induce substantial fluctuations of the absolute phase φ. For instance, a random shift of φ_n resulting from the accumulation of small changes in $\Delta\varphi_n$ that are roughly constant ($\delta_n \approx \delta_0$) over a period T_{const} (this noise of $\Delta\varphi_n$ has its main spectral component at $f_{noise} = 1/2T_{const}$) is approximately given by $\sigma_\varphi \approx \delta_0 f_r T_{const}$. The longest time period over which small changes in $\Delta\varphi_n$ of constant sign can accumulate without damping is roughly given by the response time of the phase-locked loop (including the optical unit controlling f_{ceo}), which may range from milliseconds to tens of microseconds depending on the technique used for affecting f_{ceo} (piezo-controlled alignment of resonator versus electro-optic or acousto-optic modulation of pump power; see previous section) and on the electronic part of the loop. On these time scales, phase excursions of the order of 2π cannot be excluded even with the servo loop in operation; hence reliable measurement of this parameter is an important prerequisite for optical waveform synthesis for time domain applications. Are standard diagnostics sufficient to achieve this goal?

The most "dangerous" fluctuations δ_n are those emerging at the lowest frequencies f_{noise} that cannot yet be efficiently regulated by the servo loop, i.e., at f_{noise} comparable to the bandwidth of the servo loop. At these frequencies even values of the noise amplitude δ_0 as small as tens to hundreds of microradians may result in random excursions of φ_n up to the order of π. These changes in $\Delta\varphi$ are far too small to be measured by the interferometric cross-correlator (Xu et al., 1996) in the time domain, even if this device were able to record correlation traces within a fraction of a millisecond. In the frequency domain, frequency counting offers a standard diagnostic just as correlation does in the time domain. However, even the best frequency counters can observe only the occasional loss of track resulting in cycle slips between the beat and reference signals. They are unable to detect even large ($> 2\pi$) phase excursions that can eventually be tracked and compensated by the servo loop. Tracing the error signal of the phase-locked loop in the time domain might provide some additional insight, but again, only for sufficiently slow noise components, because the carrier-slip beat signal has to be filtered to suppress the influence of pulse energy noise on the phase stabilization. Hence, additional diagnostics are required for reliable assessment of the subcycle jitter of φ_n over the entire RF spectral range of interest ($f_{noise} < f_r/2$). Only if the overall carrier envelope phase jitter [as defined by Eq. (10)] is found to be much smaller than π is the cycle-slip-free phase locking "tight" enough to provide reproducible waveforms for time domain applications.

12.4.2 Measurement of Carrier Envelope Phase Jitter with Subcycle Precision

The first technique enabling the measurement of carrier-envelope phase jitter with sub-cycle precision was recently reported and relies on the intracavity pulse energy dependence of the carrier envelope phase (Poppe et al., 2001). A comparison of the energy-noise-induced carrier envelope phase jitter $\sigma_{\varphi(W)}(T_m)$ with the servo loop for phase locking the carrier-slip beat signal in and out of operation reveals in which spectral range energy noise induced phase jitter is dominant and allows assessment of the accumulated carrier envelope phase error at which the servo loop starts to respond. This permits the estimation of σ_φ with subcycle ($\ll \pi$) precision. Figure 13 depicts the measured power spectral density $S_W(f)$ of pulse energy variations and the energy-induced carrier envelope phase jitter $\sigma_{\varphi(W)}$ [calculated from $S_W(f)$ according to Eq. (4)] for locked and unlocked carrier envelope phase in a sub- 10 fs system (consisting of a mirror dispersion controlled Ti:sapphire laser and a fiber compressor, as depicted in Fig. 11) with $f_r = 24$ MHz, in which f_{ceo} locking has been implemented by controlling the intracavity pulse energy via control of the pump power (Poppe et al., 2001).

Comparison of the carrier-slip beat signal phase locked to a local oscillator at $f_{loc} = 1$ MHz with the reference signal revealed that no cycle slips occur over the maximum gating time of 10 s of the frequency counter. This fact can be reconciled with the observed increase of $\sigma_{\varphi(W)}$ (see Fig. 13) well beyond 2π for $T_m > 100$ ms only by assuming that substantial carrier envelope phase jitter originates from effects other than pulse energy fluctuations at frequencies $f_{noise} \leq 1/(10 \text{ ms}) = 100$ Hz. In this range, a significant fraction of the pulse energy fluctuations are "rephased" by the control loop in order to compensate for a jitter from other sources, and $\sigma_{\varphi(W)}$ mirrors this jitter. The dramatic energy noise reduction below 1 Hz indicates that energy noise induced phase jitter becomes vastly dominant for measurement times $T_m > 1$ s. Thus, as a by-product of the carrier envelope offset frequency stabilization, pulse energy fluctuations are reduced to an unprecedented root mean square value of $\sigma_W \approx 0.01\%$ over the spectral range of

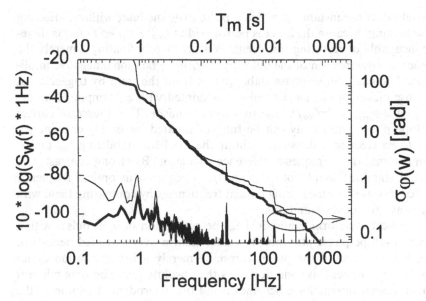

Figure 13 The lower traces show the power spectral density $Sw(f)$ of the pulse energy fluctuations at the output of the sub-10 fs Ti:sapphire laser with the servo loop interrupted (thin line) and closed (thick line). The upper traces represent the carrier envelope phase jitter $\sigma(w)$ as a function of T_m introduced by pulse energy variations, with the servo loop interrupted (thin line) and closed (thick line).

0.1 Hz to −0.1 MHz at the output of the oscillator. This value is enhanced by almost a factor of 10 at the output of the fiber used for spectral broadening/pulse compression, still resulting in a reproducibility of the *field* amplitude to within less than 0.05% (corresponding to $\sigma_W \approx 0.1\%$) in the sub-6 fs pulse train. Figure 13 shows that energy noise induced jitter clearly dominates for $T_m < 10$ ms, and the servo loop starts responding (i.e., reducing the phase jitter) at jitter levels of around 0.2 rad. This implies that the carrier envelope phase fluctuations are kept safely within ±0.3 rad by the servo loop, corresponding to a phase error of less than $\pi/10$. (The same nonlinearities employed for controlling the round-trip phase change $\Delta\varphi$ are used as a diagnostic tool for determining the subcycle jitter of φ, here.)

12.4.3 Differences in the Implementation of Optical Waveform and Frequency Synthesis

It is important to note that the degree of control discussed above can be *achieved without stabilizing f_r to a reference oscillator*. Minor variations in

f_r merely shift the instants at which a pulse exits the laser without affecting phase locking, because the latter is performed at $f_{ceo} = f_{ref} \ll f_r$ and is therefore incapable of resolving this event. With f_r "freely floating," merely the absence of synchronism of the pulse train with f_{ref} has some negative implication. The selection of phase-stable pulses from the train by triggering of the pulse picker by the local oscillator is tainted with a "sampling" uncertainly of $\Delta\varphi_{error} \leq 2\pi f_{ceo}/f_r$ due to desynchronism. This source of carrier envelope phase uncertainly can be fully eliminated by simply deriving f_{ref} from f_r via frequency division, without the need for stabilizing f_{ceo} and f_r to an external radio-frequency reference oscillator. By strong contrast, this external stabilization of both f_{ceo} and f_r is required in optical frequency synthesizers used for measuring optical frequencies by comparing them with microwave frequency standards.

Optical waveform synthesis, i.e., the generation of light pulses with a carrier envelope phase jitter much smaller than π over extended periods of time, is useful only if the pulses comprise merely a few oscillation cycles (Fig. 1). Only under this condition does the absolute phase become relevant in light–matter interactions. Schemes using the heterodyne detection of the fundamental and second harmonics of the laser spectrum usually require additional spectral broadening by means of nonlinear propagation in a waveguide. A simple analysis reveals that the overall energy-dependent contribution to a change of φ is expected to be enhanced by more than one order of magnitude in a photonic crystal fiber, compared to the slope in Figure 5, because of the extended length of the nonlinear interaction. Hence, the fiber can translate pulse energy fluctuations as small as a fraction of a percent into random shifts (jitter) of the carrier envelope phase with a characteristic amplitude as large as $\pi/2$. Once the stabilization feedback loop (including the fiber) is closed, this jitter appears at the output of the laser oscillator and may destroy the stabilization of the carrier envelope phase in the laser output. This problem can be most safely avoided by realizing few-cycle waveform synthesis at the same part of the system, where the carrier-slip beat signal for phase locking f_{ceo} to some reference signal is extracted. This can be implemented either by recompressing the pulses at the output of the fiber used for extracavity spectral broadening (Apolonski et al., 2000; Poppe et al., 2001) or by producing octave-spanning spectra directly in the modelocked laser (Morgner et al., 2000). In the former case, PCFs are unsuitable for spectral broadening because the large amount of high-order dispersion they introduce (Xu et al., 2000) cannot be easily handled by standard dispersive optics (chirped mirrors or prisms). Hence a standard telecommunications monomode fiber in conjunction with higher energy seed pulses must be used as described in the last part of the preceding section.

In conclusion, frequency and time domain applications make vastly different demands on carrier envelope phase-controlled femtosecond oscillators. Frequency and time domain applications call for maximized power in a single mode and in a single pulse, respectively, giving rise to the opposite requirements of highest possible and lowest possible repetition rate f_r, respectively. For frequency domain metrology, f_{ceo} and f_r need to be locked to at least one stabilized reference oscillator. Time domain applications tolerate f_r freely floating, with f_{ceo} being phase locked to a signal derived from f_r by frequency division. In the frequency domain, cycle-slip-free phase locking of f_{ceo} and f_r to one or more stabilized reference oscillators allows the optical frequency measurements to be made with an accuracy limited merely by that of RF reference source(s). Frequency counters provide adequate diagnostic tools for assessing fulfillment of this requirement. Cycle-slip-free operation is not sufficient for time domain applications because it tolerates a carrier envelope phase jitter comparable to or larger than 2π, which is unacceptable in phase-sensitive interactions of few-cycle light pulses with matter. Diagnostics other than frequency counting are required to measure the carrier envelope phase jitter with subcycle precision.

12.5 FEW-CYCLE-DRIVEN STRONG-FIELD INTERACTIONS, MEASUREMENT OF φ

The impact of the absolute phase on strong-field interactions remained experimentally unexplored in spite of the availability of intense 5 fs pulses at $\lambda_0 \approx 0.8\,\mu m$ for several years (Nisoli et al., 1997; Sartania et al., 1997). Recent advances in controlling φ at low pulse energies (see Sec. 12.3) are expected to trigger efforts to transfer the degree of control of φ achieved in oscillators to the output of amplifier systems. This will open the way to governing strong-field processes with controlled evolution of the electric and magnetic fields. The regime where this becomes feasible is defined by the Keldysh parameter

$$\frac{1}{\gamma} = \frac{eE_a}{2\pi f_c \sqrt{2mW_b}}$$

(where E_a is the amplitude of the linearly polarized field, e is the electron charge, m is the electron mass, f_c is the carrier frequency of the optical field, and W_b denotes the ionization potential of the atom), which allows discriminating between the perturbative ($\gamma > 1$) and the strong-field ($\gamma < 1$) regimes of nonlinear optics (Keldysh, 1965). For $\gamma < 1$, the laser field suppresses the Coulomb potential to an extent that allows the wave function of the most weakly bound electron to penetrate the barrier and reach its outer side within a fraction of the laser oscillation cycle. Consequently, the probability of tunnel ionization follows nearly adiabatically the changes of the optical field. The characteristics of the freed electron (e.g., ionization yield and angular distribution) and of the radiation resulting from its possible recombination with its parent ion are therefore expected to be sensitive to the carrier envelope offset phase if the energy of the exciting field is concentrated within only a few oscillation cycles. On the one hand, these methods could be employed for measuring the actual value of φ. On the other hand, the absolute phase can provide a unique means of coherent control in these interactions, once intense phase-controlled few-cycle pulses become available. In the remaining part of this section we give an overview of strong-field methods proposed for measuring the carrier envelope phase (Brabec and Krausz, 2000), including strong-field ionization (Christov, 1999; Dietrich et al., 2000), tunnel photoemission (Xu et al. 1996; Poppe at al., 1999), and high-order harmonic generation in gases (de Bohan et al., 1998; Tempea et al., 1999).

12.5.1 Optical Field Ionization

Within the framework of the quasi-static approximation, the instantaneous optical field ionization rate is a function of the instantaneous field strength $w(t) = w(E(t))$, and hence the integrated ionization yield can be written as

$$\rho(\tau) = \rho_0 \left(1 - \exp\left[- \int\limits_{-\infty}^{t} w(E(t'))\, dt' \right] \right) \tag{11}$$

where ρ is the ion (or free electron) density and ρ_0 the initial neutral atom density. Figure 14 shows $\rho(t)/\rho_0$ in helium exposed to 5 fs, 800 nm pulses that have a Gaussian intensity envelope and a peak intensity of $4 \times 10^{15}\,\text{W/cm}^2$ as obtained by using the computed quasi-static ionization rates of Scrinzi et al. (1999). The instantaneous ionization rate is sensitive to the carrier envelope phase, as revealed by the computations performed for $\varphi = 0$ and $\varphi = \pi/2$ (corresponding to consinusoidal and sinusoidal carrier fields, respectively). Somewhat surprisingly, the integrated ionization yield, however, is found to be independent of φ, at least in the framework of the quasi-static approximation. An investigation based on the full numerical solution of the time-dependent Schrödinger equation indicated a slight dependence of the integrated ionization yield on φ (Christov, 1999). The phase sensitivity of the integrated optical field ionization yield can be substantially enhanced either by (1) "switching off" ionization for one of the two directions of the electric field vector in a linearly polarized wave by

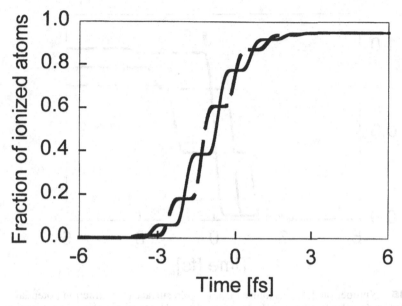

Figure 14 Temporal evolution of the relative ionization in He atoms exposed to 5 fs pulses having a peak intensity of 4×10^{15} W/cm^2 and a carrier wavelength of 800 nm. The broken line corresponds to a carrier envelope phase $\varphi = 0$ and the full line to $\varphi = \pi/2$.

inducing the process on a metal surface (photoemission) or by (2) ionizing atoms with a circularly polarized light wave and resolving the angular distribution of the photoelectrons.

Photoemission from a Metal Surface in the Tunneling Regime

The total number of photoelectrons emitted from a photocathode irradiated with p-polarized few-cycle laser pulses impinging at oblique incidence has been predicted to be sensitively dependent on the absolute phase (Poppe et al., 1999). The enhancement of the sensitivity of the overall ionization yield to the carrier envelope phase (compared to ionization of a gaseous medium by the same radiation) originates from symmetry breaking: Only half of each oscillation cycle contributes to the overall yield. In the strong-field regime, photoionization can be calculated using the quasi-static ionization theory for metal surfaces (Sommerfeld, 1967). Figure 15 depicts the predicted temporal evolution of the number of ejected photoelectrons when a 5 fs, 800 nm p-polarized pulse with a peak intensity of 2×10^{13} W/cm^2 impinges at an angle of 45° on a gold surface (ionization

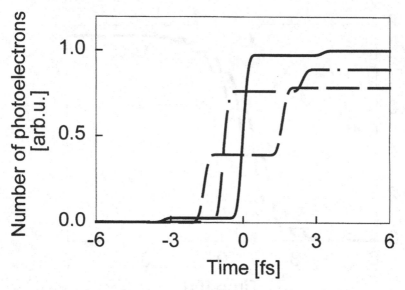

Figure 15 Number of electrons ionized from a gold surface (of extraction potential 4.9 eV) for a 5 fs pulse having a peak intensity of 2×10^{13} W/cm^2 and a carrier wavelength of 800 nm impinging at an angle of incidence of 45°. The solid curve corresponds to $\varphi = 0$, the dashed curve to $\varphi = \pi/2$, and the dot-dash curve to $\varphi = \pi$

potential $W_b = 4.9$ eV). In strong contrast with ionization in the gas phase (Fig. 14), not only the ionization dynamics but also the total number of photoelectrons left after the pulse exhibit a substantial phase dependence, which rapidly increases for decreasing pulse durations (Fig. 16). According to this theoretical prediction, a phase-dependent component of the photoemission current should be easily detectable for sub-7 fs pulses focusable to peak intensities $\geq 10^{13}$ W/cm^2. Preliminary experiments carried out with a mirror dispersion controlled Kerr lens modelocked Ti:sapphire oscillator on a gold photocathode in this parameter range resulted in photoemission, but no phase effects could be observed. A more sophisticated recent theoretical analysis revealed that the simple Sommerfeld theory tends to overestimate the phase sensitivity of the photoemission process and that shorter pulses (≤ 5 fs at 800 nm) may be required for observable phase effects (Yudin and Ivanov, 2001). Pulses with such durations are now available from a laser oscillator (Ell et al., 2001).

Optical Field Ionization with Circularly Polarized Light

Other than the ionization yield, the momentum distribution of the ejected electrons can provide information on the phase of the laser field (Dietrich

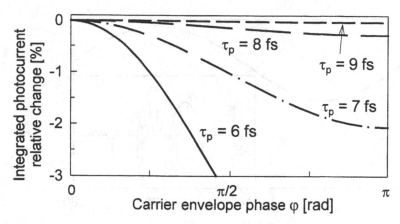

Figure 16 Intergrated photocurrent as a function of the carrier phase φ for different pulse durations.

et al., 2000). If the intense few-cycle laser pulse to which an atomic gas is exposed is circularly polarized, the direction of the photoelectron momentum associated with its drift motion will depend on the carrier envelope offset phase of the pulse. By employing circularly polarized light, electron rescattering off the parent ion is prevented, and the motion of the electrons subsequent to ionization can accurately be determined from Newton's equations. The final direction of the electrons rotates with the electric field vector. For long pulses, tunneling occurs over many optical cycles, and the electron distribution is isotropic, because ionization is equally probable for any phase. In the case of a few-cycle pulse, significant ionization occurs only on a subcycle time scale and the carrier envelope phase determines the direction of the field at the moment of ionization and consequently the direction of the freed electrons. Figure 17 shows the angular distribution of the strong-field-ionized electrons produced in He with a 4.8 fs pulse having a peak field 6×10^{10} V/m. The ionization rates were determined from a quasi-static model (Corkum, 1993), whereas the electron trajectories were determined analytically by integrating the classical equations of motion. For $\varphi = 0$, the angular distribution of the electrons peak at $\theta = 270°$; a change $\Delta\varphi$ of the carrier envelope phase will rotate the electron trajectories by the same angle, i.e., will shift the angular distribution shown in Figure 17. by $\Delta\varphi$ (Dietrich et al., 2000). At moderate peak intensities, where ionization occurs only at the peak of the pulse, amplitude fluctuations do not change the direction of the electrons, which is hence unambiguously related to the carrier envelope offset phase.

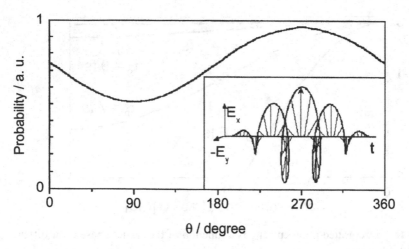

Figure 17 Angular distribution of the electrons tunneling from He atoms irradiated with circularly polarized 4.8 fs pulses at 800 nm, for an absolute phase $\varphi = 0$. The inset shows the time-dependent electric field vector rotating around the direction of propagation (Dietrich et al., 2000).

12.5.2 High Harmonic Generation

Atoms exposed to intense, linearly polarized ultrashort laser pulses in the range of $\gamma \ll 1$ (strong-field regime) emit coherent, high-order harmonics of the impinging laser light (L'Huillier and Balcou, 1993; Macklin et al., 1993). In a semiclassical approach, high harmonic radiation emerges from a three-step process (Corkum, 1993; Kulander et al., 1993; Lewenstein et al., 1994). An atom is ionized via tunnel ionization in the optical field, and the freed electron gains energy from the laser pulse. Upon reversal of the direction of the linearly polarized electric field vector, the electron is driven back to the proximity of its parent ion and, with some probability, can radiatively recombine into its original ground state, giving rise to the emission of a high-energy (XUV, soft X-ray) photon. The ultrashort rise time of the intensity envelope of femtosecond pulses is essential for this process for several reasons. It allows atoms to be exposed to high field strengths before their ground state is depleted and permits harmonic emission to coherently grow over extended lengths owing to a low background-free electron density (which unavoidably results from ionization and leads to phase velocity mismatch between the fundamental and high-harmonic pulses).

As a consequence of the above-mentioned particular features of ultrashort pulse–atom interactions, the shortest possible driver pulses in the few-cycle regime benefit high-harmonic generation in several ways. They are capable of (1) extending the emission spectrum to shorter wavelengths [down to the water window (Spielmann et al., 1997; Chang et al., 1997; Schnürer et al., 1998)], (2) enhancing the conversion efficiency (Rundquist et al., 1998; Schnürer et al., 1998; Tempea et al., 2000), and last but not least (3) generating X-ray pulses shorter than the oscillation cycle of the driver laser (Drescher et al., 2001) with the potential of durations in the attosecond regime (Schafer and Kulander, 1997; Christov et al., 1997). At the same time, few-cycle pulses also inevitably give rise to the onset of phase-sensitive features in both the temporal and spectral characteristics of the high-harmonic radiation. The highest harmonic order can be easily related to the maximum field strength by means of the cutoff law (Krause et al., 1992; Corkum, 1993; Kulander et al., 1993). As the maximum field strength becomes phase-sensitive for few-cycle pulses (as illustrated in Fig. 1), the position of the cutoff will in turn be affected by φ (de Bohan et al., 1998). Although this property originates from the single-atom response, it has been shown that the phase sensitivity of the harmonic signal is substantially preserved upon propagation over macroscopic distances (Tempea et al., 1999).

Results presented here were obtained by solving the coupled propagation equations for the fundamental and harmonic fields in one space dimension. The ionization rate is calculated from the Ammosov–Delone–Krainov theory (Ammosov et al., 1986) or by using the computed quasi-static rates of Scrinzi et al. (1999). The atomic dipole moment for HHG is calculated by means of the Lewenstein model, modified to account for nonadiabatic effects. For a detailed description of the mathematical model the reader is referred to the relevant literature (Lewenstein et al., 1994; Tempea et al., 1999).

The calculated single-atom response of He irradiated with 5 fs pulses having a carrier wavelength of 750 nm and a peak intensity of 2×10^{15} W/cm^2 shows that harmonic spectra corresponding to absolute phases $\varphi = 0$ and $\varphi = \pi/2$ of the laser are, particularly in the cutoff region, substantially different (Fig. 18a). Shorter wavelength radiation can be produced with the pulse having a consinusoidal carrier ($\varphi = 0$), because it contains a higher peak electric field (see Fig. 1). Furthermore, harmonics are resolved in the cutoff region only for $\varphi = \pi/2$, whereas the higher-energy end of the spectrum is a smooth continuum for $\varphi = \pi$. The latter feature can easily be understood by considering the temporal structure of the signal in the cutoff range, obtained by a Fourier transformation from the filtered harmonic spectrum (Fig. 19). A single half-cycle of the cosine laser pulse contributes to the generation of the cutoff harmonics, because the peak of the electric field coincides with the peak of the intensity envelope;

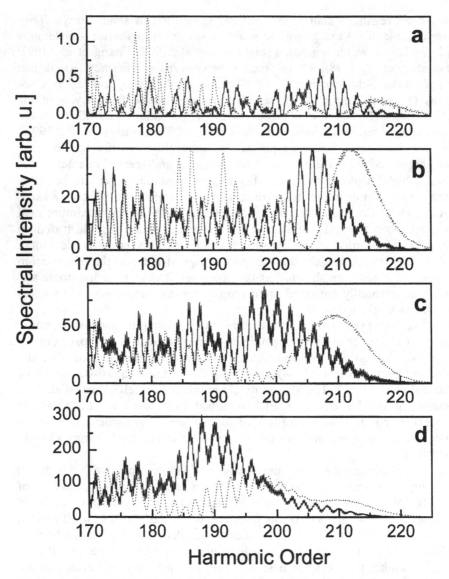

Figure 18 Spectral intensity of the harmonic radiation emitted from an 18 μm helium target at 500 torr, irradiated by 5 fs pulses having a peak intensity of 2×10^{15} W/cm^2 and carrier wavelength of 750 nm, for propagation distances of (a) 0.2 μm, (b) 5 μm, (c) 9 μm, and (d) 18 μm. Dotted curves correspond to $\varphi = 0$, and full lines to $\varphi = \pi/2$.

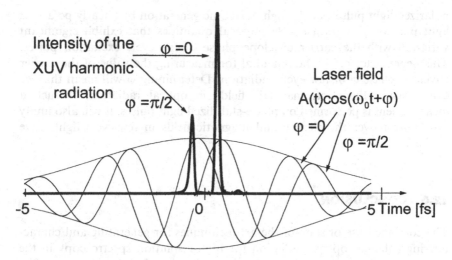

Figure 19 Fourier transform of the harmonic amplitude spectrum in the cutoff range (between orders 210 and 230) for the parameters of Figure 18d, for two different values of the carrier envelope phase. The harmonic radiation for a given order N can grow only over the coherence length L_N, for which the dephasing between the laser and the harmonic field has reached π. Propagation over distances much larger than L_N is therefore of no practical interest, if phase matching cannot be achieved. For the cutoff harmonics in this calculation, $L_{N=215} = 9\,\mu m$. The propagation distance was chosen twice as long as $L_{N=215}$ in Figure 18 in order to demonstrate that phase-sensitive features are preserved over experimentally useful target thicknesses.

consequently, a single attosecond high-harmonic burst is produced. The pulse having a sinusoidal carrier field contains two half-cycles of comparable strength that contribute equally to the generation of cutoff harmonics, leading to the appearance of two attosecond harmonic pulses separated by half the optical period of the laser field. Interference between these two pulses leads to the modulation observed in the frequency domain. This phase sensitivity is preserved in the high-harmonic emission from a macroscopic source in spite of a notable self-phase and self-amplitude modulation suffered by the few-cycle laser pulse upon propagation through an extended ionizing interaction medium (atomic gas) as revealed by Figure 18.

12.5.3 Measurement of φ

All the processes discussed in this section—optical field ionization of (1) metals by linearly polarized light pulses and (2) atoms by circularly polarized light pulses, or (3) high-harmonic generation by linearly polarized light pulses—yield measurable physical quantities that exhibit significant variation with the carrier envelope phase of few-cycle excitation pulses. This dependence offers the potential for measuring the value of the carrier envelope phase φ of few-cycle radiation. Determining φ will mean that we can measure the electromagnetic fields in optical radiation. If such a measurement is performed on phase-stabilized light pulses, it will also imply full control over the electric and magnetic fields in few-cycle light wave packets.

12.6 CONCLUSION

The combined use of state-of-the-art techniques for generating and characterizing coherent optical radiation for high-resolution spectroscopy in the frequency and time domains in the same device recently led to the emergence of the technical capability of synthesizing a vast number of narrowband optical lines at precisely known (equidistant) frequencies (optical frequency synthesis) and ultrashort optical waveforms with precisely reproducible electromagnetic field evolution. The impacts on frequency domain and time domain metrology can hardly be overestimated. Extremely complex and inflexible (usually capable of generating only one frequency) chains used during the past 25 years for frequency measurements can now be replaced by a universal optical frequency comb synthesizer that provides the long-missing simple link between optical and microwave spectral ranges. This tool makes it possible for the first time for small-scale spectroscopy laboratories to measure or synthesize any optical frequency with extreme precision. Beyond benefiting these well-established fields of optical science, phase-controlled ultrabroadband femtosecond lasers open up entirely new perspectives in vastly different areas of science and technology. For instance, the stabilized ultrabroadband comb of optical frequencies emitted may offer a new way to carry enormous amounts of information in long-distance telecommunication. Intense few-cycle light pulses with controlled and known absolute phase, once available, will allow strong-field control of electronic, atomic, and molecular as well as plasma dynamics on time scales substantially shorter than the oscillation period of visible light. The same phase-controlled few-cycle light pulses hold promise for generating attosecond bursts of X-rays and relativistic electrons for recording "movies" of sub-light-cycle microscopic dynamics they will trigger and control. Fascinating prospects open up in both science and technology.

REFERENCES

Ammosov MV, Delone NB, Krainov VP. (1986). Tunneling ionization of complex atoms and atomic ions by an alternating electromagnetic field. Sov Phys JETP 64:1191.

Apolonski A, Poppe A, Tempea G, Spielmann Ch, Udem Th, Holzwarth R, Hänsch TW, Krausz F. (2000). Controlling the phase evolution of few-cycle light pulses. Phys Rev Lett 85:740.

Baltuska A, Wei Z, Pshenichnikov MS, Wiersma DA. (1997). Optical pulse compression to 5 fs at a 1-MHz repetition rate. Opt Lett 22:102.

Bartels R, Backus S, Zeek E, Misoguti L, Vdovin G, Christov IP, Murnane MM, Kapteyn HC. (2000). Shaped-pulse optimization of coherent emission of high-harmonic soft X-rays. Nature 406:164.

Brabec T, Krausz F. (2000). Intense few-cycle laser fields: frontiers of nonlinear optics. Rev Mod Phys 72:545.

Chang Z, Rundquist A, Wang H, Murnane M, Kapteyn HC. (1997). Generation of coherent soft x-rays at 2.7 nm using high harmonics. Phys Rev Lett 79:2967.

Cheng Z, Tempea G, Brabec T, Ferencz K, Spielmann Ch, Krausz F. (1998). Generation of intense diffraction-limited white light and 4-fs pulses. In: T Elsässer, JG Fujimoto, DA Wiersma, W Zinth, eds. Ultrafast Phenomena, Vol XI. Berlin: Springer, p 8.

Christov IP. (1999). Phase-dependent loss due to nonadiabatic ionization by sub-10-fs pulses. Opt Lett 24:1425.

Christov IP, Murnane MM, Capteyn HC. (1997). High-harmonic generation of attosecond pulses in the single-cycle regime. Phys Rev Lett 78:1251.

Corkum P. (1993). Plasma perspective on strong field multiphoton ionization. Phys Rev Lett 71:1994.

De Bohan A, Antoine P, Milosevic DB, Piraux B. (1998). Phase-dependent harmonic emission with ultrashort laser pulses. Phys Rev Lett 81:1837.

Diddams SA, Jones DJ, Ye J, Cundiff ST, Hall JL, Ranka JK, Windeler RS, Holzwarth R, Udem Th, Hänsch TW. (2000). Direct link between microwave and optical frequencies with a 300 THz femtosecond laser comb. Phys Rev Lett 84:5102.

Dietrich P, Krausz F, Corkum PB. (2000). Determining the absolute carrier phase of a few-cycle laser pulse. Opt Lett 25:16.

Drescher M, Hentschel M, Kienberger R, Tempea G, Spielmann Ch, Reider GA, Corkum PB, Krausz F. (2001). X-ray pulses approaching the attosecond frontier. Science 291:1923.

Eckstein JN, Ferguson AI, Hänsch, TW. (1978). High-resolution two-photon spectroscopy with picosecond light pulses. Phys Rev Lett 40:847.

Ell R, Morgner U, Kärtner FX, Fujimoto JG, Ippen EP, Scheuer V, Angelow G, Tschudi T, Lederer MJ, Boiko A, Luther-Davies B. (2001). Generation of 5 fs pulses and octave-spanning spectra directly from a Ti:sapphire laser. Opt Lett 26:373.

Ferguson AI, Eckstein JN, Hänsch TW. (1979). Polarization spectroscopy with ultrashort light pulses. Appl Phys 18:257.

Hänsch TW. (1997). Proposal for a universal optical frequency comb synthesizer. Private communication.

Holzwarth R, Udem Th, Hänsch TW, Knight JC, Wadsworth WJ, Russel PStJ. (2000). Optical frequency synthesizer for precision spectrosopy. Phys Rev Lett 85:2264.

Iaconis C, Walmsley IA. (1998). Spectral phase interferometry for direct electric field reconstruction of ultrashort optical pulses. Opt Lett 23:792.

Jones DJ, Diddams SA, Ranka JK, Stentz A, Windeler RS, Hall JL, Cundiff ST. (2000). Carrier-envelope phase control of femtosecond modelocked lasers and direct optical frequency synthesis. Science 288:635.

Judson RS, Rabitz H. (1992). Teaching lasers to control molecules. Phys Rev Lett 68:1500.

Kane DM, Bramwell SR, Ferguson AI. (1986). FM dye lasers. Appl Phys B 39:171.

Kärtner FX, Morgner U, Ell R, Jirauschek Ch, Metzler G, Schibli TR, Chen Y, Haus HA, Ippen EP, Fujimoto J, Scheuer V, Angelow G, Tschudi T. (2000). Challenges and limitations on generating few cycle laser pulses directly from oscillators. In: T Elsaesser, S Mukamel, MM Murnane, NF Scherer, eds. Ultrafast Phenomena, Vol XII. Berlin: Springer, p 51.

Keldysh LV. (1965). Ionization in the field of a strong electromagnetic wave. Sov Phys JETP 20:1307.

Knight JC, Birks TA, Russell PStJ, Atkin DM. (1996). All-silica single-mode optical fiber with photonic crystal cladding. Opt Lett 21:1547.

Krause JL, Schafer KJ, Kulander KC. (1992). High-order harmonic generation from atoms and ions in the high intensity regime, Phys. Rev. Lett. 68:3535.

Kulander KC, Schafer KJ, Krause JL. (1993). Dynamics of short-pulse excitation, ionization and harmonic conversion. In: B Piraux, ed. Proceedings of the Workshop on Super Intense Laser Atom Physics (SILAP III), Vol 316. New York: Plenum Press, p 95.

Lewenstein M, Balcou Ph, Ivanov MYu, L'Huillier A, Corkum PB. (1994). Theory of high-harmonic generation by low frequency laser fields. Phys Rev A 49:2117.

L'Huillier A, Balcou P. (1993). High-order harmonic generation in rare gases with a 1-ps 1053-nm laser. Phys Rev Lett 70:774.

Macklin JJ, Kmetek JD, Gordon CL III. (1993). High-order harmonic generation using intense femtosecond pulses. Phys Rev Lett 70:766.

Mehendale M, Mitchell SA, Likforman JL, Vielleneuve DM, Corkum PB. (2000). A method for single shot measurement of the carrier-envelope phase of a few cycle laser pulse. Opt Lett 25:1672.

Morgner U, Kärtner, FX, Cho· S, Chen Y, Haus HA, Fujimoto JG, Ippen EP, Scheuer V, Angelow G, Tschudi T. (1999). 5.5 fs pulses from a Kerr-lens mode-locked Ti:sapphire laser. Opt Lett 24:411.

Morgner U, Ell R, Drexler W, Kärtner FX. (2001). Erzeugung und Anwendungen ultrakurzer Lichtpulse, Physik Journal 3:53.

Nisoli M, Silvestri S De, Svelto O, Szipöcs R, Ferencz K, Spielmann Ch, Sartania S, Krausz F. (1997). Compression of high-energy laser pulses below 5 fs. Opt Lett 22:522.

Poppe A, Xu L, Krausz F, Spielmann Ch. (1998). Noise characterization of sub-10-fs Ti:sapphire oscillators. IEEE J Selected Topics Quantum Electron 4:179.

Poppe A, Fürbach A, Spielmann Ch, Krausz F. (1999). Electronic on the time scale of the light oscillation period? Proc OSA Trends Opt Photon Ser 28:31.

Poppe A, Holzwarth R, Apolonski A, Tempea G, Spielmann Ch, Hänsch TW, Krausz F. (2001). Few-cycle optical waveform synthesis. Appl Phys B 72:373.

Ranka JR, Windeler RS, Stentz AJ. (2000). Visible continuum generation in air silica microstructure optical fibers with anomalous dispersion at 800 nm. Opt Lett 25:25.

Reichert J, Holzwarth R, Udem Th, Hänsch TW. (1999). Measuring the frequency of light with mode-locked lasers. Opt Commun 172:59.

Rundguist A, Durfee CG-III, Zhengu C, Herne C, Backus S, Murnane MM, Kapteyn HC. (1998). Phase-matched generation of coherent soft X-rays. Science 280:1412.

Sartania S, Cheng Z, Lenzner M, Tempea G, Spielmann Ch, Krausz F, Ferencz K. (1997). Generation of 0.1-TW 5-fs optical pulses at a 1-kHz repetition rate. Opt Lett 22:1562.

Schafer KJ, Kulander KC. (1997). High harmonic generation from ultrafast pump lasers. Phys Rev Lett 78:638.

Schnürer M, Spielmann Ch, Wobrauschek P, Streli C, Burnett NH, Kan C, Ferencz K, Koppitsch R, Cheng Z, Brabec T, Krausz F. (1998). Coherent 0.5-keV x-ray emission from helium driven by a sub-10 fs laser. Phys Rev Lett 80:3236.

Scrinzi A, Geissler M, Brabec T. (1999). Above barrier ionization. Phys Rev Lett 83:706.

Sommerfeld A. (1967). Elektronentheorie der Metalle. Berlin: Springer, p 109.

Spielmann Ch, Burnett NH, Sartania S, Koppitsch R, Schnürer M, Kan C, Lenzner M, Wobrauschek P, Krausz F. (1997). Generation of coherent x-rays in the water window using 5-fs laser pulses. Science 278:661.

Sutter DH, Steinmeyer G, Gallmann L, Matuschek N, Morier-Genoud F, Keller U, Scheuer V, Angelow G, Tschudi T. (1999). Semiconductor saturable-absorber-mirror-assisted Kerr-lens mode-locked Ti:sapphire laser producing pulses in the two-cycle regime. Opt Lett 24:631.

Telle HR, Steinmeyer G, Dunlop AE, Stenger J, Sutter DH, Keller U. (1999). Carrier-envelope offset phase control: a novel concept for absolute optical frequency measurement and ultrashort pulse generation. Appl Phys B 69:327.

Tempea G, Geissler M, Brabec T. (1999). Phase sensitivity of high-order harmonic generation with few-cycle laser pulses. J Opt Soc Am B 16:669.

Tempea G, Geissler M, Schnürer M, Brabec T. (2000). Self-phase-matched high harmonic generation. Phys Rev Lett 84:4329.

Trebino R, DeLong KW, Fittinghoff DN, Sweetser JN, Krumbügel MA, Richman BA. (1997). Measuring ultrashort laser pulses in the time-frequency domain using frequency-resolved optical gating. Rev Sci Instrum **68**:3277.

Udem Th, Reichert J, Holzwarth R, Hänsch TW. (1999a). Accurate measurement of large optical frequency differences with a modelocked laser. Opt Lett **24**:881.

Udem Th, Reichert J, Holzwarth R, Hänsch TW. (1999b). Absolute optical frequency measurement of the cesium D1 line with a modelocked laser. Phys Rev Lett **82**:3568.

Weiner AM. (2000). Femtosecond pulse shaping using spatial light modulators. Rev Sci Instrum **71**:1929.

Wineland DJ, Bergquist JC, Itano WM, Diedrich F, Weiner CS. (1989). Frequency standards in the optical spectrum, In: GF Bassani, M Inguscio, TW Hänsch, eds. The Hydrogen Atom. Berlin: Springer-Verlag.

Xu L, Spielmann Ch, Poppe A, Brabec T, Krausz F, Hänsch T. (1996). Route to phase control of ultrashort light pulses. Opt Lett **21**:2008.

Xu L, Kimmel MW, O'Shea P, Trebino R, Ranka JK, Windeler RS, Stentz AJ. (2000). Measuring the intensity and phase of ultrabroadband continuum. In: T Elsässer, S Mukamel, MM Murnane, NF Scherer, eds. Ultrafast Phenomena XII. Berlin: Springer-Verlag, p 129.

Yudin GL, Ivanov MY. (2001). Non-adiabatic tunnel ionization: looking inside a laser cycle. Phys Rev A **64**:013409.

13

Ultrahigh Bit Rate Communication Systems

Masataka Nakazawa
Tohoku University, Miyagi-ken, Japan

13.1 INTRODUCTION

In recent years, rapid progress has been made on rare-earth-doped fiber lasers because of their simplicity and wide range of oscillation wavelengths. The generation of short optical pulses from 1.55 μm erbium-doped fiber lasers is particularly attractive for high-speed optical communication, because the source can generate transform-limited picosecond to subpicosecond pulses and the oscillation wavelength in the 1.5 μm region corresponds to the minimum loss region of silica-based optical fibers.

There are two modelocking schemes for generating short pulses: passive modelocking and active modelocking. Passive modelocking is the preferred technique for generating the shortest possible pulse, and the repetition rate of the output pulse is determined by the cavity length, which is conventionally less than 100 MHz. One method by which to construct a passively modelocked fiber laser is to use a nonlinear amplifying loop mirror (NALM) [1,2]. A NALM acts as an ultrafast saturable absorber in the laser cavity by using self-phase modulation in the fiber [3]. An alternative approach to passive modelocking is to use nonlinear polarization rotation (NPR) [4,5].

A stable optical short pulse source that operates in the gigahertz region is important to the development of ultrahigh-speed optical communication. For the generation of high repetition rate pulses, the preferred technique is active modelocking. Conventional active modelocking uses an optical modulator, which is driven at the harmonic of the fundamental frequency determined by the cavity length [6,7]. In 1994, we developed a new fiber laser that we called a "regeneratively and harmonically modelocked fiber laser" [8]. This fiber laser has been used for various high-speed optical communication systems, which we describe in this chapter.

13.2 REGENERATIVELY AND HARMONICALLY MODELOCKED FIBER LASERS

13.2.1 Regeneratively and Harmonically Modelocked Fiber Laser in Free-Running Condition

With a view to realizing ultrahigh-speed optical communication, we require a stable optical source of short pulses that operates in the gigahertz region. Among many potential sources, a harmonic actively modelocked fiber laser is simple to construct and can easily produce pulses shorter than 10 ps in that region [6,7]. With conventional harmonic active modelocking, a synthesizer is used to generate a radio-frequency (RF) signal that drives the modulator. This modulation frequency is tuned to a high harmonic of the fundamental cavity frequency. If the modulation frequency from the synthesizer is not equal to a harmonic of the fundamental frequency, then the phase among the cavity modes is not locked, and eventually unstable pulsation occurs. When temperature variations and/or mechanical vibrations are applied to the cavity, the cavity length or the fundamental cavity frequency changes. This causes a mismatch between the modulation frequency and cavity modes.

A conventional harmonic actively modelocked fiber laser that operates in the gigahertz region is especially sensitive to temperature variations and/or mechanical vibrations because the cavity length (from a few meters to 100 m) is typically longer than that of other modelocked lasers. Therefore, it is possible to generate clean pulses in a short period, but it is difficult to maintain optimum operational conditions over a long period. However, this problem can be overcome with a regenerative modelocking technique [8,9].

The harmonically and regeneratively modelocked erbium-dopted fiber laser at a repetition rate of 10 GHz is shown in Figure 1. This laser consisted of a polarization-maintaining erbium-doped fiber (PM-EDF), a wavelength division multiplexing (WDM) coupler, a polarization-maintaining dispersion-shifted fiber (PM-DSF), a 15% output coupler, a polarization-dependent isolator, a lithium niobate (LiNbO$_3$) intensity modulator, and an optical bandpass filter. The pumping source was 1.48 μm laser diodes. The dispersion-shifted fiber was 180 m long, and the group velocity dispersion (GVD) was 3.3 ps/km/nm. The optical filter had a bandwidth of 2.5 nm.

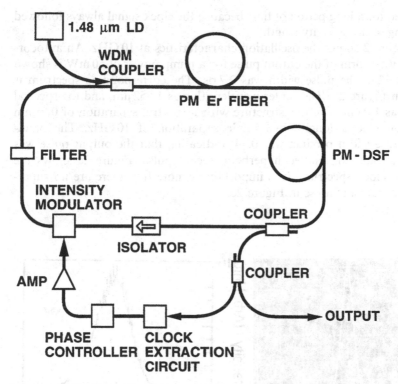

Figure 1 Configuration of a regeneratively and harmonically modelocked erbium-doped fiber laser.

All the fibers in the cavity were polarization-maintaining to prevent polarization fluctuation. The clock extraction circuit was composed of a high-speed photodetector, a high-Q dielectric filter ($Q > 1400$), and a high-gain amplifier. The center frequency of the dielectric filter was set at 10 GHz. Part of the laser output beam was coupled into the clock extraction circuit. A longitudinal self-beat signal near 10 GHz, which was one of the harmonic longitudinal modes of the cavity, was detected with the clock extraction circuit. This harmonic beat signal was amplified through an electrical amplifier and fed back to the LiNbO$_3$ intensity modulator. The phase between the pulse and the clock signal was adjusted by using a phase controller [8].

The laser started from noise and regenerative modelocking was accomplished automatically by adjusting the phase between the pulse and the clock signal. Once modelocking was achieved, stable operation continued for a long period of time because the clock signal always followed the changes in the cavity length.

Figure 2 shows the oscillation characteristics at 10 GHz. An autocorrelation waveform of the output pulse for a pump power of 50 mW is shown in Figure 2a. The pulse width was 2.7 ps. The corresponding spectrum is shown in Figure 2b. The center wavelength was 1.552 µm, and the spectral width was 1.0 nm. The fine structure with a spectral separation of 0.08 nm corresponds to a longitudinal mode separation of 10 GHz. The bandwidth–pulse width product was 0.34, indicating that the output pulse was almost a transform-limited hyperbolic secant pulse. Figure 2c shows the extracted clock spectrum. It is important to note that there are no super-modes or parasitic noise in Figure 2c.

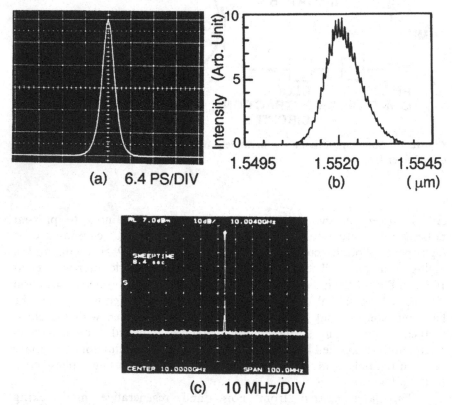

Figure 2 Oscillation characteristics of a regeneratively modelocked fiber laser at 10 GHz. (a) Autocorrelation waveform; (b) spectrum; (c) extracted clock spectrum.

13.2.2 Phase-Locked Loop Regeneratively and Harmonically Modelocked Fiber Laser

Regenerative modelocking is accomplished by feeding back the self-beat signal [8,9]. Thus, complete modelocking is achieved automatically, because an ideal feedback signal, which reflects the instantaneous frequency change between the longitudinal beats, is used as the modulation frequency even when some perturbations are applied.

The otherwise excellent regenerative modelocking technique has one drawback—the repetition rate fluctuates with time in a free-running condition when the optical path length in the cavity is not stabilized. Therefore, there is an urgent need to stabilize the repetition rate of the regeneratively modelocked fiber laser at a fixed frequency. For this purpose we developed an offset locking technique in which the repetition rate is stabilized at a fixed frequency [10]. The repetition rate stability thus obtained was better than 10^{-9}, and this value was determined mainly by the stability of the synthesizer. However, it was still not possible to lock the repetition rate to an external clock signal with a zero offset frequency. There have been several interesting reports on repetition rate stabilization of fiber lasers [11,12].

Here, we describe the first ideal phase-locked loop (PLL) operation of a fiber laser that uses a voltage-controlled regenerative modelocking technique [13]. The key to success is that the regeneratively modelocked fiber laser, whose cavity length is controlled with a PZT on which part of the fiber cavity is wound, operates as an optical voltage-controlled oscillator (OVCO).

A basic PLL consists of a self-oscillatory circuit such as a voltage-controlled x-tal oscillator (VCO), a mixer, a synthesizer (as an external clock), and a feedback control circuit. To date, in order to achieve PLL operation in a modelocked fiber laser, an external clock signal is simultaneously fed back to the laser cavity to maintain the oscillation. That is, no voltage-controlled self-oscillatory laser such as an electric VCO has been reported. However, a regeneratively modelocked fiber laser operates as an optical VCO when part of the fiber cavity is wound on a Piezoelectric transducer (PZT) and the cavity length is changed by applying a DC voltage to the PZT.

The experimental setup for the PLL fiber laser using voltage-controlled regenerative modelocking is shown in Figure 3. This laser operates as a harmonically modelocked fiber laser in which part of the fiber cavity is wound on a PZT so that the repetition rate of the regenerative laser can be varied by applying a DC voltage. That is, this oscillator is an optical self-oscillatory circuit (optical VCO) in which the repetition rate changes very stably when the DC voltage is changed. To obtain a sinusoidal harmonic beat signal between the longitudinal laser modes, part of the output beam

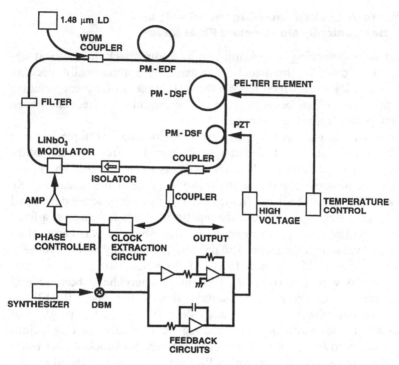

Figure 3 Experimental setup for phase-locked-loop (PLL) fiber laser using voltage-controlled regenerative modelocking.

is coupled into a clock extraction circuit that consists of a high-speed photodetector, a 10 GHz high-Q dielectric filter ($Q \sim 1400$), and a high-gain amplifier. After the phase between the pulse and the modulation peak has been adjusted, the beat signal is amplified and fed back to the LiNbO$_3$ intensity modulator, which enables regenerative modelocking to be realized.

To achieve PLL operation, a sinusoidal clock signal from the laser is mixed with a signal from a synthesizer, and its beat frequency is set at nearly zero. Then locking occurs. That is, if the laser clock signal has the same frequency as the synthesizer and the phase difference between the laser clock and the synthesizer is $\theta(t)$, the output signal from the double-balanced mixer becomes $\sin(\theta(t))$ when there is an appropriate phase bias between the laser clock and the synthesizer. In this case, the phase retardation of the laser clock is linearly converted to a voltage signal, which can be used as an error signal to detect the phase change. This signal is fed back to the PZT after proportional or integral control. It is important to note that there is no need to install a phase controller to obtain a $\sin(\theta(t))$ signal, because the regenerative laser itself is automatically phase-shifted and the phase is completely locked to the synthesizer. Once the system is locked, the laser repetition rate will accurately track that of the synthesizer by tuning the cavity length such that $\theta(t)$ is always zero.

The laser has a unidirectional ring cavity constructed with a 15 m polarization-maintaining erbium-doped fiber (PM-EDF), a WDM coupler for pumping the erbium fiber with 1.48 µm laser diodes, a 200 m polarization-maintaining dispersion-shifted fiber (PM-DSF) for solition compression, a 10% output coupler, a polarization-dependent isolator, a lithium niobate (LN) modulator, and an optical filter with a bandwidth of 1.0 nm. Part of the fiber cavity is wound on a PZT so the cavity length can be changed by applying a voltage signal. When the voltage applied to the PZT is changed, the repetition frequency change is 120 Hz/V.

The PLL circuit comprises a frequency synthesizer, a double-balanced mixer, proportional and integral feedback circuits, and high-voltage controller. Thus, by feeding this error signal back to the PZT after it has passed through a proportional and integral control circuit, the repetition rate of the laser is PLL-stabilized. The use of the PLL is similar to that in Ref. 11, but the synthesizer signal is simultaneously fed back to the laser modulator itself, which does not form an optical VCO.

The change in the repetition rate frequency with time is shown in Figure 4. Without PLL stabilization, there was a frequency difference of approximately 2–10 kHz over the course of an hour with a temperature

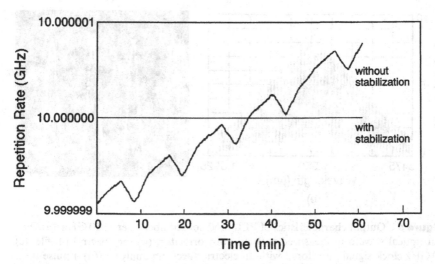

Figure 4 Change in the repetition rate frequency with time.

change of 0.5°C at around 23°C. However, when the laser was PLL-stabilized, the repetition rate was completely locked to that of the synthesizer within the stability of the synthesizer itself. The repetition rate stability is better than 10^{-9}. The resolution of the counter was better than 7×10^{-10}. Because the frequency stability of the synthesizer is the same as that of the output repetition rate, the stability of the repetition rate is mainly determined by the synthesizer itself. PLL operation occurs when the two frequencies are set below 500 Hz and the tracking range is approximately 40 kHz.

The single polarization average output was 1.5 mW, and the output pulse width was 5.7 ps. A waveform of the output monitored with an autocorrelator is shown in Figure 5a, and its spectral profile is shown in

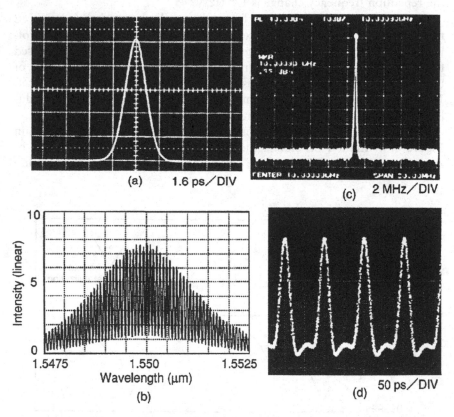

Figure 5 Output characteristics of PLL modelocked fiber laser at 10 GHz. (a) Output optical waveform measured with an autocorrelator; (b) the spectral profile; (c) 10 GHz clock signal monitored with an electric spectrum analyzer; (d) a pulse train monitored with a high-speed photodetector.

Figure 5b. A fine structure that indicates a mode separation of 10 GHz is clearly seen in Figure 5b. The spectral width of the output pulse is 0.6 nm, resulting in a pulse width–bandwidth product of 0.43. The 10 GHz clock extracted from laser output is shown in Figure 5c, and a pulse waveform monitored with a high-speed photodetector is shown in Figure 5d. As shown in Figure 5c and 5d, the amplitude of the pulse train is very stable and has no supermode noise.

This PLL regeneratively modelocked laser, which is now commercially available, is highly advantageous as a data transmission source and can be used to provide a control pulse for demultiplexing and many optical measurement techniques. Figure 6 is a photographic external view of the product. The upper housing contains all the electronics for controlling the laser, and the lower housing includes the fiber cavity optics.

13.2.3 40 GHz PLL Regeneratively and Harmonically Modelocked Fiber Laser

Recently many 40 Gbit/s optical transmission experiments have been reported because electronics at 40 Gbit/s has gradullay matured. A stable

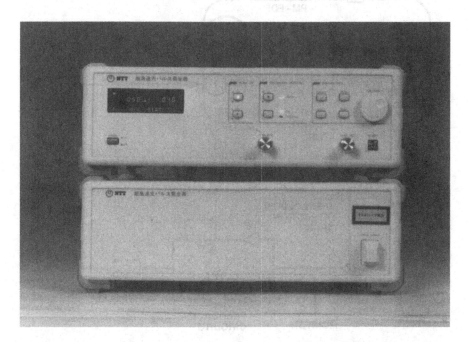

Figure 6 A photographic external view of the modelocked fiber laser.

optical pulse source at such a high repetition rate is highly desirable for future ultrafast time division multiplexed (TDM) and WDM optical transmissions. In this section, we describe the successful stable operation of a 40 GHz regeneratively modelocked fiber laser from which we obtained a 0.9 ps transform-limited short pulse. The laser was also locked to an external signal by a phase-locked loop (PLL) operation.

The experimental setup is shown in Figure 7. The laser consisted of a polarization-maintaining erbium-doped fiber (PM-EDF), a WDM coupler, a polarization-maintaining dispersion-shifted fiber (PM-DSF), a 10% output coupler, a polarization-dependent isolator, a LiNbO$_3$ (LN) intensity modulator, and an optical bandpass filter. The pumping source was 1.48 μm laser diodes, and the filter bandwidth was 5 nm. The average group velocity dispersion of the cavity was 1 ps/km/nm at a wavelength of 1.55 μm. The cavity length wass 230 m, which corresponded to a cavity mode of

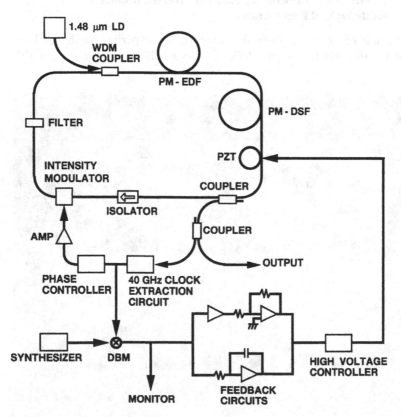

Figure 7 A 40 GHz PLL regeneratively and harmonically modelocked fiber laser.

0.9 MHz. All the fibers in the cavity were polarization-maintaining. Part of the laser output was coupled into the clock extraction circuit, which was composed of a uni-traveling-carrier (UTC) PD [14], a high-Q electric filter ($Q \sim 1200$) with a center frequency of 40 GHz, and a high-gain amplifier. The UTC-PD could operate at a high average input power of about a few tens of milliwatts. The damage threshold of an ordinary photodetector at 40 GHz is a few milliwatts. A longitudinal self-beat signal near 40 GHz was detected with the clock extraction circuit and fed back to the intensity modulator. This process accomplished regenerative modelocking. The laser started self-oscillation without the external signal generator. Once mode-locking was achieved, stable operation continued over a long period of time because the clock signal always follows the changes in cavity length.

Stable 40 GHz operation requies a higher driving voltage for the LN modulator because the modulator loss at such a high frequency is greater than that at 10–20 GHz. The half-wave voltage of the modulator we used was 4.5 V for DC operation. A power of 23 dBm was coupled into the modulator by using a high-power amplifier. Figure 8 shows typical output pulse characteristics. Figures 8a and 8b are autocorrelation waveforms. Figure 8a shows the repetitive feature of the output pulse. The interval between the two pulses is 25 ps, which corresponds to a repetition rate of 40 GHz. Figure 8b is the expansion waveform of Figure 8a and the pulse width is as short as 0.9 ps. Figure 8c is an optical spectrum. The spectrum has a symmetrical profile, and the longitudinal modes of 40 GHz (0.32 nm in the frequency domain) are clearly seen. The center wavelength is 1.552 μm, and the spectral width is 2.7 nm. Thus, the bandwidth–pulse width product ($\Delta v \Delta \tau$) is 0.3, which indicates that the output pulse is close to a transform-limited sech pulse. Figure 8d is a 1 GHz beat RF spectrum that we obtained from the difference between the 40 GHz original clock signal and a 39 GHz synthesizer signal. There was only one clock component at 1 GHz, and the supermode noise was suppressed to 70 dB down from the peak.

Figure 9a shows the pump power dependence of the output power and pulse width. The open and filled circles correspond to the output power and the pulse width, respectively. When the pump power was less than 60 mW, supermode noise appeared and no stable oscillation was obtained. When the pump power was above 60 mW, a stable pulse train was formed. The supermode noise can be suppressed by the energy stabilization mechanism that is achieved by the combination of self-phase modulation and an optical filter [15]. The output power increased linearly with a increase in the pump power, whereas the pulse width decreased monotonously. This is due to the soliton effect in the cavity. The shortest pulse width of 0.9 ps was obtained for a pump power of 110 mW. The average output power was 3.9 mW. When the pump power exceeded 110 mW, the laser became unstable and

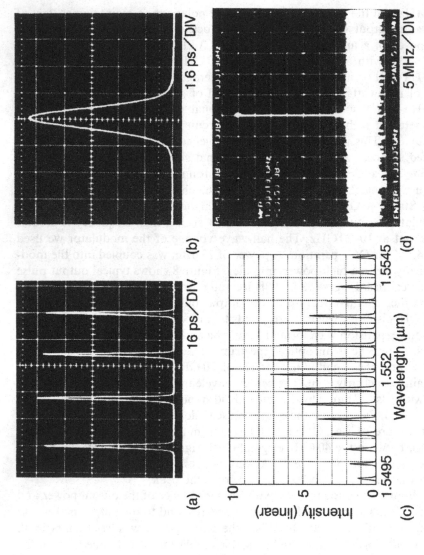

Figure 8 Output characteristics, (a, b) Autocorrelation waveforms; (c) optical spectrum; (d) clock signal at 40 GHz (frequency converted to 1 GHz).

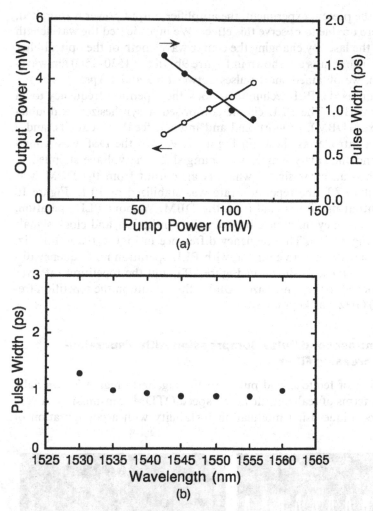

Figure 9 Pump power dependence of the output pulse and its wavelength tunability. (a) Pump power dependence of the output power and pulse width; (b) wavelength tunability.

supermode noise appeared. In this case the waveform in the correlator developed a pedestal, which indicates a deviation from the fundamental soliton. This is a result of higher order soliton formation. If greater electric power is coupled into the modulator, we may obtain shorter pulses as the result of a deeper modulation. In our experiments conducted at 10 GHz [8,16], greater electric power to the modulator resulted in shorter pulses.

However, in the present experiment, the amplifier output power was limited; hence, we were unable to observe this effect. We investigated the wavelength tunability of the laser by changing the center wavelength of the opitcal filter in the cavity. The results are shown in Figure 9b. In the 1530–1560 nm wavelength region, we obtained short pulses between 0.9 and 1.3 ps.

We again used a PLL technique to lock the repetition frequency to an external signal [13]. The PLL circuit comprised a synthesizer, a double-balanced mixer (DBM), proportional and integral feedback circuits, and a high-voltage controller as shown in Figure 7. Part of the DSF was wound on a PZT, and the cavity length was changed by the voltage supplied to the PZT. When an error signal, which is an output from the DBM, was fed back to the PZT, the repetition rate was stabilized by PLL. Figure 10 shows the output monitor signal from the DBM. Without PLL operation, there was a frequency difference between the synthesizer and clock signals as shown in Figure 10a. The frequency difference in Figure 10a is 1.3 kHz, and it changes with time. In contrast, with PLL operation no frequency difference was observed, as shown in Figure 10b, and the repetition rate was completely locked to the synthesizer. Under this condition the repetition frequency of 40 GHz was kept constant.

13.2.4 Femtosecond Pulse Compression with Dispersion-Decreasing Fiber

The generation of femtosecond pulses in the gigahertz region is especially important in terms of realizing ultrahigh-speed OTDM transmission. A very interesting technique using modulation instability with a configuration of

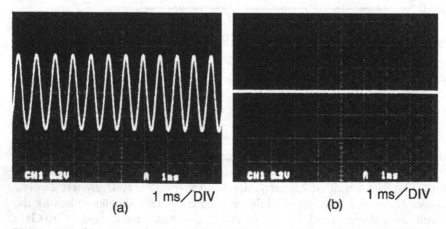

1 ms/DIV	1 ms/DIV
(a)	(b)

Figure 10 Output monitor signal from DBM (a) without and (b) with PLL.

two light beams can produce a femtosecond pulse train at a repetition rate of more than 100 GHz [17]. However, when the aim is to digitally modulate such a high-speed signal, conventional optical modulators are not applicable because they cannot respond at all in this region. Therefore, a femtosecond pulse train at a repetition rate of 10–40 GHz would seem to be the most attractive candidate for digitally modulating a signal with a conventional high-speed modulator. The method for compressing the output pulse from the fiber laser is the most advantageous for the generation of a femtosecond pulse train at 10 GHz.

We use an adiabatic soliton pulse compression in dispersion-decreasing fibers (DDFs). This technique is currently one of the most practical means for compressing high repetition rate picosecond pulses to femtosecond durations [18,19]. Such short pulse trains are needed for high-speed time division multiplexed optical communications, all-optical sampling, and all-optical switching. The mechanism of the compression is based on the principle of keeping soliton energy constant. The energy of an $N = 1$ soliton is proportional to D/τ. Here, D is the GVD and τ is the input pulse width. By keeping the $N = 1$ energy constant and decreasing the GVD, the pulse width τ of the soliton automatically shortens corresponding to the ratio of the decrease in the dispersion to the original dispersion. Here, adiabatic compression occurs when the dispersion decrease per soliton period is much smaller than the original GVD.

In all of these applications, pulses in several wavelength regimes are desirable. One problem with a conventional DDF constructed from step index fibers, however, is that the third-order dispersion (TOD) can cause the compressed pulse to form in a different wavelength regime from the input pulse. Another problem is that the compressed pulse spectrum varies significantly with input wavelength. In this section, we show that both of these problems are substantially reduced when using a dispersion-flattened DDF, which has low TOD. The reduced TOD also shortens the minimum compressed pulse width.

We first compared the compression properties of a conventional DDF and a dispersion-flattened DDF of similar lengths and dispersions. The conventional DDF was constructed from step index dispersion-shifted fiber (DSF) and had a length of 908 m. The dispersion-flattened DDF was constructed from a dispersion-flattened fiber (DFF) with a W-shaped core and a length of 1106 m. In previous work, a dispersion-flattened DDF crossing into the normal dispersion regime at the output end was used for supercontinuum (SC) generation [20]. The measured dispersions of the DDFs fitted with a five-term Sellmeier equation are shown in Figure 11. In Figure 11, D_t is the total dispersion and D_o is the dispersion of a 100 m segment cut from the output end. The profile of D_o has the most

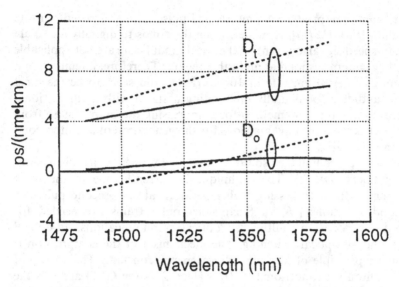

Figure 11 Dispersion characteristics of the conventional DDF (dashed lines) and dispersion-flattened DDF (solid lines). D_t is the total dispersion of the DDF. D_o is the dispersion measured in a 100 m segment cut from the output end.

important effect on the compressed pulse, and for the conventional DDF, D_o has a slope (TOD) of approximately 4.5×10^{-2} ps/(nm$^2 \cdot$ km). For the dispersion-flattened DDF, D_o has a slope that changes from 1.4×10^{-2} to 2×10^{-3} ps/(nm$^2 \cdot$ km) over the wavelength range of 1530–1565 nm. The changing slope indicates that the fourth-order dispersion (FOD) is -3.6×10^{-4} ps/(nm$^3 \cdot$ km). Both fibers had mode field diameters of approximately 6 μm. The seed pulses were generated from a 10 GHz regeneratively modelocked fiber laser emitting 3.0–3.6 ps transform-limited pulses over a wavelength range of 1533–1565 nm [8]. The pulses were amplified to average powers in excess of 140 mW by using a high-power erbium-doped fiber amplifier (HP-EDFA). The amplified pulses were launched into the DDF, and the output was characterized on a spectrometer and an autocorrelator.

The compression ratio (the ratio of input to output pulse width) in each DDF varied with input power. We compared the wavelength dependence of the compression at a power that gave a compression ratio of approximately 21 at the center of the tuning range. Figure 12a shows the output spectra from the conventional DDF as the input wavelength λ_i varied from 1533 to 1565 nm. The average power at the output end was 80–85 mW. Owing to the TOD, there are large variations in the spectra,

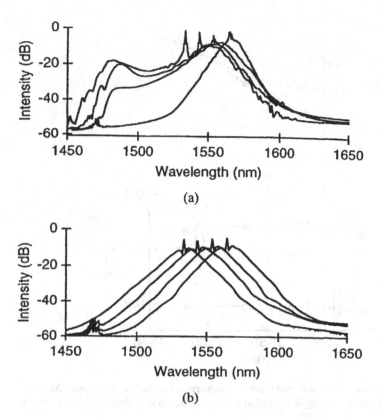

Figure 12 Output spectra of (a) conventional DDF and (b) dispersion-flattened DDF as input wavelength is tuned from 1533 to 1565 nm.

which are broadened asymmetrically. The spectral spike corresponds to uncompressed light at the input wavelength, and the compressed soliton is shifted to the longer wavelength side. Figure 12b shows the output spectra from the dispersion-flattened DDF over the same input wavelength range. The output power was between 51 and 55 mW. In contrast to the conventional DDF, similar and symmetrically broadened spectra are obtained over the entire range of input wavelengths, which is a result of reducing the effects of TOD.

Figure 13 compares the wavelength at the peak of the compressed soliton spectrum (λ_s), compressed pulse width $\Delta\tau$, and output time–bandwidth products ($\Delta\nu\,\Delta\tau$) as functions of λ_i for the conventional DDF (open symbols) and the dispersion-flattened DDF (filled symbols). For the conventional DDF, λ_s appears to reach a limit near 1550 nm, but for the dispersion-flattened DDF, λ_s and λ_i remain nearly equal. The plots of

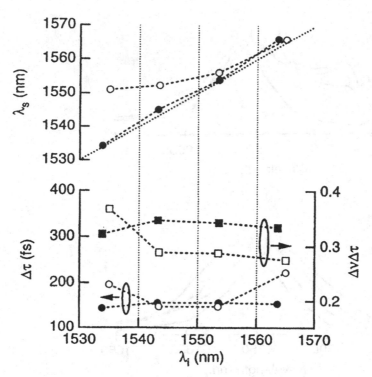

Figure 13   Center of soliton spectrum λ_s, compressed pulse width $\Delta\tau$, and $\Delta\tau\,\Delta\nu$ as functions of input wavelength λ_i. Filled symbols: Dispersion-flattened DDF. Open symbols: Conventional DDF.

$\Delta\tau$ and $\Delta\nu\,\Delta\tau$ confirm that the compression properties of the dispersion-flattened DDF are uniform with λ_i. In calculating $\Delta\nu\,\Delta\tau$, the spectral width was estimated by fitting a squared hyperbolic secant (sech^2) to the broadened portion of the spectrum. In the autocorrelations, the peak-to-pedestal ratios (PPRs) were found to be greater than 20 dB for the dispersion-flattened DDF, whereas for the conventional DDF they were approximately 17 dB for λ_i shorter than 1550 nm and greater than 20 dB for λ_i longer than 1550 nm. The pedestal widths ranged between 6 and 10 ps.

   We examined the limits to pulse compression for each fiber by measuring the compressed pulse as a function of input power. $\Delta\nu\,\Delta\tau$ and PPR as functions of output power from each DDF are plotted in Figure 14 for an input wavelength of 1553.5 nm. The degradation in PPR with power is mainly due to increased nonadiabatic higher order soliton compression effects, which increase the pulse wings. The pulse shortening reached its limit at approximately 131 fs for the conventional DDF and at approximately

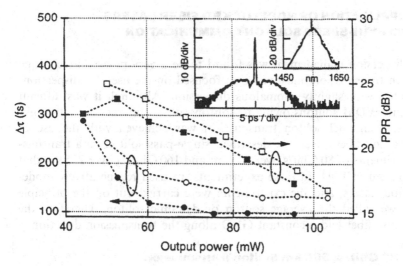

Figure 14 Δτ (left) and PPR (right) as functions of output power for conventional DDF (open symbols) and dispersion-flattened DDF (filled symbols). Input wavelength is 1553.6 nm. Inset shows autocorrelation and spectrum for 98 fs pulse from dispersion-flattened DDF. Vertical axes are intensity. Dashed curve is sech² fit to spectrum.

100 fs for the dispersion-flattened DDF. The inset shows the 98 fs compressed pulse that was generated from the dispersion-flattened DDF for an average output power of 75 mW. The PPR is approximately 21 dB. From the spectral fit (dashed lines), the bandwidth is 30.8 nm. The corresponding time–bandwidth product $\Delta v \Delta \tau$ is 0.37, which indicates that the pulse is nearly transform-limited. The compression properties of the dispersion-flattened DDF began to show a small wavelength dependence at these high compression ratios. Nevertheless, pulses shorter than 115 fs were obtained over the entire range from 1533 to 1565 nm.

A variety of mechanisms may cause the limit to Δτ observed in Figure 14. Among them are higher order dispersion, the finite nonlinear response time of fiber (Raman effect), and polarization mode dispersion (PMD). The effective lengths associated with the Raman effect were estimated to be >150 m at the DDF outputs. The PMD was measured to be < 0.1 ps/km$^{1/2}$ over the entire DDF length. Hence, higher order dispersion was expected to be the main limit to compression.

Recently, we succeeded in generating a 50 fs pulse train at 10 GHz with the use of a polarization-maintaining DF-DDF [21]. In our terabit per second OTDM transmission experiment, which is described in Section 13.4, this soliton compressor was used.

13.3 APPLICATION OF MODELOCKED FIBER LASERS TO HIGH-SPEED SOLITON COMMUNICATION

Before dispersion management was found to be a very powerful technique for soliton transmission, most research focused on the use of a dispersion-shifted fiber with slightly anomalous dispersion. Although it was difficult to transmit WDM solitons owing to soliton–soliton interaction, ultrahigh-speed single-channel soliton transmission was achieved with the use of low dispersion fibers. Here we described single-pass soliton data transmission experiments of 80 Gbit/s over 500 km and 160 Gbits/s over 225 km that were reported in 1994. In these experiments we used regeneratively mode-locked fiber lasers, and the experiments were carried out on the principle of what we called the average soliton or dynamic soliton. Therefore the transmission fiber had a constant GVD along the transmission direction.

13.3.1 80 Gbit/s, 500 km Soliton Transmission Using Uniform GVD Fiber

The experimental setup used for 80 Gbit/s, 500 km soliton transmission is shown in Figure 15 [22]. The soliton source was a regeneratively (actively) modelocked 10 GHz erbium fiber ring laser that could emit a transform-limited 3 ps soliton pulse at 1.552 μm [23]. The pulse was modulated at 10 Gbit/s

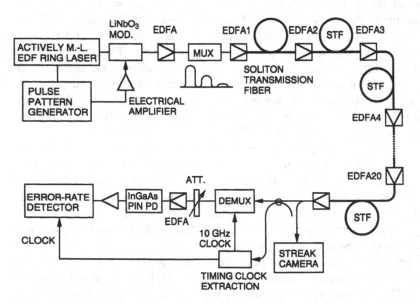

Figure 15   Experimental setup for 80 Gbit/s soliton transmission over 500 km.

with a $2^{15}-1$ PRBS using an LN intensity modulator. A planar lightwave circuit (PLC) was used as a stable optical multiplexer to obtain a 80 Gbit/s pulse train. It consisted of a three-stage Mach–Zehnder interferometer with different arm lengths that corresponded to time differences of 12.5, 25, and 50 ps. To obtain a 10 GHz clock signal easily from the transmitted 80 Gbit/s signal, 10 GHz soliton units were superimposed on each other with slightly different soliton amplitudes. This technique is also useful for reducing soliton–soliton interaction [24].

The soliton transmission fibers (STFs) were dispersion-shifted fibers with an average anomalous dispersion of 0.19 ps/(km·nm) at 1.552 µm. The average soliton period was 19.0 km, which meant that the amplifier spacing had to be shortened to as a little as 25 km. The coded pulses were amplified by EDFAs to an average soliton power level of +8.2 dBm. The average $N = 1$ soliton peak power was as high as 31.5 mW. The average fiber loss including the connector loss for one span was about 6.0 dB. A narrowband optical filter with a passband of 3 nm was installed every 50 km to stabilize the soliton train.

An 80 Gbit/s soliton data signal was demultiplexed to a 10 Gbit/s signal by using a polarization-insensitive nonlinear optical loop mirror (PI-NOLM) [25]. The demultiplexing circuit is shown in Figure 16. Part of the transmitted signal was detected with a high-speed InGaAs p–i–n photodiode, and a 10 GHz clock signal was extracted. A high signal-to-noise ratio

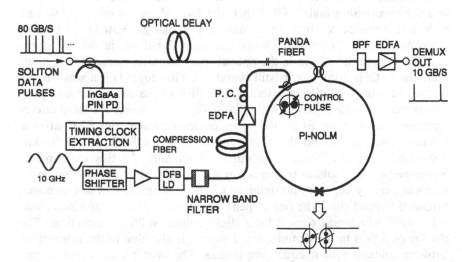

Figure 16 Optical demultiplexing circuit using a polarization-insensitive nonlinear optical loop mirror.

(SNR) clock was obtained because of the 10 GHz component resulting from our use of an unequal-amplitude soliton train. Then the sinusoidal clock signal drove the 1.533 μm DFB LD under a gain-switching condition, and the generated optical pulse was converted to a transform-limited 9 ps pulse with a combination of spectral filtering and linear compression techniques with a 500 m dispersion compensation fiber. The 9 ps pulse train at 10 GHz was amplified and coupled into a part of the loop terminal of the NOLM as a control pulse. The peak power of the control pulse was 198 mW. The NOLM consisted of a 6.0 km long dispersion-shifted polarization-maintained (PANDA) fiber, where the polarization axis of the fiber was rotated by 90° and spliced to remove the polarization mode dispersion at the middle point of the loop.

The timing between the 80 Gbit/s signal and the 10 GHz optical clock signal was tuned with an electric phase shifter and an optical delay. Thus we obtained a demultiplexed signal at 10 Gbit/s. We installed an optical band-pass filter centered at 1.552 μm to remove the control pulse at the output, and an EDFA was used to amplify the demultiplexed signal. The bit error rate (BER) of the demultiplexed signal was measured with an error rate detector.

Examples of the performance of the NOLM demultiplexer are given in Figures 17a and 17b, which show a fixed data pattern at 80 Gbit/s before demultiplexing and a demultiplexed 10 Gbit/s data pattern, respectively. The waveforms were measured with the aid of a streak camera. Because the pulse width was 2.7 ps, the waveforms do not provide precise information. However, it is clearly seen in Figure 17b that clean demultiplexing was achieved with a high SNR. When the NOLM gate is off, a ripple can be seen that was caused by the nonlinear optical bias generated by the counterpropagating 80 Gbit/s signal when the input signal was in the "1" state.

Power penalty is seen in the optical spectrum after transmission; however, the 80 GHz components still remained. This degradation was caused by a slight soliton–soliton interaction. The BER of the demultiplexed signal versus received optical power is shown in Figure 18, where the filled circles represent the baseline BER, the open circles represent the BER after a 500 km transmission, and the open squares show the BER after a 500 km transmission in a different demultiplexed channel. In this experiment, power-penalty-free soliton transmission was possible over 300 km; however, a power penalty began to accumulate when the transmission distance was extended beyond this. The power penalty after a 500 km transmission was 2.5–3.0 dB, which was caused by a slight soliton–soliton interaction. The change of 0.5 dB in the demultiplexed signals is also due to the interaction between solitons with unequal amplitudes. The inset photo shows an eye pattern of the demultiplexed 10 Gbit/s signal after 500 km transmission, which indicates a clear eye opening. A further extension of the maximum

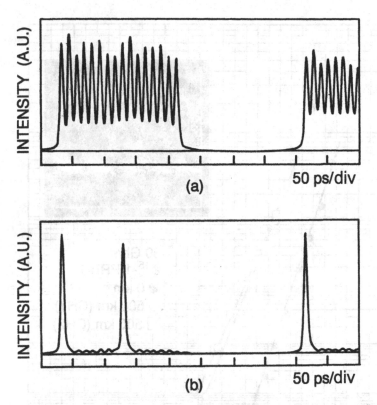

Figure 17 Demultiplexed signal at 10 Gbit/s using a PI-NOLM. (a) 80 Gbit/s soliton burst signal at input; (b) demultiplexed data signal at 10 Gbit/s.

transmission distance will be possible by properly adopting the dispersion management.

13.3.2 160 Gbit/s, 225 km Soliton Transmission

The transmission speed of 80 Gbit/s was increased to 160 Gbit/s by using polarization [26] and time domain multiplexing techniques. The experimental setup for the 160 Gbit/s, 200 km soliton transmission is similar to that of the 80 Gbit/s transmission system shown in Figure 15 [27]. To reduce soliton–soliton interaction between adjacent solitons, the 3 ps pulse from a regeneratively modelocked erbium-doped fiber laser was compressed to 1.5 ps with a soliton effect [28]. To obtain a 160 Gbit/s pulse train, polarization and time domain multiplexing techniques were implemented by employing the polarization mode dispersion of a highly birefringent fiber

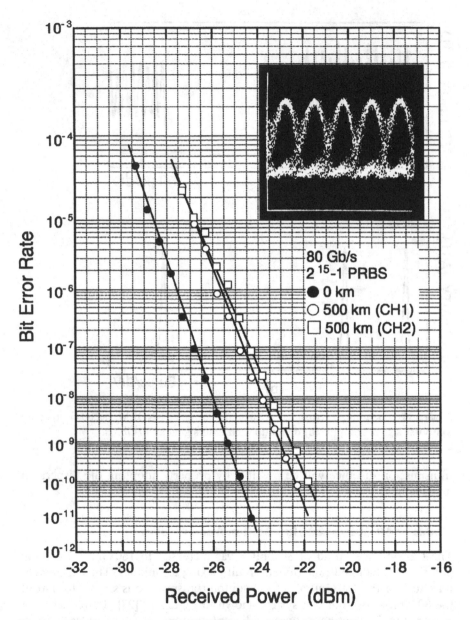

Figure 18 Bit error rate of a 10 Gbit/s signal demultiplexed from a transmitted 80 Gbit/s signal vs. received optical power. (●) Baseline; (○,□) after 500 km transmission.

[polarization-maintaining single-mode (PANDA) fiber]. This meant that the bit separation was 6.25 ps and the polarization adjacent solitons was orthogonal to reduce soliton–soliton interaction. The soliton transmission fibers were dispersion-shifted fibers with an average dispersion of 0.17 ps/(km · nm) at 1.552 μm. The average soliton period was 5.5 km, and the amplifier spacing was 25 km. This condition violates the principle of the average soliton, because the perturbation occurs within a shorter distance than the soliton period. Stable transmission over longer distances would be possible if the amplifier spacing were reduced to less than 10 km. However, this would not be practical, because an EDFA can be installed with a much longer span because of its low noise and high gain characteristics. The average $N = 1$ soliton peak power was as high as 109 mW. A narrowband filter with a passband of 5 nm was installed every 50 km to stabilize the soliton energy. In the optical demultiplexing, the 160 Gbit/s soliton data signal was first demultiplexed to 80 Gbit/s with a polarization beam splitter, then further demultiplexed to a 10 Gbit/s signal with a PI-NOLM as shown in Figure 16.

The bit error rate (BER) of the demultiplexed signal versus received power is shown in Figure 19. A BER of $<10^{-10}$ was obtained for both polarized outputs. Although there was a power penalty of 1.5–2.0 dB after a 200 km transmission, there was no indication of the existence of an error floor. Power-penalty-free soliton transmission was possible over 175 km; however, a power penalty began to accumulate when the transmission distance was extended beyond this. The power penalty deviation in the different demultiplexed signals is also due to the interaction between solitons of unequal amplitudes. The inset photograph shows a demultiplexed eye pattern after 200 km transmission, which indicates a clear eye opening. The maximum transmission distance was 225 km with a power penalty of 2.5 dB. Further extension of the maximum transmission distance can be achieved by reducing the average GVD from 0.17 to 0.01–0.02 ps/(km · nm), which is also advantageous to reduce the Gordon–Haus jitter [29].

When one uses a dispersion management technique for high-speed soliton systems, one segment of the dispersion-managed transmission line should be much shorter than the soliton period [30]. This means that we have to develop a new fiber that changes its dispersion in a short period [31] and a stable pico- to subpicosecond pulse source. If this kind of fiber becomes available in the near future, it might be very powerful for ultrahigh-speed soliton transmission over long distances. It is well known that the DM soliton has a large power margin and dispersion tolerance and is very useful for reducing soliton–soliton interaction, so a DM soliton with rapid dispersion management is advantageous not only for high-speed TDM but also for WDM soliton transmission.

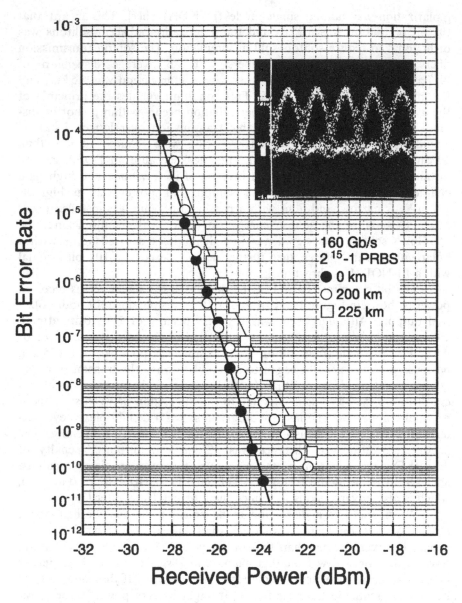

Figure 19 Bit error rate of a 10 Gbit/s signal demultiplexed from a transmitted 160 Gbit/s signal vs. received optical power. (●) Baseline; (○) after 200 km transmission; (□) after 225 km transmission.

13.3.3 Soliton In-Line Control for Ultrahigh-Speed Transmission

Principles of Soliton Transmission Control

In a linear system, the group velocity dispersion broadens the pulse width and fiber loss reduces the signal intensity. If one uses optical solitons as the signals, the dispersion problem is overcome and the loss is compensated for by EDFAs. However, the use of solitons causes new problems such as soliton–soliton interaction, Gordon–Haus jitter [29], and the accumulation of ASE noise.

Gordon–Haus jitter has a cubic dependence on the propagation distance and is a major factor in limiting soliton transmission distances [29]. The variance in the jitter is given by

$$\langle \delta t^2 \rangle = \frac{0.196(G-1)hn_2|D|L^3}{t_0 A_{\text{eff}} Z_a} \tag{1}$$

where G is amplifier gain expressed by $G = \exp(\Gamma Z_a)$, Γ is normalized fiber loss, Z_a is amplifier spacing, h is Planck's constant, n_2 is the nonlinear index, D is GVD, L is transmission distance, t_0 is the full width at half-maximum pulse width, and A_{eff} is effective area. For example, for a pulse width of 30 ps, a repeater spacing of 50 km, a loss of 0.25 dB/km, a GVD of 0.4 ps/(km · nm), and a spot size of 4 μm, the Gordon–Haus jitter is as large as 19 ps after a propagation of 10,000 km. The limit of the bit rate–distance product that provides a BER of 10^{-9} is given by

$$(RL)^3 \leq \frac{0.1372 t_0 t_w^2 R^3 A_{\text{eff}}}{h\Gamma n_2|D|} \tag{2}$$

where R is the bit rate of the system and $2t_w$ is the detector window width. For example, when $R = 10$ Gbit/s and $2t_w = 67$ ps, the maximum transmission distance L is only 7800 km. To ensure a smaller BER, the total transmission distance should be shortened.

We proposed a way to overcome the Gordon–Haus limit, ASE accumulation, and soliton–soliton interaction [32–35]. This technique is called "soliton control" or "in-line synchronous modulation" and was used to achieve a 1×10^6 km soliton transmission [32]. Soliton control is very similar to the active or passive modelocking technologies described in Section 13.2. In the active and passive modelocking of lasers, synchronous modulation is used for pulse shaping and retiming and a saturable absorber is used for noise reduction and pulse shaping. The pulse shaping and retiming techniques are very useful for soliton control. Synchronous modulation (soliton control in the time domain) enables us to retime the position of the soliton pulse that experiences jitter as a result of amplified spontaneous emission (ASE) noise. It is also possible to remove the interaction forces that inevitably occur between closely adjacent solitons. It has been proved theoretically that soliton data transmission over unlimited distances is possible with this technique because periodic synchronous modulation can reduce the ASE noise to a very low level [34]. A bandpass filter with a 0.3–0.4 nm bandwidth was also installed in the loop as a soliton control in the frequency domain to stabilize the soliton energy [33,34].

Figure 20 illustrates the principle of noise reduction through the use of synchronous modulation [33]. When modulation is applied to a soliton, it can be reshaped and retimed for transmission, but noise, including ASE noise and nonsoliton components, that has a small amplitude is modulated and disperses after transmission over a certain distance. This is because a low-amplitude signal is just a dispersive wave. After a transmission of a certain distance, the next modulator is installed. So the soliton is unchanged, and the noise remains at a very low level.

We assume sinusoidal modulation for the shaping function $f(t)$ and that its extinction ratio is $1/x$ $(x < 1)$. For example, $x = 0.01$ means an extinction ratio of 20 dB. The signal pulse shape is assumed to be a hyperbolic secant squared.

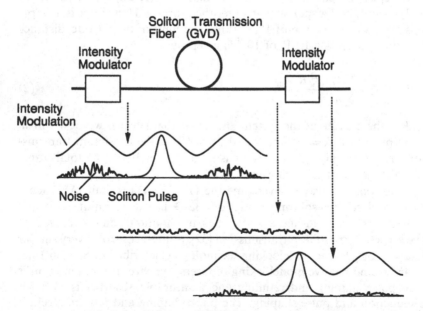

Figure 20 Principle of noise reduction by the use of synchronous modulation.

$$f(t) = x + (1 - x)\cos^2\frac{\pi t}{T} = x + (1 - x)\frac{\cos(2\pi t/T) + 1}{2} \tag{3a}$$

$$g(t) = \text{sech}^2(1.76t/\Delta t) \tag{3b}$$

where T is the period of the shaping function $(= 1/R)$. The transfer function for the noise, or noise reduction ratio, due to $f(t)$ is $T_R = (1 + x)/2$ per shaping if the sinusoidally modulated noise disperses during propagation and becomes continuous wave (cw) noise again before it reaches the next modulator.

The transmitting signal power through the modulator T_m is

$$T_m = \frac{\int_{-T/2}^{T/2} f(t)g(t)\,dt}{\int_{-T/2}^{T/2} g(t)\,dt} \cong x + \frac{1-x}{2}\left[1 + \frac{\pi^2/1.76\alpha}{\sinh(\pi^2/1.76\alpha)}\right] \tag{4}$$

where $\alpha = T/\Delta t$ and is assumed to be > 1. It is necessary to compensate for the attenuation to maintain the soliton pulse. The required excess gain is $1/T_m$. The next amplifier adds noise, but the noise is also reduced by succeeding modulators. Thus the total amount of noise from k amplifiers is written as

$$\langle N_T \rangle = \langle N \rangle \frac{1 - (T_R G_m)^k}{1 - T_R G_m} \tag{5}$$

where $\langle N \rangle$ is the ASE power from each amplifier. Here,

$$T_R G_m = \frac{(1 + x)/2}{x + \frac{1-x}{2}\left[1 + \frac{\pi^2/1.76\alpha}{\sinh(\pi^2/1.76\alpha)}\right]} < 1 \tag{6}$$

therefore, $T_R G_m$ is always less than unity, so $\langle N_T \rangle$ converges for any x value. For example, $x = 0.01$ (20 dB modulation) and $\alpha = 5$ gives $\langle N_T \rangle = 2.25 \langle N \rangle$. The Mach–Zehnder intensity modulator has a sinusoidal transmission as a function of the applied voltage, and the driving voltage is sinusoidal. Therefore, the transmission function has a steeper edge than the sinusoidal modulation that is simply assumed here. In an actual situation, this asymmetrical modulation function can remove the noise more efficiently.

Another large advantage of soliton control is that the amplifier spacing L_a can be extended to be of the order of soliton period Z_{sp}. Theory suggests that the pulse distortion is proportional to $(Z_a/Z_{\text{sp}})^2$ [36], and dispersive nonsoliton waves are increased with increases in Z_a/Z_{sp}. This result means that $(Z_a/Z_{\text{sp}})^2$ should be much smaller than unity in order to achieve ultralong-distance soliton transmission. Higher bit rate communication requires a shorter optical pulse, which shortens Z_{sp}. The ASE noise added by EDFA also limits the amplifier spacing. It has been stated that the EDFA gain G should be less than 10 dB because excess noise increases at a rate of $F(G) = (G-1)^2/G(\ln G)^2$ [37]. These difficulties can be removed through the use of soliton control, which reshapes the soliton pulse and removes dispersive nonsoliton waves as well as ASE noise [38].

40 and 80 Gbit/s Single-Channel DM Soliton Transmission over 10,000 km

In this section we describe single-channel 40 Gbit/s soliton transmission over 70,000 km and 80 Gbit/s over 10,000 km with the use of in-line synchronous modulation [39,40]. Both experiments were performed in a 250 km dispersion-shifted fiber loop, as shown in Figure 21, where the 80 Gbit/s in-line modulation scheme is given in the dotted box. The optical soliton source was a 10 GHz harmonically modelocked erbium fiber laser at 1550.5 nm [8,13]. The pulse and spectral widths for 40 Gb/s were 5 ps and 0.50 nm, respectively $(\Delta\nu\,\Delta\tau = 0.31)$, and those for 80 Gbit/s were 3.0 ps and 0.81 nm, respectively. The pulse was modulated at 10 Gbit/s with a $2^{11}-1$

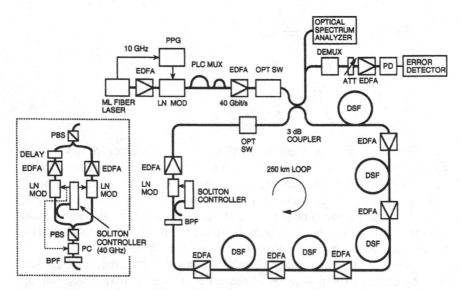

Figure 21 Experimental setup for 40 Gbit/s single-channel soliton transmission using synchronous modulation and in-line filtering. For 80 Gbit/s transmission, inline modulation was replaced with the apparatus shown in the dotted box.

PRBS using an LN intensity modulator. In order to obtain a 10 GHz clock signal from the transmitted 40 Gbit/s signal, 10 GHz soliton units were superimposed on each other with slightly different amplitudes. This technique is also useful for reducing soliton–soliton interaction. The amplifier spacing was 50 km, and the average fiber loss for the span was 12.5 dB. We used a four-segment dispersion-decreasing configuration to reduce the influence of the dispersive waves. For a 40 Gbit/s experiment, the four 12.5 km long dispersion-shifted fibers (DSFs) we used had GVDs of 0.24, 0.06, −0.04, and −0.10 ps/(km · nm). We set the average GVD at approximately 0.04 ps/(km · nm). The average launched power was +7.0 dBm, and the peak power into the first segment was 50 mW, which corresponded to an almost $N = 1$ soliton.

For the 80 Gbit/s experiment, the average launched power was + 9.5 dBm and the peak power into the first segment was 74 mW, which corresponded to an $N = 1.1$ soliton. The 80 Gbit/s transmission configuration is similar to 40 Gbit/s, 70,000 km transmission [39] except for the polarization-multiplexing technique. To achieve soliton control at 80 Gbit/s based on 40 Gbit/s electronics, we exmployed a polarization-multiplexing technique. That is, a 40 Gbit/s single-polarization signal was polarization-multiplexed and interleaved with an 80 Gbit/s TDM signal by using a 7.5 m long PANDA fiber.

After a 250 km transmission through the loop, we applied in-line soliton control using synchronous modulation and narrowband optical filtering, in which the filter bandwidth was 0.80 nm for 40 Gbit/s and 1.1 nm for 80 Gbit/s. The 40 GHz clock signal was extracted from part of the transmitted soliton pulses with an ultrahigh-speed photodetector that had a 50 GHz bandwidth and a high-Q dielectric filter. The LN modulator for soliton control was driven by the extracted 40 GHz clock. To obtain 80 Gbit/s in-line modulation, the polarization-multiplexed 80 Gbit/s signal was separated into two 40 Gbit/s signals with a polarization beam splitter (PBS), and single-polarization soliton control was employed at 40 Gbit/s. The 40 GHz clock signal was extracted from part of the transmitted soliton pulses with an ultrahigh-speed photodetector with a bandwidth of 50 GHz and a high-Q dielectric filter. One clock signal was applied to two orthogonal signals. The LN modulators for soliton control were driven by the extracted 40 GHz clock, where each modulator operated at a single polarization. Simultaneously, the amplitude level of the clock signal was fed back to a polarization controller (PC) to obtain maximum clock power. This ensured that the orthogonal channel was also automatically optimized. After the in-line modulation, two orthogonal 40 Gbit/s signals were passed through a delay unit and were reconverted to a 80 Gbit/s signal with another PBS.

The transmitted 40 Gbit/s soliton data signal was demultiplexed into a 10 Gbit/s signal by using an electroabsorption (EA) intensity modulator. The 10 GHz clock signal was extracted from the transmitted signal. Then, by adjusting the DC bias voltage and the amplitude of the 10 GHz clock signal for the EA modulator, the gate width of the EA modulator was set a 20 ps to extract a 10 Gbit/s signal from the 40 Gbit/s signal. The demultiplexed signal was detected with a 10 Gbit/s optical receiver and the BER was measured. As for the transmitted 80 Gbit/s soliton data signal, it was demultiplexed into a 10 Gbit/s signal by using a PBS (80 Gbit/s to 40 Gbit/s) and an electroabsorption (EA) modulator (40 Gbit/s to 10 Gbit/s).

Figures 22a and 22b show the measured BERs corresponding to 40 and 80 Gbit/s. In Figure 22a the open circles indicate the BER with soliton control after a transmission of 70,000 km. The filled circles show the BER without soliton control after a transmission of 4500 km, and the triangles show the BER at 0 km. When the soliton control was not applied, the maximum transmission distance was 4500 km. The BER beyond 4500 km increased rapidly with an increase in the transmission distance owing to the accumulation of the noise component and timing jitter. In contrast, when we employed the soliton control [32], the maximum transmission distance was greatly extended up to 70,000 km. When the received power was greater than -31.0 dBm, no error appeared at a 10^{-8}-order error counter setting, which indicates that the BER was less than 1×10^{-8}. The inset photograph shows an eye pattern after a 70,000 km transmission, in which the eye is clearly open.

In Figure 22b, we show eight-channel 10 Gbit/s signals, four of which were vertical and four of which were from orthogonal components. The inset photograph and figure show one of the eye patterns that was demultiplexed to 10 Gbit/s and a spectral profile of the 80 Gbit/s soliton signal after 10,000 km transmission, respectively. The solid line with open diamonds indicates the BER at 0 km. The power penalty after the 10,000 km transmission was typically 2.5 dB, and all data fell within a power penalty difference of 1.5 dB. When the received power was greater than -27.5 dBm, no error appeared at a 10^{-9}-order error counter setting, which indicates that the BER was less than 1×10^{-9}. Without the in-line control, the transmission distance was approximately 1750 km, but we were able to extend it to 10,000 km by using in-line modulation.

These results suggest that an ultrahigh bit rate well beyond 100 Gbit/s soliton can be transmitted over 10,000 km by using in-line synchronous modulation. This technique will be more effective when the single-channel bit rate is higher. The speed of the electronics may be limited to 100 Gbit/s, and this makes all-optical soliton control using high-speed optical switching

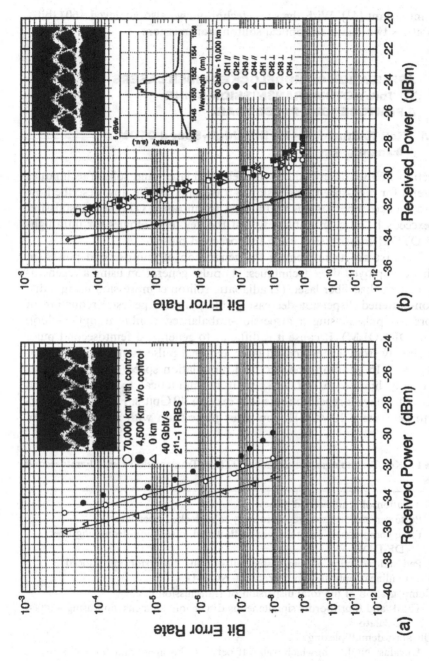

Figure 22 Bit error rate (BER) characteristics. (a) 40 Gbit/s soliton data. (○)BER after 70,000 km transmission with soliton control; (●)BER after 4,500 km transmission without soliton control; (△)BER at 0 km. (b) 80 Gbit/s TDM soliton data.

more important [41]. With the same technique we also achieved 160 Gbit/s (80 Gbit/s × two channels) optical soliton transmission.

13.4 APPLICATION OF MODELOCKED FIBER LASERS TO Tbit/s OTDM TRANSMISSION USING A FEMTOSECOND PULSE TRAIN

13.4.1 Key Technologies for Ultrahigh-Speed OTDM Transmission

In Section 13.3 we described how modelocked lasers and soliton techniques are useful for high-speed long-distance transmission. We now describe how the modelocked fiber laser is useful even for ultrafast OTDM transmission that exceeds 1 Tbit/s. In Table 1 we list the key technologies for ultrahigh-speed OTDM transmission using femtosecond pulses.

The first key technology is the generation of femtosecond pulses, in which we employed three techniques: (1) pulse generation using a regeneratively modelocked fiber laser, (2) adiabatic soliton compression using a dispersion-flattened dispersion-decreasing fiber, and (3) pedestal reduction of compressed pulses using a dispersion-imbalanced nonlinear optical loop mirror (DI-NOLM). Because it is difficult to generate a femtosecond pulse train directly from a modelocked laser, external pulse compression is commonly used. In addition, the electrical modulation speed of digital signals is currently limited to 40 Gbit/s, so it is impractical to prepare optical sources that have a repetition rate that exceeds 40 Gbit/s. Hence it is important to generate a 10–40 GHz low-jitter pulse train with a pulse width of

Table 1 Key Technologies for Ultrafast OTDM Transmission Using Femtosecond Pulses

---

1. Generation of femtosecond pulses
     Pulse generation using regeneratively modelocked fiber lasers
     Pulse compression using dispersion-flattened, dispersion-decreasing fiber
         (DF-DDF)
     Pedestal reduction of compressed pulses using dispersion-imbalanced
         nonlinear optical loop mirror (DI-NOLM)
2. Compensation of the total dispersion of transmission line
     Third- and fourth-order simultaneous dispersion compensation using a phase
         modulator
3. Ultrafast demultiplexing
     Ultrafast NOLM in which walk-off between the signal and control pulses
         is reduced

100–400 fs. Our approach is to use a regeneratively modelocked fiber laser that can emit a 3 ps, 10 GHz pulse train [8,13]. The 3 ps pulse is compressed to 50–200 fs by using adiabatic soliton compression in a dispersion-decreasing fiber [42]. The unwanted dispersive waves (pedestal components) are removed with the DI-NOLM [43].

The second key technology is compensation of the total dispersion of the transmission line. Dispersion compensation and dispersion slope compensation fibers are commonly used to compensate for second- and third-order dispersion. It is also important to minimize the fourth-order dispersion because of the large spectral width of the transmitted pulse. For this, we adopted an active dispersion compensation technique that uses a phase modulator [44].

The third key technology is ultrafast demultiplexing using a walk-off-free nonlinear optical loop mirror [45]. Because the transmitted OTDM Tbit/s signal is too fast to detect directly with a photodetector, down-conversion of the signal via demultiplexing with a NOLM, four-wave mixing, or electro-absorption (EA) modulator is required. We used a NOLM to demultiplex the 1.28 Tbit/s signal to 10 Gbit/s.

13.4.2 A Single-Channel 1.28 Tbit/s OTDM Transmission Experiment

Because we compensated for the second- and third-order dispersion, the fourth-order dispersion became relatively large and symmetrically broadened the wings of the femtosecond pulse after 100 km transmission. This effect is no longer negligible in this regime.

There are two methods of compensating for fourth-order dispersion. One is to use a grating pair and a spatial light modulator [46], and the other is to use a phase modulator [47]. Our technique uses a phase modulator with a conventional single-mode fiber (SMF) and a reverse dispersion fiber (RDF) [48]. This modified phase modulation technique enables simultaneous compensation of the third- and fourth-order dispersion and reduces the pulse broadening from 200 fs to as small as 20 fs. We found that PMD was not a serious problem as long as the input polarization was properly aligned. With this methods, we achieved for the first time single-wavelength channel 1.28 Tbit/s polarization-multiplexed (PM) OTDM transmission over 70 km.

Generation of a Femtosecond Pulse Train for OTDM Tbit/s Transmission

The experimental setup for the 1 Tbit/s OTDM experiment is shown in Figure 23. The generation of the femtosecond pulse train and its transmission

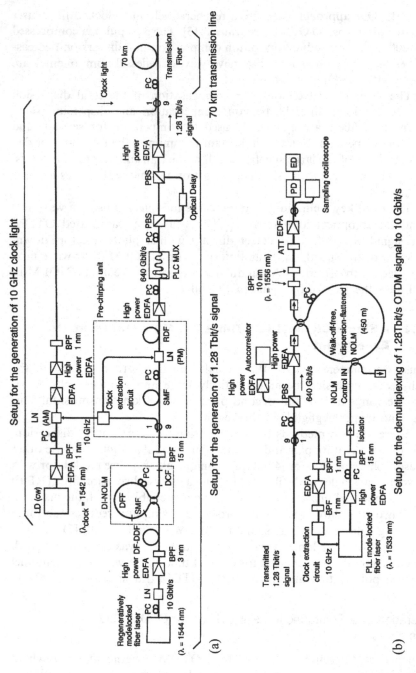

Figure 23 Experimental setup for 1.28 Tbit/s, 70 km OTDM transmission.

are shown in Figure 23a, and the corresponding demultiplexing system is shown in Figure 23b. A 3 ps, 10 GHz regeneratively and harmonically mod-eclocked fiber laser at 1.544 μm was used as the original pulse source. The advantage of this laser is its low jitter of approximately 100 fs, which is very important for realizing ultrafast multiplexing in the time domain.

The output laser pulse is intensity-modulated at 10 Gbit/s with a PRBS of $2^{15}-1$, and the pulse train was coupled into a 1.1 km long dispersion-flat-tened dispersion decreasing fiber (DF-DDF) [42]. The DDF was designed to have approximately zero dispersion and a zero dispersion slope at the out-put end for a wavelength of 1550 nm. The dispersion at the input end for a wavelength of 1550 nm was approximately 10 ps/(km·nm). This enabled us to realize adiabatic soliton compression to less than 200 fs. After the pulse compression, the signal was coupled into the DI-NOLM to reduce the spec-tral spike and pedestal of the compressed pulse [43,49]. The DI-NOLM con-sists of a 50/50 coupler, a fiber loop, and a dispersion compensation fiber (DCF) as shown in Figure 23a. The fiber loop in our experiment consisted of a 100 m long dispersion-flattened fiber and a 1 m length of SMF. The SMF was inserted at an asymmetrical position in the loop. The chirp caused in the fiber loop was compensated for by using a DCF. When a narrow-bandwidth signal is coupled into the DI-NOLM, the signal is reflected for all intensities because the effect of dispersion in the loop is negligible. By using these DI-NOLM characteristics, we can reduce the pedestal of the sig-nal pulse by more than 15 dB. The pulse width at the DI-NOLM output was 200 fs, and the peak-to-pedestal radio was greater than 30 dB.

Preparation of the Input Tbit/s Signal with Pre-Chirping for Fourth-Order Dispersion Compensation

Before optically multiplexing the signal into a 1.28 Tbit/s signal, we incorpo-rated a phase modulation technique that enabled us to compensate for third- and fourth-order dispersion. The distortion of pulse waveforms by dispersion results from the phase shift that affects each spectral component of the pulse. The frequency dependence of the phase shift (in radians) is expressed in a Taylor series as

$$\beta L = \beta_0 L + \beta_1 L(\omega - \omega_0) + \frac{\beta_2 L}{2}(\omega - \omega_0)^2 + \frac{\beta_3 L}{6}(\omega - \omega_0)^3$$
$$+ \frac{\beta_4 L}{24}(\omega - \omega_0)^4 + \cdots \tag{7}$$

where L is fiber length and ω_0 is center frequency. Second-, third-, and fourth-order dispersion arise from the terms including $\beta_2, \beta_3,$ and β_4, respectively.

Figure 24 illustrates the principle of third- and fourth-order dispersion compensation by showing calculated phase shift results. Parameters β_2 ($= -7.9 \times 10^{-3}$ ps^2/km), β_3 ($= 4.0 \times 10^{-4}$ ps^3/km), and β_4 ($= 8.6 \times 10^{-4}$ ps^4/km) used in the calculation are the same values as those for our transmission line, which includes a 70 km transmission fiber, three EDFAs, and a pre-chirping unit (dispersion compensator). Figure 24a illustrates the principle of fourth-order dispersion compensation [47]. A quartic curve expressing fourth-order dispersion can be converted to a cosine curve around ω_0 by adding an appropriate quadratic curve, that is, by adjusting the second-order dispersion of the whole transmission line. If we can apply a cosine phase change with an adequate amplitude for the spectral component of the pulse, the total phase shift has a broader linear region than the original quartic curve. That is, by broadening the linear region of the phase shift curve, we can reduce the influence of fourth-order dispersion. To apply the cosine phase modulation to the frequency domain, we linearly chirp the pulse before launching it into a phase modulator using a fiber with a large second-order dispersion. This is usually achieved by using an SMF. With this linear chirp, the cosine phase modulation in the time domain applied by the phase modulator automatically becomes the cosine phase modulation in the frequency domain.

Figure 24b shows the principle of third-order dispersion compensation [50]. In this case, by applying a sine phase modulation, we broaden the linear region of the phase shift curve compared with that of the original cubic curve expressing third-order dispersion. That is, the sine phase modulation can reduce the influence of third-order dispersion. When we apply the cosine and sine phase modulation simultaneously, third- and fourth-order dispersion are compensated for simultaneously. As shown in Figure 24c, the sum of a cosine phase modulation $\phi_c(t) = -\phi_c \cos(2\pi R_0 t)$ and a sine phase modulation $\phi_s(t) = -\phi_s \sin(2\pi R_0 t)$ can be expressed as a generalized sinusoidal phase modulation:

$$\phi(t) = -\text{sign}(\phi_c)\sqrt{\phi_c^2 + \phi_s^2} \cos\left[2\pi R_0 t - \arctan\left(\frac{\phi_s}{\phi_c}\right)\right] \qquad (8)$$

where R_0 is the repetition rate of the phase modulation. From Eq. (8), it is clear that we can compensate for not only the fourth-order dispersion but also the third-order dispersion simply by changing the amplitude and the timing of the cosine phase modulation depending on the magnitude of the third-order dispersion. Figure 24d shows the sum of the phase shift caused by the second-, third-, and fourth-order dispersion, total phase modulation, and the resulting modified phase shift. We convert the sum of the quadratic, cubic, and quartic curves so that it becomes linear around ω_0 by applying a single sinusoidal modulation.

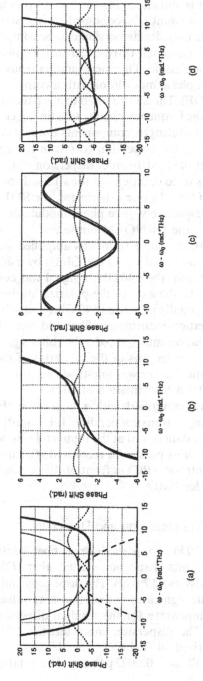

Figure 24 Principle of third- and fourth-order dispersion compensation showing calculated phase shift results. (a) Phase shift caused by second-order (dashed line) and fourth-order (thin solid line) dispersion, cosine phase modulation (dotted line), and the sum of these phase shifts (thick solid line). (b) Phase shift caused by third-order dispersion (thin solid line), sine phase modulation (dotted line), and the sum of these phase shifts (thick solid line). (c) Phase shift caused by cosine phase modulation (thin solid line), sine phase modulation (dotted line), and the sum of these phase shifts (thick solid line). (d) Phase shift caused by the sum of the second-, third-, and fourth-order dispersion (thin solid line), total phase modulation (dotted line), and the sum of these phase shifts (thick solid line).

The pre-chirping unit we used is shown in Figure 23a. In this scheme, we deliberately introduce a certain amount of second-order dispersion, in which the sum of the second- and fourth-order dispersion can be compensated for by a cosine phase modulation. The small third-order dispersion is compensated for by a sine phase modulation. This new simultaneous dispersion compensation is achieved by a phase modulator with a conventional SMF and reverse dispersion fiber (RDF). The SMF is used to chirp the input femtosecond signal, which converts the frequency change to a timing change. That is, the SMF spreads the input wavelength components linearly in time. The latter RDF gives us a second-order dispersion to create the cosine wave chirping with a combination of the fourth-order dispersion of the transmission line. This technique allows us to compensate for all the dispersion from the EDFAs and transmission fibers. In our experiment, the SMF and RDF lengths were 276 and 139 m, respectively. The phase modulator was driven by a 10 GHz clock extracted at the DI-NOLM output.

Then the pre-chirped 10 GHz pulse train was amplified with a high-power EDFA and optically multiplexed up to 640 Gbit/s by using a planar lightwave circuit (PLC). After that, the 640 Gbit/s signal was coupled with the polarization at a 45° angle to the axis of the polarization beam splitter (PBS), thus creating two orthogonally polarized 640 Gbit/s signals. Each signal was coupled two a polarization-maintaining fiber and recombined with another PBS to form a polarization-multiplexed 1.28 Tbit/s signal. The difference between the PANDA fiber lengths of the two arms of the PBS was approximately 4 m, and the optical delay was fine-tuned so that a 1.28 Tbit/s signal was clearly obtained after a transmission of 70 km.

The 1.28 Tbit/s signal was amplified with a high-power EDFA and coupled into a dispersion-managed transmission line. It is important to note that due to the pre-chirping, the data signal at the input end was no longer in the femtosecond regime, but a short pulse was recovered after transmission. We installed a polarization controller (PC) in front of the transmission fiber to compensate for the first-order PMD.

Femtosecond OTDM Signal Transmission and Detection

The transmission line was a 70 km long single-span fiber consisting of a 39.7 km long SMF, a 4.6 km long dispersion-shifted fiber (DSF), and a 25.1 km long RDF. The fiber loss was 17.6 dB. The dispersion and dispersion slope of the RDF had opposite signs and almost the same values as SMF. The DSF was included to compensate for the residual dispersion slope of the whole transmission line. The dispersion characteristics of the present transmission line are summarized in Table 2. The third-order dispersion (dispersion slope) was as small as $-0.0023\,\text{ps/nm}^2$. The total third-order

Table 2 Dispersion Characteristics of a 70 km Transmission Line

	Length (km)	Loss (dB)	Dispersion (1556 nm)			PMD ps/(km · nm)
			2nd dispersion ps/(km · nm)	3rd dispersion ps/(km · nm^2)	4th dispersion ps/(km · nm^3)	
SMF	39.7	8.5	18.994	0.0547	−0.00005	0.08
DSF	4.6	1.4	−0.041	0.0696	−0.00003	0.10
RDF	25.1	7.8	−30.033	−0.0991	−0.00101	0.02
Total	69.4	17.7	0.003 ps/nm	−0.0023 ps/nm^2	−0.02748 ps/nm^3	0.05

dispersion including three wideband EDFAs was $+0.015\,\text{ps/nm}^2$. Two advantages of using RDF for ultrafast OTDM signal transmission are that it has a lower PMD and a flatter dispersion profile than a dispersion-slope compensation fiber (DSCF). The PMD of the RDF in this experiment was $0.02\,\text{ps/km}^{1/2}$, and the total PMD was $0.05\,\text{ps/km}^{1/2}$. It is important to note that the accumulated fourth-order dispersion became as large as $-0.028\,\text{ps/nm}^3$, which is no longer negligible for the long-distance transmission of a femtosecond pulse train. This occurs because the sign of fourth-order dispersion is the same in all fibers. There are several ways to compensate for the second- and third-order dispersion, but it is quite difficult to compensate for the fourth-order dispersion. Usually, fourth-order dispersion does not cause serious problems, but when the transmission distance becomes of the order of 100 km, the value is no longer negligible. The second-order dispersion, which we deliberately introduced, was $-0.42\,\text{ps/nm}$. To compensate for all these dispersions, we use the pre-chirping technique that we previously described.

Figure 25 shows the dispersion profile of our transmission line. The profile may not look dispersion-flattened, but the vertical axis corresponds to a total dispersion of over 70 km. For convenience, a 3 dB bandwidth of the signal spectrum is plotted. Figure 26 shows how the transmitted waveform is distorted due to the higher order dispersions in our 70 km transmission line. Figure 26a shows distortion due to the third-order dispersion, in which ripples appear on the wing of the transmitted pulse. On the other hand, waveform distortion due to the fourth-order dispersion is shown in Figure 26b, in which both wings of the transmitted pulse are substantially broadened. This broadening is fatal in terabit per second OTDM trans-

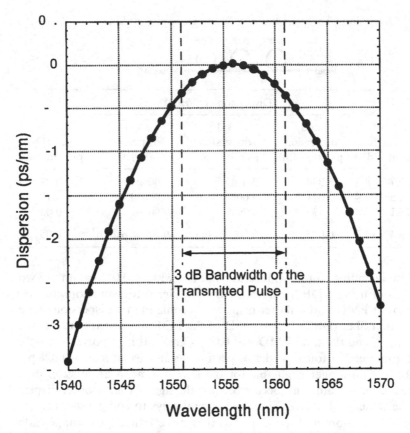

Figure 25   Dispersion profile of a 70 km transmission line.

mission because the energy is extended into an adjacent channel, resulting in bit errors. However, when we adopt the pre-chirping scheme described above, third- and fourth-order dispersion are well compensated for by a sinusoidal phase modulation and the energy leakage into the adjacent channel is quite small. This is shown in Figure 26c.

The transmitted 1.28 Tbit/s signal was converted into two 640 Gbit/s signals with a PBS and then demultiplexed into 10 Gbit/s signals using a walk-off-free, dispersion-flattened NOLM as shown in Figure 23b [45]. The control pulse was generated from a 1533 nm PLL-operated modelocked fiber laser [13]. To obtain a 10 GHz clock to drive the PLL fiber laser, we generated a 10 GHz optical clock light at a different wavelength, which was synchronized to the original 10 Gbit/s signal, and transmitted the clock light and a 640 Gbit/s TDM signal simultaneously.

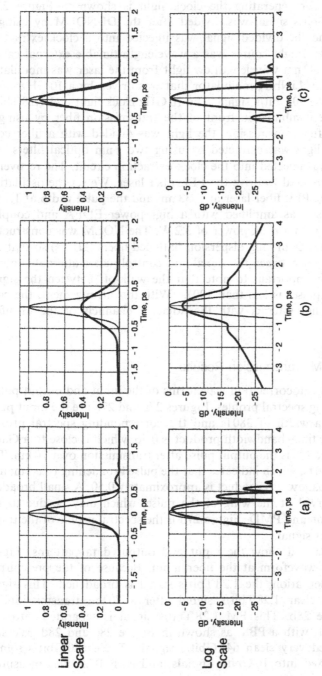

Figure 26 Pulse waveform changes after 70 km transmission. (a) Waveform distorted by third-order dispersion; (b) waveform compensated for dispersion by phase modulation. Third- and fourth-order dispersion are simultaneously compensated for. $L_{smf} = 276$ m; $L_{rdf} = 139$ m; phase modulation amplitude $= 1.21\pi$. (c) waveform distorted by fourth-order dispersion.

The setup for generating the clock light is shown in Figure 23a. Part of the 10 Gbit/s signal was divided after the DI-NOLM by using a fiber coupler, and the divided signal was injected into a clock extraction circuit. The control light source was a wavelength-tunable external cavity laser diode. A 1542 nm wavelength cw light from the laser was modulated at 10 GHz with an LiNbO$_3$ modulator that was driven by a 10 GHz clock extracted from the 10 Gbit/s signal. The 10 GHz clock light and 1.28 Tbit/s signal light were combined in front of the transmission fiber by using a fiber coupler. After transmission, the light was divided with a fiber coupler. The clock light was extracted by using two 1 nm optical filters and an EDFA and was injected into the clock extraction circuit. The recovered 10 GHz clock was used to drive the PLL fiber laser. We set the oscillation wavelength of the PLL fiber laser at 1533 nm and the pulse width at 1.1 ps. The control pulse was amplified with a high-power EDFA and coupled into the NOLM with a peak power of 3.2 W. The NOLM was constructed by connecting nine 50 m long dispersion-flattened fibers (DFFs) so that the sign of the walk-off between the signal and control pulses alternated along the NOLM. It is important to note that the walk-off between the signal and the control pulses is less than 100 fs. With this arrangement, the control pulse does not interact with the adjacent channels of the ultrafast OTDM signals.

1.28 Tbit/s OTDM Transmission Results

Figure 27 shows autocorrelation waveforms of the input and output pulses and corresponding spectral profiles. Figures 27a and 27b are the input pulse waveform with a width of 380 fs and the corresponding spectral profile, respectively. The time–bandwidth product is 0.46, which is close to a Gaussian pulse. Figure 27c is the output pulse after transmission over 70 km. The pulse width is 400 fs, which indicates that the pulse broadening is as small as 20 fs. The time–bandwidth product is approximately 0.50. A small broadening is also observed at the wing of the pulse, which may be due to the residual dispersion and PMD. Figure 27d is the spectral profile of the transmitted 1.28 Tbit/s signal.

Figures 28a–28d show the input and output data patterns. Figure 28a is the input waveform at the fiber input. Because of the pre-chirped dispersion compensation, the 1.28 Tbit/s signal is broadened. This signal changed into a clear 1.28 Tbit/s signal after a 70 km transmission, as shown in Figure 28b. This PM 1.28 Tbit/s signal is separated into two 640 Gbit/s signals with a PBS, as shown in Figure 28c and 28d. We successfully obtained very clean 640 Gbit/s signals. These 640 Gbit/s signals were demultiplexed into 10 Gbit/s signals, and each BER was measured.

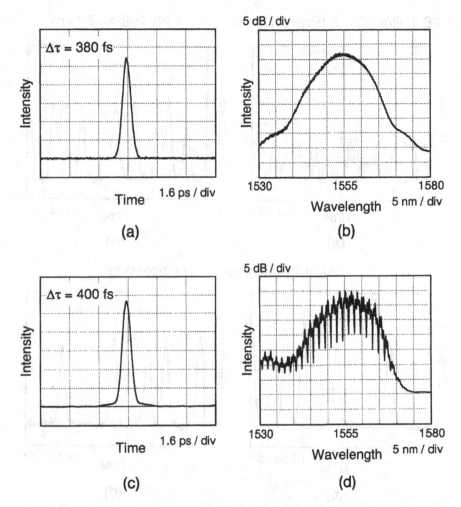

Figure 27 Autocorrelation waveforms of the input and transmitted pulses and corresponding spectral profiles. (a) Input pulse before pre-chirping; (b) corresponding input spectrum; (c) transmitted waveform over 70 km; (d) corresponding spectrum.

Figure 29 shows the received power at a BER of 1×10^{-9} for the 128 channels after a 70 km transmission. A BER of 10^{-9} was achieved for all of the channels with a received power of more than -21 dBm. The fluctuation in the received power for each channel is attributed to the small residual amplitude differences between the signal channels, which were caused by the PLC.

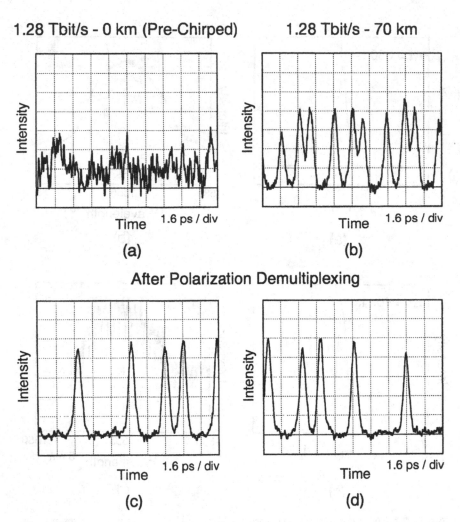

1.28 Tbit/s - 0 km (Pre-Chirped)   **1.28 Tbit/s - 70 km**

(a)   (b)

After Polarization Demultiplexing

(c)   (d)

Figure 28 Input and transmitted data patterns. (a) Input; (b) 1.28 Tbit/s OTDM signal after 70 km transmission; (c, d) polarization-demultiplexed 640 Gbit/s signals.

13.5 SUMMARY

We have described modelocked fiber lasers in the gigahertz region and their applications to high-speed optical communication such as soliton and terabit per second OTDM transmission. The fiber laser we emphasized in this chapter is a regeneratively and harmonically modelocked fiber laser that operates very stably in the repetition rate regime of 10–40 GHz. As

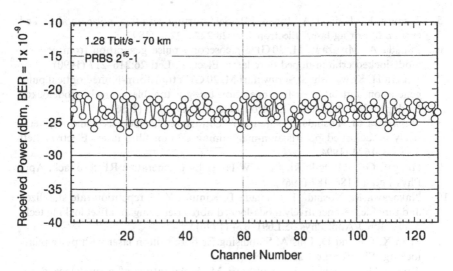

Figure 29 Received power at a BER of 1×10^{-9} after 70 km transmission vs channel number (128 channels).

for the soliton communication, such a pulse train was very advantageous, because the fiber laser could directly generate a transform-limited hyperbolic secant pulse using a soliton effect in the fiber cavity. It is also useful as an input pulse for an optical pulse compressor using the adiabatic soliton compression effect. With this technique, we were able to compress the input picosecond pulses to 50–200 fs. These compressed pulses were used for a Tbit/s OTDM transmission experiment, realizing for the first time a 1.28 Tbit/s OTDM transmission over 70 km. Because modelocked fiber lasers can easily generate a clean transform-limited short pulse train, they are applicable to a variety of advanced optical communication technologies.

REFERENCES

1. Duling IN III. Subpicosecond all-fibre erbium laser. Electron Lett 27:544–545 (1991).
2. Nakazawa M, Yoshida E, Kimura Y. Generation of 98 fs optical pulses directly from an erbium-doped fiber ring laser at 1.57 μm. Electron Lett 29:63–64 (1993).
3. Fermann ME, Haberl F, Hofer M, Hochreiter H. Nonlinear amplifying loop mirror. Opt Lett 15:752–754 (1990).
4. Matsas VJ, Newson TP, Richardson DJ, Payne DN. Selfstarting passively mode-locked fiber ring soliton laser exploiting nonlinear polarisation rotation. Electron Lett 28:1391–1393 (1992).

5. Tamura K, Haus HA, Ippen EP. Self-starting additive pulse modelocked erbium fibre ring laser. Electron Lett 28:2226–2228 (1992).
6. Takada A, Miyazawa H. 30 GHz picosecond pulse generation from actively mode-locked erbium-doped fibre laser. Electron Lett 26:216–217 (1990).
7. Takara H, Kawanishi S, Suruwatari M. 20 GHz transform-limited optical pulse generation and bit-error-free operation using a tunable, actively modelocked Er-doped fibre ring laser. Electron Lett 29:1149–1150 (1993).
8. Nakazawa M, Yoshida E, Kimura Y. Ultrastable harmonically and regeneratively modelocked polarisation-maintaining erbium fibre laser. Electron Lett 30:1603–1604 (1994).
9. Huggett GR. Mode-locking of CW lasers by regenerative RF feedback. Appl Phys Lett 13:186–187 (1968).
10. Nakazawa M, Yoshida E, Yamada E, Kimura Y. A repetition-rate stabilized and tunable, regeneratively mode-locked fibre laser using an offset locking technique. Jpn J Appl Phys 35:L691–694 (1996).
11. Shan X, Cleland D, Ellis A. Stabilising Er fibre soliton laser with pulse phase locking. Electron Lett 28:182–184 (1992).
12. Takara H, Kawanishi S, Saruwatari M. Stabilisation of a modelocked Er-doped fibre laser by suppressing the relaxation oscillation frequency component. Electron Lett 31:292–293 (1995).
13. Nakazawa M, Yoshida E, Tamura KR. Ideal phase-locked-loop (PLL) operation of a 10 GHz erbium-doped fibre laser using regenerative modelocking as an optical voltage controlled oscillator. Electron Lett 33:1318–1319 (1997).
14. Ishibashi T, Shimizu N, Kodama S, Ito H, Nagatsuma T, Furuta T. Uni-traveling-carrier photodiodes. Proc Ultrafast Electronics and Optoelectronics, UEO, 1997, pp 166–168.
15. Nakazawa M, Tamura K, Yoshida E. Supermode noise suppression in a harmonically modelocked fiber laser by selfphase modulation and spectral filtering. Electron Lett 32:461–463 (1996).
16. Yoshida E, Nakazawa M. Wavelength tunable 1.0 ps pulse generation in 1.530–1.555 µm region from PLL regeneratively modelocked fibre laser. Electron Lett 34:1753–1754 (1998).
17. Chernikov SV, Richardson DJ, Laming RI, Dianov EM, Payne DN. 70 Gbit/s fibre based source of fundamental solitons at 1550 nm. Electron Lett 28:1210–1212 (1992).
18. Chernikov SV, Mamyshev PV. Femtosecond soliton propagation in fibers with slowly decreasing dispersion. J Opt Soc Am B 8:1633–1641 (1991).
19. Nakazawa M, Yoshida E, Kubota K, Kimura Y. Generation of a 170 fs, 10 GHz transform-limited pulse train at 1.55 µm using a dispersion-decreasing, erbium-doped active soliton compressor. Electron Lett 30:2038–2040 (1994).
20. Mori K, Takara H, Kawakami S, Saruwatari M, Morioka T. Flatly broadened supercontinuum spectrum generated in a dispersion decreasing fiber with convex dispersion profile. Electron Lett 33:1806–1808 (1997).

21. Tamura KR, Nakazawa M. 54-fs, 10 GHz soliton generation from a polarization-maintaining dispersion-flattened dispersion-decreasing fiber compressor. Opt Lett 26:762–764 (2001).

22. Nakazawa M, Yoshida E, Yamada E, Suzuki K, Kitoh T, Kawachi M. 80 Gbit/s soliton data transmission over 500 km with unequal amplitude solitons for timing clock extraction. Electron Lett 30:1777–1778 (1994).

23. Nakazawa M, Yoshida E, Kimura Y. Ultrastable harmonically and regeneratively modelocked polarization-maintaining erbium fibre ring laser. Electron Lett 30:1603–1604 (1994).

24. Chu PL, Desem C. Mutual interaction between solitons of unequal amplitudes in optical fibre. Electron Lett 24:1133–1134 (1985).

25. Uchiyama K, Takara H, Kawanishi S, Morioka T, Saruwatari M. Ultrafast polarisation-independent all optical switching using a polarization diversity scheme in the nonlinear optical loop mirror. Electron Lett 28:1864–1865 (1992). See also Andrekson PA, Olsson NA, Simpson JR, Digiovanni DJ, Morton PA, Tanbun-Ek T, Logan RA, Wecht KW. 64 Gb/s all-optical demultiplexing with the nonlinear optical-loop mirror. IEEE Photon Technol Lett 4: 644–647 (1992).

26. Evangelides SG. Jr, Mollenauer LF, Gordon JP, Bergano NS. Polarization multiplexing with solitons. IEEE J Lightwave Technol 10:28–35 (1992).

27. Nakazawa M, Suzuki K, Yoshida E, Yamada E, Kitoh T, Kawachi M. 160 Gbit/s soliton data transmission over 200 km. Electron Lett 31:565–566 (1995).

28. Nakazawa M, Kurokawa K, Kubota H, Yamada E. Observation of the trapping of an optical soliton by adiabatic gain narrowing and its escape. Phys Rev Lett 65:1881–1884 (1990).

29. Gordon JP, Haus HA. Random walk of coherently amplified solitons in optical fiber transmission. Opt lett 11:665–667 (1986).

30. Liang AH, Toda H, Hasegawa A. High-speed soliton transmission in dense periodic fibers. Opt Lett 24:799–801 (1999).

31. Anis H, Berkey G, Bordogna G, Cavallari M, Charbonnier B, Evans A, Hardcastle I, Jones M, Pettitt G, Shaw B, Srikant V, Wakefield J. Continuous dispersion managed fiber for very high speed soliton systems. Proc Eur Conf Opt Commun, ECOC, 1999, pp I231–I231.

32. Nakazawa M, Yamada E, Kubota H, Suzuki K. 10 Gbit/s soliton transmission over one million kilometers. Electron Lett 27:1270–1271 (1991).

33. Kubota H, Nakazawa M. Soliton transmission control in time and frequency domains. IEEE J Quantum Electron 29:2189–2197 (1993).

34. Nakazawa M, Kubota H, Yamada E, Suzuki K. Infinite-distance soliton transmission with soliton controls in time and frequency domains. Electron Lett 28:1099–1101 (1991).

35. Mecozzi A, Moores JD, Haus HA, Lai Y. Soliton transmission control. Opt Lett 16:1841–1843 (1991).

36. Hasegawa A, Kodama Y. Guiding-center soliton. Phys Rev Lett 66:161–164 (1991).

37. Gordon JP, Mollenauer LF. Effects of fiber nonlinearities and amplifier spacing on ultra-long distance transmission. IEEE J Lightwave Technol 9:170–173 (1991).

38. Kubota H, Nakazawa M. Soliton transmission with long amplifier spacing under soliton control, Electron Lett 29:1780–1782 (1993). See also Aubin G, Montalant T, Moulu J, Nortier B, Pirio F, Thomine JB. Record amplifier span of 105 km in a soliton transmission experiment at 10 Gbit/s over 1 Mm. Electron Lett 31:217–219 (1995).

39. Suzuki K, Kubota H, Sahara A, Nakazawa M. 40 Gbit/s single channel optical soliton transmission over 70,000 km using in-line synchronous modulation and optical filtering. Electron Lett 34:98–99 (1998).

40. Nakazawa M, Suzuki K, Kubota H. Single-channel 80 Gbit/s soliton transmission over 10,000 km using in-line synchronous modulation. Electron Lett 35:162–163 (1999).

41. Bigo S, Leclerc O, Brindel P, Vendrôme G, Desurvire E, Doussière P, Ducellier T. 20 Gbit/s all-optical regenerator. Proc Opt Fiber Commun Conf, OFC, 1997, PD22.

42. Tamura KR, Nakazawa M. Femtosecond soliton generation over a 32-nm wavelength range using a dispersion-flattened dispersion-decreasing fiber. IEEE Photon Technol Lett 11:319–321 (1999).

43. Tamura K.R, Nakazawa M. Spectral-smoothing and pedestal reduction of wavelength tunable quasi-adiabatically compressed femtosecond solitons using a dispersion-flattened dispersion-imbalanced loop mirror. IEEE Photon Technol Lett 11:230–232 (1999).

44. Nakazawa M, Yamamoto T, Tamura KR. 1.28 Tbit/s—70 km OTDM transmission using third- and fourth-order simultaneous dispersion compensation with a phase modulator. Proc Eur Conf Opti Commun, ECOC, 2000, PD2.6.

45. Yamamoto T, Yoshida E, Nakazawa M. Ultrafast nonlinear optical loop mirror for demultiplexing 640 Gbit/s TDM signals. Electron Lett 34:1013–1014 (1998).

46. Chang C-C, Sardesai HP, Weiner AM. Dispersion-free fiber transmission for femtosecond pulses by use of a dispersion-compensating fiber and a programmable pulse shaper. Opt Lett 23:283–285 (1998).

47. Pelusi MD, Matsui Y, Suzuki A. Fourth-order dispersion suppression of ultrashort optical pulses by second-order dispersion and cosine phase modulation. Opt Lett 25:296–298 (2000).

48. Mukasa K, Akasaka Y, Suzuki Y, Kamiya T. Novel network fiber to manage dispersion at 1.55 µm with combination of 1.3 µm zero dispersion single mode fiber. Proc Eur Conf Opt Commun, ECOC, 1997, I127–I130.

49. Wong WS, Namiki S, Margalit M, Haus HA, Ippen EP. Self-switching of optical pulses in dispersion-imbalanced nonlinear loop mirrors. Opt Lett 22:1150–1152 (1997).

50. Pelusi MD, Matsui Y, Suzuki A. Electrooptic phase modulation of stretched 250-fs pulses for suppression of third-order fiber dispersion in transmission. IEEE Photon Technol Lett 19:1461–1463 (1999).

14
Nonlinear Microscopy with Ultrashort Pulse Lasers

Michiel Müller
University of Amsterdam, Amsterdam, The Netherlands

Jeff Squier
Colorado School of Mines, Golden, Colorado, U.S.A.

14.1 INTRODUCTION

There are many reasons for using an ultrashort pulse laser microscope, but they are all associated with the specific properties of ultrashort pulse laser systems: high peak power, short pulse duration and large spectral bandwidth. Nonlinear optics thrives on power. Ultrashort pulse lasers have the unique ability to provide high peak power at moderate average power, avoiding sample damage and permitting the implementation of nonlinear optical techniques in microscopy.

The use of nonlinear optical interactions is of itself of interest to microscopy. In particular, the nonlinear dependence of the signal on the input laser power provides inherent optical sectioning and hence three-dimensional imaging capability. In addition, by limiting the interaction to the focal region, possible detrimental out-of-focus interactions (i.e., photobleaching and photoinduced sample damage) are reduced. Also, the use of a nonlinear interaction may, as in the case of multiphoton absorption, permit the use of radiation of longer wavelengths. This reduces the losses from scattering and may be less harmful to, for instance, biological samples. Finally, some nonlinear interactions, such as harmonic generation, are electronically nonresonant with the sample transitions, eliminating energy deposition to the specimen and related photoinduced damage phenomena.

Another important reason for using nonlinear optics in microscopy is that it makes it possible to obtain spatially resolved spectroscopic measurements. Nonlinear optical spectroscopy, particularly in the form of four-wave mixing, can provide a wealth of information in both the time and frequency domains through intricate interactions with the molecular energy levels.

Some microscopic techniques utilize the brute force of the high peak power of femtosecond pulses to induce highly nonlinear phenomena. These techniques range from laser ablation to the creation of waveguides in transparent media with laser-induced breakdown, and from microsurgery to plasma spectroscopy.

Apart from the nonlinear response induced by the large peak power of femtosecond pulses, microscopic applications can profit from the extremely short pulse duration or the intrinsically large spectral bandwidth. Pump-probe experiments have been developed to spatially resolve, for instance, fluorescence lifetime distributions or the thickness of semiconductor layers. Alternatively, the spectral bandwidth has been used in chirped pulse excitation to control the fluorescence emission of fluorophores and in spectral correlation based techniques such as optical coherence tomography.

The rapid development in laser technology has provided reliable laser sources that have become more and more compact and robust and have "turnkey" operation. In addition, these laser sources now cover almost the complete optical spectrum, ranging from the ultraviolet to the far-infrared. This development is extremely important for applications of ultrashort pulse lasers in microscopy, because it permits implementing such laser systems in industrial and clinical environments. Landmarks in this regard are the first commercial applications of an all-solid-state femtosecond laser in the measurement of single- or multilayer metal film thicknesses of wafers (*Meta* PULSE, Rudolph Technology), in lithographic mask repair systems (DRS photomask repair system, Quantronix), and in the first materials processing workstation (Clark-MXR, Inc.).

This chapter is organized as follows. Section 14.2 is devoted to various applications of ultrashort lasers to microscopy. Starting with an introduction to nonlinear optical microscopy and a discussion of the now widely used multiphoton absorption technique, it proceeds with an overview of various coherent nonlinear optical microscopic techniques (second- and third-harmonic generation and coherent anti-Stokes Raman scattering). This section concludes by addressing spatial resolution in ultrashort laser microscopy.

Section 14.3 is concerned with a number of general (technical) aspects of the use of ultrashort laser pulses in high-resolution optical microscopy. This includes a discussion of a number of techniques to measure the temporal characteristics of the focal field of a high numerical aperture microscope objective. Because of the significant spectral bandwidth of ultrashort optical pulses, dispersion plays an important role in ultrashort pulse laser microscopy. The effects of dispersion on the efficiency of nonlinear optical processes in microscopy and the methods that can be used to control and precompensate for this dispersion are discussed at the end of this section.

Ultrashort pulse lasers come in all "flavors," varying in pulse duration, peak power, repetition rate, and wavelength. Section 14.4 addresses the issue of balancing the various ultrashort laser parameters for optimal use in microscopy. This is especially important in the imaging of biological specimens, where inevitably a compromise has to be made between maximum signal and minimal photoinduced damage.

It should be noted at this point that the overview presented here of the field of ultrashort pulse microscopy is in no way complete. Not only did we have to limit the number of examples from the literature, but also, the material presented here is mainly concerned with novel microscopic methods that generate contrast based on some nonlinear optical interaction requiring the use of ultrashort pulses. This admittedly limited perspective does not include, for instance, important topics such as optical coherence tomography, transient reflection measurements, or data storage applications.

14.2 NONLINEAR OPTICAL MICROSCOPY

Ultrashort pulse lasers provide unique opportunities for the implementation of nonlinear optical techniques in high-resolution microscopy. Whereas nonlinear optics thrives on peak power, ultrashort pulse lasers combine high peak power with moderate average power, limiting sample damage through processes that scale linearly with intensity.

14.2.1 Introduction

Before discussing in some detail some of the nonlinear optical techniques that have found application in microscopy we summarize some of the general characteristics of these techniques. In nonlinear optics it is customary to split the macroscopic polarization—i.e., the result of an ensemble of oscillating dipole moments—into a linear and a nonlinear part and expand it in successive orders of the electromagnetic field ($\mathbf{E}$):

$$\mathbf{P}(\mathbf{r}, t) = \mathbf{P}_L(\mathbf{r}, t) + \mathbf{P}_{NL}(\mathbf{r}, t)$$

$$= \chi_L^{(1)} \, \mathbf{E} + \chi_{NL}^{(2)} \, \mathbf{E}\mathbf{E} + \chi_{NL}^{(3)} \, \mathbf{E}\mathbf{E}\mathbf{E} + \cdots \tag{1}$$

$\chi_L^{(1)}$ and $\chi_{NL}^{(n)}$ denote the linear and nonlinear susceptibility, respectively, which relate the complex amplitude of the electromagnetic fields with the polarization. In general, for a material with dispersion and/or loss, the susceptibility is a complex tensor quantity. For example, for the third-order susceptibility (see, e.g., Ref. 1),

$$P_i = \sum_{jkl} \chi_{ijkl}^{(3)} E_j E_k E_l \tag{2}$$

where i, j, k, and l run over all Cartesian coordinates. Note that from symmetry arguments it follows immediately that even orders of the nonlinear susceptibility—$\chi_{NL}^{(2n)}$—vanish for materials that possess a center of inversion symmetry. Various approaches have been developed to relate the macroscopic susceptibility to microscopic quantities (see, e.g., Refs. 2 and 3).

Coherent Signal Generation

In nonlinear optical processes where a coherent signal field is generated, as in the case of harmonic generation or stimulated Raman, the macroscopic polarization $\mathbf{P}(\mathbf{r}, t)$ acts as a source term in the wave equation. In this case the signal intensity is proportional to the complex square of the field. For instance, third harmonic generation (THG) is related to the third-order susceptibility through*

$$\mathbf{P}(\omega_4) \propto \chi^{(3)}(\omega_4 = \omega_1 + \omega_2 + \omega_3) \mathbf{E}(\omega_1) \mathbf{E}(\omega_2) \mathbf{E}(\omega_3) \tag{3}$$

where the explicit spatial and temporal dependence has been dropped. Using $\omega_1 = \omega_2 = \omega_3 = \omega$ and $\omega_4 = 3\omega$, it follows that the THG signal intensity is proportional to the cube of input power:

$$I_{3\omega} \propto |\mathbf{P}(3\omega)|^2 \propto |\chi^{(3)}(3\omega = \omega + \omega + \omega) \mathbf{E}(\omega) \mathbf{E}(\omega) \mathbf{E}(\omega)|^2 = |\chi^{(3)}|^2 I_\omega^3 \tag{4}$$

Incoherent Signal Generation

Alternatively, for nonlinear processes that result in the absorption of energy, such as multiphoton absorption, the absorbed energy follows from [5]

$$\left\langle \frac{d}{dt} \left(\frac{\text{absorbed energy}}{\text{volume}} \right) \right\rangle_{\text{time}} = \langle \mathbf{j} \cdot \mathbf{E} \rangle \tag{5}$$

---

*The notation used here follows Ref. 4.

where the induced current $\mathbf{j}$ can be approximated by

$$\mathbf{j} = \frac{\partial \mathbf{P}}{\partial t} + \cdots \tag{6}$$

For instance, two-photon absorption is related to the third-order suscept-ibility as

$$\frac{\partial \mathbf{P}(\omega)}{\partial t} \propto \frac{\partial}{\partial t} \left[\chi^{(3)}(\omega = \omega + \omega - \omega)\mathbf{E}(\omega)\mathbf{E}(\omega)\mathbf{E}(-\omega) \right] \tag{7}$$

It follows that the rate of two-photon absorption, with a transition from initial to final state $P_{i \to f}$, is proportional to the square of the input intensity:

$$P_{i \to f} \propto \left\langle \frac{\partial \mathbf{P}}{\partial t} \mathbf{E} \right\rangle = \left\langle \frac{\partial}{\partial t} \chi^{(3)} \mathbf{EEEE} \right\rangle \propto I^2 \tag{8}$$

Optical Sectioning Capability

All nonlinear optical microscopic applications have one characteristic in common: inherent optical sectioning capability. This is a direct result of the supralinear dependence of the detected signal on input laser intensity. When the signal response depends linearly on the input intensity—as is the case for, e.g., single-photon absorption fluorescence—a uniform sample layer contributes equally to the signal whether it is in or out of focus. Thus the axial position of such a layer cannot be determined with widefield fluor-escence microscopy. The use of a confocal pinhole suppresses the contribu-tions from out-of-focus planes, permitting "optical sectioning" of the sample and thus a true axial resolution.

Nonlinear optical microscopic applications have inherent optical sec-tioning capability. Because the signal in this case depends supralinearly on the input intensity, the signal response is always primarily from in-focus contributions. Thus even without the use of a confocal pinhole, these micro-scopic techniques provide three-dimensional imaging capability. This can provide significant advantages, especially for imaging in turbid media.

14.2.2 Multiphoton Absorption Microscopy

In multiphoton absorption microscopy, image formation is based on the detection of fluorescence from molecules that have been excited through the simultaneous absorption of two or more photons. A number of distinct properties result from both the nonlinearity of the induced response and the relatively long wavelength of excitation. Since its introduction in 1990 by

Denk and coworkers [6], multiphoton absorption microscopy has found widespread application, especially in biology.

The Principle

To describe the principle of multiphoton absorption, first consider the two-photon absorption process, (Three-photon absorption will be discussed later in this section.) As shown in the energy level diagram of Figure 1, two-photon absorption relies on the simultaneous absorption of two photons in a single quantized event. In practice, "simultaneous" here means within 10^{-16} s or less. The two-step absorption process results in a quadratic dependence of the rate of absorption ($P_{i \to f}$), and thus of the induced fluorescence, on the excitation intensity:

$$P_{i \to f} \propto I_{\text{flu}} \propto \delta I_{\text{laser}}^2 \tag{9}$$

where δ [in meters to the fourth power seconds squared per photon ($\text{m}^4 \, \text{s}^2$ photon^{-1})] is the two-photon absorption crosssection. I_{flu} and I_{laser} denote the fluorescence and excitation laser intensity, respectively.

Quantum selection rules specify that one- and two-photon processes are complementary in nature. Whereas for one-photon processes the selection rule $\Delta L = \pm 1$ holds for the orbital angular momentum, the selection rule for a two-photon absorption process is $\Delta L = 0, \pm 2$ (see, e.g., Ref. 1).

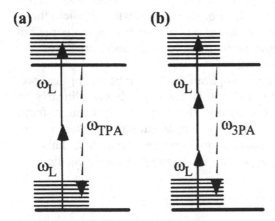

Figure 1   Energy level diagram for the (a) two-photon absorption (TPA) and (b) three-photon absorption (3PA) processes, where the thick solid lines denote the atomic or molecular ground state and electronic excited state, respectively. The thin solid lines represent vibrational and rotational energy levels. The dashed line represents fluorescence.

In practice, in most common fluorophores, which are multilevel systems with a high density of states, two-photon absorption is generally allowed within the one-photon absorption band. The two-photon absorption cross-section for these molecules is typically of the order of $10^{-58}\,\mathrm{m^4\,s^2}$ photon^{-1} [7,8]. The fact that the two-photon absorption spectra are generally different from the one-photon spectra [9], with typically a blue shift of the absorption maximum ($\lambda_{\text{two-photon}}^{\max} \leq 2\lambda_{\text{one-photon}}^{\max}$), exemplifies the different nature of the one- and two-photon transitions. It should be noted at this point that until now almost all two-photon absorption microscopic experiments have been performed with fluorophores originally designed for one-photon absorption applications. Preliminary work indicates [10] that fluorophores can be designed that exhibit a substantially increased two-photon absorption cross-section.

After excitation, intramolecular relaxation rapidly brings the fluorophore to its lowest electronically excited state. The two-photon induced fluorescence can in general not be discriminated from one-photon induced fluorescence. Thus the typical two-photon absorption characteristics are lost in the intramolecular relaxation processes of the excited state. The induced fluorescence is emitted in a random direction (following the emission pattern of an oscillating dipole).

Through anisotropy measurements, additional information on rotational motion of the molecule can be obtained. Because of the square dependence on the excitation intensity, a more strongly oriented excited state population is induced, which results in a larger time-zero anisotropy value [11]. This effect is even more pronounced in three-photon absorption.

Three-Photon Absorption Microscopy

In many ways, three-photon absorption is very similar to two-photon absorption. In this case, three photons are absorbed simultaneously in a single quantum event (Fig. 1). The induced fluorescence depends on the excitation laser intensity cubed:

$$I_{\text{flu}} \propto \sigma_3 I_{\text{laser}}^3 \tag{10}$$

where σ_3 is the three-photon absorption cross section of the order of 10^{-94} $\mathrm{m^6\,s^3}$ photon^{-1} [9]. As expected from the quantum-mechanical selection rules, the three-photon absorption spectra mimic their one-photon counterpart, in contrast to those from two-photon absorption. Although in general a relatively weak phenomenon, three-photon absorption microscopy has been demonstrated in practice (e.g., Ref. 12) and can be useful for excitation in the UV of endogenous chromophores, which would otherwise be difficult to probe with conventional optics. Note that a gradual transition from a two-photon to a three-photon absorption process has been observed in some fluorophores as a function of excitation wavelength [9].

The Properties

The main characteristics of multiphoton absorption microscopy follow from the square dependence of the fluorescence on the laser intensity and the longer wavelength of excitation. Here we summarize the main features:

1. Multiphoton absorption is an electronically resonant interaction in which energy is transferred to the specimen. It is related to an odd-power nonlinear susceptibility ($\chi^{(3)}$ and $\chi^{(5)}$ in the case of two- and three-photon absorption, respectively) and is thus permitted in all materials, either with or without a center of inversion symmetry.

2. The rate of absorption, and thus the induced fluorescence response, depends nonlinearly on the excitation intensity. This provides multiphoton absorption with inherent optical sectioning. Because only in the close vicinity of the microscope objective's focus is the excitation intensity high enough to induce the nonlinear response, the interaction is confined to the focal region only. No significant absorption occurs in out-of-focus regions. This has a number of significant consequences. First, it provides the technique with inherent three-dimensional image formation capability. There is no need for a confocal pinhole in the detection, because there is no out-of-focus fluorescence. Second, because of the inherent optical sectioning capability, descanning optics are not required, and the optical losses in the detection channel of the microscope can be minimized, maximizing the collection efficiency. Third, the fact that the absorption is strongly localized in three dimensions can be used to photoactivate certain agonists locally through photoinduced release of a "caged" compound (see, e.g., Ref. 13). Fourth, because there is no out-of-focus absorption there is also no out-of-focus photobleaching. This is especially important for three-dimensional imaging. Finally, what holds for the induced fluorescence is true also for other nonlinear induced damage effects, being limited to the focal region only. Because of the nonlinear dependence of the rate of multiphoton absorption on the excitation intensity, the induced fluorescence depends on the laser pulse duration. Assuming for simplicity laser pulses with a square temporal envelope, it follows that the amount of two-photon absorption is proportional to the inverse pulse duration, whereas three-photon

absorption is proportional to the inverse of the pulse duration squared. Thus for the same average power, shorter pulses provide more efficient excitation. This may be important when specimen damage is related to average, rather than peak, power. For optical pulses shorter than $\sim$150 fs, dispersion compensation techniques are required to optimize the pulse duration at the focal point of the microscope objective (see Sec. 14.3). Note, however, that even two-photon absorption excitation with a continuous wave laser source has been demonstrated in practice [14].

3. Multiphoton absorption permits excitation of a certain optical transition with photons of a longer wavelength than is required for a one-photon absorption event. Scattering by the specimen decreases strongly with increasing wavelength (e.g., an l^{-4} dependence for Rayleigh scattering). This permits high-resolution imaging at greater depth ($>$100 µm) in strongly scattering samples such as brain slices, intact embryos, or other primary culturetissue preparations. In addition, the longer wavelength of excitation can be used to effectively address UV transitions. In particular, UV wavelengths below $\sim$300 nm are very problematic to address with regular microscopic optics. Also, various studies indicate (see below) that excitation at longer wavelength ($>$800 nm) is beneficial for retaining cell viability.

The Instrument

Various configurations have been used in multiphoton absorption microscopy. In all cases either laser scanning or specimen scanning is required. This can be accomplished in several ways: by adaptation of a confocal microscope retaining the descanned detection channel; by use of laser scanning optics and a specific detection channel without descanning optics and optimized for fluorescence collection efficiency; or by using laser scanning optics in combination with widefield CCD detection. In general a specifically optimized microscope configuration is required to take full advantage of the various features of multiphoton absorption microscopy. A schematic of a typical two-photon microscope setup is depicted in Figure 2.

Typically, in high-resolution biological applications, power levels not exceeding $\sim$100 pJ/pulse should be used to retain cell viability. On the other hand, ultrashort lasers routinely produce power levels well over 10 nJ/pulse. Thus ample power is available for parallel excitation to speed up the image acquisition process. The first approach used in this respect was based on the use of line—rather than point—excitation, in combination with widefield CCD detection [15]. This provides real-time image acquisition rates at the expense of reduced axial resolution.

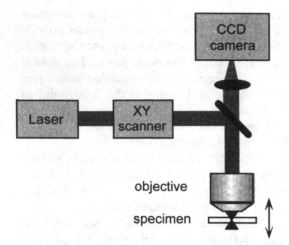

Figure 2 Generalized schematic of a two-photon absorption microscopy setup. A microscope objective focuses the laser pulses onto the specimen. The specimen is raster scanned in a point excitation mode, generally using mirror scanning for the x and y directions and specimen scanning for the z direction. The fluorescence is detected in the backscattering direction and may be imaged directly onto a CCD camera.

In another approach, the available laser power is spread out over multiple excitation spots [16,17]. In this case, the single-point optical resolution can be maintained if the spacing between the excitation spots is chosen carefully.

Biological Damage

An increasing number of studies address the issue of biological damage in multiphoton microscopy. This in itself is a complicated issue because of the large number of parameters involved (laser power, pulse energy, pulse duration, wavelength, dwell time, specimen related conditions, etc.). However, a number of general observations are consistently noted in various studies.

Photoinduced biological damage appears to be a threshold phenomenon [18,19]. The threshold is generally found to be wavelength dependent, with less damage at longer wavelength. For instance, in one study a 50% cloning efficiency of CHO cells was observed at power levels of 2.5 mW and 6 mW for two-photon excitation wavelengths of 780 and 920 nm, respectively (Ti:sapphire laser; 240 fs, 80 MHz) [19]. The precise damage mechanism is as yet unclear.

Multiphoton experiments have been performed almost exclusively with near-infrared (NIR) lasers, especially Ti:sapphire (700–1100 nm) and

Nd:YLF (1053 nm). Because water absorption increases significantly at longer wavelengths in going from the visible to the NIR region of the spectrum [20], biological damage through heating can be expected. However, careful calculations have shown that this is generally not significant under typical experimental conditions [21]. Because, in addition, biological specimens are virtually nonabsorbing in the NIR (in the absence of hemoglobin, melanin, or chlorophyll), photothermal induced damage through linear (i.e., one-photon) absorption is unlikely. This was confirmed by the observation of less biological damage for an excitation wavelength of 920 nm relative to that at 780 nm despite a sevenfold increase in water absorption in going to the longer NIR wavelength [19].

A more likely damage mechanism is through multiphoton absorption of endogenous cellular absorption, such as NAD(P)H, flavins, porphyrins, DNA, and proteins. This can in turn lead to the production of oxygen radicals. The fact that the damage process is itself multiphoton-induced is substantiated by monitoring cell viability as a function of pulse duration [19]. At shorter wavelengths such as excitation at visible wavelengths as in regular confocal microscopy, oxidative stress may also be induced through linear absorption [22].

Photobleaching

Photobleaching is a general problem in fluorescence microscopy. It limits the number of images that can be taken from a single specimen in, for instance, in vivo microscopic applications. In addition, chemical by-products generated in the photochemical reactions are considered to play a part in photo-induced biological damage. In three-dimensional imaging, two-photon absorption may provide a significant advantage over one-photon absorption techniques in terms of photobleaching. Indeed, because in the former case absorption occurs only in the focal plane, the total amount of photobleaching during three-dimensional image acquisition is significantly reduced. This is true as long as the rate of photobleaching depends only on the rate of absorption. However, recent data suggest that increased photobleaching in the case of two-photon absorption may be a general phenomenon [23]. For several fluorophores it was observed that the two-photon photobleaching rate increased with a slope of ≥ 3 with the excitation power. As a possible explanation for these observations it was proposed that two-photon photobleaching may be a two-step process, with a destructive photon interaction with the excited molecule after two-photon absorption.

Selected Applications

The selected examples presented here are chosen only to illustrate some of the specific features of multiphoton absorption microscopy. It is outside the scope of this work even to attempt a concise review of this field, which is still expanding at a staggering pace.

Squirrell et al. [22] demonstrated that multiphoton absorption microscopy can indeed have advantages in terms of cell viability relative to confocal microscopy. They showed that hamster embryos could be imaged for over 24 h without adverse effects (Fig. 3). In these experiments a Nd:YLF laser at 1047 nm and with 175 fs pulses was used. Regular confocal imaging of the same specimen for less than 8 h inhibited development of the embryos even in the absence of fluorophore excitation. For confocal and two-photon absorption microscopy the exposure level was ~280 μJ/ embryo and ~2 J/embryo, respectively. It is proposed by the authors that the biological damage induced by confocal microscopy is related to oxidative stress from the photoinduced production of H_2O_2.

Various groups have demonstrated the ability of multiphoton absorption microscopy to image deep into strongly scattering samples. A striking example of this is given in Figure 4, which shows in vivo two-photon

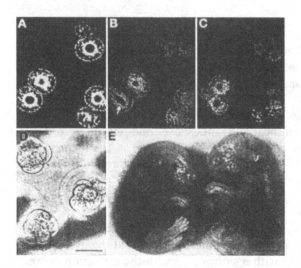

Figure 3 Fetal imaging following long-term imaging. Two-photon images of mitochondria-labeled embryos (A) at the initiation of imaging; (B) after 8.5 h of imaging, showing a mitotic spindle (arrow); and (C) after the completion of a 24 h imaging sequence. (D) After imaging, embryos were cultured in the incubator until 82 h PEA (Nomarski), at which point they were transferred to a recipient female. (E) A black-eyed fetus that developed from one of these imaged embryos is shown next to an albino uterine mate. Scale bar for (A)–(D) = 4.5 μm. Scale bar for (E) = 4.75 mm. (From Ref. 22, with permission.)

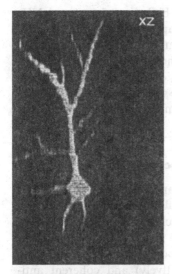

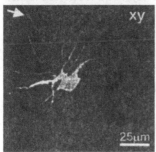

Figure 4 Three-dimensional reconstruction of a layer 2/3 pyramidal neuron. Projections onto *xz* (top) and *xy* (bottom); stacks consist of 110 sections acquired with 2 μm spacing. In the *xz* projection the surface of the cortex is toward the top. (From Ref. 24 with permission.)

absorption microscopy in the primary vibrissa cortex of anaesthetized rats [24]. Using a 100 fs, ~820 nm Ti:sapphire laser with <200 mW of power at the brain, dendrites could be resolved down to 500 μm below the pial surface.

Conclusions

Multiphoton absorption microscopy is rapidly developing into an important tool in biological research. The most striking advantages of this technique relative to, e.g., confocal microscopy are the increased penetration capability in scattering specimens and reduced biological damage at NIR excitation wavelengths (>800 nm). The scope of applications is much greater than can be covered in this section, and many important contributions had to be omitted. Also, new applications of multiphoton absorption are emerging in, for instance, the field of extremely high optical resolution through so-called 4π microscopy (e.g., Ref. 25) and the measurement of diffusion properties and quantitative concentration determination using fluorescence correlation spectroscopy (e.g., Ref. 26).

14.2.3   Coherent Nonlinear Optical Microscopy

In the previous section we considered nonlinear optical phenomena that result in absorption and subsequent emission of fluorescence. In this section we address another class of nonlinear optical techniques in which a coherent, rather than an incoherent, signal response is generated. Practical applications of these types of processes have been reported for second and third harmonic generation (SHG and THG, respectively) and coherent anti-Stokes Raman scattering (CARS).

The Principle

The processes of SHG, THG, and CARS are schematically depicted in energy level diagrams in Figure 5. All these processes represent electronically non-resonant interactions that generate a macroscopic polarization according to

$$\mathbf{P}(\omega_{\text{SHG}}) \propto \chi^{(2)}(\omega_{\text{SHG}} = 2\omega_L)\mathbf{E}(\omega_L)\mathbf{E}(\omega_L) \tag{11a}$$

$$\mathbf{P}(\omega_{\text{THG}}) \propto \chi^{(3)}(\omega_{\text{THG}} = 3\omega_L)\mathbf{E}(\omega_L)\mathbf{E}(\omega_L)\mathbf{E}(\omega_L) \tag{11b}$$

$$\mathbf{P}(\omega_{\text{CARS}}) \propto \chi^{(3)}(\omega_{\text{CARS}} = \omega_L - \omega_S + \omega_P)\mathbf{E}(\omega_L)\mathbf{E}(\omega_S)\mathbf{E}(\omega_P) \tag{11c}$$

This macroscopic polarization in turn generates a coherent signal field, the intensity of which shows a square (SHG) or cubic (THG and CARS) dependence on the input laser intensity.

A general property of all these coherent techniques is that they concern an electronically nonresonant and, for all practical purposes, instantaneous interaction with the specimen. This results in the absence of a net energy transfer to the specimen, rendering these techniques potentially nondestructive. In addition, owing to the nonlinear dependence on the input laser fields, all of these processes provide inherent sectioning capability.

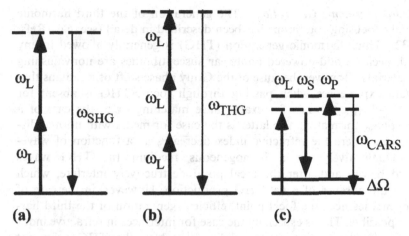

(a)   **(b)**   **(c)**

Figure 5 Energy level diagram for (a) SHG, (b) THG, and (c) CARS. The solid lines denote the atomic or molecular electronic ground state. The dashed lines represent so-called virtual states, which are not energy eigenlevels of the system but rather represent the combined energy of one of the energy eigenstates of the atom (or molecule) and of one or more photons of the radiation field [4]. The thick arrows represent input laser fields, and the thin arrows represent the coherent signal fields. The CARS process is shown to be resonant with a particular ground-state vibrational transition.

Second Harmonic Generation. Although second harmonic generation (SHG) is widely used to generate new frequencies, it has only recently been introduced as a tool for imaging nonlinear susceptibilities in various materials [27–30]. Because it is related to the second-order nonlinear susceptibility, SHG vanishes for a symmetrical distribution of dipoles. It is therefore especially sensitive to local molecular asymmetry. The general properties of SHG imaging can be summarized as follows.

1. SHG vanishes for a (locally) symmetric distribution of dipoles. Therefore, either it is confined to the surface of isotropic specimens or it is able to probe local molecular asymmetries.
2. The SHG signal is proportional to the square of both the incident laser power and the second-order nonlinear susceptibility.
3. The SHG signal is generated in the forward direction. The so-called phase anomaly of tightly focused laser beams—with a phase retardation of the focused wave relative to an unfocused plane wave—results in specific, nonisotropic, signal emission patterns [30].

Third Harmonic Generation. The generation of the third harmonic under tight focusing conditions has been described in detail (see, e.g., Refs. 4, 31, 32). Third harmonic generation (THG) is generally allowed in any material, because odd-powered nonlinear susceptibilities are nonvanishing in all materials. However, because of the Gouy phase shift of π radians that any beam experiences when passing through focus, THG is absent for $\Delta k = 3k_L - k_{THG} \leq 0$, i.e., for exact phase matching or in the case of a negative phase mismatch. The latter is the case for media with normal dispersion, i.e., where the refractive index decreases as a function of wavelength. Qualitatively, in a homogeneous medium the THG waves generated before and after the focal point destructively interfere, which results in the absence of a net THG production. However, in the case of inhomogeneities near the focal point, efficient generation of the third harmonic is possible. This is especially the case for interfaces in refractive index and/or third-order nonlinear susceptibility ($\chi^{(3)}$). Note that THG is thus not restricted to the surface of the material but rather results from the bulk of the material contained within the focal volume and the presence of interfaces or inhomogeneities therein. This is confirmed also by the absence of a back-propagating THG signal [32].

From theory it follows that THG imaging has a number of characteristics, most of which have now been verified in practice. Here we restrict ourselves to those properties that apply to microscopy in particular:

1. Under the tight focusing conditions pertinent to microscopy, efficient generation of the third harmonic occurs only near interfaces (or inhomogeneities) in refractive index and/or third-order nonlinear susceptibility ($\chi^{(3)}$). Thus, in contrast to phase microscopy, THG imaging can discern the interface between two media on the basis of a difference in nonlinear susceptibility alone. An example of this is the fact that the boundary between immersion oil and a cover glass that have been matched in refractive index explicitly is clearly imaged in THG microscopy. Note that the contrast relates to the inherent properties of the specimen and that no exogenous labeling is required. The efficiency of the THG process depends critically on the orientation of the interface relative to the optical axis [33].

2. The generation of the third harmonic is restricted to the focal region. In particular, the full width at half- maximum (FWHM) of the axial response of a THG microscope to an interface between two media with a difference in nonlinear susceptibility alone is equal to the confocal parameter ($b = kw_0^2$, where k is the wave vector and w_0 the beam waist radius) at the fundamental wavelength [32].

3. THG is a coherent phenomenon in which the third harmonic radiation is generated in the forward direction. For a linearly polarized input laser beam, the generated third harmonic is also linearly polarized in the same direction [34]. The third-order power dependence of THG on the input laser power results in an approximately inverse square dependence on the laser pulse width. Typical conversion efficiencies from fundamental to third-harmonic are in the range of 10^{-7}–10^{-9} [34], and conversion efficiencies up to 10^{-5} have been reported for specific materials [35].

4. THG imaging is a transmission mode microscopy, similar to phase contrast or DIC microscopy but with inherent three-dimensional sectioning properties. Thus, whereas phase contrast microscopy depends on accumulated phase differences along the optical pathlength, THG microscopy is sensitive to differences in specimen properties localized within the focal volume.

5. The noninvasive character of THG imaging has been demonstrated in various applications of microscopic imaging of biological specimens in vivo [34,36,37]. In addition, fading of contrast—equivalent to the bleaching of fluorescence—is absent in THG imaging applications.

Coherent Anti-Stokes Raman Scattering. Raman spectroscopy is sensitive to molecular vibrations, which in turn reflect molecular structure, composition, and intermolecular interactions. As such, Raman spectroscopy is unique in that it provides detailed intra- and intermolecular structural information and specificity at room temperature, even within extremely complex systems such as living cells. However, Raman spectroscopy suffers particularly from a low scattering cross section as well as from interference from naturally occurring luminescence, limiting its applicability to high resolution microscopy.

Coherent anti-Stokes Raman scattering (CARS) is the nonlinear optical analog of spontaneous Raman scattering. In this technique a particular Raman transition is coherently driven by two laser fields—the "laser" and "Stokes"—and is subsequently probed by a third laser field—the "probe"—generating an anti-Stokes signal field. The coherent signal is emitted in a particular direction determined by the phase-matching condition $\mathbf{k}_{AS} = \mathbf{k}_L - \mathbf{k}_S + \mathbf{k}_P$. This permits efficient suppression of background luminescence. In addition, it follows from the energy level diagram (Fig. 5) that no net energy transfer to the specimen results from the interaction. This renders CARS microscopy a mild, nondisruptive, imaging technique for biological applications and also guarantees the absence of bleaching of contrast, which is a severe problem in fluorescence-based techniques.

CARS microscopy was first introduced in the early 1980s [38,39] but found limited application at that time, probably owing to shortcomings in laser technology. The technique was recently reinvented [40,41], using novel approaches in terms of both the laser apparatus used and the application of noncollinear phase-matching geometries at high numerical aperture focusing conditions.

The general properties of CARS microscopy can be summarized as follows:

1. The CARS signal level is proportional to the product of the intensities of the input laser, Stokes, and probe fields, to the square of the particle number density, and to the square of the Raman scattering cross section.
2. CARS is a coherent process in which the signal is degenerated in a forward direction determined by the phase-matching geometry $(\mathbf{k}_{AS} = \mathbf{k}_L - \mathbf{k}_S + \mathbf{k}_P)$. The divergence of the generated emission scales directly with the lateral waist of the interaction volume.
3. The resolution in CARS microscopy is determined by the interaction volume, which is substantially reduced relative to one-photon interaction type microscopy because of the signal's cubic dependence on the laser input intensity. While sacrificing some of the attainable spatial resolution, nonlinear phase-matching configurations [41] can be implemented to enhance the detection sensitivity.
4. CARS microscopy is a form of "chemical imaging" that, through the Raman spectrum, is particularly sensitive to both molecular signature and intermolecular interactions.

Selected Applications

All coherent nonlinear optical microscopic techniques use basically the same setup, which is schematically depicted in Figure 6. The laser output is focused onto the specimen with a high numerical aperture microscope objective. The coherent signal is collected in the forward direction by a second objective (or condenser lens) and either imaged directly on a CCD camera or detected by a photomultiplier tube. The specimen is raster scanned using an XY beam scanner, and the specimen stage provides for axial scanning. In general a point-scanning mode is used, although line scanning has been demonstrated [33] to be feasible for high-power laser sources in the case of THG imaging.

Some of the specific properties of coherent nonlinear imaging are illustrated below with some examples selected from recent literature. Again it should be stressed that this is in no way a complete overview of the recent advances in this field.

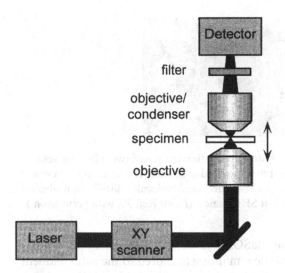

Figure 6 Schematic of a general coherent nonlinear optics imaging setup. A microscope objective focuses the ultrashort optical pulses (originating from one laser, as in the case of SHG and THG, or from two lasers, as in the case of CARS) onto the specimen. The specimen is raster scanned in a point excitation mode, generally using mirror scanning for the x and y directions and specimen scanning for the z direction. The coherent signal is collected in the forward direction with a second microscope objective or condenser lens. A filter blocks the residual light at the fundamental frequency. The image is either collected directly on a CCD camera or reconstructed after point-by-point detection with a photomultiplier tube.

Moreaux and coworkers [30] exploited the different properties of two-photon absorption and SHG by combining the two techniques simultaneously in a single microscope. In this study use was made of a special fluorophore that exhibits both a significant second-order nonlinear susceptibility and a two-photon absorption cross section ($\sim 3 \times 10^{-57}$ $m^4 s^2$ photon^{-1}). Of particular interest for this application (Fig. 7) is that SHG, in addition to its different specimen symmetry related properties, provides information complementary to that of two-photon absorption by probing different nonlinear susceptibilities ($\chi^{(2)}$ rather than $\chi^{(3)}$).

The first application of THG microscopy to live cells was reported by Squier et al. [34]. In these experiments the nondisruptiveness of the THG imaging technique was shown for high-resolution in vivo imaging of the rhizoid from the green alga *Chara* (Fig. 8). Rhizoids are tubular single cells that form the roots of the alga and have been studied widely, especially with respect to their response to gravity. Within the live cell there is strong cytoplasmic streaming to and from the rhizoid tip. The tip contains so-called

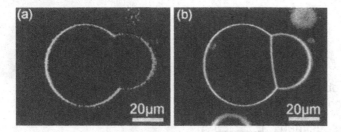

Figure 7 (a) SHG and (b) two-photon absorption images of two adhering vesicles labeled with di-6-ASPBS (equatorial slice), excited at 880 nm. The adhesion area where the membranes are fused exhibits a centrosymmetrical molecular distribution wherein two-photon absorption is allowed but SHG is not. (From Ref. 30, with permission.)

statoliths—vesicles that contain $BaSO_4$ crystals. In in vivo imaging, the statoliths show dynamic motion while remaining anchored to the actin filament network. No disruption of the cytoplasmic streaming nor any fading of contrast has been observed in more than an hour of continuous exposure, a first indication that the cell remains functional.

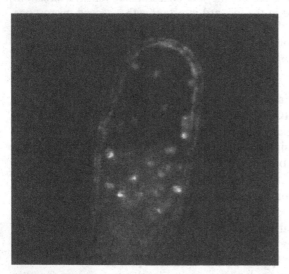

Figure 8 THG image of a single optical section of the tip of a rhizoid from green alga. The image (image size: $\sim$75 µm $\times$ 75 µm) clearly shows the cell wall and the statoliths. The persistence of the strong cytoplasmic streaming and absence of fading of contrast are indications of the nondisruptiveness of the technique even for times of continuous exposure exceeding 1 h.

In a material sciences application, Yelin et al. [35] used an OPO system (130 fs, 1500 nm, 80 MHz) to demonstrate strong phase-matched third harmonic generation in a nematic liquid crystal cell. Here the optical field can induce its own inhomogeneities in the nonlinear susceptibility by deforming the internal structure of the molecular orientation distribution (Fig. 9). Phase-matched THG conversion efficiencies of the fundamental to third harmonic as high as 10^{-5} have been observed in this case.

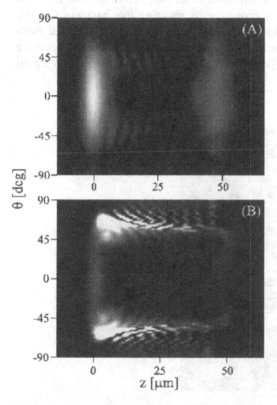

Figure 9 THG distribution in a nematic liquid crystal (LC) cell for different linear polarization states with average laser intensity of (A) 100 mW and (B) 200 mW. The strong signals for $z = 0\,\mu m$ and $z = 50\,\mu m$ correspond to THG from the glass/LC interface, which maximizes for zero angle ($\theta = 0$) between the optical field and the rubbing direction. At elevated power levels, optical field induced inhomogeneities in the susceptibility result in significant third harmonic being produced from the bulk of the LC. Conversion efficiencies of 10^{-5} from fundamental to third harmonic were observed in this case. (From Ref. 35, with permission.)

In another application, Squier and Müller [42] demonstrated the use of THG imaging to visualize the result of laser-induced breakdown in glass (Fig. 10). Using THG, the induced damage can be imaged in three dimensions, providing insight into the effects of exposure levels and duration on the process of laser-induced breakdown. It was demonstrated also that this technique has potential for reading in optical data storage, where the same laser beam at different power levels can be used both for "writing" and "reading" the data.

A number of recent experiments have demonstrated the potential of CARS microscopy in practice. Zumbusch et al. [40] demonstrated the mild imaging conditions of CARS microscopy by imaging *live* bacteria and the mitochondria in *live* HeLa cells. It has also been shown that a folded Box-CARS phase-matching configuration can be used at high numerical aperture imaging conditions, i.e., at high spatial resolution [41]. This permits signal detection with high sensitivity.

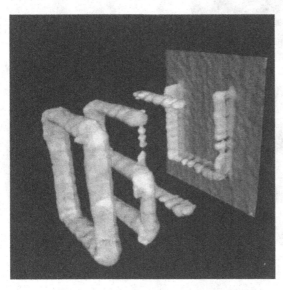

Figure 10 Three-dimensional reconstruction of a stack of axially sectioned THG images. Using laser-induced breakdown (LIB), a three-dimensional pattern is first created inside the bulk of a microscope coverglass, with the letters (which are ~20 μm wide) "written" in different axial planes approximately 19 μm apart. The induced damage is subsequently imaged using THG contrast at a laser power below threshold for LIB. Based on the THG imaging characteristics, a potential ~10^{12} bits/cm^3 data storage and retrieval can be realized. (From Ref. 42, with permission.)

14.2.4 Resolution of an Ultrashort Pulse Microscope

The resolution of any microscope in general, and that of an ultrashort optical microscope in particular, is not easily defined in general terms. Various criteria have been used in terms of either the apparent spreading of the image of an infinitely small point object or the capability to discriminate two separated point objects or, alternatively, in terms of the frequency bandpass of the optical system. A detailed description of all aspects of this subject is outside the scope of this work, so the discussion here is restricted to the effective three-dimensional size of the interaction volume in an ultrashort pulse microscope. This interaction volume is determined not only by the focusing properties of the microscope objective, i.e., by its numerical aperture, and the wavelength of the radiation but also by the order of nonlinearity of the interaction. In addition, confocal detection can effectively reduce the observation volume.

The Rayleigh Criterion

Optical resolution is traditionally defined through the Rayleigh criterion, which states that *two components of equal intensity should be considered to be just resolved when the principal intensity maximum of one coincides with the first minimum of the other* [43]. In the classical description of the focal field produced by a high numerical aperture lens, Kirchhoff, Debye, and paraxial approximation are all imposed and dimensionless optical co-ordinates are introduced:

$$v = r\frac{2\pi}{\lambda_m}\sin\alpha, \qquad u = z\frac{2\pi}{\lambda_m}\sin^2\alpha \tag{12}$$

where r, z are the radial and axial co-ordinates, respectively, and $\lambda_m = \lambda/n$ is the wavelength of the optical field in the medium with refractive index n. In this description the intensity distribution near the focal point for an optical field with a constant amplitude across its wavefront has an analytical solution in the form of the Lommel functions [43,44]. The intensity distribution of the focal field in a plane coinciding with the focal point and orthogonal to the optical axis is given by

$$I(0, v) \propto \left|\frac{2J_1(v)}{v}\right|^2 \tag{13}$$

where $J_1(v)$ is the first-order Bessel function of the first kind. Similarly, the intensity distribution along the optical axis is described by

$$I(u, 0) \propto \left(\frac{\sin(u/4)}{u/4}\right)^2 \tag{14}$$

The first node in these lateral and axial distributions is at $v_0 = 1.22\pi$ and $u_0 = 4\pi$, respectively, in optical co-ordinates. In real units the first nodes are thus found at

$$r_0 = \frac{0.61\lambda}{NA} \quad \text{and} \quad z_0 = \frac{2n\lambda}{(NA)^2} \tag{15}$$

For convenience, it is often assumed that the full width at half-maximum (FWHM) of these distributions is approximately equal to the distance to the first node, i.e., $FWHM_{lateral} \approx r_0$ and $FWHM_{axial} \approx z_0$. In fact, the FWHM is only ~85% of this value. Nevertheless, Eqs. (15) are generally used to describe the diffraction-limited focusing properties of a high numerical aperture microscope objective and subsequently those of the optical resolution through the Rayleigh criterion. Note that the imposed approximations are not strictly valid for high numerical aperture focusing. Full diffraction theory calculations [44] predict even smaller values for the FWHM of the focal field distributions than those based on classical theory (Fig. 11).

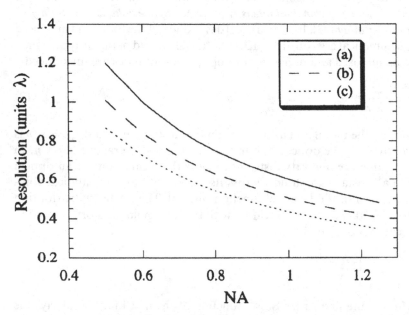

Figure 11 Lateral resolution of an optical microscope (a) as a function of numerical aperture (NA) based on the lateral position of the first node of the focal field distribution (Rayleigh criterion) or as the FWHM of the lateral focal field distribution in the focal plane from (b) classical theory [43] and (c) full diffraction theory [44].

The Nonlinear Interaction Volume

In general, the influence of a signal response that depends nonlinearly on the input intensity reduces the effective interaction volume and thus increases the resolution relative to its linear counterpart. Let us assume that the contrast in the microscope depends on the focal field intensity to the power N. To evaluate the influence of the nonlinear interaction on the effective interaction volume, consider a general Gaussian distribution

$$I(x) = \exp\left[\frac{-x^2}{2\sigma^2}\right] \tag{16}$$

for which the FWHM is at

$$\text{FWHM} = 2\sigma\sqrt{-2\ln(1/2)} \tag{17}$$

It follows that for an Nth-order process the FWHM reduces to

$$\text{FWHM} = \frac{1}{\sqrt{N}}2\sigma\sqrt{-2\ln(1/2)} \tag{18}$$

Hence, in general, the interaction volume of an Nth-order nonlinear process decreases by a factor of $\sqrt{N}$ relative to the linear interaction volume at the same optical wavelength. This is true for, for instance, four-wave mixing processes. However, in multiphoton absorption processes, excitation of the one-photon equivalent electronic transition of the fluorophore requires a longer wavelength that scales with the nonlinearity of the process as

$$\lambda^{(n)} = N\lambda^{(1)} \tag{19}$$

where (n) denotes the order of the process. Thus in multiphoton absorption processes the decrease in resolution resulting from the increase in wavelength is compensated for only in part by the decrease in the interaction volume due to the nonlinearity of the interaction. Effectively the interaction volume for an N-photon absorption process increases by a factor of $\sqrt{N}$ relative to its single-photon absorption counterpart.

The Influence of Detection

The actual resolution of the microscope is determined not only by the size of the effective interaction volume but also by the optical configuration of the detection—especially confocal detection—and the coherence properties of the signal. If the effective point spread functions of illumination and detection are denoted by psf_{ill} and psf_{det}, respectively, then the image formation for an incoherent signal is described by

$$I \propto |\text{psf}_{\text{ill}} \ \text{psf}_{\text{det}}|^2 \otimes O^2 \tag{20}$$

where I and O denote the image and object, respectively, and $\otimes$ denotes a convolution operation. For a fully coherent process, on the other hand, Eq. (20) becomes

$$I \propto |h_{\text{ill}} h_{\text{det}} \otimes O|^2 \tag{21}$$

Note that the two imaging modes become identical for a single point object [45,46].

As shown above, the effective point spread function for an Nth-order nonlinear interaction is in first approximation related to its linear counterpart through

$$\text{psf}^{(N)} = \left[\text{psf}^{(1)}\right]^N \tag{22}$$

The influence of confocal detection is to suppress out-of-focus signals relative to in-focus signals. Theoretically this results in a multiplication of the excitation point spread function with that of the detection. For an infinitely small detection pinhole and a signal wavelength that is equal to that of the fundamental, the influence of the confocal pinhole is to increase the order of the nonlinear process by one:

$$\text{psf}^{(N)}_{\text{confocal}} = \left[\text{psf}^{(1)}\right]^{N+1} \tag{23}$$

Thus confocal detection reduces the effective interaction volume.

As an example, consider an Nth-order multiphoton absorption process with an excitation wavelength of $N\lambda_{\text{exc}}$, where λ_{exc} is the corresponding single-photon absorption wavelength. Assume further that the fluorescence wavelength $\lambda_{\text{flu}} = N\lambda_{\text{flu}}/\beta$. Then the size of the effective interaction volume is described by

$$\text{FWHM} = \frac{N}{\sqrt{N + \beta^2}} 2\sigma\sqrt{-2\ln(1/2)} \tag{24}$$

Ultrashort Pulses

Ultrashort pulses intrinsically have a large optical bandwidth. For transform-limited Gaussian shaped pulses the time–bandwidth product is given by $\Delta t\, \Delta v = 0.44$. The significant bandwidth of ultrashort pulses results in dispersion while they propagate through the optical system (see also Sec. 14.3). This dispersion in turn results in a broadening and distortion of the temporal profile of the pulses. If the induced dispersion is equal for all rays through the optical system, it can be compensated for by re-chirping, retrieving the minimal pulse width at focus [47].

More important with respect to resolution in nonlinear optical microscopy is the dispersion that arises from a varying dispersion—including group delay—across the pupil of the microscope objective. This effect is directly related to spherical and chromatic aberrations of the system [48,49]. This leads to a spreading of the optical intensity both in time and in space that cannot be compensated for in any straightforward manner. To minimize these effects the imaging system should be as achromatic as possible.

The Influence of Noise

In terms of the Rayleigh criterion, two equally bright point objects can just be resolved when the maximum of the intensity distribution of one coincides with the first minimum of the other. In this case—and in the absence of noise—there is a drop of approximately 20% (depending on the shape of the distributions) in the intensity in the image of the two point objects. Whether this drop in intensity is detectable in practice depends strongly on the presence of noise. All kinds of noise—shot noise, thermal noise, readout noise, photon statistical noise, photon scattering, quantization noise, etc.—affects either the background level or the signal's standard deviation and will therefore decrease the signal-to-noise ratio (SNR) of the system. A decrease in SNR in turn decreases the visibility of the drop in intensity between the maxima as specified by the Rayleigh criterion. Thus the attainable resolution depends not only on the imaging properties of the microscope—i.e., the numerical aperture and the wavelength—but also on the SNR of the image acquisition (Fig. 12).

In the special case of a single point object—as, for instance, in the case of single-molecule fluorescence imaging—the attainable precision with which the position of the point object can be determined is in fact determined solely by the SNR of the system. In this case it is sufficient to determine the centroid position of the image intensity distribution to localize the object. This situation is analogous to the precision with which the positions of stars can be determined using relatively low resolution telescopes.

In general, a priori knowledge of the object can be used to increase the attainable resolution of the system. Deconvolution techniques can be applied to limit the influence of noise and limited spatial resolution of the

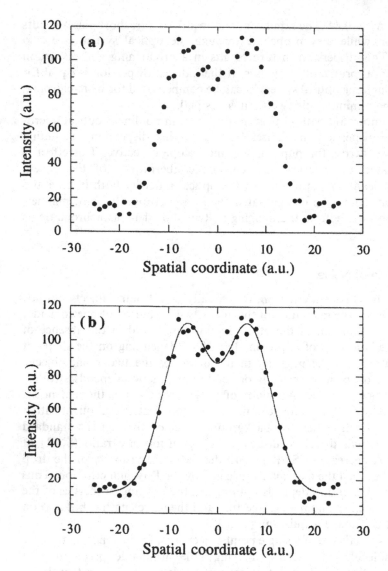

Figure 12 (a) Intensity profile of two signal point sources, separated according to the Rayleigh criterion, with Poisson noise and 10% detector noise. (b) The use of a priori knowledge permits the determination of the position of the two point sources with an accuracy determined by the SNR of the measurement only.

microscope. Alternatively one can argue that the applicability of deconvolution techniques depends strongly on the SNR of the image acquisition.

14.3 DISPERSION CONTROL IN HIGH NUMERICAL APERTURE OPTICS

14.3.1 Characterizing the Field of an Ultrashort Pulse at the Focus of a High Numerical Aperture (NA) Optic

A crucial aspect of multiphoton microscopy is characterization of the spatial and temporal characteristics of the laser field at the focus of the microscope objective. Detailed knowledge of the field ensures optimum resolution performed with the highest efficiency. In addition, quantitative intensity measurements are necessary for accurately gauging the exposure limits that a living specimen can withstand. In this section a variety of in situ methods by which the temporal characteristics of the pulse can be measured are presented. Most important, they can be performed at the full NA of the objective, which is the condition under which the system will be used for imaging.

Inherent to virtually all pulse measurement techniques is the need for an autocorrelator that is capable of sweeping a time-delayed replica of the pulse across itself. The two arms are perfectly balanced: Each pulse travels the same material path length and is subject to the identical number of reflections. This is important because ultrashort pulses are dispersed by dielectric coatings. By keeping the arms perfectly balanced, the amounts of dispersion due to material and coatings are kept identical, and a true autocorrelation results. A second important aspect of this autocorrelator geometry is the interferometric aspect. This has the advantage that the beams are fully collinear and therefore the autocorrelation measurement can be performed at the full NA of the objective.

The beams from the autocorrelator are relayed into the microscope to ensure that the autocorrelation measurement is performed at the objective focus. A material capable of generating a nonlinear intensity-dependent signal is placed at the focus. At this point a variety of materials and methods are available for the actual pulse measurement. Three of the simplest to implement are described here.

Two-Photon Absorption Autocorrelation

One of the easiest methods for recording the autocorrelation is to take a small sample of a fluorophore and perform a two-photon absorption autocorrelation (TPAA) [50]. The dye should be appropriately mounted to match the imaging conditions under which the objective will be used. Care should be taken that the correct immersion medium, coverslip thickness, and mounting medium are in place. By blocking an arm of the autocorrelator and scanning the sample in the z direction, a sectioning measurement can also be performed. Thus, spatial and temporal characteristics of the microscope can be determined.

Two-Photon Absorption in Photodiodes

A disadvantage of using a fluorophore for performing TPAA is the limited bandwidth. Typical dyes have bandwidths on the order of 30 nm, meaning that a new dye is necessary should the laser be tuned beyond this limit. A more achromatic alternative is to use the two-photon response of GaAsP photodiodes. These diodes have been shown to have a spectral response in the range of 720–950 nm that is more uniform than that of many types of doubling crystals used in typical autocorrelation measurements. The actual two-photon response for these diodes covers the range of 600–1360 nm, completely encompassing the entire tunability range of femtosecond lasers such as the Ti:sapphire lasers that are typically used in multiphoton microscopy. In addition, these diodes are extremely robust and inexpensive, making them a nearly ideal choice for autocorrelation measurements. Millard et al. [51] demonstrated how such diodes are easily implemented in the microscope and, as with TPAA, they can be used for spatial sectioning measurements as well. The autocorrelations taken with GaAsP photodiodes also show superior signal-to-noise ratios compared to TPAA measurements.

Third Harmonic Autocorrelation

A third alternative is to generate the third harmonic at the coverslip interface and use this signal to perform the autocorrelation [52]. Interferometric intensity autocorrelations generated in this manner have several advantages. First, no special materials are needed; the glass coverslip is the nonlinear medium. It is achromatic; autocorrelations using this method have been performed from 800 nm to 1500 nm [53,54]. This technique is especially useful for very short pulses, because group velocity mismatch considerations across the interface are negligible. Because this is a third-order process, the autocorrelation envelopes do not follow an 8:1 contrast as in our previous examples; rather, they are 32:1.

In each of these examples a method is described wherein the pulse duration at the focus of the objective can be measured. This information, although valuable, does in fact not provide an unambiguous and therefore complete characterization of the temporal profile of the pulse. Certain applications may require knowledge of the pulse phase, which cannot be determined by any of the measurements described to this point. Full pulse characterization requires a more sophisticated measurement such as frequency-resolved optical gating (FROG) [55]. The basic phase retrieval technique requires that the autocorrelation be spectrally resolved. This can be accomplished, for instance, by placing a spectrometer between the output of the autocorrelator and the detector. A high numerical aperture FROG technique was demonstrated recently by Fittinghoff et al. [56] and used to characterize 22 fs pulses at the focus of a 1.2 NA objective. This method makes it possible to determine the intensity profile and phase of the pulse. Such a scheme is nontrivial to implement and should be considered only if phase information is critical for the imaging application.

Radially Varying Propagation Time Delay

One factor affecting each of the aforementioned measurement techniques is radially varying time-of-flight errors [49]. When using refractive optics, it is important to note that the net group delay for those rays traveling along the center of the optical path may vary from those that are off-center. This radially varying time-of-flight error manifests itself in several ways in the autocorrelation measurement. First, the envelopes that define the interferometric autocorrelation traces deviate from the standard 8:1 or 32:1 contrast ratios. This is due to a mixing of collinear and noncollinear autocorrelation functions produced by the radially varying time-of-flight error. Second, the pulse duration also becomes a function of pupil position as a result of this phenomenon, and the autocorrelation measurement convolves this varying pulse duration information in a complicated way. The radially varying time-of-flight errors vary from objective to objective.

14.3.2   Dispersion in Microscopes

Clearly, in dealing with ultrashort pulses, dispersion of the optical glasses and dielectric coatings that the pulse passes through or reflects from is an important consideration. Important from at least two aspects: the efficiency of the multiphoton process, and the damage threshold of the specimen. Without due consideration of the microscope's dispersive properties, neither of these aspects of multiphoton imaging can be adequately addressed.

First consider the average power of the detected signal, assuming two-photon absorption fluorescence as the imaging modality. Whereas the instantaneous signal scales as the square of the excitation intensity and hence depends inversely on the square of the pulse width, the detected integrated signal power scales inversely with the pulse width to the power of unity. Thus a 1 ps pulse produces a signal that is 10 times smaller than that of a 100 fs pulse with the same energy. The typical optical path lengths found in microscopes provide sufficient dispersion that 100 fs pulses can be stretched by up to a factor of 2, reducing the excitation efficiency by the same factor of 2. For higher order processes, such as third harmonic generation, this becomes even more significant. A pulse stretched by a factor of 2 reduces the detected signal by a factor of 4.

The dispersion of typical objectives has been measured by several groups [47,57]. From these measurements, broadening of femtosecond pulses can be estimated. Consider a 100 fs pulse at 800 nm—typical pulse parameters used in multiphoton imaging. The measured group delay dispersion (GDD) ranges from 580 to 1200 fs^2 for standard objectives. The net dispersion of the microscope (including beam expanders, filters, dichroics) can then be on the order of 4500 fs^2. This amount of dispersion broadens a 100 fs pulse to ~270 fs, significantly impacting the efficiency of any nonlinear imaging process.

The dispersion produced by the microscope can be precompensated through a variety of means. The use of a grating pair or prism pair arranged to provide negative GDD is the most common means of dispersion precompensation. Gratings have the advantage of high dispersion, making the system very compact, whereas prisms have the advantage of high efficiency. A second advantage of prisms over gratings is the sign of the third-order dispersion (TOD), which is negative for prisms and positive for gratings. Thus, the higher order dispersion for a prism system is opposite to that of the glass material that is being compensated, whereas the grating's TOD adds to the TOD of the glass. Ultimately it is the residual higher order dispersion from the compressor that in fact limits the pulse width. Careful attention to the higher order terms is obviously necessary to ensure optimum compensation.

Although the prism pair is better matched with respect to balancing the higher order dispersion than a grating pair, care must also be taken with regard to the type of glass used in the prism sequence. SF10 glass is often used because it is highly dispersive and a compact compensator can be designed around it. However, as shown by Müller [46], such dispersive glasses result in high residual TOD that can limit pulse compression. Fused silica is a better glass choice when either the shortest pulse width or greatest wavelength tunability is desired. Pulses as short as 15 fs have been generated at the focus of high NA optics using this type of sequence [47]. Shorter pulses have been obtained by further balancing the residual TOD through the use of dielectric mirror coatings [58].

14.4 ULTRASHORT PULSE LASERS FOR MICROSCOPY

There has been a revolution in solid-state lasers that are applicable for microscopy that was sparked by the development of Kerr lens modelocked

Ti:sapphire oscillators. Many new lasers from different materials are available, but Ti:sapphire-based systems remain the most flexible source in terms of balancing the parameters for optimizing the image. For instance, Ti:sapphire is tunable from 700 nm to 1070 nm and is capable of producing short pulses throughout this wavelength range. The pulse widths from this system are very flexible, ranging from 10 ps (optimal for CARS imaging) down to 20 fs or less (optimal for two-photon absorption imaging with multiple labels, for instance). These systems can be frequency-doubled or used to pump an optical parametric oscillator, extending the wavelength range of the basic system from the ultraviolet to the infrared. The repetition rate of commercially available systems runs in the range of 60–100 MHz, which is fairly optimal for fluorescence imaging applications where the lifetime of the fluorophore is on the order of 1–10 ns. The average power of a Ti:sapphire system is typically 300 mW to 1 W—more than sufficient to efficiently drive one or many foci in intensity-dependent imaging applications. Thus, a state-of-the-art Ti:sapphire system can be used to optimize all the major operating parameters found in intensity-dependent imaging: wavelength, average power, pulse width, and repetition rate.

Many other lasers have demonstrated their utility in nonlinear imaging. These include Nd:glass, Cr:LiSAF, Cr:fosterite, fiber lasers, Nd:YLF, etc. Many of these are chosen based on the fact that they can be directly diode-pumped, resulting in a very compact, less expensive system. Rather than pursue the individual details of these lasers here, it is more useful to briefly discuss some general laser configurations that are equally applicable to all of the aforementioned lasers. Depending on the technique, these configurations can be used to further balance the laser parameters to give the most favorable imaging conditions.

14.4.1 Cavity Dumped Lasers

By using an acousto-optic cavity dumper, the energy per pulse can be boosted by a factor of ~10. The addition of the cavity dumper has virtually no impact on pulse widths down to at least the 20 fs level. In addition to the energy gain, cavity-dumped systems allow the user to easily adjust the repetition rate. Common cavity-dumped systems can be varied from 1 MHz to 1 kHz at the flick of a switch. This is a very nice option when the pulse repetition rate is of concern in optimizing the image.

14.4.2 Extended Cavity Lasers

The use of an extended cavity laser is an exceptionally straightforward way of increasing the pulse energy in home-built systems without the expense of a cavity dumper or an external amplifier. By simply extending the cavity length the average power is kept constant, but the energy per pulse is increased. For instance, a 400 mW laser operating at 100 MHz produces 4 nJ per pulse. When the cavity is extended and the repetition rate is reduced to 20 MHz, the average power remains at 400 mW, but the pulse energy is now 20 nJ. This combination of energy and repetition rate can be especially attractive in multifocal imaging systems.

REFERENCES

1. M Schubert, B Wilhelmi. Nonlinear Optics and Quantum Electronics. New York: Wiley, 1986.
2. C Flytzanis. Theory of nonlinear optical susceptibilities. In: H Rabin, CL Tang, eds. Quantum Electronics, Vol. I, Nonlinear Optics. New York: Academic Press, 1975, pp 9–207.
3. Y Prior. A complete expression for the third-order susceptibility ($\chi^{(3)}$). Perturbative and diagrammatic approaches. IEEE Quantum Electron 20:37–42 (1984).
4. RW Boyd. Nonlinear Optics. Boston: Academic Press, 1992.
5. H Mahr. Two-photon absorption spectroscopy. In: H Rabin, CL Tang, ed. Quantum Electronics. New York: Academic Press, 1975, pp 285–361.
6. W Denk, JH Strickler, WW Webb. Two-photon laser scanning fluorescence microscopy. Science 248:73–76 (1990).
7. C Xu, WW Webb. Measurement of two-photon excitation cross sections of molecular fluorophores with data from 690 to 1050 nm. J Opt Soc Am B 13: 481–491 (1996).
8. C Xu, RM Williams, W Zipfel, WW Webb. Multiphoton excitation cross-sections of molecular fluorophores. Bioimaging 4:198–207 (1996).
9. C Xu, W Zipfel, JB Shear, RM Williams, WW Webb. Multiphoton fluorescence excitation: new spectral windows for biological nonlinear microscopy. Proc Nat Acad Sci USA 97:10763–10768 (1996).
10. M Albota, D Beljonne, J Brédas, JE Ehrlich, J Fu, AA Heikal, SE Hess, T Kogej, MD Levin, SR Marder, D McCord-Maughon, JW Perry, H Röckel, M Rumi, G Subramanian, WW Webb, X Wu, C Xu. Design of organic molecules with large two-photon absorption cross sections. Science 281:1653–1656 (1998).
11. I Gryczynski, H Malak, JR Lakowicz. Multiphoton excitation of the DNA stains DAPI and Hoechst. Bioimaging 4:138–148 (1996).
12. SW Hell, K Bahlmann, M Schrader, A Soini, H Malak, I Gryczynski, JR Lakowicz. Three-photon excitation in fluorescence microscopy. J Biomed Opt 1:71–74 (1996).
13. W Denk. Two-photon scanning photochemical microscopy: mapping ligand-gated ion channel distributions. Proc Natl Acad Sci USA 91:6629–6633 (1994).

14. PE Hänninen, E Soini, SW Hell. Continuous wave excitation two-photon fluorescence microscopy. J Microsc 176:222–225 (1994).
15. GJ Brakenhoff, J Squier, T Norris, AC Bliton, MH Wade, B Athey. Real-time two-photon confocal microscopy using a femtosecond, amplified Ti:sapphire system. J Microsc 181:253–259 (1996).
16. AH Buist, M Müller, GJ Brakenhoff. Real time two-photon microscopy using multi point excitation. J Microsc 192:217–226 (1998).
17. J Bewersdorf, R Pick, SW Hell. Multifocal multiphoton microscopy. Opt Lett 23:655–657 (1998).
18. K König, PTC So, WW Mantulin, E Gratton. Cellular response to near-infrared femtosecond laser pulses in two-photon microscopy. Opt Lett 22:135–137 (1997).
19. K König, TW Becker, P Fischer, I Riemann, K-J Halbhuber. Pulse-length dependence of cellular response to intense near-infrared laser pulses in multiphoton microscopes. Opt Lett 24:113–115 (1999).
20. JA Curcio, CC Petty. The near infrared absorption spectrum of water. J Opt Soc Am 41:302–304 (1951).
21. A Schönle, SW Hell. Heating by absorption in the focus of an objective lens. Opt Lett 23:325–327 (1998).
22. JM Squirrell, DL Wokosin, JG White, BD Bavister. Long-term two-photon fluorescence imaging of mammalian embryos without compromising viability. Nature Biotechnol 17:763–767 (1999).
23. GH Patterson, DW Piston. Photobleaching in two-photon excitation microscopy. Biophys J 78:2159–2162 (2000).
24. K Svoboda, W Denk, D Kleinfeld, DW Tank. In vivo dendritic calcium dynamics in neocortical pyramidal neurons. Nature 385:161–165 (1997).
25. S Hell, EHK Stelzer. Fundamental improvement of resolution with a 4Pi-confocal fluorescence microscope using two-photon excitation. Opt Commun 93:277–282 (1992).
26. P Schwille, U Haupts, S Maiti, WW Webb. Molecular dynamics in living cells observed by fluorescence correlation spectroscopy with one- and two-photon excitation. Biophys J 77:2251–2265 (1999).
27. R Hellwarth, P Christensen. Nonlinear optical microscopic examination of structure in polycrystalline ZnSe. Opt Commun 12:318–322 (1974).
28. R Gauderon, PB Lukins, CJR Sheppard. Three-dimensional second-harmonic generation imaging with femtosecond laser pulses. Opt Lett 23:1209–1211 (1998).
29. PJ Campagnola, M Wei, A Lewis, LM Loew. High-resolution nonlinear optical imaging of live cells by second harmonic generation. Biophys J 77:3341–3349 (1999).
30. L Moreaux, O Sandre, M Blanchard-Desce, J Mertz. Membrane imaging by simultaneous second-harmonic generation and two-photon microscopy. Opt Lett 25:320–322 (2000).
31. JF Ward, GHC New. Optical third harmonic generation in gases by a focused laser beam. Phys Rev 185:57–72 (1969).

32. Y Barad, H Eisenberg, M Horowitz, Y Silberberg. Nonlinear scanning laser microscopy by third-harmonic generation. Appl Phys Lett 70:922–924 (1997).
33. M Müller, J Squier, KR Wilson, GJ Brakenhoff. 3D-microscopy of transparent objects using third-harmonic generation. J Microsc 191:266–274 (1998).
34. J Squier, M Müller, GJ Brakenhoff, KR Wilson. Third harmonic generation microscopy. Opt Express 3:315–324 (1998).
35. D Yelin, Y Silberberg, Y Barad, JS Patel. Phase-matched third-harmonic generation in a nematic liquid crystal cell. Phys Rev Lett 82:3046–3049 (1999).
36. D Yelin, Y Silberberg. Laser scanning third-harmonic-generation microscopy in biology. Opt Express 5:169–175 (1999).
37. AC Millard, P Wiseman, DN Fittinghoff, JA Squier, M Müller. Third-harmonic generation microscopy by use of a compact, femtosecond fiber laser source. Appl Opt 38:7393–7397 (1999).
38. MD Duncan, J Reijntjes, TJ Manuccia. Scanning coherent anti-Stokes microscope. Opt Lett 7:350–352 (1982).
39. MD Duncan. Molecular discrimination and contrast enhancement using a scanning coherent anti-Stokes Raman microscope. Opt Commun 50:307–312 (1984).
40. A Zumbusch, GR Holtom, XS Xie. Three dimensional vibrational imaging by coherent anti-Stokes Raman scattering. Phys Rev Lett 82:4142–4145 (1999).
41. M Müller, J Squier, CA de Lange, GJ Brakenhoff. CARS microscopy with folded BoxCARS phasematching. J Microsc 197:150–158 (2000).
42. J Squier, M Müller. Third harmonic generation imaging of laser-induced breakdown in glass. Appl Opt 38:5789–5794 (1999).
43. M Born, E Wolf. Principles of Optics. 6th ed. Oxford: Pergamon Press, 1980.
44. JJ Stamnes, Waves in Focal Regions. Bristol: IOP Pub, 1986.
45. JW Goodman. Introduction to Fourier Optics. New York: McGraw-Hill, 1968.
46. J Deitche, M Kempe, W Rudolph. Resolution in nonlinear laser scanning microscopy. J Microsc 174:69–73 (1994).
47. M Müller, J Squier, R Wolleschensky, U Simon, GJ Brakenhoff. Dispersion pre-compensation of 15 femtosecond optical pulses for high-numerical-aperture objectives. J Microsc 191:141–150 (1998).
48. M Kempe, U Stamm, B Wilhelmi, W Rudolph. Spatial and temporal transformation of femtosecond laser pulses by lenses and lens systems. J Opt Soc Am B 9:1158–1165 (1992).
49. Z Bor. Distortion of femtosecond laser pulses in lenses. Wave optical description. Opt Commun 94:249–258 (1992).
50. M Müller, J Squier, GJ Brakenhoff. Measurement of femtosecond pulses in the focal point of a high-numerical-aperture lens by two-photon absorption. Opt Lett 20:1038–1040 (1995).
51. AC Millard, DN Fittinghoff, JA Squier, M Muller, AL Gaeta. Using GaAsP photodiodes to characterize ultrashort pulses under high numerical aperture focusing in microscopy. J Microsc 193:179–181 (1999).
52. TYF Tsang. Optical third-harmonic generation at interfaces. Phys Rev A (Atomic, Mol Opt Phys), 52:4116–4125 (1995).

53. JA Squier, DN Fittinghoff, CPJ Barty, KR Wilson, M Muller, GJ Brakenhoff. Characterization of femtosecond pulses focused with high numerical aperture optics using interferometric surface-third-harmonic generation. Opt Commun 147:153–156 (1998).

54. D Meshulach, Y Barad, Y Silberberg. Intensity and interferometric autocorrelation measurements of ultrashort pulses by third harmonic generation. 1997.

55. R Trebino, KW DeLong, DN Fittinghoff, JN Sweetser, MA Krumbugel, BA Richman, DJ Kane. Measuring ultrashort laser pulses in the time-frequency domain using frequency-resolved optical gating. Rev Sci Instrum 68:3277–3295 (1997).

56. DN Fittinghoff, AC Millard, JA Squier, M Muller. Frequency-resolved optical gating measurement of ultrashort pulses passing through a high numerical aperture objective. IEEE J Quantum Electron 35:479–486 (1999).

57. JB Guild, C Xu, WW Webb. Measurement of group delay dispersion of high numerical aperture objective lenses using two-photon excited fluorescence. Appl Opt 36:397–401 (1997).

58. T Jasapara, W Rudolph. Characterization of sub-10-fs pulse focusing with high-numerical-aperture microscope objectives. Opt Lett 24:777–779 (1999).

15
Optical Coherence Tomography

James G. Fujimoto, Mark Brezinski, Wolfgang Drexler, Ingmar Hartl, Franz Kärtner, Xingde Li, and Uwe Morgner
Massachusetts Institute of Technology, Cambridge, Massachusetts, U.S.A.

15.1 INTRODUCTION TO OPTICAL COHERENCE TOMOGRAPHY

Optical coherence tomography (OCT) is a fundamentally new type of optical imaging modality. OCT performs high-resolution cross-sectional tomographic imaging of the internal microstructure in materials and biological systems by measuring the echo time delay and magnitude of backscattered light (Huang et al., 1991a). Image resolutions of 1–15 µm can be achieved, one to two orders of magnitude higher than conventional ultrasound. Imaging can be performed in situ and in real time. The unique features of this technology enable a broad range of research and clinical applications. OCT imaging integrates many technologies, including ultrafast optics, interferometry, fiber optics, and Fourier transform spectrometry. This chapter provides an overview of OCT technology, its background, and its applications.

Optical coherence tomography performs cross-sectional imaging by measuring the time delay and magnitude of optical echoes at different transverse positions. A cross-sectional image is acquired by performing successive rapid axial measurements while transversely scanning the incident optical beam as shown in Figure 1. The result is a two-dimensional data set that represents the optical reflection or backscattering in a cross-sectional plane through the material or tissue. Optical coherence tomography was first demonstrated in 1991 (Huang et al., 1991a). Imaging was performed in vitro in the human retina and in atherosclerotic plaque as examples of imaging in transparent, weakly scattering media as well as in highly scattering media.

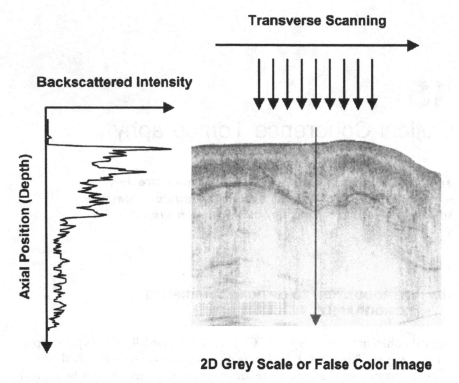

Figure 1 Optical coherence tomography generates cross-sectional images by performing measurements of the echo time delay of light at different transverse positions. The result is a two-dimensional data set that represents the backscattering in a cross-sectional plane of the tissue.

In vivo OCT imaging of the human retina was demonstrated in 1993 (Fercher et al., 1993, Swanson et al., 1993), and clinical studies in ophthalmology began in 1995 (Hee et al., 1995, Puliafito et al., 1995, 1996). Since that time, numerous applications of OCT for both materials and medical applications have emerged (Brezinski and Fujimoto, 1999; Brezinski et al., 1996; Fujimoto et al., 1995, 2000a, 2000b; Schmitt et al., 1994, 1995; Tearney et al., 1997b). At the same time, the resolution and capabilities of OCT technology have improved dramatically.

15.2 IMAGING USING LIGHT VERSUS SOUND

Imaging by OCT is analogous to ultrasound B mode imaging except that it uses light instead of sound. There are several different embodiments of OCT, but essentially OCT performs imaging by measuring the echo time delay and intensity of backscattered or backreflected light from internal microstructure in materials or tissues (Huang et al., 1991a). OCT images are two-dimensional or three-dimensional data sets that represent optical backscattering or backreflection in a cross-sectional plane or volume. When a beam of sound or light is directed onto tissue, it is backreflected or backscattered from structures that have different acoustic or optical properties as well as from boundaries between structures. The dimensions of the different structures can be determined by measuring the time it takes for a sound or light "echo" to be backreflected or backscattered from structures at varying axial distances.

In ultrasound, the axial measurement of distance or range is called A mode scanning, whereas imaging is called B mode scanning. The principal difference between ultrasound and optical imaging is the fact that the velocity of propagation of light is significantly greater than the velocity of sound. The velocity of sound in water is approximately 1500 m/s, whereas the velocity of light is approximately 3×10^8 m/s. The measurement of distances or dimensions with a resolution on the 100 μm scale, which would be typical for ultrasound, corresponds to a time resolution of approximately 100 ns. One advantage of ultrasound is that echo time delays are on the nanosecond time scale and well within the limits of electronic detection. Ultrasound technology has dramatically advanced in recent years with the availability of high-performance, low-cost digital signal processing technology. Because ultrasound imaging depends on sound waves, it requires direct contact with the material or tissue being imaged or immersion in a liquid or other medium to facilitate the transmission of sound waves. In contrast, optical imaging techniques such as OCT rely on the use of light rather than sound and can be performed without physical contact with the structure being imaged and without the need for a special transducing medium. Unlike sound, the echo time delays associated with light are extremely short. The measurement of a structure with a resolution on the 10 μm scale, which is typical in OCT, corresponds to a time resolution of approximately 30 fs. Direct electronic detection is not possible on this time scale. Thus, OCT measurements of echo time delay are based on correlation techniques that compare the backscattered or backreflected light signal to reference light traveling a known path length.

15.3 PERFORMANCE OF OCT VERSUS OTHER IMAGING TECHNOLOGIES

It is helpful to compare OCT, ultrasound, and microscopy as shown in Figure 2. The resolution of ultrasound imaging depends directly on the fre-

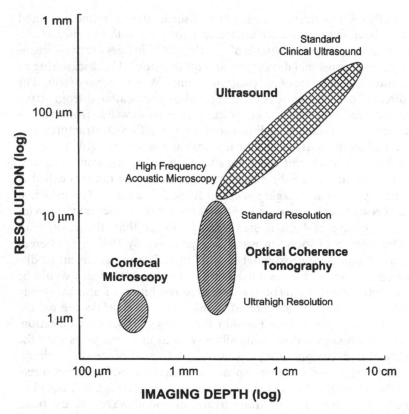

Figure 2 Resolution and penetration of ultrasound, OCT, and confocal micro-scopy. In ultrasound imaging both resolution and penetration are determined by the sound wave frequency. Higher frequencies yield higher resolution but increase ultrasonic attenuation, which limits image penetration. The image resolution in OCT ranges from 1 to 15 μm and is determined by the coherence length of the light source. In most biological tissues, image penetration is limited to 2–3 mm by attenua-tion from scattering. Confocal microscopy can have submicrometer transverse reso-lution. However, the image penetration of confocal microscopy is typically limited to a few hundred micrometers in tissues.

quency or wavelength of the sound waves (Fish, 1990; Kremkau, 1984; Zwiebel, 1992). For typical clinical ultrasound systems, sound wave frequen-cies are in the 10 MHz regime and yield spatial resolutions of up to 150 μm. Ultrasound imaging has the advantage that sound waves at this frequency are readily transmitted into most biological tissues and therefore it is possi-ble to obtain images of structures up to several tens of centimeters deep within the body. High-frequency ultrasound has been developed and investigated extensively in laboratory applications as well as some clinical applications. Resolutions of 15–20 µm and higher have been achieved with frequencies of 100 MHz. However, high-frequency ultrasound is strongly attenuated in biological tissues, and attenuation increases approximately in proportion to the frequency. Thus, high-frequency ultrasound imaging is limited to depths of only a few millimeters. The sound frequency is an important parameter in ultrasound imaging because it is possible to optimize image resolution for a given application while trading off image penetration depth.

The axial resolution in OCT is determined by the coherence length of the light source. Current OCT imaging technologies have resolutions ranging from 1 to 15 µm, approximately 10–100 times higher resolution than standard ultrasound imaging. The inherently high resolution of OCT permits the imaging of features such as tissue architectural morphology as well as cellular structure. For medical applications, this enables the diagnosis of a wide range of clinically important pathologies. OCT is ideally suited for ophthalmology because of the ease of optical access to the eye. The principal disadvantage of optical imaging is that light is highly scattered by most biological tissues. In tissues other than the eye, optical scattering limits image penetration depths to 2–3 mm. However, because OCT is an optical technology, it can be interfaced to a wide range of instruments such as endoscopes, catheters, or laparoscopes that enable the imaging of internal organ systems.

15.4 MEASURING ULTRAFAST OPTICAL ECHOES

The concept of using optical echoes to perform imaging in scattering systems such as biological tissue was first proposed by Michel Duguay in 1971 (Duguay, 1971; Duguay and Mattick, 1971). Duguay demonstrated an ultrafast optical shutter using the Kerr effect to photograph pulses of light in flight. The Kerr shutter is actuated by an intense ultrashort light pulse that induces birefringence in an optical medium (the Kerr effect) and is placed between crossed polarizers. If the induced birefringence is electronically mediated, the Kerr shutter can achieve picosecond or femtosecond resolution depending upon the gate or reference pulse duration. Because the optical signal or image that is transmitted through the Kerr shutter is detected by a slow detector (i.e., a camera), the signal is a convolution with the transmission function of the Kerr shutter. The shutter operates like a sampling gate, and the delay between the gating or reference pulse and the transient optical signal is adjusted to detect optical signals with varying echo delays. Optical scattering limits the ability to image biological tissues, and Duguay postulated that a high-speed shutter could be used to gate out unwanted scattered light and detect light echoes from the internal structure of tissue (Duguay and Mattick, 1971). Thus, ultrahigh-speed optical gating could be used to "see through" tissues and noninvasively image pathology.

The Kerr shutter requires high-intensity laser pulses to induce the Kerr effect. An alternative approach for high-speed gating is to use nonlinear processes such as harmonic generation or parametric conversion (Bruckner, 1978; Fujimoto et al., 1986; Park et al., 1981). Short pulses are used to illuminate the object or specimen being imaged, and the backscattered light is upconverted or parametrically converted by mixing with a reference pulse in a nonlinear optical crystal. The reference pulse is obtained from the same laser source delayed by a variable time delay. Nonlinear optical gating can measure the time delay and intensity of a high-speed optical signal. The time resolution is determined by the pulse duration, and the sensitivity is determined by the conversion efficiency of the nonlinear process. Optical ranging measurements have been demonstrated in biological tissues using femtosecond pulses and nonlinear intensity autocorrelation to measure structures such as the eye and in skin with axial resolutions of 15 μm (Fujimoto et al., 1986). Sensitivities of 10^{-7} can be achieved; however, these sensitivities are insufficient to image biological tissues, which have strong optical attenuation due to scattering.

15.5 LOW COHERENCE INTERFEROMETRY AND HETERODYNE DETECTION

Interferometric detection provides a powerful approach for measuring echo time delay of backreflected or backscattered light with high dynamic range and high sensitivity. These techniques are analogous to coherent optical detection in optical communications. OCT is based on a classic optical measurement technique known as low coherence interferometry or white light interferometry, first described by Sir Isaac Newton. More recently, low coherence interferometry has been used to characterize optical echoes and backscattering in optical fibers and waveguide devices (Gilgen et al., 1989; Takada et al., 1987; Youngquist et al., 1987). The first biological application of low coherence interferometry was in ophthalmological biometry for measurement of eye length (Fercher et al., 1988). Since then, related versions of this technique have been developed for noninvasive high precision biometry (Clivaz et al., 1992b; Schmitt et al., 1993). High-resolution measurements of corneal thickness in vivo were demonstrated using standard low coherence interferometry (Huang et al., 1991b).

Low coherence interferometry measures the field of the optical beam rather than its intensity. The electric and magnetic fields of an electromagnetic wave oscillate in time and space and propagate with a characteristic velocity. In vacuum, the velocity of light is $c = 3 \times 10^8$ m/s, whereas in water, biological tissues, or materials, the velocity of propagation of light is reduced from its speed in vacuum according to the index refraction n of the medium, $v = c/n$. The functional form of the oscillating electric field in a light wave is

$$E_t(t) = E_i \cos\left(2\pi v t - \frac{2\pi}{\lambda z}\right) \tag{2}$$

When two beams of light are combined, their fields may add either constructively or destructively according to the relative phase of the oscillations. Figure 3 shows a schematic diagram of a simple Michelson interferometer. The incident optical wave into the interferometer is directed onto a partially reflecting mirror or beam splitter that splits the field into two beams, one of which acts as a reference beam and the other as a measurement or signal beam. The beams travel given distances in the two arms of the interferometer. The reference beam $E_r(t)$ is reflected from a reference mirror, and the measurement or signal beam $E_s(t)$ is reflected from the biological specimen or tissue that is to be imaged. The beams then recombine and interfere at the beam splitter. The output of the interferometer is the sum of the electromagnetic fields from the reference beam reflected from the reference mirror and the measurement or signal beam reflected from the specimen or tissue. The electric field at the output of the interferometer is

$$E_O(t) \sim E_r(t) + E_s(t) \tag{3}$$

The detector measures the intensity of the output optical beam, which is proportional to the square of the electromagnetic field. For the purposes of this simple analysis, assume that the reflected measurement consists of a single reflection at a given distance rather than multiple echoes. Then, if the distance that light travels in the reference path of the interferometer is ℓ_r and the length that light travels in the measurement path, reflected from the specimen, is ℓ_s, then the intensity of the output from the interferometer will oscillate as a function of the difference in distance $\Delta\ell = \ell_r - \ell_s$ because of interference effects. The intensity as a function of E_s, E_r, and $\Delta\ell$ is given as

$$I_O(t) \sim \frac{1}{4}|E_r|^2 + \frac{1}{4}|E_s|^2 + \frac{1}{2}E_rE_s \cos\left(2\frac{2\pi}{\lambda}\Delta\ell\right) \tag{4}$$

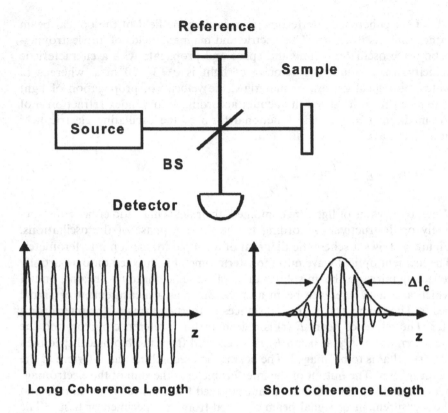

Figure 3 Optical coherence tomography measures the echo time delay of reflected light by using low coherence interferometry. The system is based on a Michelson-type interferometer. Reflections or backscattering from the object being imaged are correlated with light that travels a known reference path delay. The detection approach is sensitive to the field rather than intensity, analogous to optical hetero-dyne detection.

If the position of the reference mirror is varied, then the path length that the optical beam travels in the reference arm changes. Interference effects will be observed in the intensity output of the interferometer if the relative path lengths are changed by scanning the reference mirror as shown schematically in Figure 3. If the light is highly coherent (narrow linewidth) with a long coherence length, then interference fringes will be observed for a wide range of relative path lengths of the reference and measurement arms. However, for applications in optical ranging or optical coherence tomography, it is necessary to precisely measure absolute distance and dimensions of structures within the material or tissue. In this case, light with a short coherence length (broad bandwidth) is used. Low coherence light may be thought of as being composed of electromagnetic fields with statistical discontinuities in phase as a function of time. Alternatively, the light may be thought of as being composed of a superposition of different frequencies or wavelengths rather than a single wavelength. Low coherence light can be characterized as having statistical discontinuities in phase over a distance known as the coherence length (ℓ_c). The coherence length is a measure of the coherence and is inversely proportional to the frequency bandwidth.

When low coherence light is used as the source for the interferometer, interference is observed only when the path lengths of the reference and measurement arms are matched to within the coherence length of the light. If the path lengths differ by more than the coherence length, then the electromagnetic fields from the two beams are not correlated, and there is no interference. The interferometer effectively measures the field autocorrelation of the light. For the purposes of ranging and imaging, the coherence length of the light determines the resolution with which optical range or distance can be measured. The magnitude and echo time delays of the reflected light can be determined by scanning the reference mirror position and demodulating the interference signal from the interferometer.

15.6 RESOLUTION AND SENSITIVITY OF OPTICAL COHERENCE TOMOGRAPHY

In contrast to conventional microscopy, in OCT the mechanisms that govern the axial and transverse image resolution are independent. The axial resolution in OCT imaging is determined by the coherence length of the light source, so high axial resolution can be achieved independently of the beam focusing conditions. The coherence length is the spatial width of the field autocorrelation produced by the interferometer. The envelope of the field autocorrelation is equivalent to the Fourier transform of the power spectrum. Thus, the width of the autocorrelation function, or the axial resolution, is inversely proportional to the width of the power spectrum. For a source with a Gaussian spectral distribution, the axial resolution Δz is given as

$$\Delta z = \left(2 \ln \frac{2}{\pi} \right) \left(\frac{\lambda^2}{\Delta\lambda} \right) \tag{5}$$

where Δz and $\Delta\lambda$ are the full widths at half-maximum of the autocorrelation function and power spectrum, respectively, and λ is the source center wavelength. Because axial resolution is inversely proportional to the bandwidth of the light source, broad-bandwidth optical sources are required to achieve high axial resolution.

The transverse resolution for OCT imaging is the same as for conventional optical microscopy and is determined by the diffraction-limited focusing of an optical beam. The minimum spot size to which an optical beam may be focused is inversely proportional to the numerical aperture or the angle of focus of the beam. The transverse resolution is given as

$$\Delta x = \frac{4\lambda}{\pi} \left(\frac{f}{d}\right) \tag{6}$$

where d is the spot size on the objective lens and f is its focal length. High transverse resolution can be obtained by using a large numerical aperture and focusing the beam to a small spot size. In addition, the transverse resolution is also related to the depth of focus or the confocal parameter b, which is $2z_R$, two times the Raleigh range:

$$2z_R = \pi \, \Delta x^2 / 2\lambda \tag{7}$$

Thus, increasing the transverse resolution produces a decrease in the depth of focus, similar to that which occurs in conventional microscopy.

Figure 4 shows schematically the relationship between focused spot size and depth of field for low and high numerical aperture focusing (Fujimoto et al., 2000a). The focusing conditions define two limiting cases for performing OCT imaging. Typically, OCT imaging is performed with low numerical aperture focusing because it is desirable to have a large depth of field and use low coherence interferometry to achieve axial resolution. In this limit the confocal parameter is larger than the coherence length, $b > \Delta z$. The image resolution is determined by the coherence length in the axial dimension and the spot size in the transverse dimension. In contrast to conventional microscopy, this mode of operation achieves high axial resolution independently of the available numerical aperture. This feature is particularly powerful for applications such as retinal imaging in ophthalmological or catheter- or endoscope-based imaging, where the available numerical aperture may be limited. However, operation in the low numerical aperture limit necessarily means that the transverse resolution is limited.

Conversely, it is also possible to focus with high numerical aperture and achieve high transverse resolutions at the expense of reduced depth of focus. This operating regime is typical for conventional microscopy or confocal microscopy. Depending upon the coherence length of the light, the depth of field can be shorter than the coherence length, $b < \Delta z$. In this case

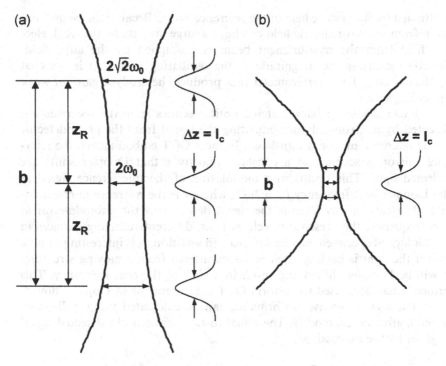

Figure 4 (a) Low and (b) high numerical aperture focusing limits of OCT. Most OCT imaging is performed with low NA focusing, where the confocal parameter is much longer than the coherence length. There is a trade-off between transverse resolution and depth of field. The high NA focusing limit achieves excellent transverse resolution with less depth of field. Low coherence detection provides more effective rejection of scattered light than confocal detection.

the depth of field can be used to differentiate backscattered or backreflected signals from different depths. This regime of operation has been referred to as optical coherence microscopy (OCM) (Izatt et al., 1994). This mode of operation can be useful for imaging scattering systems because the coherence gating effect removes the contributions from scattering in front and in back of the focal plane more effectively than confocal gating.

Optical coherence tomography can achieve high detection sensitivity because interferometry measures the field rather than the intensity of light using optical heterodyne detection. This effect can be seen qualitatively in Eq. (4), which describes the intensity of the signal at the output of the interferometer. The oscillating interference term is the result of the backscattered or backreflected electric field from the tissue (which can be very weak), multiplied by the electric field of the reference beam. Because the beam from the reference mirror electric field can have a large amplitude, the weak electric field from the measurement beam is multiplied by the large field, thereby increasing the magnitude of the oscillating term that is detected by the detector. The interferometer thus produces heterodyne gain for weak optical signals.

Backreflected or backscattered optical echoes from the specimen are detected by electronically demodulating the signal from the photodetector as the reference mirror is translated. In most OCT embodiments, the reference mirror is scanned at a constant velocity v that Doppler shifts the reflected light. This results in a modulation of the interference signal at the Doppler beat frequency $f_D = 2v/\lambda$, where v is the reference mirror velocity. By electronically filtering the signal detected by the photodetector at this frequency, the presence of echoes from different reflecting surfaces in the biological specimen may be detected. In addition, it is interesting to note that if the light is backreflected or backscattered from a moving structure, it will be Doppler shifted and result in a shift of the beat frequency. This principle has been used to perform OCT measurements of Doppler flow.

The signal-to-noise performance can be calculated using well-established mathematical models. The signal-to-noise ratio in the detected signal is given by the expression

$$\text{SNR} = 10\log\left(\frac{\eta P}{2h\nu}\text{NEB}\right) \tag{8}$$

where η is the detector quantum efficiency, $h\nu$ is the photon energy, P is the signal power, and NEB is the noise equivalent bandwidth of the electronic filter used to demodulate or detect the signal. This expression implies that the signal-to-noise ratio scales as the detected power divided by the noise equivalent bandwidth of the detection. This means that high image acquisition speeds or higher image resolutions require higher optical powers for a given signal-to-noise ratio. The performance of OCT systems varies widely according to their design and data acquisition speed requirements. However, for typical measurement parameters, sensitivities to reflected signals in the range of -90 to $-100\,\text{dB}$ can be achieved. This corresponds to the detection of backreflected or backscattered signals that are 10^{-9} or 10^{-10} of the incident optical power. This is sufficient to achieve imaging depths of 2–3 mm in most biological tissues.

Optical coherence tomography generates cross-sectional images by performing successive axial measurements of backreflected or backscattered light at different transverse positions. Figure 1 schematically illustrates one example of how optical coherence tomography is performed. A two-dimensional cross-sectional image is acquired by performing successive rapid axial measurements of optical backscattering or backreflection profiles at different transverse positions by scanning the incident optical beam. The result is a two-dimensional data set where each trace represents the magnitude of reflection or backscattering of the optical beam as a function of depth in the tissue.

A wide range of OCT scan patterns are possible, as shown in Figure 5. The most common method of OCT scanning acquires data with depth priority. However, it is also possible to acquire data with transverse priority, by detecting backreflections or backscattering at a given depth or range while transversely scanning the imaging beam. A cross-sectional image can be generated by detecting the backscattering along successive x scans for different z depths. It is also possible to perform OCT imaging in an *en face* plane (Podoleanu et al., 1997, 1998). In this case the backreflected or backscattered signals are detected at a fixed z depth, scanning along successive x and y directions. This mode of imaging is analogous to that used in confocal microscopy.

Optical coherence tomographic image data are usually acquired by computer and displayed as a two-dimensional gray scale or false color image. Figure 6 shows an example of a tomographic image of a leopard frog (*Rana pipiens*) tadpole displayed in false color scale. The vertical direction corresponds to the direction of the incident optical beam and the axial data sets. The optical beam was scanned in the transverse direction, and 200 axial measurements were performed. The backscattered signal ranges from approximately $-50\,dB$, the maximum signal, to $-100\,dB$, the sensitivity limit. Because the signal varies over five orders of magnitude, it is convenient to use the log of the signal to display the image. This expands the dynamic range of the display but results in compression of relative variations in signal.

5.7 OPTICAL COHERENCE TOMOGRAPHY SYSTEMS

Optical coherence tomography can be implemented using compact fiber-optic components and integrated with a wide range of medical instruments. Figure 7 shows a schematic of the modules in an OCT imaging system. The OCT system can be divided into the imaging engine, low coherence light source, optical delay scanner, beam delivery and probes, receiver, computer control, and image processing. The imaging engine is the heart of the OCT system. In general, the imaging engine can be any optical detection device that performs high-resolution and high-sensitivity ranging and detection of backreflected or backscattered optical echoes. Most OCT systems include

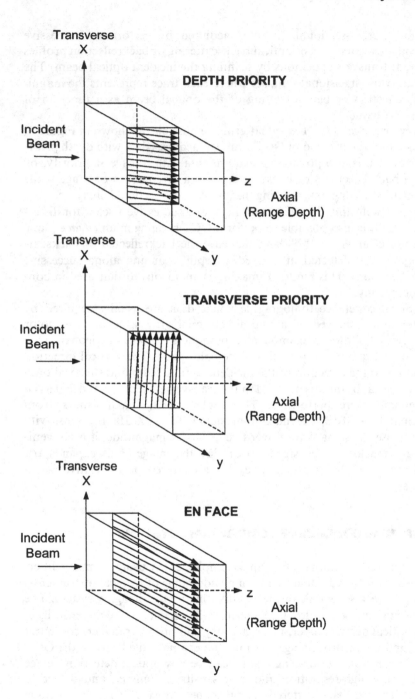

a reference delay scanning interferometer that employs a low coherence light source. There are many different embodiments of the interferometer and imaging engine for specific applications. Doppler flow imaging has been performed using imaging engines that detect the interferometric output rather than demodulating the interference fringes (Chen et al., 1997a, 1997b; Izatt et al., 1997; Kulkarni et al., 1998). Polarization-sensitive detection techniques have been demonstrated using a dual channel interferometer (de Boer et al., 1997; Everett et al., 1998; Hee et al., 1992). These techniques permit imaging of the birefringence properties of structures. Collagen and other tissues are strongly birefringent, and polarization-sensitive OCT can be a sensitive indicator of the tissues' microstructural organization. Conversely, polarization diversity interferometers can also be built using well-established polarization diversity detection methods from coherence heterodyne detection.

In the reference delay scanning embodiments of OCT systems, the optical delay scanner determines the image acquisition rate of the system. High-speed OCT imaging requires technology for high-speed optical delay scanning of the reference path length. The earliest scanning devices were constructed using galvanometers and retro reflectors (Hee et al., 1995). These systems can scan ranges of up to a centimeter or more and have the advantage of simplicity and ease of use. However, scan repetition rates are limited to approximately 100 Hz. A novel technique based on PZT scanning of optical fibers has been developed that can achieve scan ranges of a few millimeters with repetition rates of 500 Hz or more (Sergeev et al., 1997; Tearney et al., 1996b). Even higher speed scanning can be performed using an optical delay line based on a diffraction grating phase control device (Tearney et al., 1997a). This device is similar to pulse-shaping devices that are used in femtosecond optics and permits the phase and group velocity of the scanning to be independently controlled (Weiner et al., 1986). Scan rates of several thousand axial scans per second, corresponding to imaging rates of several frames per second, are possible (Tearney et al., 1997b).

Figure 5 Three scanning protocols for OCT imaging. There are numerous scanning protocols that can be used for OCT imaging depending upon the imaging engine and application. Cross-sectional images may be acquired using either depth or transverse priority scanning. In depth priority scanning, axial scans are acquired at successive transverse positions. In transverse priority scanning, the beam is scanned rapidly in the transverse direction, and light from successive axial depths (or ranges) is detected. Transverse priority scanning in two transverse dimensions can be used to perform *en face* imaging at a given depths.

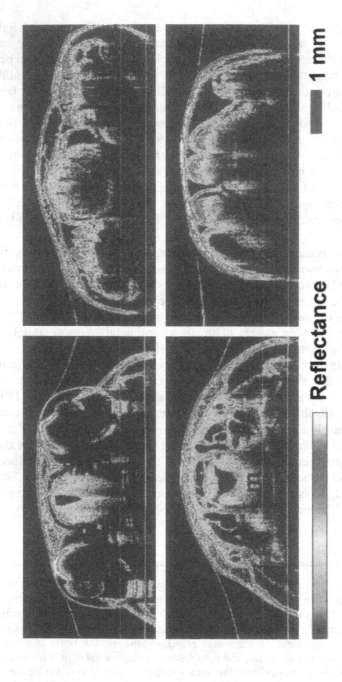

Figure 6 Optical coherence tomographic images of a *Rana pipiens* tadpole. Images are displayed in a false color, log scale. Scans were taken from the dorsal and ventral sides of the specimen. Image penetration depths of 2–3 mm are sufficient to perform whole-body imaging in these small specimens. Internal organ morphology can be visualized noninvasively and in real time. (From Boppart et al., 1996.)

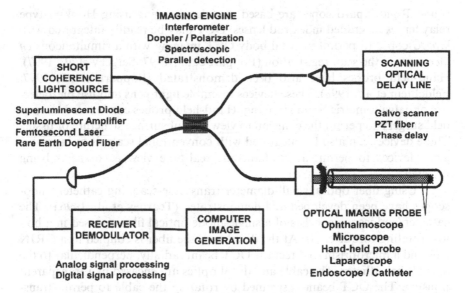

IMAGING ENGINE
Interferometer
Doppler / Polarization
Spectroscopic
Parallel detection

SHORT COHERENCE LIGHT SOURCE

Superluminescent Diode
Semiconductor Amplifier
Femtosecond Laser
Rare Earth Doped Fiber

SCANNING OPTICAL DELAY LINE

Galvo scanner
PZT fiber
Phase delay

RECEIVER DEMODULATOR

Analog signal processing
Digital signal processing

COMPUTER IMAGE GENERATION

OPTICAL IMAGING PROBE
Ophthalmoscope
Microscope
Hand-held probe
Laparoscope
Endoscope / Catheter

Figure 7 Optical coherence tomography technology. The major submodules in the OCT system include the imaging engine, light source, optical delay line scanner, image delivery probes, receiver, computer control, and image processing.

15.8 OCT IMAGING DEVICES

Because OCT imaging technology is based on fiber optics, it can be easily integrated with many standard optical diagnostic instruments including instruments for internal body imaging. OCT imaging has been integrated with a slit lamp and fundus camera for ophthalmic imaging of the retina (Hee et al., 1995; Puliafito et al., 1995). In these instruments, the OCT beam is scanned using a pair of galvanometric beam-steering mirrors and relay-imaged onto the retina. The ophthalmic OCT imaging instrument is a relatively complex design because it must allow the retina (fundus) to be viewed *en face* while showing the position of the tomographic scanning beam as well as the resulting OCT image. The ability to register OCT images with *en face* features and pathology is an important consideration for many applications. Similar principles apply for the design of OCT using low numerical aperture microscopes. Low numerical aperture microscopes have been used for imaging in vivo developmental biology specimens as well as for surgical imaging applications (Boppart et al., 1996, 1998, 1999).

Closely related beam delivery systems include forward imaging devices, which permit the delivery of a one- or two-dimensional scanned beam. Rigid laparoscopes are based on relay imaging using Hopkins-type relay lenses or graded index rod lenses. OCT may be readily integrated with laparoscopes to permit internal body OCT imaging with a simultaneous *en face* view of the scan registration (Boppart et al., 1997; Sergeev et al., 1997). Hand-held probes have also been demonstrated (Boppart et al., 1997; Feldchtein et al., 1998). These devices resemble light pens and use piezoelectric or galvanometric beam scanning. Hand-held probes can be used in open field surgery to permit the clinician to view the subsurface structure of tissue. These devices can also be integrated with conventional scalpels or laser surgical devices to permit a simultaneous, real-time view as tissue is being resected.

Using fiber optics, small-diameter transverse-scanning catheter/endoscopes have been developed and demonstrated (Tearney et al., 1996a). The catheter/endoscope consists of a single-mode optical fiber encased in a hollow rotating torque cable. At the distal end, the fiber is coupled to a GRIN lens and a microprism to direct the OCT beam radially, perpendicular to the axis of the catheter. The cable and distal optics are encased in a transparent housing. The OCT beam is scanned by rotating the cable to permit transluminal imaging in a radar-like pattern, cross-sectionally through vessels or hollow organs. Figure 8 shows a photograph of the prototype catheter. The catheter/endoscope has a diameter of 2.9 French or 1 mm, comparable to the size of a standard intravascular ultrasound catheter. This is small enough to allow imaging in a human coronary artery or imaging using the accessory port of a standard endoscope or bronchoscope.

The catheter/endoscope OCT system enables the acquisition of in vivo images of internal organ systems. In vivo imaging of the pulmonary, gastrointestinal, and urinary tracts as well as arterial imaging has been demonstrated in animals (Fujimoto et al., 1999; Tearney et al., 1997b, 1997c). Figure 9 shows an example of a catheter/endoscopic OCT image of the rabbit gastrointestinal tract. Imaging could be performed with either 256 or 512 lateral pixels, corresponding to image acquisition times of 125 ms or 250 ms, respectively. The two-dimensional image data were displayed using a polar coordinate transformation and inverse gray scale. OCT images of the in vivo rabbit esophagus permitted differentiation of the layers of the esophageal wall.

Endoscopic OCT imaging in humans can be performed by introducing OCT imaging probes into the accessory port of standard endoscopes. Figure 10 shows an example of endoscopic OCT imaging of the human esophagus (Li et al., 2000). Imaging was performed with 13 µm resolution at 1.3 µm wavelengths. These images show a representative linear scan OCT image, an endoscopic video image, and biopsy histology of normal squa-

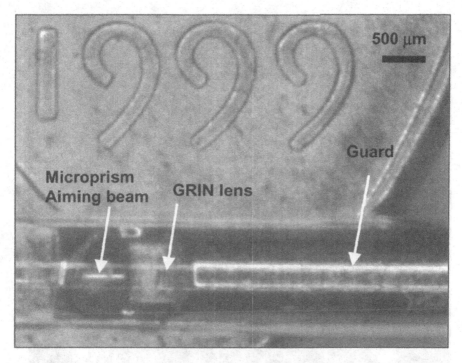

Figure 8 Photograph of prototype OCT catheter for transverse intraluminal imaging. A single-mode fiber lies within a rotating flexible speedometer cable enclosed in a protective plastic sheath. The distal end focuses the beam at 90° from the axis of the catheter. The diameter of the catheter is 2.9 French, or 1 mm.

mous epithelium compared to Barrett's esophagus, a metaplastic condition associated with increased cancer risk. The OCT image (4 mm × 2.5 mm, 512 × 256 pixels) of normal epithelium illustrates the epithelium, lamina propria, muscularis mucosa, submucosa, and muscularis propria. The OCT image of Barrett's pathology shows loss of normal structure and replacement of squamous epithelium with columnar epithelium and glandular structures characteristic of this pathology. These studies demonstrate the feasibility of performing OCT imaging of internal organ systems and suggest a range of future clinical applications. Endoscopic OCT imaging studies in patients have been reported (Bouma and Tearney, 1999; Li et al., 2000; Rollins et al., 1999; Sergeev et al., 1997). Numerous research groups are performing OCT imaging studies in patients for a wide range of clinical applications.

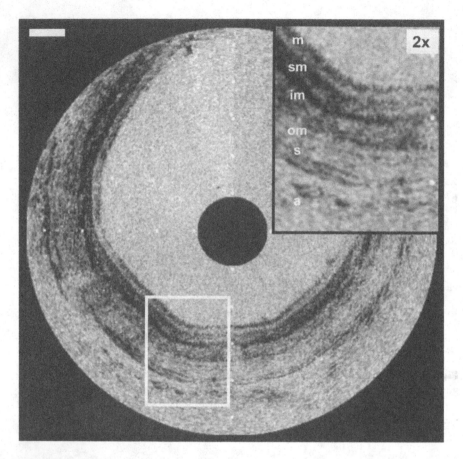

Figure 9 Optical coherence tomographic catheter/endoscope image in vivo of the esophagus of a New Zealand White rabbit. The image clearly differentiates the layers of the esophagus, including the mucosa, submucosa, inner muscularis, and outer muscularis. (From Tearney et al., 1997b.)

15.9   ULTRAHIGH RESOLUTION OPTICAL COHERENCE
TOMOGRAPHY USING FEMTOSECOND LASERS

The light source used for imaging determines the axial resolution of the OCT system. For research applications, short pulse lasers are powerful light sources for OCT imaging because they have extremely short coherence lengths and high output powers, enabling high-resolution, high-speed imaging. For clinical applications, compact superluminescent diodes or semiconductor-based light sources have been used extensively. Although these

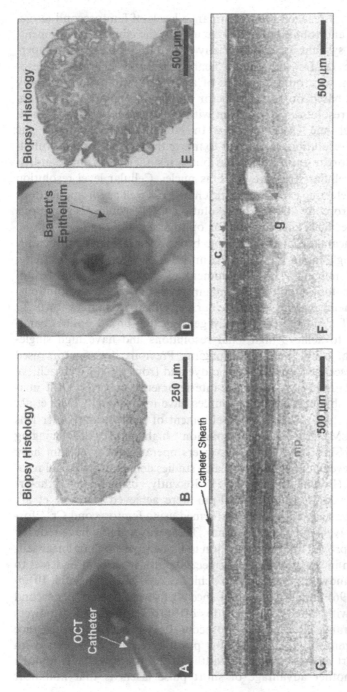

Figure 10 Clinical endoscopic OCT imaging of normal human esophagus using linear scanning. (A) Endoscopic image of normal region. (B) Biopsy histology of normal squamous epithelium. (C) OCT image of normal squamous epithelium with relatively uniform and distinct layered structures. (D) Endoscopic image of region showing pathology (finger-like projection of Barrett's epithelium). (E) Biopsy histology of Barrett's esophagus showing characteristic specialized columnar epithelium. (F) OCT image of Barrett's epithelium with disruptions of layered morphology due to multiple crypt- and gland-like structures (arrows). (From Li et al., 2000.)

sources have not yet reached the performance levels of femtosecond lasers, they are compact and robust enough to be used in the clinical environment. Ophthalmic OCT systems use commercially available compact GaAs super-luminescent diodes (SLDs) that operate near 800 nm and achieve axial reso-lutions of $\sim$10 μm with output powers of a few milliwatts. Commercially available sources based on semiconductor amplifiers operating at 1.3 μm can achieve axial resolutions of $\sim$15 μm with output powers of 15–20 mW, sufficient for real-time OCT imaging. In biomedical applications, the 10–15 μm level of resolution is sufficient to image the architectural morphol-ogy or glandular organization of tissues but is insufficient to image indivi-dual cells or subcellular structures such as nuclei. Cellular level resolution is important for detecting early neoplastic changes as well as for applications in biological microscopy. Ultrahigh-resolution OCT would also improve sensitivity and specificity of diagnosis for ophthalmic diseases.

Although incandescent sources have broad spectral bandwidths, they have very low single-mode power and cannot be used for high-sensitivity, high-speed OCT imaging. The development of high-performance femto-second laser technology enables OCT imaging at unprecedented resolutions. Ultrashort pulse femtosecond lasers are currently the best sources for high-performance OCT systems because they generate the broad optical band-widths necessary to achieve high axial resolutions and have high single-mode powers that enable real-time imaging. Previous investigators used broadband fluorescence from an organic dye and from $Ti:Al_2O_3$ to achieve resolutions of $\sim$2 μm in low- coherence interferometry; however, OCT ima-ging was not possible because these sources have low power (Clivaz et al., 1992a; Liu et al., 1993). With the development of femtosecond Kerr lens modelocked (KLM) lasers, high-resolution, high-speed OCT imaging became possible. Short pulse $Ti:Al_2O_3$ lasers operating near 800 nm have been used to achieve high resolutions. Early studies demonstrated axial reso-lutions of 4 μm (Bouma et al., 1995). Recently, using state-of-the-art $Ti:Al_2O_3$ lasers, axial resolutions of 1 μm were achieved (Drexler et al., 1999). Many studies have also been performed using femtosecond Cr^{4+}:for-sterite lasers at wavelengths near 1300 nm. These wavelengths have reduced scattering and improved image penetration in tissue. By using nonlinear self-phase modulation in optical fibers, the spectral output can be broadened to produce broad bandwidths sufficient to achieve axial resolutions of 5–10 μm (Bouma et al., 1996). Image acquisition speeds of several frames per second can be achieved with signal-to-noise ratios of 100 dB using incident powers in the 5–10 mW range. Recently, femtosecond Cr^{4+}:forsterite performance has improved significantly, and it is now possible to generate pulses, with durations as short as 14 fs and bandwidths of 250 nm directly (Chudoba et al., 2001). Another advantage of short pulse lasers is that nonlinear optical processes such as self-phase modulation, continuum generation, or Raman generation can be used to broaden optical bandwidths. Recently, image resolutions of 3 μm at a wavelength of 1200 nm were achieved using a continuum generation in a photonic crystal fiber starting with 100 fs pulses from a commercial Ti:Al$_2$O$_3$ laser (Hartl et al., 2001).

15.9.1   Ultrahigh-Resolution Imaging Using Ti:Al$_2$O$_3$ Lasers

Femtosecond Ti:Al$_2$O$_3$ lasers can now directly generate pulse durations of ~5 fs corresponding to only two optical cycles and bandwidths of up to one octave centered at 800 nm (Ell et al., 2001; Morgner et al., 1999; Sutter et al., 1999). These high-performance lasers have been made possible through the development of double-chirped mirror technology, which enables extremely broad bandwidths by linearizing the phase response (compensating higher order dispersion) in the laser. OCT systems must also be optimized to accommodate these ultrabroad bandwidths. However, unlike ultrafast femtosecond time-resolved measurements where special care must be exercised to maintain the short pulse duration, OCT measurements depend on field correlations rather than intensity correlations. Dispersion in the reference and signal paths of the interferometer must be precisely matched but need not be equal to zero. Thus, OCT systems can be implemented fiber-optically.

The axial resolution of OCT is determined by the coherence length of the light source. The interferometric signal detected by OCT is the electric field autocorrelation of the light source and is related to the Fourier transform of the power spectrum of the light source. Figure 11 shows the free-space resolution for a given optical bandwidth at a center wavelength of 800 nm as predicted by Eq. (5). A resolution of approximately 1.5 μm (or 1 μm in tissue) requires the use of a light source with ~260 nm optical bandwidth. Figure 12 shows a schematic of an ultrahigh-resolution OCT system using a KLM Ti:Al$_2$O$_3$ laser. The system is dispersion balanced and optimized for ultrabroad bandwidths. The KLM Ti:Al$_2$O$_3$ laser emits pulses as short as 5.5 fs corresponding to bandwidths of up to ~300 nm centered at 800 nm with an average power of 150 mW. The output spectrum can be shaped using a prismatic filter. The shape of the axial resolution function depends on the Fourier transform of the spectrum, so sharp edges or modulation on the spectrum give rise to wings on the axial point spread function. Figure 13 shows a comparison of the optical bandwidth and interferometric output traces determining the axial resolution of a conventional superluminescent diode (SLD) and the KLM Ti:Al$_2$O$_3$ laser (Drexler et al., 1999). The axial image resolution can be improved 5–10-fold compared to superluminescent diodes by using femtosecond lasers.

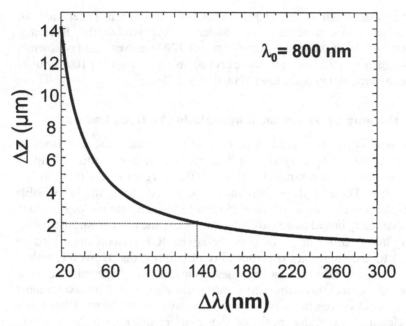

Figure 11 Plot of OCT axial resolution as a function of optical bandwidth for a center wavelength of 800 nm. Micrometer-scale axial resolution requires extremely broad optical bandwidths of a few hundred nanometers.

Group velocity dispersion causes pulse broadening in femtosecond optics, but in low coherence interferometry total dispersion does not affect the point spread function. Instead, the mismatch of dispersion between the reference and sample arms of the interferometer produces broadening of the resolution. The effect of dispersion mismatching and broadening will become more severe as the optical bandwidth and thickness of the sample are increased. In addition to degrading resolution, dispersion mismatch between the interferometer arms also degrades the peak height of the interference envelope, which reduces the system dynamic range. As shown in Figure 12, dispersion introduced by the fiber length and optics mismatch between the sample and the reference arm is balanced by using fused silica (FS) and BK7 prisms with their faces in contact. Dispersion balancing is achieved in real time by performing a fast Fourier transform analysis of the interference signal to extract the phase.

Figure 14 shows an example of an ultrahigh-resolution OCT image of human skin (Fujimoto et al., 2000a). The axial resolution is 2 μm, and the transverse resolution is 6 μm. The image dimensions are 1.1 mm × 1.8 mm,

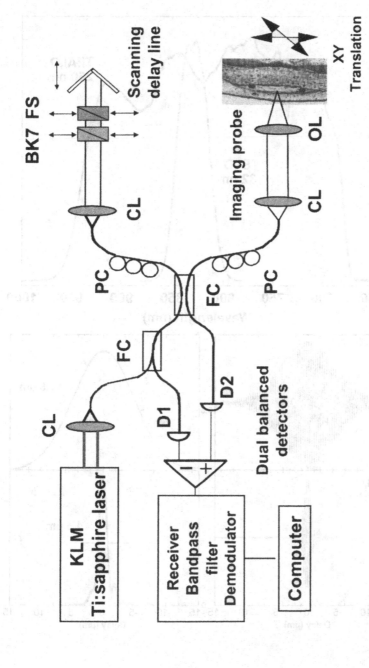

Figure 12 Ultrahigh-resolution fiber OCT system using a femtosecond Ti:Al$_2$O$_3$ laser. The interferometer is dispersion balanced and optically and electronically optimized to support broad optical bandwidths. (From Drexler et al., 1999.)

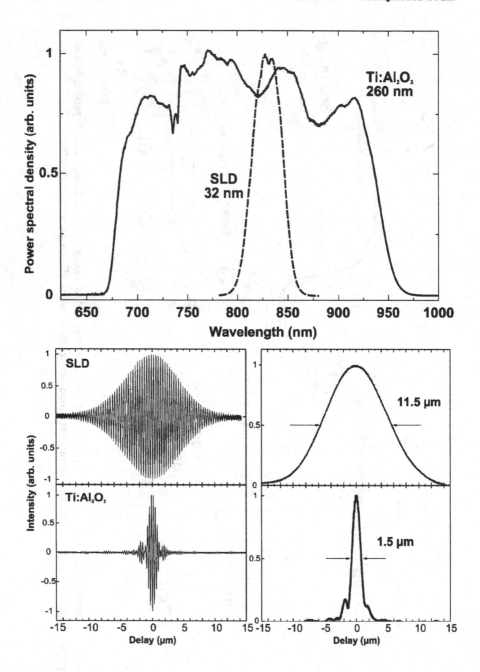

corresponding to 1100×600 pixels. Skin imaging could have important applications for the cosmetics industry. Because the axial and transverse resolutions are extremely fine, ultrahigh-resolution images can have large pixel densities. Ultrahigh-resolution images can be beyond the resolution of a standard computer monitor and must be viewed with panning and zooming.

The minimum achievable transverse resolution is given by the smallest achievable spot size on the sample. An additional constraint for high-resolution images might be the depth of focus of the beam, which is given by the confocal parameter. High numerical aperture focusing yields small spot sizes at the expense of reduced depth of field. Depth-of-field limitations may be overcome by borrowing a novel technique from ultrasound known as C mode imaging. Multiple images are acquired with the focusing set to different depths within the specimen. Each image is in focus over a depth range comparable to the confocal parameter. The in-focus regions from each of the images are selected and fused together to form a single image, which has a greatly extended depth of field. Figure 15 shows an example of C mode imaging of an African frog (*Xenopus laevis*) tadpole (Drexler et al., 1999). Imaging was performed with $\sim$1 μm axial resolution and 3 μm transverse resolution. The fused image covers an area of $0.75\,\text{mm} \times 0.5\,\text{mm}$ and consists of 1700×1000 pixels. Because of the small focal spot size, the confocal parameter was only 40 μm; however, the use of image fusion of eight images enabled imaging over a depth of 750 μm. These OCT images show in vivo subcellular features. Tissue morphology, including the neural olfactory tract and pleomorphic mesenchymal cells, can be visualized. OCT can image nuclear and intracellular morphology as well as mitotic activity. In developmental biology and genetics, imaging can be very powerful because it provides a way to track development and genetic expression using repeated measurements on a single specimen, without the need to sacrifice multiple specimens as in conventional histology.

Figure 16 shows a comparison of conventional (top) and ultrahigh-resolution (bottom) OCT images of the retina along the papillomacular axis (spanning the fovea and the optic disc) of a normal human subject (Drexler et al., 2001). The axial resolution is $\sim$3 μm and is limited by the chromatic aberration of the eye. The axial dimension of the OCT

Figure 13 Optical output spectrum (top) and interference signals and envelopes (bottom) of a femtosecond Ti:Al$_2$O$_3$ laser versus those of a superluminescent diode (SLD). An optical bandwidth of 260 nm enables a free-space resolution of 1.5 μm compared to 32 nm bandwidths and 11.5 μm resolutions using an SLD. (From Drexler et al., 1999.)

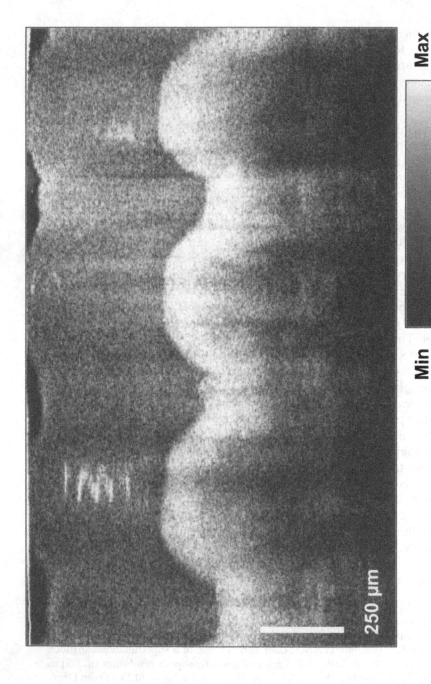

Figure 14   Ultrahigh-resolution OCT image of human skin. This image has an axial resolution of 2 μm and transverse resolution of 6 μm. The image dimensions are 1.1 mm × 1.8 mm, corresponding to 1100 × 600 pixels. Note the presence of the sweat gland and spiral duct. (From Fujimoto et al., 2000a.)

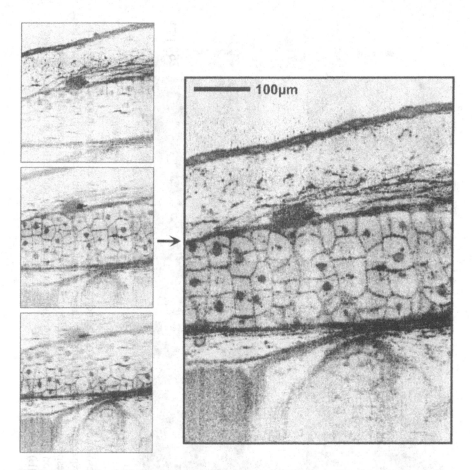

Figure 15 Ultrahigh-resolution cellular imaging. In vivo subcellular level resolution 1 μm × 3 μm (axial × transverse) 1700 × 1000 pixel image of an African frog (*Xenopus laevis*) tadpole. Images were recorded with different depths of focus and fused to construct the image shown. Multiple mesenchymal cells of various sizes and nuclear-to-cytoplasmic ratios and intracellular morphology and the mitosis of cell pairs are clearly shown. The bar is 100 μm. (From Drexler et al., 1999.)

image is normalized by the index of refraction 1.36 to convert optical delay into geometrical distances. The transverse resolution is governed by the numerical aperture of the pupil of the eye and is ~15 μm. Measurements were performed with 250–300 μW incident optical power, well within the ANSI standard for safe retinal exposure. In order to reduce the intensities associated with the short pulses generated by the Ti:Al₂O₃ laser, the

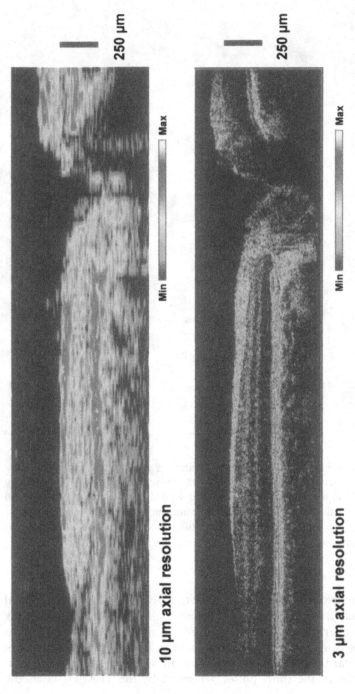

10 μm axial resolution

250 μm

Max

Min

3 μm axial resolution

250 μm

Max

Min

Figure 16 Conventional and ultrahigh-resolution OCT images of the normal human retina. The axial resolution is 10 μm (top) and 3 μm (bottom). The image has been expanded by a factor of 2 in the axial direction. The images have 100 × 180 (transverse × axial) and 600 × 725 pixel sampling, respectively. The nerve fiber layer is well differentiated and varies in thickness between the fovea and optic disc. Ultrahigh-resolution OCT enables the architectural morphology of the retina to be imaged noninvasively and in real time. (From Drexler et al., 2001.)

laser output was coupled into a 100 m length of standard optical fiber in order to dispersively broaden the pulse duration to several hundred picoseconds. Because of the high 90 MHz repetition rate, the light source can be treated as a continuous wave. Ultrahigh-resolution OCT enables visualization of the foveal and optic disc contours as well as internal architectural morphology of the retina and choroid that is not resolvable with conventional resolution OCT. The retinal nerve fiber layer (NFL) is clearly differentiated, and the variation of its thickness toward the optic disc can be observed. This figure shows that ultrahigh resolution provides significantly better image quality with the ability to resolve internal retinal architectural morphology that is difficult to observe at lower resolutions.

15.9.2 Ultrahigh-Resolution Imaging Using Nonlinear Fiber Light Sources

Although high-performance femtosecond lasers are powerful sources for optical coherence tomographic imaging, achieving broad-bandwidth performance requires double-chirped mirror technology. Recently, high nonlinearity photonic crystal, air–silica microstructure fibers as well as tapered fibers have emerged as powerful techniques for dramatically increasing nonlinearities and managing dispersion (Birks et al., 2000; Ranka et al., 2000). These fibers can generate an extremely broadband continuum using nonlinear self-phase modulation produced by standard duration femtosecond pulses. These fibers achieve enhanced nonlinearities by using air spaces adjacent to the guided mode that reduce dispersive pulse broadening by shifting the zero dispersion wavelength to shorter wavelengths. The small mode diameters also increase the effective intensity in the fiber core. Dramatic nonlinear effects can be observed in the fiber, and broadband continua extending from the visible to the near infrared, wavelengths of 390–1600 nm, may be generated.

Photonic crystal and tapered fibers enable ultrahigh-resolution OCT to be performed in a wide range of new wavelength regimes (Hartl et al., 2001). Axial image resolutions of 2.5 µm at a 1.2 µm center wavelength were recently demonstrated. Figure 17 shows a schematic of the experimental setup. A broadband continuum was generated by using 100 fs pulses from a commercially available Kerr lens modelocked $Ti:Al_2O_3$ laser (Coherent Mira) coupled into a 1 m length of microstructure fiber. The pulse energy in the fiber was only ~ 2 nJ, corresponding to an average power of 150 mW at a 75 MHz repetition rate. The dispersion of both interferometer arms was balanced by using identical materials in the two arms. Dual detection was used to cancel excess noise in the light source.

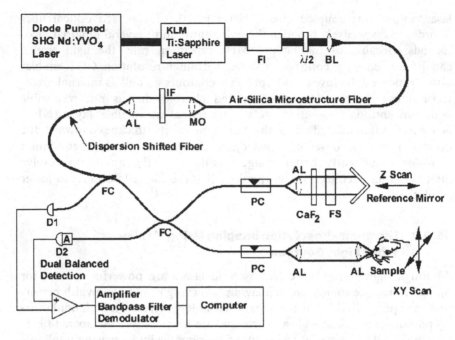

Figure 17 Ultrahigh-resolution OCT system using continuum generation in a non-linear fiber as the light source. Dual balanced detection is used to reduce excess noise. The interferometer is dispersion balanced and optically and electronically optimized to support broad optical bandwidths. (From Hartl et al., 2001.)

The spectrum of the light source at the entrance port of the second fiber coupler was measured by an optical spectrum analyzer and had a bandwidth of 450 nm centered at 1.2 µm as shown in Figure 18. The interferometric signal together with its demodulated envelope is also shown. The axial resolution was 2.5 µm in free space, corresponding to ~2 µm in tissue. This is the highest longitudinal OCT resolution demonstrated to date in this wavelength range (Hartl et al., 2001). The spectral bandwidth of the system was measured to be 370 nm by calculating the detected signal bandwidth using the Fourier transform of the interferometric signal. This reduction of the bandwidth occurs on the short-wavelength part of the spectrum and is probably caused by the spectral limitations of the InGAs photodiodes. Thus even higher resolutions should be possible if the entire bandwidth can be detected.

Imaging at longer wavelengths in the 1.2 µm wavelength range is important because these wavelengths have less scattering and better image

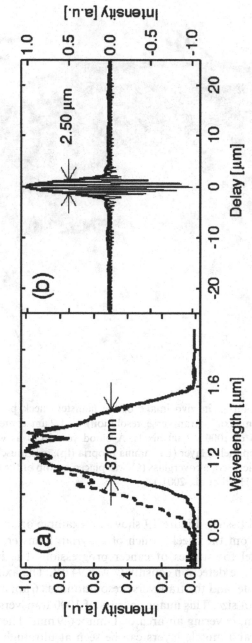

Figure 18 The bandwidth of the fiber continuum light source after spectral filtering and coupling to a single-mode fiber was 450 nm at the entrance port of the second fiber coupler. Interference fringes and demodulated output from the interferometer demonstrate an axial resolution of 2.5 μm in air. The overall system bandwidth was 370 nm as measured by Fourier transforming the interferometric signal. (From Hartl et al., 2001.)

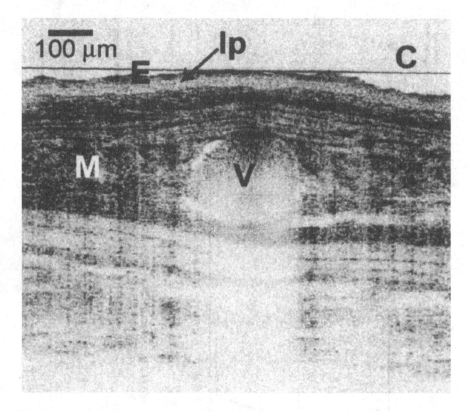

Figure 19 Ultrahigh-resolution in vivo image of the hamster cheek pouch performed with ~2 μm × 6 μm (axial × transverse resolution) at 1.2 μm center wavelength. The image consists of 1000 × 2700 pixels. A blood vessel (V) as well as a layered structure showing epithelial layer (E), lamina propria (lp), and muscle layers (M) are clearly visible. A microscope coverglass (C) was placed on top of the tissue to reduce aberrations. (From Hartl et al., 2001.)

penetration in biological tissues. Figure 19 shows an example of an in vivo image of oral mucosa from the cheek pouch of a Syrian hamster, a well-established animal model for studies of cancer progression. The incident power was 0.5 mW, and the detection sensitivity was 94 dB. The axial resolution was ~2 μm in tissue, and the transverse resolution was 6 μm as determined by the focused spot size. This image consists of 1000 transverse scans with 2700 pixels per scan, covering an area of 1 mm × 0.9 mm. The epithelium, connective tissue, and muscle layers can be seen at ultrahigh resolution. A blood vessel is also visible along with shadowing beneath the vessel arising from optical scattering in blood.

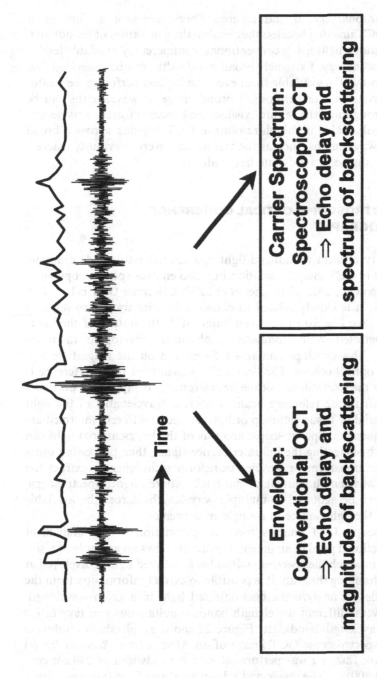

Figure 20 Amplitude versus spectroscopic imaging. Conventional OCT detects the envelope of the interferometric output that corresponds to the field envelope of the backscattered light. Spectroscopically resolved OCT can be performed by detecting and signal processing the interferometric waveform. Spectroscopic information is mapped into the interference fringe frequency. Spectroscopic OCT is analogous to Fourier transform infrared spectroscopy. Because the full fringe data are recorded, image sizes can exceed several tens of megabytes.

(in figure)

Time

Envelope:
Conventional OCT
⇒ Echo delay and
magnitude of backscattering

Carrier Spectrum:
Spectroscopic OCT
⇒ Echo delay and
spectrum of backscattering

Nonlinear photonic crystal and tapered fibers represent an important advance for OCT imaging because they enable the generation of broadband, single spatial mode light using conventional, commercially available femtosecond laser technology. Extremely broad bandwidths can be generated that exceed the bandwidths available from even the highest performance femtosecond laser systems. Furthermore, a broad range of wavelengths can be generated throughout the entire visible and near-infrared wavelength ranges. This will enable ultrahigh-resolution OCT imaging across a broad range of new wavelengths, such as the visible, that were previously inaccessible using femtosecond solid-state lasers alone.

15.10 SPECTROSCOPIC OPTICAL COHERENCE TOMOGRAPHY

The availability of ultra broadband light sources not only yields dramatic improvements in OCT image resolution but also enables spectroscopic imaging (Fujimoto et al., 2000a; Morgner et al., 2000). Because OCT is based on interferometry, it is closely related to classical Fourier transform infrared spectroscopy. (See Fig. 20.) In conventional OCT, the output of the interferometer is detected and demodulated to obtain the envelope of the interference signal. This envelope contains information on the magnitude and time delay of optical echoes. The "carrier" frequency of the interference is determined by the scan velocity of the interferometer delay, which produces a Doppler shift in the reference beam. Different wavelengths of the light from the material or tissue will map or heterodyne to different intermediate or carrier frequencies. Spectroscopic analysis of the backscattered light can be performed by detecting the full interference signal, then processing using digital Fourier transform or wavelet transform techniques to extract the frequency or wavelength content of the backscattered light. Spectroscopic information can be acquired at multiple wavelengths across the available bandwidth of the light source in a single measurement.

Spectroscopic OCT enables spectral information to be obtained at every cross-sectional pixel in an image. Figure 21 shows an example of a single axial scan in which the spectrum of the backscattered light is displayed at different depths along the scan. It is possible to extract information from the spectra in order to analyze the backreflected light in a given wavelength band or in several different wavelength bands simultaneously or take ratios of different wavelength bands, etc. Figure 22 shows an ultrahigh resolution image and a spectroscopic OCT image of an African frog (*Xenopus laevis*) tadpole in vivo. Imaging was performed with a bandwidth of 230 nm centered at about 800 nm. The image had a 1 μm axial and 5 μm transverse reso-

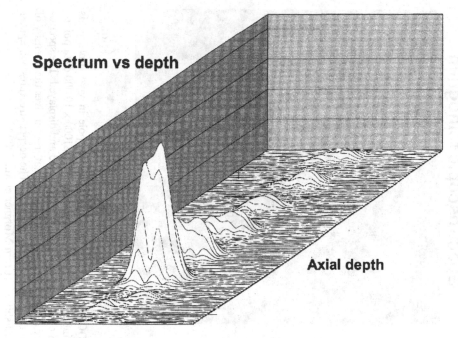

Figure 21 Spectroscopic OCT. Using a windowed Fourier transform or wavelet transform, the spectrum of the backscattered light can be determined. Spectroscopic OCT requires broadband light sources but has the advantage that the spectroscopic behavior at all wavelengths in the available bandwidth is characterized in a single measurement. Spectroscopic information is available at each axial and transverse pixel in an image. (From Fujimoto et al., 2000a.)

lution and covered an area of ~0.54 mm × 0.5 mm. The interferometric image data consist of 1000 × 10,000 data points. In this image, the spectral center of gravity of the backscattered light was extracted and displayed. Both the magnitude of the backscattered signal and the spectral center of gravity will vary, so a two-dimensional mapping is required to display the image. The spectroscopic OCT image is displayed in the Hue Saturation Luminance (HSL) color space so that the saturation gives the intensity of the signal, the hue gives the center of gravity of the spectrum, and the luminance is constant. The red hue in the spectroscopic OCT image indicates that the center of gravity of the backscattered spectrum is shifted to longer wavelengths, and a green hue indicates a shift to shorter wavelengths. The spectroscopic OCT image clearly shows that longer wavelengths of light

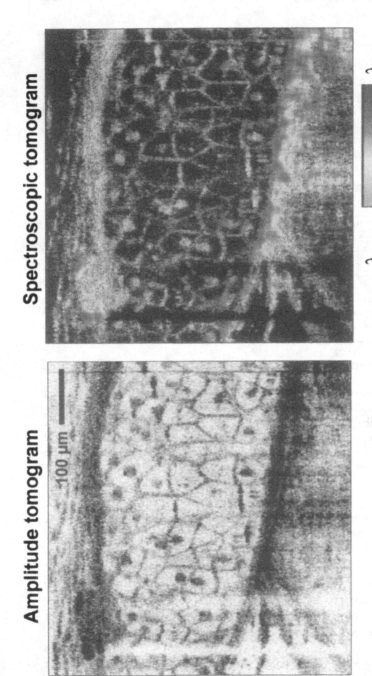

Figure 22 (a) Ultrahigh-resolution and (b) spectroscopic OCT images of a *Xenopus laevis* tadpole in vivo. Imaging is performed with 1 μm axial and 5 μm transverse resolution. The interferometric image data consist of 1000 × 10,000 data points. In the spectroscopic image (b), the spectral center of gravity of the backscattered light was extracted and displayed. The spectroscopic OCT image is displayed in the Hue Saturation Luminance (HSL) color space so that the saturation gives the intensity of the signal, the hue gives the center of gravity of the spectrum, and the luminance is constant. The deeper structures will appear with a red hue, indicating that the backscattered spectrum is red shifted. (From Morgner et al., 2000.)

penetrate deeper into tissue, with less attenuation from scattering, as expected. The melanocytes (which contain melanin) appear bright red in the spectroscopic OCT image, and the pigment layer below the cell layer is visible as a layer of red scatters. Some structures, such as the melanocytes in the upper layers, are better differentiated in the spectroscopic OCT image than in the conventional OCT image. Spectroscopic imaging promises to improve the ability of OCT to differentiate tissue types as well as to enable functional imaging.

15.11 SUMMARY

Optical coherence tomography has a number of features that make it attractive for a broad range of applications. First, imaging can be performed in situ and nondestructively. Image resolutions of 10–15 μm are possible using standard technology, and ultrahigh resolutions of ~1 μm may be achieved using state-of-the-art systems. High speed real-time imaging is possible with acquisition rates of several frames per second. OCT technology is based on fiber optics and draws upon components developed for the telecommunications industry. Image information is generated in electronic form and can be processed as well as transmitted via network.

Optical coherence tomography is a powerful imaging technology in medicine because it enables the real-time in situ visualization of tissue microstructure without the need to excise and process a specimen as in conventional biopsy and histopathology. Nonexcisional "optical biopsy" and the ability to visualize tissue morphology in real time under operator guidance can be used both for diagnostic imaging and to guide surgical intervention. In tissues other than the eye, optical scattering limits image penetration depths to 2–3 mm. However, because OCT is an optical technology, it can be interfaced to a wide range of instruments such as endoscopes, catheters, or laparoscopes, which enables the imaging of internal organ systems. OCT promises to have a powerful impact on many medical applications ranging from the screening and diagnosis of neoplasia to enabling new microsurgical and minimally invasive surgical procedures.

Optical coherence tomographic imaging with ~1 μm axial resolution is now possible using ultrabroad-bandwidth femtosecond laser technology. High-resolution OCT has important implications for early cancer diagnosis and for applications such as ophthalmic imaging. The broad bandwidths available from short pulse light sources also enable spectroscopic OCT imaging. Whereas conventional OCT only detects the envelope of the interference pattern, spectroscopic OCT is similar to Fourier transform spectroscopy, detecting and processing the full interference pattern. This permits the spectrum as well as the time delay and magnitude of optical echoes to be characterized, enabling spectroscopic cross-sectional imaging with micrometer scale resolution. Spectroscopic OCT may be used analogously to staining in histopathology, to enhance image contrast and provide better differentiation of tissue pathologies. With further development, spectroscopic techniques hold the promise of enabling functional imaging such as cross-sectional mapping of tissue oxygenation with micrometer level resolution. The ultimate availability of this ultrahigh-resolution and spectroscopic OCT technology for future clinical applications will depend on the development of low-cost compact femtosecond lasers or other ultra-broad-bandwidth light sources.

ACKNOWLEDGMENTS

During the work associated with this chapter, W. Drexler was visiting from the University of Vienna and U. Morgner and F. Kärtner were visiting from the University of Karlsruhe. Dr. Mark Brezinski is also with the Brigham and Women's Hospital. We gratefully acknowledge the contributions of Steve Boppart, Christian Chudoba, Ravi Ghanta, Pei-Lin Hsiung, Tony Ko, and Costas Pitris. We thank J. Ranka and R. Windeler from Lucent Technologies and T. A. Birks and W. J. Wadsworth from the University of Bath for their collaboration on photonic crystal and tapered fibers. W. Drexler acknowledges support from Max Kade Foundation and the Österreichische Akademie der Wissenschaften. I. Hartl acknowledges support from the BASF Aktiengesellschaft and the German National Scholarship Foundation. U. Morgner and F. Kärtner acknowledge support from the Deutsche Forschungsgemeinschaft. This research was supported in part by the U.S. Air Force Office of Scientific Research, contract F4920-98-1-0139; the Office of Naval Research Medical Free Electron Laser Program, contract N000014-97-1-1066; and the National Institutes of Health, contracts NIH-9-RO1-CA75289-04 and NIH-9-RO1-EY11289-15.

REFERENCES

Birks TA, Wadsworth WJ, Russel PSJ. (2000). Supercontinuum generation in tapered fibers. Opt Lett 25:1415–1417.
Boppart SA, Brezinski ME, Bouma BE, Tearney GJ, Fujimoto JG. (1996). Investigation of developing embryonic morphology using optical coherence tomography. Dev Biol 177:54–63.

Boppart SA, Bouma BE, Pitris C, Tearney GJ, Fujimoto JG, Brezinski ME. (1997). Forward-imaging instruments for optical coherence tomography. Opt Lett 22:1618–1620.

Boppart SA, Bouma BE, Pitris C, Southern JF, Brezinski ME, Fujimoto JG. (1998). Intraoperative assessment of microsurgery with three-dimensional optical coherence tomography. Radiology 208:81–86.

Boppart SA, Herrmann J, Pitris C, Stamper DL, Brezinski ME, Fujimoto JG. (1999). High-resolution optical coherence tomography-guided laser ablation of surgical tissue. J Surg Res 82:275–284.

Bouma BE, Tearney GJ. (1999). Power-efficient nonreciprocal interferometer and linear-scanning fiber-optic catheter for optical coherence tomography. Opt Lett 24:531–533.

Bouma B, Tearney GJ, Boppart SA, Hee MR, Brezinski ME, Fujimoto JG. (1995). High-resolution optical coherence tomographic imaging using a mode-locked Ti-Al$_2$O$_3$ laser source. Opt Lett 20:1486–1488.

Bouma BE, Tearney GJ, Bilinsky IP, Golubovic B. (1996). Self phase modulated Kerr-lens mode locked Cr:forsterite laser source for optical coherence tomography. Opt Lett 21:1839–1841.

Brezinski ME, Fujimoto JG. (1999). Optical coherence tomography: high resolution imaging in nontransparent tissue. IEEE J Selected Topics Quantum Electron 1185–92.

Brezinski ME, Tearney GJ, Bouma BE, Izatt JA, Hee MR, Swanson EA, Southern JF, Fujimoto JG. (1996). Optical coherence tomography for optical biopsy: properties and demonstration of vascular pathology. Circulation 93:1206–1213.

Bruckner AP. (1978). Picosecond light scattering measurements of cataract microstructure. Appl Opt 17:3177–3183.

Chen Z, Milner TE, Dave D, Nelson JS. (1997a). Optical Doppler tomographic imaging of fluid flow velocity in highly scattering media. Opt Lett 22:64–66.

Chen Z, Milner TE, Srinivas S, Wang X, Malekafzali A, van Gemert MJC, Nelson JS. (1997b). Noninvasive imaging of in vivo blood flow velocity using optical Doppler tomography. Opt Lett 22:1119–1121.

Chudoba C, Fujimoto JG, Ippen EP, Haus HA, Morgner U, Kärtner FX, Scheuer V, Angelow G, Tschudi T. (2001). All-solid-state Cr:forsterite laser generating 14 fs pulses at 1.3 µm. Opt Lett 26:292–294.

Clivaz X, Marquis-Weible F, Salathe RP. (1992a). Optical low coherence reflectometry with 1.9 µm spatial resolution. Electron Lett 28:1553–1554.

Clivaz X, Marquis-Weible F, Salathe RP, Novak RP, Gilgen HH. (1992b). High-resolution reflectometry in biological tissues. Opt Lett 17:4–6.

de Boer JF, Milner TE, van Gemert MJC, Nelson JS. (1997). Two-dimensional birefringence imaging in biological tissue by polarization-sensitive optical coherence tomography. Opt Lett 22:934–936.

Drexler W, Morgner U, Kaertner FX, Pitris C, Boppart SA, Li XD, Ippen EP, Fujimoto JG. (1999). In vivo ultrahigh resolution optical coherence tomography. Opt Lett 24:1221–1223.

Drexler W, Morgner U, Ghanta RK, Kärtner FX, Schuman JS, Fujimoto JG. (2001). Ultrahigh resolution ophthalmic optical coherence tomography. Nature Med 7:502–507.

Duguay MA. (1971). Light photographed in flight. Am Sci 59:551–556.

Duguay MA, Mattick AT. (1971). Ultrahigh speed photography of picosecond light pulses and echoes. Appl Opt 10:2162–2170.

Ell R, Morgner U, Kärtner FX, Fujimoto JG, Ippen EP, Scheuer V, Angelow G, Tschudi T. (2001). Generation of 5 fs pulses and octave-spanning spectra directly from a Ti:sapphire laser. Opt Lett 26:373–375.

Everett MJ, Schoenenberger K, Colston BW, Da Silva LB. (1998). Birefringence characterization of biological tissue by use of optical coherence tomography. Opt Lett 23:228–230.

Feldchtein FI, Gelikonov GV, Gelikonov VM, Iksanov RR, Kuranov RV, Sergeev AM, Gladkova ND, Ourutina MN, Warren JA, Reitze DH. (1998). In vivo OCT imaging of hard and soft tissue of the oral cavity. Opt Express 3:239–250.

Fercher AF, Hitzenberger CK, Drexler W, Kamp G, Sattmann H. (1993). In vivo optical coherence tomography. Am J Ophthalmol 116:113–114.

Fercher AF, Mengedoht K, Werner W. (1988). Eye-length measurement by interferometry with partially coherent light. Opt Lett 13:1867–1869.

Fish P. (1990). Physics and Instrumentation of Diagnostic Medical Ultrasound. New York: Wiley.

Fujimoto JG, De Silvestri S, Ippen EP, Puliafito CA, Margolis R, Oseroff A. (1986). Femtosecond optical ranging in biological systems. Opt Lett 11:150–152.

Fujimoto JG, Brezinski ME, Tearney GJ, Boppart SA, Bouma B, Hee MR, Southern JF, Swanson EA. (1995). Optical biopsy and imaging using optical coherence tomography. Nature Med 1:970–972.

Fujimoto JG, Boppart SA, Tearney GJ, Bouma BE, Pitris C, Brezinski ME. (1999). High resolution in vivo intra-arterial imaging with optical coherence tomography. Heart 82:128–133.

Fujimoto JG, Drexler WD, Morgner U, Kartner F, Ippen E. (2000a). Optical coherence tomography: high resolution imaging using echoes of light. Opt Photon News 1:24–31.

Fujimoto JG, Pitris C, Boppart SA, Brezinski ME. (2000b). Optical coherence tomography: an emerging technology for biomedical imaging and optical biopsy. Neoplasia 2:9–25.

Gilgen HH, Novak RP, Salathe RP, Hodel W, Beaud P. (1989). Submillimeter optical reflectometry. IEEE J Lightwave Technol 7:1225–1233.

Hartl I, Li XD, Chudoba C, Ghanta R, Ko T, Fujimoto JG, Ranka JK, Windeler RS, Stentz AJ. (2001). Ultrahigh resolution optical coherence tomography using continuum generation in an air-silica microstructure optical fiber. Opt Lett 26:608–610.

Hee MR, Huang D, Swanson EA, Fujimoto JG. (1992). Polarization-sensitive low-coherence reflectometer for birefringence characterization and ranging. J Opt Soc Am B 9:903–908.

Hee MR, Izatt JA, Swanson EA, Huang D, Lin CP, Schuman JS, Puliafito CA, Fujimoto JG. (1995). Optical coherence tomography of the human retina. Arch Ophthalmol 113:325–332.

Huang D, Swanson EA, Lin CP, Schuman JS, Stinson WG, Chang W, Hee MR, Flotte T, Gregory K, Puliafito CA, Fujimoto JG. (1991a). Optical coherence tomography. Science 254:1178–1181.

Huang D, Wang J, Lin CP, Puliafito CA, Fujimoto JG. (1991b). Micron-resolution ranging of cornea and anterior chamber by optical reflectometry. Lasers Surg Med 11:419–425.

Izatt JA, Hee MR, Owen GM, Swanson EA, Fujimoto JG. (1994). Optical coherence microscopy in scattering media. Opt Lett 19:590–592.

Izatt JA, Kulkami MD, Yazdanfar S, Barton JK, Welch AJ. (1997). In vivo bidirectional color Doppler flow imaging of picoliter blood volumes using optical coherence tomography. Opt Lett 22:1439–1441.

Kremkau FW. (1984). Diagnostic Ultrasound: Principles, Instrumentation, and Exercises. Philadelphia: Grune and Stratton.

Kulkarni MD, van Leeuwen TG, Yazdanfar S, Izatt JA. (1998). Velocity-estimation accuracy and frame-rate limitations in color Doppler optical coherence tomography. Opt Lett 23:1057–1059.

Li XD, Boppart SA, Van Dam J, Mashimo H, Mutinga M, Drexler W, Klein M, Pitris C, Krinsky ML, Brezinski ME, Fujimoto JG. (2000). Optical coherence tomography: advanced technology for the endoscopic imaging of Barrett's esophagus. Endoscopy 32:921–930.

Liu HH, Cheng PH, Wang J. (1993). Spatially coherent white light interferometer based on a point fluorescent source. Opt Lett 18:678–680.

Morgner U, Kartner FX, Cho SH, Chen Y, Haus HA, Fujimoto JG, Ippen EP, Scheuer V, Angelow G, Tschudi T. (1999). Sub-two-cycle pulses from a Kerr-lens mode-locked Ti:sapphire laser. Opt Lett 24:411–413.

Morgner U, Drexler W, Li X, Kaertner FX, Pitris C, Boppart SA, Ippen EP, Fujimoto JG. (2000). Spectroscopic optical coherence tomography. Opt Lett 25:111–113.

Park H, Chodorow M, Kompfner R. (1981). High resolution optical ranging system. Appl Opt 20:2389–2394.

Podoleanu AG, Dobre GM, Webb DJ, Jackson DA. (1997). Simultaneous en-face imaging of two layers in the human retina by low-coherence reflectometry. Opt Lett 22:1039–1041.

Podoleanu AG, Dobre GM, Jackson DA. (1998). En-face coherence imaging using galvanometer scanner modulation. Opt Lett 23:147–149.

Puliafito CA, Hee MR, Lin CP, Reichel E, Schuman JS, Duker JS, Izatt JA, Swanson EA, Fujimoto JG. (1995). Imaging of macular diseases with optical coherence tomography. Ophthalmology 102:217–229.

Puliafito CA, Hee MR, Schuman JS, Fujimoto JG. (1996). Optical Coherence Tomography of Ocular Diseases. Thorofare, NJ: Slack Inc.

Ranka JK, Windeler RS, Stentz AJ. (2000). Visible continuum generation in air silica microstructure optical fibers with anomalous dispersion at 800 nm. Opt Lett 25:25–27.

Rollins AM, Ung-arunyawee R, Chak A, Wong CK, Kobayashi K, Sivak MV, Izatt
    JA. (1999). Real-time in vivo imaging of human gastrointestinal ultrastructure
    by use of endoscopic optical coherence tomography with a novel efficient
    interferometer design. Opt Lett 24:1358–1360.
Schmitt JM, Knuttel A, Bonner RF. (1993). Measurement of optical-proper-
    ties of biological tissues by low-coherence reflectometry. Appl Opt 32:6032–
    6042.
Schmitt JM, Knuttel A, Yadlowsky M, Eckhaus MA. (1994). Optical coherence
    tomography of a dense tissue—statistics of attenuation and backscattering.
    Phys Med Biol 39:1705–1720.
Schmitt JM, Yadlowsky M, Bonner RF. (1995). Subsurface imaging of living skin
    with optical coherence tomography. Dermatology 191:93–98.
Sergeev AM, Gelikonov VM, Gelikonov GV, Feldchtein FI, Kuranov RV, Glad-
    kova ND, Shakhova NM, Snopova LB, Shakov AV, Kuznetzova IA, Deni-
    senko AN, Pochinko VV, Chumakov YP, Streltzova OS. (1997). In vivo
    endoscopic OCT imaging of precancer and cancer states of human mucosa.
    Opt Express 1:432.
Sutter DH, Steinmeyer G, Gallmann L, Matuschek N, Morier-Genoud F,
    Keller U, Scheuer V, Angelow G, Tschudi T. (1999). Semiconductor
    saturable-absorber mirror assisted Kerr-lens mode-locked Ti:sapphire
    laser producing pulses in the two-cycle regime. Opt Lett 24:631–
    633.
Swanson EA, Izatt JA, Hee MR, Huang D, Lin CP, Schuman JS, Puliafito CA, Fuji-
    moto JG. (1993). In vivo retinal imaging by optical coherence tomography.
    Opt Lett 18:1864–1866.
Takada K, Yokohama I, Chida K, Noda J. (1987). New measurement system for
    fault location in optical waveguide devices based on an interferometric techni-
    que. Appl Opt 26:1603–1608.
Tearney GJ, Boppart SA, Bouma BE, Brezinski ME, Weissman NJ, Southern JF,
    Fujimoto JG. (1996a). Scanning single mode catheter/endoscope for optical
    coherence tomography. Opt Lett 21:543–545.
Tearney GJ, Bouma BE, Boppart SA, Golubovic B, Swanson EA, Fujimoto JG.
    (1996b). Rapid acquisition of in vivo biological images by use of optical coher-
    ence tomography. Opt Lett 21:1408–1410.
Tearney GJ, Bouma BE, Fujimoto JG. (1997a). High-speed phase- and group-
    delay scanning with a grating-based phase control delay line. Opt Lett
    22:1811–1813.
Tearney GJ, Brezinski ME, Bouma BE, Boppart SA, Pitris C, Southern JF, Fuji-
    moto JG. (1997b). In vivo endoscopic optical biopsy with optical coherence
    tomography. Science 276:2037–2039.
Tearney GJ, Brezinski ME, Southern JF, Bouma BE, Boppart SA, Fujimoto JG.
    (1997c). Optical biopsy in human urologic tissue using optical coherence
    tomography. J Urol 157:1915–1919.
Weiner AM, Heritage JP, Kirschner EM. (1986). High resolution femtosecond pulse
    shaping. J Opt Soc Am B 5:1563–1572.

Youngquist RC, Carr S, Davies DEN. (1987). Optical coherence-domain reflectome-
try: a new optical evaluation technique. Opt Lett 12:158–160.
Zwiebel WJ. (1992). Introduction to Vascular Ultrasonography. Philadelphia: WB
Saunders.

16
Ultrafast Lasers in Ophthalmology

Ronald M. Kurtz and Melvin A. Sarayba
University of California at Irvine, Irvine, California, U.S.A.

Tibor Juhasz
University of Michigan, Ann Arbor, Michigan, U.S.A.

16.1 INTRODUCTION

16.1.1 Laser–Tissue Interaction Mechanisms

Only a few years after the development of the first lasers, ophthalmic surgical procedures were already being performed with these devices. Since then, the ability to transmit light energy to almost any ocular structure, in addition to the functional importance of vision, have continued to make the eye a favored target organ for laser surgery. Ophthalmic surgeons make use of three classes of laser–tissue interaction: photocoagulation, photoablation, and photodisruption.

Photocoagulation employs continuous wave laser light applied to absorbing tissue targets, with clinical effects mediated by primary and secondary effects of thermal damage. This technique is most widely used in the eye to treat retinal diseases such as diabetic retinopathy and macular degeneration (ETDRS, 1991). In photoablation, highly absorbing ultraviolet wavelengths are used to vaporize superficial tissues, primarily for therapeutic and refractive surgical applications in the cornea (Orndahl and Fagerholm, 1998; Leroux les Jardins et al., 1994).

Until recently, photodisruption was limited to procedures performed in the middle of the eye due to the relatively large collateral tissue damage zones associated with the available nanosecond-pulsed Nd:YAG lasers (Steinert and Puliafito, 1985). In contrast to these systems, low-energy femtosecond laser photodisruption affords micrometer precision and minimal collateral tissue damage. This combination allows femtosecond laser pulses to be used in high-precision tissue-cutting applications, offering the potential for an ideal surgical scalpel.

16.1.2 Tissue Photodisruption with Femtosecond Lasers

As in inorganic material, tissue photodisruption begins with laser-induced optical breakdown (LIOB) when a strongly focused, short-duration laser pulse generates a high-intensity electric field, leading to the formation of a mixture of free electrons and ions that constitutes the plasma state (Bloembergen, 1974). The optically generated hot plasma expands with supersonic velocity, displacing surrounding tissue (Fujimoto et al., 1985; Zysset et al., 1989; Vogel et al., 1986; Glezer et al., 1997). As the plasma expansion slows, the supersonic displacement front propagates through the tissue as a shock wave. The shock wave loses energy and velocity as it propagates, relaxing to an ordinary acoustic wave that dissipates harmlessly (Vogel et al., 1990). Adiabatic expansion of the plasma occurs on a time scale that is short in comparison with the local thermal diffusion time constant, thereby confining thermal damage. The cooling plasma vaporizes a small volume of tissue, eventually forming a cavitation bubble. The cavitation bubble consists mainly of CO_2, N_2, and H_2O, which can diffuse out from the tissue via normal mechanisms (Habib et al., 1995).

The foregoing discussion describes the general process of tissue photodisruption, but the specific features of the process depend largely on the pulse duration and focusing geometry of a particular system. In fact, initial ophthalmic procedures utilizing photodisruption were limited to just a few intraocular procedures due to the relatively large energies needed to initiate LIOB with available nanosecond pulse duration Nd: YAG lasers (Steinert and Puliafito, 1985). The resulting large shock waves and cavitation bubbles produce significant collateral tissue effects. Reducing either the focal spot size or the pulse duration of the laser decreases the threshold energy for LIOB (Fujimoto et al., 1985; Zysset et al., 1989; Vogel et al., 1986, 1990; Glezer et al., 1997). Although a small spot size is a requirement, beam delivery systems capable of scanning over large areas or volumes become impractical for f-numbers below $f/1$. On the other hand, the laser pulse duration can now be decreased by six orders of magnitude from the nanosecond to the femtosecond regime.

The LIOB threshold fluence decreases as the pulse duration falls from nanoseconds to ~ 20 fs, exhibiting a weak dependence below a few hundred femtoseconds. Consequently, the use of extremely short pulses does not greatly reduce the photodisruption threshold. The large optical bandwidth and high intensities of pulses shorther than a few hundred femtoseconds make pulse generation and amplification difficult at the required surgical energies and are prone to temporal broadening and undesirable nonlinear phenomena (self-phase modulation, self-focusing, etc.) Practical pulse durations for surgical photodisruptive lasers therefore fall in the several hundred femtosecond range (Loesel et al., 1996, 1998; Kurtz et al., 1997).

Table 1 summarizes LIOB threshold fluence and the scale of secondary photodisruptive phenomena for several pulse durations. Although these phenomena depend in detail on pulse energy, tissue elasticity, laser wavelength, and focal geometry, general statements can be made about the dependence of secondary photodisruption on pulse duration (pulse width), primarily because LIOB threshold fluence falls with pulse width. The result is that photodisruptive shock wave size (radial propagation distance to decay to an acoustic wave) and cavitation size (maximum bubble radius) for pulses in the range of 100 fs are much smaller than those generated by picosecond and nanosecond pulses. The small amount of energy deposited in tissue and the rapid (faster than 1 μs) adiabatic plasma expansion prevent significant heat transfer to surrounding tissue. Narrow collateral tissue damage zones have been demonstrated in ex vivo tissue experiments with femtosecond photodisruption (Lubatshcowski ct al., 2000).

Another important feature of femtosecond photodisruption is the deterministic nature of the optical breakdown process. In general, optical breakdown proceeds by avalanche ionization, a highly nonlinear process in which the strong electric field of the laser accelerates charge carriers that ionize other atoms, producing an "avalanche" of ionization. For picosecond or longer pulses, avalanche ionization is initiated by carriers in the material, with optical breakdown proceeding probabilistically. Femtosecond pulses are associated with electric fields high enough to produce their own seed carriers through tunnel and multiphoton ionization, producing LIOB that is very deterministic and repeatable (Du et al., 1994).

Table 1 Fluence Threshold, Shockwave Radial Extent, and Cavitation Bubble Radial Extent in Cadaver Bovine and Porcine Corneas as a Function of Laser Pulse Duration

Pulse width	Threshold fluence (J/cm^2)	R_s, radial extent of shock wave (μm)	R_c, radial extent of cavitation (μm)
150 fs	1.5	20	3–15
500 fs	1.6	20	3–15
60 ps	14	200	30–120
10 ns	185	700	300–1200

16.1.3   Femtosecond Laser Surgical Principles

Owing to their large collateral tissue effects, nanosecond ophthalmic photo-disrupters have mainly been used to create explosive tears in surgical targets far from delicate structures. The best example of such an application is posterior capsulotomy, performed to photodisrupt the opacity that develops in the optical path, behind the plastic intraocular lens, after cataract surgery (Steinert and Puliafito, 1985). In contrast, the localized effects of femtosecond photodisruption described in the previous section permit its use as a highly precise cutting tool. To be used in this manner, essentially as a remote-controlled scalpel, individual laser pulses must be placed contiguously, creating a "postage stamp" effect that results in incisional planes within the tissue. These planes can be placed in any orientation to create horizontal, vertical, or oblique incisions (Fig. 1).

Tissue targets that are transparent to the laser wavelength allow optical breakdown to occur at any depth or location without affecting tissue outside the photodisruption zone. For targets in the eye, this generally restricts laser wavelengths to the visible and near-infrared. The only limitation to creating arbitrary incision planes is that they must be written from the deepest portion of the tissue to the shallowest, because static gas bubbles that persist in tissue shadow the laser if the focus is moved to a plane below previously produced bubbles. Using a femtosecond laser with high pulse repetition rates (in the kilohertz range) and a computer-controlled scanning optical delivery system, localized microphotodisruptions can be placed in a

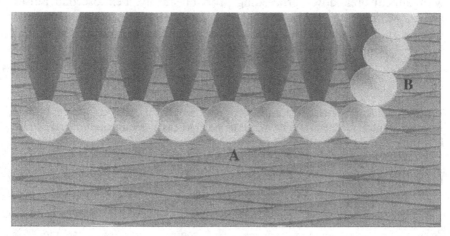

Figure 1   Individual laser pulses placed contiguously horizontally for a lamellar cut (A) and vertically or obliquely for a side-cut (B).

contiguous fashion to produce incisions of any shape to produce high-precision tissue separations. Complex shapes can also be created by intersecting these resection planes.

16.2 FEMTOSECOND LASER SURGICAL APPLICATIONS IN THE CORNEA

16.2.1 Background

The cornea presents an attractive initial target for femtosecond laser surgical applications because it is easily accessible and lacks blood vessels. The cornea is only 500–600 μm thick centrally, permitting delivery of femtosecond pulses with negligible nonlinear effects. Because photodisruptive lasers do not coagulate blood vessels, the lack of vascular structures in the normal cornea obviates the need for additional techniques to control bleeding. The Food and Drug Administration has already cleared four surgical procedures for use in the United States: hinged flap for LASIK, anterior lamellar keratoplasty (lamellar corneal transplant), intracorneal channel formation for corneal implants, and removal of a block of corneal tissue through a hinged flap incision (keratomileusis) for the correction of myopic refractive error. Several additional applications are under development.

A key factor that permits increased precision when using femtosecond lasers in corneal surgical procedures is fine control of the focal plane with respect to the corneal surface. To accomplish this, all corneal procedures use an applanation system consisting of a suction ring and a flat contact lens located at the tip of the laser delivery system. The suction ring fixates the eye, allowing the contact lens to temporarily flatten the front surface of the cornea. The flat contact lens is securely attached to the suction ring by an internal cylindrical clamp, mechanically coupling the eye to the beam delivery system. This system allows depth precision to be controlled to less than 10 μm (Fig. 2).

16.2.2 Femtosecond Laser in Refractive Surgery

Creation of Corneal Flaps for LASIK

Traditional LASIK. To understand the advantages of the femtosecond laser for creating a corneal flap, one must consider the limitations of the LASIK procedure. Introduced in the early 1990s, laser-assisted in situ keratomileusis (LASIK) has become the dominant surgical technique to correct simple refractive errors such as myopia, hyperopia, and astigmatism. Traditional LASIK uses a mechanical blade (called a microkeratome) to

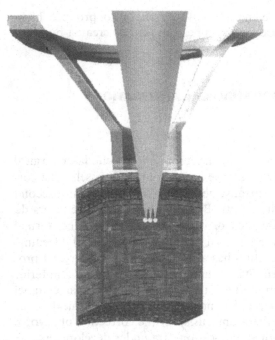

Figure 2 The contact lens temporarily flattens the cornea, allowing depth precision.

create a lamellar hinged corneal cut (called a flap), exposing internal corneal layers (stroma) for subsequent excimer laser ablation. Although current microkeratomes have attained a high level of clinical experience, outcome variability and patient anxiety remain concerns. In fact, flap creation with microkeratomes is responsible for the majority of intraoperative and post-operative LASIK complications, occurring in as many as 5% of cases (Pallikaris and Saiganos, 1994; Azar and Farah, 1998; Carr et al., 1997). Intraoperative complications most commonly involve abrasions of the epithelium (the sensitive outer layer of protective cells), "buttonholed" flaps, incomplete cuts, free flap (cut through the hinge), thin flaps, and thick flaps. Postoperative complications include flap slippage or dislocation, ingrowth or downgrowth of the epithelium into the stroma, and severe inflammation (diffuse lamellar keratitis). These events can significantly delay recovery of good visual acuity and may even lead to permanent visual loss. Even when operating normally, the accuracy and precision of the microkeratome contrasts significantly with that of the subsequent excimer laser ablation of the stroma. Microkeratome flap depth can vary significantly both between and within cuts, depending on patient factors (corneal curvature,

orbit size) as well as variations in individual instruments, blades, and operator skill. Surgeon control of microkeratomes is limited, with little flexibility
to accommodate individual surgical requirements imposed by corneal thickness, pupil location, or refractive state.

LASIK with the Femtosecond Laser. A flap is created by scanning a
pattern of laser pulses at the appropriate depth to create a resection plane
parallel to the applanated corneal surface. An arc is then scanned with pro-
gressive movement closer to the surface to create a hinged side-cut. Following creation of the flap, the suction ring is released and the applanating
contact lens removed. The flap is then elevated, and the excimer laser treatment is performed. A representative procedure sequence is illustrated in
Figure 3.

The first large clinical series using the femtosecond laser for LASIK
flaps suggests that the laser technique may offer both safety and performance advantages over current mechanical methods (Nordan et al., 2002).
No operative or postoperative complications were noted in over 200 procedures, with the only undesired events being interrupted procedures due to

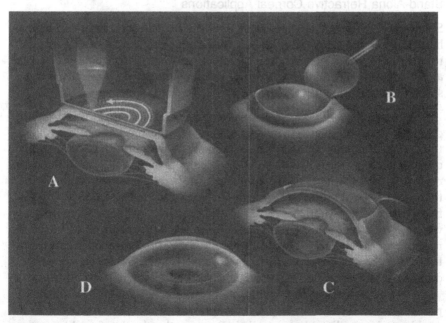

Figure 3 Schematic illustration of IntraLasik. Lamellar resection with a side-cut is
performed to create a flap (A). The flap is lifted and excimer photoablation is performed (B, C). After the procedure, the flap is replaced (D).

loss of applanation. The nature of femtosecond laser resection allows these cases to be completed *on the same surgical day*, with the same results obtained in uninterrupted cases. The absence of a delay following such an event is a significant advantage over procedures performed using traditional keratomes, for which a 3 month wait is generally recommended before a second attempt following an aborted resection. It is important to note that excimer laser nomograms developed for traditional LASIK with the microkeratome appear to apply equally well to the new all-laser procedure, suggesting that the femtosecond laser does not remove significant amounts of tissue in a single layer. This finding is consistent with ex vivo and animal model studies in which micrographs and histological evidence indicate that only small zones of tissue are removed (Vogel et al., 1990).

In contrast to a traditional mechanical microkeratome, femtosecond surgical technology allows the surgeon precise control over flap parameters that may affect clinical outcomes. These include flap thickness, diameter, hinge position and angle, and side (entry) cut angle. Future investigations will need to evaluate the clinical significance of the increased flexibility in flap architecture afforded by this new technology.

Stand-Alone Refractive Corneal Applications

Stand-alone refractive changes using only the femtosecond laser can be realized through the removal of laser-cut lenticular tissue volumes (laser keratomileusis) or via a combination of direct volumetric destruction of tissue and biomechanical changes in corneal structure induced by selective femtosecond laser treatment (intrastromal procedures.)

Femtosecond Laser Keratomileusis. Keratomileusis was initially developed and popularized in the 1980s using the mechanical microkeratome and ultimately led to the development of the current LASIK techniques following introduction of the excimer laser. In keratomileusis, removal of stromal tissue is accomplished via two lamellar resections rather than with tissue ablation by the excimer laser. In the first resection, a corneal flap is created. A second resection in the corneal bed frees stromal tissue, which is then removed manually. The corneal flap is then repositioned to restore corneal integrity. The anterior curvature of the cornea flattens owing to this tissue removal, resulting in a refractive change. When performed in a donor eye, the removed tissue can be used for transplanation.

Owing to the poor precision of the mechanical microkeratome, this procedure never gained widespread acceptance, because refractive outcomes could not be predictably controlled. Because the depth of each resection could vary by as much as 30–60 μm, the thickness of the removed lenticule (which determines the refractive effect) could not be predicted with any degree of accuracy. In contrast, the relative accuracy of femtosecond laser resections, performed during the same applanation, is approximately 1 μm, potentially making femtosecond laser keratomileusis comparable to current excimer laser techniques. Potential advantages of femtosecond laser keratomileusis over LASIK include the use of a single laser technology and higher precision with respect to the three-dimensional position and centration of the removed refractive lenticule. This may be especially true for hyperopic corrections, where excimer ablations produce smaller optical zones that require better centration.

As in LASIK, the refractive change induced by keratomileusis is determined by the shape, size, and position of the removed tissue lenticule. Clinical nomograms and models that predict clinical refractive results are already widely available for myopia and can be developed with relative case for hyperopia and astigmatism (Djotyan et al., 2001). As can be seen in Figures 4 and 5, both the top and bottom surfaces of the lenticule as well as the surface opening (either a full flap or side-port opening) are created during a single applanation. The shape of each lamellar resection can be modified not only to produce specific corrections for simple sphero-cylindrical errors but also to correct higher order aberrations diagnosed using wavefront technology.

Intrastromal Vision Correction. Refractive techniques that are completely intrastromal and use photodisruptive lasers operating at infrared wavelengths to further reduce the complications and risks of elective refractive procedures were first proposed in the mid-1980s (Fig. 6). Initial intrastromal procedures that used Nd:YAG and Nd:YLF laser pulses were not successful, possibly due to the limited precision afforded by these relatively long pulse duration nanosecond and picosecond lasers (Brown et al., 1994a, 1994b; Vogel et al., 1997).

Initial in vivo testing of femtosecond laser intrastromal procedures in rabbits was aimed at evaluating the effects of layered intrastromal femtosecond laser photodisruption on corneal transparency, as well as studying alterations in biomechanical and structural properties of the cornea as they relate to corneal thickness changes (Sletten et al., 1999). In this experimental model, 10 layers of a standard spiral pattern (3–6 mm in diameter) were placed in the central cornea using a modified delivery system. Postoperative slit lamp examination and corneal pachymetric measurements were performed at regular intervals for up to 6 months. Animals were euthanized at 1 day, 1 week, 2 weeks, 4 weeks, 2 months, 4 months, and 6 months, and their eyes enucleated for histological evaluation.

As shown in Figure 7, a mild reticular haze could be discerned with retroillumination by 2 weeks postoperatively in animals treated with the

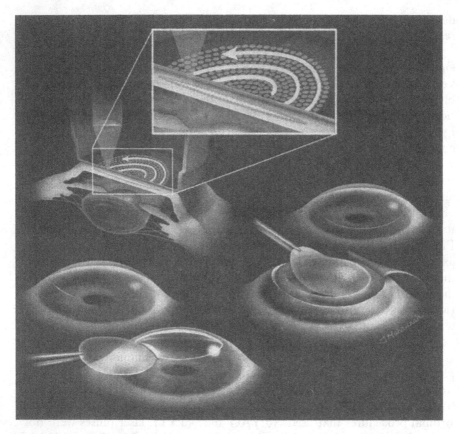

Figure 4 Femtosecond laser keratomileusis. Stromal tissue is removed by applying two lamellar resections. The flap is opened with another resection, and the tissue
is removed manually. The excised tissue can be used also be used as donor tissue for
transplantation.

10-layer pattern. No significant corneal or intraocular inflammation was
seen on slit lamp examination, despite the minimal use of topical anti-
inflammatories. Histological results suggest that photodisrupted material
remains within the corneal stroma as disorganized lamellae but does not
incite a significant inflammatory response. No damage to the corneal
endothelium was seen on histological sections.

Although initially there was a small increase in thickness, consistent
decreases in central corneal thickness were noted by 2 weeks, stabilizing at
approximately $-70 \pm 35\,\mu m$ for over 4 months. This amount of thinning
represented approximately 70% of the original 100 μm treated thickness.

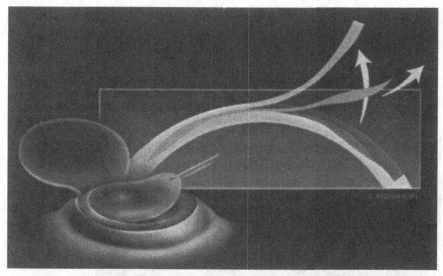

Figure 5 Femtosecond laser hyperopic keratomileusis. Top and bottom surfaces are resected to the prescribed pattern. A full flap opening is used here to remove the lenticule.

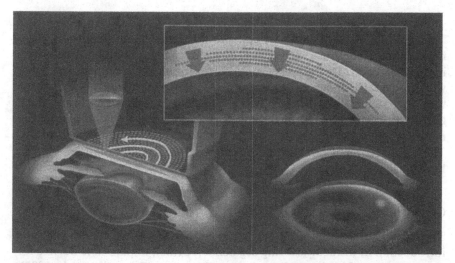

Figure 6 Intrastromal myopic refractive surgery. Laser pulse pattern is more contiguous and dense in the central cornea for more tissue photodisruption.

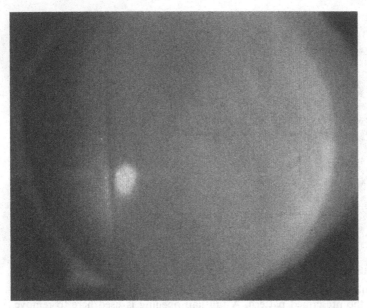

Figure 7 Intrastromal laser keratectomy in rabbit eye. Top shows preoperative and intraoperative surgeon's view. The gas bubbles generated during the procedure dissipate within several hours, leaving a mild reticular haze within the treatment volume (2 weeks).

This ratio, termed the "corneal collapse quotient," is an important parameter in the models used to predict refractive changes in humans described in a later section. By one year, some regression of the treatment was seen, with an average change in thickness of 35 μm. In contrast, rabbits treated with the excimer laser have shown a complete return to baseline corneal thickness after several months, suggesting that the regeneration of collagen in the rabbit make it a poor long-term model for human surgical results. These intrastromal procedures are now under investigation in various human clinical trials.

Channels for Corneal Implant

The creation of intracorneal tunnels has become a topic of interest since the introduction of intracorneal ring segments (ICRSs), implantable devices for the correction of low myopia without astigmatism (Schanzlin et al., 1997). ICRS technology has significant advantages over other refractive surgical techniques, including sparing of the visual axis from any surgical intervention. In addition, the potential for removing the ICRS with subsequent return to the original refractive state is particularly attractive to pre-prebyopes, who may later want to make use of their low myopia for reading. Finally, the maintenance of prolate asphericity of the cornea may improve visual quality and may allow the device to be used in therapeutic applications, such as in keratoconus (Colin et al., 2001).

Despite these advantages, ICRS technology has not gained widespread clinical acceptance since clearance by the FDA in 1996. A key factor in this trend may be technical problems associated with available mechanical devices, which can limit the safety, reproducibility, and convenience of this procedure. Although excellent intraoperative safety was demonstrated in a recent ICRS PMA cohort, approximately 1% (4 of 449) patients experienced severe operative complications. These included three corneal surface perforations from overly shallow resections and one anterior chamber perforation due to an excessively deep resection. Less severe surgical complications included limbal hemorrhages (two patients) from excessively long entry cut incisions (Schanzlin et al., 1997).

Problems with reproducibility and predictability have also been seen, with relatively large variations in intended versus achieved postoperative refractive corrections (Schanzlin et al., 1997; Holmes-Higgin et al., 1999). These have necessitated the use of refractive ranges instead of the high predictable nomograms used with other refractive procedures. Such variations in achieved refractive correction may be partly linked to difficulty in achieving reproducible tunnel resection depth and/or centration. In addition, placement of ring segments at different depths in the same eye has been noted and might be linked to the relatively high percentage of patients who experience induced cylindrical abberation (approximately 4% greater than 1 diopter). As with all procedures, adverse events, poor outcomes, and complications may be more common when laser surgery is performed by surgeons less experienced than those in the clinical trial. The implantation procedure's general awkwardness also makes it difficult to incorporate into busy refractive surgery practices. This results in limited availability to patients who might otherwise benefit from a choice of refractive procedures.

The femtosecond laser offers several advantages that could reduce complications and variations associated with traditional ICRS implantation techniques. Because laser energy is delivered optically to a precise depth, using a precalibrated and preassembled disposable contact lens, tunnel resections and entry cuts may be highly reproducible, with essentially no risk of corneal perforation. Improved tunnel resection reproducibility could also improve the effectiveness of the ICRS procedure by reducing operative variables, and the ease of the procedure may make it more available, thereby increasing patient refractive choices.

To create the tunnel and entry cut for ring implantation, the femtosecond laser first scans an annulus-shaped pattern at the surgeon-selected depth. The width of the annulus is determined by the implant size. A vertical cut is then made to the surface by scanning a line with movement toward the epithelial surface. The surgeon can select the meridional location of the entry cut as well as its length, easily allowing implantation geometries different from the traditional nasal/temporal ring orientation (Fig. 8).

Ex Vivo Results

The depth reproducibility of laser-created tunnels was evaluated in a series of resections performed in porcine eyes. As Table 2 shows, extremely high reproducibility was seen, with standard deviation of about 10 µm. Access to such precision may increase the use of this and other corneal implant technologies.

Figure 8   Channels for corneal implants. Channels are created by disc-shaped laser pattern. Implants are inserted through side cuts.

Table 2 Tunnel Resection Depth Reproducibility

Number of eyes	12
Intended tunnel resection depth	440 µm
Average measured tunnel resection depth	441 µm
Standard deviation	9.9
Percent variance	0.2%

16.2.3 Femtosecond Laser in Corneal Transplantation

In contrast to flap creation in LASIK, the goal of lamellar keratoplasty is therapeutic—namely, the replacement of diseased or damaged corneal tissue with normal tissue from a cadaver donor eye. Anterior lamellar corneal transplantation replaces superficial tissue for clinical indications such as corneal scarring, corneal ectasia, peripheral or central corneal thinning, or perforation. Although this procedure has been performed for decades using simple manual instruments and microkeratomes, most corneal surgeons continue to find these procedures technically difficult to perform with precision and reproducibility. Full thickness corneal transplants (penetrating keratoplasty) are therefore used in most clinical situations, despite the association of penetrating keratoplasty with more significant complications and longer delays to achieve functional vision. A reproducible femtosecond laser lamellar keratoplasty procedure has the potential to dramatically alter the surgical approach to anterior corneal dysfunction.

To perform lamellar keratoplasty a complete circular side-cut is used to create a free corneal cap in both the recipient and donor eyes (instead of the partial side-cut that creates a hinged corneal flap for LASIK). Once placed on the recipient bed, the donor cap is held in place by the hydrostatic pressure provided by the pumping action of the endothelial cell layer at the posterior surface of the cornea. Sutures are often used to secure positioning until healing occurs (Fig. 9). The precision provided by the femtosecond laser allows the recipient and donor corneas to be cut with high accuracy, guaranteeing proper fit of the donor corneal graft in the recipient bed.

In contrast to anterior lamellar transplantation, posterior lamellar corneal transplantation to address endothelial cell dysfunction accounts for the majority of full thickness corneal transplant procedures performed in developed countries (Barraquer 1964, 1972; Haimovici and Culbertson 1991). Several alternative methods for endothelial cell restoration have been studied. Isolated endothelial cenn transplantation has been attempted with limited success (Gospodarowicz et al., 1979; Mohay et al., 1994; McCulley et al., 1980; Jumblatt et al., 1978; Aboalchamat et al., 1999). With the development of new biomaterials, this method might become more feasible (McDonnell, 2000).

Figure 9 Femtosecond anterior lamellar keratoplasty. Free flaps are created on both donor and recipient corneas. The donor tissue is sutured to the recipient bed.

Use of a posterior stromal lamellar carrier (posterior lamellar kerato-plasty) has shown great promise recently. In this approach, replacement of the posterior stroma, Descemet's membrane, and endothelial cell layer is accomplished through a scleral pocket (Melles et al., 1998, 1999; Terry, 2000). The technique is mechanical and laborious and requires a highly skilled surgeon. Several femtosecond laser posterior lamellar keratoplasty methods have been proposed, either as described above or through other surgical approaches (Fig. 10). Potential advantages include the preservation of the original corneal surface, avoidance of extensive superficial suturing, and isolated posterior tissue transplantation. This may translate into earlier postoperative visual recovery and lower rejection rates. In addition to these advantages, the residual anterior cap of the donor may be used for an anterior lamellar keratoplasty in another patient, improving the efficiency of tissue procurement by a factor of 2.

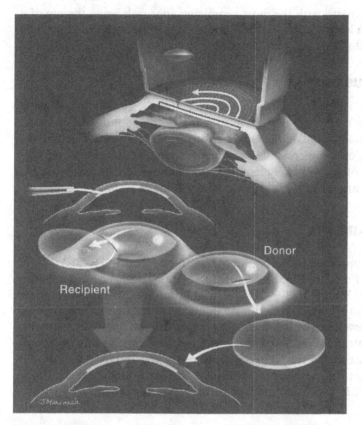

Figure 10 Femtosecond posterior lamellar keratoplasty. Endostromal tissue is removed from the donor and recipient corneas. Tissues are removed and replaced through a corneal pocked.

16.2.4 Other Ophthalmic Applications

Other obvious targets include the rest of the transparent ocular tissues (lens, capsule, and vitreous) as well as surgical procedures in translucent tissues such as the sclera. In the lens, potential applications under investigation include tissue cutting and removal for cataract and/or restoration of accommodation. Release of traction on the retina from the transparent vitreous gel has also been proposed, using the precise cutting afforded by ultrafast pulses, possibly permitting development of transpupillary noninvasive vitreoretinal procedures. Subsurface machining of incisions, channel, and other flow features in sclera may lead to novel glaucoma treatments. The important optical problem here is the overcoming of strong scattering in this nontransparent tissue, with the most promising approach being the use of a laser wavelength long enough to substantially reduce the scattering but short enough to avoid strong infrared water absorption (Sacks et al., 1999).

16.3 CONCLUSIONS

The minimally invasive, high-precision characteristics of femtosecond laser technology make it highly promising for various ophthalmic surgical procedures. A number of initial applications in the cornea are already in clinical use and/or commercial development. Continued advances in procedures and technology will likely lead to expanded use of this technology in ophthalmology and possibly in other medical fields as well.

ACKNOWLEDGMENTS

This work was supported by the National Institute of Health (1R43EY12340-01), U.S. Air Force (F29601-98-C-0146), National Science Foundation (STCPHY 8920108 and DMI-99017126), and IntraLase Corporation (Irvine, CA). The authors acknowledge a financial interest in IntraLase Corporation, a commercial developer of this technology. Drs. Juhasz and Kurtz are also employees of IntraLase. Any opinions, findings, and conclusions or recommendations expressed in this material are those of the author(s) and do not necessarily reflect the views of the National Science Foundation, National Institute of Health, U.S. Air Force, or IntraLase Corporation.

REFERENCES

Aboalchamat B, Engelmann K, Bohnke M, Eggli P, Bednarz J. (1999). Morphological and functional analysis of immortalized human corneal endothelial cells after transplantation. Exp Eye Res 69:547–553.

Azar DT, Farah SG. (1998). Laser in situ keratomileusis versus photorefractive keratectomy: an update on indications and safety. Ophthalmology 105(8):1357–1358.

Barraquer JI. (1972). Lamellar keratoplasty (special techniques). Ann Ophthalmol 4(6):437–469.

Barraquer JI. (1964). Queratomileusis para la correción de la miopía. Arch Soc Am Oftamol Optom 5:27–48.

Bloembergen N. (1974). Laser-induced electric breakdown in solids. IEEE J Quantum Electron 10:375–386.

Brown DB, O'Brien WJ, Schultz RO. (1994a). Corneal ablations produced by the neodymium doped yttrium-lithium-fluoride picosecond laser. Cornea 13:471–478.

Brown DB, O'Brien WJ, Schultz RO. (1994b). The mechanism of ablation of corneal tissue by the neodymium doped yttribum-lithium-fluoride picosecond laser. Cornea 13:479–486.

Carr JD, Stulting RD, Thompson KP, Waring GO III. (1997). Laser in-situ keratomileusis. Refract Surg 10(4):533–542.

Colin J, Cochener B, Savary G, Malet F, Holmes-Higgin D. (2001). INTACS inserts for treating keratoconus: one-year results. Ophthalmology. 108(8):1409–1414.

Djotyan GP, Kurtz RM, Fernandez DC, Juhasz T. (2001). An analytical solvable model for biomechanical response of the cornea to refractive surgery. Trans ASME 123:440–445.

Du D, Squier J, Kurtz R, Elner V, Liu, X, Guttmann G, Mourou G. (1994). Damage threshold as a function of pulse duration in biological tissue. Ultrafast Phenomena IX. Springer Ser Chem Phys 60:254.

ETDRS. (1991). Early photocoagulation for diabetic retinopathy. ETDRS Rep 9. Early Treatment Diabetic Retinopathy Study Research Group. Ophthalmology 98(5 suppl):766–785.

Fujimoto JG, Lin WZ, Ippen EP, Puliafito CA, Steinert RF. (1985). Time-resolved studies of Nd:YAG laser-induced breakdown. Invest Ophthalmol Vis Sci 26:1771–1777.

Glezer EN, Scaffer CB, Nishimura N, Mazur E. (1997). Minimally disruptive laser induced breakdown in water. Opt Lett 23:1817.

Gospodarowicz D, Greenburg G, Alvarado J. (1979). Transplantation of cultured bovine corneal endothelial cells to rabbit cornea: clinical implications for human studies. Proc Natl Acad Sci USA 76:464–468.

Habib MS, Speaker MG, Shnatter WF. (1995). Mass spectrometry analysis of the byproducts of intrastromal photorefractive keratectomy. Ophthalmol Surg Lasers 26:481–483.

Haimovici R, Culbertson WW. (1991). Optical lamellar keratoplasty using the Barraquer microkeratome. Refract Corneal Surg 7:42–45.

Holmes-Higgin D, Baker P, Burris T, Silvestrini T. (1999). Characterization of the aspheric corneal surface with the intrastromal ring segment. J Refract Surg 15:520–529.

Jumblatt MM, Maurice DM, McCulley JP. (1978). Transplantation of tissue-cultured corneal endothelium. Invest Ophthalmol Vis Sci 17(12):1135–1141.

Khodadoust AA, Arkfeld DF, Caprioli J, Sears ML. (1984). Ocular effect of neodymium-YAG laser, Am J Ophthalmol 98:144–152.

Kurtz RM, Liu X, Elner VM, Squier JA, Du D, Mourou GA. (1997). Plasma-mediated ablation in human cornea as a function of laser pulse width. Refract Surg 13:653–658.

Leroux les Jardins S, Auclin F, Roman S, Burtschy B, Leroux les Jardins J. (1994). Results of photorefractive keratectomy on 63 myopic eyes with six months minimum follow-up. J Cataract Refract Surg 20(Suppl):223–228.

Loesel FH, Niemz MH, Bille JF, Juhasz T. (1996). Laser-induced optical breakdown on hard and soft tissues and its dependence on the pulse duration: experiment and model. IEEE J Quantum Electron 32:1717–1722.

Loesel FH, Tien AC, Backus S, Kapteyn H, Murnane M, Kurtz RM, Sayegh S, Juhasz T. (1998). Effect of reduction of laser pulse width from 100 ps to 20 fs on the plasma-mediated ablation of hard and soft tissue, Proc SPIE 3565:1325–1328.

Lubatschowski H, Maatz G, Heisterkamp A, Hetzel U, Drommer W, Welling H, Ertmer W. (2000). Application of ultrashort laser pulses for intrastromal refractive surgery. Graefe's Arch Clin Exp Ophthalmol 238:33–39.

McCulley JP, Maurice DM, Schwartz BD. (1980). Corneal endothelial transplantation. Ophthalmology 87:194–201.

McDonnell PJ. (2000). Emergence of refractive surgery. Arch Ophthalmol 118:1119–1120.

Melles GR, Eggink FA, Lander F, Pels E, Rietveld FJ, Beekhuis WH, Binder PS. (1998). A surgical technique for posterior lamellar keratoplasty. Cornea 17:618–626.

Melles GR, Lander F, Beekhuis WH, Remeijer L, Binder PS. (1999a). Posterior lamellar keratoplasty for a case of pseudophakic bullous keratopathy. Am J Ophthalmol 127:340–341.

Melles GR, Lander F, Rietveld FJ, Remeijer L, Beekhuis WH, Binder PS. (1999b). A new surgical technique for deep stromal, anterior lamellar keratoplasty. Br J Ophthalmol 83:327–333.

Mohay J, Lange TM, Soltau JB, Wood TO, McLaughlin BJ. (1994). Transplantation of corneal endothelial cells using a cell carrier device. Cornea 13:173–182.

Nordan LT, Slade SG, Baker R, Juhasz T, Kurtz R. (2002, in press). Femtosecond laser flap creation: initial 6 month follow-up. J Refract Surg.

Orndahl MJ, Fagerholm PP. (1998). Treatment of corneal dystrophies with phototherapeutic keratectomy. J Refract Surg 14(2):129–135.

Pallikaris IG, Saiganos DS. (1994). Excimer laser in-situ keratomileusis and photorefractive keratectomy for correction of high myopia. J Refract Corneal Surg 10(5):498–510.

Sacks ZS, Loesel F, Durfee C, Kurtz RM, Juhasz T, Mourou. (1999). Transscleral photodisruption for the treatment of glaucoma. SPIE Proc.

Schanzlin D, Asbell PA, Burris TE, Durrie D. (1997). The intracorneal ring segments: phase 2 results for the correction of myopia. Ophthalmology 104:1067–1078.

Sletten KR, Yen KG, Sayegh S, Loesel F, Eckhoff C, Horvath C, Meunier M, Juhasz T, Kurtz RM. (1999). An in vivo model of femtosecond laser intrastromal refractive surgery. Ophthal Surg Lasers 30(9):742–749.

Steinert RF, Puliafito CA. (1985). The Nd:YAG Laser in Ophthalmology. Philadelphia: Saunders.

Terry MA. (2000). The evolution of lamellar grafting techniques over twenty-five years. Cornea 19:611–616.

Vogel A, Hentschel W, Holzfuss J, Lauterborn W. (1986). Cavitation bubble dynamics and acoustic transient generation in ocular surgery with pulsed neodymium:YAG laser. Ophthalmology 93:1259–1269.

Voegel A, Schweiger P, Freiser A, Asio MN, Birngruber R. (1990). Intraocular Nd:YAG laser surgery: light-tissue interactions, damage-range and reduction of collateral effects. IEEE Quantum Electron 26:2240–2260.

Vogel A, Gunther T, Asiyo-Vogel M, Birngruber R. (1997). Factors determining the refractive effects of intrastromal photorefractive keratectomy with the picosecond laser. J Cataract Refract Surg 23:1301–1310.

Zysset B, Fujimoto JG, Puliafito CA, Birngruber R, Deutsch TF. (1989a). Picosecond optical breakdown: tissue effects and reduction of collateral damage. Lasers Surg Med 9:193–204.

Zysset B, Fujimoto JG, Deutsch TF. (1989b). Time resolved measurements of picosecond optical breakdown. Appl Phys B 48:139–147.

Index

Printed in the United States
by Baker & Taylor Publisher Services